Painlevé Transcendents

The Riemann-Hilbert Approach

Mathematical
Surveys
and
Monographs

Volume 128

Painlevé Transcendents

The Riemann-Hilbert Approach

Athanassios S. Fokas
Alexander R. Its
Andrei A. Kapaev
Victor Yu. Novokshenov

AMERICAN MATHEMATICAL SOCIETY
Providence, Rhode Island

2000 *Mathematics Subject Classification.* Primary 34M55.

For additional information and updates on this book, visit
www.ams.org/bookpages/surv-128

Library of Congress Cataloging-in-Publication Data

Painlevé transcendents : the Riemann-Hilbert approach / Athanassios S. Fokas ... [et al.].
p. cm. — (Mathematical surveys and monographs, ISSN 0076-5376 ; v. 128)
Includes bibliographical references and index.
ISBN 0-8218-3651-X (alk. paper) | ISBN 1-4704-7556-1 (Softcover)
1. Painlevé equations. 2. Riemann-Hilbert problems. 3. Differential equations, Nonlinear–Asymptotic theory. I. Fokas, A. S., 1952– II. Series: Mathematical surveys and monographs ; no. 128.

QA372.P34 2006
515′.352–dc22 2006042772

∞ The paper used in this book is acid-free and falls within the guidelines
established to ensure permanence and durability.
Visit the AMS home page at `http://www.ams.org/`

11 10 9 8 7 6 5 4 3 2 28 27 26 25 24 23

To our wives Regina, Liza, Veronica, and Olga

Contents

Preface

This book summarizes recent developments in the theory of classical Painlevé equations based on the so-called Riemann-Hilbert, or Isomonodromy, method. The emphasis is on the use of this method for the global asymptotic analysis of the Painlevé transcendents. Specifically, we will study the *connection problem* for the Painlevé transcendents, i.e., the problem of finding explicit formulae relating the relevant asymptotic parameters at different critical points. In more detail we will present:

(i) A complete description of all types of possible asymptotic behavior of the solutions of the important particular cases of Painlevé II and III equations on the real line.
(ii) The derivation of the associated connection formulae.
(iii) A complete description of the asymptotic behavior in the complex plane; quasi-linear and nonlinear Stokes phenomena.
(iv) The distributions of zeros and poles of the solutions.

In addition, we will show how the isomonodromy approach can be used for the study of algebraic aspects of the Painlevé theory. These include the study of the symmetry properties of Painlevé equations (Bäcklund transformations) and the derivation of their elementary solutions and the solutions expressed in terms of classical special functions. In other words, although the asymptotic analysis is the primary theme of this book, several other aspects of modern Painlevé theory will be presented as well. One of our methodological goals is to stress the unique role of the Riemann-Hilbert approach as a universal tool which provides the most effective treatment of many aspects of the theory. Another methodological message is that the Riemann-Hilbert method is, essentially, the only tool available for the study of the problem which is analytically the most challenging, namely the problem of global asymptotics.

It should be emphasized that before the emergence of the Riemann-Hilbert approach, it was thought that the explicit global analysis of the solutions was possible only for the linear differential equations associated with the classical special functions of the hypergeometric type. We will show that the Riemann-Hilbert formalism brings both the *nonlinear special functions*, i.e., the Painlevé functions, and the *linear special functions*, such as Bessel and Airy functions, under the same analytic umbrella.

The Riemann-Hilbert method is based on the remarkable observation, which goes back to the classical works of Garnier and Fuchs, that the Painlevé equations describe the isomonodromy deformations of certain systems of linear differential equations with rational coefficients. This implies that solving a Painlevé equation is equivalent to solving the inverse monodromy problem, i.e., *the Riemann-Hilbert problem* for that particular linear system which is associated to the given Painlevé equation and whose monodromy data are the first integrals of the latter. What was apparently missed in the classical works is that this monodromy relation provides a powerful tool for analyzing the Painlevé transcendents. Indeed, as will be demonstrated in this book, these *Riemann-Hilbert representations* of the Painlevé functions play the same efficient role that the contour integral representations play in the theory of linear special functions.

This book is the second attempt in the literature to present systematically the theory of Painlevé equations from the Riemann-Hilbert point of view. The first attempt was made by A. Its and V. Novokshenov in 1986 in the monograph *Isomonodromic Method in the Theory of Painlevé Transcendents.* Since then the field has developed considerably. The three most important and interrelated advances are: (1) the asymptotic analysis of the Painlevé equations has been extended to the entire complex plane, (2) the Riemann-Hilbert methodology has been made rigorous, (3) the method has been enhanced by the powerful *nonlinear steepest descent* asymptotic techniques introduced in the beginning of the 1990s by P. Deift and X. Zhou. Also, many new concrete asymptotic results have been obtained and many new exciting applications, notably in random matrix theory, have appeared. In the present book we will attempt to present some of these new developments as fully as possible.

We have tried to make the book self-contained so that no prior knowledge of the Painlevé equations or of the Riemann-Hilbert problems is needed. In fact, we begin the book with a detailed introduction to the monodromy theory of systems of linear ordinary differential equations with rational coefficients and with a discussion of the associated Riemann-Hilbert problems. In this context, the Painlevé equations appear as the isomonodromy deformations of the first nontrivial examples of linear systems. This point of view will allow us to put the theory of Painlevé transcendents from the beginning into the proper analytic context. Indeed, in this approach the Painlevé equations will appear simultaneously with the Riemann-Hilbert representations for their solutions. It is these representations, i.e., the representations in terms of solutions of certain matrix Riemann-Hilbert problems, that will allow us to obtain all the results mentioned earlier.

We also present some of the recent applications of Painlevé transcendents in physics and mathematics. We hope that this, together with the lists of the asymptotic formulae obtained and discussed in the book, will make the monograph valuable, not only for the readers interested in theoretical aspects of the Painlevé equations and of the Riemann-Hilbert method, but also for the broad scientific community as a source of reference material on Painlevé transcendents. A small part of this material is included in the introduction so that the reader can quickly appreciate the type of results contained in this book.

INTRODUCTION

Painlevé Transcendents as Nonlinear Special Functions

The six classical Painlevé transcendents were introduced at the turn of the twentieth century by Paul Painlevé and his school, as the solution of a specific classification problem for second order ODEs of the type

$$u_{xx} = F(x, u, u_x),$$

where F is a function meromorphic in x and rational in u and u_x. The problem was to find all equations of this form, which have the property that their solutions are free from movable critical points, i.e., the locations of possible branch points and essential singularities of the solution do *not* depend on the initial data. The motivation for posing this problem is quite clear: the absence of movable critical points means that every solution of the equation can be meromorphically extended to the entire universal covering of a punctured complex sphere, determined only by the equation. This implies that such equations share one of the fundamental properties of linear equations.

It was shown by Painlevé and Gambier (1900, 1910), that within a Möbius transformation,

$$u \mapsto \frac{\alpha(x)u + \beta(x)}{\gamma(x)u + \delta(x)}, \qquad x \mapsto \varphi(x),$$

$$\alpha, \beta, \gamma, \delta, \varphi - \text{are meromorphic in } x,$$

there exist only fifty such equations (see monograph [**In**]). Each of them either can be integrated in terms of known functions, or can be mapped to a set of six equations, which cannot be integrated in terms of known functions. These six equations are called Painlevé equations, and their general solutions are called Painlevé functions or Painlevé transcendents. The canonical forms for the Painlevé equations are:

1. $u_{xx} = 6u^2 + x$
2. $u_{xx} = xu + 2u^3 - \alpha$
3. $u_{xx} = \dfrac{1}{u}{u_x}^2 - \dfrac{u_x}{x} + \dfrac{1}{x}(\alpha u^2 + \beta) + \gamma u^3 + \dfrac{\delta}{u}$
4. $u_{xx} = \dfrac{1}{2u}{u_x}^2 + \dfrac{3}{2}u^3 + 4xu^2 + 2(x^2 - \alpha)u + \dfrac{\beta}{u}$
5. $u_{xx} = \dfrac{3u - 1}{2u(u-1)}{u_x}^2 - \dfrac{1}{x}u_x + \dfrac{(u-1)^2}{x^2}(\alpha u + \dfrac{\beta}{u}) + \dfrac{\gamma u}{x} + \dfrac{\delta u(u+1)}{u-1}$

6. $$u_{xx} = \frac{1}{2}\left(\frac{1}{u} + \frac{1}{u-1} + \frac{1}{u-x}\right){u_x}^2 - \left(\frac{1}{x} + \frac{1}{x-1} + \frac{1}{u-x}\right)u_x + \frac{u(u-1)(u-x)}{x^2(x-1)^2}\left(\alpha + \beta\frac{x}{u^2} + \gamma\frac{x-1}{(u-1)^2} + \delta\frac{x(x-1)}{(u-x)^2}\right)$$

Here, α, β, γ, δ are complex parameters. During the rest of the twentieth century a great deal of facts about these equations were discovered: the structure of movable singularities, families of explicit solutions, their transformation properties, etc. In connection with these developments, we mention the works of N. P. Erugin, N. L. Lukashevich, A. I. Yablonsky, V. I. Gromak (see the reviews of N. P. Erugin [**Er**], and more recent works by A. S. Fokas and M. J. Ablowitz [**FoA3**], A. S. Fokas and Y. Yortsos [**FoY**], L. A. Bordag [**Bor2**], V. I. Gromak [**GroL1**], [**GroL2**]).

A new surge of interest in Painlevé functions occurred in the late 1970s after the pioneering work of M. J. Ablowitz and H. Segur [**AS**], and of B. M. McCoy, C. A. Tracy and T. T. Wu [**McTW**]. It was shown in [**AS**] that some of the Painlevé equations appear as exact ODE reductions of integrable nonlinear PDEs. Using this fact together with the powerful analytical technique of inverse scattering (and in particular of the Gelfand-Levitan-Marchenko equations), these authors were able to characterize, for the first time, classes of solutions which are not expressible in terms of elementary functions, as well as to solve a certain connection problem for this class of solutions. The paper [**McTW**] was the first rigorous study on the Painlevé connection formulae. In this work, for the first time, a complete connection formulae, in the case of a specific one-parameter family of solutions of the third Painlevé equations, was derived. This one-parameter family, as it was discovered in the earlier work of E. Barouch, B. M. McCoy, C. A. Tracy and T. T. Wu [**Bar**], plays a very important role in the theory of 2D Ising model.

At the same time, Painlevé equations began to appear in a wide range of physical applications[1]. It is now becoming clear that the Painlevé transcendents play the same role in nonlinear mathematical physics that the classical special functions, such as Airy functions, Bessel functions, etc., play in linear physics. During the last twenty to twenty-five years, great progress in the theory of Painlevé equations themselves has been achieved. Just as for their linear counterparts, it is now possible to derive explicit connection formulae for the Painlevé transcendents relating the relevant asymptotics parameters at different critical points. This fact, apparently unknown to P. Painlevé and his contemporaries, is based on the Isomonodromy Method. This method was introduced in 1980 by H. Flaschka and A.C. Newell [**FN1**], and by M. Jimbo, T. Miwa and K. Ueno [**JMU**], and it is based on the intrinsic relation of the Painlevé functions to *the monodromy theory of systems of linear ODE with rational coefficients.* It seems it was R. Garnier [**Gar1**] who first discovered this connection. Other classical references on this subject are the works by R. Fuchs [**Fu**] and by L. Schlesinger [**Sch**].

Let us outline (for details see the main text) the isomonodromy formulation of the Painlevé equations. Consider the generic case of a linear system with rational coefficients, i.e., the *Fuchsian system*,

$$\frac{d\Psi}{d\lambda} = \sum_{j=1}^{n} \frac{A_j}{\lambda - a_j}\Psi, \qquad \Psi, A_j - N \times N \text{ matrices.}$$

[1]The above mentioned paper [**Bar**] was perhaps the first example of these applications.

The *monodromy group* of this system is defined as a representation of the fundamental group of the punctured Riemann sphere (more exactly, a conjugate class of representations),

$$\rho : \pi_1(\mathbb{C}\mathrm{P}^1 \setminus \{a_1, \ldots, a_n, \infty\}) \mapsto \mathbf{GL}(N, \mathbb{C}),$$

generated by encircling the singular points $a_1, \ldots, a_n, a_\infty = \infty$:

$$\Psi(\lambda)|_{(\lambda - a_j)\mapsto(\lambda - a_j)e^{2\pi i}} = \Psi(\lambda)\mathsf{M}_j; \qquad \lambda - a_\infty \equiv \lambda^{-1},$$

$$\mathsf{M}_\infty \mathsf{M}_n \ldots \mathsf{M}_1 = I.$$

The matrices $\mathsf{M}_1, \ldots, \mathsf{M}_n, \mathsf{M}_\infty$ are called monodromy matrices, and the set $\mathfrak{m}=\{\mathsf{M}_1, \ldots, \mathsf{M}_n\}$, monodromy data. This set completely defines (up to a conjugation) the monodromy group $\mathfrak{M}$ of the above Fuchsian equation. Following [**JMU**] we will call the set $\mathbb{A} = \{a_1, \ldots, a_n; A_1, \ldots, A_n\}$ singular data of the Fuchsian system and the set $\mathbb{M} = \{a_1, \ldots, a_n; \mathsf{M}_1, \ldots, \mathsf{M}_n\} \equiv \{a_1, \ldots, a_n; \mathfrak{m}\}$ its extended monodromy data. We will also use the notations

$$\mathcal{A} \equiv \{\mathbb{A}\}, \quad \mathcal{M} \equiv \{\mathfrak{m}\}, \quad \text{and} \quad \mathcal{M}_e \equiv \{\mathbb{M}\},$$

for the sets of singular, monodromy, and extended monodromy data, respectively.

The associated *Riemann-Hilbert problem* consists of proving the existence of a Fuchsian system with given singular points and monodromy group, i.e., with the given set $\{a_1, \ldots, a_n; \mathfrak{M}\} \equiv \mathbb{M}$. More generally, one has to analyze the direct and inverse monodromy maps, i.e., the maps

$$\mathcal{A} \mapsto \mathcal{M}$$

and

$$\mathcal{M}_e \mapsto \mathcal{A}\ ,$$

respectively. This constitutes the central question of the *global* theory of Fuchsian systems.

The Riemann-Hilbert (RH) problem has a long and illustrious history. A comprehensive solution was obtained only relatively recently by A.A. Bolibruch [**Bol2**]. Rather than discussing this problem in general, we will only limit our discussions to how the monodromy problem can be used to study the Painlevé equations. In this respect we will first discuss the appearance of the Painlevé functions in the monodromy theory.

Let us study the above RH problem for the first nontrivial case, i.e., for the matrix size $N = 2$, increasing successively the number of singular points.

(1) TWO REGULAR SINGULAR POINTS, $n = 1$.
After proper gauge, scaling and conformal transformations, it can be shown that:
   (a) $\dim\mathcal{A} = \dim\mathcal{M}$.
   (b) Both the inverse and direct monodromy problems can be solved explicitly in terms of elementary functions (details - in Chapters 2 and 3).

(2) THREE REGULAR SINGULAR POINTS, $n = 2$.
In this case we again have:
   (a) $\dim\mathcal{A} = \dim\mathcal{M}$.

(b) Both the inverse and direct monodromy problems can be solved explicitly in terms of hypergeometric functions, i.e., in terms of elementary functions and of a finite number of contour integrals of these functions (details - in Chapters 2 and 3).

(3) FOUR REGULAR SINGULAR POINTS, $n = 3$.

In the generic situation, the system can be written as follows:

$$\frac{d\Psi}{d\lambda} = \left(\frac{A}{\lambda} + \frac{B}{\lambda - 1} + \frac{C}{\lambda - x}\right)\Psi \equiv \mathsf{A}(\lambda)\Psi, \tag{1}$$

with

$$\mathrm{tr}A = \mathrm{tr}B = \mathrm{tr}C = 0,$$
$$A + B + C = \mathrm{diag}\,(A + B + C).$$

Thus

$$\dim\mathcal{A} = 8.$$

In this case it can be shown that

$$\dim\mathcal{M} = 7 = \dim\mathcal{A} - 1.$$

Hence, one expects a one-parameter family of equations, i.e., a curve in the space $\mathcal{A}$, with a given monodromy group $\mathfrak{M}$. This is in fact the point where the Painlevé equations appear. Indeed, if one writes (cf. [**JMU**]) the (1,2)-th element of matrix $\mathsf{A}(\lambda)$ as

$$\mathsf{A}_{12}(\lambda) = \frac{w(\lambda - u)}{\lambda(\lambda - 1)(\lambda - x)},$$

then,

$$\mathfrak{m} \equiv \mathrm{const} \quad \Longleftrightarrow \quad u = u(x) \text{ satisfies the PVI equation.}$$

The proof of the last statement is based on the fundamental observation that the x-independence of the monodromy matrices $\mathsf{M}_1, \ldots, \mathsf{M}_3$, implies that the logarithmic derivative

$$\frac{d\Psi}{dx}\Psi^{-1}$$

is a rational function on the λ-plane which has (in the generic case) a simple pole at the point $\lambda = x$. In fact, one can see (details are given in Chapter 4) that

$$\frac{d\Psi}{dx}\Psi^{-1} = -\frac{C}{\lambda - x}.$$

Therefore, one concludes that the matrix function $\Psi \equiv \Psi(\lambda, x)$ satisfies the following overdetermined linear system (cf. system (4.2.6) of Chapter 4),

$$\begin{cases} \dfrac{\partial\Psi}{\partial\lambda} = \left(\dfrac{A}{\lambda} + \dfrac{B}{\lambda - 1} + \dfrac{C}{\lambda - x}\right)\Psi \equiv \mathsf{A}(\lambda)\Psi, \\ \dfrac{\partial\Psi}{\partial x} = -\dfrac{C}{\lambda - x}\Psi \equiv \mathsf{U}(\lambda)\Psi\,. \end{cases} \tag{2}$$

The compatibility condition of this system, i.e.,

$$\Psi_{\lambda x} = \Psi_{x\lambda},$$

implies

$$\mathsf{A}_x(\lambda) - \mathsf{U}_\lambda(\lambda) + [\mathsf{A}(\lambda), \mathsf{U}(\lambda)] = 0 \quad \text{(identically in } \lambda), \tag{3}$$

which in turn yields the following system of nonlinear ODEs on the matrices A, B, C (cf. system (4.2.8) of Chapter 4),

$$\left\{\begin{aligned} \frac{dA}{dx} &= \frac{1}{x}[C,A], \\ \frac{dB}{dx} &= \frac{1}{x-1}[C,B], \\ \frac{dC}{dx} &= -\frac{dA}{dx} - \frac{dB}{dx}. \end{aligned}\right. \tag{4}$$

Restricting the last system on its invariant orbit, which is determined by the relations

$$A+B+C = \begin{pmatrix} -\theta_\infty & 0 \\ 0 & \theta_\infty \end{pmatrix},$$

$$\text{the eigenvalues of } A,\, B,\, C \;=\; \pm\theta_0, \quad \pm\theta_1, \quad \pm\theta_3, \tag{5}$$

one arrives (details in [**JMU**], [**JM**]) at the Painlevé VI equation for the function $u(x)$, with the parameters

$$\alpha = \frac{1}{2}(2\theta_\infty - 1)^2, \quad \beta = -2\theta_0^2,$$

$$\gamma = 2\theta_1^2, \quad \delta = \frac{1}{2}(1 - 4\theta_3^2).$$

Although, it is technically nontrivial to pass from (4) to PVI, the fact that system (4) leads to a Painlevé equation can be predicted from general principles. Indeed, this system is a particular case of a *Schlesinger equation* (these equations will be discussed in detail in Chapter 4), which is known to posses the Painlevé property (which in turn follows from the meromorphic, w.r.t. x, solvability of the associated Riemann-Hilbert problem; see [**Mal**], [**Miw**]). Also the fact that the ODE for $u(x)$ is of second order can be predicted by counting the dimension of the spectral orbit (5).

According to the terminology of integrable systems (see [**AC**], [**FT**], [**NMPZ**]), the linear system (2) and the nonlinear equation (3), are the *Lax pair* and the *zero-curvature* (or *Lax*) *representation*, respectively, of the nonlinear ordinary differential equation PVI.

In order to put the other Painlevé equations into a similar context, we consider more general linear systems, namely systems with irregular singularities. The first nontrivial case is the system with only one irregular singularity of order 3 at $\lambda = \infty$:

$$\frac{d\Psi}{d\lambda} = \mathsf{A}(\lambda)\Psi, \qquad \mathsf{A}(\lambda) = A_3\lambda^2 + A_2\lambda + A_1, \qquad A_i \in \mathbf{Mat}(2,\mathbb{C})$$

(if $A_3 = 0$ the system can be solved in terms of the confluent hypergeometric functions; if $A_3 = A_2 = 0$, the system can be solved in terms of elementary functions). In the generic situation one can always reduce this system to the normal form

$$\mathsf{A}(\lambda) = -4i\lambda^2\sigma_3 + 4i\lambda \begin{pmatrix} 0 & u \\ -v & 0 \end{pmatrix} + \begin{pmatrix} a & -2w \\ -2z & -a \end{pmatrix}, \tag{6}$$

$$\sigma_3 = \begin{pmatrix} 1 & 0 \\ 0 & -1 \end{pmatrix}.$$

The corresponding *canonical solutions* $\Psi_k(\lambda)$ (details are in Chapters 1 and 2) in the neighborhood of $\lambda = \infty$ are characterized by their asymptotics

$$\Psi_k(\lambda) = (I + O(1/\lambda)) \exp\left\{-\frac{4}{3}i\lambda^3\sigma_3 - ix\lambda\sigma_3 - \nu\log\lambda\sigma_3\right\},$$

$$\frac{k-2}{3}\pi < \arg\lambda < \frac{\pi k}{3}, \qquad \lambda \to \infty,$$

$$k = 1, \ldots, 7,$$

where

$$x = ia - 2uv, \qquad \nu = vw - uz.$$

The monodromy data $\mathfrak{m}$ are given by the parameter ν and the set of *Stokes matrices* S_k defined as

$$\mathsf{S}_k = \Psi_k^{-1}\Psi_{k+1}, \qquad k = 1, \ldots, 6.$$

These matrices have a triangular structure,

$$\mathsf{S}_{2l} = \begin{pmatrix} 1 & s_{2l} \\ 0 & 1 \end{pmatrix}, \qquad \mathsf{S}_{2l+1} = \begin{pmatrix} 1 & 0 \\ s_{2l+1} & 1 \end{pmatrix},$$

and satisfy the cyclic relation

$$\mathsf{S}_1\mathsf{S}_2 \ldots \mathsf{S}_6 = e^{-2\pi i\nu\sigma_3}.$$

Therefore,

$$\dim\mathcal{M} = 4 = \dim\mathcal{A} - 1,$$

and again we expect nontrivial *isomonodromy deformations.* In fact (cf. [**FN1**], [**JMU**]), if one takes the quantity x as a parameter for the isomonodromy curve, so that u, v, z and w become functions of x, then the condition $\mathfrak{m} \equiv$ const. is equivalent to the system

$$(7) \qquad \begin{cases} w = u_x, & u_{xx} - xu - 2u^2v = 0\,, \\ z = v_x, & v_{xx} - xv - 2v^2u = 0\,. \end{cases}$$

Similarly to the Fuchsian case considered earlier, the proof of (7) is based on the fact that the x-independence of the Stokes matrices $\mathsf{S}_1, \ldots, \mathsf{S}_6$, implies that the function $\Psi \equiv \Psi_1$ satisfies the Lax pair (details in Chapter 4),

$$(8) \qquad \begin{cases} \dfrac{\partial\Psi}{\partial\lambda} = \mathsf{A}(\lambda)\Psi\,, \\ \dfrac{\partial\Psi}{\partial x} = \mathsf{U}(\lambda)\Psi\,, \end{cases}$$

where the matrix $\mathsf{A}(\lambda)$ is the one from (6), whereas the matrix $\mathsf{U}(\lambda)$ is given by the equation,

$$\mathsf{U}(\lambda) = -i\lambda\sigma_3 + i\lambda\begin{pmatrix} 0 & u \\ -v & 0 \end{pmatrix}. \tag{9}$$

This time it is straightforward to show that the associated zero-curvature relation,

$$\mathsf{A}_x(\lambda) - \mathsf{U}_\lambda(\lambda) + [\mathsf{A}(\lambda), \mathsf{U}(\lambda)] = 0 \quad \text{(identically in } \lambda\text{)}, \tag{10}$$

is equivalent to system (7).

Under the reduction,

$$u = v,$$

system (7) becomes a particular case ($\alpha = 0$) of the second Painlevé equation. The reduction $u = v$ implies the restrictions

$$s_{k+3} = -s_k, \quad \nu = 0,$$

so that

$$\mathcal{M} = \{(s_1, s_2, s_3) : s_2 - s_1 - s_3 - s_1 s_2 s_3 = 0\} \tag{11}$$

and hence

$$\dim\mathcal{M} = 2.$$

This means that for generic (s_1, s_2, s_3), one can take s_1 and s_3 as independent coordinates on $\mathcal{M}$.

In order to obtain the general Painlevé II equation it is necessary (cf. [**FN1**]) to add in (6) (under the reduction $u = v$, $w = z$) a regular singularity at $\lambda = 0$ of the form

$$-\frac{\alpha}{\lambda}\begin{pmatrix} 0 & -i \\ i & 0 \end{pmatrix}.$$

Also, there exists an elementary transformation which maps the general u–v system (7) into the general Painlevé II equation, whose parameter α is related to the first integral $\nu \equiv vu_x - uv_x$ of system (7); we will describe this map in Section 3.5 of Chapter 4.

Summarizing the above discussion, we can say that *the Painlevé transcendents mark the border beyond which both the inverse and the direct monodromy problems cannot be solved in terms of elementary functions and their contour integrals.* At the same time, this is the border beyond which the monodromy map ceases to be one-to-one.

The above relation was used actually by R. Garnier and L. Schlesinger for the solution of the inverse monodromy problem. Their scheme for the constructive solution of this inverse monodromy problem can be formulated as follows: Given the monodromy data, the coefficients of the corresponding linear system can be expressed in terms of the solution of the associated nonlinear equations of the isomonodromy deformations, i.e., in the first nontrivial cases — the Painlevé equations.

The principal question which remained open was: What are the initial data for the isomonodromy equations (in our examples for the PVI and PII equations)?

What apparently was missed in the classical works, was precisely the opposite point of view. Namely, the fact that the initial linear system and the associated RH problem can be used for solving the Painlevé equations themselves. The latter is the modern achievement, which was initiated in the works [**FN1**], [**JMU**] mentioned above, and which constitutes what is now called the *isomonodromy* or *Riemann-Hilbert method.* As an important by-product of this development, it is now possible to answer the above question in the classical scheme, namely it is possible to choose the correct initial data of the isomonodromy equations (see paper [**IN2**] and Theorem 2 below).

By the word "solving" in the previous paragraph, we mean the ability to perform a comprehensive *global* analysis of the Painlevé transcendents, i.e., to derive explicitly all relevant connection formulae. It also should be mentioned that, throughout the book, we accept the terminology, that an equation, a formula, an expression, etc. is called *explicit* if it is given in terms of a finite algebraic combination of elementary and elliptic functions, and a finite number of contour integrals (including quadratures) of these functions. In particular, all classical "linear" special functions admit explicit representations. The Painlevé transcendents do not. At the same time, the Painlevé transcendents, as we know now, admit explicit connection formulae. This fact, in our opinion, makes them as efficient in applications as the linear special functions.

In the second part of the book, two general approaches for solving the connection problem for the Painlevé equations will be presented. Both of them use the relation of the Painlevé transcendents to the RH problem. Due to J. Plemelj [**Pl**], the latter problem can be reformulated as a factorization problem (homogeneous Hilbert problem) in the theory of sectionally analytic matrix valued functions (see below). Using geometric language, this means that the Painlevé functions (and their natural generalizations), which can be called *nonlinear* special functions, can be analyzed within the theory of holomorphic vector bundles of rank > 1 over compact Riemann surfaces (of genus ≤ 1). In the same spirit, the classical *linear* special functions belong to the same theory, but for rank $= 1$ (the scalar factorization problem).

To explain in more detail the essence of the isomonodromy method let us consider the particular case, $\alpha = 0$, of the PII equation

$$u_{xx} - xu - 2u^3 = 0. \tag{12}$$

Reversing the arguments used earlier relating this equation to the isomonodromy deformations of the system (6), we can now say that the monodromy data,

$$s_1 = s_1(x, u, u_x), \qquad s_3 = s_3(x, u, u_x)$$

are the *first integrals* of equation (12). Formally this means that we have solved equation (12)! Indeed, we have found a complete set of independent first integrals. However, since these integrals are *not* explicit, the important question is: Can we make this fact an efficient tool for the study of the PII equation? It turns out that the answer is positive. Of course, one does not expect an explicit expression for the functions $s_1(x, u, w)$ and $s_3(x, u, w)$, since this would imply the possibility of expressing the second Painlevé transcendent in terms of classical special functions.

Nevertheless, one can calculate the functions s_1 and s_3 asymptotically. Assuming a certain kind of behavior of $u(x)$ as $x \to \pm\infty$, and applying an *extended version* of the classical WKB-technique to the λ-equation (2), it is possible to evaluate s_1 and s_3 asymptotically as $x \to \pm\infty$. However, since these functions are independent of x, this implies the connection formulae between the asymptotic parameters of the solution $u(x)$ at "$+\infty$" and "$-\infty$". Following this scheme, which will be explained in detail later, many global asymptotic results of the Painlevé equations have been obtained during the last twenty years (A. P. Bassom, P. A. Clarkson, M. Jimbo, A. R. Its, A. A. Kapaev, A. V. Kitaev, C. K. Law, B. M. McCoy, J. B. McLeod, V. Yu .Novokshenov, B. I. Suleimanov, Sh. Tang). Among these results are:

(a) The parametrization of all solutions by their asymptotic behaviour near critical points and the construction of the corresponding connection formulae.

(b) The global description of the asymptotic behaviour for complex x-nonlinear Stokes phenomenon.

(c) The determination of the distributions of zeros and poles.

The above scheme is based on the asymptotic analysis of the *direct monodromy problem* associated with a given Painlevé equation. It has the disadvantage that it requires the use of a priori information about the solution. An alternative approach, which remedies the above disadvantage is based on the asymptotic solution of the *inverse monodromy problem.* The starting point of this approach is, following ideas of Birkhoff [**Bir**], to consider the basic monodromy relation, $S_k = {\Psi_k}^{-1}\Psi_{k+1}$, as a jump condition for the sectionally analytic function $Y(\lambda)$:

$$(13) \qquad Y(\lambda) = \Psi_k(\lambda)e^{\theta(\lambda)\sigma_3} \equiv Y_k(\lambda), \qquad \theta(\lambda) = \frac{4i}{3}\lambda^3 + ix\lambda,$$

$$\frac{\pi}{3}(k-2) + \frac{\pi}{6} \le \arg\lambda \le \frac{\pi}{3}k - \frac{\pi}{6}, \quad k = 1, \dots, 6,$$

on the anti-Stokes rays, $\arg\lambda = \pi k/3 - \pi/6$, $k = 1, \dots, 6$. This gives rise to a special type of oscillatory Riemann-Hilbert problem, which is now understood as a *factorization or jump problem*:

$$(14) \qquad Y_{k+1}(\lambda) = Y_k(\lambda)e^{-\theta(\lambda)\sigma_3}S_k e^{\theta(\lambda)\sigma_3},$$

$$\arg\lambda = \frac{\pi k}{3} - \frac{\pi}{6}, \qquad k = 1, \dots, 6,$$

$$\lim_{\lambda\to\infty} Y(\lambda) = I.$$

(The last normalization equation is actually the reason for the introduction of the function $Y(\lambda)$.)

The problem (14) is depicted in Figure I. The corresponding solution $u(x, s_1, s_3)$ of the second Painlevé equation (12) can be determined from $Y(\lambda)$ by the equation

$$(15) \qquad u(x; s_1, s_3) = 2\lim_{\lambda\to\infty}(\lambda Y_{12}(\lambda)).$$

The problem (14), as well as the Fuchsian inverse monodromy problem (which also can be reformulated as a factorization problem — see Chapter 3), are particular cases of a *Riemann-Hilbert factorization problem.* This problem involves finding an analytic (matrix valued) function having prescribed jumps across a given contour. Following the tradition developed in mathematical physics, it is such more general

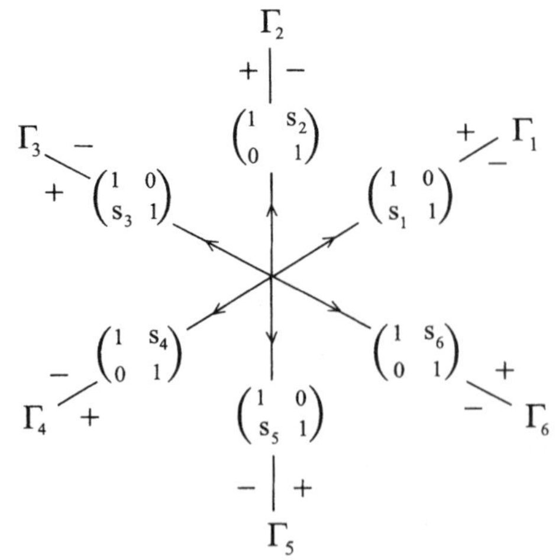

FIGURE I. The contour and jump matrices for the RH problem (14)

factorization problems (and not just the inverse monodromy Fuchsian and/or the Birkhoff problems) that we will call *Riemann-Hilbert problems.*[2]

The investigation of the Cauchy problem for the Painlevé II equation based on the analysis of the RH problem (14), as well as the investigation of its discontinuous generalization for $\alpha \neq 0$ (complete Painlevé II equation) was initiated by A. S. Fokas and M. J. Ablowitz in [**FoA3**], [**FoA1**]. The rigorous theory was developed by A. S. Fokas and X. Zhou in [**FoZ**] for PII and PIV, and was later applied to PI, PIII and PV in [**FoMA**]. In the works of Fokas, Mugan, Ablowitz and Zhou (see the review paper [**FoI1**]) the first rigorous proofs were given (based on the isomonodromy method) that all solutions of PI – PV possess the Painlevé-property. Furthermore, it was shown that if the monodromy data satisfy explicit constraints, then the solutions are free from poles (on the real line).

The last basic step was taken in the works of P. Deift and X. Zhou [**DeZ1**], [**DeZ2**], who introduced an elegant scheme of the asymptotic analysis of this type of oscillatory factorization problems; their method can be thought of as a *nonlinear steepest descent method.* The asymptotic method of Deift and Zhou, in complete analogy with the classical method, exploiting the analytic structure of the relevant jump matrices deforms the original contours of the RH problem to contours where the relevant oscillatory factors become exponentially small as $x \to \infty$. In this way, the original RH problem reduces to a collection of local RH problems associated with the relevant saddle points. However, the noncommutative nature of the RH problem requires the development of several totally new and rather sophisticated technical ideas, which, in particular, allow us to solve the local RH problems in closed form. The remarkable fact is that the final result of the analysis is as efficient as the asymptotic evaluation of the classical oscillatory integrals.

It is worth mentioning that the first step towards the direct asymptotic analysis of oscillatory RH problems was made in 1973 by S. V. Manakov [**Mn**] in connection

[2]It should be mentioned that in the theory of boundary value problems for analytic functions, the problem of reconstructing a function from its jumps across a curve is sometimes called the "Hilbert boundary-value problem".

with the theory of integrable PDEs. Manakov's approach was further developed in [**I2**], [**I3**], where certain aspects of the nonlinear steepest descent method (e.g. the analysis and the explicit solution of some basic local RH problems) already existed.

Asymptotic results

In order to give a flavor of the type of results which are contained in this book, we first present the complete asymptotic description of the purely imaginary solutions of the PII equation, x real, $\alpha = 0$, as $x \to \pm\infty$.[3]

THEOREM 1. *An arbitrary purely imaginary solution $u(x)$ of the second Painlevé equation (12), has the following asymptotic behavior as $x \to -\infty$:*

$$u(x) = i(-x)^{-1/4} d \sin\left\{\frac{2}{3}(-x)^{3/2} + \frac{3}{4}d^2 \log(-x) + \varphi\right\} + o((-x)^{-1/4}). \tag{16}$$

Here the constants $d > 0$ and $\varphi \in \mathbb{R}$ (mod 2π) can be any real numbers; they determine the solution $u(x)$ uniquely and hence form a set $\mathbf{a}_- \equiv (d, \varphi)$ of asymptotic parameters at $-\infty$.

Let

$$\Delta(d,\varphi) \equiv \frac{3}{2}d^2 \log 2 - \frac{\pi}{4} - \arg\Gamma\left(i\frac{d^2}{2}\right) - \varphi,$$

where $\Gamma(z)$ denotes Euler's gamma function. The behavior of $u(x) \equiv u(x; d, \varphi)$ as $x \to +\infty$ depends on the value of $\Delta(d,\varphi)$. If (generic case)

$$\Delta(d,\varphi) \neq 0 \quad (\text{mod } \pi), \tag{17}$$

then

$$u(x) = \sigma i\sqrt{\frac{x}{2}} + \sigma i(2x)^{-1/4}\rho\cos\left\{\frac{2\sqrt{2}}{3}x^{3/2} - \frac{3}{2}\rho^2 \log x + \theta\right\} + o(x^{-1/4}), \tag{18}$$

$$\text{as } x \to +\infty,$$

where $\sigma = \pm$, $\rho > 0$, and $\theta \in \mathbb{R}$ (mod 2π). If instead,

$$\Delta(d,\varphi) = 0 \quad (\text{mod } \pi), \tag{19}$$

then

$$u(x) = \sigma i\frac{\rho}{2\sqrt{\pi}}x^{-1/4}e^{-(2/3)x^{3/2}}(1 + o(1)), \quad x \to +\infty, \tag{20}$$

where $\sigma = \pm$ and $\rho > 0$. The set $\mathbf{a}_+$ of asymptotic parameters at $+\infty$ is the triple (ρ, θ, σ) in the generic situation and the pair (ρ, σ) in the special case.

The following explicit connection formulae characterize the map $\mathbf{a}_- \mapsto \mathbf{a}_+$:

$$\begin{aligned} \rho^2 &= d^2 - \frac{1}{\pi}\log 2(e^{\pi d^2} - 1)^{1/2}|\sin\Delta(d,\varphi)|, \\ \theta &= -\frac{3\pi}{4} - \frac{7}{2}\rho^2 \log 2 + \arg\Gamma(i\rho^2) + \arg\left(1 + (e^{\pi d^2} - 1)e^{2i\Delta(d,\varphi)}\right), \\ \sigma &= -\text{sign}\,(\sin\Delta(d,\varphi)), \end{aligned} \tag{21}$$

if $\Delta(d,\varphi) \neq 0$ (mod π), and

[3]By letting $u \to iu$ in (12), the results stated below are transformed to the complete asymptotic description of the real solutions of the equation $u_{xx} - xu + 2u^3 = 0$, x real, as $x \to \pm\infty$.

$$\rho = (e^{\pi d^2} - 1)^{1/2}, \quad \sigma = (-1)^n \tag{22}$$

if $\Delta(d, \varphi) = n\pi$.

Here are some remarks about Theorem 1.

- A detailed proof of the theorem is given in Part II of the book.
- Connection formulae for the one-parameter family of solutions (19) were first obtained in 1977 in [**AS**].
- Connection formulae for the generic case were first obtained in 1986 in [**IK1**]; some technical gaps in the original proof were filled later in the paper [**IFoK**]. Another proof of the theorem based on the nonlinear steepest descent method was obtained in [**DeZ2**].
- Some parts of the theorem can be obtained without using the Riemann-Hilbert formalism. This is true for the existence, for a given pair (d, φ), of a solution $u(x)$ with the asymptotics (16) (see [**Abd**]), or with the asymptotics (18) and (20). Since these are *local* statements, they do not reflect the integrability of the second Painlevé equation. The *global* fact that formulae (16) and (18)–(20) describe *all* possible types of asymptotic behavior of the purely imaginary solutions of the second Painlevé equation (12) as $x \to \pm\infty$, can also be proved without appealing to the Riemann-Hilbert representation (see [**JoKr1**]), but one has to make use of the integrability of equation (12); indeed, the arguments of [**JoKr1**] exploit in an essential way the Painlevé property.

  Without using the Riemann-Hilbert formalism, it does not seem feasible to obtain the parts of the theorem concerning the bifurcation condition (17) and the connection formulae (21)–(22).
- In the limit of small d the Ablowitz-Segur connection formulae (19), (22) become the classical Airy connection formulae.

We would like to emphasize that it is precisely the connection formulae that make the Airy functions so useful in applications. Theorem 1 shows that in the nonlinear case of PII, the associated connection formulae, though more complicated, are still explicit. In the Tables 1 and 2 below, we present similar results for PIII (the relevant linear counterparts are the Bessel functions).

Because u is imaginary, there exists only one independent monodromy datum $s_1 = -\bar{s}_3 = s$. As we will see in Chapter 9, the proof of Theorem 1, also provides the explicit representations of the monodromy parameter s in terms of the asymptotic parameters d, φ and ρ, θ associated with the "nonlinear" Stokes rays, $\arg x = \pi, 0$. For instance, the formula $s = s(d, \varphi)$ reads as follows,

$$s = (e^{\pi d^2} - 1)^{1/2} \exp\{\Delta(d, \varphi)\}. \tag{23}$$

It is important to emphasize that it is also possible to obtain similar results for the remaining Stokes lines, i.e., for the rays $\arg x = \pm\pi/3$ and $\arg x = \pm 2\pi/3$ (see paper [**K5**] by A. A. Kapaev). The formulae presented in Theorem 1 and the formulae of Kapaev show that the asymptotic behavior of PII on the Stokes lines is characterized by elementary (trigonometric) functions.

TABLE 1. Connection formulas (oscillating sector)

Bessel functions	Third Painlevé functions
$u_{xx} + \frac{1}{x}u_x + u = 0$	$u_{xx} + \frac{1}{x}u_x + \sin u = 0$
$u(x;s): \quad (sJ_0(x))$ $u(0;s) = s$ $u_x(0;s) = 0$	$u(x;s):$ $u(0;s) = s$ $u_x(0;s) = 0$
$x \to +\infty$ $u(\|s) \cong \alpha x^{-1/2}\cos(x+\varphi)$ $\alpha = s\sqrt{\frac{2}{\pi}}$ $\varphi = -\frac{\pi}{4}$	$x \to +\infty$ (V.Yu. Novokshenov [**N2**]) $u(\|s) \cong 2\pi k + \alpha x^{-1/2}\cos(x - \frac{\alpha^2}{16}\log x + \varphi)$ $\alpha^2 = -\frac{16}{\pi}\log\|\cos\frac{s}{2}\|$ $\varphi = -\frac{\alpha^2}{8}\log 2 - \arg\Gamma(-\frac{i\alpha^2}{16}) + \frac{\pi}{4}$ $k = \mathit{entier}\,\{\frac{1}{2\pi}[\pi - s]\}$

TABLE 2. Connection formulas (decreasing sector)

Bessel functions	Third Painlevé functions
$u_{xx} + \frac{1}{x}u_x - u = 0$	$u_{xx} + \frac{1}{x}u_x - \sinh u = 0$
$u(x\|\alpha): \quad (\alpha K_0(x))$ $x \to +\infty$ $u(0\|\alpha) \cong \alpha\sqrt{\pi}2x^{-1/2}e^{-x}$	$u(x\|\alpha):$ $x \to +\infty$ $u(0\|\alpha) \cong \alpha\sqrt{\pi}2x^{-1/2}e^{-x}$
$x \to 0$ $u(0\|\alpha) \cong r\log x + s$ $r = -\alpha$ $s = -r(\log 2 + \gamma)$	$x \to 0$ (B.McCoy, T.Wu, C.A.Tracy [**McTW**]) $u(0\|\alpha) \cong r\log x + s$ $\alpha = -\frac{4}{\pi}\sin\frac{\pi r}{4}$ $e^s = 2^{-3r}\Gamma^2(\frac{1}{2} - \frac{r}{4})\Gamma^{-2}(\frac{1}{2} + \frac{r}{4})$

It turns out that if x is not on the Stokes lines, then the asymptotic behavior is more complicated and is characterized by elliptic functions. It is shown by V. Yu. Novokshenov [**N1**] and A. A. Kapaev [**K5**] (see also Chapters 7 and 8 of the book) that for $u(x;s_1,s_3)$ the asymptotics at infinity in the sectors $\frac{\pi}{3}j < \arg x < \frac{\pi}{3}(j+1)$,

$j = 0, 1, \ldots, 5$, is given by the following formula:

$$\begin{aligned}(24)\ u(x) &= e^{\frac{2\pi i}{3}n}|x|^{1/2}(\xi_1 - \xi_2) \times \\ &\mathrm{sn}\left\{(\xi_1+\xi_2)\left(e^{i\varphi+\frac{2\pi i}{3}n}\frac{2}{3}|x|^{3/2} + \chi_{n,k}\right)\Big|\frac{\xi_1-\xi_2}{\xi_1+\xi_2}\right\} + O(|x|^{-1}),\end{aligned}$$

where

$$\frac{\pi}{6}(7-k) - \frac{2\pi}{3}n < \varphi < \frac{\pi}{6}(9-k) - \frac{2\pi}{3}n, \qquad |x| \to \infty, \qquad x \in \{\arg x = \varphi\} \cap D_\varepsilon,$$

$$n = 0, \pm 1, \ldots, \qquad k = 1, 3,$$

and D_ε is the complement to the union of the ε-neighborhoods of all the poles of the elliptic function appearing in (24). The module parameters, $\xi_{1,2} \equiv \xi_{1,2}(\varphi)$, of the elliptic sine are given by the equations (Boutroux equations)

$$(25)\ \ {\xi_1}^2 + {\xi_2}^2 = -\frac{1}{2}e^{i\varphi+\frac{2\pi i}{3}n}, \qquad \mathrm{Im}\left(\int_{\xi_2}^{\xi_1}, \int_{-\xi_1}^{\xi_1}\right)\sqrt{(\xi^2-\xi_1^2)(\xi^2-\xi_2^2)}d\xi = 0\,.$$

The connection formulae between the asymptotic phases $\chi_{n,k}$, and the monodromy data s, are given by

$$(26) \qquad \chi_{n,k} = \chi_{n,k}(s_1, s_3) = \frac{\omega_2}{2\pi i}(-1)^{\frac{k+1}{2}}\log s_{k+2n} + \frac{\omega_1}{2\pi i}\log(1 - s_{1+2n}s_{3+2n}),$$

$$\omega_{1,2} = \left(i\int_{-\xi_1}^{\xi_1}, 2i\int_{\xi_2}^{\xi_1}\right)\frac{d\xi}{\sqrt{(\xi^2-\xi_1^2)(\xi^2-\xi_2^2)}},$$

where the monodromy data satisfy

$$s_{k+3} = -s_k, \qquad s_1 - s_2 + s_3 + s_1s_2s_3 = 0, \qquad -\overline{s_3} = s_1 \equiv s.$$

The elliptic asymptotics of this type for the Painlevé I equation, was first obtained by P. Boutroux in 1913 [**Bou**], but without the phase connection formulae.

Formulae (24-26) are actually valid for the general complex valued solution of the Painlevé II equation (without the restriction that u is purely imaginary). Also, in [**K5**] the general case of the second Painlevé equation is considered (without the restriction $\alpha = 0$). The complete connection formulae for the elliptic asymptotics of Painlevé I equation are given in papers [**Ki7**], [**KK2**].

The connection formulae valid for $\arg x = 0, \pm\pi/3, \pm 2\pi/3, \pi$, together with the formulae (26) for the complex phases $\chi_{n,k}$, provide an explicit description of the *nonlinear Stokes phenomenon* for the second Painlevé equation. Indeed, knowing the asymptotic behaviour of the solution $u(x)$ on one of the rays, one can calculate the corresponding monodromy parameter s, and hence obtain complete information about the asymptotic behaviour of $u(x)$ everywhere on the complex x-plane.

The connection formulae can be used to fill the gap of the classical approach to the inverse monodromy problem mentioned earlier, namely these formulae specify the proper initial conditions for the isomonodromy equations [**IN2**].

THEOREM 2. *Consider the RH problem, i.e., the inverse monodromy problem (IMP) for the system (6),*

$$\{S_k, \nu, x\} \to \{u, v, w, z, x\}$$

and assume that

(27) $$s_{k+3} = -s_k, \quad \nu = 0,$$

(28) $$-\overline{s_3} = s_1 \equiv s.$$

Then, for all real x the IMP is uniquely solvable and the solution is given by

$$v = u = u(x,s),\ y = w = u_x(x,s),$$

where $u(x,s)$ is the solution of the second Painlevé equation,

$$u_{xx} = xu + 2u^3$$

specified by the following asymptotic condition ("initial" data):

$$\begin{aligned} u(x) &= i(-x)^{-1/4} d \sin\left\{\frac{2}{3}(-x)^{3/2} + \frac{3}{4}d^2\log(-x) + \varphi\right\} \\ &+ o((-x)^{-1/4}), \quad x \to -\infty, \end{aligned} \tag{29}$$

where (cf. (23))

(30) $$d^2 = \frac{1}{\pi}\log(1+|s|^2), \quad d > 0,$$

(31) $$\varphi = \frac{3}{2}d^2\log 2 - \frac{\pi}{4} - \arg\Gamma(i\frac{d^2}{2}) - \arg s.$$

The analogous result for the Fuchsian system (1), i.e., an explicit expression of the initial data for the sixth Painlevé equation in terms of the given monodromy data, was obtained for the generic case by M. Jimbo [**J1**]. Important special cases, not covered by Jimbo's results, were obtained in the recent works of B. Dubrovin and M. Mazzocco [**DM**] and D. Guzzetti [**Guz1**].

If instead of (28), the restriction

(32) $$\overline{s_3} = s_1 \equiv s$$

is imposed, then results similar to the statement of Theorem 2 are valid but only for $x < 0$, $|x|$ sufficiently large, and $|s| < 1$.

If in addition to the restriction (32), the conditions

(33) $$|s| > 1, \quad \mathrm{Re}s \neq 0,$$

hold, then the Painlevé transcendent $u(x,s)$ has sequences $\{x_n{}^{\pm}\}$ of real poles which accumulate at $x = \pm\infty$. The asymptotic behaviour of $x_n{}^{\pm}$ as $n \to \infty$ can also be described explicitly in terms of s [**K1**], [**KN**]:

$$\begin{aligned} (x_n^+)^{3/2} &= \frac{3}{\sqrt{2}}\pi n - \frac{3}{2\sqrt{2}}\mu^+\log\frac{3}{2\sqrt{2}}\pi n - \frac{21}{4\sqrt{2}}\mu^+\log 2 - \frac{3}{2\sqrt{2}}\arg\Gamma(1/2 - i\mu^+) \\ &- \frac{3}{2\sqrt{2}}\arg\frac{1+s^2}{1-|s|} + o(1), \end{aligned} \tag{34}$$

$$n \to \infty,\ \mu^+ = \frac{1}{\pi}\log\frac{2|\mathrm{Re}s|}{|s|^2-1}$$

and

$$(-x_n^-)^{3/2} = 3\pi n - \frac{3}{2}\mu^-\log 3\pi n - \frac{9}{2}\mu^-\log 2 + \frac{3}{2}\arg\Gamma(1/2 + i\mu^-)$$

(35)
$$-\frac{3\pi}{4}(1-(-1)^{n+n_0})+\frac{3}{2}\arg s+o(1),$$
$$n\to\infty,\ \mu^- = \frac{1}{2}\pi\log(|s|^2-1).$$

It is shown in [**IN2**] that the poles of $u(x,s)$ give the values of x for which the RH problem for the system (6) is not solvable. Equations (34) and (35) give the asymptotic distribution of these values on the real axis in the case that the monodromy data satisfy the restrictions (32) and (33). At the same time, these equations give, in closed form, the asymptotic distribution of the real values of x for which the holomorphic bundle over $\mathbb{C}\mathrm{P}^1$ defined by the associated Stokes cocycle (see paper [**Si1**]),
$$S_k(\lambda) = \mathrm{e}^{\{-i\frac{4}{3}\lambda^3-ix\lambda\}\sigma_3}S_k\mathrm{e}^{\{i\frac{4}{3}\lambda^3+ix\lambda\}\sigma_3}$$
is not trivial.

When (33) is replaced by,

(36)
$$|s|=1,\quad \mathrm{Re}s=0,$$
i.e., when $s=\pm i$, then for real x the solution becomes smooth. It decays exponentially as $x\to+\infty$, and it grows as $\mp\sqrt{-x}$ as $x\to-\infty$. This particular (but still transcendental) solution, which was first singled out and studied by S. P. Hastings and J. B. McLeod [**HasM**], plays now an important role in the theory of random matrices (it determines the so-called Tracy-Widom distributions; see the applications to random matrices discussed later in the introduction).

Regarding asymptotic results we also mention the following:

(a) The asymptotics of particular families of solutions of Painlevé equations can be investigated rigorously by linear integral equations of the Gel'fand-Levitan-Marchenko type (P. A. Clarkson, J. B. McLeod, [**ClM**]).
(b) Important facts (but not explicit connection formulae) about the complex asymptotics of the Painlevé functions can be obtained without using the isomonodromy method (N. Joshi and M. D. Kruskal [**JoKr1**]).
(c) An improved version of the direct monodromy asymptotic approach has been developed by Bassom, Clarkson, Law and McLeod in [**BCLM**].
(d) The direct monodromy asymptotic approach outlined above can be thought of as the ODE version of the Zakharov-Manakov scheme [**ZM**], which was developed for integrable PDEs.
(e) An alternative approach to the study of the elliptic asymptotics for the Painlevé equations, which uses a combination of the isomonodromy, algebrogeometric, and Bogolyubov-Whitham techniques, has been developed by S. P. Novikov and P. Grinevich in [**GN**].
(f) The asymptotics of some important special families of the third and fifth Painlevé equations (and their higher order generalizations) have been studied via the Fredholm operator techniques by C. Tracy and H. Widom [**TW3**], [**TW5**], and by E. Basor [**Bas**]. This approach, in particular, allows one to evaluate exactly the important constants of integration appearing in the applications of the Painlevé transcendents.
(g) A study of the various connection formulae for the first, third, fourth and fifth Painlevé equations via the direct isomonodromy approach has been carried out by the following authors: for Painlevé I, by A. A. Kapaev and

A. V. Kitaev [**K2**], [**KK2**], [**Ki7**]; for Painlevé III, by V. Yu. Novokshenov [**N2**], [**N3**], A. V. Kitaev [**Ki3**], [**Ki4**], A. I. Bobenko and A. R. Its [**BobI**], V. Yu. Novokshenov and A. G. Shagalov [**N5**], [**N6**], and by A. V. Kitaev and A. H. Vartanian [**KiV**]; for Painlevé IV, by A. V. Kitaev [**Ki2**], [**Ki5**], A. A. Kapaev [**K6**], and by A. R. Its and A. A. Kapaev [**IK2**]; and for Painlevé V, by B. M. McCoy and Sh. Tang [**McT1**], [**McT2**], and by F. V. Andreev and A. V. Kitaev [**AnK2**].

(h) The asymptotics and connection formulae for the sixth Painlevé equation were evaluated via the isomonodromy approach for the generic case by M. Jimbo [**J1**], and for important special cases, not covered by Jimbo's results, by B. Dubrovin and M. Mazzocco [**DM**] and D. Guzzetti [**Guz1**].

(i) The first connection formulae for specific families of the Painlevé transcendents were obtained in [**AS**] (the one-parameter family (19) of the solutions of the second Painlevé equation presented in Theorem 1) and in [**McTW**] (a one-parameter family of solutions of the third Painlevé equation arising in the 2D Ising model and presented in Table 2). The authors of [**AS**] used the results of [**ZM**] and the fact (first discovered by the authors of [**AS**]) that the second Painlevé equation is a self-similar reduction of the KdV equation. The Ablowitz-Segur connection formulae were rigorously justified later in the works [**Su2**] and [**ClM**]. The work of B. M. McCoy, C. A. Tracy and T. T. Wu [**McTW**] is actually the first rigorous work on the Painlevé connection formulae. Remarkably, it was done before the discovery of the Riemann-Hilbert formalism for Painlevé equations. However, there is an important specific technical point used by the above authors for the one-parameter family considered in [**McTW**]; this family admits a Fredholm determinant representation, which, in a sense, is a "shadow" of the Riemann-Hilbert formalism.

(j) Among the pioneering studies of the connection formulae for Painlevé equations, the works of J. W. Miles [**Mil1**], [**Mil2**] must be mentioned. These are the works where the view on the Painlevé transcendents as the *nonlinear special functions* was stressed, perhaps, for the first time.

Applications

We conclude our brief review of the modern theory of Painlevé functions, by mentioning some of their typical physical applications. The first group of these applications is related to the fact that all reductions of integrable nonlinear PDEs, and also some ODE reductions of some nonintegrable PDEs, are Painlevé equations. For example, ODE reductions of Korteweg-de Vries equation yield PI and PII, of nonlinear Schrödinger equation — PIV, of Ernst equation – PVI, of sine-Gordon equation — PIII, etc. Some of these ODEs play a crucial role in the asymptotic analysis of the corresponding PDEs. For example, it was shown by A. S. Fokas, D. Kaup and C. Menyuk [**FoM**], [**KaM**] that a certain important asymptotic limit of the PDEs describing stimulated Raman scattering is provided by a particular class of solutions of PIII.

Another impressive field of applications is exactly solvable models in statistical physics and in quantum field theory. Among field-theoretical problems which can be solved in terms of the Painlevé transcendents we mention the following:

1. The two-point correlation function for the two-dimensional Ising model (E.Barouch, B.M. McCoy, C.A. Tracy, T.T. Wu [**Bar**], [**WTMTB**], [**McTW**]).
2. The two-point correlation functions at zero temperature for the one-dimensional impenetrable Bose-gas (M. Jimbo, T. Miwa, Y. Mari, M. Sato [**JMMS**]).
3. The two-point correlation functions at zero temperature for the one-dimensional isotropic XY-model (B. M. McCoy, J.H.H. Perk, R.E. Shrock, H.G. Vaidya, C.A. Tracy [**McPS**], [**VT**]).
4. The correlation functions in some topological field theories (C. Vafa, S. Cecotti [**CeVa1**], B. Dubrovin [**D**]).
5. The partition function of 2D quantum gravity (E. Brézin, V. Kazakov [**BreKaz**], M. Douglas, S. Shenker [**DoS**], D. Gross, A. Migdal [**GrMi**]).

A very interesting (as well as "visual") application of the modern theory of Painlevé equations is the differential geometry of surfaces (see the monograph of A. I. Bobenko and U. Eitner [**BobEi2**]). More recent applications include:

(1) Orthogonal polynomials and random matrices.
(2) Random permutations, growth processes, last passage percolation.
(3) Number theory (the distribution of zeros of zeta-function).
(4) String Theory.

An intriguing appearance of the second Painlevé function occurs in recent studies of the classical Hele-Shaw problem (A. S. Fokas and S. Tanveer [**FoT**]), where it seems that for the first time the properties of the Painlevé function on the *complex plane* were used. Another exciting example, which is also related to the Hele-Shaw problem, is the appearance of the Painlevé functions in the analysis of the normal matrix ensembles (see the series of works of A. Zabrodin, P. Wiegmann and their collaborators [**TBAZ**] and references therein).

In what follows we give a brief outline of some specific applications.

THREE-DIMENSIONAL WAVE COLLAPSE AND PII.

The so-called strong wave collapse in the three-dimensional nonlinear Schrödinger equation,

$$iU_t + \frac{1}{2} \triangle U + |U|^2 U = 0, \quad x \in \mathbb{R}^3 \tag{37}$$

(condition for the existing of collapse: $H \equiv \int_{R^3}(| \bigtriangledown U|^2 - |U|^4)dx < 0$) was first analyzed in detail by V.E. Zakharov, E.A. Kuznetsov, and S.L. Musher [**ZK**]. In particular, they discovered that the corresponding transition mode can be described in terms of the second Painlevé transcendent. The analysis of [**ZK**] had been completed by V. Yu. Novokshenov who calculated the uniform asymptotics of the corresponding solution of equation (37) using the modern theory of PII. Below we present this result following Chapter 15 of the monograph [**IN1**].

Suppose that collapse occurs at the origin $x = 0$, at $t = t_0 > 0$. Suppose also, that $U = U(r,t)$, $r = \sqrt{x_1^2 + x_2^2 + x_3^2}$. The aim is to find an asymptotic expansion for $U(r,t)$ uniform in r as $t \to t_0$. As usual, we have an inner, an outer and a transition regions.

THE INNER EXPANSION

$$U_{in} = \exp\left\{i\tau^{-1/5}\Phi(\xi)\right\}\cdot\left[\tau^{-3/5}f(\xi) + \tau^{-1/2}\alpha(\xi)\cos\theta(\xi,\tau) + \dots\right],$$
$$\tau \to 0,$$
$$\xi = \frac{r}{l\tau^{2/5}},\ \tau = t_0 - t > 0,\ \Phi(\xi) = \frac{3}{5}l^2\xi_c - \frac{1}{5}l^2\xi,$$
$$f(\xi) = \frac{l}{5}\sqrt{3(\xi_c^2 - \xi^2)},\quad \xi < \xi_c,$$
$$\Theta(\xi,\tau) = \tau^{-1/5}(\xi_c - \xi)^{3/2}\left(-\frac{4l^2}{5\sqrt{3}}\sqrt{2\xi_c}\right)$$
$$-\frac{C^2}{\sqrt{6\xi_c}}\log\tau + \frac{5C^2}{4\sqrt{\xi_C}}\sqrt{6}\log(\xi_c - \xi) + \beta_0 + \dots$$
$$\alpha(\xi) = \frac{C}{2}(\xi_c - \xi)^{-1/4} + \dots,\quad \xi \to \xi_c - 0.$$

The above expansion implies the parameters : $(l,\ \xi_c,\ C,\ \beta_0)$.

THE OUTER EXPANSION

$$U_{out} = \exp\left\{i\tau^{-1/5}\Phi(\xi)\right\}\cdot\left[\tau^{-1/2}\tilde{\alpha}(\xi)\sin\tilde{\theta}(\xi,\tau) + \dots\right],$$
$$\tau \to 0,\quad \tau > 0,\quad \xi > \xi_c,$$
$$\tilde{\Theta}(\xi,\tau) = \frac{4l^2}{5\sqrt{3}}\sqrt{\xi_c}(\xi - \xi_c)^{3/2}$$
$$-\frac{\tilde{C}^2}{4\sqrt{3\xi_c}}\log\tau + \frac{5\tilde{C}^2}{8\sqrt{\xi_c}}\sqrt{3}\log(\xi_c - \xi) + \tilde{\beta}_0 + \dots$$
$$\alpha(\xi) = \frac{\tilde{C}}{2}(\xi - \xi_c)^{-1/4} + \dots,\quad \xi \to \xi_c + 0.$$

The above expansion implies the parameters : $(l,\ \xi_c,\ \tilde{C},\ \tilde{\beta}_0)$.

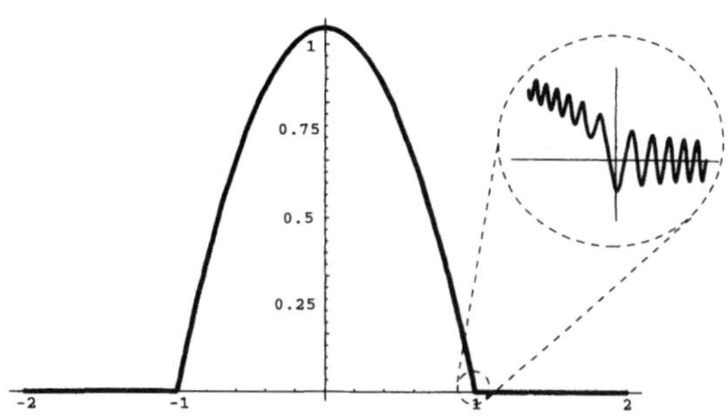

FIGURE II. The matching of the core parabolic profile with the zero tail by the transition mode expressed in terms of a solution of PII

TRANSITION FORMULAE

$$U_{trans} \cong \tau^{-8/15}u(\zeta)\exp\left\{i\tau^{-1/5}\Phi(\xi)\right\},$$

(38)
$$\frac{6}{25}\xi_c l^2\zeta u + \frac{1}{2l^2}u_{\zeta\zeta} + u^3 = 0,$$

$$\xi = \xi_c + \tau^{2/15}\zeta, \quad \text{transition region}: \quad \xi_c - \xi = O(\tau^{2/15-\epsilon}).$$

Introducing the new variables,

$$\zeta = -\left(\frac{12l^4}{25}\xi_c\right)\eta^{-1/3}, \quad u = \left(\frac{12l}{25}\xi_c\right)^{1/3} v(\zeta),$$

we can rewrite (37) in the canonical form of the second Painlevé equation,

$$v_{\eta\eta} = \eta v - 2v^3.$$

The known connection formulae for PII (use Theorem 1 as well as the transformation $u \to v = iu$, which changes the sign before the nonlinearity) allow us to connect the inner and outer parameters,

$$(C,\ \beta_0) \leftrightarrow (\tilde{C},\ \tilde{\beta}_0). \tag{39}$$

For example, denoting

$$b \doteq C\sqrt{\frac{5}{2}}(3\xi_c)^{-1/4}, \quad a \doteq \tilde{C}\sqrt{\frac{5}{2}}(\xi_c)^{-1/4},$$

we find

$$b^2 = \frac{3a^2}{4\sqrt{2}} - \frac{1}{2\pi\sqrt{2}}\log 2\sinh\frac{\pi a^2}{2} -$$

$$\frac{1}{\pi\sqrt{2}}\log\left|2\sin\left\{\frac{3}{2}a^2\log 2 + \frac{\pi}{4} - \arg\Gamma(\frac{ia^2}{2}) - \tilde{\beta}_0 + \frac{a^2}{4}\log(\frac{12l^4}{25}\xi_c)\right\}\right|.$$

Finally, using the conservation laws,

$$E = \int_{R^3}|U|^2 dx, \quad H = \int_{R^3}(|\nabla U|^2 - |U|^4)dx,$$

one can determine the remaining parameters $l, \xi_c, \tilde{C}, \tilde{\beta}_0$, through E and H, i.e., through the initial data $U_0(x)$.

Bound states of the elliptic sine-Gordon equation and PIII

Singular solutions of the 2D elliptic sine-Gordon equation

$$u_{xx} + u_{yy} = \sin(u) - f, \quad f = \text{const}, \tag{40}$$

which admit point-like singularities of the type

$$u \approx \rho\ \log r + \sigma + o(1), \qquad r^2 = x^2 + y^2 \to 0, \tag{41}$$

have important applications in condensed-matter physics, as models of defects. Such solutions are associated with supercurrent inhomogeneities near the corners of large area Josephson junctions [**Sha1**], as well as with patterns generated by dust particles in liquid crystals located in a rotating magnetic field [**Mi, Sha2**]. Moreover, equation (40) is an approximation [**Fri**] of the anisotropic Ginzburg-Landau equation, thus it represents a simple model for stationary phase defects in this equation.

The physical solutions should have *finite energy E*,

$$E = \int\left[\frac{1}{2}u_x^2 + \frac{1}{2}u_y^2 - \cos(u) + \cos(u_0) - f(u - u_0)\right]dxdy, \tag{42}$$

in analogy with the bound states in the scalar field theories. Here, u_0 represents a homogeneous asymptotic state $u \to u_0$, $r \to \infty$. However, it is well-known that in a

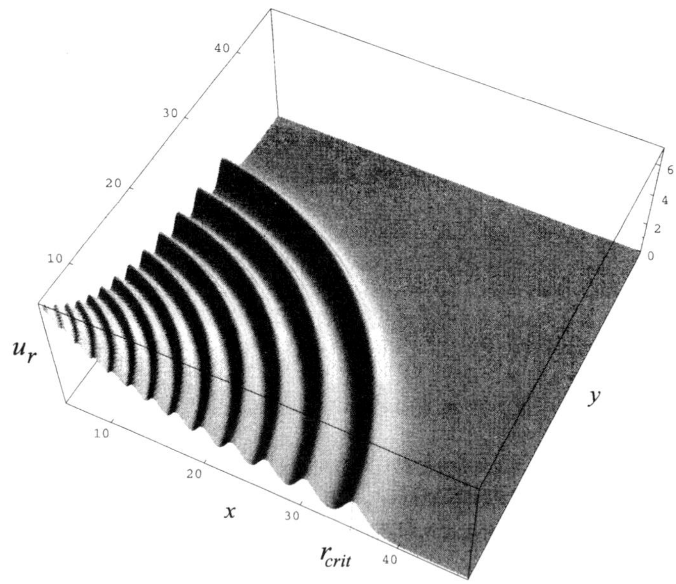

FIGURE III. The Josephson current intensity at the corner of a large rectangular junction.

class of smooth solutions, equation (40) has no bound states in $\mathbb{R}^2$. Nevertheless, if we admit solutions with point-like singularity satisfying (41), then global solutions with finite energy, i.e., bound states, do exist. But now, we have to exclude a small disk with radius $a \ll 1$ around the singular point, where a is the "cut-off" radius defined by the physical applicability of the continual model (40) [**L**].

In the force free case $f = 0$, bound state solutions with one singular point were found by computer simulation in [**Sha1, Sha2**]. They are radially symmetric, and they vanish exponentially at infinity $r \to \infty$. In addition, they have oscillatory radial structure up to a certain critical distance $r = r_{crit}$, where the exponentially decreasing mode comes to effect (see Figure).

This structure was interpreted in [**Sha1, Mi**] as a system of ring-like solitons (quasi-plane kinks of equation (40) at large distances), with common center at the singular point. Qualitatively, the picture does not depend on the driving force f up to a certain critical value f_{crit}. It was also found numerically [**Sha1, Sha2**] that for $f > f_{crit}$, the stationary radial-symmetric solutions of (40) become unstable and give rise to new dynamical patterns which satisfy the nonstationary equation

$$-u_t + u_{xx} + u_{yy} = \sin(u) - f.$$

The detailed treatment of the above radially symmetric structure was given by V. Novokshenov and A. Shagalov in [**N6**]. In the symmetric case the sine-Gordon equation (40) takes the form

$$u_{rr} + \frac{1}{r}u_r = \sin(u) - f. \tag{43}$$

For the case $f = 0$, the equation (43) reduces to a special case of the PIII equation. The solution of PIII describing the bound-state solutions can be characterized by the isomonodromy method as a limit case of the monodromy data manifold (see

Chapter 14 for details). At the same time, this solution can be described by its asymptotics both at the origin and at the infinity

$$u(r) = \begin{cases} \rho \ln x + \sigma + o(r), & r \to 0, \\ 2\pi\left(k + \frac{1}{2}\right) - \gamma\sqrt{\frac{2}{\pi}} r^{-\frac{1}{2}} \mathrm{e}^{-r}\left(1 + o(1)\right), & r \to +\infty, \end{cases} \tag{44}$$

where the constants γ, ρ and σ are related by the constraints

$$\gamma = 2\sinh\frac{\pi\rho}{4}, \qquad \sigma = -3\rho\log 2 - 4\arg\Gamma\left(\frac{1}{2} - \frac{i\rho}{4}\right),$$

and the integer k is defined by

$$k = \left[\frac{1}{2\pi}\left\{\pi - \sigma + 3\rho\ln 2 - 4\arg\Gamma\left(\frac{1}{2} + \frac{i\rho}{4}\right)\right\}\right].$$

Due to the explicit connection formula (44), one can estimate the bound state radius r_{crit}, which is, roughly speaking, the coordinate of the most distant annular soliton of the target-like radial structure

$$r_{crit} = \frac{\pi}{4}\rho - \frac{1}{2}\log\rho + O(1), \quad \rho \to \infty.$$

This above analysis provides a guideline for the study of bound-state solutions to the nonintegrable equation (43). Namely, for large ρ and $|f| \le 1$, it is possible to derive a critical value $f_c(\rho) \approx \mathrm{const}\cdot\rho^{-1}$, which governs the existence and stability of the bound-state solutions. For large area Josephson junctions and for nematic liquid crystals in a rotating magnetic field, these critical values were found in [**N6**]; they are in a good correlation with computer simulations and experimental data [**Sha1, Sha2**].

Regarding the sinh-Gordon equation, i.e., equation (43) with sine replaced by sinh, we note that bound-state solutions vanishing exponentially at infinity were studied in detail in [**McTW**]. However, these solutions correspond to the infinite-energy solutions of equation (43) under the map $u \to iu$, transforming sinh-Gordon into sine-Gordon equation. On the other hand, the connection formula for the amplitude $\gamma = \gamma(\rho)$ in (44) is correctly reproduced under this map.

Another important application of PIII (and actually somewhat similar to the one considered above) is the occurrence of the sinh-Painlevé III, as well as of some of its generalizations, in the theory of polyelectrolytes [**TW4**]. In this application, the principal role is played by the connection formulae derived in [**McTW**], [**Ki4**], and in [**TW3**].

Random matrices and random permutations

One of the basic random matrix models is the so-called *Unitary Ensemble* which is defined (see [**Meh**]) as the set of $N\times N$ Hermitian matrices $M = \{M_{ij}\}$, equipped with the unitary invariant probability measure given by the equation

$$\mu_N(dM) = Z_N^{-1} e^{-NTrV(M)} dM, \tag{45}$$

where

$$V(z) = \sum_{j=1}^{2m} t_j z^j, \quad t_{2m} > 0,$$

$$dM = \prod_{j<k} (d\,\mathrm{Re}\; M_{jk} d\,\mathrm{Im}\; M_{jk}) \prod_j dM_{jj},$$

and Z_N is the normalization constant, i.e.,

$$Z_N = \int e^{-NTrV(M)} dM. \tag{46}$$

An important particular case, the *Gaussian Unitary Ensemble GUE*, is characterized by the choice

$$V(z) = z^2.$$

The central question of random matrix theory is the analysis of the statistical properties of the ordered eigenvalues,

$$z_1(M) < z_2(M) < \cdots < z_N(M),$$

as $N \to \infty$. For instance, an important question is the computation of the probability $E_N(J)$ of finding no eigenvalues in the interval J. Another interesting question is whether there exist constants a and δ such that

$$(z_N - a)N^\delta$$

converges in distribution to a nontrivial distribution function $F(x)$, i.e.

$$\lim_{N\to\infty} \text{Prob}\left(z_N \le a + \frac{x}{N^\delta}\right) = F(x); \tag{47}$$

if such constants exist, what is this limiting distribution?

Consider the GUE case and choose the scaling $z \to \gamma_N z$, so that the expected number of eigenvalues per unit interval is 1. Then one can show (see [**Meh**], [**TW1**] or [**De1**] for more detail) that the limit

$$E(J) = \lim_{N\to\infty} E_N(\gamma_N J),$$

exists for every Borel subset J of $\mathbb{R}$. Furthermore, this limit is given by the formula

$$E(J) = \det(1 - K_{\text{sine}}), \tag{48}$$

where K_{sine} is the (trace class) operator with kernel,

$$K_{\text{sine}}(z, z') = \frac{\sin \pi(z - z')}{\pi(z - z')}$$

acting on $L^2(J, dz)$. In 1980 M. Jimbo, T. Miwa, Y. Môri, and M. Sato [**JMMS**] expressed this Fredholm determinant in terms of the solution of a system of integrable differential equations describing the isomonodromy deformations of a certain linear matrix ODE with rational coefficients. In the simplest case of a single interval, $J = (0, x)$, the differential equation is a particular case of the fifth Painlevé equation. In this case, the Fredholm determinant (48) is given by

$$\det(1 - K_{\sin}) = \exp\left(\int_0^{\pi x} \frac{\sigma(t)}{t} dt\right), \tag{49}$$

where $\sigma(x)$ is the solution of the so-called sigma (or Hirota) form of PV (see, e.g. [**JMMS**]),

$$(x\sigma'')^2 + 4(\sigma - x\sigma' - (\sigma')^2)(\sigma - x\sigma') = 0, \tag{50}$$

which is characterized by the initial condition

$$\sigma(x) \sim -\frac{x}{\pi}, \quad x \to 0. \tag{51}$$

The Riemann-Hilbert integrability of equation (50), suggests the possibility of the evaluation of the large x asymptotics of the probability $E(J)$. Indeed, the following asymptotic expansion is valid:

$$\log E(J) \sim -\frac{\pi^2 x^2}{8} - \frac{1}{4}\log\frac{\pi x}{2} + c + \frac{c_1}{x} + \frac{c_2}{x^2} + \dots, \tag{52}$$

where

$$c = \frac{1}{12}\log 2 + 3\zeta'(-1), \tag{53}$$

and the constants c_1, c_2, ..., can be computed, in principle, in terms of integrals of Bessel functions. The first two terms of this expansion were found in 1973 by des Cloizeaux and Mehta [**dCM**], whereas the full expansion, including equation (53) for the constant term, were derived in 1976 by Dyson [**Dy2**]. The rigorous proof of the asymptotic series (52) was obtained (using the Riemann-Hilbert approach) in [**DeIZ**], but without the explicit evaluation of the constant term c. The analysis of [**DeIZ**] was preceded by [**Wi1**], where a rigorous derivation of the first term of the asymptotics was suggested, and by [**Su1**] where the leading term of the asymptotics of the corresponding fifth Painlevé transcendent was rigorously evaluated. The proof of Dyson's formula (53) for the constant c was obtained in 2003 by I. V. Krasovsky [**Kra**] and, independently, by T. Ehrhardt [**Eh**]. The asymptotic representations (51) and (52), constitute an example of connection formulae for a Painlevé equation. The ability of deriving such formulas makes equation (49) an efficient tool for the calculation of the gap probability $E(J)$.

A Fredholm determinant representation, similar to (49), also exists for the limiting distribution function $F(x)$ defined by relation (47). More precisely, one can show (see [**TW2**] and references therein) that in the case of GUE, $a = \sqrt{2}$, $\delta = 2/3$, and

$$\text{Prob}\left(z_N \le \sqrt{2} + \frac{\sqrt{2}x}{N^{2/3}}\right) \to F(x) = \det(1 - K_{\text{airy}}), \tag{54}$$

where K_{airy} is the (trace class) operator with kernel,

$$K_{\text{airy}}(z, z') = \frac{\text{Ai}(z)\text{Ai}'(z') - \text{Ai}'(z)\text{Ai}(z')}{z - z'} \tag{55}$$

acting on $L^2((x, \infty), dz)$. The notations $\text{Ai}(z)$ and $\text{Ai}'(z)$ are used for the classical Airy function and its derivative, respectively. It was established in [**TW2**] by Tracy and Widom that the Airy kernel determinant (55) can also be evaluated in terms of a Painlevé transcendence - this time Painlevé II:

$$\begin{aligned} F(x) &\equiv \det(1 - K_{\text{airy}}) \\ &= \exp\left(\int_x^\infty (x-y)u^2(y)dy\right), \end{aligned} \tag{56}$$

where $u(x)$ is the solution of the second Painlevé equation,

$$u_{xx} = xu + 2u^3,$$

uniquely specified by the boundary condition,

$$u(x) = \frac{1}{2\sqrt{\pi}} x^{-1/4} \exp\left\{-\frac{2}{3}x^{3/2}\right\}(1 + o(1)), \quad x \to +\infty. \tag{57}$$

This particular solution of the second Painlevé equation was first singled out and studied by S. P. Hastings and J. B. McLeod in [**HasM**], where it is shown that the solution $u(x)$ is smooth for all real x, and its behavior as $x \to -\infty$ is given be the equation,

$$u(x) = \sqrt{-\frac{x}{2}}\left(1 + O\left(\frac{1}{x^2}\right)\right), \quad x \to -\infty. \tag{58}$$

This is again an example of an explicit connection formula for a Painlevé equation which again makes the representation (56) an efficient tool for evaluating the distribution function $F(x)$. The solution $u(x)$ was studied extensively in the later works [**Su2**], [**K5**], [**DeZ2**], and [**K10**]. A detailed analysis of this solution, which includes its behavior for complex x, and the generalization of this solution to the case that the parameter α in PII is nonzero, is presented in Chapter 11 of this book.

The importance of the results presented above is based on the *universality* of the distribution functions $E(J)$ and $F(x)$. Namely, the fact is that equations (48) and (54) are valid for an arbitrary generic choice of the polynomial $V(z)$ in the measure (45) which defines the random matrix model (provided that the number $\sqrt{2}$ in (54) is replaced by a proper value of the constant a, which, of course, depends on $V(z)$). This statement, which constitutes the *Dyson conjecture*, has been proven for different classes of the potential $V(z)$, (see [**PS**], [**BlI1**], [**DeKMVZ**]; see also the review article [**TW7**], and the monograph [**De1**] for more details and for the history of the subject). We note that the Riemann-Hilbert approach has played a crucial role in the rigorous analysis of the random matrix universality. We also mention that, as it was shown by A. Soshnikov [**So**], the Tracy-Widom edge scaling limit (47), is also valid for the so-called Wigner ensembles (where the matrix elements on and above the diagonal are i.i.d. random variables). This result is particularly important because the corresponding measures are not unitary invariant, and therefore the "integrable" techniques, such us Fredholm determinant and Riemann-Hilbert methods, are not applicable.

The Hastings-McLeod Painlevé solution and the related hierarchy of integrable ODEs, appear in yet another universality issue. Namely, these functions describe the so-called *double scaling* limits of the random matrix distribution functions at the neighborhoods of certain critical values of the coefficients $t_1, t_2, \ldots$ of the polynomial $V(z)$ (see [**BlI2**], [**BlE**], and [**CK**]; see also [**BlI2**] and [**BlE**] for the history of this problem and basic references).

All above results are based on the *orthogonal polynomial method* in the random matrix theory, i.e., on the fact that the eigenvalue correlations for finite N can be expressed in terms of orthogonal polynomials on the line with respect to the weight $e^{-NV(z)}$, where the function $V(z)$ is the same as in (45) (see [**Dy1**]; see also [**Meh**], [**TW1**] or [**De1**]). This implies the equivalence of the asymptotic analysis of random matrices with the asymptotic investigation of orthogonal polynomials. All the results mentioned above have their counterparts in the theory of orthogonal polynomials, and they provide solutions to some of the most basic long-standing questions of this theory. In fact, the majority of the works cited above deals with the relevant problems of the theory of orthogonal polynomials. The random matrix statements follow as simple corollaries. In other words, *the theory of orthogonal polynomials, provides yet another important field of applications of modern Painlevé theory.*

Painlevé equations do not only appear in the large N limit and in the investigation of universality, but they also occur before the large N limit is taken. Actually, *all* basic distribution functions can be expressed in terms of solutions of the various Painlevé equations. However, for finite N, these are *classical solutions*, unlike the "genuine" Painlevé transcendents which appear at $N = \infty$. For instance, as it was first shown in [**FoIK3**] (see also [**Ma**]), in the case of a quartic potential, $V(z) = t_2 z^2 + t_4 z^4$, the partition function Z_N, as a function of the ratio $t_2/\sqrt{t_4}$, is given in terms of a special solution of the fourth Painlevé equation which can be expressed in the parabolic cylinder functions. We refer the reader to the works of M. Adler and P. van Moerbeke [**AvM**], P. Forrester and N. Witte [**ForW1**], [**ForW2**], and G. Tracy and H. Widom [**TW6**] for more details on this aspect of the random matrix theory (in these papers, the universal Painlevé functions appearing at $N = \infty$ limit are also discussed).

As it is becoming increasingly evident, the limiting distributions of random matrix theory, and hence the related Painlevé transcendents, represent new limit laws in probability theory, which have numerous applications in number theory, combinatorics, statistical physics, growth processes, solid state physics, and statistics. We will now describe one of these applications, which concerns random permutations.

Let S_n be the symmetric group of degree n, i.e., the group of permutations σ of the numbers 1, 2, ..., n. If $1 \le m_1 < m_2 < \cdots < m_k \le n$, then $\sigma(m_1), \sigma(m_2), \ldots, \sigma(m_k)$ is an increasing subsequence of σ of length k provided that

$$\sigma(m_1) < \sigma(m_2) < \cdots < \sigma(m_k).$$

Let $l_n(\sigma)$ be the length of the longest increasing subsequence in the permutation σ. Then,

$$\lim_{N\to\infty} \text{Prob}\left(l_n \le 2\sqrt{N} + xN^{1/6}\right) = F(x), \tag{59}$$

where $F(x)$ is the Tracy-Widom distribution function of the random matrix theory (54), (56). This remarkable fact, proven in the celebrated paper [**BDJ**], led to an avalanche of activity in random permutations and related areas of combinatorics, representation theory and the theory of percolations and growing processes. The interested reader can find more about these exciting developments, in the review articles [**TW7**] and [**De2**]. For most recent results and references - see [**BKMM**].

2D QUANTUM GRAVITY AND PI

Consider the partition function of the random matrix model (cf. (46)),

$$Z_N(t) = \int e^{-NTrV(M)} dM, \tag{60}$$

corresponding to the quartic potential,

$$V(z) \equiv V(z;t) = \frac{1}{2}z^2 + tz^4. \tag{61}$$

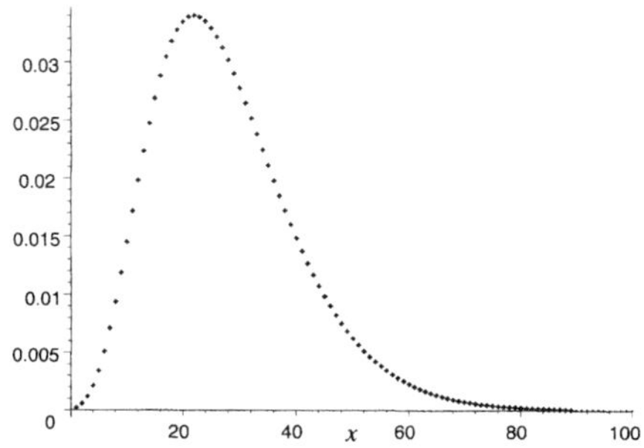

FIGURE IV. The density of the Tracy - Widom distribution

It was shown in the seminal paper of D. Bessis, C. Itzykson, and J. B. Zuber [**BIZ**], that the integral (60) (which is well defined for all $t \geq 0$) admits the following formal expansion[4] over N^{-2},

$$\log \frac{Z_N(t)}{Z_N(0)} \sim \sum_{g=0}^{\infty} N^{2-2g} E_g(t), \tag{62}$$

where

$$E_g(t) = \sum_{n\geq 1} (-t)^n \frac{\kappa_g(n)}{n!}. \tag{63}$$

The coefficients κ_n of the series (62) have a profound topological meaning, namely

$$\kappa_g(n) = \#\{\text{connected maps of genus } g \text{ with } n \text{ 4-valent vertices}\}. \tag{64}$$

See [**EM**] and also the survey [**FGZ**] for details.

The series (62) is a convergent series; moreover, by analyzing the explicit formulae that exist for $E_g(t)$ for small values of the genus g, one can suggest (see e.g. [**FGZ**]) that the radius of convergence of series (62) is equal to $1/48$ for all g, and that the point

$$t = -\frac{1}{48},$$

is a singular point for all the functions $E_g(t)$. The basic idea of E. Brézin and V. Kazakov [**BreKaz**], M. Douglas and S. Shenker [**DoS**], and D. Gross and A. Migdal [**GrMi**], is to associate the free energy functional $\mathcal{F}(x)$ of two-dimensional quantum gravity (for definitions see [**FGZ**]), with the second term in the following double-scaling expansion:

$$\log \frac{Z_N(t)}{Z_N(0)} = N^2 \mathcal{F}^0 + \mathcal{F}(x) + \dots, \tag{65}$$

$$N \to \infty, \quad t = -\frac{1}{48} - c_o N^{-4/5} x.$$

To be more precise, one has to perform an analytic continuation of the function $Z_N(t)$ to the negative values of t (which exists) and then take the limit indicated. The parameter $-x$ has the meaning of a renormalized string coupling constant. A

[4]The rigorous proof of the fact that the formal series in the right-hand side of equation (62) represents the large N asymptotics of its left-hand side was obtained only in 2003 by N. Ercolani and K. McLaughlin [**EM**].

detailed explanation of this terminology and for the reason that this particular limit should be taken as a definition of 2D quantum gravity, can be found in [**FGZ**]. Here, we only note that the double scaling limit (65) is supposed to make all the terms of series (62) of the same order. This fact, the assumption that the asymptotics (62) is valid for $-1/48 < t < 0$, and the topological meaning of the numbers $\kappa_g(n)$, imply that the limit (65) can be thought of as a "Riemann sum approximation" to the formal functional integral of the string theory.

The remarkable fact, discovered in [**BreKaz**], [**DoS**] and [**GrMi**], is that the limiting function $\mathcal{F}(x)$ satisfies (with the proper choice of the numerical coefficient $c_0 > 0$) the relation

$$\mathcal{F}''(x) = -u(x), \tag{66}$$

where $u(x)$ is the solution of the first Painlevé equation,

$$u_{xx} = 6u^2 + x, \tag{67}$$

uniquely specified by the asymptotic condition,

$$u(x) \sim \sqrt{\frac{-x}{6}} + \sum_{l=1}^{\infty} c_l(-x)^{1/2-5l/2}, \tag{68}$$

$$x \to \infty, \quad -\frac{\pi}{5} < \arg x < \frac{7\pi}{5}. \tag{69}$$

Strictly speaking, it was shown in [**BreKaz**], [**DoS**] and [**GrMi**] only that $u(x)$ belongs to a one-parameter family of solutions of PI with the asymptotics (68) as $x \to -\infty$. The characterization (68 - 69) was obtained in the works [**Da**] and [**FoIK1**] (see also [**FoIK2**], [**FoIK3**], [**FoIK4**], [**Mo**], and [**Mor**]). In [**FoIK1**], an alternative characterization of $u(x)$ was also found, using the earlier results of the global asymptotic analysis of PI obtained by Kapaev [**K2**] together with the Riemann-Hilbert approach: On the boundaries of the sector (69), the solution $u(x)$ exhibits the following oscillatory behavior,

$$\begin{aligned} u(x) &= \sqrt{\frac{-x}{6}} + \sum_{l=1}^{n} c_l(-x)^{1/2-5l/2} + O(x^{-2-5n/2}) \\ &-is(-x)^{-1/8}\exp\left\{-\frac{8}{5}\left(\frac{3}{2}\right)^{\frac{1}{4}}(-x)^{5/4}\right\}\Big(1+o(1)\Big), \end{aligned} \tag{70}$$

as

$$x \to \infty, \quad \arg x = \arg(-x) + \pi = \frac{7\pi}{5},$$

and

$$\begin{aligned} u(x) &= \sqrt{\frac{-x}{6}} + \sum_{l=1}^{n} c_l(-x)^{1/2-5l/2} + O(x^{-2-5n/2}) \\ &+s(-x)^{-1/8}\exp\left\{\frac{8}{5}\left(\frac{3}{2}\right)^{\frac{1}{4}}(-x)^{5/4}\right\}\Big(1+o(1)\Big), \end{aligned} \tag{71}$$

as

$$x \to \infty, \quad \arg x = \arg(-x) + \pi = -\frac{\pi}{5},$$

where

$$s = \frac{1}{\sqrt{8\pi}}\left(\frac{2}{3}\right)^{1/8}. \tag{72}$$

Each of the two asymptotics (70) and (71) characterizes $u(x)$ uniquely.

It is also worth noticing that the Riemann-Hilbert method in the theory of matrix models and orthogonal polynomials, mentioned in the previous subsection, was first introduced in [**FoIK1**]–[**FoIK4**].

A formal evaluation of the parameter s, as the coefficient of the exponentially decaying term in the asymptotics of $u(x)$ as $x \to -\infty$, was also performed in [**SY**].

The solution $u(x)$ is actually the classical *tri-tronquée* solution of the first Painlevé equation, which was already singled out and studied by P. Boutroux in [**Bou**]. Equations (70) - (72) give an *explicit* evaluation of the *Stokes phenomenon* for this solution, and they constitute a modern achievement ([**K2**], [**KK2**], [**GN**], [**K11**]), based on the isomonodromy method. In Chapter 11 we will discuss the issue of Stokes phenomenon in great detail for the case of PII.

Equation (66) has also an interesting combinatorial application. In fact, (66) suggests that the singularities of the functions $E_g(t)$ in (63) at $t = -1/48$ are of the following power form:

$$E_g(t) \sim e_g\left(t + \frac{1}{48}\right)^{5/2(1-g)} + \text{constant}. \tag{73}$$

Indeed, if we assume (73) and the validity of (62) for $-1/48 < t < 0$ (the absence of the Stokes phenomenon for the partition function $Z_N(t)$!) then by a direct formal calculations we find (cf. [**FGZ**])

$$\log\frac{Z_N(t)}{Z_N(0)} - N^2\,\text{const} \sim \sum_{g=0}^{\infty} N^{2-2g} e_g\left(t + \frac{1}{48}\right)^{5/2(1-g)}$$

$$= \sum_{g=0}^{\infty} N^{2-2g} e_g\left(-c_0 x N^{-4/5}\right)^{5/2(1-g)} = \sum_{g=0}^{\infty} \hat{c}_g(-x)^{5/2-5g/2}.$$

The second derivative of the last series with respect to x is of the form of the series (68). This, in view of (66), can be considered as a formal confirmation of the estimate (73) (see the survey [**FGZ**] and references therein for more details). In turn, the estimate (73), following a standard reasoning, implies the following estimate for the combinatorial numbers (64):

$$\kappa_g(n) \sim Cn!\frac{48^n}{n^{1+5/2(1-g)}}. \tag{74}$$

The above arguments are, of course, formal. The status of estimate (73), and hence of (74), remains that of a conjecture. Nevertheless, the current level of development of the Riemann-Hilbert techniques, and the experience with other combinatorial problems e.g. in random permutations [**BDJ**], suggest that all the gaps in the above construction will be soon filled. It should be mentioned, that the estimate (73) is also predicted by the continuum string theory (Liouville gravity, see [**FGZ**]).

A GUIDANCE TO THE BOOK AND TO THE RELATED LITERATURE

The book consists of three parts (with a continuous numeration of chapters) and of two appendices. Each part has a short introduction, which provides a brief outline of the material discussed in this part. In what follows, we present a more extended overview of the content of the book.

Part 1 consists of Chapters 1 – 6, and it has two goals. The first goal is to develop the general theory of linear systems of ODEs. This is carried out in the first four chapters, which are essentially based on the monographs of W. Balser [**Bal**], E. A. Coddington and N. Levinson [**CodL**], Y. Sibuya [**Si2**], and W. Wasow [**W1**], as well as on the papers of M. Jimbo, T. Miwa, and K. Ueno [**JMU**], [**JM**]. The second goal is to introduce (following [**I1**] and [**I4**] as the "blueprint" outlined earlier) the both Painlevé functions and the "linear" special functions of hypergeometric type, as an intrinsic element of the monodromy theory of linear ODEs. This allows us to place the theory of Painlevé equations, from the very beginning, in the context of the Riemann-Hilbert method. We believe that this point of view on the theory of special functions has not been presented in the existing textbook literature; at least, it has not been put forward to the extent we do. The closest to our methodology (though still rather different), are the recent monographs [**IwKSY**] and [**SL**], as well as the isomonodromy approach to special functions initiated in [**IKK**] and further developed in [**Ki8**]. At the same time, in order to present the Riemann-Hilbert method as fully as possible we were obliged to limit our presentation of more conventional approaches to Painlevé equations. In particular, the classification problem mentioned at the beginning of the introduction and several important issues concerning the general properties of Painlevé functions are not addressed in this book. These issues are well covered in the existing literature, and we refer the interested reader to the classical book of Ince [**In**], to the recent monographs [**GroLS**], [**Cos**], and to the original articles [**JoK**] and [**JoKr2**]. We also recommend the proceedings volume [**Con**], as a good source of the classical as well as modern results concerning the Painlevé equations together with many interesting applications.

Chapters 3, 4, and 5 are the central chapters of this part of the book. This is where the general Riemann-Hilbert formalism is introduced (Chapter 3), the general theory of isomonodromy deformations (following mostly [**JMU**]) is presented (Chapter 4), the Painlevé equations are introduced and the specific Riemann-Hilbert formalism for integrating each of the six Painlevé transcendents (following mostly [**FoZ**] and [**FoMZ**]) is developed (Chapters 4 and 5).

As we have already indicated, we have attempted to make the book self-contained; therefore, in Chapter 3 we present and discuss (following very closely [**Zh**], [**DeZ4**], and [**DeZ5**]) the general theory of Riemann-Hilbert factorization and the related singular integral operators. The main analytical statement of Chapter 3 is meromorphic solvability of the Riemann-Hilbert problem related to the second Painlevé equation. The relevant proof is presented in **Appendix A** and it is based on the classical Birkhoff-Grothendieck theorem with a parameter, whose detailed proof is presented in **Appendix B**. These two appendices are based on the work [**BolIK**], and they relate parts of the book to the theory of holomorphic vector bundles. However, we do not develop further this aspect of modern theory of isomonodromy deformations. The interested reader can find more on this and other important geometrical aspects of the monodromy theory in the works of P. Boalch [**Bo3**], B. Malgrange [**Mal**], T. Miwa [**Miw**], P. Palmer [**Pa**], L. Mason, M. A. Singer, and N. Woodhouse [**Mas**]. It should be emphasized that the invariant geometric language of vector bundles is unavoidable for the rigorous analysis of the inverse monodromy problem and isomonodromy deformations in the case of general linear systems. Bearing this in mind, we believe that, although we do not use vector bundles directly in our study (in the case of Painlevé equations it is sufficient to use the "elementary" language of classical complex analysis), Appendices A and B prepare the interested reader to the more advanced texts on the general Riemann-Hilbert-Birkhoff problems.

An alternative approach to the inverse monodromy problems is based on the analytic Fredholm theory of the relevant singular integral operators. This theory is outlined in Chapter 3. It is used then in Chapter 5 (where we follow [**FoZ**] and [**FoMZ**]) to analyze the global solvability of the Riemann-Hilbert problems associated to the Painlevé I, II, III, IV, and V equations, and hence to establish the global solvability of these equations (see also [**BS**] where a similar technique is applied to the isomonodromy deformations of general linear systems). Painlevé VI is rather different than the other five Painlevé equations (although the relevant Riemann-Hilbert formalism for this equation was developed [**Mu**]). In the analysis of the sixth Painlevé equation a very important role is played by certain nontrivial algebraic properties of the space of its monodromy data. For the modern theory and for applications of the sixth Painlevé equation, we refer the reader to the pioneering work of M. Jimbo [**J1**] and to the more recent works of P. Boalch [**Bo4**], B. Dubrovin and M. Mazzocco [**DM**], D. Guzzetti [**Guz1**] - [**Guz2**], and A. Kitaev [**Ki9**] - [**Ki10**]. We will indicate later (see the description of Part III below) some of the specific features of the asymptotic analysis of the sixth Painlevé equation, which distinguish this equation from the other Painlevé functions.

In the last chapter of the first part we introduce the notion of *Bäcklund transformation* which maps solutions of a Painlevé equation into solutions of the same equation but with different parameters. These transformations play an important role in the theory of integrable systems. A large class of Bäcklund transformations for the Painlevé equations, is generated by the so-called Schlesinger transformations of the associated Riemann-Hilbert problems (see [**FoA2**], [**FoY**], [**Ki6**], [**MuFo**]). In Chapter 6 we construct sets of Bäcklund transformations for Painlevé II–IV equations (Painlevé I does not have a parameter). An important application of the Bäcklund transformations is that they generate the classical solutions of the Painlevé equations.

The theory of Bäcklund transformations has beautiful algebraic and algebrogeometrical ramifications, which we do not discuss in this book. We refer the reader to the works of A. P. Bassom, P. A. Clarkson and A. C. Hicks [**BCH**], P. Clarkson, N. Joshi and A. Pickering [**ClJP**], H. Sakai [**S**], to the monograph of M. Noumi [**No2**], and to the references in Chapter 6. Closely related to the algebrogeometrical approach to Painlevé equations are their Hamiltonian theory and the modern proofs of the irreducibility of the Painlevé equations. The main references on these aspects of the theory are the works of P. Boalch [**Bo1**]–[**Bo2**], H. Flaschka and A. Newell [**FN2**], J. Harnad [**Ha1**], J. Harnad and M. Wisse [**Ha2**], D. Korotkin and

H. Samtleben [**Ko3**], I. Krichever [**Kri1**] , K. Okamoto [**Ok1**]–[**Ok6**], M. Ugaglia [**U**], H. Umemura and H. Watanabe [**Um1**]–[**UmW**], [**Wa1**], [**Wa2**].

Bäcklund transformations are also closely related to the most recent direction in the theory of Painlevé equations, namely to the theory of the *discrete Painlevé equations* (for which, unfortunately, we have no room in our book). In fact, "half" of the discrete Painlevé equations are just the chains of Bäcklund transformations of the usual continuous Painlevé equations. For instance, one of the first discrete Painlevé equations, which was identified as the discrete Painlevé I equation, is the Bäcklund chain of the Painlevé IV equation (see [**FoIK3**], [**FoIK2**]). We also note that the PI and PII double scaling limits of the random matrix theory and the PII double scaling limit of random permutations, which were discussed above, are so far the only nontrivial examples for which the global asymptotic analysis (via a proper version of the Riemann-Hilbert method) of discrete Painlevé functions was carried out (d-PI in the case of [**FoIK3**] and [**BlI2**], and d-PII in the case of [**BDJ**]).

The other "half" of the discrete Painlevé equations consists of the so-called q-discrete equations. Their theory is not part of the continuous theory. They too are the isomonodromy (Schlesinger) transformations, but of *difference* linear equations, i.e., *both* equations of their Lax pairs are discrete. This is an extremely interesting area, which is in the very beginning of its development. We refer the reader to the works of M. Jimbo and H. Sakai [**JS**], A. Borodin [**Br**] and I. Krichever [**Kri2**], for more details on the q-Lax pairs and the setting of the q-monodromy problem.

For more details on the discrete Painlevé equations see the survey [**GrNR**]. We also refer the reader to the beginning of Chapter 6 where we briefly review some of the algebraic aspects of the Painlevé theory.

Part 2 which consists of Chapters 7–11, is the main part of the book. Taking the second Painlevé equation as a case study, we present in full detail the two major approaches to the asymptotic analysis of Painlevé equations via the Riemann-Hilbert method. The first scheme is based on the asymptotic solution of the *direct monodromy problem* for the associated linear λ-equation and uses the fact that the monodromy data form a complete set of first integrals of the Painlevé equations. This approach is presented in Chapter 7, Sections 1-3 of Chapter 9 and in Chapter 10. The basic technical tool of this approach is an extended version of the complex WKB method which was first suggested in [**IPe**], [**I3**], [**N3**], and [**IN1**] and which was further used and substantially developed in the works of A. Kapaev [**K1**]–[**K9**], A. Kitaev [**Ki2**]–[**Ki7**], F. Andreev and A. Kitaev [**AnK2**], and A. Kitaev and A. Vartanian [**KiV**]. A similar approach was also used by B. McCoy and Sh. Tang [**McT1**]–[**McT2**]. The key element of the technique is the introduction of a set of complex WKB solutions of the λ-equation. These solutions can be glued with the canonical solutions at the irregular singular points, and, at the same time, can be connected to each other through the relevant complex turning points (with the help of explicitly solvable model systems arising at the turning points). This allows an asymptotic evaluation of the corresponding Stokes matrices. Since the latter are *integrals of motion*, their asymptotic evaluation yields in fact the parametrization of the asymptotics of the solution of the Painlevé equation in different directions and near different critical points by the *same* parameters — the constant values of the entries of the Stokes matrices. This, in turn, provides the connection formulae. In its simplest form, this scheme is presented in Sections 1-3 of Chapter 9 (where we mostly follow the original method of [**IN1**]). Chapters 7 and 10 use a more advanced version of the WKB scheme which was developed by A. Kapaev in [**K7**]. Somewhat similar to this version, although still different in several important points, is the WKB method of A. P. Bassom, P.A. Clarkson, C. K. Law and J. B. McLeod [**BCLM**].

A crucial methodological element of the above direct monodromy approach, is the necessity of (at least some) a priori information about the local asymptotic behavior of the solution. Hence the connection formulae obtained by this approach must be rigorously justified. This issue can be resolved with the help of an elegant method based on the Brouwer fixed point theorem suggested by A. Kitaev [**Ki1**] (see Subsection 6.1 of Chapter 7; see also the discussion of the justification problem at the end of Section 1 of Chapter 9).

Chapters 8, Sections 4 and 5 of Chapter 9 and Chapter 11 present the second scheme of the asymptotic analysis of Painlevé equations. This scheme is based on the asymptotic solution of the *inverse monodromy problem* for the associated linear λ-equation, and it uses the nonlinear steepest descent method of P. Deift and X. Zhou for oscillatory Riemann-Hilbert problems [**DeZ1**]. This method was outlined earlier in the introduction; its important advantage is that it does not rely on any sort of a priori information, and therefore the rigorous proofs of the asymptotics and connection formulae are obtained in a much more straightforward way than in the case of the first scheme. At the same time, in many situations, and especially when one is confident about the asymptotic form of the solution, the formal derivation of the relevant connection formulae (with its a posteriori justification), is often easier via the first scheme. In particular, this is the case when one deals with the *generic* types of asymptotic behavior, such as the ones considered in Chapters 7 - 8 and Chapters 9 - 10. In order to help the reader to appreciate both schemes, in Chapter 8 we present the steepest descent derivation of the complex asymptotics, which was already obtained via the WKB technique in Chapter 7, while in Sections 4, 5 of Chapter 9 we rederive the WKB results of Sections 1–3 of the same chapter. However, in the case of the *special separatrix families* of Painlevé transcendents, the Deift-Zhou method is often better than the WKB scheme. We demonstrate this in Chapter 11 where, with the help of the nonlinear steepest descent approach, we describe rigorously and in full detail the quasi-linear Stokes phenomenon associated with the so-called *tronquée* solutions of the second Painlevé equation (this is methodologically similar with the contour integral techniques of the linear theory). One can also observe that the steepest descent approach, although technically more involved in the generic complex case, reveals in a most elegant way all the nontrivial elements of the asymptotics as the result of the proper deformation of the jump contours of the relevant Riemann-Hilbert problems.

The results obtained in the second part of the book give a complete description of the leading terms of the asymptotic behavior for complex $x \to \infty$ and the relevant connection formulae for all the generic families of solutions in the special case of the second Painlevé equation with the parameter $\alpha = 0$ (Chapters 7 - 10). For the general case of the second Painlevé equation with $\alpha \neq 0$, we only describe the quasi-linear Stokes phenomenon exhibited by the separatrix families of solutions. These families play a significant role in random matrix and string theory applications [**CKV**], [**Se**].

Part 3 consists of Chapters 12–16 where we study the important particular case of the third Painlevé equation which appears as a self-similar reduction of the sine-Gordon equation. From the methodological point of view, the analysis of the third Painlevé equation adds an important new component to the scheme developed in the preceding part. Indeed, unlike the second Painlevé equation, the third Painlevé equation has two critical point, $x = 0$ and $x = \infty$. This means that the connection problem for this equation, in addition to the nonlinear Stokes phenomenon near $x = \infty$, includes the so-called central connection problem, i.e., the evaluation of the relation between the asymptotic parameters at $x = 0$ and the asymptotic parameters at $x = \infty$. In fact, it is the central connection problem for the SG-PIII, or, in other words, the asymptotic behavior of the solution of the *Cauchy problem*, which is the question we consider in this part of the book.

Unlike the limit $x \to \infty$, whose analysis is very similar to the one performed in the case of the second Painlevé equation, the evaluation of the PIII monodromy data at the limit $x \to 0$ does not require the WKB arguments or the steepest descent constructions: The key observation is that, as $x \to 0$, the corresponding λ-equation and the Riemann-Hilbert problem can be, with the help of a proper rescaling, solved explicitly (see Chapter 14 for more details). This is a dramatic difference, between the investigation of the Riemann-Hilbert problem at a critical point which is an essential singularity of the Painlevé equation (the point $x = \infty$ in the case of PII and PIII), and the investigation of the Riemann-Hilbert problem at the point which is a branch singularity of the equation (the point $x = 0$ in the case of PIII). The evaluation of the relevant monodromy data in terms of the Cauchy data at the branch points was first performed, in the cases of the sixth and fifth Painlevé equations, by M. Jimbo in [**J1**]. A similar technique (together with the WKB analysis at the limit $x \to \infty$) was then used in the papers [**N2**]–[**N3**], [**Ki3**]–[**Ki4**], and [**KiV**] and in [**McT1**]–[**McT2**], [**AnK2**], for the third and fifth Painlevé equations, respectively. It should be emphasized that the papers [**N2**]–[**N3**] were the first works where the isomonodromy approach was applied to a Painlevé equation which is not the sixth Painlevé equation (which was done in [**J1**]). Also, the works [**McT1**]–[**McT2**] and [**AnK2**] were the works where the analysis of [**J1**], for the fifth Painlevé equation was completed by evaluating the monodromy data at $x = \infty$ and hence producing the Painlevé V connection formulae. An extended version of the method of [**J1**] has also been used in the recent studies of the sixth Painlevé equation in the papers [**DM**], [**Guz1**]–[**Guz2**], and [**Bo4**].

A detailed exposition of the asymptotic solution of the generalized Cauchy problem for the SG-PIII equation is presented in Chapters 13 - 16. In these chapters we discuss the smooth (for real x) two-parameter families of solutions (Chapter 14), the important separatrix solutions (Chapter 13 and Chapter 14), and the two-parameter singular (for real x) solutions (Chapter 15). The latter analysis adds yet another new component to the scheme of Part 2; namely it presents the evaluation (in terms of the Cauchy data) of the limiting distribution of the poles of the solutions of the Painlevé equations. In the last chapter of this part, the asymptotic behavior of the solution of the Cauchy problem in the complex plane is evaluated. The important new feature of this analysis is that unlike the PII case, the setting of the Riemann-Hilbert problem *depends on the value of* $\arg x$. It is important to notice that this feature is also shared with the fifth and sixth Painlevé equations and, in fact, with the isomonodromy deformations of the general linear systems.

In the third part of the book we also present (see Chapter 12) the sets of rational and algebraic solutions of the general PIII equation, presenting more details than those presented for the analogous case of the second Painlevé equation (Part 2 of the book).

In the third part of the book we mostly follow the original works of V. Novokshenov and A. Shagalov [**N2**]–[**N6**], and some of the parts of the monograph [**IN1**]. It should be mentioned that the third part of the book is mainly intended for the prospective "users" of the Painlevé functions. Therefore, we emphasize the qualitative properties and the physical meaning of the connection formulae derived, as well as the physical meaning of the third Painlevé equation itself. Also, for the same reason (and in contrast to Part 2), when we obtain the relevant asymptotic results, we focus on the formal derivation rather than on the details of the rigorous proofs. The proofs are often only sketched, with the understanding that the interested reader could complete them using the relevant material of the preceding parts.

The Parts 2 and 3, although methodologically related to the general Part 1, can be read independently of the latter by a reader who is already familiar with the general monodromy theory of linear systems.

ACKNOWLEDGEMENT

The work on this book was supported in part by the National Science Foundation, Grants Nos. DMS-9501559, DMS-9801608, and DMS-0099812, by the Russian Foundation of Basic Research, Grants Nos. 02-01-00144, 04-01-00190, and 04-01-14020 and by the Engineering and Physical Sciences Research Council.

Part 1

Riemann-Hilbert Problem, Isomonodromy Method and Special Functions

In the following six chapters we discuss our understanding of the modern theory of Isomonodromy Deformations. We start with an overview of the classical theory of linear systems of ODEs with rational coefficients in the complex plane. The monodromy group, i.e., the group of linear transformations over canonical paths circling the singular points, has been the subject of study by the classics, starting with Jacobi and Weierstrass. The intriguing problem of reconstructing an ODE from its monodromy group had been attempted by a number of mathematicians. This problem was formulated as the Hilbert's twenty-first problem and was finally solved by A. Bolibruch in 1989.

The above insight naturally leads to the Riemann-Hilbert (RH) problem which is equivalent to the inverse monodromy problem. This is again classical material dealing with the analytical behavior of the fundamental solutions of an ODE in the complex plane. The set of conjugation matrices of the RH problem form a natural representation of the monodromy group. We also discuss the issue of RH solvability.

The general facts of the first two chapters prepare the reader for the introduction of the key idea of Isomonodromic Deformations: the monodromy group is fixed, whereas the poles of ODEs rational coefficients are free. This immediately yields a set of nonlinear equations for the residues at the poles considered as functions of the poles' positions. These are known as the Schlesinger equations, and they are integrable in the sense of Frobenius as well as of modern soliton theory. Indeed, there exists an additional linear operator involving deformation parameter which commutes with the original ODE.

Finally the Painlevé equations appear as isomonodromic deformations of particular two by two matrix linear ODEs. It is interesting to notice that traditional special functions can be considered as isomonodromic deformations of triangular linear systems. Hence one can define special functions, such as hypergeometric function, via isomonodromic deformations of triangular two by two matrix linear ODEs.

In the last two chapters, the rigorous Riemann-Hilbert formalism for the Painlevé I, II, III, IV, and V equations is developed together with its application to the theory of Bäcklund transformations.

CHAPTER 1

Systems of Linear Ordinary Differential Equations with Rational Coefficients. Elements of the General Theory

0. Introduction. The basic notations and facts

The following notations are used throughout the book:

- $\mathbb{N} = \{1, 2, \dots\}$ – the set of natural numbers;
- $\mathbb{Z} = \{\dots, -2, -1, 0, 1, 2, \dots\}$ – the set of integers;
- $\mathbb{C}$ – the complex plane;
- $\overline{\Omega}$ – the closure of a set $\Omega \in \mathbb{C}$;
- $\mathbb{C}P^1$ – the one-dimensional complex projective space; it is isomorphic to the compactified complex plane, or the Riemann sphere $\overline{\mathbb{C}}$;
- $\mathrm{Re}\ \lambda$, $\mathrm{Im}\ \lambda$ – real and imaginary parts of a complex number λ;
- $|\lambda|$ and $\arg \lambda$ – modulus and argument of a complex number λ;
- ξ – local parameter near the point $\lambda_0 \in \mathbb{C}P^1$. (This specific usage of the letter ξ is restricted to Part 1 only.);
- D_{λ_0} – disk with a center at the point λ_0;
- γ – path in a domain of the Riemann sphere; γ_{ab} – path, beginning at the point a and ending at the point b; two different paths γ_{ab} and $\tilde{\gamma}_{ab}$ are called homotopic, if they can be deformed continuously into each other;
- $\widetilde{X}$ – the universal covering of the Riemann surface X;
- I – the $N \times N$ unit matrix, its elements are given by the Kroneker symbol,

$$I_{ij} = \delta_{ij} = \begin{cases} 1, & i = j, \\ 0, & i \neq j; \end{cases}$$

- $[A, B] := AB - BA$ – matrix commutator;
- diagonal matrix, $D = \mathrm{diag}\,(d_1, \dots, d_N)$ is a matrix with diagonal elements equal to $D_{ii} = d_i$, and all other elements zero: $D_{ij} = 0$, $i \neq j$;
- off-diagonal matrix Y is a matrix with all diagonal elements equal to zero: $Y_{ii} = 0$, $i = 1, \dots, N$;
- $\mathrm{diag}\,(A) = \mathrm{diag}\,(A_{11}, \dots, A_{NN})$ is the diagonal part of the matrix A, i.e., $A - \mathrm{diag}\,(A)$ is an off-diagonal matrix;
- $\|A\|$ — a norm of a matrix A;
- $\mathrm{trace}\ (A) \equiv \mathrm{trace}\ A$ – trace of the square matrix A;
- $\det(A) \equiv \det A$ – determinant of the square matrix A;
- singular matrix is a matrix with zero determinant;
- $\mathbf{GL}(N, \mathbb{C})$ – the set of all $N \times N$ nonsingular complex matrices;
- $\mathbf{Mat}(N, \mathbb{C})$ – the set of all $N \times N$ complex matrices;

- $P^{-1}AP = \Lambda$ — the normal Jordan form of a matrix A; if Λ is diagonal, then it is called eigenvalue matrix, whereas matrix P is called eigenvector matrix of A.
- Pauli matrices

$$\sigma_1 = \begin{pmatrix} 0 & 1 \\ 1 & 0 \end{pmatrix}, \quad \sigma_2 = \begin{pmatrix} 0 & -i \\ i & 0 \end{pmatrix}, \quad \sigma_3 = \begin{pmatrix} 1 & 0 \\ 0 & -1 \end{pmatrix}.$$

- We shall call a matrix-valued function $f(\lambda)$ simply a "function $f(\lambda)$" if it is clear from a context that we are talking about a matrix-valued object;
- PI, PII, PIII, PIV, PV, and PVI - will denote Painlevé I, Painlevé II, Painlevé III, Painlevé IV, Painlevé V, and Painlevé VI equations, respectively;
- RH - is a shorthand notation for Riemann-Hilbert problem.

The following basic notations and facts concerning asymptotic expansions are used throughout the book. Proofs can be found, e.g., in [**Fe1**], [**Ol**], or [**W1**].

- Let Ω be a set on the complex x-plane such that $0 \in \overline{\Omega}$. Suppose that the matrix- (or vector-) valued functions $f(x)$ and $g(x)$ are defined on $\Omega \setminus \{0\}$. Then, the asymptotic relation

$$f(x) = O\Big(g(x)\Big), \quad x \in \Omega,$$

  means that there exists a positive constant C such that $\|f(x)\| \leq C\|g(x)\|$ for all $x \in \Omega$.
- Let Ω, $f(x)$, and $g(x)$ be as above. Then, the asymptotic relation

$$f(x) = O\Big(g(x)\Big), \quad x \to 0, \quad x \in \Omega,$$

  means that there exists a positive constants C and ρ such that $\|f(x)\| \leq C\|g(x)\|$ for all $x \in \Omega$, $0 < |x| \leq \rho$.
- Let Ω, $f(x)$, and $g(x)$ be as above. Then, the asymptotic relation

$$f(x) = o\Big(g(x)\Big), \quad x \to 0, \quad x \in \Omega,$$

  means that for every $\epsilon > 0$ there exists $\rho > 0$ such that $\|f(x)\| \leq \epsilon\|g(x)\|$ for all $x \in \Omega$, $0 < |x| \leq \rho$.
- Let Ω, $f(x)$, and $g(x)$ be as above, and, in addition, assume that $f(x)$ and $g(x)$ are scalar functions and that $g(x) \neq 0$. Then, the asymptotic relation

$$f(x) \sim g(x), \quad x \to 0,$$

  means that

$$\lim_{x \to 0} \frac{f(x)}{g(x)} = 1.$$

- The asymptotic relation,

$$f_1(x) = f_2(x) + O\Big(g(x)\Big), \quad x \to 0, \quad x \in \Omega,$$

  means that $f_1(x) - f_2(x) = O\Big(g(x)\Big), \quad x \to 0, \quad x \in \Omega.$

Similarly, with $O\Big(g(x)\Big)$ replaced by $o\Big(g(x)\Big)$.

- Let Ω be an open set such that $0 \in \overline{\Omega}$. Suppose that the matrix- (or vector-) valued functions $f(x)$ and $g(x)$ are defined on $\Omega \setminus \{0\}$. Then the asymptotic relation

$$f(x) = O\Big(g(x)\Big), \quad x \to 0, \quad x \in \Omega,$$

is said to be valid uniformly on Ω if

$$f(x) = O\Big(g(x)\Big), \quad x \to 0, \quad x \in \Omega_1,$$

holds for every compact set Ω_1 such that $0 \in \Omega_1$ and $\Omega_1 \subset \Omega \cup \{0\}$. Similarly if $O\Big(g(x)\Big)$ is replaced by $o\Big(g(x)\Big)$. Unless otherwise indicated, for open sets Ω, the asymptotic relations

$$f(x) = O\Big(g(x)\Big), \quad x \to 0, \quad x \in \Omega,$$

and

$$f(x) = o\Big(g(x)\Big), \quad x \to 0, \quad x \in \Omega,$$

will always be assumed to be valid uniformly on Ω.

- Let

$$F_f(x) := \sum_{k=0}^{\infty} F_k x^k$$

be a formal power series. We say that $F_f(x)$ is the asymptotic series (or expansion) of the function $F(x)$ at $x = 0$, and we will write,

$$F(x) \sim F_f(x), \quad x \to 0, \quad x \in \Omega, \tag{1.0.1}$$

if for every integer $m > 0$, the following asymptotic relation takes place:

$$F(x) = \sum_{k=0}^{m-1} F_k x^k + O\Big(x^m\Big), \quad x \to 0, \quad x \in \Omega.$$

Let Ω be the sectorial domain,

$$\Omega = \Big\{0 < |x| < \rho, \ \alpha < \arg x < \beta\Big\}.$$

Then, in more detail, equation (1.0.1) means that for any positive numbers $m \in \mathbb{N}$, $\rho_1 < \rho$, and $\alpha < \alpha_1 < \beta_1 < \beta$, there exists a positive constant $C_{m,\rho_1,\alpha_1,\beta_1}$ such that the inequality,

$$\Bigg\|F(x) - \sum_{k=0}^{m-1} F_k x^k\Bigg\| < C_{m,\rho_1,\alpha_1,\beta_1}|x|^m \tag{1.0.2}$$

holds for every x satisfying $0 < |x| \le \rho_1$ and $\alpha_1 \le \arg x \le \beta_1$.

- Suppose that Ω is a sectorial domain as defined above and assume that $F(x)$ is holomorphic in Ω. Then, the asymptotic relation (1.0.1) can be integrated and differentiated term by term with respect to x.
- Suppose that Ω is a sectorial domain as defined above. Suppose also that the function $F(x)$ depends holomorphically on the set of extra parameters $t \equiv (t_1, t_2, \dots, t_p)$, $\ t \in \mathcal{D}$, where $\mathcal{D}$ is a polydisk (in addition to the holomorphicity with respect to the basic variable x). Suppose also that all the coefficients F_k of the series $F_f(x)$ are holomorphic functions on

the polydisk $\mathcal{D}$. Assume, in addition, that the asymptotic relation (1.0.1) holds *uniformly* with respect to t, i.e., for every compact subset $\mathcal{D}_1 \subset \mathcal{D}$, the inequality (1.0.2) holds for all $t \in \mathcal{D}_1$ with the *same* constant $C_{m,\rho_1,\alpha_1,\beta_1} \equiv C_{m,\rho_1,\alpha_1,\beta_1}(\mathcal{D}_1)$. Then, the asymptotic relation (1.0.1) can be integrated and differentiated term by term with respect to each t_j (in addition to the same properties with respect to the basic variable x).

- Let $\{g_k(x)\}_{k=0}^{\infty}$ be a sequence of functions such that

$$g_{k+1}(x) = o\Big(g_k(x)\Big), \quad x \to 0, \quad x \in \Omega,$$

and Ω is a sectorial domain (e.g., $g_k(x) = x^{\frac{k}{2}}$). Then the notion of the asymptotic series (1.0.1) can be generalized as follows: We will say that

$$F(x) \sim \sum_{k=0}^{\infty} F_k g_k(x), \quad x \to 0, \quad x \in \Omega,$$

if for every integer $m > 0$, we have

$$F(x) = \sum_{k=0}^{m-1} F_k g_k(x) + O\Big(g_m(x)\Big), \quad x \to 0, \quad x \in \Omega.$$

- All the definitions concerning the asymptotic behavior of a function $f(x)$ as $x \to x_0 \neq 0$ or as $x \to \infty$ are obtained from the relevant definitions concerning the asymptotic behavior of the function $\tilde{f}(y) \equiv f(x(y))$ as $y \to 0$ and $y := x - x_0$ or $y := x^{-1}$, respectively. For instance, the asymptotic relation,

$$F(x) \sim \sum_{k=1}^{\infty} F_k x^{-k}, \quad x \to \infty, \quad \alpha < \arg x < \beta, \tag{1.0.3}$$

means that for any positive numbers $m \in \mathbb{N}$, $0 < \rho$, and $\alpha < \alpha_1 < \beta_1 < \beta$, there exists a positive constant $C_{m,\rho,\alpha_1,\beta_1}$, such that the inequality,

$$\left\| F(x) - \sum_{k=0}^{m-1} F_k x^{-k} \right\| < C_{m,\rho,\alpha_1,\beta_1} |x|^{-m}, \tag{1.0.4}$$

holds for every x satisfying $|x| \geq \rho$ and $\alpha_1 \leq \arg x \leq \beta_1$. The asymptotic relation (1.0.3) can be integrated and differentiated with respect to x.

The following basic notations and facts concerning the theory of linear ODEs are used throughout the book. Proofs can be found, e.g., in [**Bal**], [**CodL**], [**Si2**], or [**W1**].

- A first order linear ODE is a differential equation of the form

$$\frac{d\Psi}{d\lambda} = \mathsf{A}(\lambda)\Psi, \tag{1.0.5}$$

where $\mathsf{A}(\lambda)$ is a given $N \times N$ ($N > 1$) matrix-valued function of the complex variable λ holomorphic on a domain $\Omega \subset \mathbb{C}$. The "unknown variable" $\Psi(\lambda)$ is also supposed to be an $N \times N$ matrix-valued function.

Each column of the matrix function $\Psi(\lambda)$ satisfies the system of linear differential (scalar) equations,

$$\frac{dy_j}{d\lambda} = \sum_{k=1}^{N} a_{jk}(\lambda) y_k, \quad j = 1, \ldots, N,$$

where $a_{jk}(\lambda)$ denote the entries of the matrix $\mathsf{A}(\lambda)$. In reference to (1.0.5), we shall use the following terms as equivalent: "differential equation", "matrix differential equation", "system of differential equations", or simply "equation", "system".

- For every point $\lambda_0 \in \Omega$, there is a unique *fundamental solution* $\Psi(\lambda)$ of equation (1.0.5), which is holomorphic at λ_0 and satisfies the normalization relation,

$$\Psi(\lambda_0) = I.$$

Any other solution $\Phi(\lambda)$ of (1.0.5), holomorphic at λ_0, is given by the equation,

$$\Phi(\lambda) = \Psi(\lambda) C,$$

where C is a constant matrix.

- Let $\Psi(\lambda)$ be a solution of equation (1.0.5). Then (Liouville's formula),

(1.0.6) $$\frac{d \ln \det \Psi(\lambda)}{d\lambda} = \operatorname{trace} \mathsf{A}(\lambda).$$

In particular, the following implication is true,

$$\operatorname{trace} \mathsf{A}(\lambda) \equiv 0 \quad \Longleftrightarrow \quad \det \Psi(\lambda) \equiv \text{constant}.$$

- Let $g(\lambda)$ denote a holomorphic and invertible (i.e., $\det g(\lambda) \neq 0$ for all λ) matrix-valued function on Ω. Then the two linear systems,

$$\frac{d\Psi}{d\lambda} = \mathsf{A}(\lambda)\Psi,$$

and

$$\frac{d\Phi}{d\lambda} = \mathsf{B}(\lambda)\Phi,$$

are called *gauge equivalent* if

(1.0.7) $$\mathsf{B}(\lambda) = g(\lambda)\mathsf{A}(\lambda)g^{-1}(\lambda) + \frac{dg(\lambda)}{d\lambda} g^{-1}(\lambda).$$

Here, $g \to g^{-1}$ means matrix inversion. In terms of the solutions, the gauge transformation (1.0.7) reads as

$$\Psi(\lambda) \to \Phi(\lambda) = g(\lambda)\Psi(\lambda).$$

Note, that, in virtue of Liouville's formula (1.0.6), any given linear system (1.0.5) can be easily transformed to one whose coefficient matrix has identically zero trace. Indeed, taking in (1.0.7) the gauge matrix

$$g(\lambda) = \exp\Big(-\frac{1}{N}\int^{\lambda} \operatorname{trace} \mathsf{A}(\mu)\, d\mu\Big)\, I,$$

we obtain that

$$\mathsf{B}(\lambda) = \mathsf{A}(\lambda) - \frac{1}{N} \operatorname{trace} \mathsf{A}(\lambda)\, I,$$

and hence $\operatorname{trace} \mathsf{B}(\lambda) \equiv 0$.

1. The local theory

1.1. Classification of singular points. Let $\mathsf{A}(\lambda)$ be an $N \times N$ $(N > 1)$ matrix-valued rational function of the complex variable λ. Consider in the complex λ-plane the first order linear ODE,

$$\frac{d\Psi}{d\lambda} = \mathsf{A}(\lambda)\Psi, \tag{1.1.1}$$

for an $N \times N$ matrix-valued function $\Psi(\lambda)$. The local analytic theory of systems of linear ordinary differential equations (ODEs) with rational coefficients deals with the behavior of the solutions of equation (1.1.1) near a given point $\lambda_0 \in \mathbb{C}P^1$. This behavior depends on which of the following three cases occurs:

- The coefficient matrix $\mathsf{A}(\lambda)$ is holomorphic at the point λ_0, and, if $\lambda_0 = \infty$, it has a zero at λ_0 of no less than second order.
- The coefficient matrix $\mathsf{A}(\lambda)$ has a simple pole at the point λ_0, if $\lambda_0 \neq \infty$, or, if $\lambda_0 = \infty$, it has a simple zero at λ_0.
- The coefficient matrix $\mathsf{A}(\lambda)$ has a multiple pole at the point λ_0, if $\lambda_0 \neq \infty$, or, if $\lambda_0 = \infty$, it does not vanish (but might have a pole) at λ_0.

Indeed in the first case the fundamental solution of (1.1.1) is always holomorphic at λ_0. In the second and third cases, the fundamental solution has in general a singularity at λ_0. Moreover, in the second case the possible growth of $\Psi(\lambda)$ as $\lambda \to \lambda_0$ is polynomial, while in the third case, in general, $\Psi(\lambda)$ has a more complicated exponential-polynomial behavior. Furthermore, in the second and the third cases, the fundamental solution of equation (1.1.1), in general, has a branch point at λ_0. Later in this section, we will formulate and discuss the classical theorems which specify exactly the types of local analytic behavior indicated above.

In the first case, the point λ_0 is usually referred to as a *regular point* of equation (1.1.1). In the second case, equation (1.1.1) is said to be *Fuchsian at* λ_0, and the point λ_0 is called a *Fuchsian singular point.* Finally, in the third case, the point λ_0 is called a *non-Fuchsian singular point* of equation (1.1.1). In terms of the Taylor and Laurent series of the coefficient matrix $\mathsf{A}(\lambda)$ we have the following:

- λ_0 is a regular point of (1.1.1) iff $\mathsf{A}(\lambda)$ has at λ_0 a Taylor expansion of the form,

$$\mathsf{A}(\lambda) = \sum_{k=0}^{\infty} A_k(\lambda - \lambda_0)^k, \quad \lambda_0 \neq \infty, \tag{1.1.2}$$

$$\mathsf{A}(\lambda) = -\sum_{k=-\infty}^{-2} A_{-k-2}\lambda^k, \quad \lambda_0 = \infty.$$

- λ_0 is a Fuchsian singular point of (1.1.1) iff $\mathsf{A}(\lambda)$ has at λ_0 a Laurent expansion of the form,

$$\mathsf{A}(\lambda) = \sum_{k=-1}^{\infty} A_{k+1}(\lambda - \lambda_0)^k, \quad \lambda_0 \neq \infty, \tag{1.1.3}$$

$$\mathsf{A}(\lambda) = -\sum_{k=-\infty}^{-1} A_{-k-1}\lambda^k, \quad \lambda_0 = \infty.$$

$$A_0 \neq 0.$$

- λ_0 is a non-Fuchsian singular point of (1.1.1) iff $\mathsf{A}(\lambda)$ has at λ_0 a Laurent expansion of the form,

$$\mathsf{A}(\lambda) = \sum_{k=-r-1}^{\infty} A_{k+1}(\lambda - \lambda_0)^k, \quad \lambda_0 \neq \infty, \tag{1.1.4}$$

$$\mathsf{A}(\lambda) = -\sum_{k=-\infty}^{r-1} A_{-k-1}\lambda^k, \quad \lambda_0 = \infty,$$

$$A_{-r} \neq 0, \quad r > 0.$$

Introducing the local variable,

$$\xi = \begin{cases} \lambda - \lambda_0, & \text{if } \lambda_0 \neq \infty, \\ 1/\lambda, & \text{if } \lambda_0 = \infty, \end{cases} \tag{1.1.5}$$

and the disk D_{λ_0} centered at λ_0,

$$D_{\lambda_0} = \begin{cases} \{\lambda \in \mathbb{C} \colon |\lambda - \lambda_0| < \rho\}, \quad 0 < \rho \leq \infty, \quad \text{if} \quad \lambda_0 \neq \infty, \\ \{\lambda \in \mathbb{C} \colon |\lambda| > \rho\} \cup \{\infty\}, \quad 0 \leq \rho < \infty, \quad \text{if} \quad \lambda_0 = \infty, \end{cases} \tag{1.1.6}$$

we can reformulate the above classification in terms of the differential form $\mathsf{A}(\lambda)\, d\lambda$ in the following compact way:

- λ_0 is a regular point of (1.1.1) iff the differential form $\mathsf{A}(\lambda)\, d\lambda$ is holomorphic at λ_0, i.e., if there exists a disk D_{λ_0} such that

$$\mathsf{A}(\lambda)\, d\lambda = \Big(\sum_{k=0}^{\infty} A_k \xi^k\Big)\, d\xi, \quad \lambda \in D_{\lambda_0}. \tag{1.1.7}$$

- λ_0 is a Fuchsian singular point of (1.1.1) iff the differential form $\mathsf{A}(\lambda)\, d\lambda$ has a simple pole at λ_0, i.e., if there exists a disk D_{λ_0} such that

$$\mathsf{A}(\lambda)\, d\lambda = \Big(\sum_{k=-1}^{\infty} A_{k+1} \xi^k\Big)\, d\xi, \quad \lambda \in D_{\lambda_0} \setminus \{\lambda_0\}, \tag{1.1.8}$$

$$A_0 \neq 0.$$

- λ_0 is a non-Fuchsian singular point of (1.1.1) iff the differential form $\mathsf{A}(\lambda)\, d\lambda$ has a multiple pole at λ_0, i.e., if there exists a disk D_{λ_0} such that

$$\mathsf{A}(\lambda)\, d\lambda = \Big(\sum_{k=-r-1}^{\infty} A_{k+1} \xi^k\Big)\, d\xi, \quad \lambda \in D_{\lambda_0} \setminus \{\lambda_0\}, \tag{1.1.9}$$

$$A_{-r} \neq 0, \quad r > 0.$$

We also note that in the non-Fuchsian case the positive number r is called the *Poincaré rank* of the singularity.

We now present a detailed discussion of the behavior of the solutions of equation (1.1.1) near its regular, Fuchsian, and non-Fuchsian points.

Throughout the rest of this section we shall use the notations ξ and D_{λ_0} introduced by equations (1.1.5) and (1.1.6), respectively.

1.2. A regular point. The principal analytic fact of the local theory of equation (1.1.1) near its regular point is given by the following theorem.

THEOREM 1.1. *Let* $\mathsf{A}(\lambda)$ *be a* $N \times N$ *matrix-valued function holomorphic in the disk* D_{λ_0}, *and let* Ψ_0 *be a constant* $N \times N$ *matrix. Then there exists a unique solution* $\Psi(\lambda)$ *of equation (*1.1.1*) holomorphic in* D_{λ_0} *and such that* $\Psi(\lambda_0) = \Psi_0$.

Sketch of the Proof.

We represent the matrix $\mathsf{A}(\lambda)$ by its Taylor expansion (1.1.2), and we look for a solution $\Psi(\lambda)$ in the form of a similar power series,

$$\Psi(\lambda) = \sum_{k=0}^{\infty} \Psi_k \xi^k. \tag{1.1.10}$$

Substituting equation (1.1.10) into (1.1.1) it follows that for this series to be a *formal solution* of equation (1.1.1) its coefficients must satisfy the following recursive equation:

$$(k+1)\Psi_{k+1} = \sum_{m=0}^{k} A_{k-m}\Psi_m, \quad k = 0, 1, 2, \ldots \tag{1.1.11}$$

This relation determines the coefficients Ψ_k uniquely via the given initial matrix Ψ_0 and hence the uniqueness part of the theorem is proven. To prove the existence part, it is enough to show that the series (1.1.10), with the coefficients Ψ_k determined recurrently by (1.1.11), converges in D_{λ_0}. A proof that this is indeed the case (which follows from the convergence of series (1.1.2)) can be found e.g. in **[Si2]**, **[Fr]**. □

We note that the maximal radius of convergence of the series (1.1.10) coincides with the distance between the point λ_0 and the closest pole of $\mathsf{A}(\lambda)$.

Given a regular point λ_0, we shall call the Taylor series (1.1.10) with its coefficients defined according to (1.1.11) *the germ of a solution of equation (*1.1.1*) at the point* λ_0. We shall also use the term "germ" for the analytic function $\Psi(\lambda)$ which is defined by the Taylor series (1.1.10) in any disk D_{λ_0} whose closure belongs to the domain of analyticity of $\mathsf{A}(\lambda)$.

Suppose that in (1.1.10) $\det \Psi_0 \neq 0$. Then we will call the function $\Psi(\lambda)$, as well as its Taylor series, *the germ of a fundamental solution of equation (*1.1.1*) at the point* λ_0.

1.3. A Fuchsian singular point. Suppose that $\lambda_0 \in \mathbb{CP}^1$ is a Fuchsian singular point of equation (1.1.1), i.e., λ_0 is a simple pole of $\mathsf{A}(\lambda)\,d\lambda$. Before formulating the analog of Theorem 1.1 let us first discuss the formal solution of equation (1.1.1) in this case.

We represent the matrix $\mathsf{A}(\lambda)$ by its Laurent expansion (1.1.3), so that (cf. (1.1.8))

$$\mathsf{A}(\lambda) = \Big(\sum_{k=-1}^{\infty} A_{k+1}\xi^k\Big)\frac{d\xi}{d\lambda}. \tag{1.1.12}$$

If we look for a formal solution $\Psi(\lambda)$ in the form of a similar power series,

$$\Psi(\lambda) = \sum_{k=-1}^{\infty} \Psi_{k+1}\xi^k, \tag{1.1.13}$$

we find the constraint,

$$-\Psi_0 = A_0\Psi_0,$$

on the "initial" matrix Ψ_0; thus the series (1.1.13) cannot characterize the generic behavior of the solutions of equation (1.1.1) at its Fuchsian point. In order to motivate the correct ansatz that must replace (1.1.13), let us consider the model situation,

$$\mathsf{A}(\lambda) = \frac{A_0}{\lambda - \lambda_0}, \quad \lambda_0 \in \mathbb{C}.$$

In this case equation (1.1.1) becomes

$$\frac{d\Psi}{d\lambda} = \frac{A_0}{\lambda - \lambda_0}\Psi \tag{1.1.14}$$

and can be solved explicitly. Indeed we have

$$\Psi(\lambda) = (\lambda - \lambda_0)^{A_0}. \tag{1.1.15}$$

Let Λ_0 be the canonical Jordan form of A_0,

$$P^{-1}A_0P = \Lambda_0, \quad \det P \neq 0. \tag{1.1.16}$$

Then (1.1.15) can be rewritten as

$$\Psi(\lambda) = P(\lambda - \lambda_0)^{\Lambda_0}P^{-1}.$$

Dropping the right nonsingular constant matrix multiplier P^{-1} and using the local variable ξ, we arrive at the following form of the fundamental solution of equation (1.1.14),

$$\Psi(\lambda) = P\xi^{\Lambda_0}. \tag{1.1.17}$$

This formula indicates that the solutions of (1.1.1) near the simple pole of $\mathsf{A}(\lambda)$ have more complicated behavior than the one indicated in (1.1.13); indeed, in general, λ_0 is a branch point of algebraic or of logarithmic type of the function (1.1.17).

Equation (1.1.17) suggests the following ansatz for the fundamental solution of equation (1.1.1) in the case that λ_0 is a simple pole of $\mathsf{A}(\lambda)\,d\lambda$:

$$\Psi(\lambda) = P\Big(\sum_{k=0}^{\infty} \Psi_k\xi^k\Big)\xi^{\Lambda_0}, \quad \Psi_0 = I \tag{1.1.18}$$

(cf. (1.1.10)). Substituting the expansions (1.1.12) and (1.1.18) into (1.1.1) and equating the coefficients of ξ^k, we obtain the following analog of the recursive equation (1.1.11):

$$k\Psi_k + [\Psi_k, \Lambda_0] = B_k, \quad B_k = \sum_{m=0}^{k-1} P^{-1}A_{k-m}P\Psi_m, \quad k = 1, 2, \ldots, \tag{1.1.19}$$

where brackets denote the usual matrix commutator.

Let us assume that the leading coefficient A_0 of the Laurent expansion (1.1.12) is a diagonalizable matrix. Then, one of the following two cases can occur.

The nonresonant (generic) case: The eigenvalues of the matrix A_0 are distinct modulo a nonzero integer, i.e.,

$$(\Lambda_0)_{ij} = \alpha_i\delta_{ij}, \quad \alpha_i - \alpha_k \notin \mathbb{Z}\setminus\{0\} \quad \forall i, k\colon\ i \neq k. \tag{1.1.20}$$

Then the coefficients Ψ_k, $k \geq 1$, of the expansion (1.1.18) are determined recursively by (1.1.19),

$$(\Psi_k)_{ij} = \frac{(B_k)_{ij}}{k + \alpha_j - \alpha_i}, \quad k = 1, 2, \ldots. \tag{1.1.21}$$

Thus equation (1.1.1), where $\mathsf{A}(\lambda)$ is given by equation (1.1.12) admits the formal solution given by the power series (1.1.18).

The resonant (general) case: the matrix A_0 is diagonalizable, but it has eigenvalues which coincide modulo a nonzero integer. In this case the ansatz (1.1.18) must be modified as follows:

$$\Psi(\lambda) = P\Big(\sum_{k=0}^{\infty} \Psi_k \xi^k\Big)\xi^{\Lambda_0}\xi^{\mathsf{N}}, \tag{1.1.22}$$

where N denotes a (nilpotent) matrix which satisfies the condition

$$(\mathsf{N})_{ij} = 0 \quad \text{unless} \quad \alpha_i - \alpha_j \in \mathbb{N}, \quad i, j = 1, \ldots, N. \tag{1.1.23}$$

Substituting the series (1.1.12) and (1.1.22) into equation (1.1.1), we find the following formal series identity (cf. (1.1.19)),

$$\sum_{k=0}^{\infty} \xi^k (k\Psi_k + [\Psi_k, \Lambda_0]) + \sum_{k=0}^{\infty} \xi^k \Psi_k (\xi^{\Lambda_0}\mathsf{N}\xi^{-\Lambda_0}) = \sum_{k=0}^{\infty} \xi^k B_k,$$

$$B_k = \sum_{m=0}^{k-1} P^{-1} A_{k-m} P \Psi_m. \tag{1.1.24}$$

We observe that the condition (1.1.23) implies that the product $\xi^{\Lambda_0}\mathsf{N}\xi^{-\Lambda_0}$ is a matrix polynomial in ξ. Therefore, equation (1.1.24) yields a recursive relation between the coefficients Ψ_k. It can be shown that this relation determines, although not uniquely, all the coefficients Ψ_k, $k \geq 1$ as well as all the nontrivial entries of the matrix N.

We skip the general resonant case [**Si2**], [**W1**] and consider instead the simplified situation when only one pair of eigenvalues of A_0 coincides modulo a nonzero integer. In fact, this is the only case we will actually need in the main part of the book. Let α_p, α_q be the above mentioned singular pair of eigenvalues of A_0, i.e.,

$$\alpha_p - \alpha_q = l \in \mathbb{N},$$

while all other pairs of α_j are distinct modulo a nonzero integer. Then in general the matrix N will have only one nontrivial entry (p, q),

$$(\mathsf{N})_{ij} = \mathcal{N}\delta_{pi}\delta_{qj},$$

where the constant $\mathcal{N}$ will be determined later. This implies that the entry (i, j) of the matrix equation (1.1.24) can be written as

$$\sum_{k=0}^{\infty} \xi^k (k + \alpha_j - \alpha_i)(\Psi_k)_{ij} + \xi^l \mathcal{N}\delta_{ip}\delta_{qj}$$

$$= -\sum_{k=l+1}^{\infty} \xi^k (\Psi_{k-l})_{ip}\mathcal{N}\delta_{qj} + \sum_{k=0}^{\infty} \xi^k (B_k)_{ij}. \tag{1.1.25}$$

(We recall that we have choosen $\Psi_0 = I$.) Equating in (1.1.25) the coefficients of ξ^k for $k < l$, we obtain equation (1.1.19) and hence determine via (1.1.21) all the coefficients Ψ_k with $k < l$ (note that for $k < l$ the denominator of (1.1.21) does *not* vanish). If $k = l$, all the entries $(i, j) \neq (p, q)$ of the coefficient Ψ_l are determined again by (1.1.21) while the (p, q) entry remains undetermined. In fact, since $\alpha_p - \alpha_q = l$, the term containing the entry $(\Psi_l)_{pq}$ in the left-hand side of (1.1.25) vanishes. For $k = l$, equation (1.1.25) with $i = p$, $j = q$, becomes an equation for $\mathcal{N}$,

$$\mathcal{N} = (B_l)_{pq}. \tag{1.1.26}$$

For all $k > l$, we have $k + \alpha_j - \alpha_i \neq 0$, for all i and j. Therefore, all the coefficients Ψ_k, $k > l$, can be determined recursively from (1.1.25) according to the formula,

$$(\Psi_k)_{ij} = \frac{(B_k)_{ij} - (\Psi_{k-l})_{ip}\mathcal{N}\delta_{qj}}{k + \alpha_j - \alpha_i}, \quad k = l+1, l+2, \ldots \tag{1.1.27}$$

We note that the entry $(\Psi_l)_{pq}$ remains a free parameter of the expansion (1.1.22).

The above discussion concerns a diagonalizable $N \times N$ matrix A_0. Consider now the case when the Jordan form Λ_0 of the matrix A_0 is an elementary Jordan cell,

$$\Lambda_0 = \alpha I + J, \quad J_{ij} = \delta_{i+1,j}, \quad J^N = 0; \tag{1.1.28}$$

this is the general case of a nondiagonal Jordan form for 2×2 matrices. We also resume our original ansatz (1.1.18) for the formal fundamental solution $\Psi(\lambda)$. Therefore, the basic recursive equation (1.1.19) holds, and, in virtue of (1.1.28), it can be rewritten as

$$\Psi_0 = I, \quad k(\Psi_k)_{i,j} + (\Psi_k)_{i,j-1} - (\Psi_k)_{i+1,j} = (B_k)_{i,j}, \quad k \geq 1, \tag{1.1.29}$$

where we put $(\Psi_k)_{i,0} = (\Psi_k)_{N+1,j} = 0$. Equation (1.1.29) allows us to compute the entries of Ψ_k recursively starting from the left lower corner $(\Psi_k)_{N,1}$,

$$(\Psi_k)_{N,1} = \frac{1}{k}(B_k)_{N,1},$$

and then moving to the next entry to the right or to the entry above,

$$(\Psi_k)_{N-i,1} = \frac{1}{k}\big((B_k)_{N-i,1} + (\Psi_k)_{N-i+1,1}\big), \quad i = 1, \ldots, N-1,$$

$$(\Psi_k)_{N,j} = \frac{1}{k}\big((B_k)_{N,j} - (\Psi_k)_{N,j-1}\big), \quad j = 2, \ldots, N,$$

$$(\Psi_k)_{i-1,j+1} = \frac{1}{k}\big((B_k)_{i-1,j+1} - (\Psi_k)_{i-1,j} + (\Psi_k)_{i,j+1}\big).$$

Expansions (1.1.18) and (1.1.22) were constructed as formal series. One of the main results of the local theory of linear ODEs is that these series are in fact convergent series. Indeed the following fundamental theorems are valid (see [**Bal**], [**CodL**], [**Si2**], or [**W1**]).

THEOREM 1.2 (nonresonant case). *Let $A(\lambda)$ be a $N \times N$ matrix-valued function holomorphic in the punctured disk $D_{\lambda_0} \setminus \{\lambda_0\}$, and let $\lambda_0 \in \mathbb{CP}^1$ be a simple pole of $A(\lambda)\,d\lambda$. Suppose also that the Jordan form Λ_0 of the residue $A_0 = res_{\lambda=\lambda_0}\{A(\lambda)\,d\lambda\}$ is either a diagonal matrix with entries distinct modulo a nonzero integer (generic case), i.e.,*

$$(\Lambda_0)_{ij} = \alpha_i\delta_{ij}, \quad \alpha_i - \alpha_k \notin \mathbb{Z} \setminus \{0\} \quad \forall i, k\colon\ i \neq k, \tag{1.1.30}$$

or, it is given by an elementary Jordan cell, i.e.,

$$\Lambda_0 = \alpha I + J, \quad J_{ij} = \delta_{i+1,j}, \quad J^N = 0. \tag{1.1.31}$$

Then equation (1.1.1) has a fundamental solution $\Psi(\lambda)$ of the form

$$\Psi(\lambda) = \hat{\Psi}(\lambda)\xi^{\Lambda_0}, \tag{1.1.32}$$

where the matrix function $\hat{\Psi}(\lambda)$ is holomorphic and invertible in D_{λ_0}, and

$$\Lambda_0 = [\hat{\Psi}(\lambda_0)]^{-1}A_0\hat{\Psi}(\lambda_0).$$

(In other words, the series in (1.1.18) is a convergent series.) Moreover, the function $\hat{\Psi}(\lambda)$ in (1.1.32) is uniquely determined by the choice of the matrix $P \equiv \hat{\Psi}(\lambda_0)$.

THEOREM 1.3 (resonant case). *Let $\mathsf{A}(\lambda)$ be a $N \times N$ matrix-valued function holomorphic in the punctured disk $D_{\lambda_0} \setminus \{\lambda_0\}$, and let $\lambda_0 \in \mathbb{CP}^1$ be a simple pole of $\mathsf{A}(\lambda)\,d\lambda$. Suppose also that the Jordan form Λ_0 of the residue $A_0 = res_{\lambda=\lambda_0}\{\mathsf{A}(\lambda)\,d\lambda\}$ is a diagonal matrix, i.e.,*

$$(\Lambda_0)_{ij} = \alpha_i\delta_{ij}. \tag{1.1.33}$$

Then equation (1.1.1) has a fundamental solution $\Psi(\lambda)$ of the form

$$\Psi(\lambda) = \hat{\Psi}(\lambda)\xi^{\Lambda_0}\xi^{\mathsf{N}}, \tag{1.1.34}$$

where N is a constant (nilpotent) matrix which satisfies the condition

$$(\mathsf{N})_{ij} = 0 \quad \textit{unless} \quad \alpha_i - \alpha_j \in \mathbb{N}, \quad i,j = 1,\dots,N. \tag{1.1.35}$$

The matrix function $\hat{\Psi}(\lambda)$ is holomorphic and invertible in D_{λ_0}, and

$$\Lambda_0 = [\hat{\Psi}(\lambda_0)]^{-1}A_0\hat{\Psi}(\lambda_0).$$

(In other words, the series in (1.1.22) is a convergent series.)

REMARK 1.1. Because of the presence of the factors,

$$\xi^{\Lambda_0} \equiv e^{\Lambda_0 \ln \xi}, \quad \text{and} \quad \xi^{\mathsf{N}} \equiv e^{\mathsf{N}\ln \xi}, \tag{1.1.36}$$

the function $\Psi(\lambda)$ in the theorems above is defined on the universal covering, $\widetilde{D_{\lambda_0} \setminus \{\lambda_0\}}$, of the punctured disk $D_{\lambda_0} \setminus \{\lambda_0\}$. This means that in concrete calculations involving $\Psi(\lambda)$, the branch of the $\ln\xi$ in (1.1.36) must be specified.

It is important to note that in all cases analyzed above the fundamental solution can be rewritten in the following universal form:

$$\Psi(\lambda) = \Big(\sum_{k=0}^{\infty}\Psi_k\xi^k\Big)\xi^{\Delta}\xi^{\mathcal{E}}, \quad \det\Psi_0 \neq 0, \tag{1.1.37}$$

where Δ is a diagonal integer valued matrix, $\mathcal{E}$ is a matrix whose eigenvalues e_j satisfy the inequalities,

$$0 \le \text{Re } e_j < 1,$$

and the series $\left(\sum_{k=0}^{\infty}\Psi_k\xi^k\right)$ converges in D_{λ_0}. It actually can be shown (see e.g. [**Bol1**]) that the representation (1.1.37) holds in the general case, i.e., no assumptions for the structure of the Jordan form Λ_0 of the matrix residue A_0 are needed. We also note that, for the 2×2 systems (1.1.1), Theorems 1.2 and 1.3 cover, in fact, all the possible types of the Jordan form Λ_0.

DEFINITION 1.1. We shall call the matrix Λ_0 in the nonresonant case, the pair $\{\Lambda_0, \mathsf{N}\}$ in the resonant case, and the matrix $\mathcal{E}$ in the most general case, the *formal monodromy exponents* associated to the Fuchsian singular point λ_0.

REMARK 1.2. If λ_0 is a pole of $\mathsf{A}(\lambda)\,d\lambda$, and all the solutions of (1.1.1) have no more than power-like growth in the neighborhood of λ_0, then the point λ_0 is called a *regular singular point* of equation (1.1.1). The above results imply that every simple pole of $\mathsf{A}(\lambda)\,d\lambda$, i.e., every Fuchsian singular point, is a regular singular point. (It is important to notice that the reverse statement is not true; see [**Bol1**] for this important distinction.)

1.4. An irregular singular point. The Stokes phenomenon. Let $\lambda_0 \in \mathbb{C}P^1$ be a non-Fuchsian singular point of equation (1.1.1), i.e., let equation (1.1.1) have the following local form

$$\frac{d\Psi}{d\xi} = \Big(\sum_{k=-r-1}^{\infty} A_{k+1}\xi^k\Big)\Psi, \quad \lambda \in D_{\lambda_0} \setminus \{\lambda_0\}, \quad r > 0, \tag{1.1.38}$$

where ξ and D_{λ_0} are the local variable and the disk centered at λ_0 respectively (see (1.1.5) and (1.1.6)). Assume the following

Main assumption: *The coefficient A_{-r} of the leading order singularity in the Laurent series in* (1.1.38) *has distinct eigenvalues.*

This means that we only discuss the *generic case* when the Jordan form Λ_{-r} of the matrix A_{-r} is a diagonal matrix with distinct entries,

$$\begin{gathered} P^{-1}A_{-r}P = \Lambda_{-r}, \quad \det P \neq 0, \\ \Lambda_{-r} = \operatorname{diag}(\alpha_1, \dots, \alpha_N), \quad \alpha_i \neq \alpha_j, \quad i \neq j. \end{gathered} \tag{1.1.39}$$

DEFINITION 1.2. A non-Fuchsian singular point λ_0 will also be called an *irregular singular point.*

The *first* main difference between Fuchsian and non-Fuchsian cases is the difference in the structures of the respective formal solutions. Indeed, the relevant model equation,

$$\frac{d\Psi}{d\lambda} = \frac{A_{-r}}{\xi^{r+1}}\Psi,$$

has the function

$$\Psi(\xi) = P\exp\left\{\frac{\Lambda_{-r}}{-r}\xi^{-r}\right\}$$

as its fundamental solution. This shows that a formal series satisfying (1.1.38) involves an exponential factor, which indicates the presence of an essential singularity. Actually, the following statement is true.

PROPOSITION 1.1. [**JMU**], [**W1**] *Under the generic conditions* (1.1.39), *equation* (1.1.38) *has a unique* formal *fundamental solution*

$$\begin{aligned} \Psi_f(\lambda) &= P\Big(\sum_{k=0}^{\infty}\Psi_k\xi^k\Big)\xi^{\Lambda_0}\exp\left\{\frac{\Lambda_{-r}}{-r}\xi^{-r} + \dots + \frac{\Lambda_{-1}}{-1}\xi^{-1}\right\} \\ &\equiv P\Big(\sum_{k=0}^{\infty}\Psi_k\xi^k\Big)e^{\Lambda(\xi)}, \quad \Psi_0 = I, \end{aligned} \tag{1.1.40}$$

where

$$\Lambda(\xi) = \sum_{k=-r}^{-1}\frac{\Lambda_k}{k}\xi^k + \Lambda_0\ln\xi, \tag{1.1.41}$$

all the matrices Λ_k, $k=-r,\dots,0$, *are diagonal, and the matrix* Λ_{-r} *is defined in* (1.1.39). *All the coefficients* Ψ_k *as well as the diagonal exponent* $\Lambda(\xi)$ *in* (1.1.40) *can be determined recursively, via the Laurent series* (1.1.38), *as polynomials of the matrix coefficients* A_k, $k\geq -r$.

PROOF. It is convenient to rewrite (1.1.40) in the following equivalent form,

$$\Psi_f(\lambda)=P\Big(\sum_{k=0}^{\infty}Y_k\xi^k\Big)e^{\Delta(\xi)},\quad Y_0=I, \tag{1.1.42}$$

where all the matrices Y_k, $k>0$ are off-diagonal, and the diagonal matrix $\Delta(\xi)$ is given by the series

$$\Delta(\xi)=\Lambda(\xi)+\sum_{k=1}^{\infty}\frac{\Lambda_k}{k}\xi^k=\sum_{\substack{k=-r\\k\neq 0}}^{\infty}\frac{\Lambda_k}{k}\xi^k+\Lambda_0\ln\xi.$$

Note that

$$\frac{d\Delta(\xi)}{d\xi}=\sum_{k=-r-1}^{\infty}\Lambda_{k+1}\xi^k.$$

We need to show that both the collection of off diagonal matrices Y_k, $k>0$ and the collection of diagonal matrices Λ_k, $k>-r$, can be defined uniquely via the matrix coefficients A_k, $k\geq -r$ of the Laurent series (1.1.38), provided that the diagonal form Λ_{-r} of the leading coefficient A_{-r} and the gauge matrix P in (1.1.39) are fixed.

Substituting the expansion (1.1.42) into equation (1.1.38) and equating coefficients of the same powers of ξ yields the following system for Λ_k and Y_k:

$$\Lambda_{-r+k}+[Y_k,\Lambda_{-r}]=F_{-r+k},\quad k>0, \tag{1.1.43}$$

where F_{-r+k} are defined by the equations

$$F_{-r+1}=P^{-1}A_{-r+1}P, \tag{1.1.44}$$

$$F_{-r+k}=P^{-1}A_{-r+k}P+\sum_{m=1}^{k-1}\big(P^{-1}A_{-r+k-m}PY_m-Y_m\Lambda_{-r+k-m}\big)$$

$$+H_{-r+k},$$

$$H_{-r+k}=\begin{cases}0, & k=2,\dots,r,\\ -(k-r)Y_{-r+k}, & k=r+1,r+2,\dots\end{cases}$$

Equations (1.1.43) determine both the diagonal matrices Λ_{-r+k} and the off-diagonal matrices Y_k uniquely in terms of F_{-r+k}. Indeed we have

$$\Lambda_{-r+k}=\operatorname{diag}(F_{-r+k}),\quad (Y_k)_{i,j}=\frac{(F_{-r+k})_{i,j}}{\alpha_j-\alpha_i}.$$

Because the matrices F_{-r+k} involve the matrices $\Lambda_{-r+1},\dots,\Lambda_{-r+k-1}$, the matrices $Y_1,\dots,Y_{k-1}$, and the given matrices P and A_k, the algebraic system (1.1.43) recursively determines the coefficients Λ_k and Y_k of the expansion (1.1.42). The last statement of the proposition follows from the polynomial structure of the right-hand sides of equations (1.1.43) and (1.1.44). □

The *second* main difference between Fuchsian and non-Fuchsian cases is that typically the series in (1.1.40) does not converge. This makes the connection between a formal solution of (1.1.1) and genuine solutions to this equation quite subtle. We shall now discuss in detail this important point.

One may conjecture that the formal solution $\Psi_f(\lambda)$ (1.1.42) represents the *asymptotics* of a genuine fundamental solution of (1.1.1) as $\lambda \to \lambda_0$. This is indeed the case provided λ belongs to *one* of certain *sectors* of D_{λ_0}. In order to understand how these sectors can be singled out let us discuss the question of uniqueness of the genuine fundamental solution $\Psi(\lambda)$ assuming that

$$\Psi(\lambda) \sim \Psi_f(\lambda), \quad \lambda \to \lambda_0, \quad \lambda \in \Omega, \tag{1.1.45}$$

where the sector Ω is defined as

$$\Omega = \{\xi \in \mathbb{C}: \quad 0 < |\xi| < \rho, \quad \theta_1 < \arg \xi < \theta_2\}.$$

Suppose $\tilde{\Psi}(\lambda)$ is another solution of (1.1.1) satisfying the same asymptotic condition (1.1.45). The matrix functions $\Psi(\lambda)$ and $\tilde{\Psi}(\lambda)$ satisfy the same linear ODE; moreover, in virtue of the asymptotics (1.1.45) they are the fundamental solutions of this equation. Hence their matrix ratio

$$C \equiv \Psi^{-1}(\lambda)\tilde{\Psi}(\lambda) \tag{1.1.46}$$

is well defined and does not depend on λ. Thus

$$C = \lim_{\lambda \to \lambda_0} \Psi^{-1}(\lambda)\tilde{\Psi}(\lambda), \quad \lambda \in \Omega. \tag{1.1.47}$$

Since the asymptotic equation (1.1.45) is satisfied by both the Ψ-functions, equation (1.1.47) can be rewritten as

$$C = \lim_{\xi \to 0} e^{-\Lambda(\xi)}\{I + O(\xi)\}e^{\Lambda(\xi)}, \quad \xi \in \Omega,$$

or, componentwise,

$$C_{ij} = \lim_{\xi \to 0} e^{(\Lambda(\xi))_{jj} - (\Lambda(\xi))_{ii}}\{\delta_{ij} + O(\xi)\}, \quad \xi \in \Omega. \tag{1.1.48}$$

Hence in order to be able to conclude that $C = I$, which would imply the uniqueness of the solution (1.1.45), we need the sector Ω to contain at least one ray

$$\mathrm{Re}\,\{(\alpha_j - \alpha_i)\xi^{-r}\} = 0, \tag{1.1.49}$$

for each pair (i, j), $i < j$. Indeed, in this case for each pair (i, j), $i \neq j$, we can choose a path terminating at $\xi = 0$ along which

$$\mathrm{Re}\ \{(\Lambda(\xi))_{jj} - (\Lambda(\xi))_{ii}\} < 0, \tag{1.1.50}$$

and hence the exponential term in (1.1.48) decays. Assuming that ξ in (1.1.48) belongs to this path, we have that

$$C_{ij} = 0, \quad \forall i \neq j.$$

This, together with the equations

$$C_{jj} = 1, \quad \forall j = 1, \dots, N,$$

(which follow from (1.1.48)), yield

$$C = I.$$

Since each difference $\alpha_j - \alpha_i \neq 0$, it follows that there are $2r$ rays emanating from $\xi = 0$ which satisfy (1.1.49). They are given by the equations,

$$\arg\xi = \frac{1}{r}\arg(\alpha_j - \alpha_i) + \frac{\pi}{r}\left(n + \frac{1}{2}\right), \quad n = 0, 1, \ldots, 2r - 1. \tag{1.1.51}$$

These rays are called *Stokes rays*; we shall use the notation $\ell_n^{(i,j)}$ for the Stokes rays corresponding to the pair (i, j), $i < j$, i.e.,

$$\ell_n^{(i,j)} = \left\{\xi\colon |\xi| < \rho, \quad \arg\xi = \frac{1}{r}\arg(\alpha_j - \alpha_i) + \frac{\pi}{r}\left(n + \frac{1}{2}\right)\right\}, \tag{1.1.52}$$

$$n = 0, 1, \ldots, 2r - 1, \quad i, j = 1, \ldots, N, \quad i < j.$$

The sector Ω which contains exactly one Stokes ray from the family $\{\ell_n^{(i,j)}\}_{n=0}^{2r-1}$ for each pair (i, j), $i < j$, will be referred to as a *Stokes sector* at the point $\lambda = \lambda_0$. The above arguments show that a genuine fundamental solution $\Psi(\lambda)$ is uniquely determined by the asymptotic condition (1.1.45), provided this asymptotic condition is specified in the Stokes sector Ω.

It is remarkable that the Stokes sectors are exactly the sectors where one can prove an existence result; this can be achieved using a variant of the usual integral Volterra equation technique but employing different contours for the each of the matrix elements in the solution (cf. [**Fe1**]).

In summary, the basic analytic fact concerning the behavior of the solutions of equation (1.1.1) near an irregular singular point is given by the following classical theorem [**W1**], [**Si2**], [**Fe1**].

THEOREM 1.4. *Let $A(\lambda)$ be a $N \times N$ matrix-valued function holomorphic in the punctured disk $D_{\lambda_0} \setminus \{\lambda_0\}$, and let $\lambda_0 \in \mathbb{CP}^1$ be a multiple pole of $A(\lambda)\,d\lambda$ of order $r + 1$, $r > 0$,*

$$A(\lambda)\,d\lambda = \Big(\sum_{k=-r-1}^{\infty} A_{k+1}\xi^k\Big)\,d\xi, \quad \lambda \in D_{\lambda_0} \setminus \{\lambda_0\}, \tag{1.1.53}$$

$$r > 0, \quad A_{-r} \neq 0$$

(we recall the definitions (1.1.5) and (1.1.6) of the local parameter ξ and the disk D_{λ_0} respectively). Suppose that the leading coefficient A_{-r} of the Laurent series (1.1.53) has distinct eigenvalues (generic case). Suppose also that the sector

$$\Omega = \{\xi \in \mathbb{C}\colon \quad 0 < |\xi| < \rho, \quad \theta_1 < \arg\xi < \theta_2\} \subset D_{\lambda_0} \tag{1.1.54}$$

is a Stokes sector at the point λ_0. Then in Ω there exists a unique fundamental solution $\Psi(\lambda)$ of equation (1.1.1) satisfying the asymptotic condition,

$$\Psi(\lambda) \sim \Psi_f(\lambda), \quad \lambda \to \lambda_0, \quad \lambda \in \Omega, \tag{1.1.55}$$

where $\Psi_f(\lambda)$ is the formal series (1.1.40), and the principal branch of $\ln\xi$ in (1.1.41) is chosen.

REMARK 1.3. The exact meaning of the asymptotic relation (1.1.55) is as follows (cf. (1.0.2)). For any positive numbers $m \in \mathbb{N}$, $\rho_1 < \rho$ and $\epsilon < (\theta_2 - \theta_1)/2$, there exists a positive constant $C_{m,\rho_1,\epsilon}$ such that the inequality,

$$\Big\|P^{-1}\Psi(\lambda)e^{-\Lambda(\xi)} - \sum_{k=0}^{m-1}\Psi_k\xi^k\Big\| < C_{m,\rho_1,\epsilon}|\xi|^m,$$

holds for every λ satisfying $0 < |\xi(\lambda)| \le \rho_1$ and $\theta_1 + \epsilon \le \arg\xi(\lambda) \le \theta_2 - \epsilon$. If neither θ_1 nor θ_2 is a Stokes ray, then the inequality above can be extended for all $\theta_1 \le \arg\xi \le \theta_2$.

REMARK 1.4. In the general case, when the eigenvalues of the leading coefficient A_{-r} of the Laurent series (1.1.53) are not all distinct (and the matrix A_{-r} is not necessarily diagonalizable), one must replace ξ by

$$\xi \mapsto \xi^{1/p}, \quad p \in \mathbb{N},$$

in the asymptotic form (1.1.40) of the solution $\Psi(\lambda)$.

The reader can find all details concerning the general non-Fuchsian case in the monograph [**W1**]. However, in our study of the Painlevé equations the general case will not be needed.

CANONICAL SOLUTIONS

For sufficiently small $\delta > 0$ any sector of the form

$$\Omega = \{\xi \in \mathbb{C}\colon\ 0 < |\xi| < \rho,\ \theta - \delta < \arg\xi < \theta + \frac{\pi}{r}\} \tag{1.1.56}$$

is a Stokes sector[1]. Also, if

$$\Omega = \{\xi \in \mathbb{C}\colon\ 0 < |\xi| < \rho,\ \theta_1 < \arg\xi < \theta_2\},$$
$$\left(\theta_2 - \theta_1 = \frac{\pi}{r} + \delta\right)$$

is a Stokes sector, then so is each of the sectors

$$\begin{aligned} \Omega_n &= e^{i\frac{\pi}{r}(n-1)}\Omega \\ &\equiv \{\xi \in \mathbb{C}\colon\ 0 < |\xi| < \rho,\ \theta_1 + \frac{\pi}{r}(n-1) < \arg\xi < \theta_2 + \frac{\pi}{r}(n-1)\}, \\ & n = 1, \dots, 2r, \\ \Omega_1 &= \Omega_{2r+1} = \Omega. \end{aligned} \tag{1.1.57}$$

Thus the punctured disk $D_{\lambda_0} \setminus \{\lambda_0\}$ can be always covered by $2r$ Stokes sectors.

Let $\{\Omega_n\}_{n=1}^{2r+1}$ be the collection of $2r+1$ Stokes sectors (1.1.57) which cover the punctured neighborhood $D_{\lambda_0} \setminus \{\lambda_0\}$ of the point λ_0. According to Theorem 1.4, in each sector Ω_n, there exists a unique solution $\Psi_n(\lambda)$ of equation (1.1.1), which satisfies the asymptotic condition,

$$\Psi_n(\lambda) \sim \Psi_f(\lambda), \quad \lambda \to \lambda_0, \quad \lambda \in \Omega_n, \tag{1.1.58}$$
$$n = 1, \dots, 2r+1.$$

The functions $\Psi_n(\lambda)$ are called *canonical solutions* of the equation (1.1.1) at the irregular point λ_0. Observe that, although $\Omega_1 = \Omega_{2r+1}$, in general, the associated

[1] Indeed, the obvious inclusion,

$$\ell_0^{(j,k)} \subset \cup_n e^{i\frac{\pi}{r}(n-1)}\Omega$$

implies that for each pair (j,k) there exists the number n, $0 \le n \le 2r-1$ such that $\ell_n^{(j,k)} \subset \Omega$. For sufficiently small δ, this number is unique. In other words, for every pair (j,k), $j < k$ there exists a positive $\delta^{(j,k)}$ such that $\delta < \delta^{(j,k)}$ implies that Ω contains exactly one Stokes ray from the family $\{\ell_n^{j,k}\}$. Put $\delta_0 := \min\{\delta^{(j,k)}\}$. Then for all $\delta < \delta_0$, the sector Ω is a Stokes sector.

canonical solutions are different, $\Psi_1(\lambda) \neq \Psi_{2r+1}(\lambda)$. In fact, because of (1.1.42), we have

$$\Psi_{2r+1}(\lambda) = \Psi_1(\lambda)e^{2\pi i\Lambda_0}. \tag{1.1.59}$$

Stokes matrices

Given $n \in \{1, \ldots, 2r\}$, the functions $\Psi_n(\lambda)$ and $\Psi_{n+1}(\lambda)$ are fundamental solutions of the same linear ODE. Therefore, similarly to the matrix C in (1.1.46), the matrices

$$\mathrm{S}_n \equiv \Psi_n^{-1}(\lambda)\Psi_{n+1}(\lambda), \quad n = 1, \ldots, 2r, \tag{1.1.60}$$

do not depend on λ. These matrices are called *Stokes matrices*, and their nontrivial entries are called *Stokes multipliers* [2]

REMARK 1.5. Equation (1.1.60) implies that each canonical solution $\Psi_n(\lambda)$ admits an analytic continuation on the universal covering, $\widetilde{D_{\lambda_0}\setminus\{\lambda_0\}}$, of the punctured disk $D_{\lambda_0} \setminus \{\lambda_0\}$.

Unlike the matrix C in (1.1.46), the Stokes matrices are in general *not trivial.* Indeed, the intersection of the Stokes sectors $\Omega_n \cap \Omega_{n+1}$, *does not contain any Stokes ray.* Therefore, the arguments that led to the equation $C = I$ fail in the case of the Stokes matrices. These arguments, in fact, imply a *triangularity property* of S_n which we will now explain.

Denote

$$\delta\Lambda_{ij}(\lambda) := \mathrm{Re}\,\{(\Lambda(\xi))_{jj} - (\Lambda(\xi))_{ii}\}.$$

Since $\Omega_n \cap \Omega_{n+1}$ does not contain any Stokes ray, it follows that for any closed subsector $K \subset \Omega_n \cap \Omega_{n+1}$ there exists a sufficiently small $\rho_1 \leq \rho$ (ρ is the radius of D_{λ_0}) such that throughout the set,

$$K_{\rho_1} := K \cap \{|\xi| < \rho_1\}$$
$$= \{\xi \in \mathbb{C}\colon\ 0 < |\xi| \leq \rho_1,\ \theta_1 + \frac{\pi}{r}n + \epsilon \leq \arg\xi \leq \theta_2 + \frac{\pi}{r}(n-1) - \epsilon\},$$

and for each fixed choice of the pair (i, j), $i \neq j$ the quantity $\delta\Lambda_{ij}$ in K_{ρ_1} is either always positive or always negative. Thus there exists a fixed dominance ordering,

$$\mathrm{Re}\,(\Lambda(\xi))_{\sigma_1\sigma_1} > \mathrm{Re}\,(\Lambda(\xi))_{\sigma_2\sigma_2} > \cdots > \mathrm{Re}\,(\Lambda(\xi))_{\sigma_N\sigma_N},$$

for all $\lambda \in K_{\rho_1}$ and for some permutation $(\sigma_1, \sigma_2, \ldots, \sigma_N)$ of $1, 2, \ldots, N$. It is worth noticing that the permutation σ does not depend on the choice of the subsector K (again, due to the absence of the Stokes ray in $\Omega_n \cap \Omega_{n+1}$). Similarly to (1.1.48) (see also Remark 1.3), we have

$$(\mathrm{S}_n)_{ij} = \lim_{\xi\to 0} e^{(\Lambda(\xi))_{jj} - (\Lambda(\xi))_{ii}}\{\delta_{ij} + O(\xi)\}, \quad \xi \in K_{\rho_1}. \tag{1.1.61}$$

Therefore,

$$(\mathrm{S}_n)_{jj} = 1, \quad j = 1, \ldots, N, \tag{1.1.62}$$

[2] Sometimes, what we just defined as Stokes matrices are in fact called Stokes multipliers, and no special term is used for the matrix entries of the Stokes matrices. We refer the reader to monograph **[Bal]** for more on the issue of these definitions and also on more detailed analysis of the algebraic structure of the Stokes matrices-multipliers.

and

$$(\mathsf{S}_n)_{\sigma_i\sigma_j} = 0, \quad i < j, \quad i,j = 1,\dots,N. \tag{1.1.63}$$

The remaining $\frac{N(N-1)}{2}$ entries of the matrix S_n cannot be deduced from equation (1.1.61), and in the generic situation these entries are nonzero.

The Stokes phenomenon

The fact that, in general, $\mathsf{S}_n \neq I$, gives rise to the so-called *Stokes phenomenon*: Two *different* fundamental solutions of equation (1.1.1), namely the solutions

$$\Psi_n \quad \text{and} \quad \Psi_{n+1} = \Psi_n S_n,$$

have the *same* asymptotic expansion,

$$\Psi_n, \Psi_{n+1} \sim \Psi_f, \quad \lambda \to \lambda_0, \quad \lambda \in \Omega_n \cap \Omega_{n+1}.$$

In other words, if the sector Ω is sufficiently narrow and properly located, then the asymptotics (1.1.45) does *not* define the solution $\Psi(\lambda)$ uniquely.

Given the set $\{\Omega_n\}_{n=1}^{2r+1}$ of the Stokes sectors, the collection of the data,

$$\mathbf{SPh}_{\lambda_0} := \{\Lambda_{-r}, \Lambda_{-r+1}, \dots, \Lambda_{-1}, \Lambda_0; \mathsf{S}_1, \mathsf{S}_2, \dots, \mathsf{S}_{2r}\}, \tag{1.1.64}$$

is determined uniquely via the coefficient matrix $\mathsf{A}(\lambda)$ provided that the transition matrix P in (1.1.39) (and hence the formal series Ψ_f) is fixed. Following [**Si2**], we call the set (1.1.64) *the Stokes phenomenon of equation* (1.1.1) *corresponding to the irregular point* λ_0.

The Stokes phenomenon $\mathbf{SPh}_{\lambda_0}$ characterizes completely the local behavior of the solutions of a linear system near an irregular singular point and at the same time it determines the system itself up to a local holomorphic gauge equivalence. Indeed the following theorem is valid.

THEOREM 1.5. *Suppose that the linear systems,*

$$\frac{d\Psi}{d\lambda} = \mathsf{A}(\lambda)\Psi$$

and

$$\frac{d\Phi}{d\lambda} = \mathsf{B}(\lambda)\Phi$$

have the same point $\lambda_0 \in \mathbb{C}\mathrm{P}^1$ as their irregular singular point of the same Poincaré rank $r > 0$. Moreover, suppose that

$$\mathbf{SPh}_{\lambda_0}(\mathsf{A}) = \mathbf{SPh}_{\lambda_0}(\mathsf{B}). \tag{1.1.65}$$

Then the above two systems are locally gauge equivalent, i.e.,

$$\mathsf{B}(\lambda) = g(\lambda)\mathsf{A}(\lambda)g^{-1}(\lambda) + \frac{dg(\lambda)}{d\lambda}g^{-1}(\lambda),$$

or equivalently,

$$\Phi(\lambda) = g(\lambda)\Psi(\lambda),$$

for some matrix-valued function $g(\lambda)$ which is holomorphic and invertible in the disk D_{λ_0}.

PROOF. Let $\{\Psi_n(\lambda)\}_{n=1}^{2r+1}$ and $\{\Phi_n(\lambda)\}_{n=1}^{2r+1}$ denote the sets of canonical solutions of the Ψ-system and Φ-system respectively (defined for the *same* set of canonical sectors Ω_n). Let

$$g(\lambda) := \Phi_1(\lambda)\Psi_1^{-1}(\lambda), \quad \lambda \in \Omega_1.$$

From the asymptotic condition (1.1.58) it follows that

$$\det \Psi_n(\lambda) \neq 0 \quad \text{and} \quad \det \Phi_n(\lambda) \neq 0, \quad \forall \lambda \in \Omega_n,$$

$$n = 1, 2, \ldots, 2r+1,$$

which implies that the function $g(\lambda)$ is holomorphic and invertible in the sector Ω_1. In order to prove the theorem we must show that $g(\lambda)$ admits an analytic continuation as an invertible and holomorphic matrix function in the whole disk D_{λ_0}.

In virtue of equation (1.1.60), the function $\Psi_1(\lambda)$ admits an analytic continuation, $\tilde{\Psi}_1(\lambda)$, along the chain of sectors $\Omega_1, \Omega_2, \ldots, \Omega_{2r+1}$ so that,

$$\tilde{\Psi}_1(\lambda) = \Psi_n(\lambda)\mathsf{S}_{n-1}^{-1}\mathsf{S}_{n-2}^{-1}\cdots\mathsf{S}_1^{-1}, \quad \lambda \in \Omega_n, \tag{1.1.66}$$

$$n = 2, \ldots, 2r+1,$$

(cf. also Remark 1.5). Because of (1.1.65), precisely the same equations are valid for the analytic continuation, $\tilde{\Phi}_1(\lambda)$ of the function $\Phi_1(\lambda)$. This implies that the function $g(\lambda)$ admits the analytic continuation $\tilde{g}(\lambda)$ along the chain of sectors $\Omega_1, \Omega_2, \ldots, \Omega_{2r+1}$. Moreover,

$$\tilde{g}(\lambda) = \Phi_n(\lambda)\Psi_n^{-1}(\lambda), \quad \lambda \in \Omega_n \quad n = 1, 2, \ldots, 2r+1. \tag{1.1.67}$$

Next we observe that

$$\tilde{g}|_{\Omega_{2r+1}} = g. \tag{1.1.68}$$

Indeed taking into account (1.1.59) and the equation $\Lambda_0(\mathsf{A}) = \Lambda_0(\mathsf{B})$, we find from (1.1.67) that

$$\tilde{g}|_{\Omega_{2r+1}}(\lambda) = \Phi_{2r+1}(\lambda)\Psi_{2r+1}^{-1}(\lambda) = \Phi_1(\lambda)\Psi_1^{-1}(\lambda),$$

which yields (1.1.68).

Equation (1.1.68) implies that the function $g(\lambda)$ admits an analytic continuation in the punctured disk $D_{\lambda_0} \setminus \{\lambda_0\}$. Also using (1.1.67) and the equations (condition (1.1.65) again),

$$\Lambda_k(\mathsf{A}) = \Lambda_k(\mathsf{B}), \quad k = -r, -r+1, \ldots, -1, 0,$$

we conclude that

$$g(\lambda) \to O(1),$$

as $\lambda \to \lambda_0$ in the *whole* neighborhood, $D_{\lambda_0} \setminus \{\lambda_0\}$, of the point λ_0. Hence this point is a removable singularity, and the proof is completed. □

REMARK 1.6. Theorems 1.2 and 1.3 imply that in the Fuchsian case two linear systems are locally equivalent if the systems have the same formal monodromy exponents at the point λ_0. In the nonresonant case, this means that the matrix residues at λ_0 of both the systems have the same Jordan form Λ_0 for both systems.

2. The global theory

2.1. The monodromy theorem. Suppose that the points

$$a_\nu \in \mathbb{CP}^1, \quad \nu = 1, \dots, m,$$

are the poles of the coefficient matrix, $A(\lambda)$, of the linear ODE (1.1.1). Let

$$\gamma = \gamma(t) \in \mathbb{CP}^1 \setminus \{a_\nu\}_{\nu=1}^m, \quad 0 \le t \le 1$$

be a curve on $\mathbb{CP}^1 \setminus \{a_\nu\}_{\nu=1}^m$, i.e.,

$$\gamma\colon\ [0,1] \mapsto \mathbb{CP}^1 \setminus \{a_\nu\}_{\nu=1}^m \tag{1.2.1}$$

is a continuous map. The global theory of solutions of (1.1.1) is based on the following fundamental result.

THEOREM 1.6 (monodromy theorem). *Let*

$$\Psi(\lambda) \equiv \sum_{k=0}^{\infty} \Psi_k \xi^k,$$

be the germ of a solution of equation (1.1.1) at the point $a = \gamma(0)$ (cf. Subsection 1.2 above). Then $\Psi(\lambda)$ can be analytically continued along γ to the point $b = \gamma(1)$. Moreover, this analytic continuation depends only on the homotopy class of the curve γ.

SKETCH OF THE PROOF.
We can always choose a finite partition,

$$0 = t_0 < t_1 < \cdots < t_{p-1} < t_p = 1, \tag{1.2.2}$$

of the interval $[0,1]$ in such a way that there exists a chain,

$$D_0,\ D_1,\ \dots,\ D_{p-1},\ D_p, \tag{1.2.3}$$

of the disks which are centered at the points $\gamma(0), \gamma(t_1), \dots, \gamma(t_{p-1}), \gamma(t_p)$, respectively, and which satisfy the following conditions,

$$\overline{D}_j \subset \mathbb{CP}^1 \setminus \{a_\nu\}_{\nu=1}^m, \quad j = 0, 1, \dots, p, \tag{1.2.4}$$

(the $\overline{D}$ means the closure of the set D),

$$\gamma\Big([t_j, t_{j+1}]\Big) \subset D_j \cap D_{j+1}, \tag{1.2.5}$$

and

$$D_j \cap D_{j+1} \ne \emptyset, \quad j = 0, \dots, p-1. \tag{1.2.6}$$

(Condition (1.2.6) is, of course, a consequence of condition (1.2.5).) Because of the condition (1.2.4) and Theorem 1.1, in every disk D_j, $j > 0$, there exists a fundamental solution $\Phi_j(\lambda)$ of equation (1.1.1). Also, we put

$$\Phi_0(\lambda) \equiv \Psi(\lambda).$$

Because of the condition (1.2.6), and since the functions $\Phi_j(\lambda)$ and $\Phi_{j+1}(\lambda)$ are solutions of the same linear ODE and $\det \Phi_j \ne 0$ for $j > 0$, it follows that all the matrix ratios,

$$C_j := \Phi_{j+1}^{-1}(\lambda)\Phi_j(\lambda), \quad \lambda \in D_j \cap D_{j+1}, \quad j = 0, \dots, p-1,$$

are constant (invertible for $j > 0$) matrices (cf. (1.1.46) and (1.1.60)). Put

$$\Psi_0(\lambda) := \Phi_0(\lambda)(= \Psi(\lambda)), \quad \lambda \in D_0,$$
$$\Psi_j(\lambda) := \Phi_j(\lambda) C_{j-1} C_{j-2} \dots C_0, \quad \lambda \in D_j,$$
$$j = 1, \dots, p.$$

By construction,

$$\Psi_{j+1}(\lambda) = \Psi_j(\lambda), \quad \forall \lambda \in D_j \cap D_{j+1}, \quad j = 0, \dots, p-1,$$

and hence the chain of germs,

$$\Psi_0(= \Psi), \ \Psi_1, \ \dots, \ \Psi_p,$$

provides an analytic continuation of $\Psi(\lambda)$ to the point $b = \gamma(1)$ along the curve γ. This proves the existence part of the theorem's statement.

More work is needed in order to see that the germ $\Psi_p(\lambda)$ depends only on the homotopy class of the curve γ. The proof of this fundamental fact is rather long. It is indeed one of the first genuinely nontrivial technical matters of the analytical theory of linear ODEs; see [**Si2**] for details. The proof can be shortened, however, at the expense of using the advanced language and results of the general theory of spaces of sheafs. This version of the proof is presented e.g. in the monograph of O. Forster [**Fr**]. Briefly, the idea of this elegant approach is the following.

Let $\mathcal{F}$ be the sheaf of germs of holomorphic functions on the Riemann surface,

$$X \equiv \mathbb{C}\mathrm{P}^1 \setminus \{a_\nu\}_{\nu=1}^m.$$

Consider the space,

$$|\mathcal{F}| := \bigcup_{\lambda_0 \in X} \mathcal{F}_{\lambda_0},$$

i.e., the space of all pairs (λ_0, ψ), where $\lambda_0 \in X$ and $\psi(\lambda) = \sum c_n (\lambda - \lambda_0)^n$ is a germ of a holomorphic function at the point λ_0. The space $|\mathcal{F}|$ can be equipped with a natural projection,

$$\pi \colon |\mathcal{F}| \mapsto X, \quad \pi\Big((\lambda_0, \psi)\Big) = \lambda_0, \tag{1.2.7}$$

and a natural topology so that $|\mathcal{F}|$ becomes a Hausdorff space and π acts as a nonramified covering map (for more details see [**Fr**]).[3] Having the curve γ on X and the germ $\Psi(\lambda)$, the existence of the analytic continuation of Ψ along γ (which was proved above) can be reformulated as the existence of a *lift* $\hat{\gamma}$ of the curve γ to the covering space $|\mathcal{F}|$, with the initial point $(\gamma(0), \Psi)$:

$$\hat{\gamma}(0) = \Psi, \quad \pi \circ \hat{\gamma} = \gamma.$$

The remaining statement of the theorem is then a straightforward application of the following two general properties of lifts (see again [**Fr**]).

LEMMA 1.1 (uniqueness of the lift). *Let X, Y be Hausdorff spaces, and let $\pi \colon Y \mapsto X$ be a nonramified covering map. Suppose that $\gamma \colon [0,1] \mapsto X$ is a continuous map and $\hat{\gamma}_1$, $\hat{\gamma}_2$ are the associated lifts such that $\hat{\gamma}_1(0) = \hat{\gamma}_2(0)$. Then $\hat{\gamma}_1 = \hat{\gamma}_2$.*

---

[3] A neighborhood of the element $(\lambda_0, \psi) \in |\mathcal{F}|$ is defined as the set $\bigcup_{\mu \in U} (\mu, \psi_\mu)$, where U is a disk centered at λ_0, and ψ_μ denotes the Taylor series $\psi(\lambda) = \sum c_n (\lambda - \lambda_0)^n$ re-expanded as the Taylor series over powers of $\lambda - \mu$. The Hausdorffness of $|\mathcal{F}|$ is a direct consequence of the uniqueness property of analytic functions.

LEMMA 1.2 (lifting of homotopic curves). *Let X, Y be Hausdorff spaces, and let $\pi\colon Y \mapsto X$ be a nonramified covering map. Suppose that a, b belong to X, and $\hat{a} \in Y$, $\pi(\hat{a}) = a$. Suppose also that there exists a continuous map $A\colon [0,1]\times[0,1] \mapsto X$ such that:*

- *$A(0,s) = a$ and $A(1,s) = b$ for all $s \in [0,1]$.*
- *Each curve, $\gamma_s(t) := A(t,s)$, can be lifted to a curve $\hat{\gamma}_s$ with the initial point at $\hat{a}$.*

Then the curves $\hat{\gamma}_0$ and $\hat{\gamma}_1$ have the same end point and are homotopic.

Take $X = \mathbb{CP}^1 \backslash \{a_\nu\}_{\nu=1}^n$ and (1.2.7) as a covering map π. Then from Lemma 1.1 it follows that the analytic continuation of the germ Ψ of the solution of the linear equation (1.1.1) along a given curve γ on X does not depend on the choice of the partition (1.2.2) or on the choice of the chain of disks (1.2.3). Also, having two homotopic curves, γ_1 and γ_2, consider the continuous map $A\colon [0,1]\times[0,1] \mapsto X$ which realizes the homotopy between the curves. Since it was proven that the germ Ψ can be analytically continued along *any* curve on X starting at $a = \gamma_{1,2}(0)$, it follows that the conditions of Lemma 1.2 are satisfied for the function $A(s,t)$ if one takes $\hat{a} = (a, \Psi)$. The lemma then implies that the results of analytic continuation of the germ Ψ to the point $b = \gamma_{1,2}(1)$ along the curves γ_1 and γ_2 are the same. This completes the proof of the theorem. □

Choose a point

$$a_0 \in X \equiv \mathbb{CP}^1 \setminus \{a_\nu\}_{\nu=1}^m,$$

and consider a simply connected domain $\hat{X}$ obtained by cutting the Riemann sphere $\mathbb{CP}^1$ along the segments $[a_0, a_\nu]$, $\nu = 1, \dots, m$. *An immediate consequence of Theorem 1.6 is that the germ $\Psi(\lambda)$ can be analytically continued as a single-valued holomorphic function, which we also denote by $\Psi(\lambda)$, on the whole $\hat{X}$.*

In fact, the domain of analyticity for the solution $\Psi(\lambda)$ can be extended even further. To this end, let us recall (for more details see [**Fr**]) the notion of universal covering of a topological space – in our case – $\mathbb{CP}^1 \setminus \{a_\nu\}_{\nu=1}^n$. Choose as above a point

$$a_0 \in X \equiv \mathbb{CP}^1 \setminus \{a_\nu\}_{\nu=1}^m.$$

Consider the set $\mathcal{X}$ of all pairs, (a, γ_{a_0a}), were $a \in X$ and γ_{a_0a} is a curve passing from a_0 to a. By defining the equivalence relation "$\sim$" on $\mathcal{X}$, which identifies the pairs (a, γ_{a_0a}) and (b, β_{a_0b}) if $a = b$ as well as the curves γ_{a_0a} and β_{a_0b} are homotopic, we arrive to the simply connected Riemann surface

$$\widetilde{X} := \mathcal{X}/\sim.$$

The Riemann surface $\widetilde{X}$ is called the *universal covering* of X. The topology and complex structure on $\widetilde{X}$ are introduced as follows: A neighborhood $\widetilde{U}_{\tilde{a}}$ of the point $\tilde{a} := (a, \gamma_{a_0a})$ is defined as the set $\widetilde{U}_{\tilde{a}} = \{\widetilde{(b, \gamma_{a_0a}\cdot\gamma_{ab})}\colon b \in U_a\}$, where U_a is an open disk centered at a, and γ_{ab} lies entirely in U_a. The symbol $\gamma_{a_0a}\cdot\gamma_{ab}$ means the product of two curves (first pass from a_0 to a along γ_{a_0a} and then from a to b along γ_{ab}). The local homeomorphism, $\phi\colon \widetilde{U}_{\tilde{a}} \mapsto \mathbb{C}$, is defined as the restriction on $\widetilde{U}_{\tilde{a}}$ of the natural projection,

$$\pi_0\colon \widetilde{X} \mapsto X, \quad \pi_0\Big(\widetilde{(a, \gamma_{a_0a})}\Big) = a. \tag{1.2.8}$$

In other words, the complex structure on $\widetilde{X}$ is defined in such a way that the projection $\lambda \in X$ of the point $\widetilde{\lambda} \in \widetilde{X}$ can be taking as a local complex coordinate on $\widetilde{X}$, and the natural projection (1.2.8) becomes a nonramified holomorphic covering map. This in turn leads to the convention of identifying the points of $\widetilde{X}$ with their projections on X. We shall also accept this convention, i.e., we will use the letters λ, a, b, etc. for both points on X and points on its universal covering $\widetilde{X}$.

We note that the universal covering $\widetilde{X}$ can be thought of as an infinite collection of the copies of the domain $\hat{X}$ cross-glued together along the sides of the cuts, $[a_0, a_\nu]$, $\nu = 1, \dots, m$.

The following corollary is now a direct consequence of Theorem 1.6 and of the above definition of the universal covering $\widetilde{\mathbb{C}\mathrm{P}^1} \setminus \{a_\nu\}_{\nu=1}^m$.

COROLLARY 1.1. *Let $\Psi(\lambda)$ be a germ of solution of equation (1.1.1) or, more generally, any solution of (1.1.1) defined on an open subset of $\mathbb{C}\mathrm{P}^1 \setminus \{a_\nu\}_{\nu=1}^m$. Then $\Psi(\lambda)$ can be analytically continued on the whole universal covering, $\widetilde{\mathbb{C}\mathrm{P}^1} \setminus \{a_\nu\}_{\nu=1}^m$, of the punctured Riemann sphere, $\mathbb{C}\mathrm{P}^1 \setminus \{a_\nu\}_{\nu=1}^m$.*

Unless otherwise indicated, we will not distinguish between the germ $\Psi(\lambda)$ and the analytic function $\Psi(\lambda)$ which the germ defines on the universal covering $\widetilde{\mathbb{C}\mathrm{P}^1} \setminus \{a_\nu\}_{\nu=1}^m$.

REMARK 1.7. Different choices of the base point a_0 when constructing $\hat{X}$ lead to different branches of the multivalued function $\Psi(\lambda)$ on X. Different choices of a_0 when constructing $\widetilde{X}$ lead to isomorphic Riemann surfaces.

2.2. Monodromy operator. The notion of monodromy operator is another important consequence of the monodromy Theorem 1.6. Recall (cf. (1.2.1)) that the map

$$\gamma\colon [0,1] \mapsto \mathbb{C}\mathrm{P}^1 \setminus \{a_\nu\}_{\nu=1}^m \tag{1.2.9}$$

is called a *curve* if $\gamma(t)$ is a continuous function. A curve γ which goes from the point

$$a = \gamma(0)$$

to the point

$$b = \gamma(1),$$

will be denoted by γ_{ab}, and it will be called a *path from the point a to the point b*. We also recall the standard definition of the product, $\gamma_{ab} \cdot \gamma_{bc}$, of two curves, γ_{ab} and γ_{bc} (note that $\gamma_{ab}(1) = \gamma_{bc}(0)$), i.e.,

$$[\gamma_{ab} \cdot \gamma_{bc}](t) = \begin{cases} \gamma_{ab}(2t), & t \in \left[0; \frac{1}{2}\right], \\ \gamma_{bc}(2t-1), & t \in \left[\frac{1}{2}; 1\right]. \end{cases} \tag{1.2.10}$$

The important fact is that the homotopy class of $\gamma_{ab} \cdot \gamma_{bc}$ is determined only by the homotopy classes of γ_{ab} and γ_{bc} (see e.g. [**Fr**]).

Now, given a path γ_{ab} on $CP^1 \setminus \{a_1, a_2, \dots, a_m\}$ from the point a to the point b, the *monodromy operator* $\tau_{\gamma_{ab}}$ is a map,

$$\tau_{\gamma_{ab}} : \mathbf{Mat}(N, \mathbb{C}) \mapsto \mathbf{Mat}(N, \mathbb{C}),$$

defined as,

$$\tau_{\gamma_{ab}}(A) = \Psi^A_{\gamma_{ab}}(b),$$

where $\Psi^A_{\gamma_{ab}}(\lambda)$ is the analytic continuation along the curve γ_{ab} of the germ of solution of (1.1.1) satisfying $\Psi^A_{\gamma_{ab}}(a) = A$.

In virtue of Theorem 1.6, the operator $\tau_{\gamma_{ab}}$ depends on the homotopy class of γ_{ab} only. In the next proposition we state the other basic properties of $\tau_{\gamma_{ab}}$.

PROPOSITION 1.2. *The following equations are valid:*

- $\tau_{\gamma_{ab}\cdot\gamma_{bc}} = \tau_{\gamma_{bc}} \circ \tau_{\gamma_{ab}}$,
- $\tau_{\gamma_{ab}}(AB) = \tau_{\gamma_{ab}}(A)B, \quad A, B \in \mathbf{Mat}(N, \mathbb{C})$,
- $\tau_{\gamma_{ab}}(A) = N(\gamma_{ab})A, \quad N(\gamma_{ab}) \equiv \tau_{\gamma_{ab}}(I) \in \mathbf{GL}(N, \mathbb{C})$.

This proposition follows directly from Theorem 1.6, the uniqueness of the solution of the Cauchy problem of equation (1.1.1) (see Theorem 1.1), and the linearity of equation (1.1.1).

CHAPTER 2

Monodromy Theory and Special Functions

0. Monodromy group

0.1. Monodromy data for a system of linear ODEs. As before, $\mathsf{A}(\lambda)$ denotes an $N\times N$ ($N>1$) matrix-valued rational function, and $\{a_\nu\}_{\nu=1}^m$, $a_\nu\in\mathbb{C}\mathrm{P}^1$, denotes the set of its poles. Our principal object of analysis is the first order linear ODE (1.1.1) for an $N\times N$ matrix-valued function $\Psi(\lambda)$.

In the previous chapter we described in detail the basic analytic properties of the solutions of (1.1.1). Based on these results we can now introduce the following sets of local characteristics of equation (1.1.1) which govern the behavior of its solutions near singular points:

- The formal monodromy exponents, i.e., the matrices

$$\Lambda_0 \quad \text{or} \quad \{\Lambda_0, \mathrm{N}\}, \tag{2.0.1}$$

  at a Fuchsian singularity in the nonresonant and resonant cases respectively (see Theorems 1.2 and 1.3).
- The Stokes phenomenon, i.e., the set

$$\mathbf{SPh}=\{\Lambda_{-r},\Lambda_{-r+1},\dots,\Lambda_{-1},\Lambda_0;\mathrm{S}_1,\mathrm{S}_2,\dots,\mathrm{S}_{2r}\}, \tag{2.0.2}$$

  at an irregular singularity (see Theorem 1.4 and equations (1.1.60) and (1.1.64)).

Indeed, it was shown (see Theorem 1.5 and Remark 1.6) that the data above determine the system (1.1.1) up to the local holomorphic gauge equivalence.

We have also established the fundamental fact, Theorem 1.6, that linear ODEs admit *global* fundamental solutions, i.e., any local solution of (1.1.1) can be analytically continued on the whole universal covering of the punctured Riemann sphere, $\mathbb{C}\mathrm{P}^1\setminus\{a_\nu\}_{\nu=1}^m$. This in turn allows us "to glue" the local data (2.0.1) and (2.0.2), associated with each of the singular point a_ν, to a finite set of *global monodromy data*. These data will determine the system (1.1.1) uniquely. The goal of this section is a detailed description of these monodromy data.

Denote by $A_{-r_\nu}^{(\nu)}$ the leading coefficient of the Laurent series,

$$\mathsf{A}(\lambda)=\Big(\sum_{k=-r_\nu-1}^{\infty}A_{k+1}^{(\nu)}\xi_\nu^k\Big)\frac{d\xi_\nu}{d\lambda},\quad \lambda\in D_{a_\nu}\setminus\{a_\nu\}, \tag{2.0.3}$$

$$A_{-r_\nu}^{(\nu)}\neq 0,\quad r_\nu\geq 0,$$

of the coefficient matrix $\mathsf{A}(\lambda)$ of equation (1.1.1) at the singular point a_ν (cf. (1.1.8), (1.1.9)). Recall that we use the notation ξ_ν for the local variable at the point a_ν, i.e., (cf. (1.1.5))

$$\xi_\nu = \begin{cases} \lambda - a_\nu, & \text{if } a_\nu \neq \infty, \\ 1/\lambda, & \text{if } a_\nu = \infty. \end{cases} \tag{2.0.4}$$

Note that the point a_ν is a Fuchsian singular point if $r_\nu = 0$, and it is an irregular point if $r_\nu > 0$.

Let us fix at each singular point a_ν the pair,

$$\{\Lambda^{(\nu)}_{-r_\nu}, P_\nu\}, \tag{2.0.5}$$

where $\Lambda^{(\nu)}_{-r_\nu}$ is the Jordan form of the leading coefficient, $A^{\nu}_{-r_\nu}$, of the Laurent series (2.0.3), and P_ν is the corresponding transition matrix,

$$P_\nu^{-1} A^{(\nu)}_{-r_\nu} P_\nu = \Lambda^{(\nu)}_{-r_\nu}, \quad \det P_\nu \neq 0. \tag{2.0.6}$$

In addition, if a_ν is an irregular point, we fix the set

$$\{\Omega^{(\nu)}_n\}_{n=1}^{2r_\nu+1}, \tag{2.0.7}$$

of Stokes sectors at a_ν (see (1.1.56), (1.1.57)). We will also assume the following restrictions on the matrices $\Lambda^{(\nu)}_{-r_\nu}$, $\nu = 1, \ldots, m$ (cf. (1.1.30) and (1.1.39)):

- If $r_\nu = 0$, i.e., a_ν is a Fuchsian point, then $\Lambda^{(\nu)}_0$ is either an elementary Jordan cell,

$$\Lambda^{(\nu)}_0 = \alpha I + J, \quad J_{ij} = \delta_{i+1,j}, \quad J^N = 0, \tag{2.0.8}$$

  or a diagonal matrix with entries distinct modulo a nonzero integer[1].

$$(\Lambda^{(\nu)}_0)_{ij} = \alpha^{(\nu)}_i \delta_{ij}, \quad \alpha^{(\nu)}_i - \alpha^{(\nu)}_j \notin \mathbb{Z} \setminus \{0\} \quad \forall i,j\colon\ i \neq j. \tag{2.0.9}$$

- If $r_\nu > 0$, i.e., a_ν is an irregular singular point, then $\Lambda^{(\nu)}_{-r_\nu}$ is a diagonal matrix with the distinct entries,

$$(\Lambda^{(\nu)}_{-r_\nu})_{ij} = \alpha^{(\nu)}_i \delta_{ij}, \quad \alpha^{(\nu)}_i \neq \alpha^{(\nu)}_j, \quad i \neq j. \tag{2.0.10}$$

Finally, having chosen a base point

$$a_0 \in X \equiv \mathbb{C}\mathrm{P}^1 \setminus \{a_\nu\}_{\nu=1}^m,$$

we fix the simply connected domain $\hat{X}$ and the universal covering $\widetilde{X}$ together with the branches of multivalued on X functions,

$$\xi_\nu^{\Lambda^{(\nu)}}, \quad \nu = 1, \ldots, m.$$

By virtue of the basic theorems of the local theory, i.e., Theorems 1.2 and 1.4, we can define the following sets of canonical solutions at each point a_ν:

- If a_ν is a Fuchsian point, then the canonical set consists of a single canonical solution $\Psi^{(\nu)}(\lambda)$ defined by the equation (cf. (1.1.32)),

$$\Psi^{(\nu)}(\lambda) = P_\nu \left(I + \sum_{j=1}^{\infty} \Psi^{(\nu)}_j \xi_\nu^j \right) \xi_\nu^{\Lambda^{(\nu)}_0}, \tag{2.0.11}$$

$$\lambda \in \hat{D}_{a_\nu} \equiv D_{a_\nu} \cap \hat{X}.$$

[1]We impose these conditions to avoid tedious technicalities in our exposition of the general theory. In some concrete Fuchsian examples later in the text, however, we will actually lift the restrictions on the matrix $\Lambda^{(\nu)}_0$.

- If a_ν is an irregular singular point of the Poincaré rank $r_\nu > 0$, then the canonical set consists of $2r_\nu + 1$ canonical solutions $\Psi_n^{(\nu)}(\lambda)$, $n = 1, \ldots, 2r_\nu + 1$, defined by the asymptotic conditions (cf. (1.1.58) and (1.1.40)),

$$\Psi_n^{(\nu)}(\lambda) \sim P_\nu \left(I + \sum_{j=1}^{\infty} \Psi_j^{(\nu)} \xi_\nu^j \right) e^{\Lambda^{(\nu)}(\xi_\nu)}, \tag{2.0.12}$$

$$\lambda \in \Omega_n^{(\nu)}, \quad \lambda \to a_\nu, \quad n = 1, \ldots, 2r_\nu + 1,$$

where $\Lambda^{(\nu)}(\xi)$ is the sum

$$\Lambda^{(\nu)}(\xi) = \sum_{k=-r_\nu}^{-1} \frac{\Lambda_k^{(\nu)}}{k} \xi^k + \Lambda_0^{(\nu)} \ln \xi, \tag{2.0.13}$$

all the matrices $\Lambda_k^{(\nu)}$, $k = -r_\nu, \ldots, 0$, are diagonal, and the matrix $\Lambda_{-r_\nu}^{(\nu)}$ is defined in (2.0.6). Also, according to (1.1.59),

$$\Psi_{2r_\nu+1}^{(\nu)}(\lambda) = \Psi_1^{(\nu)}(\lambda) e^{2\pi i \Lambda_0^{(\nu)}}. \tag{2.0.14}$$

The matrix coefficients $\Psi_k^{(\nu)}$ of the series above, as well as the matrix exponents $\Lambda^{(\nu)}(\xi)$ are polynomial functions of the coefficients of equation (1.1.1). In fact, they all can be explicitly evaluated via the recursive equations (1.1.21), (1.1.29), (1.1.43) and (1.1.44).

The *local monodromy data* corresponding to the singular point a_ν are (cf. (2.0.1) and (2.0.2)):

- the formal monodromy exponent

$$\Lambda_0^{(\nu)}, \tag{2.0.15}$$

if a_ν is a Fuchsian singular point and
- the Stokes phenomenon

$$\mathbf{SPh}^{(\nu)} = \{\Lambda_{-r_\nu}^{(\nu)}, \Lambda_{-r_\nu+1}^{(\nu)}, \ldots, \Lambda_{-1}^{(\nu)}, \Lambda_0^{(\nu)}; \mathrm{S}_1^{(\nu)}, \mathrm{S}_2^{(\nu)}, \ldots, \mathrm{S}_{2r_\nu}^{(\nu)}\}, \tag{2.0.16}$$

if a_ν is an irregular singular point of the Poincaré rank $r_\nu > 0$.

We recall that the Stokes matrices $\mathrm{S}_n^{(\nu)}$ are the constant matrices defined by the equation (cf. (1.1.60))

$$\Psi_{n+1}^{(\nu)}(\lambda) = \Psi_n^{(\nu)}(\lambda) \mathrm{S}_n^{(\nu)}, \quad \lambda \in \Omega_n^{(\nu)} \cap \Omega_{n+1}^{(\nu)}, \quad n = 1, \ldots, 2r_\nu. \tag{2.0.17}$$

Fix a matrix $\Psi_0 \in \mathbf{GL}(N, \mathbb{C})$, and let $\Psi(\lambda)$ be the fundamental solution of (1.1.1) determined by the pair $\{a_0, \Psi_0\}$, i.e.,

$$\Psi(a_0) = \Psi_0.$$

By virtue of the basic theorem of the global theory, i.e., Theorem 1.6, each of the functions $\Psi(\lambda)$, $\Psi^{(\nu)}(\lambda)$, and $\Psi_n^{(\nu)}(\lambda)$ admits analytic continuation on the domains $\hat{X}$ and $\widetilde{X}$. Moreover, each of them is a solution of the same linear ODE. Hence, they differ from each other by constant right matrix multipliers. This allows us to introduce the *connection matrices* E_ν as the constant matrices defined by the equations

$$\Psi(\lambda) = \Psi^{(\nu)}(\lambda) \mathsf{E}_\nu, \tag{2.0.18}$$

if a_ν is a Fuchsian point, and

$$\Psi(\lambda) = \Psi_1^{(\nu)}(\lambda)\mathsf{E}_\nu, \tag{2.0.19}$$

if a_ν is an irregular singular point. We also note that the Stokes equations (2.0.17) are valid on the full domains $\hat{X}$ and $\widetilde{X}$. The *global monodromy data*, or the *extended monodromy data*, of equation (1.1.1) are defined as the set

$$\mathbb{M} := \{a_1, \dots, a_m;\ \Lambda_0^{(1)}, \dots, \Lambda_0^{(p)};\ \mathbf{SPh}^{(p+1)}, \dots, \mathbf{SPh}^{(m)};$$

$$\mathsf{E}_1, \dots, \mathsf{E}_m\}, \tag{2.0.20}$$

where we have ordered the singular points a_ν in such a way that the Fuchsian singular points (if there are any) are the points $a_1, \dots, a_p$, $0 \le p \le m$. If $p = 0$ $(p = m)$, then the second (third) subset in (2.0.20) is absent.

It should be emphasized that, for a general linear system, the Stokes and the connection matrices form the part of the monodromy data which are *transcendental* functions of the coefficients of the system. This motivates the introduction of the *essential monodromy data* or simply the *monodromy data* as the subset

$$\mathfrak{m} = \{\Lambda_0^{(\nu)}; \mathsf{S}_n^{(\mu)}; \mathsf{E}_\nu\}_{\mu=p+1,\dots,m;n=1,\dots,2r_\mu}^{\nu=1,\dots,m}. \tag{2.0.21}$$

The reason for including the formal monodromy exponents $\Lambda_0^{(\nu)}$ in the monodromy data (which are elementary functions of the coefficients of the system) will become clear latter on; see Remark 2.3 in the end of Section 1.4 below.

REMARK 2.1. Admissible transformations of the normalization matrix Ψ_0 or of any of the transition matrices P_ν, induce some simple transformations of the monodromy data of the system (1.1.1) which leave the system itself invariant. In the generic case, these transformations are given by the following formulae:

(1) (the transformation $\Psi_0 \to \Psi_0 C^{-1}$)

$$\mathsf{E}_\nu \to \mathsf{E}_\nu C, \tag{2.0.22}$$

(2) (the transformation $P_\nu \to P_\nu D^{-1}$)

$$\mathsf{S}_n^{(\nu)} \to D\mathsf{S}_n^{(\nu)}D^{-1}, \tag{2.0.23}$$

$$\mathsf{E}_\nu \to D\mathsf{E}_\nu, \tag{2.0.24}$$

(3) (the transformation $P_\nu \to P_\nu \Pi^{(\alpha_1,\alpha_2)^{-1}}$)

$$\Lambda_k^{(\nu)} \to \Pi^{(\alpha_1,\alpha_2)}\Lambda_k^{(\nu)}\Pi^{(\alpha_1,\alpha_2)^{-1}}, \tag{2.0.25}$$

$$\mathsf{S}_n^{(\nu)} \to \Pi^{(\alpha_1,\alpha_2)}\mathsf{S}_n^{(\nu)}\Pi^{(\alpha_1,\alpha_2)^{-1}}, \tag{2.0.26}$$

$$\mathsf{E}_\nu \to \Pi^{(\alpha_1,\alpha_2)}\mathsf{E}_\nu. \tag{2.0.27}$$

Here, C denotes an arbitrary invertible matrix, D is an arbitrary invertible diagonal matrix, and $\Pi^{(\alpha_1,\alpha_2)}$ denotes the permutation matrix obtained from the identical matrix by switching the α_1 and α_2 columns. For each point a_ν, the matrices D and $\Pi^{(\alpha_1,\alpha_2)}$ can be chosen independently. When counting the dimension of the sets $\mathbb{M}$ and $\mathfrak{m}$ of the monodromy data for a given linear system, the transformations (2.0.22 -2.0.27) should be taken into account. Also, in every concrete example, these transformations can be used to eliminate (factor out) nonessential parameters from the set $\mathfrak{m}$.

REMARK 2.2. The classical theorems concerning the analytic dependence of the solutions of linear equations on the coefficients of the equations and initial data, imply the important fact that all canonical solutions $\Psi_n^{(\nu)}(\lambda)$, and hence all monodromy data, can be chosen to be (locally) holomorphic functions of the entries of the matrix coefficients of the rational function $A(\lambda)$ and of the points a_ν. Moreover, both the convergent series representation (2.0.11) and the asymptotics (2.0.12) can be differentiated termwise with respect to the parameters indicated (as well as with respect to the variable λ).

The extended monodromy data $\mathbb{M}$ characterize completely the global behavior of the solutions of a linear system and, simultaneously, determine uniquely the system itself. In fact, in Subsection 1.5 of this Chapter (see Proposition 2.2) we will prove that if two linear systems have the same set of singular points and the same monodromy data, then they coincide. For this proof we need one more important general concept from the global theory of linear ODEs — the concept of the *monodromy group* of equation (1.1.1).

0.2. The definition of the monodromy group. The monodromy group of the linear system (1.1.1) is defined as a representation of the fundamental group $\pi_1(\mathbb{C}P^1 \setminus \{a_\nu\}_{\nu=1}^m)$, which can be obtained via solutions of the system as follows:

Fix a base point $a_0 \in \mathbb{C}P^1 \setminus \{a_\nu\}_{\nu=1}^m$ and realize π_1 as the collection of loops,

$$\{\gamma = \gamma(t), \quad 0 \leq t \leq 1, \quad \gamma(0) = \gamma(1) = a_0\},$$

in the usual way[2]. Fix a matrix $\Psi_0 \in \mathbf{GL}(N, \mathbb{C})$. Now let $\Psi(\lambda)$ be the germ of fundamental solution of (1.1.1) with $\Psi(a_0) = \Psi_0$. Under an analytic continuation along a loop $\gamma \in \pi_1$, the germ $\Psi(\lambda)$ is transformed to another germ $\Psi_\gamma(\lambda)$. Both the matrix functions $\Psi(\lambda)$ and $\Psi_\gamma(\lambda)$ are fundamental solutions of the same linear ODE (1.1.1). Therefore, there exists an invertible constant matrix M such that

$$\Psi_\gamma(\lambda) = \Psi(\lambda)M.$$

We note that the matrix M can be also defined as

$$M = \Psi_0^{-1}\Psi_\gamma(a_0). \tag{2.0.28}$$

With the point a_0 and the matrix Ψ_0 fixed, the matrix M can of course depend on the loop γ (or rather on its homotopy class),

$$M \equiv M(\gamma).$$

PROPOSITION 2.1. *The correspondence,*

$$\pi_1 \ni \gamma \mapsto M^{-1}(\gamma) \in \mathbf{GL}(N, \mathbb{C}), \tag{2.0.29}$$

is a linear representation of $\pi_1(\mathbb{C}P^1 \setminus \{a_\nu\}_{\nu=1}^n)$.

PROOF. Equation (2.0.28) can be rewritten as

$$M = \Psi_0^{-1}\tau_\gamma(\Psi_0), \tag{2.0.30}$$

where τ_γ is the monodromy operator introduced in the end of the previous chapter. Assuming that γ, $\beta \in \pi_1$, and taking into account the basic properties of the

[2]Unless otherwise indicated, we will always identify curves and their homotopy classes, and will use the same notations for both objects.

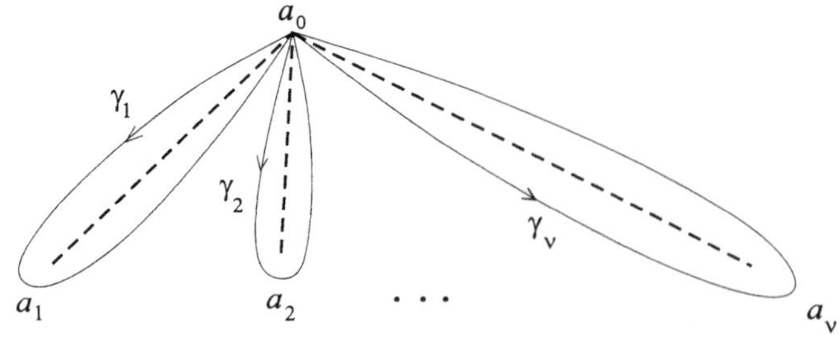

FIGURE 2.1. Generators of the fundamental group π_1 and the cuts $[a_0, a_\nu]$

monodromy operator (see Proposition 1.2) we find

$$\begin{aligned}\mathsf{M}(\gamma \cdot \beta) &= \Psi_0^{-1}\tau_{\gamma\cdot\beta}(\Psi_0) = \Psi_0^{-1}\tau_\beta(\tau_\gamma(\Psi_0)) \\ &= \Psi_0^{-1}\tau_\beta(\Psi_0\mathsf{M}(\gamma)) = \Psi_0^{-1}\tau_\beta(\Psi_0)\mathsf{M}(\gamma) = \mathsf{M}(\beta)\mathsf{M}(\gamma).\end{aligned}$$

The last equation, together with the obvious relation,

$$\mathsf{M}(\{a_0\}) = I,$$

proves the proposition. □

A representation (2.0.29) is called a *monodromy representation* of the system (1.1.1). Given $\mathsf{A}(\lambda)$, the subgroup

$$\mathfrak{M} \equiv \{\mathsf{M}(\gamma),\ \gamma \in \pi_1\} \subset \mathbf{GL}(N, \mathbb{C})$$

is called the *monodromy group* of system (1.1.1).

The above terminology of course involves a slight abuse of notation: The group $\mathfrak{M}=\mathfrak{M}(a_0, \Psi_0)$ depends on the choice of a_0 and Ψ_0. However, all these groups are isomorphic via a conjugation, e.g.,

$$\mathfrak{M}(a_0, \Psi_0) = \Psi_0^{-1}\mathfrak{M}(a_0, I)\Psi_0,$$

and when we say monodromy group we really mean the equivalence class of monodromy group modulo conjugation. At the same time, in every particular calculation involving the monodromy group the pair $\{a_0, \Psi_0\}$ should be specified. In fact, it is often convenient (see examples below) to choose the base point a_0 as one of the singular points a_k.

Let $\gamma_1, \gamma_2, \dots, \gamma_m$ denote the usual set of generators of the fundamental group π_1, i.e., the homotopy class γ_ν corresponds to the "small" loop around the point a_ν passing in the positive direction (see Figure 2.1).

Then the matrices

$$\mathsf{M}_\nu \equiv \mathsf{M}(\gamma_\nu), \quad \nu = 1, 2, \dots, m,$$

form a set of generators for the monodromy group $\mathfrak{M}$. They are usually called *monodromy matrices.*

The homotopy relation,

$$\gamma_1\gamma_2\cdots\gamma_m \sim \{a_0\},$$

implies that the generators M_ν satisfy the *cyclic* equation,

$$\mathsf{M}_m \dots \mathsf{M}_2 \mathsf{M}_1 = I. \tag{2.0.31}$$

The importance of the monodromy group is that it is an *invariant* global characteristic of the solutions of equation (1.1.1). Indeed, the monodromy group $\mathfrak{M}$ (more exactly, its conjugation class) depends only on the coefficient matrix $\mathsf{A}(\lambda)$. However, in a number of important questions of the theory of linear ODEs (including the main topic of this book) it is necessary to know the relation of the monodromy group with the local characteristics of the solutions introduced in the previous section. Indeed we need to evaluate each of the monodromy matrices M_ν in terms of the set of monodromy data $\mathbb{M}$ as defined in (2.0.20).

0.3. Evaluation in terms of monodromy data. Fuchsian equations. Suppose that the rational matrix function $\mathsf{A}(\lambda)$ has only simple poles. Then all singular points of equation (1.1.1) are Fuchsian, and the equation itself is called *Fuchsian equation.* Note that Fuchsian equations represent the generic case of systems of linear ordinary differential equations with rational coefficients.

Assuming that $a_m = \infty$, a Fuchsian equation with m (regular) singular points can be written as

$$\frac{d\Psi(\lambda)}{d\lambda} = \sum_{k=1}^{m-1} \frac{A_k}{\lambda - a_k} \Psi(\lambda). \tag{2.0.32}$$

Observe, that if $\sum A_k = 0$, then the infinity is actually a regular point and the equation has in fact $m-1$ singular points located on the affine part of $\mathbb{C}P^1$.

Consider the Fuchsian system (2.0.32). Let, as before, M_ν, $\Lambda_0^{(\nu)}$, E_ν, denote the corresponding monodromy matrices, the formal monodromy exponents, and the connection matrices, respectively. Our goal in this subsection is to show that, with the proper choice of the branches of the multi-valued functions $\xi_\nu^{\Lambda_0^{(\nu)}}$ appearing in the definitions of canonical solutions, the following explicit formula is valid:

$$\mathsf{M}_\nu = \mathsf{E}_\nu^{-1} e^{2\pi i \Lambda_0^{(\nu)}} \mathsf{E}_\nu, \quad \nu = 1, \dots, m. \tag{2.0.33}$$

To derive (2.0.33) we first notice that equations (2.0.18) and (2.0.11) imply that in some neighborhood of the point a_ν,

$$\Psi(\lambda) = P_\nu \left(I + \sum_{j=1}^{\infty} \Psi_j^{(\nu)} (\lambda - a_\nu)^j \right) (\lambda - a_\nu)^{\Lambda_0^{(\nu)}} \mathsf{E}_\nu, \tag{2.0.34}$$

$$\lambda - a_\nu \to \lambda^{-1}, \quad \text{if} \quad \nu = m.$$

Here, as in the previous Subsection, $\Psi(\lambda)$ is the fundamental solution of (2.0.32) determined by the normalization pair $\{a_0, \Psi_0\}$, i.e.,

$$\Psi(a_0) = \Psi_0.$$

Let us now take a point b from the neighborhood of the point a_ν where the canonical representation (2.0.34) of the solution $\Psi(\lambda)$ holds. In addition, let δ_{bb} be the positively oriented circle centered at a_ν and passing through b. The loop γ_ν can be represented as the product,

$$\gamma_\nu = \gamma_{a_0 b} \cdot \delta_{bb} \cdot \gamma_{b a_0}.$$

(see Figure 2.2). Therefore, taking into account the first property of the monodromy

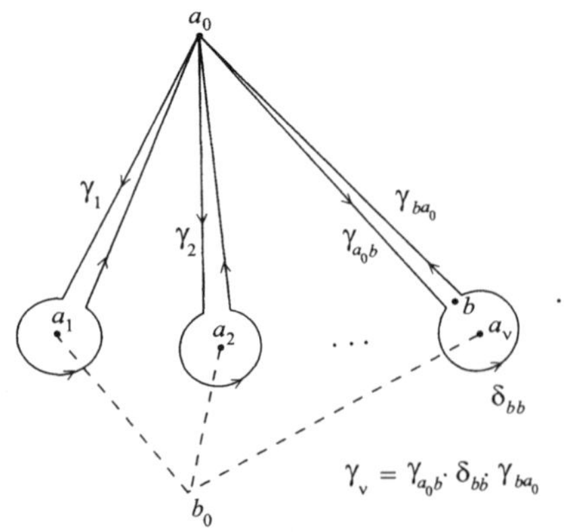

FIGURE 2.2. Generators of the fundamental group π_1 and the cuts $[b_0, a_\nu]$

operator (see Proposition 1.2), we have

$$\mathsf{M}_\nu = \Psi_0^{-1}\Psi_{\gamma_\nu}(a_0) = \Psi_0^{-1}\tau_{\gamma_\nu}(\Psi_0)$$

$$= \Psi_0^{-1}\tau_{\gamma_{ba_0}}(\tau_{\delta_{bb}}(\tau_{\gamma_{a_0b}}(\Psi_0))) = \Psi_0^{-1}\tau_{\gamma_{ba_0}}(\tau_{\delta_{bb}}(\Psi_b)), \tag{2.0.35}$$

where

$$\Psi_b \equiv \tau_{\gamma_{a_0b}}(\Psi_0). \tag{2.0.36}$$

Note that in fact $\Psi_b = \Psi(b)$, where for $\Psi(\lambda)$ one can use formula (2.0.34). More precisely, we will assume that the branch of the power function in (2.0.34) is chosen in such a way that the equation

$$\Psi_b = P_\nu\left(I + \sum_{j=1}^{\infty}\Psi_j^{(\nu)}(b-a_\nu)^j\right)(b-a_\nu)^{\Lambda_0^{(\nu)}}\mathsf{E}_\nu$$

is valid. From the representation (2.0.34) it follows then that

$$\tau_{\delta_{bb}}(\Psi_b) = \Psi_b\mathsf{E}_\nu^{-1}e^{2\pi i\Lambda_0^{(\nu)}}\mathsf{E}_\nu, \tag{2.0.37}$$

and hence equation (2.0.35) can be rewritten (using the second property of τ_γ) as

$$\mathsf{M}_\nu = \Psi_0^{-1}\tau_{\gamma_{ba_0}}(\Psi_b)\mathsf{E}_\nu^{-1}e^{2\pi i\Lambda_0^{(\nu)}}\mathsf{E}_\nu.$$

Noticing (cf. (2.0.36)) that

$$\tau_{\gamma_{ba_0}}(\Psi_b) = \Psi_0,$$

formula (2.0.33) follows.

0.4. Evaluation in terms of monodromy data. Non-Fuchsian case. Let a_ν, $\nu \in \{1,\dots,m\}$ be an irregular singular point of equation (1.1.1). This means that the differential form $\mathsf{A}(\lambda)\,d\lambda$ has a pole of order $r_\nu + 1 \geq 2$ at $\lambda = a_\nu$. Denote, as before, with $\mathsf{S}_n^{(\nu)}$ the corresponding Stokes matrices relating the canonical solutions, and with E_ν the connection matrix between the basic solution $\Psi(\lambda)$ and the canonical solution $\Psi_1^{(\nu)}(\lambda)$ (see (2.0.19), (2.0.17), and (2.0.12)).

The evaluation of the monodromy matrices performed in the previous subsection, can be easily modified for the case of a irregular singular point. Indeed assuming that the intermediate point b belongs to the Stokes sector $\Omega_1^{(\nu)}$, let us calculate the action of the monodromy operator $\tau_{\delta_{bb}}$ on $\Psi_b = \Psi(b)$. By definition (see Subsection 2.2 of the previous chapter),

$$\tau_{\delta_{bb}}(\Psi_b) = \Psi_{\delta_{bb}}(b),$$

where $\Psi_{\delta_{bb}}(\lambda)$ is the analytic continuation of the solution $\Psi(\lambda)|_{\Omega_1^{(\nu)}}$ along the curve δ_{bb}. Observe now the series of relations,

$$\Psi(\lambda) = \Psi_2^{(\nu)}(\lambda)\mathsf{S}_1^{(\nu)-1}\mathsf{E}_\nu, \quad \Psi(\lambda) = \Psi_3^{(\nu)}(\lambda)\mathsf{S}_2^{(\nu)-1}\mathsf{S}_1^{(\nu)-1}\mathsf{E}_\nu,$$

$$\Psi(\lambda) = \Psi_n^{(\nu)}(\lambda)\mathsf{S}_{n-1}^{(\nu)-1}\mathsf{S}_{n-2}^{(\nu)-1}\dots\mathsf{S}_1^{(\nu)-1}\mathsf{E}_\nu, \quad n \geq 3,$$

which in fact perform the analytic continuation of the solution $\Psi(\lambda)|_{\Omega_1^{(\nu)}}$ along the curve δ_{bb}. Therefore,

$$\Psi_{\delta_{bb}}(\lambda) = \Psi_{2r_\nu+1}^{(\nu)}(\lambda)\mathsf{S}_{2r_\nu}^{(\nu)-1}\dots\mathsf{S}_1^{(\nu)-1}\mathsf{E}_\nu.$$

Taking into account equation (2.0.14), we conclude that

$$\tau_{\delta_{bb}}(\Psi_b) = \Psi_{\delta_{bb}}(\lambda) = \Psi_1^{(\nu)}(b)e^{2\pi i\Lambda_0^{(\nu)}}\mathsf{S}_{2r_\nu}^{(\nu)-1}\dots\mathsf{S}_1^{(\nu)-1}\mathsf{E}_\nu$$

$$= \Psi(b)\mathsf{E}_\nu^{-1}e^{2\pi i\Lambda_0^{(\nu)}}\mathsf{S}_{2r_\nu}^{(\nu)-1}\dots\mathsf{S}_1^{(\nu)-1}\mathsf{E}_\nu,$$

and hence equation (2.0.37) is replaced by the relation

$$\tau_{\delta_{bb}}(\Psi_b) = \Psi_b\mathsf{E}_\nu^{-1}e^{2\pi i\Lambda_0^{(\nu)}}\mathsf{S}_{2r_\nu}^{(\nu)-1}\dots\mathsf{S}_1^{(\nu)-1}\mathsf{E}_\nu. \tag{2.0.38}$$

This in turn implies that formula (2.0.33) is transformed into

$$\mathsf{M}_\nu = \mathsf{E}_\nu^{-1}e^{2\pi i\Lambda_0^{(\nu)}}\mathsf{S}_{2r_\nu}^{(\nu)-1}\dots\mathsf{S}_1^{(\nu)-1}\mathsf{E}_\nu. \tag{2.0.39}$$

REMARK 2.3. Equations (2.0.39) and (2.0.33) explain the reason for including in the essential monodromy data m the formal monodromy exponents $\Lambda_0^{(\nu)}$: they are needed in order to make the essential set m completely responsible for the principal global characteristic of a linear system — its monodromy group.

0.5. The direct and inverse monodromy problems. The Riemann-Hilbert and Riemann-Hilbert-Birkhoff problems. Denote

$$\mathcal{A} \equiv \{\mathsf{A}(\lambda)\},$$

the set of all rational matrices with the given number m of poles and given multiplicities r at each pole. Also denote,

$$\mathcal{M} \equiv \{\mathfrak{m}\},$$

and

$$\mathcal{M}_e \equiv \{\mathbb{M}\},$$

the sets of all possible corresponding monodromy and extended monodromy data, respectively. We are now ready to prove the following simple but fundamental fact.

PROPOSITION 2.2. *The map,*

$$\mathcal{A} \ni \mathsf{A}(\lambda) \mapsto \mathbb{M} \in \mathcal{M}_e, \tag{2.0.40}$$

is one-to-one.

PROOF. Let $A(\lambda)$, $B(\lambda) \in \mathcal{A}$ and suppose that

$$\mathbb{M}\Big(A(\lambda)\Big) = \mathbb{M}\Big(B(\lambda)\Big). \tag{2.0.41}$$

Let $\Psi(\lambda)$ and $\Phi(\lambda)$ be the basic fundamental solutions of the equations

$$\frac{d\Psi}{d\lambda} = A(\lambda)\Psi \tag{2.0.42}$$

and

$$\frac{d\Phi}{d\lambda} = B(\lambda)\Phi, \tag{2.0.43}$$

respectively. Consider the matrix ratio,

$$g(\lambda) := \Phi(\lambda)\Psi^{-1}(\lambda).$$

The matrix function $g(\lambda)$ shares with the functions $\Psi(\lambda)$ and $\Phi(\lambda)$ their principal analytic property: the function $g(\lambda)$ is an analytic function on the universal covering $\widetilde{X}$, $X = \mathbb{C}\mathrm{P}^1 \setminus \{a_\nu\}_{\nu=1}^m$. Moreover, in virtue of the previous two subsections, relation (2.0.41) implies that equations (2.0.42) and (2.0.43) have exactly the same monodromy groups, or that the functions $\Psi(\lambda)$ and $\Phi(\lambda)$ gain exactly the same right matrix multipliers M_ν when circling around the singular points. Therefore, the function $g(\lambda)$ is, in fact, a single valued analytic function on the base space, i.e., on the punctured Riemann sphere $X = \mathbb{C}\mathrm{P}^1 \setminus \{a_\nu\}_{\nu=1}^m$. Relation (2.0.41) means even more. Indeed, it tells us that at every singular point a_ν, the singularities of the functions $\Psi(\lambda)$ and $\Phi(\lambda)$ cancel out in the product $\Phi(\lambda)\Psi^{-1}(\lambda)$ (compare with the proof of the *local* version of this proposition — Theorem 1.5, in Chapter 1). Hence each point a_ν is a removable singularity of the function $g(\lambda)$. Taking into account that we assume that both the functions $\Psi(\lambda)$ and $\Phi(\lambda)$ are normalized by the same pair $\{a_0, \Psi_0\}$, we finally conclude that $g(\lambda) \equiv I$. In other words,

$$\Psi(\lambda) \equiv \Phi(\lambda),$$

which in turn yields the identity $A(\lambda) \equiv B(\lambda)$. □

Proposition 2.2 establishes the key property of the extended monodromy data $\mathbb{M}$, which have already been announced and which states that the extended monodromy data $\mathbb{M}$ determine the linear system (1.1.1) uniquely.

The principal focus of the global theory of system of linear ODEs with rational coefficients is the analysis of the direct,

$$\mathcal{A} \mapsto \mathcal{M}, \tag{2.0.44}$$

and the inverse,

$$\mathcal{M}_e \mapsto \mathcal{A},$$

monodromy maps associated with equation (1.1.1). Specifically, the *Riemann-Hilbert problem*, also known as Hilbert's twenty-first problem, consists of proving the existence of a Fuchsian system with the given set $\{a\}$ of (regular) singular points and the given monodromy group $\mathfrak{M}$. A natural generalization of the Riemann-Hilbert problem for non-Fuchsian equations is due to Birkhoff, and it consists of proving the existence of a linear ODE of the form (1.1.1) with the given set $\{a\}$ of singular points, the given monodromy group $\mathfrak{M}$, and the given Stokes phenomenon **SPh** at every irregular singular point.

REMARK 2.4. Proposition 2.2 states that the map

$$\mathcal{A} \mapsto \mathcal{M}_e$$

is one-to-one. This is, of course, not (in general) true for the *direct monodromy map* (2.0.44). In fact, it is the pre-images of the single points under the direct monodromy map which are of the principal interest in this book. We will discuss this issue in full detail in Chapter 4.

At this juncture we will temporarily stop the discussion of the general monodromy theory. In the next section we will illustrate the general theory by considering the first nontrivial examples of 2×2 linear systems. Simultaneously, this will prepare us for the introduction of the major object of our study, the Painlevé equations. We will resume the discussion of the general theory in Chapters 3 and 4.

1. Two by two systems. Special functions

We start with the 2×2 matrix ODE containing regular singular points only, i.e., with systems of Fuchsian type. We will consider cases with two, three and four singular points, respectively.

1.1. Two Fuchsian points. Using an appropriate transformation belonging to the group $PGL(2, \mathbb{C})$ of automorphisms of $\mathbb{C}\mathrm{P}^1$, we can always make the singular points to be 0 and ∞, so that the system in question is

$$\frac{d\Psi}{d\lambda} = \frac{A_0}{\lambda}\Psi. \tag{2.1.1}$$

A fundamental solution $\Psi(\lambda)$ is

$$\Psi(\lambda) = \lambda^{A_0}.$$

The corresponding monodromy group is commutative, since it is generated by only one monodromy matrix,

$$\mathrm{M}_0 = \left[\mathrm{M}_\infty\right]^{-1} = e^{2\pi i A_0}. \tag{2.1.2}$$

Equation (2.1.2) constitutes an explicit solution of the direct monodromy problems for the system (2.1.1). The inverse monodromy problem is uniquely solvable by the equation

$$A_0 = \frac{1}{2\pi i}\log \mathrm{M}_0.$$

1.2. Three Fuchsian points. *Hypergeometric function.* Up to an automorphism of $\mathbb{C}\mathrm{P}^1$, the singular points can be placed at 0, 1 and ∞, thus we have the system

$$\frac{d\Psi}{d\lambda} = \left[\frac{A_0}{\lambda} + \frac{A_1}{\lambda - 1}\right]\Psi. \tag{2.1.3}$$

We can also assume that the system is traceless, and, assume for simplicity, that the matrices A_0, A_1 and $A_\infty = -A_0 - A_1$ have the nonzero eigenvalues $\pm\alpha$, $\pm\beta$

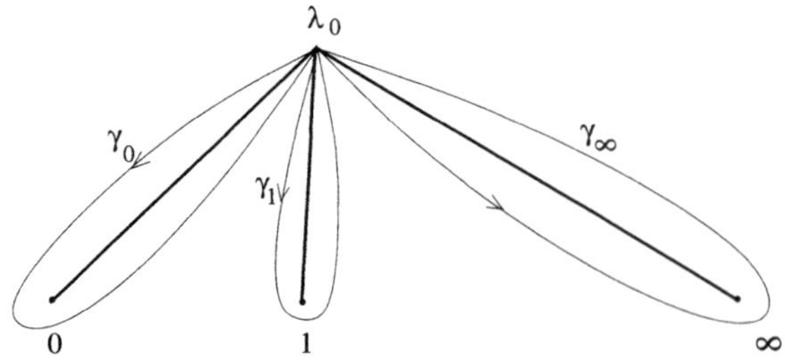

FIGURE 2.3. Basis of cycles associated with hypergeometric equation

and $\pm\gamma$, respectively. Using a constant gauge transformation, we normalize the equation at infinity,

$$A_\infty = \begin{pmatrix} \gamma & 0 \\ 0 & -\gamma \end{pmatrix}. \tag{2.1.4}$$

It is convenient to parametrize the constant matrices A_0 and A_1 by four constants α, β, γ and u,

$$A_0 = \begin{pmatrix} z & -u(z-\alpha) \\ \frac{1}{u}(z+\alpha) & -z \end{pmatrix}, \quad A_1 = \begin{pmatrix} -z-\gamma & u(z-\alpha) \\ -\frac{1}{u}(z+\alpha) & z+\gamma \end{pmatrix}, \tag{2.1.5}$$

where $z = \frac{\beta^2-\alpha^2-\gamma^2}{2\gamma}$.

Let us proceed now to the analysis of the manifold $\mathcal{M}$ of the *monodromy data* of the system (2.1.3), (2.1.5). Choose the base point a_0, branch cuts $[a_0, 0]$, $[a_0, 1]$, $[a_0, \infty)$ and the loops γ_0, γ_1 and γ_∞ as shown on Figure 2.3. The relevant monodromy data are exhausted by the monodromy matrices M_0, M_1, M_∞, satisfying the equations

$$\begin{aligned} &\det \mathrm{M}_\nu = 1, \quad \nu = 0, 1, \infty, \\ &\mathrm{M}_\infty \mathrm{M}_1 \mathrm{M}_0 = I, \\ &\mathrm{M}_\infty = \text{diag } \mathrm{M}_\infty \equiv \begin{pmatrix} e^{2\pi i\gamma} & 0 \\ 0 & e^{-2\pi i\gamma} \end{pmatrix}. \end{aligned} \tag{2.1.6}$$

The first equation in (2.1.6) is a consequence of the gauge condition, $\mathrm{Tr}\, \mathsf{A}(\lambda) = 0$, the second equation in (2.1.6) is the cyclic constraint (see (2.0.31)), and the last equation fixes the conjugacy class of the monodromy group as the group of the solution $\Psi_\infty(\lambda)$ determined by its asymptotics at infinity,

$$\Psi_\infty(\lambda) = \big(I + O(\lambda^{-1})\big)\lambda^{-\gamma\sigma_3}, \quad \lambda \to \infty,$$

where $\lambda^{-\gamma\sigma_3}$ is defined on the plane cut as above. Equations (2.1.6) imply that the number of independent parameters for $\mathcal{M}$ equals 4, i.e.,

$$\dim \mathcal{M} = 4.$$

Therefore, $\dim\mathcal{M} = \dim\mathcal{A}$, and one expects a one-to-one correspondence between 2×2 systems with three regular singular points and their monodromy data. In fact, this correspondence can be represented in *closed* form using *higher transcendental functions*. Indeed, the system (2.1.3), (2.1.5) is equivalent to the hypergeometric equation: Assume that $(\alpha+\gamma)^2 \neq \beta^2$, i.e., $z \neq \alpha$ (if this condition is not satisfied, the subsequent computation can be performed after the transformation $\Psi \mapsto \sigma_1\Psi$, where $\sigma_1 = \left(\begin{smallmatrix} 0 & 1 \\ 1 & 0 \end{smallmatrix}\right)$, which changes the sign $\alpha \mapsto -\alpha$). The transformation

$$\Psi(\lambda) = V(\lambda)\Phi(\lambda),$$

where the gauge matrix $V(\lambda)$ is defined by

$$V(\lambda) = \left(\frac{u(z-\alpha)}{\lambda(1-\lambda)}\right)^{\sigma_3/2}\sigma_3 \begin{pmatrix} 1 & 0 \\ -\frac{z+\frac{1}{2}}{\lambda} - \frac{z+\gamma-\frac{1}{2}}{1-\lambda} & 1 \end{pmatrix}, \quad \sigma_3 = \begin{pmatrix} 1 & \\ & -1 \end{pmatrix},$$

yields the equation

$$\frac{d\Phi}{d\lambda} = \begin{pmatrix} 0 & 1 \\ Q(\lambda) & 0 \end{pmatrix}\Phi, \tag{2.1.7}$$

with

$$Q(\lambda) = \frac{\alpha^2 - \frac{1}{4}}{\lambda^2} + \frac{\beta^2 - \frac{1}{4}}{(\lambda-1)^2} - \frac{\alpha^2+\beta^2-\gamma(\gamma-1)-\frac{1}{2}}{\lambda(\lambda-1)}.$$

The matrix equation (2.1.7) implies the scalar second order equation for the first row entries of Φ,

$$\frac{d^2\Phi_1}{d\lambda^2} = Q(\lambda)\Phi_1, \quad \Phi_2 = \frac{d\Phi_1}{d\lambda}. \tag{2.1.8}$$

The first of the equations (2.1.8) is one of the forms of the hypergeometric equation. Namely, the function

$$y = \lambda^{\alpha-\frac{1}{2}}(1-\lambda)^{\beta-\frac{1}{2}}\Phi_1, \tag{2.1.9}$$

satisfies the canonical hypergeometric equation; see [**BE**].

$$\lambda(1-\lambda)\frac{d^2y}{d\lambda^2} + \big(c-(a+b+1)\lambda\big)\frac{dy}{d\lambda} - aby = 0, \tag{2.1.10}$$

with the parameters

$$a = -\alpha-\beta+\gamma, \quad b = -\alpha-\beta-\gamma+1, \quad c = 1-2\alpha,$$

or

$$\alpha = \frac{1-c}{2}, \quad \beta = \frac{c-a-b}{2}, \quad \gamma = \frac{a-b+1}{2}.$$

If $2\alpha \notin \mathbb{Z}$, then two linearly independent solutions $y_{01}(\lambda)$ and $y_{02}(\lambda)$ of this equation, which form the row vector $Y(\lambda) = (y_{01}, y_{02})$, are expressed in terms of the hypergeometric series as follows:

$$y_{01}(\lambda) = F(a,b,c;\lambda), \quad y_{02}(\lambda) = \lambda^{1-c}F(a-c+1, b-c+1, 2-c;\lambda),$$

where

$$F(a,b,c;\lambda) \equiv {}_2F_1(a,b,c,\lambda) := \sum_{n=0}^{\infty}\frac{(a)_n(b)_n}{n!(c)_n}\lambda^n, \tag{2.1.11}$$

$$(a)_0 = 1, \quad (a)_n = \frac{\Gamma(a+n)}{\Gamma(a)} = a(a+1)\cdots(a+n-1), \quad n = 1, 2, 3, \ldots$$

In the generic case, when $2\alpha, 2\beta, 2\gamma \notin \mathbb{Z}$, their analytic continuation due to Barnes *integral representation* and Kummer's relations, is given by the following formulae (see e.g. [**BE**]):

$$Y(\lambda) = \big(y_{01}(\lambda), y_{02}(\lambda)\big), \quad |\lambda| < 1, \tag{2.1.12}$$

$$Y(\lambda) = \big(y_{11}(\lambda), y_{12}(\lambda)\big)G_1, \tag{2.1.13}$$

$$G_1 = \begin{pmatrix} \dfrac{\Gamma(c)\Gamma(c-a-b)}{\Gamma(c-a)\Gamma(c-b)} & \dfrac{\Gamma(2-c)\Gamma(c-a-b)}{\Gamma(1-a)\Gamma(1-b)} \\ \dfrac{\Gamma(c)\Gamma(a+b-c)}{\Gamma(a)\Gamma(b)} & \dfrac{\Gamma(2-c)\Gamma(a+b-c)}{\Gamma(a-c+1)\Gamma(b-c+1)} \end{pmatrix},$$

$$y_{11}(\lambda) = F(a, b, a+b-c+1; 1-\lambda),$$
$$y_{12}(\lambda) = \lambda^{1-c}(1-\lambda)^{c-a-b}F(1-a, 1-b, c-a-b+1; 1-\lambda),$$
$$|1-\lambda| < 1, \quad \arg(1-\lambda) \in (-\pi, \pi),$$

$$Y(\lambda) = \big(y_{\infty 1}(\lambda), y_{\infty 2}(\lambda)\big)G_2, \tag{2.1.14}$$

$$G_2 = \begin{pmatrix} \dfrac{\Gamma(c)\Gamma(b-a)}{\Gamma(b)\Gamma(c-a)} & \dfrac{\Gamma(2-c)\Gamma(b-a)}{\Gamma(1-a)\Gamma(b-c+1)} \\ \dfrac{\Gamma(c)\Gamma(a-b)}{\Gamma(a)\Gamma(c-b)} & \dfrac{\Gamma(2-c)\Gamma(a-b)}{\Gamma(a-c+1)\Gamma(1-b)} \end{pmatrix},$$

$$y_{\infty 1}(\lambda) = (-\lambda)^{-a}F(a, a-c+1, a-b+1, \lambda^{-1}),$$
$$y_{\infty 2}(\lambda) = (-\lambda)^{-b}F(b, b-c+1, b-a+1, \lambda^{-1}),$$
$$|\lambda| > 1, \quad \arg(-\lambda) \in (-\pi, \pi).$$

These expressions imply that the function $\Psi(\lambda)$ defined by

$$\Psi(\lambda) = V(\lambda)\begin{pmatrix} \lambda^{\frac{1}{2}-\alpha}(1-\lambda)^{\frac{1}{2}-\beta}Y(\lambda) \\ \frac{d}{d\lambda}\left(\lambda^{\frac{1}{2}-\alpha}(1-\lambda)^{\frac{1}{2}-\beta}Y(\lambda)\right) \end{pmatrix} \tag{2.1.15}$$
$$\times \frac{e^{-i\pi\alpha}}{\sqrt{2\gamma-1}}G_2^{-1}\sigma_3(2\gamma-1)^{\sigma_3/2}e^{i\pi\gamma\sigma_3}(u(z-\alpha))^{-\sigma_3/2},$$

is characterized by the following asymptotic behavior as λ approaches one of the singular points:

$$\lambda \to 0: \qquad \Psi(\lambda) = \begin{pmatrix} 1 & u(z-\alpha) \\ 1/u & z+\alpha \end{pmatrix}(I + O(\lambda))\lambda^{\alpha\sigma_3}\,\mathrm{E}_0,$$

$$\lambda \to 1: \qquad \Psi(\lambda) = \begin{pmatrix} 1 & u(z-\alpha) \\ \frac{z+\beta+\gamma}{u(z-\alpha)} & z-\beta+\gamma \end{pmatrix}(I + O(\lambda-1))(1-\lambda)^{\beta\sigma_3}\mathrm{E}_1,$$

$$\lambda \to \infty: \qquad \Psi(\lambda) = \big(I + O(\lambda^{-1})\big)\lambda^{-\gamma\sigma_3},$$

where

$$\mathsf{E}_0 = \frac{e^{-i\pi\alpha}}{\sqrt{2\gamma-1}}(u(z-\alpha))^{\sigma_3/2}\sigma_1 G_2^{-1}\sigma_3(2\gamma-1)^{\sigma_3/2}e^{i\pi\gamma\sigma_3}(u(z-\alpha))^{-\sigma_3/2},$$

$$\mathsf{E}_1 = \frac{e^{-i\pi\alpha}}{\sqrt{2\gamma-1}}(u(z-\alpha))^{\sigma_3/2}\sigma_1 G_1 G_2^{-1}\sigma_3(2\gamma-1)^{\sigma_3/2}e^{i\pi\gamma\sigma_3}(u(z-\alpha))^{-\sigma_3/2},$$

and therefore coincides with $\Psi_\infty(\lambda)$ whose monodromy group we are looking for. The relevant monodromy matrices are given by (2.0.33),

$$\mathsf{M}_0 = \mathsf{E}_0^{-1}e^{2\pi i\alpha\sigma_3}\mathsf{E}_0$$

$$= i\begin{pmatrix} \dfrac{e^{-2\pi i\gamma}\cos 2\pi\alpha - \cos 2\pi\beta}{\sin 2\pi\gamma} & \dfrac{2\pi\Gamma(2\gamma)^2 e^{2\pi i\gamma}}{u(z-\alpha)\Gamma(-\alpha-\beta+\gamma)\Gamma(\alpha-\beta+\gamma)\Gamma(-\alpha+\beta+\gamma)\Gamma(\alpha+\beta+\gamma)} \\ -\dfrac{2\pi u(z-\alpha)\Gamma(1-2\gamma)^2 e^{-2\pi i\gamma}}{\Gamma(1+\alpha+\beta-\gamma)\Gamma(1+\alpha-\beta-\gamma)\Gamma(1-\alpha-\beta-\gamma)\Gamma(1-\alpha+\beta-\gamma)} & -\dfrac{e^{2\pi i\gamma}\cos 2\pi\alpha - \cos 2\pi\beta}{\sin 2\pi\gamma} \end{pmatrix},$$

$$\mathsf{M}_1 = \mathsf{E}_1^{-1}e^{2\pi i\beta\sigma_3}\mathsf{E}_1 = \frac{2\beta(1-2\gamma)}{\cos^2\pi\alpha}(u(z-\alpha))^{\sigma_3/2}e^{-i\pi\gamma\sigma_3}(2\gamma-1)^{-\sigma_3/2}\sigma_3$$

$$\times\begin{pmatrix} \dfrac{\cos\pi(-\beta+\gamma)\Gamma(-2\beta)\Gamma(1-2\gamma)}{\Gamma(1+\alpha-\beta-\gamma)\Gamma(1-\alpha-\beta-\gamma)} & -\dfrac{\cos\pi(\beta+\gamma)\Gamma(2\beta)\Gamma(1-2\gamma)}{\Gamma(1+\alpha+\beta-\gamma)\Gamma(1-\alpha+\beta-\gamma)} \\ -\dfrac{\cos\pi(\beta+\gamma)\Gamma(-2\beta)\Gamma(2\gamma-1)}{\Gamma(\alpha-\beta+\gamma)\Gamma(-\alpha-\beta+\gamma)} & \dfrac{\cos\pi(-\beta+\gamma)\Gamma(2\beta)\Gamma(2\gamma-1)}{\Gamma(\alpha+\beta+\gamma)\Gamma(-\alpha+\beta+\gamma)} \end{pmatrix}$$

$$\times\begin{pmatrix} \dfrac{\cos\pi(-\beta+\gamma)\Gamma(2\beta)\Gamma(2\gamma-1)e^{-2\pi i\beta}}{\Gamma(\alpha+\beta+\gamma)\Gamma(-\alpha+\beta+\gamma)} & \dfrac{\cos\pi(\beta+\gamma)\Gamma(2\beta)\Gamma(1-2\gamma)e^{-2\pi i\beta}}{\Gamma(1+\alpha+\beta-\gamma)\Gamma(1-\alpha+\beta-\gamma)} \\ \dfrac{\cos\pi(\beta+\gamma)\Gamma(-2\beta)\Gamma(2\gamma-1)e^{2\pi i\beta}}{\Gamma(\alpha-\beta+\gamma)\Gamma(-\alpha-\beta+\gamma)} & \dfrac{\cos\pi(-\beta+\gamma)\Gamma(-2\beta)\Gamma(1-2\gamma)e^{2\pi i\beta}}{\Gamma(1+\alpha-\beta-\gamma)\Gamma(1-\alpha-\beta-\gamma)} \end{pmatrix}$$

$$\times\sigma_3(2\gamma-1)^{\sigma_3/2}e^{i\pi\gamma\sigma_3}(u(z-\alpha))^{-\sigma_3/2},$$

$$\mathsf{M}_\infty = e^{2\pi i\gamma\sigma_3}.$$

1.3. Four Fuchsian points. Using the proper automorphism of $\mathbb{C}P^1$ and gauge transformations, we can always assume that the singular points of the equation are 0, 1, ∞ and x. Thus we consider the 2×2 system

$$\frac{d\Psi}{d\lambda} = \left[\frac{A_0}{\lambda} + \frac{A_1}{\lambda-1} + \frac{A_2}{\lambda-x}\right]\Psi, \tag{2.1.16}$$

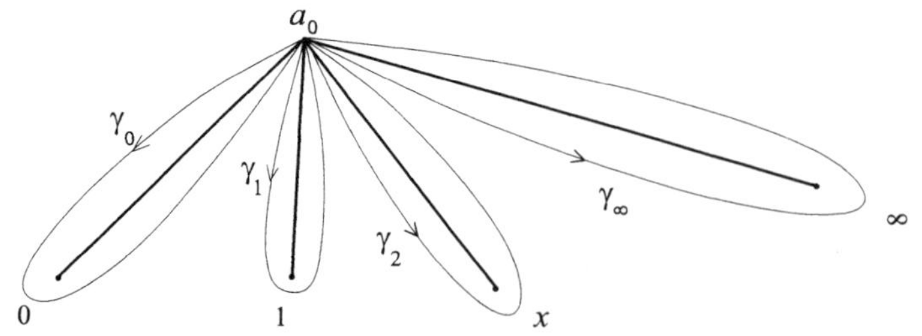

FIGURE 2.4. Basis of cycles associated with PVI equation

where, without loss of generality, the coefficient matrices A_ν, $\nu = 0, 1, 2$, are traceless, and the system is diagonal at $\lambda = \infty$, i.e.,

$$\text{Tr } A_\nu = 0, \quad \nu = 0, 1, 2, \tag{2.1.17}$$

$$A_\infty = -A_0 - A_1 - A_2 = \begin{pmatrix} \delta & 0 \\ 0 & -\delta \end{pmatrix}. \tag{2.1.18}$$

We use the notations $\pm\alpha$, $\pm\beta$, $\pm\gamma$ for the eigenvalues of A_0, A_1 and A_2, respectively, and assume that

$$2\alpha,\ 2\beta,\ 2\gamma \notin \mathbb{Z}.$$

Equations (2.1.17) imply that the complex dimension of the manifold $\mathcal{A}$ of the singular data equals 8 (7 complex numbers parametrize the coefficient matrices A_ν and 1 parameter x marks the location of the fourth singular point),

$$\dim \mathcal{A} = 8.$$

Consider now the corresponding monodromy group $\mathfrak{M}$. Choose the base point a_0, branch cuts $[a_0, 0]$, $[a_0, 1]$, $[a_0, x]$, $[a_0, \infty]$ and the loops γ_0, γ_1, γ_2 and γ_∞ as shown on Figure 2.4. The relevant monodromy data are exhausted by the monodromy matrices M_ν, $\nu = 0, 1, 2, \infty$, which give the $SL(2, \mathbb{C})$ representation of the respective loops γ_j. Just like the previous case, we have the equations

$$\det \mathsf{M}_\nu = 1, \quad \nu = 0, 1, 2, \infty,$$

$$\mathsf{M}_\infty \mathsf{M}_2 \mathsf{M}_1 \mathsf{M}_0 = I, \tag{2.1.19}$$

$$\mathsf{M}_\infty = \begin{pmatrix} e^{2\pi i\delta} & 0 \\ 0 & e^{-2\pi i\delta} \end{pmatrix}. \tag{2.1.20}$$

The first equation is a consequence of the gauge condition, Tr $\mathsf{A}(\lambda) = 0$, equation (2.1.19) is the cyclic constraint (see (2.0.31)), and equation (2.1.20) fixes the conjugacy class of the monodromy group as the group of the solution $\Psi_\infty(\lambda)$ determined by its asymptotics at infinity,

$$\Psi_\infty(\lambda) = \big(I + O(\lambda^{-1})\big)\lambda^{-\delta\sigma_3}, \quad \lambda \to \infty,$$

where $\lambda^{-\delta\sigma_3}$ is defined on the plane cut as above.

Equations (2.1.19) imply that the complex dimension of the manifold of the monodromy data of the system (2.1.16) equals 7, so that

$$\dim\mathcal{M} = \dim\mathcal{A} - 1. \tag{2.1.21}$$

This fact underlines the principal difference between the cases of three and four singularities. The monodromy map,

$$\mathcal{A} \mapsto \mathcal{M}$$

is no more a one-to-one correspondence; given a monodromy data, i.e., a point $\mathfrak{M} \in \mathrm{M}$, one expects a one-parameter family of equations (2.1.16), i.e., a curve in the space $\mathcal{A}$ with given monodromy group $\mathfrak{M}$. Therefore, for the first time we face the problem of analyzing the *isomonodromy deformations* of system (2.1.16). As we will see in Chapter 4, this is precisely the moment that Painlevé VI equation is born!

MONODROMY SURFACE AND PARAMETRIZATION OF THE MONODROMY DATA

We conclude our presentation of the general theory of the 2×2 systems with four regular singular points by indicating the explicit parametrizations of their singular and monodromy data suggested by M. Jimbo in [**J1**] following Iwasaki [**Iw**]. We will also clarify the relation of the matrix equation (2.1.16) with the so-called Heun equation.

Modulo an overall conjugation, in generic position, the monodromy matrices M_ν, $\nu = 0, 1, 2$, can be parametrized by the invariants of the adjoint action $a_\nu, t_{\mu\nu}$ or by the complex "angles" $\alpha_\nu, \sigma_{\mu\nu}$, defined as follows:

$$\begin{aligned} a_\nu &= \mathrm{Tr}\ \mathrm{M}_\nu = 2\cos 2\pi\alpha_\nu, \quad \nu = 0,1,2,3, \\ t_{\mu\nu} &= \mathrm{Tr}\ \mathrm{M}_\mu \mathrm{M}_\nu = 2\cos 2\pi\sigma_{\mu\nu}, \quad \mu,\nu = 0,1,2. \end{aligned} \tag{2.1.22}$$

Thus $\alpha_0 = \alpha$, $\alpha_1 = \beta$, $\alpha_2 = \gamma$, $\alpha_3 = \delta$ are identified with the local monodromy exponents.

Assume that there is a pair of indices $\mu = i, \nu = j$ such that the corresponding invariant

$$t_{ij} \neq \pm 2. \tag{2.1.23}$$

Let (i, j, k) be a cyclic permutation of $(2, 1, 0)$. Fix a square root of $t_{ij}^2 - 4$ and introduce the notations

$$r_{ij} = \sqrt{t_{ij}^2 - 4}, \quad \lambda_{ij}^{\pm} = \frac{t_{ij} \pm r_{ij}}{2}. \tag{2.1.24}$$

Due to (2.1.23), $r_{ij} \neq 0$, and the eigenvalues $\lambda_{ij}^{\pm}$ of the product $\mathrm{M}_i\mathrm{M}_j$ are distinct, $\lambda_{ij}^{+} \neq \lambda_{ij}^{-}$. Thus there exists a matrix $P \in SL(2,\mathbb{C})$ which diagonalizes this product,

$$P(\mathrm{M}_i\mathrm{M}_j)P^{-1} = \mathrm{diag}\,(\lambda_{ij}^{+}, \lambda_{ij}^{-}). \tag{2.1.25}$$

Put

$$PM_iP^{-1} = U = \begin{pmatrix} u_{11} & u_{12} \\ u_{21} & u_{22} \end{pmatrix},$$

$$PM_jP^{-1} = V = \begin{pmatrix} v_{11} & v_{12} \\ v_{21} & v_{22} \end{pmatrix},$$

$$PM_kP^{-1} = W = \begin{pmatrix} w_{11} & w_{12} \\ w_{21} & w_{22} \end{pmatrix}. \tag{2.1.26}$$

Then the above diagonalization of M_iM_j reads

$$UV = \operatorname{diag}(\lambda_{ij}^+, \lambda_{ij}^-); \tag{2.1.27}$$

the conditions $\operatorname{Tr} W = \operatorname{Tr} M_k = a_k$ and $\operatorname{Tr}(UVW) = \operatorname{Tr} M_\infty^{-1} = a_3$ yield

$$\begin{cases} w_{11} + w_{22} = a_k, \\ \lambda_{ij}^+ w_{11} + \lambda_{ij}^- w_{22} = a_3. \end{cases} \tag{2.1.28}$$

The system (2.1.28) determines the diagonal part of W,

$$w_{11} = \frac{a_3 - \lambda_{ij}^- a_k}{r_{ij}}, \qquad w_{22} = -\frac{a_3 - \lambda_{ij}^+ a_k}{r_{ij}}, \tag{2.1.29}$$

where the equality $\lambda_{ij}^+ - \lambda_{ij}^- = r_{ij}$ is used. To determine the diagonal entries of the matrices U and V, we apply conditions $U = \operatorname{diag}(\lambda_{ij}^+, \lambda_{ij}^-)V^{-1}$, the $\operatorname{Tr} U = \operatorname{Tr} M_i = a_i$ and $\operatorname{Tr} V = \operatorname{Tr} M_j = a_j$, which give

$$u_{11} = \lambda_{ij}^+ v_{22}, \quad u_{22} = \lambda_{ij}^- v_{11},$$

$$u_{11} + u_{22} = a_i, \quad v_{11} + v_{22} = a_j. \tag{2.1.30}$$

Thus

$$v_{11} = -\frac{a_i - \lambda_{ij}^+ a_j}{r_{ij}}, \qquad v_{22} = \frac{a_i - \lambda_{ij}^- a_j}{r_{ij}}, \tag{2.1.31}$$

and

$$u_{11} = -\frac{a_j - \lambda_{ij}^+ a_i}{r_{ij}}, \qquad u_{22} = \frac{a_j - \lambda_{ij}^- a_i}{r_{ij}}. \tag{2.1.32}$$

Applying (2.1.29) to the condition $\det W = \det M_k = 1$, we have

$$w_{12}w_{21} = -\frac{p(t_{ij}, a_k, a_3)}{t_{ij}^2 - 4}, \tag{2.1.33}$$

where the polynomial $p(a, b, c)$ is defined by

$$p(a, b, c) = a^2 + b^2 + c^2 - abc - 4. \tag{2.1.34}$$

Similarly, applying (2.1.31) to the condition $\det V = \det M_j = 1$, we find

$$v_{12}v_{21} = -\frac{p(t_{ij}, a_i, a_j)}{t_{ij}^2 - 4}. \tag{2.1.35}$$

The conditions $\mathrm{Tr}\,(WU) = \mathrm{Tr}\ \mathsf{M}_k\mathsf{M}_i = t_{ki}$ and $\mathrm{Tr}\,(VW) = \mathrm{Tr}\ \mathsf{M}_j\mathsf{M}_k = t_{jk}$ with $U = \mathrm{diag}\,(\lambda_{ij}^+, \lambda_{ij}^-)V^{-1}$ yield

$$
\begin{aligned}
&w_{12}v_{21} + w_{21}v_{12} = t_{jk} - w_{11}v_{11} - w_{22}v_{22},\\
&\lambda_{ij}^- w_{12}v_{21} + \lambda_{ij}^+ w_{21}v_{12} = -t_{ki} + \lambda_{ij}^+ w_{11}v_{22} + \lambda_{ij}^- w_{22}v_{11}.
\end{aligned} \tag{2.1.36}
$$

Solving this system for the products $w_{12}v_{21}$ and $w_{21}v_{12}$, and using in the resulting expression (2.1.29) and (2.1.31), we find

$$
w_{12}v_{21} = \frac{1}{r_{ij}^2}(y_{ij}^-\lambda_{ij}^+ + y_{ji}^+), \quad w_{21}v_{12} = -\frac{1}{r_{ij}^2}(y_{ij}^+\lambda_{ij}^- + y_{ji}^-), \tag{2.1.37}
$$

where

$$
y_{ij}^\pm = r_{ij}t_{jk} \pm (a_k a_i + a_j a_3). \tag{2.1.38}
$$

Thus we can evaluate the product $w_{12}w_{21}v_{12}v_{21}$ in two different ways, namely, we have the following identity,

$$
\begin{aligned}
0 &= (w_{12}w_{21})(v_{12}v_{21}) - (w_{12}v_{21})(w_{21}v_{12})\\
&= \frac{1}{(t_{ij}^2-4)^2}\Big[p(t_{ij}, a_k, a_3)p(t_{ij}, a_i, a_j)\\
&+\big(\lambda_{ij}^+(r_{ij}t_{jk} - a_k a_i - a_j a_3) + r_{ij}t_{ki} + a_j a_k + a_i a_3\big)\\
&\times\big(\lambda_{ij}^-(r_{ij}t_{jk} + a_k a_i + a_j a_3) + r_{ij}t_{ki} - a_j a_k - a_i a_3\big)\Big] = \frac{F(t_{\mu\nu}, a_\nu)}{(t_{ij}^2-4)},
\end{aligned} \tag{2.1.39}
$$

where

$$
\begin{aligned}
F(t_{\mu\nu}, a_\nu) &= t_{ij}t_{jk}t_{ki} + t_{ij}^2 + t_{jk}^2 + t_{ki}^2\\
&-t_{ij}(a_i a_j + a_k a_3) - t_{jk}(a_j a_k + a_i a_3) - t_{ki}(a_k a_i + a_j a_3)\\
&+a_i^2 + a_j^2 + a_k^2 + a_3^2 + a_i a_j a_k a_3 - 4.
\end{aligned} \tag{2.1.40}
$$

Equation

$$
F(t_{\mu\nu}, a_\nu) = 0 \tag{2.1.41}
$$

describes a complex two-dimensional cubic in $t_{\mu\nu}$ surface in the space of the monodromy data. According to Iwasaki [**Iw**], the surface (2.1.41) has singular points if and only if

$$
q(a_\nu)\prod_{l=0}^{3}(a_l^2 - 4) = 0, \tag{2.1.42}
$$

with a polynomial $q(a_\nu)$ defined by

$$
q(a_\nu) = \prod_{\varepsilon_0\varepsilon_1\varepsilon_2=1}(\varepsilon_0 a_0 + \varepsilon_1 a_1 + \varepsilon_2 a_2 + a_3) - \prod_{i=0}^{2}(a_i a_3 - a_j a_k),
$$

where $\varepsilon_l \in \{\pm 1\}$, $l = 0, 1, 2$. We call the values of the parameters a_ν satisfying (2.1.42) the singular parameters. According to [**Iw**], the polynomials $p(t_{ij}, a_k, a_3)$ and $p(t_{ij}, a_i, a_j)$ may have common roots iff $q(a_\nu) = 0$, and therefore, for any singular parameter, the monodromy surface (2.1.41) contains at most 64 singular points. Outside the singular points, the monodromy matrices can be parametrized by the invariants $t_{\mu\nu}$. We present such a parametrization assuming $p(t_{ij}, a_k, a_3) \neq 0$. The inequality $p(t_{ij}, a_i, a_j) \neq 0$ can be used in a similar way.

Consider an open subset of the monodromy data satisfying the inequality

$$p(t_{ij}, a_k, a_3) \neq 0. \tag{2.1.43}$$

In this subset, according to (2.1.33), $w_{12}w_{21} \neq 0$ and, without loss of generality,

$$w_{21} = 1. \tag{2.1.44}$$

Using (2.1.33), (2.1.37) and (2.1.27), we obtain

$$\begin{aligned} w_{12} &= -\frac{p(t_{ij}, a_k, a_3)}{t_{ij}^2 - 4}, \\ v_{12} &= -\frac{y_{ij}^+\lambda_{ij}^- + y_{ji}^-}{t_{ij}^2 - 4}, \quad v_{21} = -\frac{y_{ij}^-\lambda_{ij}^+ + y_{ji}^+}{p(t_{ij}, a_k, a_3)}, \\ u_{12} &= \frac{y_{ij}^+ + y_{ji}^-\lambda_{ij}^+}{t_{ij}^2 - 4}, \quad u_{21} = \frac{y_{ij}^- + y_{ji}^+\lambda_{ij}^-}{p(t_{ij}, a_k, a_3)}. \end{aligned} \tag{2.1.45}$$

Since we have the matrices U, V, W defined in (2.1.26), to compute the monodromy matrices M_ν, $\nu = 0, 1, 2$, it is enough to find the conjugation matrix P. This can be archived using the cyclic relation; see (2.1.19). Of course, the result depends on the above choice of the cyclic permutation (i, j, k) of the indices $(2, 1, 0)$. Indeed, for the identical permutation,

(2.1.46) for $(i, j, k) = (2, 1, 0)$:

$$PM_\infty^{-1}P^{-1} = UVW = \begin{pmatrix} \lambda_{ij}^+ & \\ & \lambda_{ij}^- \end{pmatrix} \begin{pmatrix} \frac{a_3 - \lambda_{ij}^- a_k}{r_{ij}} & -\frac{p(t_{ij}, a_k, a_3)}{r_{ij}^2} \\ 1 & -\frac{a_3 - \lambda_{ij}^+ a_k}{r_{ij}} \end{pmatrix}.$$

Here, we use (2.1.27). Thus, due to normalization $\mathrm{M}_\infty = e^{2\pi i \delta \sigma_3}$ (see (2.1.20)), the matrix P diagonalizes the product $\operatorname{diag}(\lambda_{ij}^+, \lambda_{ij}^-)W$ and therefore is determined modulo a right multiplication by a diagonal matrix D:

(2.1.47) $(i, j, k) = (2, 1, 0)$:

$$P = P_0 := \begin{pmatrix} a_3\lambda_{ij}^- - a_k + e^{-2\pi i \delta} r_{ij} & a_3\lambda_{ij}^- - a_k + e^{2\pi i \delta} r_{ij} \\ r_{ij}\lambda_{ij}^- & r_{ij}\lambda_{ij}^- \end{pmatrix} D.$$

Similarly,

(2.1.48) for $(i, j, k) = (1, 0, 2)$:

$$PM_\infty^{-1}P^{-1} = WUV = \begin{pmatrix} \frac{a_3 - \lambda_{ij}^- a_k}{r_{ij}} & -\frac{p(t_{ij}, a_k, a_3)}{r_{ij}^2} \\ 1 & -\frac{a_3 - \lambda_{ij}^+ a_k}{r_{ij}} \end{pmatrix} \begin{pmatrix} \lambda_{ij}^+ & \\ & \lambda_{ij}^- \end{pmatrix},$$

and therefore the conjugation matrix P is given by the following expression:

$$P = P_1 := \begin{pmatrix} \lambda_{ij}^- & \\ & \lambda_{ij}^+ \end{pmatrix} P_0. \tag{2.1.49}$$

In the remaining case,

$$(i,j,k) = (0,2,1): \quad PM_\infty^{-1}P^{-1} = VWU, \tag{2.1.50}$$

and

$$P = P_2 := U^{-1}P_0. \tag{2.1.51}$$

Thus we have constructed a parametrization of the monodromy data by the invariants $t_{\mu\nu}$ outside the singular points of the surface (2.1.41). At the singular points of this surface, it is necessary to use one or more additional complex parameters. In other words, the monodromy data for the 2×2 matrix linear ODE with four Fuchsian points can be parametrized by the points of the surface (2.1.41) with one or more copies of $\mathbb{C}P^1$ attached to its singular points. The precise description of this object however is still not completed.

PARAMETRIZATION OF THE SINGULARITY DATA AND THE HEUN EQUATION

Let us turn to the singularity data, i.e., to a parametrization of the matrices A_ν in the equation (2.1.16). Let

$$A_0 = \begin{pmatrix} z_0 & -u(z_0-\alpha) \\ \frac{1}{u}(z_0+\alpha) & -z_0 \end{pmatrix}, \quad A_1 = \begin{pmatrix} z_1 & -v(z_1-\beta) \\ \frac{1}{v}(z_1+\beta) & -z_1 \end{pmatrix},$$

$$A_2 = \begin{pmatrix} z_2 & -w(z_2-\gamma) \\ \frac{1}{w}(z_2+\gamma) & -z_2 \end{pmatrix}, \tag{2.1.52}$$

where six parameters u, v, w and z_ν, $\nu = 0,1,2$, $z_0 + z_1 + z_2 = -\delta$, satisfy two constraints,

$$\begin{aligned} &u(z_0-\alpha) + v(z_1-\beta) + w(z_2-\gamma) = 0, \\ &\frac{1}{u}(z_0+\alpha) + \frac{1}{v}(z_1+\beta) + \frac{1}{w}(z_2+\gamma) = 0. \end{aligned} \tag{2.1.53}$$

Introduce the notations

$$a_3 := \big(A(\lambda)\big)_{11}, \quad a_+ := \big(A(\lambda)\big)_{12} = \frac{P_1(\lambda)}{\lambda(\lambda-1)(\lambda-x)}, \quad a_- := \big(A(\lambda)\big)_{21}. \tag{2.1.54}$$

Due to the first of the constraints (2.1.53),

$$P_1(\lambda) = -\lambda\big[(x+1)(A_0)_{12} + x(A_1)_{12} + (A_2)_{12}\big] + x(A_0)_{12}$$

is a first-degree polynomial in λ with the only zero at the point λ_+,

$$\lambda_+ = \frac{xu(z_0-\alpha)}{(x+1)u(z_0-\alpha) + xv(z_1-\beta) + w(z_2-\gamma)}.$$

Let us apply to (2.1.16), (2.1.52) the gauge transformation

$$\Psi(\lambda) = V(\lambda)\Phi(\lambda), \tag{2.1.55}$$

where

$$V = a_+^{\sigma_3/2}\begin{pmatrix} 1 & 0 \\ -a_3 + \frac{a_+'}{2a_+} & 1 \end{pmatrix}.$$

The new matrix function $\Phi(\lambda)$ satisfies the equation

$$\frac{d\Phi}{d\lambda} = \begin{pmatrix} 0 & 1 \\ Q(\lambda) & 0 \end{pmatrix}\Phi, \tag{2.1.56}$$

where

$$Q(\lambda) = a_3^2 + a_+a_- + a_3' - a_3\frac{a_+'}{a_+} - \frac{1}{2}\left(\frac{a_+'}{a_+}\right)' + \frac{1}{4}\left(\frac{a_+'}{a_+}\right)^2$$

is a rational function of λ with the double poles at 0, 1, x and λ_+. The latter singular point is apparent since it is the point of holomorphicity for $\Psi(\lambda)$, thus the singularity of the function $\Phi(\lambda)$ at λ_+ is described by the gauge matrix $V^{-1}(\lambda)$. This apparent singularity $\lambda = \lambda_+$ disappears when it approaches one of the actual singularities, i.e., when the parameters of (2.1.16), (2.1.52) satisfy either one of the conditions:

$$\lambda_+ = 0: \qquad z_0 = \alpha, \tag{2.1.57}$$

$$\lambda_+ = 1: \qquad z_1 = \beta, \tag{2.1.58}$$

$$\lambda_+ = x: \qquad z_2 = \gamma, \tag{2.1.59}$$

$$\lambda_+ = \infty: \qquad (x+1)u(z_0 - \alpha) + xv(z_1 - \beta) + w(z_2 - \gamma) = 0. \tag{2.1.60}$$

For instance, if (2.1.57) takes place, then equations (2.1.53) simplify to

$$v(z_1 - \beta) + w(z_2 - \gamma) = 0, \qquad \frac{2\alpha}{u} + \frac{z_1 + \beta}{v} + \frac{z_2 + \gamma}{w} = 0.$$

Thus, without loss of generality,

$$A_0 = \begin{pmatrix} \alpha & 0 \\ \varkappa & -\alpha \end{pmatrix}, \quad A_1 = \begin{pmatrix} z & -v(z-\beta) \\ \frac{1}{v}(z+\beta) & -z \end{pmatrix},$$

$$A_2 = \begin{pmatrix} -\alpha - \delta - z & v(z-\beta) \\ -\varkappa - \frac{1}{v}(z+\beta) & \alpha + \delta + z \end{pmatrix}, \quad \varkappa = \frac{2\alpha}{u}, \tag{2.1.61}$$

are parametrized by 6 complex parameters. Now, the matrix equation (2.1.56) with (2.1.61) is equivalent to the second-order scalar equation for the entries of the first row,

$$\frac{d^2\Phi_1}{d\lambda^2} = Q(\lambda)\Phi_1, \quad \Phi_2 = \frac{d\Phi_1}{d\lambda},$$

where

$$Q(\lambda) = \frac{\alpha(\alpha-1)}{\lambda^2} + \frac{\beta^2 - \frac{1}{4}}{(\lambda-1)^2} + \frac{\gamma^2 - \frac{1}{4}}{(\lambda - x)^2} + \frac{R}{\lambda} + \frac{S}{\lambda - 1} + \frac{T}{\lambda - x}, \tag{2.1.62}$$

with

$$R = -\alpha(2z+1) + \tfrac{x-1}{x}\varkappa v(z-\beta) + \tfrac{2}{x}\alpha(\alpha+\delta+z-\tfrac{1}{2}),$$
$$S = \alpha(2z+1) - \tfrac{x}{x-1}\varkappa v(z-\beta) - 2\tfrac{z^2-\beta^2}{x-1} + \tfrac{2z+1}{x-1}(\alpha+\delta+z-\tfrac{1}{2}),$$
$$T = \tfrac{2x-1}{x(x-1)}\varkappa v(z-\beta) + 2\tfrac{z^2-\beta^2}{x-1} - \left(\tfrac{2\alpha}{x} + \tfrac{2z+1}{x-1}\right)(\alpha+\delta+z-\tfrac{1}{2}),$$
$$R+S+T=0.$$

The change of the dependable variable

$$\Phi_1 = \lambda^{\alpha}(\lambda-1)^{\frac{1}{2}-\beta}(\lambda-x)^{\frac{1}{2}-\gamma}y$$

yields the canonical Heun equation for the function $y(\lambda)$,

$$\frac{d^2y}{d\lambda^2} + \left(\frac{c}{\lambda} + \frac{d}{\lambda-1} + \frac{e}{\lambda-x}\right)\frac{dy}{d\lambda} + \frac{ab\lambda-q}{\lambda(\lambda-1)(\lambda-x)}y = 0,$$
$$a,b,c,d,e,q = const, \quad c+d+e = a+b+1, \tag{2.1.63}$$

where the parameters are given by

$$c = 2\alpha, \quad d = 1-2\beta, \quad e = 1-2\gamma,$$

$$ab = Rx + S(x-1) + 2\alpha(1-\beta-\gamma) + \tfrac{1}{2}(1-2\beta)(1-2\gamma),$$

$$q = xR + x\alpha(1-2\beta) + \alpha(1-2\gamma).$$

Heun equation is an extremely interesting object of study on its own. We refer the reader to the recent monograph [**SL**] for more on the theory of Heun equation and its relation to the Painlevé theory.

1.4. One irregular singular point of Poincaré rank 1. Our next example is 2×2 systems with only one irregular singular point of Poincaré rank 1 and no other singular points. Each such a system can be reduced (using automorphisms of $\mathbb{C}P^1$, gauge transformations and a proper scaling transformation, $\lambda \mapsto a\lambda$) to the model system

$$\frac{d\Psi}{d\lambda} = \mathsf{A}\Psi, \tag{2.1.64}$$

with the constant diagonal traceless coefficient matrix,

$$\mathsf{A} = \sigma_3 = \begin{pmatrix} 1 & 0 \\ 0 & -1 \end{pmatrix}.$$

The system (2.1.64) has the explicit solution

$$\Psi(\lambda) = e^{\lambda\sigma_3},$$

that indicates the triviality of the monodromy data set:

$$\Lambda_0^{(\infty)} = 0, \quad \mathsf{S}_1^{(\infty)} = \mathsf{S}_2^{(\infty)} = I.$$

1.5. One irregular singular point of Poincaré rank 2. *Parabolic cylinder functions.* Let the 2×2 system have one singular point of Poincaré rank 2 and let its leading matrix coefficient be diagonalizable. Then, using the automorphism of $\mathbb{C}P^1$ and a proper gauge transformation, the singular point can be placed at $\lambda = \infty$ and the system can be transformed into the following normal form,

$$\frac{d\Psi}{d\lambda} = \left[\frac{\lambda}{2}\sigma_3 + \begin{pmatrix} 0 & \alpha \\ \beta & 0 \end{pmatrix}\right]\Psi, \tag{2.1.65}$$

so that

$$\dim \mathcal{A} = 2.$$

Using the procedure explained in Chapter 1 (see (1.1)), the structure of the essential singularity of $\Psi(\lambda)$ at infinity is described by the formal series

$$\Psi_f(\lambda) = \Big(I + \sum_{k=1}^{\infty} \lambda^{-k}\Psi_k\Big)e^{(\frac{\lambda^2}{4} + \alpha\beta \ln \lambda)\sigma_3}. \tag{2.1.66}$$

Since $r_\infty = 2$, there are five canonical sectors

$$\Omega_n = \Big\{\lambda \in \mathbb{C}\colon \ \arg \lambda \in \big(\frac{\pi}{4}(2n-3), \frac{\pi}{4}(2n+1)\big)\Big\}. \quad n = 1, \ldots, 5. \tag{2.1.67}$$

The canonical solutions $\Psi_n(\lambda)$ of equation (2.1.65) are specified by the asymptotic condition

$$\Psi_n(\lambda) \simeq \Psi_f(\lambda), \quad \lambda \to \infty, \quad \lambda \in \Omega_n, \tag{2.1.68}$$

where the principal branch of the logarithm is chosen.

The set of the monodromy data contains four Stokes matrices S_n and the formal monodromy exponent Λ_0,

$$S_n = \Psi_n^{-1}(\lambda)\Psi_{n+1}(\lambda), \quad n = 1, \ldots, 4, \quad \Lambda_0 = \alpha\beta\sigma_3. \tag{2.1.69}$$

These data are subject to the constraints described for a general system in Chapter 1.

Triangularity conditions

The Stokes matrices S_n have the following triangular structure:

$$S_{2l-1} = \begin{pmatrix} 1 & s_{2l-1} \\ 0 & 1 \end{pmatrix}, \quad S_{2l} = \begin{pmatrix} 1 & 0 \\ s_{2l} & 1 \end{pmatrix}. \tag{2.1.70}$$

Cyclic relation

Since the coefficients of system (2.1.65) have no singularities in $\mathbb{C}$, each solution $\Psi_n(\lambda)$ is a single-valued entire function of $\lambda \in \mathbb{C}$, i.e.,

$$\tau\big(\Psi_n(\lambda)\big) = \Psi_n(\lambda), \tag{2.1.71}$$

where τ is the monodromy operator corresponding to the unit circle. On the other hand, the asymptotics (2.1.66), (2.1.68) and the definition of the canonical sectors (2.1.67), imply that

$$\Psi_5(\lambda) = \Psi_1(\lambda)e^{2\pi i\alpha\beta\sigma_3}. \tag{2.1.72}$$

From the definition of the Stokes matrices it follows that

$$\Psi_5(\lambda) = \tau\big(\Psi_1(\lambda)\big)S_1S_2S_3S_4. \tag{2.1.73}$$

Combining equations (2.1.71) with $n = 1$, (2.1.72) and (2.1.73) together, we find the cyclic relation

$$S_1S_2S_3S_4 = e^{2\pi i\alpha\beta\sigma_3}. \tag{2.1.74}$$

Symmetry relation

Besides (2.1.71), equation (2.1.65) admits a symmetry w.r.t. the reflection $\lambda \mapsto -\lambda$. Namely, if $\Psi(\lambda)$ is a solution of (2.1.65), the function

$$\Phi(\lambda) = \sigma_3\Psi(e^{i\pi}\lambda)\sigma_3 \tag{2.1.75}$$

is also a solution of equation (2.1.65). Applying the symmetry (2.1.75) to the canonical solutions, we find

$$\sigma_3\Psi_{k+2}(e^{i\pi}\lambda)\sigma_3 = \Psi_k(\lambda)e^{i\pi\alpha\beta\sigma_3}. \tag{2.1.76}$$

The latter equation implies the relations

$$S_{k+2} = e^{-i\pi\alpha\beta\sigma_3}\sigma_3S_k\sigma_3e^{i\pi\alpha\beta\sigma_3}, \tag{2.1.77}$$

or, equivalently,

$$s_{2l+1} = -s_{2l-1}e^{-2\pi i\alpha\beta}, \quad s_{2l+2} = -s_{2l}e^{2\pi i\alpha\beta}. \tag{2.1.78}$$

Using (2.1.77) in (2.1.74), we find

$$\left(S_1S_2e^{-i\pi\alpha\beta\sigma_3}\sigma_3\right)^2 = I,$$

which contains only one nontrivial equation, namely

$$1 + s_1s_2 = e^{2\pi i\alpha\beta}. \tag{2.1.79}$$

This means that the manifold of the monodromy data of the system (2.1.65),

$$\mathcal{M} = \Big\{S_1, S_2, S_3, S_4, \Lambda_0\Big\} \equiv \{s_1, s_2, s_3, s_4, \alpha\beta\},$$

is parametrized by two complex parameters. For these two local parameters of certain charts of the monodromy manifold, one can choose one of the Stokes multipliers, e.g., s_1 or s_2, and the formal monodromy exponent $\alpha\beta$. (For instance, if $\alpha\beta \in \mathbb{Z}$, then (2.1.79) implies that either s_1 or s_2 is trivial. The remaining Stokes multiplier, i.e., s_2 or s_1, respectively, parametrizes this special chart).

Similarly, with the case of three Fuchsian singular points, the dimensions of the singularity and monodromy manifolds coincide,

$$\dim\mathcal{A} = \dim\mathcal{M} = 2.$$

Furthermore, as with the case of three Fuchsian singular points, the correspondence between these two manifolds can be described in a closed form since equation (2.1.65) is solvable in terms of the contour integrals, namely in terms of special functions of hypergeometric type. Indeed, the matrix equation (2.1.65) is equivalent to the system of scalar equations for the entries $\Psi_{ij}(\lambda)$,

$$\frac{d^2\Psi_{1j}}{d\lambda^2} + \left(\nu + \tfrac{1}{2} - \tfrac{\lambda^2}{4}\right)\Psi_{1j} = 0, \qquad (2.1.80)$$
$$\Psi_{2j} = \dot{\Psi}_{1j}, \quad j = 1, 2, \quad \nu = -1 - \alpha\beta,$$

where we have introduced the notation

$$\dot{\Psi} = \frac{1}{\alpha}\left(\frac{d\Psi}{d\lambda} - \frac{\lambda}{2}\Psi\right). \qquad (2.1.81)$$

Equation (2.1.80) is called the *parabolic cylinder equation*, and its solutions $D_\nu(\lambda)$ and $D_{-\nu-1}(i\lambda)$ are called the parabolic cylinder functions. The solution $D_\nu(\lambda)$ is fixed by the asymptotic condition,

$$D_\nu(\lambda) \sim \lambda^\nu e^{-\frac{\lambda^2}{4}}, \quad \lambda \to +\infty.$$

By construction, the matrix function

$$\mathbf{D}(\lambda) = \begin{pmatrix} D_{-\nu-1}(i\lambda) & D_\nu(\lambda) \\ \dot{D}_{-\nu-1}(i\lambda) & \dot{D}_\nu(\lambda) \end{pmatrix} \qquad (2.1.82)$$

is the fundamental solution of (2.1.65). As with the hypergeometric equation, the remarkable fact is that the function $D_\nu(\lambda)$ admits an *integral representation* which allows us to obtain the asymptotic expansions for $\mathbf{D}(\lambda)$ as $\lambda \to \infty$ along any direction on the complex plane (see e.g. [**BE**]). Taking into account these classical asymptotic formulae, one can represent the canonical solutions $\Psi_{-1}(\lambda)$, $\Psi_0(\lambda)$ and $\Psi_1(\lambda)$ as the products

$$\Psi_{-1}(\lambda) = \mathbf{D}(\lambda)\begin{pmatrix} e^{-i\frac{\pi}{2}\alpha\beta} & \frac{\sqrt{2\pi}}{\beta\Gamma(\alpha\beta)}e^{i\frac{\pi}{2}\alpha\beta} \\ 0 & -\alpha \end{pmatrix},$$

$$\Psi_0(\lambda) = \mathbf{D}(\lambda)\begin{pmatrix} e^{-i\frac{\pi}{2}\alpha\beta} & 0 \\ 0 & -\alpha \end{pmatrix},$$

$$\Psi_1(\lambda) = \mathbf{D}(\lambda)\begin{pmatrix} e^{-i\frac{\pi}{2}\alpha\beta} & 0 \\ -\frac{i\sqrt{2\pi}}{\Gamma(-\alpha\beta)} & -\alpha \end{pmatrix},$$

In turn, these formulae allow one to compute the Stokes matrices S_{-1} and S_0:

$$S_{-1} = \begin{pmatrix} 1 & \frac{\sqrt{2\pi}}{\beta\Gamma(\alpha\beta)}e^{i\pi\alpha\beta} \\ 0 & 1 \end{pmatrix}, \quad S_0 = \begin{pmatrix} 1 & 0 \\ -\frac{i\sqrt{2\pi}}{\alpha\Gamma(-\alpha\beta)} & 1 \end{pmatrix}. \qquad (2.1.83)$$

Formulae (2.1.83) constitute an explicit solution of the direct monodromy problem in the case $\alpha\beta \notin \mathbb{Z}$ (if $\alpha\beta \in \mathbb{Z}$, it is enough to use the limiting form of these expressions). In fact, the map of the singularity manifold on either of the charts

$$\mathcal{A} = \{\alpha, \beta\} \longmapsto \mathcal{M}_j = \{s_j, \alpha\beta\}, \quad j = 1, 2,$$

is given by the following equation:

$$\begin{aligned} (2.1.84) \qquad & s_1 = \frac{\sqrt{2\pi}\, e^{-\pi i\alpha\beta}}{\beta\Gamma(\alpha\beta)}, \quad s_2 = -\frac{i\sqrt{2\pi}\, e^{2\pi i\alpha\beta}}{\alpha\Gamma(-\alpha\beta)}, \quad \alpha\beta \notin \mathbb{Z}, \\ & s_1 = \alpha(-1)^n \frac{\sqrt{2\pi}}{n!}, \quad s_2 = 0, \quad \alpha\beta = n \in \mathbb{N}, \\ & s_1 = 0, \quad s_2 = i\beta\frac{\sqrt{2\pi}}{n!}, \quad -\alpha\beta = n \in \mathbb{N}, \\ & s_1 = \alpha\sqrt{2\pi}, \quad s_2 = i\beta\sqrt{2\pi}, \quad \alpha\beta = 0. \end{aligned}$$

The triviality of one of the Stokes multipliers for the integer formal monodromy $\alpha\beta$, reflects the fact that one of the columns of the function $\Psi(\lambda)$ is described by the Hermite polynomial of the n-th degree.

The inverse map

$$\mathcal{M}_j = \{s_j, \alpha\beta\} \longmapsto \mathcal{A} = \{\alpha, \beta\}, \quad j = 1, 2,$$

can be described in closed form using the inversion of (2.1.84),

$$\begin{aligned} & \alpha = \frac{s_1 e^{\pi i\alpha\beta}}{\sqrt{2\pi}}\Gamma(1+\alpha\beta), \quad \beta = \frac{\sqrt{2\pi}\, e^{-\pi i\alpha\beta}}{s_1\Gamma(\alpha\beta)}, \quad \alpha\beta \notin \mathbb{Z}, \\ & \alpha = s_1(-1)^n \frac{n!}{\sqrt{2\pi}}, \quad \alpha\beta = n \in \mathbb{N}, \\ & \beta = -is_2\frac{n!}{\sqrt{2\pi}}, \quad -\alpha\beta = n \in \mathbb{N}, \\ (2.1.85) \qquad & \alpha = \frac{s_1}{\sqrt{2\pi}}, \quad \beta = -\frac{is_2}{\sqrt{2\pi}}, \quad \alpha\beta = 0. \end{aligned}$$

In other words, similarly with the case of three Fuchsian singular points, the direct and inverse monodromy problems for the 2×2 system with only one irregular singularity of the Poincaré rank 2, are solved in terms of classical transcendental functions.

1.6. One irregular singular point of Poincaré rank 3. Let the 2×2 system has one singularity of Poincaré rank 3 and let the leading order coefficient matrix be diagonalizable. Then, using the proper automorphism of $\mathbb{C}P^1$ and a gauge transformation, this matrix equation can be reduced to the following normal form,

$$(2.1.86) \qquad \frac{d\Psi}{d\lambda} = \left[-4i\lambda^2\sigma_3 + 4i\lambda\begin{pmatrix} 0 & u \\ -v & 0 \end{pmatrix} + \begin{pmatrix} a & -2w \\ -2z & -a \end{pmatrix}\right]\Psi,$$

so that

$$(2.1.87) \qquad \mathcal{A} = \{u, v, w, y, a\}, \quad \dim\mathcal{A} = 5.$$

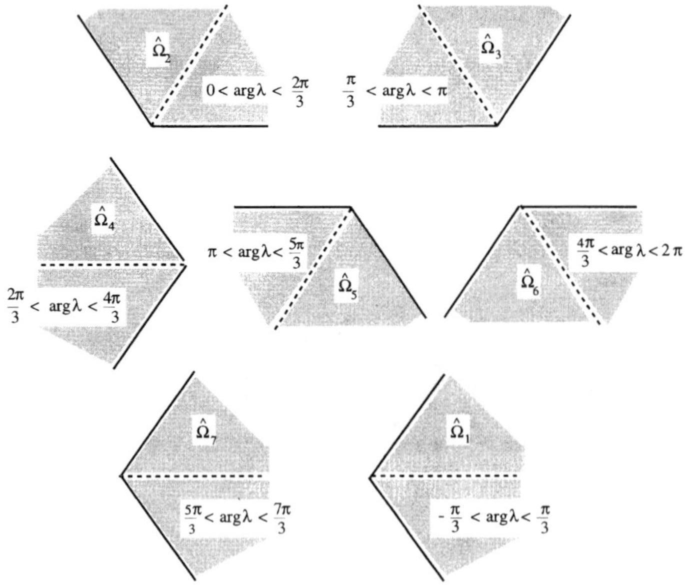

FIGURE 2.5. In each $\hat\Omega_k$ the dashed line indicates the unique ray where Re $\lambda^3 = 0$

Using the procedure explained in Chapter 3, we find the formal expansion of $\Psi(\lambda)$ for the system (2.1.86) as $\lambda \to \infty$,

$$\Psi_f(\lambda) = \Big(I + \sum_{k=1}^{\infty} \Psi_k \lambda^{-k}\Big) e^{-\frac{4i}{3}\lambda^3\sigma_3 - ix\lambda\sigma_3 - \nu \ln \lambda\,\sigma_3}, \tag{2.1.88}$$

where

$$x = ia - 2uv, \qquad \nu = vw - uz. \tag{2.1.89}$$

The canonical solutions $\Psi_n(\lambda)$ of (2.1.86) are fixed uniquely by the asymptotic conditions

$$\Psi_n(\lambda) \simeq \Psi_f(\lambda), \quad \lambda \to \infty, \quad \lambda \in \Omega_n, \quad n \in \mathbb{Z},$$

where the canonical sectors Ω_n are given by

$$\Omega_n = \Big\{\lambda \in \mathbb{C}\colon \ \arg\lambda \in \big(\frac{\pi}{3}(n-2), \frac{\pi}{3}n\big)\Big\}, \tag{2.1.90}$$

as indicated in Figure 2.5, and the principal branch of the logarithm is chosen. Note that

$$\Omega_7 = \Omega_1,$$

and

$$\Psi_7(\lambda) = \Psi_1(\lambda) e^{-2\pi i \nu \sigma_3}.$$

As in the previous case, the only point that is responsible for the nontrivial global behavior of the solution of system (2.1.86) is $\lambda = \infty$. The manifold $\mathcal{M}$ consists of six Stokes matrices,

$$S_n = \Psi_n^{-1}(\lambda)\Psi_{n+1}(\lambda), \quad n = 1, \dots, 6, \tag{2.1.91}$$

and the formal monodromy exponent

$$\Lambda_0 = -\nu\sigma_3.$$

Repeating the arguments that led us to the equations (2.1.70) and (2.1.74), we obtain the following *triangularity and cyclic constraints* on $\{S_k\}$:

$$\text{(2.1.92)} \qquad S_{2l+1} = \begin{pmatrix} 1 & 0 \\ s_{2l+1} & 1 \end{pmatrix}, \quad S_{2l} = \begin{pmatrix} 1 & s_{2l} \\ 0 & 1 \end{pmatrix},$$

$$\text{(2.1.93)} \qquad S_1S_2S_3S_4S_5S_6 = e^{-2\pi i\nu\sigma_3}.$$

These constraints imply the three scalar equations

$$\text{(2.1.94)} \qquad \begin{aligned} &1 + s_1s_2 = (1 + s_4s_5)e^{2\pi i\nu}, \quad 1 + s_2s_3 = (1 + s_5s_6)e^{-2\pi i\nu}, \\ &s_1 + s_3 + s_1s_2s_3 = -s_5e^{2\pi i\nu}, \end{aligned}$$

on the seven scalar monodromy parameters $\{s_1, s_2, s_3, s_4, s_5, s_6, \nu\}$. For instance, if $s_1 + s_3 + s_1s_2s_3 \neq 0$, then equations (2.1.94) allow one to compute s_4, s_5 and s_6 in terms of the remaining parameters s_1, s_2, s_3 and ν. Thus we conclude that the dimension of the monodromy manifold $\mathcal{M}$ at a generic point equals 4, and as in the case of four Fuchsian points,

$$\dim\mathcal{M} = \dim\mathcal{A} - 1.$$

This indicates the necessity to study the *isomonodromy deformations* of the system (2.1.86). As we will see in Chapter 4, these deformations are governed by the Painlevé II equation.

We conclude our analysis of system (2.1.86) by pointing out two of its important reductions.

1) Triangular reduction (*Airy integrals*)
Let $u = w = 0$. In this case, from (2.1.89), $x = ia$ and $\nu = 0$, equation (2.1.86) is lower triangular and therefore solvable by quadratures. Its fundamental solution can also be chosen in a lower triangular form,

$$\text{(2.1.95)} \qquad \Psi_k(\lambda) = \begin{pmatrix} 1 & 0 \\ -\int_{\lambda_k}^{\lambda}(4iv\zeta + 2z)e^{2(\theta(\lambda)-\theta(\zeta))}\,d\zeta & 1 \end{pmatrix} e^{-\theta(\lambda)\sigma_3}.$$

Here $\theta(\lambda) = \frac{4i}{3}\lambda^3 + ix\lambda$, and λ_k is the point of normalization for $\Psi_k(\lambda)$. Choosing λ_k by $\lambda_k = e^{i\frac{\pi}{2}+i\frac{2\pi}{3}n}\infty$, $n \in \{0, 1, -1\}$, we find that (2.1.95) defines the canonical solutions of (2.1.86). Namely, their asymptotics as

$\lambda \to \infty$ is given by the canonical exponentials in the indicated sectors,

$$\lambda_1 = e^{-i\frac{\pi}{6}}\infty: \quad \Psi_1(\lambda) = \big(I + O(\lambda^{-1})\big)e^{-i(\frac{4}{3}\lambda^3 + x\lambda)\sigma_3},$$

(2.1.96) $$\arg\lambda \in \left(-\tfrac{2\pi}{3}, \tfrac{\pi}{3}\right) = \Omega_0 \cup \Omega_1;$$

$$\lambda_3 = i\infty: \quad \Psi_3(\lambda) = \big(I + O(\lambda^{-1})\big)e^{-i(\frac{4}{3}\lambda^3 + x\lambda)\sigma_3},$$

(2.1.97) $$\arg\lambda \in (0, \pi) = \Omega_2 \cup \Omega_3;$$

$$\lambda_5 = e^{i\frac{7\pi}{6}}\infty: \quad \Psi_5(\lambda) = \big(I + O(\lambda^{-1})\big)e^{-i(\frac{4}{3}\lambda^3 + x\lambda)\sigma_3},$$

(2.1.98) $$\arg\lambda \in \left(\tfrac{2\pi}{3}, \tfrac{5\pi}{6}\right) = \Omega_4 \cup \Omega_5.$$

In the lower triangular reduction, all the even Stoke multipliers are trivial, $s_{2k} = 0$, since the fundamental solutions (2.1.96)–(2.1.98) differ from each other by the right lower triangular matrix multipliers,

(2.1.99) $$\Psi_k^{-1}(\lambda)\Psi_j(\lambda) = \begin{pmatrix} 1 & 0 \\ \int_{\lambda_k}^{\lambda_j}(4iv\zeta + 2z)e^{-2\theta(\zeta)}\,d\zeta & 1 \end{pmatrix}.$$

Observing that the functions

(2.1.100) $$g_{kj}(x) = \int_{\lambda_k}^{\lambda_j} e^{-i\frac{8}{3}\zeta^3 - i2x\zeta}\,d\zeta$$

satisfy the classical Airy equation, $g''_{jk} - xg_{jk} = 0$, where the prime means derivative w.r.t. x, we conclude that, in the lower triangular reduction, the Stokes multipliers are described explicitly in terms of the Airy functions and their derivatives; these are uniquely determined by equation (2.1.100) with the particular choice of the points λ_k and λ_j described in (2.1.96)–(2.1.98),

$$\begin{aligned} s_1(x) &= 2zg_{13} - 2vg'_{13}, \\ s_3(x) &= 2zg_{35} - 2vg'_{35}, \\ s_5(x) &= 2zg_{51} - 2vg'_{51}, \\ s_1 &+ s_3 + s_5 = 0. \end{aligned}$$

2) SPECIAL REDUCTION (*Airy functions*)

This reduction is described by the equations $a = w = 0$, $y = -\frac{1}{2u}$. In this case, definitions (2.1.89) imply that $x = -2uv$ and $\nu = \frac{1}{2}$. We will show that equation (2.1.86) is exactly solvable in terms of the Airy functions. Indeed, equation (2.1.86) now reads,

(2.1.101) $$\frac{d\Psi}{d\lambda} = \begin{pmatrix} -4i\lambda^2 & 4iu\lambda \\ -4iv\lambda + \frac{1}{u} & 4i\lambda^2 \end{pmatrix}\Psi,$$

which, after the change of the independent variable, $\lambda^2 = \zeta$, becomes

$$\frac{d\Psi}{d\zeta} = \begin{pmatrix} -2i\sqrt{\zeta} & 2iu \\ -2iv + \frac{1}{2u\sqrt{\zeta}} & 2i\sqrt{\zeta} \end{pmatrix}\Psi.$$

Observing that the equivalent scalar equation for the entries of the first row,

$$\frac{d^2\Psi_{1j}}{d\zeta^2} = (-4\zeta - 2x)\Psi_{1j}, \quad j = 1, 2,$$

is solvable in terms of Airy functions, we find the final expression for $\Psi(\lambda)$,

$$\Psi(\lambda) = \begin{pmatrix} 1 & 0 \\ \frac{\lambda}{u} & i\frac{2^{-1/3}}{u} \end{pmatrix} \begin{pmatrix} v_1\big(-2^{2/3}(\lambda^2 + \frac{x}{2})\big) & v_2\big(-2^{2/3}(\lambda^2 + \frac{x}{2})\big) \\ v_1'\big(-2^{2/3}(\lambda^2 + \frac{x}{2})\big) & v_2'\big(-2^{2/3}(\lambda^2 + \frac{x}{2})\big) \end{pmatrix}, \tag{2.1.102}$$

where $v_j(\xi)$ and $v_j'(\xi)$, $j = 1, 2$, are independent solutions of the Airy equation, $v_{\xi\xi} - \xi v = 0$, and their derivatives. In fact, $v_j(\xi)$ can be taken as linear combinations of the Airy integrals (2.1.100) (replacing in the latter x by ξ); moreover, one can ensure the asymptotics

$$\begin{aligned} &v_1(\xi) = \xi^{-1/4} e^{\frac{2}{3}\xi^{3/2}}\big(1 + O(\xi^{-3/2})\big), \\ &v_2(\xi) = \xi^{-1/4} e^{-\frac{2}{3}\xi^{3/2}}\big(1 + O(\xi^{-3/2})\big), \\ &\xi \to \infty, \quad \arg\xi \in \omega_{1,2}, \end{aligned}$$

where $\omega_{1,2}$ are certain sectors of the complex ξ-plane, we find the asymptotics of the matrix function $\Psi(\lambda)$ defined by (2.1.102),

$$\Psi(\lambda) = \big(I + O(\lambda^{-1})\big) e^{-i\frac{4}{3}\lambda^3\sigma_3 - ix\lambda\sigma_3 - \frac{1}{2}\ln\lambda\sigma_3} \begin{pmatrix} 1 & \\ & \frac{2}{u} \end{pmatrix} e^{-i\frac{\pi}{4}} 2^{-\frac{1}{6}},$$

as $\lambda \to \infty$, $\arg\lambda = -\frac{\pi}{2} + \frac{1}{2}\arg\xi$. Taking into account the well-known connection formulae for the Airy functions [**BE**], we can compute the Stokes matrices S_k:

$$S_{2k-1} = \begin{pmatrix} 1 & 0 \\ -\frac{2i}{u} & 1 \end{pmatrix}, \quad S_{2k} = \begin{pmatrix} 1 & -\frac{iu}{2} \\ 0 & 1 \end{pmatrix}. \tag{2.1.103}$$

It is interesting to notice that the same Airy functions are showing up in both the reductions. However, in the first, i.e., triangular reduction, the Airy integrals describe the Stokes matrices while in the second reduction they describe the canonical solutions.

CHAPTER 3

Inverse Monodromy Problem and Riemann-Hilbert Factorization

0. Introduction

In the general case, the solution of system (1.1.1) cannot be achieved via an explicit formulae involving a finite number of contour integrals[1] of elementary functions. Hence another, more general analytic tool is needed in order to investigate the general case of the inverse monodromy problem for system (1.1.1). This tool exists and it is based on the idea of treating the basic monodromy relations (2.0.17) and (2.0.18) as *jump conditions* for the sectionally analytic function consisting of the collection $\{\Psi_k^{(\nu)}(\lambda)\}$ of canonical solutions. This idea places the inverse monodromy problem into the context of the analytic factorization of matrix-valued functions, i.e., in the context of *Riemann-Hilbert factorization problems*. Thus, it brings into the monodromy theory the arsenal of the methods of singular integral equations and holomorphic vector bundles.

In this chapter we will present the basic facts concerning Riemann-Hilbert factorization problems. Then, using the first nontrivial 2×2 examples, we will show how the Riemann-Hilbert factorization problems can be used to analyze the Riemann-Hilbert inverse monodromy problems. *To avoid confusion of terminology, we will follow from now on the usual convention and use the term "Riemann-Hilbert (or RH) problem" for any factorization problem, and not only for the one related to the inverse monodromy problem of the theory of Fuchsian systems.*

1. Riemann-Hilbert problems

A Riemann-Hilbert (or RH) problem, roughly speaking, is the problem of finding an analytic function in the complex plane having a prescribed jump across a curve. More precisely, suppose Γ is an oriented contour in the complex plane λ. The contour Γ might have points of self-intersection, and it might have more than one connected component. Typical contours appearing in applications are depicted in Figure 3.1. The orientation defines, in the usual way, the $+$ and the $-$ sides of Γ.

[1]Strictly speaking, this is the general belief, not a theorem! In fact, in 1914 Whittaker [**Wit**] conjectured that the Heun equation, which was discussed in the previous chapter, is the simplest Fuchsian equation whose solution cannot be represented by a contour integral. To the best of our knowledge, the validity of this conjecture remains open. (We thank V. Kuznetsov for informing us about the reference [**Wit**].)

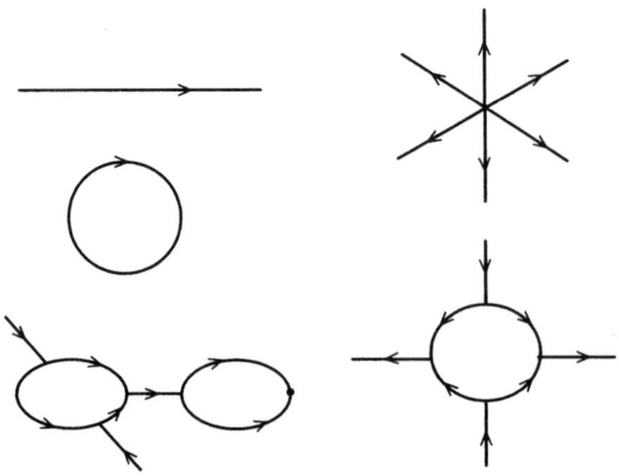

FIGURE 3.1. Typical RH contours

Suppose, in addition, that we are given a map G from Γ into a set of $N \times N$ invertible matrices. A *Riemann-Hilbert (RH) problem* defined by the pair (Γ, G), consists of finding an $N \times N$ matrix-valued function $Y = Y(\lambda)$ with the following properties:

- Y is analytic in $\mathbb{C} \setminus \Gamma$,
- The limit $Y_-(\lambda)$ of Y from the minus side of Γ and the limit $Y_+(\lambda)$ from the plus side of Γ are related for $\lambda \in \Gamma$ by
$$Y_+(\lambda) = Y_-(\lambda)G(\lambda),$$
- $Y(\lambda)$ tends to the identity matrix I as $\lambda \to \infty$.

The precise sense in which these limits exist, and also the sense in which $Y(\lambda) \to I$ as $\lambda \to \infty$ are technical matters that should be specified for each given pair (Γ, G). Later in this section we will discuss typical rigorous settings of the RH problems in some details. However, before we do this, let us single out the two basic facts of the general Riemann-Hilbert theory.

Consider the scalar, i.e., $N = 1$ case. In this case the original, multiplicative jump condition can be rewritten in the following additive form,

$$\log Y_+(\lambda) = \log Y_-(\lambda) + \log G(\lambda).$$

Observe now that an additive jump relation, $y_+(\lambda) = y_-(\lambda) + g(\lambda)$, can be resolved by means of the following contour integral (the Plemelj-Sokhotskii formula),

$$y(\lambda) = \frac{1}{2\pi i}\int_\Gamma \frac{g(\mu)}{\mu - \lambda}d\mu.$$

Therefore, in scalar case, or more generally in the *Abelian* case when

$$[G(\lambda_1), G(\lambda_2)] \equiv G(\lambda_1)G(\lambda_2) - G(\lambda_2)G(\lambda_1) = 0, \quad \forall \lambda_{1,2} \in \Gamma,$$

the solution of the RH problem admits an explicit integral representation,

$$Y(\lambda) = \exp\left\{\frac{1}{2\pi i}\int_\Gamma \frac{\log G(\mu)}{\mu - \lambda}d\mu\right\}. \tag{3.1.1}$$

Of course, there is the subtle point of how to treat this equation if the problem has a nonzero index, i.e., if $\Delta \log G|_\Gamma \neq 0$. Still, formula (3.1.1), after a proper modification for the case of nonzero index (see e.g.[**Ga**]), yields a contour integral representation for the solution of the original problem. Furthermore, in typical concrete situations, the integral in (3.1.1) can be evaluated in closed form or in terms of known elementary or special functions.

In the general, *non-Abelian* matrix case, the formula (3.1.1) fails. A generic matrix RH problem cannot (apparently!) be solved explicitly in terms of contour integrals. However, it can be reduced to the analysis of a linear singular integral equation. Indeed the following representation for the solution $Y(\lambda)$ holds.

$$Y(\lambda) = I + \frac{1}{2\pi i}\int_\Gamma \frac{\rho(\mu)(G(\mu)-I)}{\mu-\lambda}d\mu, \quad \lambda \notin \Gamma, \tag{3.1.2}$$

where $\rho(\lambda) \equiv Y_-(\lambda)$ satisfies the singular integral equation,

$$\begin{aligned}\rho(\lambda) &= I + \frac{1}{2\pi i}\int_\Gamma \frac{\rho(\mu)(G(\mu)-I)}{\mu-\lambda_-}d\mu \\ &\equiv I + [C_-(\rho(G-I))](\lambda), \quad \lambda\in\Gamma.\end{aligned} \tag{3.1.3}$$

We use notation $C_\pm$ for the Cauchy operators related to the contour Γ, i.e.,

$$[C_\pm f](\lambda) = \lim_{\lambda'\to\lambda_\pm}\frac{1}{2\pi i}\int_\Gamma \frac{f(\mu)}{\mu-\lambda'}d\mu, \quad \lambda\in\Gamma, \tag{3.1.4}$$

where $\lambda' \to \lambda_\pm$ denotes the nontangential limit from the $\pm$ side of Γ respectively.

The derivation of equations (3.1.2) and (3.1.3) is a relatively simple matter. It is especially simple in the case when Γ is a smooth simple closed arc oriented, say, counterclockwise and the jump matrix $G(\lambda)$ is smooth too. In this case the set $\mathbb{C}P^1\setminus\Gamma$ is the disjoint union of the two open connected components which we will denote Ω_+ and Ω_- assuming that $\infty\in\Omega_-$. This fact allows the following specification of the setting of the RH problem:

- The functions $Y|_{\Omega_+}$ and $Y|_{\Omega_-}$ are analytic in Ω_+ and Ω_- and continuous in $\overline{\Omega}_+$ and $\overline{\Omega}_-$, respectively.
- The pointwise limit $Y_-(\lambda)$ of Y from the minus side of Γ and the pointwise limit $Y_+(\lambda)$ from the plus side of Γ, which exist and are continuous by virtue of the previous condition, are related for $\lambda\in\Gamma$ by

$$Y_+(\lambda) = Y_-(\lambda)G(\lambda). \tag{3.1.5}$$

- $Y(\infty) = I$.

Suppose now that $Y(\lambda)$ is the solution of the RH problem just formulated. Let us show that the function

$$\rho(\lambda) := Y_-(\lambda), \quad \lambda\in\Gamma$$

satisfies the integral equation (3.1.3). Indeed, by the jump relation (3.1.5) and Cauchy theorem we have that

$$\begin{aligned}\frac{1}{2\pi i}\int_\Gamma \frac{\rho(\mu)(G(\mu)-I)}{\mu-\lambda_-}d\mu &\equiv \frac{1}{2\pi i}\int_\Gamma \frac{Y_-(\mu)(G(\mu)-I)}{\mu-\lambda_-}d\mu \\ &= \frac{1}{2\pi i}\int_\Gamma \frac{Y_-(\mu)G(\mu)}{\mu-\lambda_-}d\mu - \frac{1}{2\pi i}\int_\Gamma \frac{Y_-(\mu)}{\mu-\lambda_-}d\mu \\ &= \frac{1}{2\pi i}\int_\Gamma \frac{Y_+(\mu)}{\mu-\lambda_-}d\mu - \frac{1}{2\pi i}\int_\Gamma \frac{Y_-(\mu)}{\mu-\lambda_-}d\mu \\ &= 0 + Y_-(\lambda) - I \equiv \rho(\lambda) - I,\end{aligned}$$

and the integral equation (3.1.3) follows. Conversely, suppose that $\rho(\lambda)$ is the (continuous) solution of the integral equation (3.1.3). Define $Y(\lambda)$ by (3.1.2). Then from the integral equation, it follows that $Y_-(\lambda) = \rho(\lambda)$. Also, from the Plemelj-Sokhotskii formula we obtain

$$Y_+(\lambda) - Y_-(\lambda) = \rho(\lambda)\big(G(\lambda) - I\big).$$

Therefore, $Y_+(\lambda) - Y_-(\lambda) = Y_-(\lambda)\big(G(\lambda) - I\big)$ and the jump relation (3.1.5) is satisfied. The other two properties of $Y(\lambda)$, which make it the solution of the RH problem, are the direct consequences of our assumptions about the pair (Γ, G) and the classical properties of the Cauchy integral.

The rigorous treatment of equations (3.1.2) and (3.1.3) in the case of more general pairs (Γ, G) requires an advanced L^p-theory of the Cauchy operators and the Riemann-Hilbert problems. In what follows we collect some of the basic facts of this theory assuming that the contour Γ is a finite union of smooth curves in $\mathbb{C}P^1$ with the finite number of self-intersections and such that $\mathbb{C}P^1 \setminus \Gamma$ has a finite number of connected components. Although it is not the most general admissible case, it well covers all the needs of the Painlevé theory discussed in this book. The proofs as well as more details concerning the facts listed below can be found in the references [**CG**], [**LiS**], [**Bea**], [**BDT**], [**Zh**], [**DeZ4**], and [**DeZ5**]. In our presentation we closely follow the papers [**Zh**], [**DeZ4**], and [**DeZ5**].

- Let $f \in L^p(\Gamma, |d\lambda|)$, $1 \le p < \infty$. Then $[C_\pm f](\lambda)$, as defined in (3.1.4), exists for $\lambda \in \Gamma$ a.e.
- Let $f \in L^p(\Gamma, |d\lambda|)$, $1 < p < \infty$. Then

(3.1.6) $$\|C_\pm f\|\|_{L^p(\Gamma)} \le c_p \|f\|\|_{L^p(\Gamma)}$$

  for some constant c_p. In other words, the Cauchy operators C_+ and C_- are *bounded* in the space $L^p(\Gamma)$ for all $1 < p < \infty$ and, in particular, they are bounded in $L^2(\Gamma)$.
- As operators in $L^p(\Gamma)$, $1 < p < \infty$, the Cauchy operators $C_\pm$ satisfy the relation (the Plemelj-Sokhotskii formulae for L^p),

(3.1.7) $$C_\pm = \pm\frac{1}{2}\mathbf{1} - \frac{1}{2}H,$$

  and (hence)

(3.1.8) $$C_+ - C_- = \mathbf{1}.$$

  Here, $\mathbf{1}$ denotes the identical operator in L^p and H is the Hilbert transform,

$$[Hf](\lambda) = v.p.\int_\Gamma \frac{f(\mu)}{\lambda - \mu}\frac{d\mu}{\pi i}.$$

- Fix $1 < p < \infty$. We will say that a pair of $L^p(\Gamma)$-functions $F_\pm(\lambda) \in \partial\,\mathrm{Ran}\; C$, if there is a (unique) function $f \in L^p(\Gamma)$ such that

$$F_\pm(\lambda) = [C_\pm f](\lambda), \quad \lambda \in \Gamma.$$

  Assume also that the jump matrix $G(\lambda)$ satisfies the conditions,

(3.1.9) $$G,\, G^{-1} \in L^\infty(\Gamma),$$

  and

(3.1.10) $$G - I \in L^p(\Gamma).$$

Then, the rigorous setting of the RH problem determined by the pair (Γ, G), which we will call the L^p-RH problem, consists in finding a pair of $N \times N$ matrix-valued function $Y_\pm(\lambda) \in L^p(\Gamma)$ with the following properties:

(i) $Y_\pm - I \in \partial\,\mathrm{Ran}\, C$.

(ii) $Y_+(\lambda) = Y_-(\lambda)G(\lambda), \quad \lambda \in \Gamma$.

The condition $Y_\pm - I \in \partial\,\mathrm{Ran}\, C$, implies that the functions $Y_\pm(\lambda)$ admit the representation,

$$Y_\pm(\lambda) = I + \frac{1}{2\pi i}\int_\Gamma \frac{h(\mu)}{\mu - \lambda_\pm} d\mu, \qquad h \in L^p(\Gamma),$$

and hence they are the L^p-boundary values of the analytic function

$$Y(\lambda) = I + \frac{1}{2\pi i}\int_\Gamma \frac{h(\mu)}{\mu - \lambda} d\mu \qquad \lambda \in \mathbb{C} \setminus \Gamma.$$

The function $Y(\lambda)$ is called the *extension* of $Y_\pm$, and it solves the RH problem in our original setting with the following specifications of the limits involved.

1) Y is analytic in $\mathbb{C} \setminus \Gamma$.

2) For a.e. $\lambda \in \Gamma$, the function Y has the nontangential limits $Y_\pm(\lambda)$ which belong to the space $L^p(\Gamma)$ and satisfy the jump relation

(3.1.11) $$Y_+(\lambda) = Y_-(\lambda)G(\lambda), \quad (a.e.\, \lambda \in \Gamma).$$

3) $Y(\lambda) \to I$ as $\lambda \to \infty$ in any nontangential direction.

- The L^p-RH problem (**i**)–(**ii**) is equivalent to the integral equation (3.1.3). Indeed, suppose that $Y(\lambda)$ is the solution of the L^p-RH problem (**i**)–(**ii**) then $Y_\pm(z) - I = [C_\pm h](\lambda)$, where $h \in L^p$, and, moreover,

(3.1.12) $$h(\lambda) = Y_+(\lambda) - Y_-(\lambda) = Y_-(\lambda)(G(\lambda) - I).$$

Define

$$R(\lambda) := \frac{1}{2\pi i}\int_\Gamma \frac{Y_-(\mu)(G(\mu) - I)}{\mu - \lambda} d\mu \qquad \lambda \in \mathbb{C} \setminus \Gamma.$$

Using (3.1.8) and (3.1.12), we have

(3.1.13) $$R_\pm(\lambda) = \left(\frac{1}{2\pi i}\int_\Gamma \frac{Y_-(\mu)(G(\mu) - I)}{\mu - \lambda_-} d\mu - Y_-(\lambda)\right) + Y_\pm(\lambda).$$

At the same time, by virtue of (3.1.12),

$$R_\pm(\lambda) = [C_\pm(Y_-(G - I))](\lambda) = [C_\pm h](\lambda) = Y_\pm(z) - I,$$

and thus relation (3.1.13) is transformed into the integral equation (3.1.3) for the function $\rho(\lambda) \equiv Y_-(\lambda)$. Conversely, suppose that $\rho - I \in L^p(\Gamma)$ and $\rho(\lambda)$ solves equation (3.1.3). Define $Y(\lambda)$ by (3.1.2). Then equation (3.1.3) implies that $Y_-(\lambda) = \rho(\lambda)$ and also that $Y_\pm - I \in \partial\,\mathrm{Ran}\, C$. Moreover, from the Plemelj-Sokhotskii formula (3.1.8) we have

$$Y_+(\lambda) - Y_-(\lambda) = \rho(\lambda)(G(\lambda) - I) = Y_-(\lambda)(G(\lambda) - I),$$

and hence $Y(\lambda)$ solves the L^p-RH problem (**i**)–(**ii**).

- Let

$$G(\lambda) = \big(I - W^-(\lambda)\big)^{-1}\big(I + W^+(\lambda)\big) \tag{3.1.14}$$

be a factorization of the jump matrix $G(\lambda)$ with

$$\big(I \pm W^\pm(\lambda)\big)^{\pm 1} - I \in L^\infty(\Gamma) \cap L^p(\Gamma). \tag{3.1.15}$$

Define the Cauchy-type operator,

$$[C_w h](\lambda) := [C_+(hW^-)](\lambda) + [C_-(hW^+)](\lambda). \tag{3.1.16}$$

Then, one can replace the integral equation (3.1.3) by the equation

$$\rho(\lambda) = I + [C_w \rho](\lambda), \tag{3.1.17}$$

and preserve, with the proper modifications, all the above indicated relations of the RH problem. Namely, if $Y(\lambda)$ is the solution of the L^p-RH problem (**i** – **ii**), then the function $\rho(\lambda)$ defined by

$$\rho(\lambda) = Y_+(\lambda)\big(I + W^+(\lambda)\big)^{-1} = Y_-(\lambda)\big(I - W^-(\lambda)\big)^{-1}$$

solves the integral equation (3.1.17). Conversely, if $\rho - I \in L^p(\Gamma)$ and $\rho(\lambda)$ solves the equation (3.1.17), then the formula

$$Y(\lambda) = I + \frac{1}{2\pi i}\int_\Gamma \frac{\rho(\mu)\big(W^+(\mu) + W^-(\mu)\big)}{\mu - \lambda}\,d\mu, \tag{3.1.18}$$

defines the solution of the L^p-RH problem (**i** – **ii**). Also, arguments similar to the ones we used above, yield the following proposition.

PROPOSITION 3.1. *$\rho \in L^p(\Gamma)$ solves the homogeneous integral equation*

$$\rho(\lambda) = [C_w \rho](\lambda)$$

iff $Y_\pm(\lambda) = \rho(\lambda)\big(I \pm W^\pm(\lambda)\big)$ solves the homogeneous L^p-RH problem,

$$Y_+(\lambda) = Y_-(\lambda)G(\lambda), \quad Y_\pm \in \partial\, Ran\, C.$$

In particular, the operator $I - C_w$ has zero kernel iff the homogeneous L^p-RH problem has only the trivial solution.

The jump matrices appearing in the theory of Painlevé equations, which we are concerned with in this book, have very special analytic properties. In fact, these properties are the following:

(1) The jump contour Γ can be represented as a finite union of smooth components in such a way that the restriction of the jump matrix $G(\lambda)$ on any smooth component admits an analytic continuation in a neighborhood of the component.
(2) $G(\lambda) \to I$ along any infinite branch of Γ; moreover, the limit is taken exponentially fast.
(3) $\det G(\lambda) \equiv 1$.
(4) Let A be a point of self-intersection, or a *node* point, of the contour Γ. Let Γ_k, $k = 1, \ldots, q$ denote the smooth components of Γ intersecting at A. We assume that Γ_k are numbered counterclockwise. Let $G_k(\lambda)$ denote the restriction of the jump matrix $G(\lambda)$ on the component Γ_k. By property 1, each $G_k(\lambda)$ is analytic in a neighborhood of the point A; we denote

this neighborhood Ω_A. Then, the following *cyclic*, or *trivial monodromy* condition, holds at every node point A:

$$G_1^{\pm 1}(\lambda)G_2^{\pm 1}(\lambda)\dots G_q^{\pm 1}(\lambda) = I, \quad \forall\lambda \in \Omega_A, \tag{3.1.19}$$

where in $G_k^{\pm 1}$ the sign "+" is chosen if Γ_k is oriented outwards from A, and the sign "−" is chosen if Γ_k is oriented inwards toward A. We will say that this property constitutes the *smoothness* of the jump matrix $G(\lambda)$ (cf. [**BDT**]).

The smoothness of the jump matrix $G(\lambda)$ allows us to pose the Riemann-Hilbert problem determined by the pair (Γ, G) in the following classical "smooth" way. Denote Ω_k, $k = 1,\dots,p$ the connected components of $\mathbb{C}P^1 \setminus \Gamma$. The RH problem consists of finding an $N \times N$ piecewise analytic matrix-valued function $Y(\lambda)$ with the following properties:

- For each k, the function $Y|_{\Omega_k}$ is analytic in Ω_k and continuous in $\overline{\Omega}_k$.
- The pointwise limit $Y_-(\lambda)$ of Y from the minus side of Γ and the pointwise limit $Y_+(\lambda)$ from the plus side of Γ, which exist and are continuous by virtue of the previous condition, are related for $\lambda \in \Gamma$, by

$$Y_+(\lambda) = Y_-(\lambda)G(\lambda). \tag{3.1.20}$$

- At $\lambda = \infty$ the function $Y(\lambda)$ satisfies the asymptotic condition,

$$Y(\lambda) = I + O\left(\frac{1}{\lambda}\right), \quad \lambda \to \infty. \tag{3.1.21}$$

All the RH problems in the rest of the book will be either directly presented in this form or they can be easily transformed to it. From now on, unless otherwise specified, when talking about RH problems we will be assuming the above setting.

Observe that the RH problem in the "smooth" formulation inherits from the L^p-theory the relation to the integral equations (3.1.3) and (3.1.17) given by the formulae (3.1.2) and (3.1.18), respectively. It also should be noted that, under the smoothness assumptions on the jump matrix $G(\lambda)$, the L^p-solvability of the Riemann-Hilbert problem implies the classical "smooth" solvability.

The cyclic equation (3.1.19) implies the following important property of the solution $Y(\lambda)$ of the RH problem at a node point A. Let ω_k, $k = 1,\dots,q$, be the sectors forming in the neighborhood Ω_A by the sectors Γ_k. Then, each of the functions $Y|_{\omega_k}$, $k = 1,\dots,q$, is in fact analytic in the whole neighborhood Ω_A. This, together with the local analyticity of the jump matrix $G(\lambda)$ (G-matrix property 1 above), implies that the jump contour Γ can be always locally *deformed.* In fact, one can deform the finite parts of the contour in an arbitrary way. The only condition which must be preserved is that the infinite branches of the deformed contour must be consistent with the asymptotic condition $G(\lambda) \to I$ as $\lambda \to \infty$ (G matrix property 2 above). *This flexibility of deforming the contour of the RH problem plays a crucial role in our future use of the RH formalism in the analysis of the Painlevé equations.*

The smoothness condition (3.1.19) has yet another important consequence. Observe that, by appropriately augmenting the jump contour Γ and/or re-orienting some of its components, we can always assume that the contour Γ is *complete.* This means (cf. [**DeZ5**], [**Zh**]) that the set $\mathbb{C}P^1 \setminus \Gamma$ is a disjoint union of two, possibly disconnected, open regions Ω_+ and Ω_-, and Γ can be viewed as the positively oriented boundary for Ω_+ and also as the negatively oriented boundary of Ω_-.

Then, the absence of monodromy at every node point means that the jump matrix $G(\lambda)$ admits the factorization (3.1.14) with the matrices $W^{\pm}(\lambda)$ such that:

(1) $\det(I \pm W^{\pm})$ does not vanish.
(2) The function $W^{+}(\lambda)$, $\lambda \in \Gamma$ and the function $W^{-}(\lambda)$, $\lambda \in \Gamma$, are smooth if they are considered as functions on the boundary of Ω_{+} and on the boundary of Ω_{-}, respectively.
(3) $W^{-} = 0$ away from some neighborhoods of the node points.

In turn, this fact (and the triviality of the index of the $\det G(\lambda)$) implies that the associated Cauchy operator $\mathbf{1} - C_w$, which is defined in (3.1.16), is a *Fredholm operator*; moreover, its Fredholm index is zero ([**Zh**]). Indeed, the following *analytic Fredholm alternative* takes place.

PROPOSITION 3.2 ([**Zh**]). *If $W^{\pm}(\lambda) \equiv W^{\pm}(\lambda; x)$ depend on a parameter x analytically, then either $(\mathbf{1} - C_w)^{-1}$ is meromorphic in x, or $\mathbf{1} - C_w$ is invertible for no x.*

Propositions 3.2 and Proposition 3.1 (taken together with the fact that the operator $\mathbf{1} - C_w$ is Fredholm with zero index) lead to the following two corollaries which form the basis for the global analysis of Painlevé equations which will be developed in Chapter 5.

COROLLARY 3.1. *If $W^{\pm}(\lambda) \equiv W^{\pm}(\lambda; x)$ depend on a parameter x analytically, then either the solution $Y(\lambda; x)$ of the associated RH problem is meromorphic in x, or the problem is solvable for no x.*

COROLLARY 3.2 (The vanishing lemma). *The RH problem determined by a pair (Γ, G) has a solution, iff the corresponding homogeneous RH problem has only the trivial solution.*

We note that in the smooth setting, the homogeneous RH problem is the RH problem whose asymptotic condition (3.1.21) is replaced by the condition

$$Y(\lambda) = O\left(\frac{1}{\lambda}\right), \quad \lambda \to \infty. \tag{3.1.22}$$

We complete our review of the Riemann-Hilbert theory by addressing two more general points.

The first point is concerned with the uniqueness of the solution of a Riemann-Hilbert problem. This is a much simpler question than the existence issue. Indeed, we first note that the determinant of the function $Y(\lambda)$ satisfies the following *scalar* RH problem.

- For each k, the function $\det Y|_{\Omega_k}$ is analytic in Ω_k and continuous in $\overline{\Omega}_k$.
- The pointwise limit $\det Y_{-}(\lambda)$ of $\det Y$ from the minus side of Γ and the pointwise limit $\det Y_{+}(\lambda)$ from the plus side of Γ, which exist and are continuous by virtue of the previous condition, are related for $\lambda \in \Gamma$ by

$$\det Y_{+}(\lambda) = \det Y_{-}(\lambda) \det G(\lambda) \equiv \det Y_{-}(\lambda). \tag{3.1.23}$$

- At $\lambda = \infty$ the function $\det Y(\lambda)$ satisfies the asymptotic condition,

$$\det Y(\lambda) = 1 + O\left(\frac{1}{\lambda}\right), \quad \lambda \to \infty. \tag{3.1.24}$$

By virtue of the Liouville theorem we immediately conclude that

$$\det Y(\lambda) \equiv 1. \tag{3.1.25}$$

Suppose now that $\tilde{Y}(\lambda)$ is another solution of the same RH problem. Put

$$X(\lambda) := \tilde{Y}(\lambda)Y^{-1}(\lambda).$$

Since on the contour Γ we have

$$\begin{aligned} X_+(\lambda) = \tilde{Y}_+(\lambda)Y_+^{-1}(\lambda) &= \tilde{Y}_-(\lambda)G(\lambda)G^{-1}(\lambda)Y_-^{-1}(\lambda) \\ &= \tilde{Y}_-(\lambda)Y_-^{-1}(\lambda) = X_-(\lambda), \end{aligned}$$

the function $X(\lambda)$ has no jumps and hence is analytic everywhere on $\mathbb{C}$. In addition,

$$X(\lambda) \to I, \quad \lambda \to \infty.$$

Therefore, applying once again the Liouville theorem we arrive at the identity

$$X(\lambda) \equiv I \Longleftrightarrow \tilde{Y}(\lambda) \equiv Y(\lambda),$$

and the uniqueness of the solution of the RH problem follows.

It should be noted that the proof of uniqueness in the more general L^p-case needs some extra analytic work. We again refer the interested reader to [**Zh**], [**DeZ4**], and [**DeZ5**] for the details. Also, the unimodularity of the jump matrix $G(\lambda)$ is not really important. What is needed is the triviality of the index of the determinant of $G(\lambda)$, i.e.,

$$\Delta \log \det G|_\Gamma = 0.$$

Indeed, under this condition one can use the explicit formula

$$y(\lambda) = \exp\left\{\frac{1}{2\pi i}\int_\Gamma \frac{\log\det G(\mu)}{\mu - \lambda}d\mu\right\} \tag{3.1.26}$$

to factorize the determinant of $G(\lambda)$ and hence to reduce the RH problem to the unimodular case.

The second general point is the possibility of using an alternative to the Fredholm method represented by Corollaries 3.1 and 3.2 for analyzing the solvability issue. This approach utilizes the connection of the RH problems with the theory of holomorphic vector bundles. The basic technical ideas are the following: (i) use specific analytic properties of the jump matrix $G(\lambda)$, to construct an explicit solution of model Riemann-Hilbert problems near $\lambda = \infty$ and near the node points; (ii) use the uniform boundedness of the Cauchy operator in some natural Banach spaces of holomorphic functions, to obtain local factorizations (these constitute the essence of the so-called *Sibuya theorem*); (iii) use the *Birkhoff-Grothendieck theorem* concerning the factorization of matrix functions analytic in an annular domain, in order to extend the local factorization results to the global solution of the RH problem. The main references concerning the vector bundle method are the monographs [**Si2**], [**AnoBol**] and [**ArnIl**] and the survey article [**Bol1**]. In fact, in the theory of Painlevé equations, we need a generalization of the steps mentioned above to the case involving a parameter. The relevant generalizations were obtained in [**Mal**] and [**Bol4**], and they allow us to prove the meromorphic solvability of the RH problems related to Painlevé equations, under very general assumptions on the Riemann-Hilbert data. We will present, following [**BolIK**], an elementary version of the holomorphic bundle approach in much more details later in this chapter and in Appendices A and B.

2. The Fuchsian Systems. Hilbert's twenty-first problem.

Reflecting on Riemann-Hilbert problems, one can easily single out the following trivial case. Suppose that Γ is a Jordan closed curve, i.e., the λ-plane is divided by Γ into two open connected sets — the interior domain Ω_+ and the exterior domain Ω_- of Γ, while the map G is a constant map, i.e., $G(\lambda) \equiv G_0, \forall\lambda$. Then the problem can be solved immediately:

$$Y(\lambda) = \begin{cases} I & \text{for} \quad \lambda \in \Omega_-, \\ G_0^{-1} & \text{for} \quad \lambda \in \Omega_+. \end{cases}$$

The easiest way to make the problem nontrivial is to pass from the constant map G (jump matrix) to a *piecewise constant.* It is interesting that the RH problem arising in this way, is precisely the kind of factorization problem which appeared in the classical works of Plemelj devoted to the solution of the inverse monodromy problem for Fuchsian systems. This was the first problem to be called "The Riemann-Hilbert Problem".

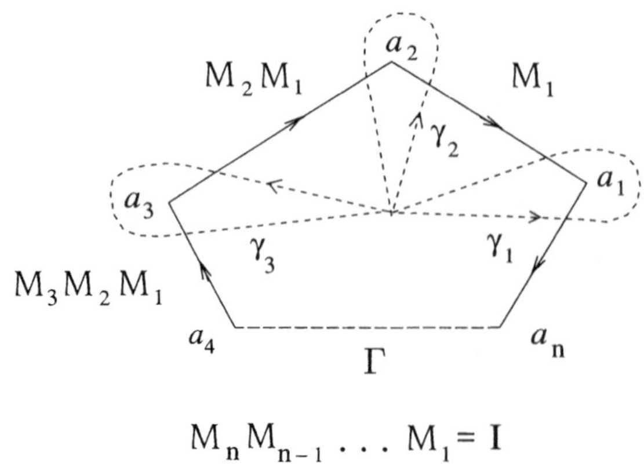

FIGURE 3.2. Polygon contour for Fuchsian equation (3.2.31)

Let the contour Γ be a polygonal line (see Figure 3.2),

$$\Gamma = [a_1, a_2] \cup [a_2, a_3] \cup \cdots \cup [a_{n-1}, a_n] \cup [a_n, a_1],$$

and assume that the jump matrix $G(\lambda)$ is defined as a piecewise constant matrix-valued function,

$$G(\lambda) = \mathrm{M}_k\mathrm{M}_{k-1}\dots\mathrm{M}_1, \quad \lambda \in (a_k, a_{k+1}), \tag{3.2.27}$$

where $\{\mathrm{M}_1, \mathrm{M}_2, \dots, \mathrm{M}_n\}$ is a given set of nonsingular constant matrices. The limits $Y_\pm(\lambda)$, $\lambda \in \Gamma \setminus \{a_k\}$ are taken in the usual pointwise sense, the behavior of $Y(\lambda)$ at the end points a_k is specified by,

$$Y(\lambda)(\lambda - a_k)^\kappa \to 0, \quad \lambda \to a_k, \quad 0 \leq \kappa < 1, \tag{3.2.28}$$

and the asymptotic condition at ∞ is replaced by the following more relaxed condition:

- $\lambda^D Y(\lambda) \to I$ as $\lambda \to \infty$, for some integer diagonal matrix D.

With these specifications and using the theory of singular integral equations, the corresponding RH problem can be shown (Plemelj, 1908) to have a solution for any choice of the nonsingular matrices $\mathsf{M}_1, \dots, \mathsf{M}_n$, satisfying only the constraint

$$\mathsf{M}_n \mathsf{M}_{n-1} \dots \mathsf{M}_1 = I \qquad \text{(the cyclic relation)}.$$

(the solution is unique if it happens that $D = 0$). The principal property of the function $Y(\lambda)$ in this case is that the matrix product,

$$\Psi(\lambda) \equiv (\lambda - a_1)^D Y(\lambda),$$

satisfies the following linear differential equation with respect to λ,

$$\frac{d\Psi(\lambda)}{d\lambda} = A(\lambda)\Psi(\lambda), \tag{3.2.29}$$

whose coefficient matrix $A(\lambda)$ is a rational function with poles at points $a_1, \dots, a_n$, and whose *monodromy group* is generated (see Figure 3.2) by the matrices $\mathsf{M}_1, \dots, \mathsf{M}_n$, i.e.,

$$\tau_{\gamma_k}(Y)(\lambda) = Y(\lambda)\mathsf{M}_k, \tag{3.2.30}$$

where τ_γ denotes the operator of analytic continuation along the loop γ.

The proof of equation (3.2.29) is based on the classical Liouville theorem and on the λ-independence of the matrices M_k. The main step of this proof is the observation that the λ-independence of the matrices M_k implies that the logarithmic derivative $\Psi_\lambda \Psi^{-1}$ is well defined as a rational function. This simple observation plays a fundamental role in the whole monodromy theory. We will explain this in more detail later. This idea will play a central role in Chapter 4 (see also [**Sh**]).

In the generic situation, the matrix $D = 0$ and all the poles of $A(\lambda)$ are simple, i.e., equation (3.2.29) becomes a *Fuchsian equation*,

$$\frac{d\Psi(\lambda)}{d\lambda} = \sum_{k=1}^{n} \frac{A_k}{\lambda - a_k}\Psi(\lambda), \quad \sum_{k=1}^{n} A_k = 0. \tag{3.2.31}$$

As it was mentioned in the introduction, the question if there always exists a Fuchsian system with given singular points and given monodromy group, constitutes Hilbert's twenty-first problem.

Equations (3.2.29) and (3.2.31) establish a relation between the Riemann-Hilbert problems with piecewise constant jump matrices, and the theory of systems of linear ODEs with rational coefficients. This relation was used by J. Plemelj in his "almost" solution of Hilbert's twenty-first problem. Indeed, having established the solvability of the RH problem above, and hence the existence of the linear system (3.2.29), Plemelj then shows that the system can be modified to the one which has the same monodromy group and the *simple* poles at all points a_k except perhaps one. He also claimed that for the remaining point the system can also be reduced to a Fuchsian one. In Plemelj's work there is no rigorous proof of this last claim to the general case. Actually, it can be shown [**ArnIl**] that this claim is indeed true if one of the monodromy matrices M_k can be diagonalized. However, in the general case, Plemelj's proof fails, and the complete solution of Hilbert's twenty-first problem, which incidentally ought to have in general a negative solution, was obtained only in 1990 by A. A. Bolibruch via a rather sophisticated analysis of the holomorphic vector bundles associated with the monodromy groups of Fuchsian equations (3.2.31).

In the next section, we will consider the first nontrivial case of Hilbert's twenty-first problem, namely the inverse monodromy problem for a 2×2 linear system with four simple poles, and we will explain in more detail the appearance and the use of Riemann-Hilbert factorization problems in the theory of Fuchsian systems. In particular, we will explain in more detail the derivation of the linear differential equation (3.2.29) and the monodromy relation (3.2.30) for the solutions of the Riemann-Hilbert problems with piecewise constant jump matrices.

3. Two by two systems

3.1. Four Fuchsian points. The first nontrivial case of the Fuchsian inverse monodromy problem, i.e., the one which does not admit a solution via contour integrals, corresponds to the 2×2 Fuchsian system (2.1.16) with four singular points. Let M_k, $k = 0, 1, 2, \infty$, be the corresponding monodromy matrices introduced in Subsection 2.3. We recall that the matrices M_k are defined via the monodromy relation,

$$\tau_{\gamma_k}(\Psi(\lambda)) = \Psi(\lambda)\mathsf{M}_k, \dots, k = 0, 1, 2, \infty, \tag{3.3.32}$$

where

$$\Psi(\lambda) \equiv \Psi^{(\infty)}(\lambda),$$

is the canonical solution at ∞ specified by the asymptotic condition

$$\Psi(\lambda) = \big(I + O(\lambda^{-1})\big)\lambda^{-\delta\sigma_3}, \quad \lambda \to \infty, \quad -\pi < \arg\lambda < \pi, \tag{3.3.33}$$

and the loops γ_0, γ_1, γ_2, and γ_∞ are indicated in Figure 3.3.

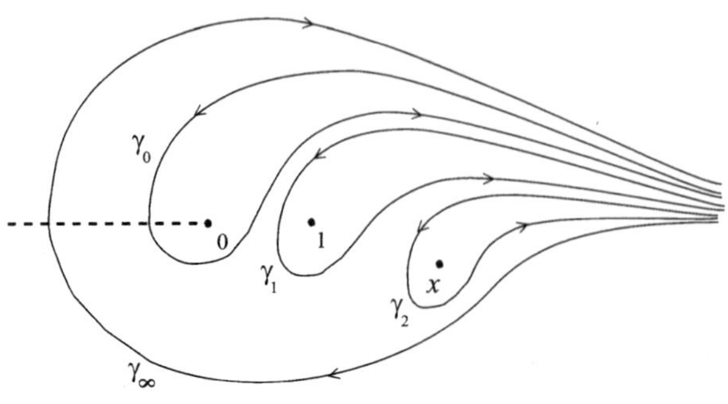

FIGURE 3.3. Monodromy group for the Fuchsian system

In (3.3.33) the matrix σ_3 is, as usual, the diagonal Pauli matrix,

$$\sigma_3 = \begin{pmatrix} 1 & 0 \\ 0 & -1 \end{pmatrix},$$

and the complex parameter δ is defined in (2.1.17), so that $\pm\delta$ are the eigenvalues of the matrix residue at infinity, A_∞, which is normalized to be diagonal,

$$A_\infty \equiv -A_0 - A_1 - A_2 = \begin{pmatrix} \delta & 0 \\ 0 & -\delta \end{pmatrix}.$$

The monodromy matrices M_k are characterized by the relations,

$$\det \mathsf{M}_\nu = 1, \quad \nu = 0, 1, 2, \infty, \tag{3.3.34}$$
$$\mathsf{M}_\infty \mathsf{M}_2 \mathsf{M}_1 \mathsf{M}_0 = I,$$

$$\mathsf{M}_\infty = \begin{pmatrix} e^{2\pi i\delta} & 0 \\ 0 & e^{-2\pi i\delta} \end{pmatrix}. \tag{3.3.35}$$

We recall that the first relation is the consequence of the gauge condition, Tr $\mathsf{A}(\lambda) = 0$, and that equation (3.3.35) is the cyclic constraint (cf. 2.0.31) which follows from the homotopy equation,

$$\gamma_\infty \gamma_0 \gamma_1 \gamma_2 \sim \{\infty\}.$$

We also recall that we use the notations $\pm\alpha$, $\pm\beta$, $\pm\gamma$ for the eigenvalues of A_0, A_1 and A_2, respectively, and assume the generic condition

$$2\alpha,\ 2\beta,\ 2\gamma \notin \mathbb{Z}.$$

In particular, this means (cf. the general formulae of Subsection 1.3) that the monodromy matrices M_k admit the following representations:

$$\mathsf{M}_0 = \mathsf{E}_0^{-1} e^{2\pi i\alpha\sigma_3} \mathsf{E}_0 \equiv \mathsf{E}_0^{-1} \begin{pmatrix} e^{2\pi i\alpha} & 0 \\ 0 & e^{-2\pi i\alpha} \end{pmatrix} \mathsf{E}_0, \tag{3.3.36}$$

$$\mathsf{M}_1 = \mathsf{E}_1^{-1} e^{2\pi i\beta\sigma_3} \mathsf{E}_1 \equiv \mathsf{E}_1^{-1} \begin{pmatrix} e^{2\pi i\beta} & 0 \\ 0 & e^{-2\pi i\beta} \end{pmatrix} \mathsf{E}_1, \tag{3.3.37}$$

$$\mathsf{M}_2 = \mathsf{E}_2^{-1} e^{2\pi i\gamma\sigma_3} \mathsf{E}_2 \equiv \mathsf{E}_2^{-1} \begin{pmatrix} e^{2\pi i\gamma} & 0 \\ 0 & e^{-2\pi i\gamma} \end{pmatrix} \mathsf{E}_2, \tag{3.3.38}$$

and, as has already been indicated,

$$\mathsf{M}_\infty = e^{2\pi i\delta\sigma_3} \equiv \begin{pmatrix} e^{2\pi i\delta} & 0 \\ 0 & e^{-2\pi i\delta} \end{pmatrix}. \tag{3.3.39}$$

Our goal now is to formulate the RH problem whose solution would yield the solution of the inverse monodromy problem for system (2.1.16). To this end, let us introduce the polygonal line,

$$\Gamma = (-\infty, 0] \cup [0, 1] \cup [1, x],$$

oriented from left to right and let us think about the canonical solution $\Psi(\lambda)$ as the holomorphic function defined on $\mathbb{C} \setminus \Gamma$.

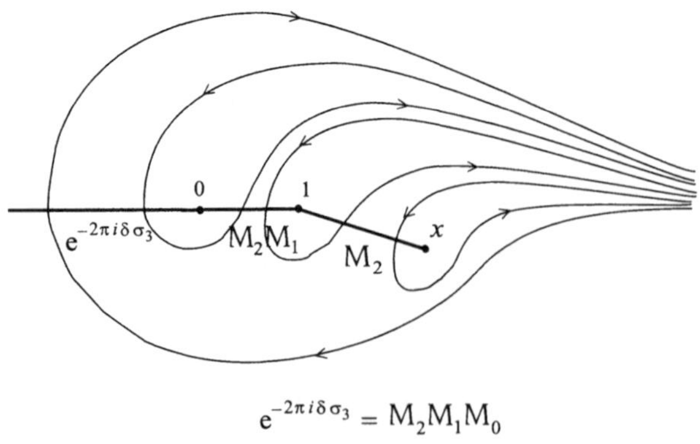

FIGURE 3.4. Jumps of the function Ψ

Using Figure 3.4 it is easy to check that the boundary values $\Psi_\pm(\lambda)$ of the function $\Psi(\lambda)$, which are well defined for all $\lambda \in \Gamma \setminus \{0, 1, x\}$, satisfy the following jump relations,

$$\Psi_+(\lambda) = \Psi_-(\lambda)\mathsf{S}(\lambda), \quad \lambda \in \Gamma \setminus \{0, 1, x\}, \tag{3.3.40}$$

where $\mathsf{S}(\lambda)$ is a piecewise constant matrix-valued function $\mathsf{S}(\lambda)$ defined by the equations

$$\mathsf{S}(\lambda) = \mathsf{M}_2\mathsf{M}_1\mathsf{M}_0 \equiv e^{-2\pi i\delta\sigma_3}, \quad \lambda \in (-\infty, 0), \tag{3.3.41}$$

$$\mathsf{S}(\lambda) = \mathsf{M}_2\mathsf{M}_1, \quad \lambda \in (0, 1), \tag{3.3.42}$$

$$\mathsf{S}(\lambda) = \mathsf{M}_2, \quad \lambda \in (1, x). \tag{3.3.43}$$

The jump condition (3.3.40) should be supplemented by the characterization of the behavior of the function $\Psi(\lambda)$ at the points of discontinuity of $\mathsf{S}(\lambda)$, i.e., at the points 0, 1, x, and ∞. From the general facts of the local theory of Fuchsian systems (see e.g. (2.0.11) and (2.0.18)) we conclude that the following representations at the points 0, 1, x, take place:

- neighborhood of $\lambda = 0$:

$$\Psi(\lambda) = \hat{\Psi}_0(\lambda)\lambda^{\alpha\sigma_3}\mathsf{E}_0 \times \begin{cases} I & \text{for} \quad \operatorname{Im}\lambda > 0, \\ \mathsf{M}_1^{-1}\mathsf{M}_2^{-1} & \text{for} \quad \operatorname{Im}\lambda < 0; \end{cases} \tag{3.3.44}$$

- neighborhood of $\lambda = 1$:

$$\Psi(\lambda) = \hat{\Psi}_1(\lambda)(\lambda - 1)^{\beta\sigma_3}\mathsf{E}_1 \times \begin{cases} I & \text{for} \quad \operatorname{Im}\lambda > 0, \\ \mathsf{M}_2^{-1} & \text{for} \quad \operatorname{Im}\lambda < 0; \end{cases} \tag{3.3.45}$$

- neighborhood of $\lambda = x$:

$$\Psi(\lambda) = \hat{\Psi}_2(\lambda)(\lambda - x)^{\gamma\sigma_3}\mathsf{E}_2. \tag{3.3.46}$$

In these equations, the matrix factors $\hat{\Psi}_0(\lambda)$, $\hat{\Psi}_1(\lambda)$, and $\hat{\Psi}_0(\lambda)$ represent the functions which are holomorphic and invertible at $\lambda = 0$, $\lambda = 1$, and $\lambda = x$, respectively. The branches of the multi-valued functions λ^{α}, $(\lambda-1)^{\beta}$, and $(\lambda-x)^{\gamma}$ are fixed by the conditions $-\pi < \arg\lambda < \pi$, $-\pi < \arg(\lambda-1) < \pi$, and $-\pi < \arg(\lambda - x) < \pi$, respectively. We also note that the asymptotic relation (3.3.33) can be also reformulated as a representation of $\Psi(\lambda)$ at the point $\lambda = \infty$ in a form similar to (3.3.44) – (3.3.46). Indeed, we can say that

$$\Psi(\lambda) = \hat{\Psi}_{\infty}(\lambda)\lambda^{-\delta\sigma_3}, \tag{3.3.47}$$

where $\hat{\Psi}_{\infty}(\lambda)$ is a holomorphic at $\lambda = \infty$ function satisfying the normalization equation,

$$\hat{\Psi}_{\infty}(\infty) = I, \tag{3.3.48}$$

and the branch of the function $\lambda^{-\delta}$ is fixed by the condition $-\pi < \arg\lambda < \pi$.

Put

$$Y(\lambda) := \Psi(\lambda)\lambda^{\delta\sigma_3}.$$

Then, the properties (3.3.40)–(3.3.48) of the canonical solution $\Psi(\lambda)$ can be reformulated in terms of the function $Y(\lambda)$ as follows:

(a) $Y(\lambda)$ is holomorphic in $\mathbb{C}P^1 \setminus \Gamma_0$, where

$$\Gamma_0 = [0,1] \cup [1,x] \equiv \Gamma \setminus (-\infty, 0), \tag{3.3.49}$$

(note the absence of a jump across the negative half-line).

(b) The function

$$\Psi(\lambda) := Y(\lambda)\lambda^{-\delta\sigma_3} \tag{3.3.50}$$

has branch singularities at the points 0, 1, and x which are characterized by the equations (3.3.44)–(3.3.46).

(c) The $\pm$ - limits of $Y(\lambda)$ on the set $\Gamma_0 \setminus \{0,1,x\}$ exist and satisfy the jump equation,

$$Y_+(\lambda) = Y_-(\lambda)G_0(\lambda) \quad \lambda \in \Gamma_0 \setminus \{0,1,x\}, \tag{3.3.51}$$

where

$$G_0(\lambda) = \lambda^{-\delta\sigma_3}\mathsf{S}(\lambda)\lambda^{\delta\sigma_3}, \tag{3.3.52}$$

and the piecewise constant matrix $\mathsf{S}(\lambda)$ is defined by the equations (3.3.42) and (3.3.43) above.

(d) At $\lambda = \infty$, the function $Y(\lambda)$ is normalized by the equation

$$Y(\infty) = I, \tag{3.3.53}$$

(note that because of the property (a) the function $Y(\lambda)$ is analytic at $\lambda = \infty$).

In other words, the function $Y(\lambda)$ solves the RH problem determined by the pair $(\Gamma_0, G_0(\lambda))$. This problem is depicted in Figure 3.5.

The inverse monodromy problem for the Fuchsian system (2.1.16) can now be formulated as the question of existence of the solution of the 2×2 RH problem whose contour and jump matrix are defined by equations (3.3.49) and (3.3.52), respectively (assuming that the matrices M_k, $k = 0,1,2$ and the complex parameters δ and x are given). Before discussing the answer to this question, let us analyze the formal setting of the RH problem in more detail.

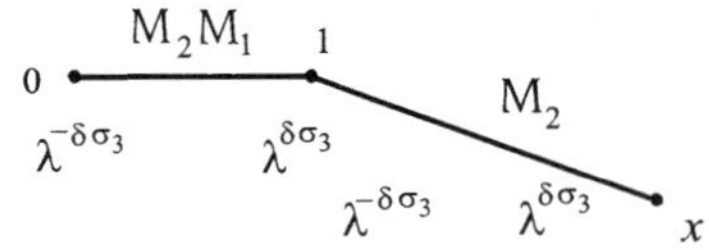

FIGURE 3.5. Four point Fuchsian RH problem

First we note that the jump relation (3.3.51) is consistent with the local behavior at the points of discontinuity prescribed by the equations (3.3.44)–(3.3.46). In fact, in the generic situation which is the one considered here, the latter is a direct corollary of the former. This can be seen as follows.

Consider a small neighborhood D_0 of the point $\lambda = 0$ and define the function (cf. (3.3.44))

$$\Psi_0(\lambda) := \lambda^{\alpha\sigma_3}\mathsf{E}_0 \times \begin{cases} I & \text{for} \quad \lambda \in D_0, \quad \operatorname{Im}\lambda > 0, \\ \mathsf{M}_1^{-1}\mathsf{M}_2^{-1} & \text{for} \quad \lambda \in D_0, \quad \operatorname{Im}\lambda < 0. \end{cases}$$

Observe that

$$\Psi_{0+}(\lambda) = \lambda_+^{\alpha\sigma_3}\mathsf{E}_0,$$

while

$$\Psi_{0-}(\lambda) = \lambda_-^{\alpha\sigma_3}\mathsf{E}_0\mathsf{M}_1^{-1}\mathsf{M}_2^{-1}.$$

If $\lambda > 0$, we have that $\lambda_+^{\alpha\sigma_3} = \lambda_-^{\alpha\sigma_3}$, and hence

$$\Psi_{0+}(\lambda) = \Psi_{0-}(\lambda)\mathsf{M}_2\mathsf{M}_1, \quad \lambda \in D_0, \quad \lambda > 0. \tag{3.3.54}$$

At the same time, if $\lambda < 0$, then

$$\lambda_+^{\alpha\sigma_3} = \lambda_-^{\alpha\sigma_3}e^{2\pi i\alpha\sigma_3},$$

and we have, in virtue of the characteristic equations (3.3.36) and (3.3.35),

$$\begin{aligned}\Psi_{0+}(\lambda) &= \Psi_{0-}(\lambda)\mathsf{M}_2\mathsf{M}_1\mathsf{E}_0^{-1}e^{2\pi i\alpha\sigma_3}\mathsf{E}_0 \\ &= \Psi_{0-}(\lambda)\mathsf{M}_2\mathsf{M}_1\mathsf{M}_0 \\ &= \Psi_{0-}(\lambda)e^{-2\pi i\delta\sigma_3}, \quad \lambda \in D_0, \quad \lambda > 0. \end{aligned} \tag{3.3.55}$$

Equations (3.3.54) and (3.3.55) show that the model function $\Psi_0(\lambda)$ and the function $\Psi(\lambda)$ satisfy in the disk D_0 *exactly* the same jump condition across the segment of the real line. Therefore, we conclude that the function $\hat{\Psi}_0(\lambda)$ *defined* by the equation

$$\hat{\Psi}_0(\lambda) := \Psi(\lambda)\Psi_0^{-1}(\lambda), \tag{3.3.56}$$

is holomorphic in the punched disk $D_0 \setminus \{0\}$.

Similar considerations, with similar results, apply in the neighborhoods of the other two points, i.e., the points $\lambda = 1$ and $\lambda = x$. Therefore, in order to claim that the representations (3.3.44) – (3.3.46) are the consequences of the jump condition (3.3.51), we must now show that: (i) the points $\lambda = 0$, 1, and x are removable singularities of the respective "hat"-functions, and (ii) that the latter have nonzero determinants.

Strictly speaking, the point (i) does not follow only from the jump relation (3.3.51). One needs to assume some a prior restriction on the behaviour of the solution $Y(\lambda)$ of the RH problem at the points of the discontinuity of the jump matrix. It is easy to see, that to ensure that the functions $\hat{\Psi}_0(\lambda)$, $\hat{\Psi}_1(\lambda)$, and $\hat{\Psi}_2(\lambda)$ are holomorphic at $\lambda = 0$, $\lambda = 1$, and $\lambda = x$, respectively, it is enough to demand that

$$\Psi(\lambda) = O\left(|\lambda|^{-\mathrm{Re}\,\alpha}\right), \quad \lambda \sim 0, \tag{3.3.57}$$

$$\Psi(\lambda) = O\left(|\lambda - 1|^{-\mathrm{Re}\,\beta}\right), \quad \lambda \sim 1, \tag{3.3.58}$$

$$\Psi(\lambda) = O\left(|\lambda - x|^{-\mathrm{Re}\,\gamma}\right), \quad \lambda \sim x, \tag{3.3.59}$$

assuming, in addition, that

$$0 \le \mathrm{Re}\,\alpha < \frac{1}{2}, \quad 0 \le \mathrm{Re}\,\beta < \frac{1}{2}, \quad \text{and} \quad 0 \le \mathrm{Re}\,\gamma < \frac{1}{2}. \tag{3.3.60}$$

The point (ii) above, i.e., the invertibility of the matrices $\hat{\Psi}_k(\lambda)$, $k = 0, 1, 2$, after their analyticity has been proven, is a trivial matter. Indeed, since $\det G(\lambda) \equiv 1$, we immediately conclude that the scalar function $\det Y(\lambda)$ has no jumps, and it is holomorphic everywhere on the complex plane λ. Moreover, in virtue of the normalization condition (3.3.53) we conclude that $\det Y(\infty) = 1$ and hence

$$\det Y(\lambda) \equiv 1; \tag{3.3.61}$$

hence the $\det \Psi(\lambda)$, as well as the determinants of the three functions $\hat{\Psi}_k(\lambda)$, $k = 0, 1, 2$, satisfy the same condition.

The RH problem (3.3.49)–(3.3.53) has a unique solution:

PROPOSITION 3.3. *The solution $Y(\lambda)$ of the problem (3.3.49)–(3.3.53) is unique.*

PROOF. Suppose $\tilde{Y}(\lambda)$ is another function satisfying (a)–(d). Put

$$X(\lambda) := \tilde{Y}(\lambda)Y^{-1}(\lambda).$$

Due to (3.3.61) and the fact that both $\tilde{Y}(\lambda)$ and $Y(\lambda)$ have the same jump matrices and the same behaviour at the singular points 0, 1, and x, the following is true:

(a_1) $X(\lambda)$ is holomorphic in $\mathbb{C}P^1 \setminus \Gamma_0$.

(b_1) $X(\lambda)$ is holomorphic in the neighborhood of 0, 1, and x.

(c_1) The $\pm$-limits of $X(\lambda)$ on Γ_0 satisfy the jump equations,

$$X_+(\lambda) = X_-(\lambda) \quad \lambda \in \Gamma_0 \setminus \{0, 1, x\}.$$

(d_1) At $\lambda = \infty$, the function $X(\lambda)$ satisfies the equation

$$X(\infty) = I.$$

Therefore, using Liouville's theorem, we conclude that $X(\lambda) \equiv 1$, and hence

$$\tilde{Y}(\lambda) \equiv Y(\lambda).$$

□

In virtue of the general restrictions (3.3.34), the independent data of the above RH problem consist of the two unimodular matrices, M_1 and M_2, and of the complex parameter δ. The matrix M_0 is uniquely determined by the equation (3.3.34), while the parameters α, β and γ are uniquely determined by the inequalities (3.3.60) and the fact that the expressions $\exp(\pm 2\pi i\alpha)$, $\exp(\pm 2\pi i\beta)$, and $\exp(\pm 2\pi i\gamma)$ represents the eigenvalues of the monodromy matrices M_0, M_1 and M_2, respectively (see (3.3.36)–(3.3.38)). The connection matrices E_k, $k = 0, 1, 2$ are defined (up to the left diagonal multipliers — which is exactly what is needed) by the equations (3.3.36)–(3.3.38). An explicit parametrization of the monodromy data has been given in Section 2.3 of Chapter 2.

Let us now address the question of the existence of the solution of the RH problem (3.3.49)–(3.3.53). We first consider the special case of the triangular reduction, $\mathsf{S}_{21}(\lambda) \equiv 0$.

Triangular case

Assume that

$$\mathsf{M}_1 = \begin{pmatrix} e^{2\pi i\beta} & s_1 \\ 0 & e^{-2\pi i\beta} \end{pmatrix} \quad \text{and} \quad \mathsf{M}_2 = \begin{pmatrix} e^{2\pi i\gamma} & s_2 \\ 0 & e^{-2\pi i\gamma} \end{pmatrix}. \tag{3.3.62}$$

Accordingly,

$$\mathsf{M}_2\mathsf{M}_1 = \begin{pmatrix} e^{2\pi i(\gamma+\beta)} & s_1 e^{2\pi i\gamma} + s_2 e^{-2\pi i\beta} \\ 0 & e^{-2\pi i(\gamma+\beta)} \end{pmatrix}, \tag{3.3.63}$$

and the jump matrix of the problem is given by the equation

$$G_0(\lambda) = \begin{cases} \begin{pmatrix} e^{2\pi i(\gamma+\beta)} & s_3\lambda^{-2\delta} \\ 0 & e^{-2\pi i(\gamma+\beta)} \end{pmatrix} & \text{for} \quad \lambda \in (0,1), \\ \begin{pmatrix} e^{2\pi i\gamma} & s_2\lambda^{-2\delta} \\ 0 & e^{-2\pi i\gamma} \end{pmatrix} & \text{for} \quad \lambda \in (1,x), \end{cases} \tag{3.3.64}$$

where we have used the notation,

$$s_3 := s_1 e^{2\pi i\gamma} + s_2 e^{-2\pi i\beta}.$$

The RH problem defined on the contour Γ_0 by the jump matrix $G_0(\lambda)$ given in (3.3.64) is not Abelian yet. However, it can be transformed to an Abelian one by factorizing the diagonal part of the matrix $G_0(\lambda)$.

Define in the domain $\mathbb{C} \setminus \Gamma_0$ the analytic function

$$\Delta(\lambda) = \left(\frac{\lambda - 1}{\lambda}\right)^{\gamma+\beta} \left(\frac{\lambda - x}{\lambda - 1}\right)^{\gamma}, \tag{3.3.65}$$

fixed by the asymptotic relation

$$\Delta(\lambda) \to 1, \quad \text{as} \quad \lambda \to \infty.$$

Note that $\Delta(\lambda)$ satisfies the following jump condition on Γ_0:

$$\Delta_+(\lambda) = \Delta_-(\lambda) \times \begin{cases} e^{2\pi i(\gamma+\beta)} & \text{for} \quad \lambda \in (0,1), \\ e^{2\pi i\gamma} & \text{for} \quad \lambda \in (1,x), \end{cases}$$

Put

$$\Phi(\lambda) := Y(\lambda)\Delta^{-\sigma_3}(\lambda).$$

Then, in terms of the function $\Phi(\lambda)$ the jump equation (3.3.51) takes the form

$$\Phi_+(\lambda) = \Phi_-(\lambda)G_\Phi(\lambda), \quad \lambda \in \Gamma_0 \setminus \{0.1, x\}, \tag{3.3.66}$$

where

$$G_\Phi(\lambda) = \begin{cases} \begin{pmatrix} 1 & s_3\lambda^{-2\delta}\Delta_-(\lambda)\Delta_+(\lambda) \\ 0 & 1 \end{pmatrix} & \text{for} \quad \lambda \in (0,1), \\ \begin{pmatrix} 1 & s_2\lambda^{-2\delta}\Delta_-(\lambda)\Delta_+(\lambda) \\ 0 & 1 \end{pmatrix} & \text{for} \quad \lambda \in (1,x), \end{cases} \tag{3.3.67}$$

which indeed yields an Abelian Riemann-Hilbert problem.

To solve the RH problem (3.3.66) we can look for $\Phi(\lambda)$ in the form

$$\Phi(\lambda) = \begin{pmatrix} 1 & \phi(\lambda) \\ 0 & 1 \end{pmatrix}. \tag{3.3.68}$$

This ansatz transforms the original multiplicative jump condition for the matrix function $\Phi(\lambda)$ into the *additive* jump condition for the scalar function $\phi(\lambda)$,

$$\phi_+(\lambda) = \phi_-(\lambda) + g(\lambda), \quad \lambda \in \Gamma_0, \tag{3.3.69}$$

supplemented by the asymptotic requirement,

$$\phi(\lambda) \to 0, \quad \lambda \to \infty.$$

In (3.3.69), the function $g(\lambda)$ is defined by the equations,

$$g(\lambda) = \begin{cases} s_3\lambda^{-2\delta}\Delta_-(\lambda)\Delta_+(\lambda) & \text{for} \quad \lambda \in (0,1), \\ s_2\lambda^{-2\delta}\Delta_-(\lambda)\Delta_+(\lambda) & \text{for} \quad \lambda \in (1,x). \end{cases}$$

The additive problem (3.3.69) can be solved via the Cauchy integral:

$$\phi(\lambda) = \frac{1}{2\pi i}\int_{\Gamma_0} g(\mu)\frac{d\mu}{\mu - \lambda} \tag{3.3.70}$$

$$= \frac{1}{2\pi i}\left\{s_3\int_0^1 + s_2\int_1^x\right\}\mu^{-2\delta}\Delta_-(\mu)\Delta_+(\mu)\frac{d\mu}{\mu-\lambda}$$

$$= \frac{s_3}{2\pi i}\int_0^1 \mu^{2\alpha}(1-\mu)^{2\beta}(x-\mu)^{2\gamma}\frac{d\mu}{\mu-\lambda}$$

$$+\frac{s_2}{2\pi i}\int_1^x \mu^{2\alpha}(\mu-1)^{2\beta}(x-\mu)^{2\gamma}\frac{d\mu}{\mu-\lambda}, \tag{3.3.71}$$

where we have taken into account that, in the triangular case, the cyclic relation (3.3.34), implies that

$$\alpha + \delta + \beta + \gamma = 0.$$

In summary, we arrive at the following theorem.

THEOREM 3.1. *Suppose that the matrices* M_1 *and* M_2 *are triangular, i.e., they are given by equations (*3.3.62*). Suppose also that the parameters* β, γ, *and* δ *satisfy the conditions (*3.3.60*), where* $\alpha := -\delta - \beta - \gamma$. *Then there exists the unique matrix function* $Y(\lambda, x)$ *solving the RH problem (*3.3.49*)–(*3.3.53*) for all* $x \in \mathbb{C} \setminus \{0,1\}$. *More precisely, the following takes place (as always,* $\tilde{X}$ *means the universal covering of the set* X*):*

(a) $Y(\lambda, x)$ *is holomorphic for all* $\lambda \in \mathbb{C}P^1 \setminus \Gamma_0$ *and for all* $x \in \mathbb{C} \setminus \{0,1\}$.

(b) *Denote by* Γ_0^ϵ *the polygonal line* Γ_0 *without* ϵ*-neighborhoods at the points* 0, 1, *and* x. *Then, the limits*

$$Y_\pm(\lambda, x) \equiv \lim_{\lambda' \to \lambda_\pm} Y(\lambda', x)$$

exist for each $\lambda \in \Gamma_0 \setminus \{0,1,x\}$ *and* $x \in \mathbb{C} \setminus \{0,1\}$, *and they are continuous functions on* Γ_0^ϵ.

(c) *The limits* $Y_\pm(\lambda)$ *satisfy the jump equation,*

(3.3.72)
$$Y_+(\lambda) = Y_-(\lambda) G_0(\lambda) \quad \lambda \in \Gamma_0 \setminus \{0,1,x\},$$

where

(3.3.73)
$$G_0(\lambda) = \lambda^{-\delta\sigma_3} \mathsf{S}(\lambda) \lambda^{\delta\sigma_3},$$

and the piecewise constant matrix $\mathsf{S}(\lambda)$ *is defined by the equations (*3.3.42*) and (*3.3.43*) above.*

(d) *The function*

(3.3.74)
$$\Psi(\lambda) := Y(\lambda) \lambda^{-\delta\sigma_3}$$

has branch singularities at the points 0, 1, x, *and* ∞ *which are characterized by the equations (*3.3.44*)–(*3.3.48*).*

(e) *Let* $Y^0(\lambda)$ *denote the restriction of the function* $Y(\lambda)$ *on a neighborhood of* $\lambda = \infty$. *Then, for every* $x \in \mathbb{C} \setminus \{0,1\}$ *the function* $Y^0(\lambda)$ *admits an analytic continuation on the whole universal covering of* $\mathbb{C} \setminus \{0,1,x\}$. *Moreover, as a function of* x, *the function* $Y^0(\lambda)$ *is an analytic function on the universal covering of* $\mathbb{C} \setminus \{0,1\}$.

(f) *Let* $\hat{\Psi}_\nu(\lambda)$, $\nu = 0, 1, x, \infty$ *be the holomorphic matrix multipliers appearing in the left-hand sides of (*3.3.44*)–(*3.3.48*). Then the Taylor series of* $\hat{\Psi}_\nu(\lambda)$ *at the respective points are differentiable with respect to* $x \neq 0, 1$. *Moreover, the (matrix) coefficients of the series are analytic functions of* $x \in \tilde{\mathbb{C}} \setminus \{0,1\}$.

(g) $\det Y(\lambda, x) \equiv 1$.

(h) **(bonus)** *The function* $Y(\lambda, x)$ *can be described by the following explicit formula:*

(3.3.75)
$$Y(\lambda) = \begin{pmatrix} 1 & y(\lambda) \\ 0 & 1 \end{pmatrix} \Delta^{-\sigma_3}(\lambda),$$

where $\Delta(\lambda)$ *is defined in* (3.3.65), *and*

$$
\begin{aligned}
y(\lambda) &= \frac{1}{2\pi i}\left\{s_3\int_0^1 + s_2\int_1^x\right\}\mu^{-2\delta}\Delta_-(\mu)\Delta_+(\mu)\frac{d\mu}{\mu-\lambda} \\
&= \frac{s_3}{2\pi i}\int_0^1 \mu^{2\alpha}(1-\mu)^{2\beta}(x-\mu)^{2\gamma}\frac{d\mu}{\mu-\lambda} \\
&\quad + \frac{s_2}{2\pi i}\int_1^x \mu^{2\alpha}(\mu-1)^{2\beta}(x-\mu)^{2\gamma}\frac{d\mu}{\mu-\lambda}.
\end{aligned}
\tag{3.3.76}
$$

PROOF. We only need to comment on the statements concerning the analyticity with respect to the parameter x. These are the direct consequences of the explicit formula (3.3.75). We skip the details. They are based on standard (see e.g. [**BE**]) analysis of the hypergeometric type contour integrals appearing in (3.3.76). □

REMARK 3.1. It is worth noticing that the representation (3.3.68), (3.3.70) is merely equation (3.1.1) for the triangular matrix $G_\Phi(\lambda)$ defined by the equation,

$$
G_\Phi(\lambda) = \begin{pmatrix} 1 & g(\lambda) \\ 0 & 1 \end{pmatrix}.
$$

Indeed,

$$
G(\mu) = \begin{pmatrix} 1 & g(\mu) \\ 0 & 1 \end{pmatrix} \implies \ln G(\mu) = \begin{pmatrix} 0 & g(\lambda) \\ 0 & 0 \end{pmatrix}
$$

$$
\implies \int_{\Gamma_0}\frac{\ln G(\mu)}{\mu-\lambda}d\mu = \begin{pmatrix} 0 & \int_{\Gamma_0} g(\mu)\frac{d\mu}{\mu-\lambda} \\ 0 & 0 \end{pmatrix}
$$

$$
\implies \Phi(\lambda) \equiv \exp\left\{\frac{1}{2\pi i}\int_{\Gamma_0}\frac{\ln G(\mu)}{\mu-\lambda}d\mu\right\} = \begin{pmatrix} 1 & \frac{1}{2\pi i}\int_{\Gamma_0} g(\mu)\frac{d\mu}{\mu-\lambda} \\ 0 & 1 \end{pmatrix}.
$$

REMARK 3.2. In the triangular case, the restrictions (3.3.60) can be dropped. Indeed, one only needs to replace the segments of integrations in (3.3.76) by proper loops — a standard operation in the classical theory of special functions (cf. [**BE**]).

The explicit formulae (3.3.75)–(3.3.76) can be used to evaluate explicitly all the matrix objects participating in the local representations (3.3.44)–(3.3.47). Let us show, for example, how this can be done in the case of equation (3.3.44) representing the function $\Psi(\lambda)$ in the neighborhood of the point $\lambda = 0$.

Assume that λ lies near zero in the λ-plane cut along the contour Γ_0. Then, the following representation for the function $y(\lambda)$ is valid

$$
y(\lambda) = \hat{y}(\lambda) - \frac{s_3 e^{2\pi i\delta}}{2i\sin 2\pi\alpha}[\lambda^{-2\delta}]_0\Delta^2(\lambda),
\tag{3.3.77}
$$

where $\hat{y}(\lambda)$ is analytic at $\lambda = 0$. In fact, as it follows from (3.3.76),

$$\hat{y}(\lambda) = \frac{1}{2\pi i}\left(s_2 - s_3\frac{\sin 2\pi(\alpha+\beta)}{\sin 2\pi\alpha}\right)\int_1^x \mu^{-2\delta}\Delta_+(\mu)\Delta_-(\mu)\frac{d\mu}{\mu-\lambda}$$
$$+\frac{s_3}{2\pi i}\frac{\sin 2\pi\delta}{\sin 2\pi\alpha}\int_x^{\infty}\mu^{-2\delta}\Delta^2(\mu)\frac{d\mu}{\mu-\lambda}$$
$$+\frac{s_3}{4\pi}\frac{e^{2\pi i\delta}}{\sin 2\pi\alpha}\int_C [\mu^{-2\delta}]_0\Delta^2(\mu)\frac{d\mu}{\mu-\lambda}, \tag{3.3.78}$$

where C is the loop emerging at $+\infty$ and going in clockwise direction around the point λ and the polygonal line,

$$\Gamma_1 := [0,1]\cup[1,x]\cup[x,+\infty).$$

In the equations (3.3.77) and (3.3.78), the notation $[\lambda^{\delta}]_0$ is used for the analytic in the domain $\mathbb{C}\setminus\Gamma_1$ branch of the associated multivalued function which is fixed by the condition,

$$\arg\lambda = \pi, \quad \text{for} \quad \lambda < 0.$$

We note, that this branch is related with the previously introduced branch λ^{δ}, which is analytic on $\mathbb{C}$ cut along the negative half-line, as follows: if $\lambda < 0$, then

$$\lambda_+^{\delta} = [\lambda^{\delta}]_0, \quad \text{and} \quad \lambda_-^{\delta} = [\lambda^{\delta}]_0 e^{-2\pi i\delta}. \tag{3.3.79}$$

Equation (3.3.77) implies that the function $\Psi(\lambda)$ in the neighborhood of $\lambda = 0$ satisfies

$$\Psi(\lambda) = \begin{pmatrix} 1 & \hat{y}(\lambda) - \frac{s_3 e^{2\pi i\delta}}{2i\sin 2\pi\alpha}[\lambda^{-2\delta}]_0\Delta^2(\lambda) \\ 0 & 1 \end{pmatrix}\lambda^{-\delta\sigma_3}\Delta^{\sigma_3}(\lambda)$$

$$= \begin{pmatrix} 1 & \hat{y}(\lambda) \\ 0 & 1 \end{pmatrix}\lambda^{-\delta\sigma_3}\Delta^{\sigma_3}\begin{pmatrix} 1 & -\frac{s_3 e^{2\pi i\delta}}{2i\sin 2\pi\alpha}[\lambda^{-2\delta}]_0\lambda^{2\delta} \\ 0 & 1 \end{pmatrix}$$

$$= \begin{pmatrix} 1 & \hat{y}(\lambda) \\ 0 & 1 \end{pmatrix}\hat{\Delta}^{\sigma_3}(\lambda)\lambda^{-\delta\sigma_3}[\lambda^{-(\gamma+\beta)\sigma}]_0\begin{pmatrix} 1 & -\frac{s_3 e^{2\pi i\delta}}{2i\sin 2\pi\alpha}[\lambda^{-2\delta}]_0\lambda^{2\delta} \\ 0 & 1 \end{pmatrix},$$

where $\hat{\Delta}(\lambda)$ is the analytic at $\lambda = 0$ part of $\Delta(\lambda)$,

$$\hat{\Delta}(\lambda) = (\lambda-1)^{\gamma+\beta}\left(\frac{\lambda-x}{\lambda-1}\right)^{\gamma} \equiv e^{i\pi(\gamma+\beta)}x^{\gamma}\left(1+\sum_{j=1}^{\infty}\Delta_k\lambda^k\right). \tag{3.3.80}$$

Using (3.3.79), the final representation of the function $\Psi(\lambda)$ in the neighborhood of $\lambda = 0$ can be written in the form,

$$\Psi(\lambda) = \begin{pmatrix} 1 & \hat{y}(\lambda) \\ 0 & 1 \end{pmatrix} \hat{\Delta}^{\sigma_3}(\lambda)\lambda^{\alpha\sigma_3} \begin{pmatrix} 1 & -\frac{s_3 e^{2\pi i\delta}}{2i\sin 2\pi\alpha} \\ 0 & 1 \end{pmatrix}$$

$$\times \begin{cases} I & \text{for} \quad \operatorname{Im}\lambda > 0, \\ \begin{pmatrix} e^{-2\pi i(\gamma+\beta)} & -s_3 \\ 0 & e^{2\pi i(\gamma+\beta)} \end{pmatrix} & \text{for} \quad \operatorname{Im}\lambda < 0. \end{cases} \tag{3.3.81}$$

Comparing equation (3.3.81) with the general equation (3.3.44), yields the following formulae for the corresponding holomorphic multiplier $\hat{\Psi}_0(\lambda)$ and connection matrix E_0,

$$\hat{\Psi}_0(\lambda) = \begin{pmatrix} 1 & \hat{y}(\lambda) \\ 0 & 1 \end{pmatrix} \hat{\Delta}^{\sigma_3}(\lambda), \tag{3.3.82}$$

$$\mathsf{E}_0 = \begin{pmatrix} 1 & -\frac{s_3 e^{2\pi i\delta}}{2i\sin 2\pi\alpha} \\ 0 & 1 \end{pmatrix}. \tag{3.3.83}$$

It follows from (3.3.81) that

$$\mathsf{M}_1^{-1}\mathsf{M}_2^{-1} = \begin{pmatrix} e^{-2\pi i(\gamma+\beta)} & -s_3 \\ 0 & e^{2\pi i(\gamma+\beta)} \end{pmatrix},$$

which, of course, is consistent with the definition of the matrices M_1 and M_2 (see (3.3.62), (3.3.63)). Also, formula (3.3.83) for the connection matrix at $\lambda = 0$, is consistent with the general equation (3.3.36) for the corresponding monodromy matrix.

General case

Consider now the general case of the RH problem (3.3.49)–(3.3.53). More precisely, we will assume, as usual, that the matrices M_ν, $\nu = 0, 1, 2$, satisfy, in addition to the general restrictions (3.3.34), (3.3.35), the generic assumption of semisimplicity manifested by equations (3.3.36)–(3.3.38). The analogue of the existence Theorem 3.1 for the general case is the following.

THEOREM 3.2. *Let $\mathfrak{m} = \{\mathsf{M}_0, \mathsf{M}_1, \mathsf{M}_2\}$ be the given monodromy data, i.e., the collection of three 2×2 diagonalizable unimodular matrices satisfying (3.3.34), (3.3.35). Let us denote by $e^{\pm 2\pi i\alpha}$, $e^{\pm 2\pi i\beta}$, and $e^{\pm 2\pi i\gamma}$ the eigenvalues of the matrices M_0, M_1, and M_2, respectively, where we assume that the exponents α, β, γ*

satisfy the inequalities (3.3.60). *Then there exists the countable set* $X_{\mathfrak{m}} = \{x_j \equiv x_j(\mathfrak{m})\} \subset \widetilde{\mathbb{C}} \setminus \{0,1\}$, *having the points* $0, 1, \infty$ *as its only accumulation points, and the unique matrix function* $Y(\lambda, x)$ *solving the RH problem* (3.3.49) – (3.3.53) *for all* $x \notin \overline{X_{\mathfrak{m}}}$. *More precisely, the following takes place:*

(a) $Y(\lambda, x)$ *is holomorphic for all* $\lambda \in \mathbb{C}P^1 \setminus \Gamma_0$ *and for all* $x \notin \overline{X_{\mathfrak{m}}}$.

(b) *Denote by* Γ_0^ϵ *the polygonal line* Γ_0 *without* ϵ*-neighborhoods at the points* 0, 1, *and* x. *Then, the limits*

$$Y_\pm(\lambda, x) \equiv \lim_{\lambda' \to \lambda_\pm} Y(\lambda', x)$$

exist for each $\lambda \in \Gamma_0 \setminus \{0, 1, x\}$ *and* $x \notin \overline{X_{\mathfrak{m}}}$, *and they are continuous functions on* Γ_0^ϵ.

(c) *The limits* $Y_\pm(\lambda)$ *satisfy the jump equation,*

$$Y_+(\lambda) = Y_-(\lambda) G_0(\lambda) \quad \lambda \in \Gamma_0 \setminus \{0, 1, x\}, \tag{3.3.84}$$

where

$$G_0(\lambda) = \lambda^{-\delta\sigma_3} \mathsf{S}(\lambda) \lambda^{\delta\sigma_3}, \tag{3.3.85}$$

and the piecewise constant matrix $\mathsf{S}(\lambda)$ *is defined by the equations* (3.3.42) *and* (3.3.43).

(d) *The function*

$$\Psi(\lambda) := Y(\lambda) \lambda^{-\delta\sigma_3} \tag{3.3.86}$$

has branch singularities at the points 0, 1, x, *and* ∞ *which are characterized by the equations* (3.3.44)–(3.3.48).

(e) *Let* $Y^0(\lambda)$ *denote the restriction of the function* $Y(\lambda)$ *on a neighborhood of* $\lambda = \infty$. *Then, for every* $x \notin \overline{X_{\mathfrak{m}}}$ *the function* $Y^0(\lambda)$ *admits an analytic continuation on the whole universal covering of* $\mathbb{C} \setminus \{0, 1, x\}$. *Moreover, as a function of* x, *the function* $Y^0(\lambda)$ *is a meromorphic function on the universal covering of* $\mathbb{C} \setminus \{0, 1\}$, *and the set* $X_{\mathfrak{m}}$ *is the set of its poles.*

(f) *Let* $\hat{\Psi}_\nu(\lambda)$, $\nu = 0, 1, x, \infty$ *be the holomorphic matrix multipliers appearing in the left-hand sides of* (3.3.44)–(3.3.48). *Then the Taylor series of* $\hat{\Psi}_\nu(\lambda)$ *at the above points are differentiable with respect to* $x \notin \overline{X_{\mathfrak{m}}}$. *Moreover, the (matrix) coefficients of the series are meromorphic functions of* $x \in \widetilde{\mathbb{C}} \setminus \{0, 1\}$, *and the set* $X_{\mathfrak{m}}$ *is the set of their poles.*

(g) $\det Y(\lambda, x) \equiv 1$.

Since we do not have the advantage of an explicit formulae for the (perspective) solution $Y(\lambda)$, the proof of this theorem is highly nontrivial. It can be achieved by using the Fredholm analysis of the associated singular integral equation or by applying methods of holomorphic vector bundles based on the generalized Birkhoff-Grothendieck theorem in the case of a factorization problem involving parameters. In the framework of the RH approach to Painlevé equations, Theorem 3.2 is related to the sixth Painlevé equation which we will not discuss in this book, and hence we will skip the proof of this theorem, which can be, in principal, extracted from the general works [**Bol4**], [**Mal**], and [**Miw**][2]. Also, later in this section a similar theorem will be discussed in more detail in connection with the first nontrivial case of the inverse monodromy problem for a linear system with an irregular singularity.

---

[2] In an equivalent formulation, Theorem 3.2 has been recently proven in [**IIwS**].

This problem will be related to the second Painlevé equation — a major object of our study.

We conclude this subsection by showing how Theorem 3.2 can be used for solving the inverse monodromy problem (the RH problem in its original meaning!), for generic 2×2 Fuchsian system with four singular points (2.1.16).

Let $\mathfrak{m} = \{\mathsf{M}_0, \mathsf{M}_1, \mathsf{M}_2\}$ and $x \notin \overline{X_{\mathfrak{m}}}$ be the given extended monodromy data and let $Y(\lambda)$ and $\Psi(\lambda) = Y(\lambda)\lambda^{-\delta\sigma_3}$ be the corresponding solution and its Ψ-function (see (3.3.86)) of the RH problem (3.3.49)–(3.3.53). Put

$$\mathsf{A}(\lambda) := \frac{d\Psi(\lambda)}{d\lambda}\Psi^{-1}(\lambda). \tag{3.3.87}$$

The principal observation is that the matrix $\mathsf{A}(\lambda)$ is holomorphic in $\mathbb{C} \setminus \{0, 1, x\}$. Indeed, the function $\mathsf{A}(\lambda)$ is certainly holomorphic for all $\lambda \in \mathbb{C} \setminus \Gamma$, where

$$\Gamma = (-\infty, 0] \cup [0, 1] \cup [1, x],$$

since the functions $\Psi(\lambda)$ and $\Psi^{-1}(\lambda)$ are holomorphic. At the same time, using the fact that the jump matrix $S(\lambda)$, which governs the jumps of the function $\Psi(\lambda)$ (see (3.3.40)), is constant along the open segments $(-\infty, 0)$, $(0, 1)$, and $(1, x)$, we conclude that the function $\mathsf{A}(\lambda)$ has no jumps at all across Γ,

$$\mathsf{A}_+(\lambda) = \mathsf{A}_-(\lambda), \quad \forall \lambda \in \Gamma \setminus \{0, 1, x\},$$

and the analyticity of $\mathsf{A}(\lambda)$ in $\mathbb{C} \setminus \{0, 1, x\}$ follows. Next we notice that the function $\mathsf{A}(\lambda)$ is in fact a rational function and the points $\lambda = 0, 1, x$ are its simple poles. To see this it is enough to insert the local representations (3.3.44)–(3.3.47) into the right-hand side of (3.3.87) and use again the fact that all the matrices E_ν and M_ν are constant. Actually, we have that

$$\mathsf{A}(\lambda) = \hat{\Psi}_0(0)\frac{\alpha\sigma_3}{\lambda}\hat{\Psi}_0^{-1}(0) + O(1), \quad \lambda \to 0,$$

$$\mathsf{A}(\lambda) = \hat{\Psi}_1(1)\frac{\beta\sigma_3}{\lambda - 1}\hat{\Psi}_1^{-1}(1) + O(1), \quad \lambda \to 1,$$

$$\mathsf{A}(\lambda) = \hat{\Psi}_2(x)\frac{\gamma\sigma_3}{\lambda - x}\hat{\Psi}_2^{-1}(x) + O(1), \quad \lambda \to x,$$

$$\mathsf{A}(\lambda) = -\frac{\delta\sigma_3}{\lambda} + O\left(\frac{1}{\lambda^2}\right), \quad \lambda \to \infty,$$

(in the last relation we have also taken into account the normalization equation (3.3.48)) and in virtue of the Liouville theorem we conclude

$$\mathsf{A}(\lambda) = \frac{A_0}{\lambda} + \frac{A_1}{\lambda - 1} + \frac{A_2}{\lambda - x}, \tag{3.3.88}$$

where the matrix residues are given by the equations

$$A_0 = \alpha P_0\sigma_3 P_0^{-1}, \quad A_1 = \beta P_1\sigma_3 P_1^{-1}, \quad A_2 = \gamma P_2\sigma_3 P_2^{-1}, \tag{3.3.89}$$

and they satisfy the normalization condition,

$$A_0 + A_1 + A_2 = -\delta\sigma_3. \tag{3.3.90}$$

In (3.3.89) we have used the notations (cf. (2.0.34)),

$$P_0 = \hat{\Psi}_0(0), \quad P_1 = \hat{\Psi}_1(1), \quad \text{and} \quad P_2 = \hat{\Psi}_2(x).$$

After establishing the rational structure of $\mathsf{A}(\lambda)$, we can reinterpret (3.3.87) as implying that the matrix function $\Psi(\lambda)$ is a fundamental solution of the Fuchsian linear system of the type (2.1.16),

$$\frac{d\Psi}{d\lambda} = \Big(\frac{A_0}{\lambda} + \frac{A_1}{\lambda - 1} + \frac{A_2}{\lambda - x}\Big)\Psi. \tag{3.3.91}$$

In virtue of equations (3.3.47) and (3.3.48) we conclude that the function $\Psi(\lambda)$ is in fact the canonical at $\lambda = \infty$ solution of (3.3.91), i.e.,

$$\Psi(\lambda) \equiv \Psi^{(\infty)}(\lambda).$$

Moreover, using Figure 3.4 we can "reinterpret back" the jump condition (3.3.40) as the monodromy relation (3.3.32) which completes the solution of the inverse monodromy problem for system (2.1.16) for the given set of its extended monodromy data $\mathfrak{m} = \{\mathsf{M}_0, \mathsf{M}_1, \mathsf{M}_2\}$ and $x \notin \overline{X_{\mathfrak{m}}}$. The corresponding inverse monodromy map is described by equations (3.3.89) which define the matrix coefficients of system (3.3.91) as meromorphic functions of $x \in \widetilde{\mathbb{C}} \setminus \{0, 1\}$ with the set $X_{\mathfrak{m}}$ as the set of their poles.

REMARK 3.3. The points of the singular set $X_{\mathfrak{m}}$ are exactly the points where the solution of the inverse monodromy problem for system (2.1.16) for given $\mathfrak{m} = \{\mathsf{M}_0, \mathsf{M}_1, \mathsf{M}_2\}$ and **under the restrictions** (3.3.60) on the formal monodromy exponents, ceases to exist. However, if we drop this restriction, then the existence of the Fuchsian system (2.1.16) having $\mathfrak{m}$ as its monodromy data can be proven for any $x \in \mathbb{C} \setminus \{0, 1\}$. Indeed, by using proper Schlesinger transformation one can always remove a pole from the coefficients of the system at the expense of shifting one of the exponents, i.e., the numbers α, β, γ, or δ, by an integer (see [**Bol2**]). In fact, in the 2×2 case, which we are concerned with, one can even drop the semisimplicity condition on the monodromy matrices and prove the solvability of the inverse monodromy problem for any given monodromy matrices and, actually for any number of singular points. As usual, we refer the reader to the monograph [**AnoBol**] for more on the general issues related to the inverse monodromy problems for Fuchsian systems.

3.2. One irregular singular point of Poincaré rank 3. The first nontrivial case of the inverse monodromy problem in the presence of irregular singularities (Riemann-Hilbert-Birkhoff problem) is provided by the 2×2 system (2.1.86). To build the relevant Riemann-Hilbert formalism we have to use, according to our basic idea, the principal monodromy relation, which is now equation (2.1.91), as a jump condition for the sectionally analytic function formed by the collection $\{\Psi_n(\lambda)\}$ of canonical solutions. More precisely, let us do the following.

In addition to the canonical sectors Ω_n, which are defined in (2.1.90), we introduce the following six sectors:

$$\Omega_n^0 = \Big\{\lambda \in \mathbb{C} \setminus \{0\} \colon \ \frac{\pi}{6}(2n-3) < \arg\lambda < \frac{\pi}{6}(2n-1)\Big\}, \quad n = 1, 2, \dots, 6. \tag{3.3.92}$$

Note that

$$\overline{\Omega_n^0} \subset \Omega_n, \quad \mathbb{C} = \cup_{n=1}^{6} \overline{\Omega_n^0},$$

and

$$\Omega_k^0 \cap \Omega_j^0 = \emptyset, \quad k \neq j.$$

Therefore, we can define the sectionally analytic function $Y(\lambda)$ by the equations,

$$Y(\lambda) = \Psi_n(\lambda)e^{\theta(\lambda)\sigma_3}, \quad \lambda \in \Omega_n^0, \quad n = 1, \dots, 6, \tag{3.3.93}$$

where

$$\theta(\lambda) = \frac{4i}{3}\lambda^3 + ix\lambda + \nu \ln \lambda,$$

and the branch of the logarithm is fixed by the condition,

$$-\frac{\pi}{6} < \arg \lambda < \frac{11\pi}{6}.$$

We will also use the natural notation $Y_n(\lambda)$ for the restriction of the function $Y(\lambda)$ on the sector Ω_n^0:

$$Y_n = Y|_{\Omega_n^0}.$$

Denote by

$$\Gamma_n \equiv \Big\{\lambda : \arg \lambda = \frac{\pi}{6}(2n-1)\Big\}, \quad n = 1, \dots, 6,$$

the common boundary of the sectors Ω_n^0 and Ω_{n+1}^0. Assume that the rays Γ_n are oriented towards infinity. Then the characteristic analytic properties of the canonical solutions $\Psi_n(\lambda)$ can be reformulated in terms of the function $Y(\lambda)$ as follows:

(a) $Y(\lambda)$ is holomorphic in $\mathbb{C} \setminus \cup_{n=1}^6 \Gamma_n$.

(b) For every n, the function

$$Y_n^0(\lambda) := Y_n(\lambda)e^{-\nu \ln \lambda \sigma_3} \tag{3.3.94}$$

has a continuous extension on $\overline{\Omega_n^0}$.

(c) The $\pm$-limits of $Y(\lambda)$ on the rays Γ_n (which exist because of the previous statement) satisfy the jump equations,

$$Y_+(\lambda) = Y_-(\lambda)e^{-\theta(\lambda)\sigma_3}S_n e^{\theta(\lambda)\sigma_3}, \quad \lambda \in \Gamma_n \setminus \{0\}, \quad n = 1, \dots, 5, \tag{3.3.95}$$

$$Y_+(\lambda) = Y_-(\lambda)e^{-\theta_-(\lambda)\sigma_3}S_6 e^{\theta_-(\lambda)\sigma_3}, \quad \lambda \in \Gamma_6 \setminus \{0\}.$$

(d) The behavior of $Y(\lambda)$ at $\lambda = \infty$ is described by the asymptotic condition,

$$Y(\lambda) \to I, \quad \lambda \to \infty \tag{3.3.96}$$

which, more precisely, means the existence of the positive constant C such that

$$\|Y(\lambda) - I\| < \frac{C}{|\lambda|}, \quad |\lambda| > 1. \tag{3.3.97}$$

In other words, the function $Y(\lambda)$ solves the Riemann-Hilbert problem determined by the pair (Γ, G), where

$$\Gamma = \cup_{n=1}^6 \Gamma_n, \tag{3.3.98}$$

$$G(\lambda) = e^{-\theta(\lambda)\sigma_3}S_n e^{\theta(\lambda)\sigma_3}, \quad \lambda \in \Gamma_n, \quad n = 1, \dots, 5, \tag{3.3.99}$$

$$G(\lambda) = e^{-\theta_-(\lambda)\sigma_3}S_6 e^{\theta_-(\lambda)\sigma_3}, \quad \lambda \in \Gamma_6,$$

and the specification of the meanings of the "$\pm$" and $\lambda \to \infty$ limits are given by condition (b) and by estimate (3.3.97), respectively. The problem is depicted in Figure 3.6.

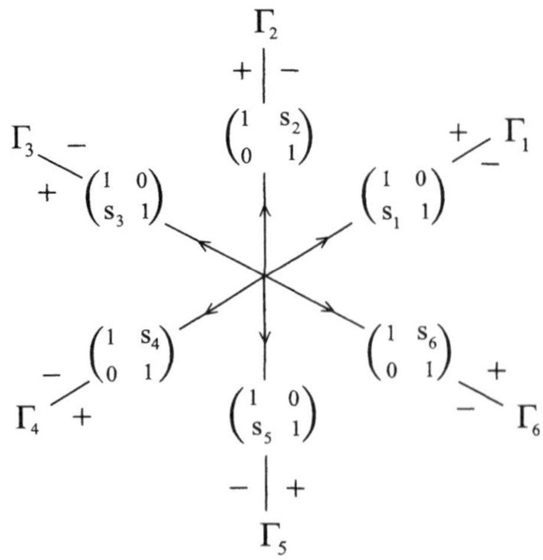

FIGURE 3.6. Poincaré rank 3 oscillatory RH problem

The inverse monodromy problem can be now formulated as the question of existence of the solution of the 2×2 RH problem whose contour and jump matrix are defined by equation (3.3.98) and (3.3.99), respectively, assuming that the matrices S_n and the complex parameters ν and x (which determine the exponent $\theta(\lambda)$) are given. Before discussing the answer to this question, let us analyze the formal setting of this RH problem.

First we note, that the specific triangular structure of the Stokes matrices S_n (see (2.1.92) implies the equations,

$$G(\lambda) = \begin{pmatrix} 1 & 0 \\ s_n e^{2\theta(\lambda)} & 1 \end{pmatrix}, \quad \lambda \in \Gamma_n, \tag{3.3.100}$$

when n is odd and

$$G(\lambda) = \begin{pmatrix} 1 & s_n e^{-2\theta(\lambda)} \\ 0 & 1 \end{pmatrix}, \quad \lambda \in \Gamma_n, \tag{3.3.101}$$

when n is even. These equations in turn imply that $G(\lambda)$ tends to the identity matrix I exponentially fast as $\lambda \to \infty$, $\lambda \in \Gamma$. This makes the jump condition of the RH problem (3.3.98)–(3.3.99) consistent with its asymptotic condition at $\lambda = \infty$.

Our next observation concerns the role of the second constraint on the monodromy data, i.e., the cyclic relation (2.1.93). We will show that it ensures the consistence of the jump condition (c) and condition (b) which, in addition to providing the meaning of the "$\pm$" limits, specifies the behavior of the solution $Y(\lambda)$ at the intersection point $\lambda = 0$. In fact, in terms of the functions $Y_n^0(\lambda)$, the jump condition (c) can be rewritten as the set of equations

$$Y_{n+1}^0(\lambda) = Y_n^0(\lambda) e^{-\theta^0(\lambda)\sigma_3} S_n e^{\theta^0(\lambda)\sigma_3}, \quad \lambda \in \Gamma_n \setminus \{0\}, \quad n = 1, \ldots, 5, \tag{3.3.102}$$

$$Y_1^0(\lambda) = Y_6^0(\lambda)e^{-\theta^0(\lambda)\sigma_3}S_6 e^{2\pi i\nu\sigma_3}e^{\theta^0(\lambda)\sigma_3}, \quad \lambda \in \Gamma_6 \setminus \{0\},$$

where

$$\theta^0(\lambda) \equiv \frac{4i}{3}\lambda^3 + ix\lambda,$$

and we have taken into account that

$$\theta_-(\lambda) = \theta_+(\lambda) + 2\pi i\nu, \quad \lambda \in \Gamma_6.$$

In virtue of condition (b), we can put in (3.3.102) $\lambda = 0$ and arrive at the relations

$$Y_{n+1}^0(0) = Y_n^0(0)S_n, \quad n = 1, \dots, 5, \tag{3.3.103}$$
$$Y_1^0(0) = Y_6^0(0)S_6 e^{2\pi i\nu\sigma_3}$$

which imply

$$Y_1^0(0) = Y_1^0(0)S_1 \dots S_6 e^{2\pi i\nu\sigma_3}. \tag{3.3.104}$$

Observe that conditions (a)-(d) yield the invertibility of matrix $Y_1^0(0)$. Indeed, together with the fact that $\det G(\lambda) \equiv 1$, they imply the following properties of the scalar function $\det Y(\lambda)$:

(a$_0$) $\det Y(\lambda)$ is holomorphic in $\mathbb{C} \setminus \cup_{n=1}^6 \Gamma_n$.
(b$_0$) For every n, the function $\det Y_n(\lambda)$ has a continuous extension on $\overline{\Omega_n^0}$.
(c$_0$) The $\pm$-limits of $\det Y(\lambda)$ on the rays Γ_n satisfy the jump equations,

$$\det Y_+(\lambda) = \det Y_-(\lambda) \quad \lambda \in \Gamma_n \setminus \{0\}, \quad n = 1, \dots, 6.$$

(d$_0$) The behavior of $\det Y(\lambda)$ at $\lambda = \infty$ is described by the asymptotic relation,

$$\det Y(\lambda) \to 1, \quad \lambda \to \infty. \tag{3.3.105}$$

In virtue of Liouville's theorem, properties (a$_0$)-(d$_0$), lead to the identity,

$$\det Y(\lambda) \equiv 1, \tag{3.3.106}$$

which in turn allows us to cancel matrix $Y_1^0(0)$ in equation (3.3.104) and conclude that this equation is equivalent to the cyclic constraint,

$$S_1 \dots S_6 = e^{-2\pi i\nu\sigma_3}.$$

Two more important intrinsic properties of the RH problem (3.3.98)–(3.3.99) should be indicated.

PROPOSITION 3.4. *The solution $Y(\lambda)$ of the above problem is unique.*

PROOF. Suppose $\tilde{Y}(\lambda)$ is another function satisfying (a)-(d). Put

$$X(\lambda) := \tilde{Y}(\lambda)Y^{-1}(\lambda).$$

Due to (3.3.106) and the fact that both $\tilde{Y}(\lambda)$ and $Y(\lambda)$ have the same jump matrices the following is true:

(a$_1$) $X(\lambda)$ is holomorphic in $\mathbb{C} \setminus \cup_{n=1}^6 \Gamma_n$.
(b$_1$) For every n, the function $X_n(\lambda)$ has a continuous extension on $\overline{\Omega_n^0}$.
(c$_1$) The $\pm$-limits of $X(\lambda)$ on the rays Γ_n satisfy the jump equations,

$$X_+(\lambda) = X_-(\lambda) \quad \lambda \in \Gamma_n \setminus \{0\}, \quad n = 1, \dots, 6.$$

(d$_1$) The behavior of $X(\lambda)$ at $\lambda = \infty$ is described by the asymptotic condition,

$$X(\lambda) \to I, \quad \lambda \to \infty.$$

Therefore, using Liouville's theorem again, we conclude that $X(\lambda) \equiv 1$, and hence

$$\tilde{Y}(\lambda) \equiv Y(\lambda).$$

□

PROPOSITION 3.5. *Each of the functions $Y_n^0(\lambda)$ defined in (3.3.94) admits an analytic continuation on the whole complex plane, i.e., they are in fact entire functions.*

PROOF. Let τ_δ^a be the operator of analytic continuation along a circle δ centered at $\lambda = 0$ and positively oriented. Note that, since the jump matrices in (3.3.102) are holomorphic in the neighborhoods of the jump contour (they are in fact entire functions), each $Y_n^0(\lambda)$ *a priori* admits an analytic continuation on the universal covering of $\mathbb{C} \setminus \{0\}$. Therefore, to prove the proposition it is enough to show that

$$\tau_\delta^a\left(Y_1^0(\lambda)\right) = Y_1^0(\lambda). \tag{3.3.107}$$

The action of the operator τ_δ^a on $Y_1^0(\lambda)$ can be evaluated by using the jump relations (3.3.102). Indeed we have,

$$\begin{aligned} Y_1^0(\lambda) &\mapsto Y_2^0(\lambda)e^{-\theta^0(\lambda)\sigma_3}S_1^{-1}e^{\theta^0(\lambda)\sigma_3} \mapsto Y_3^0(\lambda)e^{-\theta^0(\lambda)\sigma_3}S_2^{-1}S_1^{-1}e^{\theta^0(\lambda)\sigma_3} \\ &\mapsto \dots \mapsto Y_6^0(\lambda)e^{-\theta^0(\lambda)\sigma_3}S_5^{-1}\dots S_1^{-1}e^{\theta^0(\lambda)\sigma_3} \\ &\mapsto Y_1^0(\lambda)e^{-\theta^0(\lambda)\sigma_3}e^{-2\pi i\nu\sigma_3}S_6^{-1}S_5^{-1}\dots S_1^{-1}e^{\theta^0(\lambda)\sigma_3} \equiv \tau_\delta^a\left(Y_1^0(\lambda)\right), \end{aligned}$$

and (3.3.107) follows from the cyclic equation. □

Let us now address the question of the existence of the solution of the RH problem (3.3.98)–(3.3.99). We first consider the special case of the triangular reduction, $s_{2k+1} = 0$. The results of the end of Section 1.6, indicates that the RH problem under the triangular restriction is always solvable; moreover, the solution can be found explicitly in terms of contour integrals (2.1.100). Let us show this fact again but using, this time, the Riemann-Hilbert formalism only.

TRIANGULAR CASE

The principal observation is that under the triangular reduction, i.e., assuming that

$$s_1 = s_3 = s_5 = 0, \quad \nu = 0, \quad s_2 + s_4 + s_6 = 0, \tag{3.3.108}$$

the RH problem becomes *Abelian* and hence can be solved explicitly in terms of contour integrals. Indeed, since in this case the matrix $G(\lambda)$ has the form,

$$G(\lambda) = \begin{pmatrix} 1 & s\,e^{-2\theta(\lambda)} \\ 0 & 1 \end{pmatrix}, \tag{3.3.109}$$

we can repeat the arguments used in the previous subsection for solving the problem (3.3.66) and obtain the following formulae for the solution $Y(\lambda)$:

$$Y(\lambda) = \begin{pmatrix} 1 & y(\lambda) \\ 0 & 1 \end{pmatrix}, \tag{3.3.110}$$

where

$$
\begin{aligned}
y(\lambda) &= \frac{1}{2\pi i}\int_{\Gamma} s\, e^{-2\theta(\mu)}\frac{d\mu}{\mu-\lambda} \\
&= \frac{1}{2\pi i}\left\{s_2\int_{\Gamma_2} + s_4\int_{\Gamma_4} + s_6\int_{\Gamma_6}\right\} e^{-\frac{8i}{3}\mu^3 - 2ix\mu}\frac{d\mu}{\mu-\lambda},
\end{aligned}
\tag{3.3.111}
$$

and in the last equation we have taken into account conditions (3.3.108). It is important to notice that the last condition in (3.3.108) allows us to rewrite the sum of the three integrals in (3.3.111) as follows,

$$
y(\lambda) = \frac{1}{2\pi i}\left\{s_2\int_{\infty e^{-\frac{5\pi i}{6}}}^{i\infty} + s_6\int_{\infty e^{-\frac{5\pi i}{6}}}^{\infty e^{-\frac{\pi i}{6}}}\right\} e^{-\frac{8i}{3}\mu^3 - 2ix\mu}\frac{d\mu}{\mu-\lambda}.
\tag{3.3.112}
$$

In this form, the function $y(\lambda)$ can be easily continued to an entire function from any sector Ω_n^0 by a proper deformation of the contours of integration. In summary, we arrive at the following theorem.

THEOREM 3.3. *Suppose that the set* $s \equiv (s_1, \ldots, s_6; \nu)$ *satisfies conditions (*3.3.108*). Then there exists the unique matrix function* $Y(\lambda, x)$ *solving the RH problem (*3.3.98*)–(*3.3.99*) for all* $x \in \mathbb{C}$. *More precisely, the following takes place:*

(a) $Y(\lambda, x)$ *is holomorphic in* $\left(\mathbb{C} \setminus \cup_{n=1}^{6}\Gamma_n\right) \times \mathbb{C}$.

(b) *For every* n, *the restriction* $Y_n(\lambda, x) \equiv Y(\lambda, x)|_{\Omega_n^0 \times \mathbb{C}}$ *has a continuous extension on* $\overline{\Omega_n^0} \times \mathbb{C}$. *In particular, the limits*

$$
Y_{\pm}(\lambda, x) \equiv \lim_{\lambda' \to \lambda_{\pm}} Y(\lambda', x)
$$

exist for each $\lambda \in \Gamma$ *and* $x \in \mathbb{C}$, *and they are continuous functions on* $\Gamma \times \mathbb{C}$.

(c) *The limits* $Y_{\pm}(\lambda, x)$ *satisfy the jump equation,*

$$
Y_+(\lambda, x) = Y_-(\lambda, x) e^{-\theta(\lambda,x)\sigma_3} S_n e^{\theta(\lambda,x)\sigma_3}, \quad \lambda \in \Gamma_n,
\tag{3.3.113}
$$

$$
n = 1, \ldots, 6, \quad \theta(\lambda, x) = \frac{4i}{3}\lambda^3 + ix\lambda.
$$

(d) *Each of the functions*

$$
Y_n(\lambda, x), \quad n = 1, \ldots, 6,
$$

admits an analytic continuation on the whole complex λ-*plane, i.e., they are in fact entire functions of both* λ *and* x.

(e) *The asymptotic condition (*3.3.97*) at* $\lambda = \infty$ *extends to the full asymptotic series*

$$
Y(\lambda, x) \sim I + \sum_{k=1}^{\infty}\frac{\Psi_k(x)}{\lambda^k}, \quad \lambda \to \infty,
\tag{3.3.114}
$$

which is differentiable with respect to λ *and* x. *The coefficient functions* $\Psi_k(x)$ *are entire in* x.

(f) $\det Y(\lambda, x) \equiv 1$.

(g) *(***bonus***) The function* $Y(\lambda, x)$ *can be described by the following explicit formula,*

$$Y(\lambda,x)=\begin{pmatrix}1 & y(\lambda,x)\\ 0 & 1\end{pmatrix}, \tag{3.3.115}$$

where

$$y(\lambda,x)=\frac{1}{2\pi i}\left\{s_2\int_{\infty e^{-\frac{5\pi i}{6}}}^{i\infty}+s_6\int_{\infty e^{-\frac{5\pi i}{6}}}^{\infty e^{-\frac{\pi i}{6}}}\right\}e^{-\frac{8i}{3}\mu^3-2ix\mu}\frac{d\mu}{\mu-\lambda}.$$

PROOF. We only need to comment on the statements concerning the analyticity with respect to the parameter x and on the existence of the full asymptotic series (3.3.114). These are the direct consequences of the explicit formula (3.3.115). In particular, for the coefficients $\Psi_k(x)$ of the series (3.3.114) we have the following integral representations,

$$\Psi_k(x)=\begin{pmatrix}1 & u_k(x)\\ 0 & 1\end{pmatrix}, \tag{3.3.116}$$

where

$$u_k(x)=-\frac{1}{2\pi i}\left\{s_2\int_{\infty e^{-\frac{5\pi i}{6}}}^{i\infty}+s_6\int_{\infty e^{-\frac{5\pi i}{6}}}^{\infty e^{-\frac{\pi i}{6}}}\right\}e^{-\frac{8i}{3}\lambda^3-2ix\lambda}\lambda^{k-1}\,d\lambda,\quad k=1,2,\dots$$

□

GENERAL CASE

Let us now address the general case of the RH problem (3.3.98)–(3.3.99), i.e. we assume that the Stokes matrices $\{S_n\}$ are arbitrary matrices satisfying the basic triangular and cyclic constraints only (see Subsection 2.6):

$$S_{2l+1}=\begin{pmatrix}1 & 0\\ s_{2l+1} & 1\end{pmatrix},\quad S_{2l}=\begin{pmatrix}1 & s_{2l}\\ 0 & 1\end{pmatrix}, \tag{3.3.117}$$

$$S_1S_2S_3S_4S_5S_6=e^{-2\pi i\nu\sigma_3}. \tag{3.3.118}$$

We also note the scalar form of these constraints:

$$\begin{aligned}&1+s_1s_2=(1+s_4s_5)e^{2\pi i\nu},\quad 1+s_2s_3=(1+s_5s_6)e^{-2\pi i\nu},\\ &s_1+s_3+s_1s_2s_3=-s_5e^{2\pi i\nu}.\end{aligned} \tag{3.3.119}$$

Since, as in the general case of four Fuchsian singular points, we no longer have an explicit formula for $Y(\lambda)$, the existence of the solution $Y(\lambda)$ of the RH problem and its "good" analytic properties with respect to the parameter x are now quite nontrivial analytic facts. They can be established by using techniques based on the Fredholm analysis of the associated singular integral equation or by applying methods of holomorphic vector bundles based on the generalized Birkhoff-Grothendieck theorem. A precise statement concerning the solution $Y(\lambda) \equiv Y(\lambda, x)$, whose proof is presented in Appendix A, reads as follows.

THEOREM 3.4. *Let $s \equiv (\mathrm{S}_1, \mathrm{S}_2, \dots, \mathrm{S}_6; \nu)$ be the given monodromy data, i.e., the complex number ν and the set of six 2×2 matrices satisfying equations (3.3.117) and (3.3.118). Then there exists the countable set $X_s = \{x_j \equiv x_j(s)\} \subset \mathbb{C}$, $\quad x_j \to \infty$ and the unique matrix function $Y(\lambda, x)$ solving the RH problem (3.3.98)–(3.3.99) for all $x \in \mathbb{C}\setminus X_s$. More precisely, the following takes place:*

(a) *$Y(\lambda, x)$ is holomorphic in $\left(\mathbb{C}\setminus \cup_{n=1}^{6}\Gamma_n\right)\times(\mathbb{C}\setminus X_s)$.*

(b) *Let, as before, $Y_n(\lambda, x)$ be the restriction of the function $Y(\lambda, x)$ on the set $\Omega_n^0\times(\mathbb{C}\setminus X_s)$. Then, for every n, the function (cf. (3.3.94))*

$$Y_n^0(\lambda, x) := Y_n(\lambda, x)e^{-\nu\ln\lambda\sigma_3}, \tag{3.3.120}$$

has a continuous extension on $\overline{\Omega_n^0}\times(\mathbb{C}\setminus X_s)$. In particular, the limits

$$Y_{\pm}(\lambda, x) \equiv \lim_{\lambda'\to\lambda_{\pm}} Y(\lambda', x)$$

exist for each $\lambda\in\Gamma\setminus\{0\}$ and $x\in\mathbb{C}\setminus X_s$, and the products,

$$Y_{\pm}(\lambda, x)e^{-\nu\ln\lambda\sigma_3},$$

are continuous functions on $\Gamma\times(\mathbb{C}\setminus X_s)$.

(c) *The limits $Y_{\pm}(\lambda, x)$ satisfy the jump equation,*

$$Y_+(\lambda, x) = Y_-(\lambda, x)e^{-\theta(\lambda,x)\sigma_3}\mathrm{S}_n e^{\theta(\lambda,x)\sigma_3}, \quad \lambda\in\Gamma_n\setminus\{0\}, \tag{3.3.121}$$

$$n = 1, \dots, 5,$$

$$Y_+(\lambda, x) = Y_-(\lambda, x)e^{-\theta_-(\lambda,x)\sigma_3}\mathrm{S}_n e^{\theta_-(\lambda,x)\sigma_3}, \quad \lambda\in\Gamma_6\setminus\{0\},$$

where

$$\theta(\lambda, x) = \frac{4i}{3}\lambda^3 + ix\lambda + \nu\ln\lambda,$$

and the branch of logarithm is fixed, as before, by the condition,

$$-\frac{\pi}{6} < \arg\lambda < \frac{11\pi}{6}.$$

(d) *Each of the functions $Y_n^0(\lambda, x)$ defined in (3.3.120) admits an analytic continuation on the whole complex λ-plane, and they become holomorphic in $\mathbb{C}\times(\mathbb{C}\setminus X_s)$ and meromorphic along $\mathbb{C}\times X_s$ (i.e., they have poles at $x_j\in X_s$ and the coefficients of the corresponding Laurent series are entire functions of λ).*

(e) *The asymptotic condition (3.3.97) at $\lambda=\infty$ extends to the full asymptotic series*

$$Y(\lambda, x) \sim I + \sum_{k=1}^{\infty}\frac{\Psi_k(x)}{\lambda^k}, \quad \lambda\to\infty, \tag{3.3.122}$$

which is differentiable with respect to λ and x. The coefficient functions $\Psi_k(x)$ are meromorphic in x and have the set X_s as the set of their poles. The asymptotics (3.3.122) is uniform, i.e., for any positive number N and for any compact subset $\mathbb{K}\in\mathbb{C}\setminus X_s$, there exists a constant $C>0$ such that

$$\left\| Y(\lambda, x) - I - \sum_{k=1}^{N} \frac{\Psi_k(x)}{\lambda^k} \right\| < \frac{C}{\lambda^{N+1}}, \tag{3.3.123}$$

for all $|\lambda| > 1$ *and* $x \in \mathbb{K}$.

(f) $\det Y(\lambda, x) \equiv 1$.

REMARK 3.4. The x-meromorphicity is a new analytic feature of the function $Y(\lambda, x)$ in the general non-Abelian case; we recall that in the Abelian case (3.3.108), the function $Y(\lambda, x)$ is entire with respect to x.

Theorem 3.4 provides us with the principal tool of solving the Riemann-Hilbert-Birkhoff inverse monodromy problem for the generic 2×2 system (2.1.86). Indeed, suppose that $(\mathrm{S}_1, \mathrm{S}_2, \dots, \mathrm{S}_6; \nu)$, and $x \in X_s$ are the given monodromy data and let $Y(\lambda)$ be the corresponding solution of the RH problem (3.3.98)–(3.3.99) (we have temporarily suppressed the indication of dependence on the variable x). Inverting the arguments that led us to the formulation of the RH problem, we now *define* the functions $\Psi_n(\lambda)$, $n = 1, 2, \dots, 7$ by the equations (cf. (3.3.93))

$$\Psi_n(\lambda) = Y(\lambda) e^{-\theta(\lambda)\sigma_3}, \quad \lambda \in \Omega_n^0, \quad n = 1, \dots, 6, \tag{3.3.124}$$

$$\Psi_7(\lambda) = \Psi_1(\lambda) e^{-2\pi i \nu \sigma_3}, \quad \lambda \in \Omega_7^0 = \Omega_1^0$$

where, as always,

$$\theta(\lambda) = \frac{4i}{3}\lambda^3 + ix\lambda + \nu \ln \lambda,$$

and the branch of the logarithm is fixed by the condition,

$$-\frac{\pi}{6} < \arg \lambda < \frac{11\pi}{6}.$$

Because of the statement (d), each $\Psi_n(\lambda)$ is in fact an entire function of λ. Moreover, in view of (3.3.122) and (3.3.121) their behavior in the neighborhood of $\lambda = \infty$ is completely characterized by the asymptotic condition,

$$\Psi_n(\lambda) \sim \left(I + \sum_{n=1}^{\infty} \frac{\Psi_k}{\lambda^k} \right) e^{\left(-\frac{4}{3} i \lambda^3 \sigma_3 - ix\lambda\sigma_3 - \nu \ln \lambda \sigma_3\right)}, \tag{3.3.125}$$

$$\lambda \to \infty, \quad \frac{\pi}{6}(2n-3) \le \arg \lambda \le \frac{\pi}{6}(2n-1),$$

and by the set of jump relations,

$$\Psi_{n+1}(\lambda) = \Psi_n(\lambda) \mathrm{S}_n, \quad n = 1, 2, \dots, 6. \tag{3.3.126}$$

We claim that these functions form a collection of the canonical solutions for a certain linear system of type (2.1.86). The proof of this claim is rather simple. Nevertheless, it involves arguments which are central for the theory of special functions developed in the book. Here is the proof.

Put

$$\mathrm{A}(\lambda) := \frac{d\Psi_1(\lambda)}{d\lambda} \Psi_1^{-1}(\lambda). \tag{3.3.127}$$

The matrix $\mathsf{A}(\lambda)$ is an entire function of λ. Moreover, since the matrices S_n do not depend on λ, we have that for every n, as $\lambda \to \infty$ and $\lambda \in \overline{\Omega_n^0}$, the following asymptotic relation takes place

$$\mathsf{A}(\lambda) = \frac{d[\Psi_1(\lambda)\mathsf{S}_1 \dots \mathsf{S}_{n-1}]}{d\lambda} [\Psi_1(\lambda)\mathsf{S}_1 \dots \mathsf{S}_{n-1}]^{-1}$$

$$\sim \left(I + \sum_{k=1}^{\infty} \frac{\Psi_k}{\lambda^k}\right) \left(-4i\lambda^2\sigma_3 - ix\sigma_3 - \frac{\nu}{\lambda}\sigma_3\right) \left(I + \sum_{k=1}^{\infty} \frac{\Psi_k}{\lambda^k}\right)^{-1}$$

$$- \sum_{k=1}^{\infty} \frac{k\Psi_k}{\lambda^{k+1}} \left(I + \sum_{k=1}^{\infty} \frac{\Psi_k}{\lambda^k}\right)^{-1}. \tag{3.3.128}$$

This means that in the full neighborhood of $\lambda = \infty$ the following estimate holds

$$\mathsf{A}(\lambda) = A_2\lambda^2 + A_1\lambda + A_0 + O\left(\frac{1}{\lambda}\right), \quad \lambda \to \infty,$$

where

$$A_2 = -4i\sigma_3, \quad A_1 = 4i\,[\sigma_3, \Psi_1], \tag{3.3.129}$$

$$A_0 = 4i\,[\sigma_3, \Psi_2] - ix\sigma_3 - 4i\,[\sigma_3, \Psi_1]\,\Psi_1, \tag{3.3.130}$$

and, as usual, the brackets denote matrix commutator. In virtue of the Liouville theorem we conclude that in fact the exact equation,

$$\mathsf{A}(\lambda) = A_2\lambda^2 + A_1\lambda + A_0, \tag{3.3.131}$$

takes place. Note that the matrix A_1 is diagonal free, and trace $A_0 = 0$. Therefore, the functions $\Psi_n(\lambda)$ are the fundamental solutions of a linear system of the type (2.1.86). Moreover, in virtue of (3.3.125) and (3.3.126), the corresponding monodromy data coincide with the given set s.

REMARK 3.5. It is worth noticing that the tracelessness of the matrix $\mathsf{A}(\lambda)$ follows from the unimodularity of the matrices $\Psi_n(\lambda)$.

REMARK 3.6. The points of the singular set X_s are exactly the points where the solution of the Riemann-Hilbert-Birkhoff problem for system (2.1.86) ceases to exist. We will discuss this issue in more detail later on.

The parameters u, v, a, w, z of the coefficient matrix (3.3.131) (see (2.1.86)) are defined by the equations,

$$u = 2(\Psi_1)_{12}, \quad v = 2(\Psi_1)_{21}, \quad a = -ix - 2iuv, \tag{3.3.132}$$

and

$$w = 2iu(\Psi_1)_{12} - 4i(\Psi_2)_{12}, \quad z = -2iv(\Psi_1)_{21} - 4i(\Psi_2)_{21}, \tag{3.3.133}$$

where Ψ_1 and Ψ_2 are the first two matrix coefficients of the asymptotic series (3.3.122). Reviving the x-dependence of the solution of the RH problem (3.3.98)–(3.3.99) we conclude that u, v, a, w, z are meromorphic functions of x with the set X_s as the set of their poles.

The functions $u \equiv u(x; s)$, $v \equiv v(x; s)$, $w \equiv w(x; s)$, and $z \equiv z(x; s)$ defined by equations (3.3.132) and (3.3.133), as well as the functions $u \equiv u(x; \mathfrak{m})$, $v \equiv v(x; \mathfrak{m})$, $w \equiv w(x; \mathfrak{m})$, and $z_\nu \equiv z_\nu(x; \mathfrak{m})$ defined by equations (3.3.89) and (2.1.52), have a profound meaning: *they describe the isomonodromy deformations of the respective linear systems*: system (2.1.86) in the former case and system (2.1.16) in the latter. Indeed, given the set s, equations (3.3.132) and (3.3.133) describe a one-parameter family of systems (2.1.86) having the same set — the set s in fact, as their Stokes matrices. Similarly, equations (3.3.89) and (2.1.52) describe a one-parameter family of 2×2 Fuchsian systems with four singular points and given monodromy group $\mathfrak{M}$.

In the next chapter we will develop some of the basic elements of the general theory of isomonodromy deformations of the systems of linear ODEs with rational coefficients. The main objective of the next chapter will be *to introduce the Painlevé equations* as the differential equations describing the first nontrivial isomonodromy deformations of linear systems of the first nontrivial, i.e., 2×2 matrix size. The formulae (3.3.132)–(3.3.133) and (3.3.89)–(2.1.52) can be then interpreted as the representations of the general solutions of the Painlevé II and Painlevé VI equations, respectively.

The theory developed in the next chapter will become in Chapter 5 the *isomonodromy method* of solving the Painlevé equations. As it was indicated in the introduction, this method plays the role of the nonlinear analog of the Fourier-Laplace method. Similarly, the *Riemann-Hilbert representations* (3.3.132)–(3.3.133) and (3.3.89)–(2.1.52) are the nonlinear analogs of contour integral representations. Then in the remaining chapters of the book we will demonstrate that the Riemann-Hilbert representation is as efficient for the study of Painlevé equations, as the contour integrals are in the linear theory.

CHAPTER 4

Isomonodromy Deformations. The Painlevé Equations

0. General setting

The theory of isomonodromy deformations studies families of the linear ODEs with rational coefficients,

$$\frac{d\Psi}{d\lambda} = \mathsf{A}(\lambda)\Psi, \tag{4.0.1}$$

which share the same set of essential monodromy data (2.0.21). As it is stated in the Introduction, the isomonodromy deformations play a central role in our treatment of special functions and especially in our treatment of the "nonlinear special functions", i.e., of the Painlevé transcendents.

To introduce the notion of isomonodromy deformation, we assume that the coefficient matrix $\mathsf{A}(\lambda)$ depends holomorphically on a certain additional parameter t (scalar or vector):

$$\mathsf{A}(\lambda) = \mathsf{A}(\lambda, t) \quad t = (t_1, t_2, \dots, t_q), \quad q \geq 1, \quad t \in \mathcal{V}. \tag{4.0.2}$$

Here $\mathcal{V}$ denotes an open set in $\mathbb{C}^q$.

DEFINITION 4.1. We call the holomorphic family (4.0.2) an admissible deformation of equation (4.0.1) if the following conditions are satisfied.

(1) The number m of the singular points of (4.0.1) does not depend on t, i.e.,

$$a_\nu(t) \neq a_\mu(t), \quad \forall t \in \mathcal{V}, \quad \nu \neq \mu.$$

(the Poincaré rank of each singular point is, of course, t-invariant) Moreover, we assume that there exists a collection of disks $\{D_\nu\}_{\nu=1}^m$ on the complex plane λ such that
   - $a_\nu(t) \in D_\nu, \quad \forall t \in \mathcal{V}$;
   - $\overline{D}_\nu \cap \overline{D}_\mu = \emptyset, \quad \nu \neq \mu$.

(2) The type of Jordan form of the leading coefficient of the Laurent series of the matrices $\mathsf{A}(\lambda, t)$ at each singular point $a_\nu(t)$ does not depend on $t \in \mathcal{V}$. In particular, if $a_\nu(t)$ is an irregular singular point, then the leading coefficient $A^\nu_{-r_\nu}(t)$ has distinct eigenvalues for all $t \in \mathcal{V}$.

(3) At each irregular singular point $a_\nu(t)$, $r_\nu > 0$, the set of canonical sectors, $\{\Omega^\nu_n\}_{n=1}^{2r_\nu+1}$, can be chosen in such a way that its image under the map, $\lambda \mapsto \lambda - a_\nu(t)$, is the same for all $t \in \mathcal{V}$.

(4) Canonical solutions can be chosen in such a way that they are holomorphic with respect to t, and for the canonical solutions near an irregular point the asymptotic conditions (2.0.12) hold uniformly with respect to $t \in \mathcal{V}$.

REMARK 4.1. We note that if equation (4.0.1) is generic, then *any* holomorphic deformation is admissible for a sufficiently small domain $\mathcal{V}$. Indeed, the only nontrivial condition, i.e., condition (4), is guaranteed by the classical theorems concerning the analytic dependence of the solutions of linear equations on the coefficients of the equations and initial data (cf. Remark 2.2 of Chapter 2). Also condition (4) implies that all the coefficients of the series (2.0.11) and (2.0.12) are holomorphic functions of t and these series can be differentiated term by term.

It immediately follows from the above definition that if (4.0.2) is an admissible deformation of (4.0.1), then one can define the set of monodromy data in such a (not unique) way that all the Stokes matrices and connection matrices are holomorphic functions of $t \in \mathcal{V}$.

DEFINITION 4.2. We call an admissible deformation (4.0.2) of equation (4.0.1) an isomonodromy deformation if, in addition to the properties indicated in Definition 4.1, a set of canonical solutions can be chosen in such a way that the formal monodromy exponents Λ_0^ν at all singular points (or $\{\Lambda_0^\nu, \mathsf{N}^\nu\}$ in the Fuchsian resonance case), the Stokes matrices $\mathsf{S}_n^{(\nu)}$ at all irregular singular points, and all the connection matrices E_ν, are independent on $t \in \mathcal{V}$.

Note that the general formulae (2.0.33) and (2.0.39) which relate the monodromy matrices M_ν with the monodromy data imply that the matrices M_ν (and hence the whole monodromy group $\mathfrak{M}$) are invariant under the isomonodromy deformations.

The fundamental fact about isomonodromy deformations is that they can be described by certain systems of *nonlinear* differential equations. In order to describe these equations we will make the following assumptions.

(i) At each Fuchsian singular point, the coefficient matrix $\mathsf{A}(\lambda)$ is nonresonant (this is the generic situation).

(ii) One of the singular points, say $a_m(t)$, coincides with infinity for all $t \in \mathcal{V}$, i.e.,

$$a_m(t) \equiv \infty.$$

(iii) The leading coefficient of the Laurent series of $\mathsf{A}(\lambda)$ at $\lambda = \infty$ is diagonal for all $t \in \mathcal{V}$, i.e., (see the notations introduced in equation (2.0.6))

$$P_m(t) \equiv I.$$

(iv) The basic fundamental solution $\Psi(\lambda, t)$, i.e., the solution with respect to which the t-independent monodromy data are defined, coincides with the canonical solution $\Psi^{(m)}(\lambda, t)$ in the case that $a_m \equiv \infty$ is a Fuchsian singular point, and with $\Psi_1^{(m)}(\lambda, t)$, in the case that $a_m \equiv \infty$ is an irregular singular point. We will use the same notation for both of these cases, namely

$$\Psi(\lambda, t) \equiv \Psi^{(\infty)}(\lambda, t).$$

(v) The base point a_0, which is used for defining the fundamental group of $X \equiv \mathbb{C}\mathrm{P}^1 \setminus \{a_\nu\}_{\nu=1}^m$, the domain $\hat{X}$ and the universal covering $\widetilde{X}$, belongs to the neighborhood of $a_m = \infty$; furthermore, if infinity is an irregular singular point, it lies in the first canonical sector. In other words, a_0 lies in the domain of normalization of the basic solution $\Psi(\lambda)$.

REMARK 4.2. Using appropriate t-dependent conformal mappings and gauge transformations, it is always possible to satisfy the assumptions (ii)-(v). The assumption (i) is made in order to simplify some of the calculations below; it is possible to derive monodromy preserving nonlinear equations *without* imposing this assumption; see [**Bol3**].

Consider the logarithmic derivative,

$$\Theta(\lambda, t) = d\Psi \cdot \Psi^{-1}, \tag{4.0.3}$$

where $d\Psi$ is defined by

$$d\Psi = \sum_j \frac{\partial \Psi}{\partial t_j} dt_j.$$

The differential form $\Theta(\lambda, t)$ as a function of λ is well defined on the universal covering $\widetilde{X}$, $X = \mathbb{C}\mathrm{P}^1 \setminus \{a_\nu\}_{\nu=1}^m$. Also, *since all the monodromy matrices M_ν do not depend on t, the differential form $\Theta(\lambda, t)$ is a single-valued analytic function on the base space, i.e., on the punctured Riemann sphere $X = \mathbb{C}\mathrm{P}^1 \setminus \{a_\nu\}_{\nu=1}^m$.* Indeed, since $d\mathsf{M}_\nu = 0$, it follows that for every basic loop γ_ν the action of the monodromy operator τ_{γ_ν} on $\Theta(\lambda, t)$ is trivial,

$$\begin{aligned}
\tau_{\gamma_\nu}[\Theta(\lambda)] &\equiv \Theta_{\gamma_\nu}(\lambda) = d\Psi_{\gamma_\nu}(\lambda) \cdot \Psi_{\gamma_\nu}^{-1}(\lambda) \\
= d\Big(\Psi(\lambda)\mathsf{M}_\nu\Big) \cdot \Big(\Psi(\lambda)\mathsf{M}_\nu\Big)^{-1} &= \Big(d\Psi(\lambda)\Big)\mathsf{M}_\nu \mathsf{M}_\nu^{-1}\Psi^{-1}(\lambda) \\
&= d\Psi(\lambda) \cdot \Psi^{-1}(\lambda) = \Theta(\lambda).
\end{aligned}$$

In order to analyze the behavior of $\Theta(\lambda, t)$ near its isolated singularities, i.e., near the points $\lambda = a_\nu$, $\nu = 1, \ldots, m$, we will make use of the fact that not only the monodromy matrices, but also all the formal monodromy exponents $\Lambda_0^{(\nu)}$, the Stokes matrices $\mathsf{S}_n^{(\nu)}$, and the connection matrices E_ν, are t-independent.

Consider a neighborhood D_{a_ν} of one of the finite singular points, a_ν, $\nu < m$. If a_ν is a Fuchsian point, then the function $\Psi(\lambda)$ can be represented in the form (cf. (2.0.11)),

$$\Psi(\lambda) = P_\nu \hat{\Psi}^{(\nu)}(\lambda) \xi_\nu^{\Lambda_0^{(\nu)}} \mathsf{E}_\nu, \quad \xi_\nu = \lambda - a_\nu, \tag{4.0.4}$$

where

$$\hat{\Psi}^{(\nu)}(\lambda) = \left(I + \sum_{j=1}^{\infty} \Psi_j^{(\nu)} \xi_\nu^j\right), \quad \lambda \in D_{a_\nu}, \tag{4.0.5}$$

and the series in the right side converges uniformly with respect to $t \in \mathcal{V}$ (for convenience of notation we have suppressed the t-dependence). If a_ν is a non-Fuchsian singularity characterized by the Stokes phenomenon,

$$\mathbf{SPh}_\nu = \{\Lambda_{-r_\nu}^{(\nu)}, \Lambda_{-r_\nu+1}^{(\nu)}, \ldots, \Lambda_{-1}^{(\nu)}, \Lambda_0^{(\nu)}; \mathsf{S}_1^{(\nu)}, \mathsf{S}_2^{(\nu)}, \ldots, \mathsf{S}_{2r_\nu}^{(\nu)}\},$$

then the function $\Psi(\lambda)$ has the following asymptotic representation in D_{a_ν} (cf. (2.0.12)),

$$\begin{gathered}
\Psi(\lambda) \sim P_\nu \hat{\Psi}^{(\nu)}(\lambda) e^{\Lambda^{(\nu)}(\xi_\nu)} \mathsf{E}_\nu, \\
\lambda \in \Omega_1^{(\nu)}, \quad \lambda \to a_\nu,
\end{gathered}$$

$$\Psi(\lambda) \sim P_\nu \hat{\Psi}^{(\nu)}(\lambda) e^{\Lambda^{(\nu)}(\xi_\nu)} \mathsf{S}_{n-1}^{(\nu)-1} \mathsf{S}_{n-2}^{(\nu)-1} \cdots \mathsf{S}_1^{(\nu)-1} \mathsf{E}_\nu, \tag{4.0.6}$$

$$\lambda \in \Omega_n^{(\nu)}, \quad \lambda \to a_\nu, \quad n = 2, \ldots, 2r_\nu + 1,$$

where $\Lambda^{(\nu)}(\xi)$, $\{\Omega_n^{(\nu)}\}$, and $\hat{\Psi}^{(\nu)}(\lambda)$ are defined as follows:

$$\Lambda^{(\nu)}(\xi) = \sum_{k=-r_\nu}^{-1} \frac{\Lambda_k^{(\nu)}}{k} \xi^k + \Lambda_0^{(\nu)} \ln \xi,$$

$\{\Omega_n^{(\nu)}\}$ is the set of the canonical sectors associated with the irregular point a_ν, and $\hat{\Psi}^\nu(\lambda)$ denote the formal power series,

$$\hat{\Psi}^{(\nu)}(\lambda) = \left(I + \sum_{j=1}^{\infty} \Psi_j^{(\nu)} \xi_\nu^j \right). \tag{4.0.7}$$

We note that: (a) The series $\hat{\Psi}^{(\nu)}(\lambda)$ is the same in each sector $\Omega_n^{(\nu)}$. (b) Both the convergent series representation (4.0.4)–(4.0.5) and the asymptotic series representation (4.0.6)–(4.0.7), can be differentiated with respect to t term by term.

Equations (4.0.4)–(4.0.7), together with the t-independence of the matrices $\Lambda_0^{(\nu)}$, $\mathsf{S}_n^{(\nu)}$, and E_ν, imply that even for an irregular point a_ν, the differential form $\Theta(\lambda)$ exhibits the following power-like asymptotic behavior *in the full neighborhood* D_ν,

$$\Theta(\lambda) = \sum_{k=1}^{r_\nu+1} (\lambda - a_\nu)^{-k} \Theta_k^{(\nu)} + O(1), \tag{4.0.8}$$

$$\lambda \in D_\nu, \quad \lambda \to a_\nu.$$

Indeed, suppose that a_ν is an irregular point. Then the matrices $\Lambda_0^{(\nu)}$, $\mathsf{S}_n^{(\nu)}$, E_ν, are t-independent, and hence in each sector $\Omega_n^{(\nu)}$, $n = 1, \ldots, 2r_\nu$,

$$\Theta(\lambda) \sim d\Big(P_\nu \hat{\Psi}^{(\nu)}(\lambda) e^{\Lambda^{(\nu)}(\xi_\nu)}\Big) \cdot \Big(P_\nu \hat{\Psi}^{(\nu)}(\lambda) e^{\Lambda^{(\nu)}(\xi_\nu)}\Big)^{-1}$$
$$= P_\nu \hat{\Psi}^{(\nu)}(\lambda) \cdot d\Lambda^{(\nu)}(\lambda) \cdot \Big[P_\nu \hat{\Psi}^{(\nu)}(\lambda)\Big]^{-1} + O(1),$$
$$\lambda \to a_\nu,$$

which implies the power-like behavior indicated in (4.0.8). Moreover, we see that the coefficient differential forms $\Theta_k^{(\nu)} \equiv \Theta_k^{(\nu)}(t)$ are uniquely determined by

$$\sum_{k=1}^{r_\nu+1} (\lambda - a_\nu)^{-k} \Theta_k^{(\nu)}$$

$$= P_\nu \hat{\Psi}^{(\nu)}(\lambda) \cdot d\Lambda^{(\nu)}(\lambda) \cdot \Big[P_\nu \hat{\Psi}^{(\nu)}(\lambda)\Big]^{-1} \quad \big(\bmod\ (\lambda - a_\nu)^0\big), \tag{4.0.9}$$

where

$$d\Lambda^{(\nu)}(\lambda) = -\sum_{k=0}^{r_\nu} (\lambda - a_\nu)^{-k-1} \Lambda_{-k}^{(\nu)}\, da_\nu(t)$$
$$-\sum_{k=1}^{r_\nu} \frac{(\lambda - a_\nu)^{-k}}{k}\, d\Lambda_{-k}^{(\nu)}(t).$$

In the case that a_ν is a Fuchsian point, one must put $r_\nu = 0$ and drop the second sum in the above formula for $d\Lambda^{(\nu)}(\lambda)$.

Let us now consider the neighborhood D_∞ of the singular point $a_m = \infty$. If a_m is a Fuchsian singular point, we find ($P_\infty = I$!)

$$\Psi(\lambda) = \hat{\Psi}^{(\infty)}(\lambda)\lambda^{-\Lambda_0^{(\infty)}},$$

where

$$\hat{\Psi}^{(\infty)}(\lambda) = \left(I + \sum_{j=1}^{\infty} \Psi_j^{(\infty)}\lambda^{-j}\right).$$

This implies the asymptotic relation,

$$\Theta(\lambda) = O\left(\frac{1}{\lambda}\right), \quad \lambda \in D_\infty, \quad \lambda \to \infty.$$

If a_m is an irregular singular point characterized by the Stokes phenomenon,

$$\mathbf{SPh}_\infty = \{\Lambda_{-r_\infty}^{(\infty)}, \Lambda_{-r_\infty+1}^{(\infty)}, \dots, \Lambda_{-1}^{(\infty)}, \Lambda_0^{(\infty)}; S_1^{(\infty)}, S_2^{(\infty)}, \dots, S_{2r_\infty}^{(\infty)}\},$$

then (cf. (4.0.6) and (4.0.7))

$$\Psi(\lambda) \sim \hat{\Psi}^{(\infty)}(\lambda)e^{\Lambda^{(\infty)}(\lambda)},$$
$$\lambda \in \Omega_1^{(\infty)}, \quad \lambda \to \infty,$$

$$\Psi(\lambda) \sim \hat{\Psi}^{(\infty)}(\lambda)e^{\Lambda^{(\infty)}(\lambda)}S_{n-1}^{(\infty)-1}S_{n-2}^{(\infty)-1}\cdots S_1^{(\infty)-1}, \tag{4.0.10}$$
$$\lambda \in \Omega_n^{(\infty)}, \quad \lambda \to \infty, \quad n = 2, \dots, 2r_\infty + 1,$$

where $\Lambda^{(\infty)}(\lambda)$, $\{\Omega_n^{(\infty)}$, and $\hat{\Psi}^{(\infty)}(\lambda)$ are defined as follows:

$$\Lambda^{(\infty)}(\xi) = -\sum_{k=1}^{r_\infty} \frac{\Lambda_{-k}^{(\infty)}}{k}\lambda^k - \Lambda_0^{(\infty)}\ln\lambda,$$

$\{\Omega_n^{(\infty)}\}$ is the set of the canonical sectors associated with the irregular point ∞, and $\hat{\Psi}^\infty(\lambda)$ denotes the formal power series,

$$\hat{\Psi}^{(\infty)}(\lambda) = \left(I + \sum_{j=1}^{\infty} \Psi_j^{(\infty)}\lambda^{-j}\right). \tag{4.0.11}$$

In this case, the same arguments as in the case a_ν, $\nu < m$, yield the asymptotic formula (cf. (4.0.8) and (4.0.9)),

$$\Theta(\lambda) = \sum_{k=0}^{r_\infty} \lambda^k \Theta_k^{(\infty)} + O\left(\frac{1}{\lambda}\right),$$
$$\lambda \in D_\infty, \quad \lambda \to \infty;$$

furthermore, the coefficient differential forms $\Theta_k^{(\infty)} \equiv \Theta_k^{(\infty)}(t)$, are uniquely determined by

$$\sum_{k=0}^{r_\infty} \lambda^k \Theta_k^{(\infty)}$$
$$= \hat{\Psi}^{(\infty)}(\lambda) \cdot d\Lambda^{(\infty)}(\lambda) \cdot \left[\hat{\Psi}^{(\infty)}(\lambda)\right]^{-1} \quad (\bmod \lambda^{-1}), \tag{4.0.12}$$

where

$$d\Lambda^{(\infty)}(\lambda) = -\sum_{k=1}^{r_\infty} \frac{\lambda^k}{k} d\Lambda^{(\infty)}_{-k}(t).$$

The above analysis is summarized in the following statement.

PROPOSITION 4.1. *The differential form $\Theta(\lambda, t)$ defined in (4.0.3), is a rational matrix-valued function of the variable λ. Its poles are located at the singular points a_ν, $\nu = 1, \dots, m-1, \infty$, and the corresponding decomposition over associated principal parts is given by*

$$\Theta(\lambda) = \Theta^{(\infty)}(\lambda) + \sum_{\nu=1}^{m-1} \Theta^{(\nu)}(\lambda), \tag{4.0.13}$$

where

$$\Theta^{(\nu)}(\lambda) \equiv \sum_{k=1}^{r_\nu+1} (\lambda - a_\nu)^{-k} \Theta_k^{(\nu)}, \quad \nu = 1, \dots, m-1, \tag{4.0.14}$$

are defined by equation (4.0.9), and for convenience of notation we have suppressed the t-dependence. If $\lambda = \infty$ is a Fuchsian point, then the term $\Theta^{(\infty)}(\lambda)$ is identically zero. If $\lambda = \infty$ is an irregular singular point, then

$$\Theta^{(\infty)}(\lambda) \equiv \sum_{k=0}^{r_\infty} \lambda^k \Theta_k^{(\infty)}, \tag{4.0.15}$$

where the right-hand side is determined by equation (4.0.12).

It follows from the recursive equations (1.1.19) and (1.1.43) of Chapter 1, that all the products $P_\nu \Psi_k^{(\nu)} P_\nu^{-1}$, where $\Psi_k^{(\nu)}$ are the coefficients in the series $\hat{\Psi}^{(\nu)}(\lambda)$, are defined uniquely and explicitly in terms of the coefficients of equation (4.0.1). Therefore, $\Theta(\lambda)$ *is determined uniquely and explicitly by equations (4.0.9) and (4.0.12) as a differential matrix-valued form on the manifold $\mathcal{A}$ of the linear systems having the given number m of singular points and the given Poincaré rank r_ν at each point.* In other words,

$$\Theta(\lambda) \equiv \Theta(\lambda; \{A_k^{(\nu)}\}, \{a_\nu\}), \tag{4.0.16}$$

where the matrices,

$$A_k^{(\nu)}, \quad \nu = 1, \dots, m-1, \infty, \quad k = 0, -1, \dots, -r_\nu,$$

$$A_k^{(\infty)}, \quad k = -1, \dots, -r_\infty, \quad r_\infty > 0,$$

are defined as the coefficients of the decomposition of the rational function $\mathsf{A}(\lambda)$ over its principal parts,

$$\mathsf{A}(\lambda) = \mathsf{A}^{(\infty)}(\lambda) + \sum_{\nu=1}^{m-1} \mathsf{A}^{(\nu)}(\lambda), \tag{4.0.17}$$

$$\mathsf{A}^{(\nu)}(\lambda) = \sum_{k=1}^{r_\nu+1} (\lambda - a_\nu)^{-k} A_{-k+1}^{(\nu)}, \quad \nu = 1, \dots, m-1,$$

$$\mathsf{A}^{(\infty)}(\lambda) = -\sum_{k=0}^{r_\infty-1} \lambda^k A_{-k-1}^{(\infty)}, \quad r_\infty > 0,$$

$$\mathsf{A}^{(\infty)}(\lambda) \equiv 0, \quad r_\infty = 0.$$

Equations (4.0.1) and (4.0.3) imply that $\Psi(\lambda, t)$, in addition to the basic λ-equation

$$\frac{\partial \Psi}{\partial \lambda} = \mathsf{A}(\lambda)\Psi, \tag{4.0.18}$$

it also satisfies an auxiliary linear system with respect to the parameter t:

$$d\Psi = \Theta(\lambda)\Psi. \tag{4.0.19}$$

Using the representation

$$\Theta(\lambda) = \sum_j \Theta_j(\lambda)dt_j,$$

equation (4.0.19) takes the form

$$\frac{\partial \Psi}{\partial t_j} = \Theta_j(\lambda)\Psi.$$

Cross-differentiating the *overdetermined* system (4.0.18) and (4.0.19), we find the following *compatibility condition*:

$$d\mathsf{A} = \frac{\partial \Theta}{\partial \lambda} + [\Theta, \mathsf{A}] \quad \text{(identically in } \lambda), \tag{4.0.20}$$

or

$$\frac{\partial \mathsf{A}}{\partial t_j} = \frac{\partial \Theta_j}{\partial \lambda} + [\Theta_j, \mathsf{A}] \quad \text{(identically in } \lambda). \tag{4.0.21}$$

Both sides of equation (4.0.20) (or (4.0.21)) are rational functions in λ. Equating the corresponding principal parts, we obtain a finite system of (nonlinear) differential equations for the matrix coefficients $A_k^{(\nu)} \equiv A_k^{(\nu)}(t)$. We will call this system (as well as its compact form (4.0.20)) the *isomonodromy deformation equations.*

THEOREM 4.1. **[JMU]** *Suppose that an admissible deformation,*

$$\mathsf{A}(\lambda) = \mathsf{A}(\lambda, t),$$

*of the linear system (*4.0.1*), satisfies the conditions (i)–(iii) above. Assume also that the relevant monodromy data are defined via the fundamental solution $\Psi(\lambda)$ discussed in condition (iv). Then, this deformation is an isomonodromy deformation only if the deformation equations(*4.0.20*) hold. Conversely, suppose that*

$$\mathsf{A}(\lambda) = \mathsf{A}(\lambda, t),$$

*is a rational matrix function of the form (*4.0.17*) which satisfies equations (*4.0.20*). Then it defines an isomonodromy deformation of (*4.0.17*) (if again the monodromy data are introduced via the fundamental solution $\Psi(\lambda)$ discussed in condition (iv) above).*

PROOF. We have already proven the necessity of equation (4.0.20) for the t-independence of monodromy data. Therefore, we only need to prove that it is sufficient to have equation (4.0.20) in order for the deformation to be isomonodromic.

Suppose that the rational matrix function $\mathsf{A}(\lambda) = \mathsf{A}(\lambda, t)$ of the type (4.0.17) satisfies equation (4.0.20). As it has already been indicated in Remark 4.1, since the function $\mathsf{A}(\lambda, t)$ is holomorphic, it determines locally an admissible deformation.

Let $\Psi_n^{(\nu)}(\lambda) \equiv \Psi_n^{(\nu)}(\lambda, t)$ be a canonical solution of the corresponding linear equation (4.0.1). Introduce the matrix-valued differential form,

$$W(\lambda) := d\Psi_n^{(\nu)} - \Theta(\lambda)\Psi_n^{\nu},$$

and consider its λ-derivative. Taking into account equation (4.0.20), we have

$$\frac{\partial W}{\partial \lambda} = d\left(\frac{\partial \Psi_n^{(\nu)}}{\partial \lambda}\right) - \frac{\partial \Theta}{\partial \lambda}\Psi_n^{(\nu)} - \Theta\frac{\partial \Psi_n^{(\nu)}}{\partial \lambda} = d\big(A\Psi_n^{(\nu)}\big) - \frac{\partial \Theta}{\partial \lambda}\Psi_n^{(\nu)} - \Theta A\Psi_n^{(\nu)}$$

$$= Ad\Psi_n^{(\nu)} + \big(dA - \frac{\partial \Theta}{\partial \lambda} - \Theta A\big)\Psi_n^{(\nu)} = Ad\Psi_n^{(\nu)} - A\Theta\Psi_n^{(\nu)} = A(\lambda)W.$$

In other words, the function $W(\lambda)$ satisfies the same linear equation as the function $\Psi_n^{(\nu)}(\lambda)$. Therefore,

$$W(\lambda) = \Psi_n^{(\nu)}(\lambda)C, \tag{4.0.22}$$

for some λ-independent matrix function $C \equiv C(t)$ (a matrix-valued differential form, to be exact). Our claim is that the gauge factor P_ν in the definition of the canonical solution $\Psi_n^{(\nu)}(\lambda)$ (see (2.0.11) and (2.0.12)), can always be fixed in such a way that $C(t) \equiv 0$. To see this, let us recall that the rational differential form $\Theta(\lambda, t) \equiv \Theta\big(\lambda; \{A_k^{(\nu)}(t)\}, \{a_\nu(t)\}\big)$ is defined according to the formulae (4.0.13)–(4.0.15), (4.0.9), and (4.0.12). This, in particular, implies (in the case $a_\nu \neq a_m = \infty$) the following asymptotic formula,

$$C = d\Lambda_0^{(\nu)}\ln(\lambda - a_\nu) + e^{-\Lambda^{(\nu)}(\lambda)}\Big(P_\nu^{-1}dP_\nu - \theta_\nu\Big)e^{\Lambda^{(\nu)}(\lambda)} \tag{4.0.23}$$

$$+e^{-\Lambda^{(\nu)}(\lambda)}\Big(0\big((\lambda - a_\nu)\big)\Big)e^{\Lambda^{(\nu)}(\lambda)}, \quad \lambda \in \Omega_n^{(\nu)}, \quad \lambda \to a_\nu,$$

where θ_ν denote the constant (with respect to λ) differential form defined by the equation

$$\theta_\nu = \Psi_1^\nu da_\nu + P_\nu^{-1}\Big(\sum_{\mu\neq\nu}\Theta^{(\mu)}(a_\nu) + \Theta^{(\infty)}(a_\nu)\Big)P_\nu. \tag{4.0.24}$$

(If $\lambda = a_\nu$ is a Fuchsian point, then the restriction $\lambda \in \Omega_n^{(\nu)}$ can be dropped). Taking into account that the differential form C does not depend on λ and choosing, when analyzing the ij-entry of the matrix C in the case $r_\nu > 0$ and $i \neq j$, the path towards a_ν where $\mathrm{Re}\big(\Lambda_{ii}^{(\nu)}(\lambda) - \Lambda_{jj}^{(\nu)}(\lambda)\big) > 0$, we conclude that all off-diagonal entries of C are zero. Note that this is exactly the same arguments used in Chapter 1 to prove the uniqueness of the canonical solutions at irregular singular points. If $\lambda = a_\nu$ is a Fuchsian point ($r_\infty = 0$), then the conclusion $C_{ij} \equiv 0$, $i \neq j$ follows immediately from the technical assumption made about the nonresonant condition satisfied at each Fuchsian point. Also, from the diagonal part of equation (4.0.23) we obtain

$$d\Lambda_0^{(\nu)} = 0,$$

and

$$(C)_{\mathrm{diag}} = \Big(P_\nu^{-1}dP_\nu - \theta_\nu\Big)_{\mathrm{diag}}.$$

By changing P_ν to $\tilde{P}_\nu := P_\nu D$, where D is a nonsingular diagonal matrix (cf. Remark 2.1 of Chapter 2) subject to the condition,

$$dDD^{-1} = -\Big(P_\nu^{-1}dP_\nu - \theta_\nu\Big)_{\mathrm{diag}}, \tag{4.0.25}$$

we obtain $C \equiv 0$.

It should be noted that if the deformation parameter t is one-dimensional ($q = 1$ in (4.0.2)), then equation (4.0.25) is easily solved (the r.h.s. of (4.0.25) is independent on D). In the case that t is multidimensional, the question of compatibility of (4.0.25) arises. A positive answer to this question (or, rather to an equivalent question) is obtained in [**JMU**]. In applications to Painlevé equations, the parameter t is always one-dimensional, therefore we will skip the proof (which is nontrivial!) of the compatibility of (4.0.25) in the general case. Also, due to the normalization condition at $\lambda = \infty$, the analog of the asymptotic relation (4.0.23) in the case of $a_\nu = a_m = \infty$, becomes

$$C = -d\Lambda_0^{(\infty)} \ln\lambda + e^{-\Lambda^{(\infty)}(\lambda)} O\left(\frac{\ln\lambda}{\lambda}\right) e^{\Lambda^{(\infty)}(\lambda)}, \tag{4.0.26}$$

$$\lambda \in \Omega_n^{(\infty)}, \quad \lambda \to \infty.$$

Hence, we have the equations,

$$d\Lambda_0^{(\infty)} = 0, \quad \text{and} \quad C = 0.$$

Having established that $C \equiv 0$ in equation (4.0.22), it follows that every canonical solution $\Psi_n^{(\nu)}(\lambda)$ satisfies the *same* dynamical equation,

$$d\Psi_n^{(\nu)}(\lambda) = \Theta(\lambda)\Psi_n^{(\nu)}(\lambda). \tag{4.0.27}$$

Therefore, in particular, we have

$$\begin{aligned} d\Psi_n^{(\nu)}\Psi_n^{(\nu)-1} &= \Theta = d\Psi_{n+1}^{(\nu)}\Psi_{n+1}^{(\nu)-1} \\ &= d\Psi_n^{(\nu)}\Psi_n^{(\nu)-1} + \Psi_n^{(\nu)} dS_n^{(\nu)} S_n^{(\nu)-1}\Psi_n^{(\nu)-1}, \end{aligned}$$

and the equation, $dS_n^{(\nu)} = 0$, follows. Similar arguments show that $dE_\nu = 0$. This, together with the already established fact that all $\Lambda_0^{(\nu)}$ are constant in t, completes the proof of the theorem. □

REMARK 4.3. Equations (4.0.27) yield

$$dP_\nu P_\nu^{-1} = \hat{\theta}_\nu, \tag{4.0.28}$$

where

$$\begin{aligned} \hat{\theta}_\nu &:= P_\nu \theta_\nu P_\nu^{-1} \\ &= P_\nu \Psi_1^\nu P_\nu^{-1} da_\nu + \sum_{\mu \neq \nu} \Theta^{(\mu)}(a_\nu) + \Theta^{(\infty)}(a_\nu) \\ &\equiv \hat{\theta}_\nu\big(\{A_k^{(\nu)}(t)\}, \{a_\nu(t)\}\big). \end{aligned}$$

In other words, we have proven the implication,

$$\left(P_\nu^{-1} dP_\nu - \theta_\nu\right)_{\text{diag}} = 0 \iff dP_\nu P_\nu^{-1} = \hat{\theta}_\nu.$$

The isomonodromy deformation equation (4.0.20) is itself an overdetermined (if $t = (t_1, \ldots, t_q)$, $q > 1$) system of *nonlinear* differential equations for the matrices $A_k^{(\nu)}$. It is shown in [JMU] that this system is integrable in the sense of Frobenius, and that the largest independent set of deformation parameters t can always be chosen as the set of singular points $a_1, \ldots, a_{m-1}$, together with matrix elements of diagonal matrices $\Lambda_k^{(\nu)}$, $k = 1, \ldots, r_\nu > 0$, $\nu = 1, \ldots, m-1, \infty$. Therefore, in what follows we will always assume

$$t \equiv \left\{a_1, \ldots, a_{m-1};\ \{\Lambda_k^{(\nu)}\}, \quad k > 0\right\}; \tag{4.0.29}$$

although, as already mentioned, in the theory of Painlevé equations the actual number of deformation parameters is always reducible to just one.

In Section 3 we will specify equation (4.0.20) for our basic 2×2 examples introduced in the previous chapter; as we will soon see, they are related to Painlevé equations.

REMARK 4.4. If we drop the nonresonance condition, then equation (4.0.20) does not describe all possible isomonodromy deformations. We refer the reader to the papers [**Bol3**] and [**KaV**] for details on this issue.

The major idea of modern theory of isomonodromy deformations, is the idea that the system

$$\begin{cases} \frac{\partial \Psi}{\partial \lambda} &= \mathsf{A}(\lambda)\Psi, \\ d\Psi &= \Theta(\lambda)\Psi, \end{cases} \tag{4.0.30}$$

can be considered as a *Lax representation* ($\Theta - \mathsf{A}$ or Lax-pair) for the nonlinear equation (4.0.20). This makes the theory of isomonodromy deformations part of the well-developed by now theory of integrable systems (Soliton Theory).

The use of the Lax pair (4.0.30) in the analysis of the solutions of the isomonodromy deformation equations (4.0.20) is based on the fact that, by the very nature of equation (4.0.20), the monodromy data of the λ-equation in (4.0.30), i.e., the collection (cf. (2.0.21)),

$$\mathfrak{m} = \{\Lambda_0^{(\nu)}; \mathsf{S}_n^{(\nu)}; \mathsf{E}_\nu\}_{\nu=1,\dots,m;n=1,\dots,2r_\nu}$$

form a *complete set of first integrals of the system (*4.0.20*)*. Hence the question of the integration of the nonlinear equation (4.0.20), is reduced to the analysis of the direct and inverse monodromy maps of the linear system (4.0.1). The principal goal of this book is to show that this point of view can be made, with the help of the Riemann-Hilbert factorization technique and the complex WKB analysis, an efficient tool for studying the isomonodromy deformation equations and, in particular, the Painlevé equations. In fact, we will provide some indication of the usefulness of this *Isomonodromy Riemann-Hilbert method* in Section 3 of this chapter.

We conclude the general part of this chapter by specifying the isomonodromy deformation equations (4.0.20) for the case of an arbitrary Fuchsian system.

1. The Fuchsian case. The Schlesinger equations

Consider a Fuchsian system, i.e.,

$$\frac{d\Psi}{d\lambda} = \mathsf{A}(\lambda)\Psi, \quad \mathsf{A}(\lambda) = \sum_{\nu=1}^{n} \frac{A_\nu}{\lambda - a_\nu}, \tag{4.1.1}$$

and assume (cf. condition (iii) from the previous section) that

$$A_\infty \equiv \sum_{\nu=1}^{n} A_\nu = \text{(diagonal matrix)}. \tag{4.1.2}$$

The isomonodromy deformations associated with this system were first considered in the works of Schlesinger [**Sch**], who chose the points $a_1, \dots, a_n$ as the only deformation parameters (cf. (4.0.29))

$$t \equiv \{a_1, \dots, a_n\}.$$

The form $\Theta(\lambda)$ is given by

$$\Theta(\lambda) = \sum_{\nu=1}^{n} \Theta^{(\nu)}(\lambda),$$

where according to (4.0.9) and (2.0.6),

$$\begin{aligned}\Theta^{(\nu)}(\lambda) &= -\frac{1}{\lambda - a_\nu} P_\nu \cdot \Lambda_0^{(\nu)} \left[P_\nu\right]^{-1} da_\nu \\ &= -\frac{A_\nu}{\lambda - a_\nu} da_\nu.\end{aligned}$$

Thus, equation (4.0.19) becomes

$$d\Psi = -\sum_{\nu=1}^{n} \frac{A_\nu \Psi}{\lambda - a_\nu} da_\nu,$$

or

$$\frac{\partial \Psi}{\partial a_\nu} = -\frac{A_\nu}{\lambda - a_\nu} \Psi, \quad \nu = 1, \dots, n.$$

Hence the isomonodromy deformation equation (4.0.20) is equivalent to the system

$$dA_\nu = \sum_{\substack{\mu \neq \nu \\ \mu=1}}^{n} [A_\mu, A_\nu] \frac{da_\mu - da_\nu}{a_\mu - a_\nu}, \qquad \nu = 1, \dots, n.$$

The component form of this system,

$$\text{(4.1.3)} \qquad \begin{cases} \dfrac{\partial A_\nu}{\partial a_\nu} = -\displaystyle\sum_{\substack{\mu \neq \nu \\ \mu=1}}^{n} \frac{[A_\mu, A_\nu]}{a_\mu - a_\nu}, & \nu = 1, \dots, n, \\ \dfrac{\partial A_\nu}{\partial a_\mu} = \dfrac{[A_\mu, A_\nu]}{a_\mu - a_\nu}, & \mu \neq \nu = 1, \dots, n, \end{cases}$$

is known as the *Schlesinger Equations*. Since these equations are the compatibility conditions of the equations

$$\text{(4.1.4)} \qquad \begin{cases} \dfrac{\partial \Psi}{\partial \lambda} = -\displaystyle\sum_{\nu=1}^{n} \frac{A_\nu}{\lambda - a_\nu} \Psi, \\ \dfrac{\partial \Psi}{\partial a_\nu} = -\dfrac{A_\nu}{\lambda - a_\nu} \Psi, \end{cases}$$

equations (4.1.4) form a *Lax pair for the Schlesinger Equations.* The meaning of the nonlinear system (4.1.3) is that it describes the *isomonodromy deformations* of the λ-equation in (4.1.4).

The system (4.1.3) has the following explicit first integrals:

- The spectrum of A_ν, i.e., the formal monodromy exponents $\Lambda_0^{(\nu)}$, $\nu = 1, \dots, n$.
- The sum

$$\sum_{\nu=1}^{n} A_\nu = \text{const}.$$

This sum is merely A_∞, which was assumed to be diagonal (basic normalizing condition). In fact,

$$A_\infty = \Lambda_0^{(\infty)} \quad - \text{ formal monodromy at infinity}$$

The collection $\{\Lambda_0^{(\nu)}\}$ represents the "trivial" part of the full set of the first integrals of (4.1.3) formed by the monodromy data of the λ-equation (4.1.1),

$$\mathfrak{m} = \{\Lambda_0^{(\nu)}; \mathsf{E}_\nu\}_{\nu=1,\dots,n}.$$

It is worth noticing, that in the (generic) Fuchsian case the complete set of the first integrals of (4.1.3) can be described (up to the integer shifts in $\Lambda_0^{(\nu)}$) as the set of the monodromy matrices of (4.1.1),

$$\tilde{\mathfrak{m}} = \{\mathsf{M}_\nu\}_{\nu=1,\dots,n}.$$

The relation between the sets $\mathfrak{m}$ and $\tilde{\mathfrak{m}}$ is expressed through the basic equation,

$$\mathsf{M}_\nu = \mathsf{E}_\nu^{-1} e^{2\pi i \Lambda_0^{(\nu)}} \mathsf{E}_\nu, \quad \nu = 1, \dots, n.$$

(Once again, we recall that the connection matrices E_ν are defined up to a left diagonal matrix multiplier.)

The direct monodromy map,

$$\{A_1, \dots, A_n; a_1, \dots, a_n\} \to \{\mathsf{M}_1, \dots, \mathsf{M}_n\}, \tag{4.1.5}$$

provides the system (4.1.3) with the complete set of its first integrals. The inverse monodromy map,

$$\{\mathsf{M}_1, \dots, \mathsf{M}_n; a_1, \dots, a_n\} \to \{A_1, \dots, A_n; a_1, \dots, a_n\}, \tag{4.1.6}$$

provides the system (4.1.3) with its general solution. More precisely, we have the following *Riemann-Hilbert representation* for the solutions of system (4.1.3).

Let

$$\Psi(\lambda) \equiv \Psi(\lambda; a_1, \dots, a_n; \mathsf{M}_1, \dots, \mathsf{M}_n),$$

be the solution of the Fuchsian RH problem determined by the set $\{\mathsf{M}_1, \dots, \mathsf{M}_n\}$ and the polygonal line $[\infty, a_1] \cup [a_1, a_2] \cup \dots \cup [a_n, \infty]$ (see Section 2 of the previous chapter). Note that in the Riemann-Hilbert setting the points $a_\nu, \nu = 1, \dots, n$ are the points of discontinuity of the corresponding jump matrix $G(\lambda)$ (see (3.2.27)). Denote by $\hat{\Psi}_\nu(\lambda)$ the holomorphic matrix multiplier participating in the equation (cf. (2.0.34),

$$\Psi(\lambda) = \hat{\Psi}_\nu(\lambda)(z - a_\nu)^{\Lambda_0^{(\nu)}} \mathsf{E}_\nu,$$

which represents the function $\Psi(\lambda)$ in the neighborhood of the point $\lambda = a_\nu$. Then, the solution of the system (4.1.3) corresponding to the given set $\{\mathsf{M}_1, \dots, \mathsf{M}_n\}$ of its first integrals, can be represented by the equations,

$$A_\nu \equiv A_\nu(a_1, \dots, a_n; \mathsf{M}_1, \dots, \mathsf{M}_n) = P_\nu \Lambda_0^{(\nu)} P_\nu^{-1}, \quad \nu = 1, 2, \dots, n, \tag{4.1.7}$$

where

$$P_\nu := \hat{\Psi}_\nu(a_\nu). \tag{4.1.8}$$

Once again, we emphasize that one of the principal achievements of the modern theory of integrable systems is that it makes the abstract and transcendental formulae (4.1.7), (and the analogous formulae for the non-Fuchsian case) an efficient tool for analyzing the isomonodromy deformation equations. This will be indicated in the next section, although the most important applications of the Riemann-Hilbert representation occurs in the asymptotic analysis of the Painlevé equations (see the second and third parts of the book).

2. Isomonodromy deformations for two by two systems: The Painlevé equations

We will discuss the isomonodromy deformations of 2×2 systems following the pattern introduced in Chapter 3, i.e., successively increasing the number of the singularities in the finite complex plane.

2.1. Two Fuchsian points. This case is trivial. The Lax pair is

$$\begin{cases} \dfrac{\partial \Psi}{\partial \lambda} &= \dfrac{\Lambda_0}{\lambda - a}\Psi, \\ \dfrac{\partial \Psi}{\partial a} &= -\dfrac{\Lambda_0}{\lambda - a}\Psi, \end{cases}$$

and the isomonodromy deformation equations degenerate to the tautological statement

$$\frac{\partial \Lambda_0}{\partial a} = 0.$$

2.2. Three Fuchsian points. This case corresponds to $n = 2$. The basic λ-equation is

$$\frac{\partial \Psi}{\partial \lambda} = \left[\frac{A_1}{\lambda - a_1} + \frac{A_2}{\lambda - a_2}\right]\Psi, \tag{4.2.1}$$

where we assume

$$A_1 + A_2 = \Lambda_0^{(\infty)} \equiv \text{const}.$$

The Schlesinger Equations (4.1.3) reduce to the system

$$\begin{cases} \dfrac{\partial A_1}{\partial a_1} &= \dfrac{[A_2, A_1]}{a_1 - a_2}, \\ \dfrac{\partial A_1}{\partial a_2} &= -\dfrac{\partial A_1}{\partial a_1}, \\ \dfrac{\partial A_2}{\partial a_\nu} &= -\dfrac{\partial A_1}{\partial a_\nu}, \qquad \nu = 1, 2. \end{cases} \tag{4.2.2}$$

This system can be easily solved in closed form:

$$\begin{aligned} &A_1(a_1, a_2) = (a_2 - a_1)^{\Lambda_0^{(\infty)}} \cdot A \cdot (a_2 - a_1)^{-\Lambda_0^{(\infty)}}, \\ &A_2 = \Lambda_0^{(\infty)} - A_1, \end{aligned} \tag{4.2.3}$$

where A is a constant matrix.

Equations (4.2.3) can also be derived by transforming the system (4.2.1) into the canonical system (2.1.3). This can be achieved by using an automorphism of

$\mathbb{C}P^1$ such that

$$\begin{aligned} a_1 &\mapsto 0 \\ a_2 &\mapsto 1 \\ \infty &\mapsto \infty \end{aligned}$$

Such a map is given by

$$\lambda \mapsto w = \frac{1}{a_2 - a_1}(\lambda - a_1). \tag{4.2.4}$$

However, this transformation violates the normalization condition at infinity:

$$\Big(I + O(\lambda^{-1})\Big) \cdot \lambda^{\Lambda_0^{(\infty)}} \mapsto \Big(I + O(w^{-1})\Big) \cdot w^{\Lambda_0^{(\infty)}} \cdot (a_2 - a_1)^{\Lambda_0^{(\infty)}}.$$

Therefore, in order to preserve the monodromy group of the equation we need to perform simultaneously a gauge transformation, namely to introduce the function $\Phi(w)$,

$$\Phi(w) = (a_2 - a_1)^{-\Lambda_0^{(\infty)}} \cdot \Psi(\lambda(w)).$$

Rewriting equation (4.2.1) in terms of $\Phi(w)$ we find

$$\frac{\partial \Phi}{\partial w} = \Big[\frac{A}{w} + \frac{B}{w-1}\Big]\Phi, \tag{4.2.5}$$

where

$$\begin{aligned} A &= (a_2 - a_1)^{-\Lambda_0^{(\infty)}} A_1 (a_2 - a_1)^{\Lambda_0^{(\infty)}}, \\ B &= (a_2 - a_1)^{-\Lambda_0^{(\infty)}} A_2 (a_2 - a_1)^{\Lambda_0^{(\infty)}}, \\ A + B &= \Lambda_0^{(\infty)}. \end{aligned}$$

Equation (4.2.5) is the canonical form of a system with three Fuchsian points. It was shown in Chapter 3 that A and B are determined uniquely in terms of the corresponding monodromy group. The latter does not depend on a_1 and a_2 (the deformations induced by (4.2.2) are the isomonodromy ones !). Therefore, A and B do not depend on a_1 and a_2. This implies (4.2.3).

We emphasize that this alternative derivation of formula (4.2.3) is based on the analysis of the λ-equation (4.2.1) and on the associated inverse monodromy problem. In other words, in order to solve equations (4.2.2) we use the *Lax representation* of these equations. In fact, this is an elementary example of the approach which, as it was indicated in the end of Section 2, we are going to use throughout the book.

2.3. Four Fuchsian points. Hypergeometric equation and Painlevé VI equation. A generic 2×2 system with four regular points can be reduced to the normal form (2.1.16). Correspondingly, the three-point Schlesinger equations (of a 2×2 matrix) can be reduced to the nonlinear differential system whose independent variable is the deformation parameter x — the location of the forth (movable) singular point. We will refer to this nonlinear system as the *Reduced Schlesinger equations (RSE)*. In this subsection we will relate these equations with the Painlevé VI equation.

For the system (2.1.16) the Lax pair has the form

$$\text{(4.2.6)} \qquad \begin{cases} \dfrac{\partial \Psi}{\partial \lambda} &= \Big(\dfrac{A_0}{\lambda} + \dfrac{A_1}{\lambda - 1} + \dfrac{A_2}{\lambda - x}\Big)\Psi, \\ \dfrac{\partial \Psi}{\partial x} &= -\dfrac{A_2}{\lambda - x}\Psi \end{cases}$$

where, as usual, we assume

$$\operatorname{Tr} A_0 = \operatorname{Tr} A_1 = \operatorname{Tr} A_2 = 0,$$

$$-A_0 - A_1 - A_2 \equiv A_\infty = \begin{pmatrix} \delta & 0 \\ 0 & -\delta \end{pmatrix},$$

and the eigenvalues of A_0, A_1, and A_2, are

$$\pm\alpha, \quad \pm\beta, \quad \text{and} \quad \pm\gamma,$$

respectively. The quantities α, β, γ, δ do not depend on x. They are the first integrals of the RSE, which, by definition, is the compatibility condition of (4.2.6):

$$\frac{\partial}{\partial x}\left(\frac{A_0}{\lambda} + \frac{A_1}{\lambda - 1} + \frac{A_2}{\lambda - x}\right) - \frac{\partial}{\partial \lambda}\left(-\frac{A_2}{\lambda - x}\right)$$

$$\text{(4.2.7)} \qquad = \left[-\frac{A_2}{\lambda - x}, \frac{A_0}{\lambda} + \frac{A_1}{\lambda - 1} + \frac{A_2}{\lambda - x}\right].$$

Equating the principal parts of the rational functions appearing in both sides of equation (4.2.7), we find the closed system of ODEs

$$\text{(4.2.8)} \qquad \begin{cases} \dfrac{dA_0}{dx} = \dfrac{1}{x}[A_2, A_0], \\ \dfrac{dA_1}{dx} = \dfrac{1}{x - 1}[A_2, A_1], \\ \dfrac{dA_2}{dx} = -\dfrac{dA_0}{dx} - \dfrac{dA_1}{dx}. \end{cases}$$

An implementation of the general Fuchsian Riemann-Hilbert scheme presented in the end of Section 2 to the case of four singular points, yields the following Riemann-Hilbert representation of the solutions of the system (4.2.8).

Let

$$Y(\lambda) \equiv Y(\lambda; x; \mathrm{M}_1, \mathrm{M}_2, \delta),$$

be the solution of the four point Fuchsian RH problem (3.3.49)–(3.3.53), and let $\hat{\Psi}_0(\lambda)$, and $\hat{\Psi}_1(\lambda)$ be the holomorphic matrix multipliers appearing in the left-hand sides of equations (3.3.44) and (3.3.45) representing the local behavior of the function $\Psi(\lambda) \equiv Y(\lambda)\lambda^{-\delta\sigma_3}$ at the points $\lambda = 0$ and $\lambda = 1$, respectively. Then, the solution of the system (4.2.8) corresponding to the given set $\{\mathrm{M}_1, \mathrm{M}_2, \delta\}$ of first integrals, can be represented by the equations (cf. (4.1.7), (4.1.8)),

$$\text{(4.2.9)} \qquad A_0 \equiv A_0(x; \mathrm{M}_1, \mathrm{M}_2, \delta) = \alpha P_0 \sigma_3 P_0^{-1},$$

$$\text{(4.2.10)} \qquad A_1 \equiv A_1(x; \mathrm{M}_1, \mathrm{M}_2, \delta) = \beta P_1 \sigma_3 P_1^{-1},$$

and

$$A_2 = -A_0 - A_1 - \delta\sigma_3,$$

where

$$P_0 = \hat{\Psi}_0(0), \quad P_1 = \hat{\Psi}_1(1). \tag{4.2.11}$$

The system (4.2.8) represents a collection of differential relations for the entries of matrices A_k. The nontrivial part of this collection is equivalent to a single second order scalar differential equation, in fact, to the Painlevé VI equation. Skipping the technical details, which can be found in [**JMU**] and [**JM**], we indicate the final result.

Define the function $u(x)$ as the zero of the 12-entry of the coefficient matrix

$$\mathsf{A}(\lambda) = \frac{A_0}{\lambda} + \frac{A_1}{\lambda - 1} + \frac{A_2}{\lambda - x};$$

that is,

$$\mathsf{A}_{12}(\lambda, x) \equiv \frac{w(x)(\lambda - u(x))}{\lambda(\lambda - 1)(\lambda - x)} \tag{4.2.12}$$

(note that in our normalization, $A_0 + A_1 + A_2$ is a diagonal matrix). Then, the function $u(x)$ satisfies the Painlevé VI equation,

$$u_{xx} = \frac{1}{2}\left(\frac{1}{u} + \frac{1}{u-1} + \frac{1}{u-x}\right){u_x}^2 - \left(\frac{1}{x} + \frac{1}{x-1} + \frac{1}{u-x}\right)u_x$$

$$+\frac{u(u-1)(u-x)}{x^2(x-1)^2}\left(\alpha_1 + \beta_1\frac{x}{u^2} + \gamma_1\frac{x-1}{(u-1)^2} + \delta_1\frac{x(x-1)}{(u-x)^2}\right), \tag{4.2.13}$$

where the parameters α_1, β_1, γ_1, and δ_1 are related to our α, β, γ, and δ through the formulae,

$$\alpha_1 = \frac{1}{2}(2\delta - 1)^2, \quad \beta_1 = -2\alpha^2,$$

$$\gamma_1 = 2\beta^2, \quad \delta_1 = \frac{1}{2}(1 - 4\gamma^2).$$

Equations (4.2.9)–(4.2.11), together with (4.2.12) provide the *Riemann-Hilbert representation for the sixth Painlevé transcendent.*

Triangular case

Assume that

$$A_0 = \begin{pmatrix} \alpha & u \\ 0 & -\alpha \end{pmatrix}, \quad \text{and} \quad A_1 = \begin{pmatrix} \beta & v \\ 0 & -\beta \end{pmatrix}. \tag{4.2.14}$$

Then,

$$A_2 = \begin{pmatrix} -\delta & 0 \\ 0 & \delta \end{pmatrix} - A_0 - A_1 = \begin{pmatrix} -\delta - \alpha - \beta & -u - v \\ 0 & \delta + \alpha + \beta \end{pmatrix} \equiv \begin{pmatrix} \gamma & -u - v \\ 0 & -\gamma \end{pmatrix},$$

and the system (4.2.8) takes the form of the following system of two linear ODEs,

$$\left\{\begin{aligned} u_x &= \frac{2\gamma}{x}u + \frac{2\alpha}{x}(u+v), \\ v_x &= \frac{2\gamma}{x-1}u + \frac{2\beta}{x-1}(u+v). \end{aligned}\right. \tag{4.2.15}$$

Solving the first equation for v,

$$v = xu_x\frac{1}{2\alpha} - u\frac{\gamma+\alpha}{\alpha}, \tag{4.2.16}$$

and substituting this expression into the second equation, we find a single second order ODE for the function $u(x)$,

$$u_{xx} + u_x\left(\frac{1-2\alpha-2\gamma}{x} - \frac{2\beta+2\gamma}{x-1}\right) + u\left(\frac{4\alpha\gamma+4\gamma^2+4\beta\gamma}{x(x-1)}\right) = 0. \tag{4.2.17}$$

The scalar equation (4.2.17) has three regular singular points located at 0, 1, and ∞. According to the analytic theory of scalar ODEs, this equation must be reducible to the hypergeometric equation. Indeed, it is easy to see that (4.2.17) can be rewritten in the canonical hypergeometric form,

$$x(1-x)u_{xx} + [c-(a+b+1)x]u_x - abu = 0, \tag{4.2.18}$$

with

$$\begin{aligned} a &= -2\alpha-2\beta-2\gamma \equiv 2\delta, \\ b &= -2\gamma, \\ c &= 1-2\alpha-2\gamma. \end{aligned}$$

It is important to emphasize, that not only have we obtained the hypergeometric equation as an Abelian reduction of the four point Schlesinger equation, but, simultaneously, we have solved it! Indeed, the above Riemann-Hilbert representation for the solutions of the Schlesinger system (4.2.8) and the analysis of the triangular RH problem presented in the previous chapter, provide us with the contour integral representation of the general solution of equation (4.2.18). This integral representation follows directly from (3.3.82) which was obtained in Subsection 3.1 of the previous chapter. In fact, formula (3.3.82) implies that the matrix P_0 participating in the representation (cf. (4.2.9)–(4.2.11))

$$A_0 = \alpha P_0\sigma_3P_0^{-1},$$

is given by the equation,

$$P_0 = \begin{pmatrix} 1 & \hat{y}(0) \\ 0 & 1 \end{pmatrix}\hat{\Delta}^{\sigma_3}(0),$$

so that,

$$A_0 = \begin{pmatrix} \alpha & -2\alpha\hat{y}(0) \\ 0 & -\alpha \end{pmatrix},$$

and

$$u(x) = -2\alpha\hat{y}(0).$$

Taking into account the explicit formulae (3.3.76) for the function $\hat{y}(\lambda)$, we arrive at the classical integral representation for the general solution (s_2 and s_3 are arbitrary constants!) of the hypergeometric equation (4.2.18),

$$u(x) = \frac{\alpha i}{\pi}\left(s_2 - s_3\frac{\sin 2\pi(\alpha+\beta)}{\sin 2\pi\alpha}\right)\int_1^x \mu^{-2\delta-1}\Delta_+(\mu)\Delta_-(\mu)d\mu$$

$$+\frac{\alpha i s_3}{\pi}\frac{\sin 2\pi\delta}{\sin 2\pi\alpha}\int_x^\infty \mu^{-2\delta-1}\Delta^2(\mu)d\mu$$

$$-\frac{\alpha s_3}{2\pi}\frac{e^{2\pi i\delta}}{\sin 2\pi\alpha}\int_C \mu_0^{-2\delta-1}\Delta^2(\mu)d\mu. \tag{4.2.19}$$

There is an alternative way of using the Riemann-Hilbert framework for solving the four point Schlesinger system (4.2.8) under the triangular reduction (4.2.14). Indeed, instead of exploiting the explicit solvability of the corresponding Abelian RH problem, one can use the explicit solvability of the corresponding basic λ-equation (2.1.16). In other words, instead of using the *inverse* monodromy problem associated with the Schlesinger system (4.2.8), one can use the corresponding *direct* monodromy problem. Let us outline this approach.

Under the condition (4.2.14), the 2×2 differential system (2.1.16) becomes triangular and can be written in the form

$$\frac{d\Psi}{d\lambda} = \Big(a(\lambda)\sigma_3 + b(\lambda)\sigma_+\Big)\Psi, \tag{4.2.20}$$

where

$$a(\lambda) := \frac{\alpha}{\lambda} + \frac{\beta}{\lambda-1} + \frac{\gamma}{\lambda-x},$$

$$b(\lambda) := u\frac{-x}{\lambda(\lambda-1)} + v\frac{1-x}{(\lambda-1)(\lambda-x)},$$

and

$$\sigma_+ := \begin{pmatrix} 0 & 1 \\ 0 & 0 \end{pmatrix}.$$

By virtue of the gauge transformation,

$$\Psi(\lambda) \to \Phi(\lambda) = e^{-\int a(\lambda)d\lambda\sigma_3}\Psi(\lambda) \equiv \kappa^{-\sigma_3}(\lambda)\Psi(\lambda), \tag{4.2.21}$$

where

$$\kappa(\lambda) = \lambda^\alpha(\lambda-1)^\beta(\lambda-x)^\gamma,$$

the system (4.2.20) is transformed to the matrix equation,

$$\frac{d\Phi}{d\lambda} = b(\lambda)\kappa^{-2}(\lambda)\sigma_+\Phi. \tag{4.2.22}$$

The coefficient matrix of equation (4.2.22) belongs to the same commutative subalgebra for all values of λ, hence the equation (4.2.22) can be explicitly integrated.

Indeed we have,

$$\Phi(\lambda) = \exp\Big(\int_{\infty}^{\lambda} b(\mu)\kappa^{-2}(\mu)d\mu\sigma_+\Big)C$$

$$= \Big(I + \int_{\infty}^{\lambda} b(\mu)\kappa^{-2}(\mu)d\mu\sigma_+\Big)C, \tag{4.2.23}$$

where C is an arbitrary nonsingular constant matrix, and we have made use of the fact that the matrix σ_+ is nilpotent.

Combining (4.2.21) and (4.2.23), one can easily derive the contour integral representation for the canonical solution $\Psi(\lambda) \equiv \Psi^{(\infty)}(\lambda)$ of the linear equation (4.2.20). This in turn leads to explicit formulae for the monodromy matrices $\mathsf{M}_1, \mathsf{M}_2$ in terms of linear combinations of u and v where the coefficients of these linear combinations involve certain integrals of $\kappa^{-2}(\lambda)/\lambda(\lambda - x)$ and $\kappa^{-2}(\lambda)/(\lambda - 1)(\lambda - x)$. Setting

$$\mathsf{M}_1 \equiv \text{const}, \quad \mathsf{M}_2 \equiv \text{const},$$

and solving the resulting linear systems for u and v, we will arrive again at the classical contour integrals for the hypergeometric function $u(x)$. We skip the details since they are straightforward.

REMARK 4.5. We want to attract the reader's attention to the remarkable fact that the hypergeometric function has appeared in our narrative twice: first - in Chapter 2, as the solution of the basic linear equation associated with the three point Fuchsian monodromy problem; second - now, as the solution of the Abelian reduction of the isomonodromy deformation equations associated with the four point Fuchsian monodromy problem. In the former case, the hypergeometric function appeared as a function of the basic complex parameter λ, and the existence of an associated integral representation had no methodological explanation in the framework of the Riemann-Hilbert approach. Indeed, in the three point case, both the direct and the inverse monodromy problems are non-Abelian. In this chapter the hypergeometric function appears as a function of the deformation parameter x, and simultaneously the relevant integral representation also appears. Moreover, this integral representation now has a simple explanation — it is the direct consequence of the commutativity of the relevant RH problem (which is the triangular reduction of the general four point Fuchsian problem).

2.4. One irregular singular point of rank ≤ 2. The generic normal forms given by (2.1.64) and (2.1.65), do not contain any deformation parameters (the monodromy map is one-to-one!). Thus these two cases are similar to the first two Fuchsian cases discussed earlier. The isomonodromy deformation equations can be easily solved in closed form, by reducing the λ-equations to the corresponding normal forms. We leave the details to the reader.

2.5. One irregular singular point of rank 3: Airy equation and Painlevé II equation. This case is similar to the case of the four Fuchsian points discussed above. The corresponding normal system, i.e., equation (2.1.86) does contain a deformation parameter x defined in (2.1.89) (x is a nontrivial entry of the matrix Λ_1). Therefore, in contrast to the case of an irregular singular point of Poincaré rank ≤ 2, the isomonodromy deformations cannot be eliminated by simple conformal, gauge, or scaling transformations, and the reduced isomonodromy deformation equations are nontrivial.

Applying to the system (2.1.86) the general constructions of Section 1, we find

$$\Theta(\lambda) \equiv \Theta^{(\infty)}(\lambda) = \mathsf{U}(\lambda)dx,$$

where $\mathsf{U}(\lambda)$ is a polynomial of first degree in λ defined by the equations (see (4.0.12) and (2.1.88))

$$\begin{aligned}
\mathsf{U}(\lambda) &= \hat{\Psi}^{(\infty)}(\lambda)\cdot\frac{d\Lambda^{(\infty)}(\lambda)}{dx}\Big[\hat{\Psi}^{(\infty)}(\lambda)\Big]^{-1} \pmod{\lambda^{-1}}\\
&= \Psi_f(\lambda)\cdot\frac{d}{dx}\Big\{-\frac{4}{3}i\lambda^3\sigma_3 - ix\lambda\sigma_3 - \nu\sigma_3\log\lambda\Big\}\Psi_f(\lambda) \pmod{\lambda^{-1}}\\
&= \Big[I+\frac{1}{\lambda}\Psi_1+\dots\Big]\cdot(-i\lambda\sigma_3)\Big[I-\frac{1}{\lambda}\Psi_1+\dots\Big] \pmod{\lambda^{-1}}\\
&= -i\lambda\sigma_3 + i[\sigma_3,\Psi_1].
\end{aligned}$$

Substituting the asymptotic expansion (2.1.88) into the basic equation (2.1.86) and equating the terms of order λ^{-1}, we find the identity

$$[\sigma_3,\Psi_1] = \begin{pmatrix} 0 & u \\ -v & 0 \end{pmatrix}.$$

This allows one to express $\mathsf{U}(\lambda)$ in terms of the coefficients of the $\mathsf{A}(\lambda)$ matrix:

$$\mathsf{U}(\lambda) = -i\lambda\sigma_3 + i\begin{pmatrix} 0 & u \\ -v & 0 \end{pmatrix}.$$

The Lax pair takes the form

$$\text{(4.2.24)}\quad \begin{cases}
\dfrac{\partial\Psi}{\partial\lambda} = \left[-4i\lambda^2\sigma_3 + 4i\lambda\begin{pmatrix} 0 & u \\ -v & 0 \end{pmatrix} + \begin{pmatrix} -ix-2iuv & -2w \\ -2z & ix+2iuv \end{pmatrix}\right]\Psi,\\[2ex]
\dfrac{\partial\Psi}{\partial x} = \left[-i\lambda\sigma_3 + i\begin{pmatrix} 0 & u \\ -v & 0 \end{pmatrix}\right]\Psi.
\end{cases}$$

The associated reduced isomonodromy deformation equation is

$$\text{(4.2.25)}\quad \frac{\partial\mathsf{A}(\lambda)}{\partial x} - \frac{\partial\mathsf{U}(\lambda)}{\partial\lambda} = [\mathsf{U}(\lambda),\mathsf{A}(\lambda)],$$

where

$$\mathsf{A}(\lambda) = -4i\lambda^2\sigma_3 + 4i\lambda\begin{pmatrix} 0 & u \\ -v & 0 \end{pmatrix} + \begin{pmatrix} a & -2w \\ -2z & -a \end{pmatrix}$$

$$\mathsf{U}(\lambda) = -i\lambda\sigma_3 + i\lambda\begin{pmatrix} 0 & u \\ -v & 0 \end{pmatrix}. \tag{4.2.26}$$

Substituting (4.2.26) into (4.2.25) and equating the terms of the same order in λ, we arrive at the following system of ordinary differential equations for the quantities $u,\ v,\ z,\ w$,

$$z = v_x, \quad w = u_x, \tag{4.2.27}$$

$$\begin{cases} u_{xx} &= xu + 2u^2v, \\ v_{xx} &= xv + 2v^2u. \end{cases} \tag{4.2.28}$$

The system (4.2.27)–(4.2.28) represents the isomonodromy differential equation (4.2.25) on the following manifold of the systems (2.1.86),

$$\mathcal{A} = \{u,\ v,\ w,\ z,\ a = -ix - 2iuv\}.$$

Equations (3.3.132), (3.3.133) obtained in Chapter 3 provide the *Riemann-Hilbert representation* for the solutions of the nonlinear system (4.2.27)–(4.2.28) in terms of its first integrals — the monodromy data of the linear equation (2.1.86). In fact, the general Theorem 4.1 implies the following result.

THEOREM 4.2. *Suppose $s \equiv (S_1, S_2, \ldots, S_6; \nu)$ is the given set of the monodromy data, i.e., the complex number ν and the set of six 2×2 matrices satisfying equations (3.3.117) and (3.3.118). Let*

$$Y(\lambda) \equiv Y(\lambda, x; s)$$

be the solution of the Riemann-Hilbert-Birkhoff problem (3.3.98)–(3.3.99) given by Theorem 3.4. Let also $\Psi_1 \equiv \Psi_1(x;s)$ be the first coefficient of the asymptotic expansion (3.3.122) of the function $Y(\lambda)$ at $\lambda = \infty$. Then, the formulae

$$u(x) = 2(\Psi_1)_{12}, \quad v(x) = 2(\Psi_1)_{21}, \tag{4.2.29}$$

describe the solutions of system (4.2.27)–(4.2.28). Moreover, any solution of (4.2.27)–(4.2.28) admits the representation (4.2.29).

PROOF. At the end of Chapter 3 we have proven that equations (4.2.29) represent an isomonodromy deformation of the linear equation (2.1.86). Therefore, by the direct statement of Theorem 4.1 we conclude that the functions $u(x)$ and $v(x)$ must satisfy the relevant isomonodromy deformation equations which for the linear system (2.1.86) are equivalent to the nonlinear system (4.2.27)–(4.2.28). Conversely, suppose that the functions $u(x)$, $v(x)$, $w(x)$, and $z(x)$ satisfy (4.2.27)–(4.2.28) in the neighborhood of a point $x = x_0$. Then by the second statement of Theorem 4.1 we conclude that the formula

$$A(\lambda, x) := -4i\lambda^2\sigma_3 + 4i\lambda \begin{pmatrix} 0 & u(x) \\ -v(x) & 0 \end{pmatrix}$$
$$+ \begin{pmatrix} -ix - 2iu(x)v(x) & -2w(x) \\ -2z(x) & ix + 2iu(x)v(x) \end{pmatrix},$$

describes the isomonodromy deformations of equation (2.1.86). Moreover, the corresponding canonical solutions $\Psi_n(\lambda)$ provide a (unique) solution of the Riemann-Hilbert problem (3.3.98)–(3.3.99) posed for the data s determined by the Stokes matrices corresponding to the coefficient matrix $A(\lambda, x_0)$. The representation (4.2.29) for the functions $u(x)$ and $v(x)$ follows from the uniqueness of the solution of the Riemann-Hilbert problem. □

Let

(4.2.30) $$\mathcal{M} = \{(S_1, S_2, \dots, S_6; \nu) : \text{ the } S_n \text{ satisfy (2.1.92), (2.1.93)}\}$$

be the set of monodromy data of equation (2.1.86). Because of the uniqueness of the local solution of the Cauchy problem for system (4.2.28) and the uniqueness of the solution of the Riemann-Hilbert problem, we arrive at the following two important corollaries of Theorem 4.2.

COROLLARY 4.1 (The Painlevé property). *Suppose that C^2-functions $u(x)$ and $v(x)$ satisfy the isomonodromy deformation equations (4.2.28). Then they admit a meromorphic continuation into the whole complex x-plane.*

COROLLARY 4.2 (Monodromy parametrization of the solution set). *The map,*

$$\{solutions\ of\ (4.2.28)\} \mapsto \mathcal{M},$$

is a bijection; hence, the notation

$$u(x) \equiv u(x; s), \quad v(x) \equiv v(x; s)$$

for the solutions of (4.2.28) is justified. Given $s \in \mathcal{M}$, both the functions $u(x; s)$ and $v(x; s)$ are meromorphic functions, and the set X_s (see Theorem 3.4) is the set of their (first order) poles.

The nonlinear system (4.2.28) is equivalent to a pair of second Painlevé equations. To see this, we first notice that the combination,

$$vu_x - uv_x,$$

is obviously a first integral of system (4.2.28). In fact, taking into account (2.1.89), we see that this combination is just the formal monodromy parameter ν of the linear equation (2.1.86) at $\lambda = \infty$, i.e.,

(4.2.31) $$\nu = (vw - uz) = vu_x - uv_x.$$

Let $p(x)$ and $q(x)$ denote the logarithmic derivatives of $u(x)$ and $v(x)$, respectively,

(4.2.32) $$p = \frac{u_x}{u}, \quad q = \frac{v_x}{v}.$$

Put

$$u_+(x) := 2^{-1/3}p(-2^{-1/3}x), \quad \text{and} \quad u_-(x) := 2^{-1/3}q(-2^{-1/3}x). \tag{4.2.33}$$

Denote by $PII^{(\alpha)}$ the second Painlevé equation with the parameter α. By direct calculation, one can check that the functions $u_\pm(x)$ satisfy the second Painlevé equations $PII^{(\pm\nu-1/2)}$, in other words,

$$u_{\pm xx} = 2u_\pm^3 + xu_\pm + \frac{1}{2} \mp \nu. \tag{4.2.34}$$

In a more direct way, the system (4.2.28) is related to the equation XXXIV in the Painlevé-Gambier list. Indeed, the product

$$w = uv,$$

solves the Painlevé XXXIV equation,

$$w_{xx} = \frac{w_x^2}{2w} + 4w^2 + 2xw - \frac{\nu^2}{2w}. \tag{4.2.35}$$

The functions p, q, and w are related to each other by the following set of formulae:

$$w = \frac{1}{2}(p_x + p^2 - x) = \frac{1}{2}(q_x + q^2 - x), \quad w_x = 2pw - \nu = 2qw + \nu,$$

$$p = \frac{w_x + \nu}{2w}, \quad p_x = x + 2w - p^2,$$

$$q = \frac{w_x - \nu}{2w}, \quad q_x = x + 2w - q^2,$$

$$q = p - \frac{2\nu}{p_x + p^2 - x}, \quad p = q + \frac{2\nu}{q_x + q^2 - x},$$

$$p_x + p^2 = q_x + q^2.$$

These formulae establish the equivalence of Painlevé XXXIV equation to Painlevé II equation, and they actually define the so-called Bäcklund transformations between these two equations (we will discuss Bäcklund transformations in Chapter 6).

REMARK 4.6. Theorem 4.2 provides the proper Riemann-Hilbert representations for Painlevé XXXIV and Painlevé II equations, and furthermore it establishes their Painlevé property. There is also an alternative Lax pair and hence an alternative Riemann-Hilbert representation for the second Painlevé equation. Indeed, one can check that the second Painlevé equation,

$$u_{xx} = 2u^3 + xu - \alpha,$$

is equivalent to the compatibility condition,

$$\mathsf{A}_x(\lambda) - \mathsf{U}_\lambda(\lambda) + [\mathsf{A}(\lambda), \mathsf{U}(\lambda)] = 0,$$

of the overdetermined linear system,

$$\Psi_\lambda = \mathsf{A}(\lambda)\Psi, \quad \Psi_x = \mathsf{U}(\lambda)\Psi, \tag{4.2.36}$$

where

$$\mathsf{A}(\lambda) = -i(4\lambda^2 + x + 2u^2)\sigma_3 - \left(4u\lambda + \frac{\alpha}{\lambda}\right)\sigma_2 - 2u_x\sigma_1,$$

and

$$\mathsf{U}(\lambda) = -i\lambda\sigma_3 - u\sigma_2,$$

and, as usual, $\sigma_{1,2,3}$ are the Pauli matrices,

$$\sigma_1 = \begin{pmatrix} 0 & 1 \\ 1 & 0 \end{pmatrix}, \quad \sigma_2 = \begin{pmatrix} 0 & -i \\ i & 0 \end{pmatrix}, \quad \sigma_3 = \begin{pmatrix} 1 & 0 \\ 0 & 1 \end{pmatrix}.$$

This Lax pair was suggested by H. Flaschka and A. Newell in [**FN1**]. The Lax pair (4.2.24) was found by M. Jimbo and T. Miwa [**JM**] (although, in an equivalent form it was already known to Garnier). In the rest of the book, when studying the second Painlevé equation, we will mostly use the Flaschka-Newell Lax pair. The associated Riemann-Hilbert problem will be discussed in detail in Chapters 5 and 11.

Theorem 4.2 suggests two fundamental ways of implementing of the Riemann-Hilbert Isomonodromy method for investigating system (4.2.28). The first approach directly exploits the fact that the monodromy data $s \equiv s(u, v, w, z, x)$ of the linear equation (2.1.86) form a complete set of first integrals of the nonlinear system (4.2.28) and hence the (transcendental!) equation

$$s = s(u, v, w, z, x) \equiv \text{constant},$$

determines implicitly the general solution of (4.2.28). The idea is then to use the *direct monodromy problem* for the linear equation (2.1.86), to extract useful information about the (transcendental!) function $s(u, v, w, z, x)$. For instance, one can try to find it asymptotically, assuming certain characteristic behaviour of $u(x)$ and $v(x)$ as $x \to \infty$. The second approach is to use the *inverse monodromy problem* for (2.1.86), i.e., the Riemann-Hilbert problem (3.3.98)–(3.3.99). In fact, we have already taken advantage of the second method by using Theorem 3.4 to extract useful information about the global properties of the solutions of (4.2.28); see Corollaries 4.1 and 4.2. In the following chapters, we will use both approaches for the concrete analysis of analytic and asymptotic properties of Painlevé functions.

REMARK 4.7. Using Theorem 3.2, we can prove the analogs of Corollaries 4.1 and 4.2 for the nonlinear system (4.2.8), i.e., for the four point Schlesinger equations. In particular, we can obtain the Painlevé property for Painlevé VI equation. In Chapter 5 we will use the relevant Riemann-Hilbert representations to establish the Painlevé property for the rest of Painlevé equations (following [**FoZ**]). The Painlevé property for the general Schlesinger equations has been proven by B. Malgrange [**Mal**] and T. Miwa [**Miw**], and by B. Malgrange [**Mal**], for the general isomonodromy equations (4.0.20), i.e., in the presence of the irregular singular points (see also [**Pa**] and [**Mas**]).

REMARK 4.8. The alternative methods for proving the Painlevé property for Painlevé equations, which do not use any relationship with the Riemann-Hilbert problems, were suggested in [**C2**] and [**JoKr2**]. However, it seems that it is very difficult, if not possible at all, to prove the Painlevé property for the general Schlesinger systems (and to perform the global asymptotic analysis of their solutions) without using the Riemann-Hilbert apparatus.

We complete this section by discussing two basic reductions of the nonlinear system (4.2.28), namely, the triangular reduction and the reduction to the special case of Painlevé II equation corresponding to zero α.

TRIANGULAR REDUCTION

Under the reduction

$$v = z = 0, \tag{4.2.37}$$

the system (4.2.28) transforms to the single linear ODE

$$u_{xx} = xu. \tag{4.2.38}$$

This is the classical Airy equation. Simultaneously, since (cf. 3.3.108),

$$s_1 = s_3 = s_5 = 0, \quad \nu = 0, \quad s_2 + s_4 + s_6 = 0,$$

the monodromy data becomes triangular; furthermore, the Riemann-Hilbert problem (3.3.98)–(3.3.99) admits, as seen in Chapter 3, an explicit solution in terms of contour integrals. The solution is given by equations (3.3.110) and (3.3.111), and it yields, via (3.3.116), the classical integral representations for the general solution of the Airy equation (4.2.38).

Similar to the Fuchsian four point case considered in the previous section, we can also proceed with the direct monodromy problem approach: the linear equation (2.1.86), under the constraint (4.2.37) becomes triangular and hence explicitly solvable.

SPECIAL CASE OF PAINLEVÉ II EQUATION

Under the reduction relation,

$$u = v, \quad (w = z), \tag{4.2.39}$$

the system (4.2.28) becomes a particular case ($\alpha = 0$) of the second Painlevé equation,

$$u_{xx} = xu + 2u^3. \tag{4.2.40}$$

The reduction (4.2.39) implies that

$$\nu = 0,$$

and it is equivalent to the symmetry equation,

$$\sigma_2 \mathsf{A}(-\lambda)\sigma_2 = \mathsf{A}(\lambda),$$

on the coefficient matrix of the linear system (2.1.86). This, in turn, yields the following symmetry relations between the canonical solutions of (2.1.86),

$$\sigma_2 \Psi_{n+3}(-\lambda)\sigma_2 = \Psi_n(\lambda).$$

This implies the relation

$$\mathsf{S}_{n+3} = \sigma_2 \mathsf{S}_n \sigma_2, \quad \sigma_2 = \begin{pmatrix} 0 & -i \\ i & 0 \end{pmatrix}, \tag{4.2.41}$$

for the Stokes matrices. In terms of the nontrivial matrix entries s_n of S_n, equation (4.2.41) reads

$$s_{n+3} = -s_n, \quad n = 1, 2, 3, \tag{4.2.42}$$

and, as a result, the cyclic equation (2.1.93) transforms into the single scalar relation,

$$s_1 - s_2 + s_3 + s_1 s_2 s_3 = 0. \tag{4.2.43}$$

This in turn implies that the set $\mathcal{M}$ of monodromy data for the second Painlevé equation (4.2.40), becomes the two-dimensional algebraic manifold,

$$\mathcal{M}_{p2}^{(0)} = \{s \equiv (s_1, s_2, s_3) \in \mathbb{C}^3 : s_1 - s_2 + s_3 + s_1 s_2 s_3 = 0\}.$$

From Corollaries 4.1 and 4.2 we deduce the following statements concerning the Painlevé equation (4.2.40).

COROLLARY 4.3 (The Painlevé property). *Suppose that a C^2-functions $u(x)$ satisfies the second Painlevé equation (4.2.40). Then it admits a meromorphic continuation into the whole complex x-plane.*

COROLLARY 4.4 (Monodromy parametrization of the Painlevé set). *The map,*

$$\{solutions\ of\ PII^{(0)}\} \mapsto \mathcal{M}_{p2}^{(0)},$$

is a bijection; hence, the notation

$$u(x) \equiv u(x; s),$$

for the solutions of the Painlevé equation is justified. Given $s \in \mathcal{M}_{p2}^{(0)}$, the functions $u(x;s)$ are meromorphic, and the set X_s (see Theorem 3.4) is the set of its (first order) poles.

We note that under the reduction

$$u = v$$

the Garnier-Jimbo-Miwa Lax pair (4.2.24), coincides with the Flaschka-Newell Lax pair (4.2.36) with $\alpha = 0$.

In many applications it is important to single out the solutions of the second Painlevé equation (4.2.40) which satisfy the relation,

$$u(x) = \overline{u}(\overline{x}). \tag{4.2.44}$$

In other words, we want $u(x)$ to be real, for real x. The reduction (4.2.44) is equivalent to the symmetry equation,

$$\sigma_1 \overline{\mathrm{A}}(\overline{\lambda})\sigma_1 = \mathrm{A}(\lambda),$$

on the coefficient matrix of the linear system (2.1.86). This, in turn, yields the following symmetry relations between the canonical solutions of (2.1.86),

$$\sigma_1 \overline{\Psi}_1(\overline{\lambda})\sigma_1 = \Psi_1(\lambda), \quad \sigma_1 \overline{\Psi}_2(\overline{\lambda})\sigma_1 = \Psi_6(\lambda),$$

$$\sigma_1 \overline{\Psi}_3(\overline{\lambda})\sigma_1 = \Psi_5(\lambda).$$

This implies the relations

$$\mathrm{S}_6 = \Psi_6^{-1}\Psi_1 = \sigma_1 \overline{\Psi}_2^{-1} \overline{\Psi}_1 \sigma_1 = \sigma_1 \overline{\mathrm{S}}_1^{-1} \sigma_1 \tag{4.2.45}$$

and

$$S_5 = \Psi_5^{-1}\Psi_6 = \sigma_1\overline{\Psi}_3^{-1}\overline{\Psi}_2\sigma_1 = \sigma_1\overline{S}_2^{-1}\sigma_1, \tag{4.2.46}$$

$$\sigma_1 = \begin{pmatrix} 0 & 1 \\ 1 & 0 \end{pmatrix},$$

for the Stokes matrices. In terms of the nontrivial matrix entries s_n of S_n, equations (4.2.45) and (4.2.46) read

$$s_3 = -s_6 = \overline{s}_1, \quad s_2 = -s_5 = \overline{s}_2. \tag{4.2.47}$$

Similarly, replacing in the above formulae the matrix σ_1 with the matrix σ_2, we find that the "purely imaginary" reduction,

$$u(x) = -\overline{u}(\overline{x}), \tag{4.2.48}$$

is equivalent to the following symmetry relations for the Stokes multipliers,

$$s_3 = -s_6 = -\overline{s}_1, \quad s_2 = -s_5 = -\overline{s}_2. \tag{4.2.49}$$

CHAPTER 5

The Isomonodromy Method

In this chapter we will introduce the main ideas and techniques needed for the implementation of the isomonodromy method. For pedagogical reasons we will adopt the following approach: First we will discuss in detail all the necessary steps needed for the solution of Painlevé II. Then we will formulate these steps in general terms. Finally, we will apply these steps to the solution of the other Painlevé equations.

0. The second Painlevé transcendent

The Painlevé II equation,

$$u_{xx} = xu + 2u^3 - \alpha, \tag{5.0.1}$$

follows from the compatibility condition (commutativity of the partial derivatives, $\Psi_{\lambda x} \equiv \Psi_{x\lambda}$)

$$\mathsf{A}_x - \mathsf{U}_\lambda + [\mathsf{A}, \mathsf{U}] = 0 \quad \Rightarrow \quad \begin{cases} u_x = v, \\ v_x = xu + 2u^3 - \alpha, \end{cases} \tag{5.0.2}$$

of the following Lax pair,

$$\frac{\partial \Psi}{\partial \lambda} = \mathsf{A}\Psi,$$

$$\mathsf{A}(\lambda, x) = -i(4\lambda^2 + x + 2u^2)\sigma_3 + \left(4\lambda u + \frac{\alpha}{\lambda}\right)\sigma_1 - 2v\sigma_2, \tag{5.0.3}$$

$$\frac{\partial \Psi}{\partial x} = \mathsf{U}\Psi, \qquad \mathsf{U}(\lambda, x) = -i\lambda\sigma_3 + u\sigma_1. \tag{5.0.4}$$

Here $\Psi(\lambda, x)$ is a 2×2 complex-valued matrix function of $\lambda \in \mathbb{C}$, $x \in \mathbb{C}$; the matrices σ_1, σ_2, σ_3 are the usual Pauli matrices and α is a constant. The linear ODE (5.0.3) defines the 2×2 matrix $\Psi(\lambda, x)$ as a function of the complex variable λ. Equation (5.0.4) describes the dependence of Ψ on the complex variable x.

The fundamental construction of the isomonodromy method involves the following:

> *Use appropriate solutions of equation (*5.0.3*) to formulate a Riemann-Hilbert problem in the complex λ-plane for some sectionally holomorphic function $Y(\lambda, x)$*

where Y is related to Ψ through some simple transformations. This RH problem is usually specified by two complex constants, referred to as the monodromy data. Thus $Y(\lambda, x)$ can be obtained by solving a linear integral equation uniquely specified by the two monodromy data. It turns out that $u(x)$ is simply related to $Y(\lambda, x)$. Thus $u(x)$ is also parametrized (through the solution of a linear integral equation) by the two monodromy data.

We now give details of the above construction, with emphasis on the correct definition of the canonical solutions of (5.0.3)–(5.0.4) and on the description of the relevant monodromy data. We note that even if a function $\Psi(\lambda)$ (depending parametrically on x) satisfies (5.0.3) (with u and v satisfying (5.0.2)), one still has to solve (5.0.4). Indeed, the difference $\mathsf{R} = \Psi_x - \mathsf{U}\Psi$ satisfies the equation

$$\mathsf{R}_\lambda = (\mathsf{A}\Psi)_x - \mathsf{U}_\lambda\Psi - \mathsf{U}\mathsf{A}\Psi = \mathsf{A}\Psi_x + \big(\mathsf{A}_x - \mathsf{U}_\lambda - \mathsf{U}\mathsf{A}\big)\Psi = \mathsf{A}\mathsf{R},$$

and therefore R solves (5.0.3). This implies that there exists a constant in λ function $P(x)$, such that $\mathsf{R} = \Psi P(x)$. Now the function $\tilde{\Psi}$,

$$\tilde{\Psi} = \Psi Q, \quad Q_x = -PQ, \quad P = \Psi^{-1}\mathsf{R} = \Psi^{-1}\Psi_x - \Psi^{-1}\mathsf{U}\Psi, \tag{5.0.5}$$

satisfies both equations of the Lax pair (5.0.3)–(5.0.4).

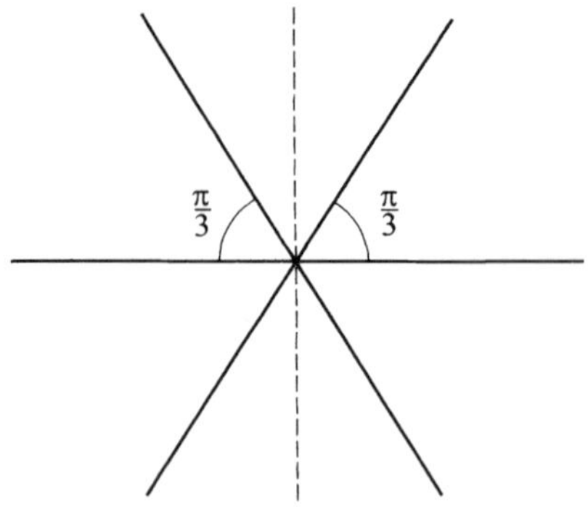

FIGURE 5.1. Stokes lines for PII

The linear ODE (5.0.3) possesses two singular points in the complex λ-plane: a Fuchsian singular point at $\lambda = 0$ and an irregular singular point at $\lambda = \infty$.

The dominant behavior of the solution $\Psi(\lambda, x)$ of equation (5.0.3) in the neighborhood of $\lambda = \infty$ is given by $\exp[-i(\frac{4}{3}\lambda^3 + x\lambda)\sigma_3]$. This implies that the Stokes curves asymptote to the rays

$$\arg(i\lambda^3) = \frac{\pi}{2} + k\pi, \qquad k \in \mathbb{Z}.$$

Thus the Stokes rays are given by $\arg\lambda = k\pi/3$, $k \in \mathbb{Z}$; see Figure 5.1.

The associated Stokes sectors $\Omega_1, \Omega_2, \ldots, \Omega_6$ are defined by $\arg\lambda$ in the following open sectors:

$$\big(-\tfrac{\pi}{3}, \tfrac{\pi}{3}\big), \big(0, \tfrac{2\pi}{3}\big), \big(\tfrac{\pi}{3}, \pi\big), \big(\tfrac{2\pi}{3}, \tfrac{4\pi}{3}\big), \big(\pi, \tfrac{5\pi}{3}\big), \big(\tfrac{4\pi}{3}, 2\pi\big).$$

We introduce the associated functions Ψ_k, $k = 1, \ldots, 6$, which are the unimodular solutions of (5.0.3) defined by the requirement that in the corresponding sectors Ω_k, they have the canonical asymptotic behavior

$$\Psi_k(\lambda, x) = \big(I + O(\lambda^{-1})\big)e^{-i(\frac{4}{3}\lambda^3 + x\lambda)\sigma_3}, \qquad \lambda \to \infty \quad \text{within} \quad \Omega_k. \tag{5.0.6}$$

The solutions $\Psi_k(\lambda, x)$ of (5.0.3) also satisfy the equation (5.0.4), provided that the compatibility condition (5.0.2) holds true. Indeed, using (5.0.3) we find the following two terms asymptotics of $\Psi_k(\lambda, x)$ as $\lambda \to \infty$, $\lambda \in \Omega_k$,

$$\Psi_k(\lambda, x) = \Big(I + \lambda^{-1}\big(-\tfrac{i}{2}D\sigma_3 + \tfrac{1}{2}u\sigma_2\big) + O(\lambda^{-2})\Big)e^{-i(\frac{4}{3}\lambda^3 + x\lambda)\sigma_3}, \tag{5.0.7}$$

where $D = v^2 - u^4 - xu^2 + 2\alpha u$. The matrix $P(x)$ defined in (5.0.5) has the following asymptotics as $\lambda \to \infty$,

$$P(x) = \lambda^{-1} e^{i(\frac{4}{3}\lambda^3 + x\lambda)\sigma_3} C(x)\big(I + O(\lambda^{-1})\big) e^{-i(\frac{4}{3}\lambda^3 + x\lambda)\sigma_3},$$

and since it is independent of λ it must be identically zero.

Associated with the Fuchsian singular point $\lambda = 0$ the solution $\Psi^{(0)}(\lambda, x)$, which is a unimodular solution of (5.0.3), is valid in the entire finite complex plane and has the following behavior as $\lambda \to 0$ (see Chapter 1, Theorems 1.2 and 1.3):

$$\text{(5.0.8)} \quad \Psi^{(0)}(\lambda, x) = \frac{1}{\sqrt{2}}(I - i\sigma_2)\hat{\Psi}^{(0)}(\lambda, x)\lambda^{\alpha\sigma_3}\lambda^{\mathsf{N}} q^{\sigma_3},$$

$$\hat{\Psi}^{(0)}(\lambda, x) = I + O(\lambda), \quad \sigma_3\hat{\Psi}(-\lambda, x)\sigma_3 = \hat{\Psi}(\lambda, x), \quad q = \exp\Big\{\int^x u\, dx\Big\},$$

where the (nilpotent) matrix $\mathsf{N} = j_+\sigma_+ + j_-\sigma_-$, with $j_+ j_- = 0$; moreover,[1]

$$j_+ = 0 \qquad \text{unless} \qquad \tfrac{1}{2} + \alpha \in \mathbb{N},$$

and

$$j_- = 0 \qquad \text{unless} \qquad \tfrac{1}{2} - \alpha \in \mathbb{N}.$$

The matrix $Q \equiv q^{\sigma_3}$ in (5.0.8) is the matrix Q from (5.0.5) which ensures that $\Psi^{(0)}(\lambda, x)$ solves both equations of the Lax pair (5.0.3)–(5.0.4).

REMARK 5.1. The symmetry property of $\hat{\Psi}^{(0)}(\lambda, x)$, $\hat{\Psi}(\lambda, x) = \sigma_3\hat{\Psi}(-\lambda, x)\sigma_3$, follows from the form of its Taylor expansion,

$$\hat{\Psi}^{(0)}(\lambda, x) = I + \sum_{m=1}^{\infty}\Big[(a_m\sigma_1 + b_m\sigma_2)\lambda^{2m-1} + (c_m I + d_m\sigma_3)\lambda^{2m}\Big].$$

Actually, this property originates from the symmetry of (5.0.3) with respect to the transformation $\lambda \mapsto -\lambda$, $\sigma_1 \mathsf{A}(-\lambda, x)\sigma_1 = -\mathsf{A}(\lambda, x)$. This implies that if $\Psi(\lambda, x)$ is a solution, then $\sigma_1\Psi(-\lambda, x)\sigma_1$ is also a solution.

The canonical solutions $\Psi_k(\lambda, x)$ defined in (5.0.6) are determined uniquely. However, the function $\Psi^{(0)}(\lambda, x)$ is defined up to a right constant diagonal unimodular multiplier, d^{σ_3}, due to the choice of the integration constant in (5.0.8). Furthermore, in the resonant case $\alpha - \frac{1}{2} \in \mathbb{Z}$, the solution $\Psi^{(0)}$ is determined up to a right multiplication with a constant triangular matrix which has 1 as its diagonal entries (cf. Chapter 1).

We introduce the Stokes matrices S_k which relate the canonical solutions Ψ_k,

$$\Psi_{k+1} = \Psi_k S_k, \quad k = 1, \ldots, 6; \qquad \Psi_7 := \Psi_1,$$

[1] It is interesting that, in contrast to the general case, the value $\alpha \in \mathbb{Z}$ is not resonant. Indeed, if $\alpha = 0$, then the Fuchsian singularity disappears, and any solution $\Psi(\lambda, x)$ of (5.0.3) is an entire function. The solution $\Psi(\lambda, x)$ for $\alpha \in \mathbb{Z} \setminus \{0\}$ can be related to a solution with $\alpha = 0$ by a rational gauge transformation (Schlesinger transformation, see Chapter 6), thus it does not have a logarithmic singularity.

where

$$\mathsf{S}_1 = \begin{pmatrix} 1 & 0 \\ s_1 & 1 \end{pmatrix}, \quad \mathsf{S}_2 = \begin{pmatrix} 1 & s_2 \\ 0 & 1 \end{pmatrix}, \quad \mathsf{S}_3 = \begin{pmatrix} 1 & 0 \\ s_3 & 1 \end{pmatrix},$$

$$\mathsf{S}_4 = \begin{pmatrix} 1 & s_4 \\ 0 & 1 \end{pmatrix}, \quad \mathsf{S}_5 = \begin{pmatrix} 1 & 0 \\ s_5 & 1 \end{pmatrix}, \quad \mathsf{S}_6 = \begin{pmatrix} 1 & s_6 \\ 0 & 1 \end{pmatrix}.$$

To describe the monodromy around the origin, we introduce the matrix M using the equation (see Remark 5.1)

$$\sigma_1 \Psi^{(0)}(\mathrm{e}^{i\pi}\lambda, x)\sigma_1 = \Psi^{(0)}(\lambda, x) M. \tag{5.0.9}$$

The matrix M can be expressed in terms of the parameters α and $j_\pm$,

$$M = \mathrm{e}^{i\pi\alpha\sigma_3} i\sigma_2 + \pi \mathrm{e}^{i\pi\alpha\sigma_3} q^{-2\sigma_3} i\sigma_3 \begin{pmatrix} j_+ & \\ & j_- \end{pmatrix}. \tag{5.0.10}$$

Finally, we introduce the connection matrix E between two singular points $\lambda = 0$ and $\lambda = \infty$, as the matrix relating Ψ_1 and $\Psi^{(0)}$,

$$\Psi_1 = \Psi^{(0)}\mathsf{E}, \quad \mathsf{E} = \begin{pmatrix} a & b \\ c & d \end{pmatrix}, \quad \det(\mathsf{E}) = ad - bc = 1.$$

The ambiguity in the definition of $\Psi^{(0)}(\lambda, x)$ implies that E can always be multiplied by a left constant diagonal unimodular matrix d^{σ_3} and, if $\alpha - \frac{1}{2} \in \mathbb{Z}$, it can also be multiplied by a left triangular matrix.

We must now show that all the entries of these matrices can be expressed in terms of only two complex constants. In this respect, we will first establish the fact that $s_{k+3} = s_k$, i.e.,

$$s_4 = s_1, \quad s_5 = s_2, \quad s_6 = s_3. \tag{5.0.11}$$

Equations (5.0.11) are a consequence of the fact that if $\Psi(\lambda, x)$ is a solution of (5.0.3), then $\sigma_1\Psi(\mathrm{e}^{i\pi}\lambda, x)\sigma_1$ is also a solution of (5.0.3); see Remark 5.1. The solutions $\Psi_k(\lambda, x)$ and $\sigma_1\Psi_{k+3}(\lambda\mathrm{e}^{-i\pi}, x)\sigma_1$ have the same asymptotic behavior as $\lambda \to \infty$ in Ω_k, thus

$$\Psi_k(\lambda\mathrm{e}^{i\pi}, x) = \sigma_1\Psi_{k+3}(\lambda, x)\sigma_1, \quad k = 1, 2, 3. \tag{5.0.12}$$

Equations (5.0.12) imply

$$\mathsf{S}_{k+3} = \sigma_1\mathsf{S}_k\sigma_1, \quad k = 1, 2, 3,$$

which are equations (5.0.11).

The Stokes multipliers s_1, s_2, s_3 belong to a 2-dimensional complex surface defined by the single equation

$$s_1 + s_2 + s_3 + s_1 s_2 s_3 = -2i \sin \alpha\pi. \tag{5.0.13}$$

This equation follows from the cyclic relation (see Chapter 2)

$$\mathsf{S}_1\mathsf{S}_2\mathsf{S}_3 = \mathsf{E}^{-1}\sigma_1 M^{-1}\mathsf{E}\sigma_1. \tag{5.0.14}$$

In order to derive (5.0.14) we note that, by the symmetry relations and the definitions,

$$\begin{aligned}\Psi_1(\lambda,x) &= \sigma_1\Psi_4(e^{i\pi}\lambda,x)\sigma_1 = \sigma_1\Psi_1(e^{i\pi}\lambda,x)S_1S_2S_3\sigma_1\\ &= \sigma_1\Psi^{(0)}(e^{i\pi}\lambda,x)\mathsf{E}S_1S_2S_3\sigma_1 = \sigma_1\Psi^{(0)}(e^{i\pi}\lambda,x)\sigma_1\,\sigma_1\mathsf{E}S_1S_2S_3\sigma_1\\ &= \Psi^{(0)}(\lambda,x)M\sigma_1\mathsf{E}S_1S_2S_3\sigma_1 = \Psi_1(\lambda,x)\mathsf{E}^{-1}M\sigma_1\mathsf{E}S_1S_2S_3\sigma_1.\end{aligned}$$

Comparing the initial and the final expressions in the sequence, we conclude that

$$\mathsf{E}^{-1}M\sigma_1\mathsf{E}S_1S_2S_3\sigma_1 = I,$$

which is equation (5.0.14). Because

$$M\sigma_1 = \begin{pmatrix} e^{i\pi\alpha} & J_+ \\ J_- & -e^{-i\pi\alpha}\end{pmatrix}, \quad J_\pm = \pm i\pi j_\pm e^{\pm i\pi\alpha}q^{\mp 2}, \quad J_+J_- = 0, \tag{5.0.15}$$

the matrix E^{-1} diagonalizes the product $\mathsf{E}S_1S_2S_3\sigma_1$ for $\alpha-\frac{1}{2}\notin\mathbb{Z}$ since in this case $J_+ = J_- = 0$.

REMARK 5.2. The parameters $J_\pm$ are multiplied by $q^{\mp 2}$ and therefore they are determined only up to an arbitrary constant factor.

A left multiplication of (5.0.14) by E yields a homogeneous system of linear algebraic equations for the entries of E,

$$\begin{aligned}&a(1+s_2s_3)+b(s_1+s_3+s_1s_2s_3-e^{-i\pi\alpha})-dJ_+=0,\\ &-bJ_-+c(1+s_2s_3)+d(s_1+s_3+s_1s_2s_3+e^{i\pi\alpha})=0,\\ &a(s_2-e^{-i\pi\alpha})+b(1+s_1s_2)-cJ_+=0,\\ &-aJ_-+c(s_2+e^{i\pi\alpha})+d(1+s_1s_2)=0.\end{aligned}\tag{5.0.16}$$

The triviality of the determinant to this system (which is the necessary condition of its nontrivial solvability), coincides with (5.0.13). Actually, if the condition (5.0.13) holds true, then the rank of the system (5.0.16) does not exceed 2.

The system (5.0.16) together with the condition $\det\mathsf{E}=1$, determine the connection matrix E, modulo the above ambiguity, as a function of the Stokes multipliers,

$$\alpha-\frac{1}{2}\notin\mathbb{Z}:$$

$$1+s_1s_2\neq 0:\quad \mathsf{E} = \begin{pmatrix} d & 0\\ 0 & d^{-1}\end{pmatrix}\begin{pmatrix} 1 & \dfrac{e^{-i\pi\alpha}-s_2}{1+s_1s_2}\\ -\dfrac{1+s_1s_2}{2\cos\alpha\pi} & \dfrac{e^{i\pi\alpha}+s_2}{2\cos\alpha\pi}\end{pmatrix}, \quad d\neq 0,$$

$$1+s_2s_3\neq 0:\quad \mathsf{E} = \begin{pmatrix} d & 0\\ 0 & d^{-1}\end{pmatrix}\begin{pmatrix} \dfrac{s_2+e^{i\pi\alpha}}{1+s_2s_3} & 1\\ \dfrac{s_2-e^{-i\pi\alpha}}{2\cos\alpha\pi} & \dfrac{1+s_2s_3}{2\cos\alpha\pi}\end{pmatrix}, \quad d\neq 0,$$

$$1+s_1s_2 = 1+s_2s_3 = 0:\quad \begin{cases} s_2 = -e^{i\pi\alpha} & \Rightarrow\quad \mathsf{E} = d^{\sigma_3}i\sigma_2,\\ s_2 = e^{-i\pi\alpha} & \Rightarrow\quad \mathsf{E} = d^{\sigma_3}.\end{cases}\tag{5.0.17}$$

$$\alpha = n - \frac{1}{2}, \quad n \in \mathbb{N}:$$

$$(5.0.18)\qquad \begin{aligned} 1 + s_1 s_2 \neq 0: \quad \mathsf{E} &= \begin{pmatrix} 1 & p \\ 0 & 1 \end{pmatrix} \Big(-\frac{J_+}{1+s_1s_2}\Big)^{\frac{\sigma_3}{2}} \begin{pmatrix} 0 & 1 \\ -1 & -\dfrac{s_2 + e^{i\pi\alpha}}{1+s_1s_2} \end{pmatrix}, \\ 1 + s_2 s_3 \neq 0: \quad \mathsf{E} &= \begin{pmatrix} 1 & p \\ 0 & 1 \end{pmatrix} \Big(\frac{J_+}{1+s_2s_3}\Big)^{\frac{\sigma_3}{2}} \begin{pmatrix} 1 & 0 \\ \dfrac{s_2 - e^{-i\pi\alpha}}{1+s_2s_3} & 1 \end{pmatrix}, \end{aligned}$$

$$\alpha = \frac{1}{2} - n, \quad n \in \mathbb{N}:$$

$$(5.0.19)\qquad \begin{aligned} 1 + s_1 s_2 \neq 0: \quad \mathsf{E} &= \begin{pmatrix} 1 & 0 \\ p & 1 \end{pmatrix} \Big(\frac{J_-}{1+s_1s_2}\Big)^{-\frac{\sigma_3}{2}} \begin{pmatrix} 1 & \dfrac{s_2 - e^{-i\pi\alpha}}{1+s_1s_2} \\ 0 & 1 \end{pmatrix}, \\ 1 + s_2 s_3 \neq 0: \quad \mathsf{E} &= \begin{pmatrix} 1 & 0 \\ p & 1 \end{pmatrix} \Big(\frac{J_-}{1+s_2s_3}\Big)^{-\frac{\sigma_3}{2}} \begin{pmatrix} \dfrac{s_2 + e^{i\pi\alpha}}{1+s_2s_3} & 1 \\ -1 & 0 \end{pmatrix}. \end{aligned}$$

For a half-integer α, the equations $1 + s_1s_2 = 1 + s_2s_3 = 0$ and (5.0.13), imply the equalities

$$(5.0.20)\qquad \alpha = \frac{1}{2} + n, \quad s_1 = s_2 = s_3 = i(-1)^{n+1}, \quad n \in \mathbb{Z}, \quad J_+ = J_- = 0,$$

and provide no restriction to the connection matrix E. Taking into account the ambiguity in the definition of E, it follows that E contains a single free parameter. This parameter, c, can be specified using the Gauss LU decomposition,

$$(5.0.21)\qquad \begin{aligned} \alpha = \frac{1}{2} - n, \quad n \in \mathbb{N}: \quad \mathsf{E} &= \begin{pmatrix} 1 & 0 \\ p & 1 \end{pmatrix} \begin{pmatrix} d & 0 \\ & d^{-1} \end{pmatrix} \begin{pmatrix} 1 & c \\ 0 & 1 \end{pmatrix}, \\ \alpha = n - \frac{1}{2}, \quad n \in \mathbb{N}: \quad \mathsf{E} &= \begin{pmatrix} 1 & p \\ 0 & 1 \end{pmatrix} \begin{pmatrix} d & \\ & d^{-1} \end{pmatrix} \begin{pmatrix} 1 & 0 \\ c & 1 \end{pmatrix}, \end{aligned}$$

where p and d are arbitrary.

We point out that equations (5.0.20) specify the points of the monodromy surface $s_1 + s_2 + s_3 + s_1s_2s_3 = -2i \sin \pi\alpha$ (5.0.13) where the gradient vanishes. Thus we obtain the description of the set of the monodromy data of equation (5.0.3), as the union of the 2-dimensional complex surface (5.0.13) in a 3-dimensional complex space with the coordinates (s_1, s_2, s_3), and of the complex line $\mathbb{CP}^1$ associated to the special point (5.0.20) of this surface parametrized by a parameter involved in the connection matrix E.

The collection of the above results is usually referred to as the direct problem. These results are summarized in Proposition 5.1.

PROPOSITION 5.1 (The direct problem for the Flaschka-Newell Lax pair). *Let $u(x)$, $v(x)$ satisfy (5.0.2),*

$$u_x = v, \quad v_x = 2u^3 + xu - \alpha. \tag{5.0.22}$$

Let $\Psi_k(\lambda, x)$, $k = 1, \ldots, 6$, be the unimodular solutions of the linear ODE

$$\frac{\partial \Psi}{\partial \lambda} = \Big\{-i(4\lambda^2 + x + 2u^2)\sigma_3 + \Big(4\lambda u + \frac{\alpha}{\lambda}\Big)\sigma_1 - 2v\sigma_2\Big\}\Psi,$$

which are uniquely specified by the asymptotic conditions

$$\Psi_k(\lambda, x) = \big(I + O(\lambda^{-1})\big)\,\mathrm{e}^{-i(\frac{4}{3}\lambda^3 + x\lambda)\sigma_3}, \quad \lambda \to \infty, \quad \lambda \in \Omega_k,$$

where Ω_1, $\Omega_2, \ldots, \Omega_6$, are the open sectors of the complex λ-plane defined by

$$\Omega_k = \Big\{\lambda \in \mathbb{C}: \quad \arg\lambda \in \big(-\tfrac{2\pi}{3} + \tfrac{\pi}{3}k, \tfrac{\pi}{3}k\big)\Big\}. \tag{5.0.23}$$

Then the functions $\Psi_k(\lambda, x)$ are related by the equations

$$\Psi_{k+1}(\lambda, x) = \Psi_k(\lambda, x)\mathsf{S}_k, \qquad k = 1, \ldots, 5;$$

$$\Psi_1(\lambda, x) = \Psi_6(\lambda \mathrm{e}^{2i\pi}, x)\mathsf{S}_6,$$

where the constant matrices S_k, $k = 1, \ldots, 6$, are defined by

$$\mathsf{S}_1 = \begin{pmatrix} 1 & 0 \\ s_1 & 1 \end{pmatrix}, \quad \mathsf{S}_2 = \begin{pmatrix} 1 & s_2 \\ 0 & 1 \end{pmatrix}, \quad \mathsf{S}_3 = \begin{pmatrix} 1 & 0 \\ s_3 & 1 \end{pmatrix},$$

$$\mathsf{S}_{k+1} = \sigma_1 \mathsf{S}_k \sigma_1, \quad k = 1, 2, 3,$$

and the constants s_1, s_2, s_3 satisfy

$$s_1 + s_2 + s_3 + s_1 s_2 s_3 = -2i \sin \alpha\pi. \tag{5.0.24}$$

This equation describes a 2-dimensional complex surface in $\mathbb{C}^3$ with the coordinates (s_1, s_2, s_3). For $\alpha = \frac{1}{2} + n$, $n \in \mathbb{Z}$, this surface has a unique special point, $s_1 = s_2 = s_3 = i(-1)^{n+1}$. The complete set of the monodromy data consists of the union of the surface (5.0.24) and the space $\mathbb{CP}^1$ associated to the special point of the surface.

Given α, and (s_1, s_2, s_3) belonging to the above surface, let the matrix E be defined by one of the relevant equations (5.0.17)–(5.0.21). Then the unimodular matrix $\Psi^{(0)}(\lambda, x)$ defined by

$$\Psi_1(\lambda, x) = \Psi^{(0)}(\lambda, x)\mathsf{E},$$

has the asymptotics (5.0.8)

$$\Psi^{(0)}(\lambda, x) = \frac{1}{\sqrt{2}}(I - i\sigma_2)\hat{\Psi}^{(0)}(\lambda, x)\lambda^{\alpha\sigma_3}\lambda^{\mathsf{N}}q^{\sigma_3},$$

where $\hat{\Psi}^{(0)}(\lambda, x)$ is holomorphic at $\lambda = 0$.

Proposition 5.1 shows that if $u(x)$ solves Painlevé II, then there exist unimodular functions $\Psi_1(\lambda, x), \ldots, \Psi_6(\lambda, x)$ which have the following properties: (a) are holomorphic in the entire finite complex λ-plane except at $\lambda = 0$; (b) have a well defined behavior as $\lambda \to \infty$ in certain sectors; (c) are related by constant matrices which involve only two complex constants; (d) have a well defined behavior as $\lambda \to 0$.

Following the procedure introduced in Chapter 3, it can be shown that these properties can be used to formulate a RH problem for a sectionally holomorphic function $Y(\lambda, x)$ in terms of the two complex constants s_1 and s_2. The function Y has different representations in different regions of the complex λ-plane: $Y = Y_0$ in the finite complex λ-plane, and $Y = Y_k$ in the sector Ω_k away from $\lambda = 0$ where Y_0 and Y_k are defined by

$$\Psi^{(0)}(\lambda, x) = Y_0 \mathrm{e}^{-i(\frac{4}{3}\lambda^3 + x\lambda)} \lambda^{\alpha\sigma_3} \lambda^{\mathrm{N}} q^{\sigma_3},$$

(5.0.25)

$$\Psi_k(\lambda, x) = Y_k(\lambda, x) \mathrm{e}^{-i(\frac{4}{3}\lambda^3 + x\lambda)\sigma_3}, \quad k = 1, \ldots, 6.$$

Equations (5.0.4) and (5.0.25b) imply that the function Y satisfies

$$\frac{\partial Y}{\partial x} + i\lambda[\sigma_3, Y] = u\sigma_1 Y, \quad \lambda \neq 0, \tag{5.0.26}$$

where [,] denotes the usual matrix commutator. Substituting

$$Y = I + \frac{Y_{-1}}{\lambda} + O\left(\frac{1}{\lambda^2}\right), \quad \lambda \to \infty$$

into equation (5.0.26) we find

$$u(x) = 2i(Y_{-1}(x))_{12} \tag{5.0.27}$$

where $(Y_{-1})_{12}$ denotes the (1,2) entry of the 2×2 matrix Y_{-1}.

In summary, if $u(x)$ satisfies the Painlevé II equation, then the 2×2 matrix $Y(\lambda, x)$ defined by equations (5.0.25), where $\Psi^{(0)}$ and Ψ_k are appropriate solutions of (5.0.3), satisfies a certain RH problem. Furthermore, $u(x) = 2i(Y_{-1})_{12}$ where $Y_{-1}(x)$ is the $O(\frac{1}{\lambda})$ term of the large λ asymptotic expansion of $Y(\lambda, x)$. One might expect that the *inverse* is also true, namely one might expect that: if $Y(\lambda, x)$ is *defined* as the solution of the above RH problem for some given complex constants s_1 and s_2, and if $u(x)$ is *defined* by (5.0.27), then $u(x)$ solves Painlevé II. This is indeed the case as will be proved in Theorem 5.1.

THEOREM 5.1 (The Solution of Painlevé II). *Let three complex constants $\{s_1, s_2, s_3\}$ satisfy the algebraic equation*

$$s_1 + s_2 + s_3 + s_1 s_2 s_3 = -2i \sin \alpha\pi. \tag{5.0.28}$$

Let the unimodular 2×2 complex-valued matrix function $Y(\lambda, x)$, $\lambda \in \mathbb{C}$, $x \in \mathbb{C}$, has the following representation in the complex λ-plane:

$$Y(\lambda, x) = \begin{cases} Y_0(\lambda, x), & \text{for } \lambda \text{ in the finite complex plane,} \\ Y_k(\lambda, x), & \text{for } \lambda \neq 0 \text{ and } \lambda \text{ in } \Omega_k, \quad k = 1, \ldots, 6, \end{cases}$$

where $\Omega_1, \ldots, \Omega_6$ are the sectors defined in (5.0.23) *and depicted in Figure* 5.2. *Let*

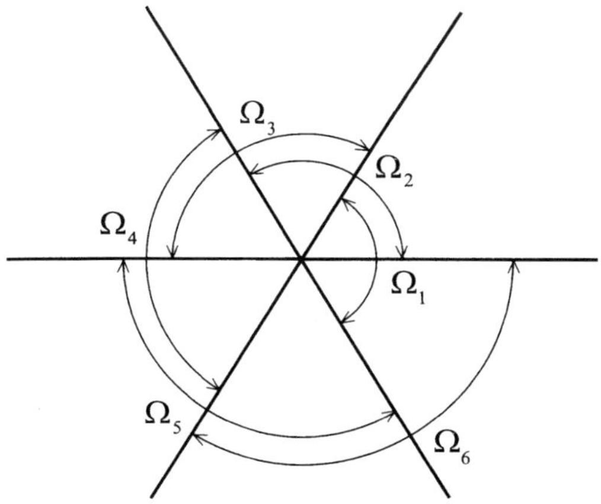

FIGURE 5.2. The Stokes sectors associated with Painlevé II

$Y(\lambda, x)$ satisfy the following Riemann-Hilbert problem:

$$Y_{k+1}(\lambda, x) = Y_k(\lambda, x) \mathrm{e}^{-i(\frac{4}{3}\lambda^3 + x\lambda)\sigma_3} \mathsf{S}_k \mathrm{e}^{i(\frac{4}{3}\lambda^3 + x\lambda)\sigma_3}, \tag{5.0.29}$$
$$k = 1, \dots, 6; \quad Y_7 = Y_1;$$

$$Y_1(\lambda, x) = Y_0(\lambda, x) \mathrm{e}^{-i(\frac{4}{3}\lambda^3 + x\lambda)\sigma_3} \lambda^{\alpha\sigma_3} \lambda^{\mathsf{N}} q^{\sigma_3} \mathsf{E} \mathrm{e}^{i(\frac{4}{3}\lambda^3 + x\lambda)\sigma_3}; \tag{5.0.30}$$

$$Y_1(\lambda, x) = I + O\left(\frac{1}{\lambda}\right), \quad \lambda \to \infty; \tag{5.0.31}$$

where the constant matrices $\mathsf{S}_1, \dots, \mathsf{S}_6$ are defined in terms of $\{s_1, s_2, s_3\}$ by

$$\mathsf{S}_1 = \begin{pmatrix} 1 & 0 \\ s_1 & 1 \end{pmatrix}, \quad \mathsf{S}_2 = \begin{pmatrix} 1 & s_2 \\ 0 & 1 \end{pmatrix}, \quad \mathsf{S}_3 = \begin{pmatrix} 1 & 0 \\ s_3 & 1 \end{pmatrix},$$

$$\mathsf{S}_{k+1} = \sigma_1 \mathsf{S}_k \sigma_1, \quad k = 1, 2, 3,$$

and E is defined by an appropriate formula of (5.0.17)–(5.0.21). Let $u(x)$ be defined by

$$u(x) = 2i \lim_{\lambda \to \infty} \lambda (Y_1(\lambda, x))_{12}, \tag{5.0.32}$$

where the subscript 12 denotes the $(1,2)$ entry of the matrix Y_1. Then:

(a) The above Riemann-Hilbert problem defines a unique 2×2 matrix $Y(\lambda, x)$ which is a meromorphic function of x.

(b) The function $u(x)$ is a meromorphic function of x and it satisfies the Painlevé II equation.

PROOF. The proof consists of two parts: (a) We will show that it is possible to choose a contour in the complex λ-plane such that on this contour the global "jump" matrix of the above Riemann-Hilbert problem is smooth and it decays

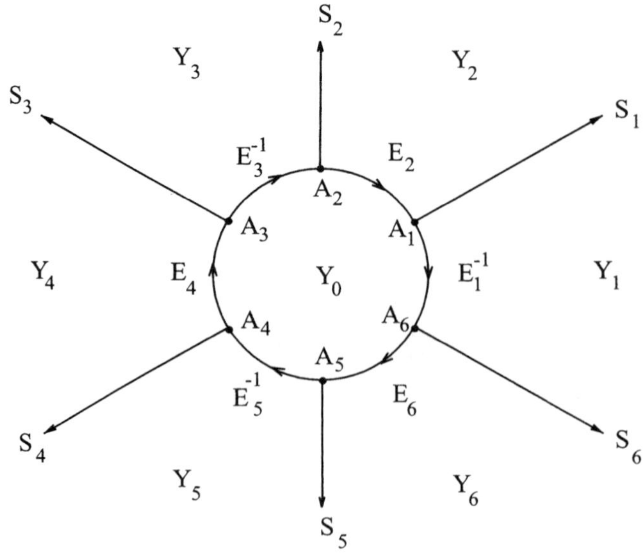

FIGURE 5.3. The Riemann-Hilbert Problem for Painlevé II

exponentially for large λ. Then, the general theory of Riemann-Hilbert problems presented in Chapter 3 implies that $Y(\lambda, x)$ is a meromorphic function of x (see Corollary 3.1).

(b) We will show that if $\Psi^{(0)}$ and Ψ_k are defined in terms of Y_0 and Y_k by equations (5.0.25), and if $u(x)$ is defined by (5.0.32), then $\Psi^{(0)}$ and Ψ_k satisfy equations (5.0.3) and (5.0.4). This implies that $u(x)$ solves Painlevé II.

(a) For convenience we choose the contour L consisting of the union of the unit circle and of the anti-Stokes rays extending from the unit circle to infinity; see Figure 5.3. The anti-Stokes rays are the rays $\arg[i\lambda^3] = k\pi$, $k \in \mathbb{Z}$, i.e., $\arg\lambda = -\frac{\pi}{6}, \frac{\pi}{6}, \frac{3\pi}{6}, \frac{5\pi}{6}, \frac{7\pi}{6}, \frac{9\pi}{6}$.

Let

$$\theta(\lambda, x) = -i\left(\frac{4}{3}\lambda^3 + x\lambda\right)\sigma_3.$$

Equations (5.0.29) and (5.0.30) imply

$$Y_2 = Y_1 e^{\theta} S_1 e^{-\theta}, \quad Y_2 = Y_3 e^{\theta} S_2^{-1} e^{-\theta}, \quad Y_4 = Y_3 e^{\theta} S_3 e^{-\theta},$$

$$Y_4 = Y_5 e^{\theta} S_4^{-1} e^{-\theta}, \quad Y_6 = Y_5 e^{\theta} S_5 e^{-\theta}, \quad Y_6 = Y_1 e^{\theta} S_6^{-1} e^{-\theta},$$

$$Y_0 = Y_1 e^{\theta} f_1^{-1} e^{-\theta}, \quad Y_2 = Y_0 e^{\theta} f_2 e^{-\theta}, \quad Y_0 = Y_3 e^{\theta} f_3^{-1} e^{-\theta},$$

$$Y_4 = Y_0 e^{\theta} f_4 e^{-\theta}, \quad Y_0 = Y_5 e^{\theta} f_5^{-1} e^{-\theta}, \quad Y_6 = Y_0 e^{\theta} f_6 e^{-\theta},$$

$$f_k = \lambda^{\alpha\sigma_3}\lambda^{\mathsf{N}} q^{\sigma_3} \mathsf{E} \mathsf{S}_1 \ldots \mathsf{S}_{k-1}, \quad k = 1, \ldots, 6, \quad \mathsf{S}_0 = I.$$

These equations define the following "jump" condition for the matrix $Y(\lambda, x)$,

$$Y_+(\lambda, x) = Y_-(\lambda, x) e^{\theta(\lambda,x)} V(\lambda) e^{-\theta(\lambda,x)}, \quad \lambda \in L,$$

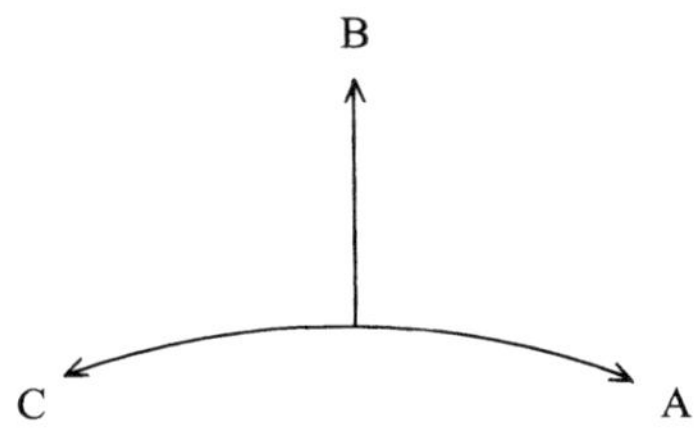

FIGURE 5.4. An intersection point

where the contour L is depicted in Figure 5.3 and $V(\lambda) = \mathrm{S}_1$ on $\arg\lambda = \frac{\pi}{6} \cap \{|\lambda| > 1\}$; $V(\lambda) = \mathrm{S}_2^{-1}$ on $\arg\lambda = \frac{\pi}{2} \cap \{|\lambda| > 1\}$, etc., see Figure 5.3.

We note that the jump matrix $\mathrm{e}^{\theta}V(\lambda)\mathrm{e}^{-\theta}$ decays exponentially for large λ. Also at the intersection points $A_1, A_2, \ldots, A_6$, appropriate cyclic products of the relevant jump matrices equal the unit matrix. Indeed, it can be verified that if the jump matrices A, B, C occur as in Figure 5.4 then $ABC = I$:

$$f_1\mathrm{S}_1 f_2^{-1} = I \text{ near } A_1; \quad f_2\mathrm{S}_2 f_3^{-1} = I \text{ near } A_2;$$

$$f_3\mathrm{S}_3 f_4^{-1} = I \text{ near } A_3; \quad f_4\mathrm{S}_4 f_5^{-1} = I \text{ near } A_4;$$

$$f_5\mathrm{S}_5 f_6^{-1} = I \text{ near } A_5; \quad f_6\mathrm{S}_6 f_1^{-1} = I \text{ near } A_6.$$

The conditions at $A_1, \ldots, A_5$, follow from the definition of f_k, $k = 1, \ldots, 6$. The condition at A_6 is a consequence of the fact that the matrix E satisfies the equation (5.0.14). For instance, if $\alpha - \frac{1}{2} \notin \mathbb{Z}$, using the relation $\mathrm{S}_1\mathrm{S}_2\mathrm{S}_3\sigma_1 = \sigma_1\mathrm{S}_4\mathrm{S}_5\mathrm{S}_6$, the square of the above equation yields

$$\mathsf{E}^{-1}\mathrm{e}^{2i\pi\alpha\sigma_3}\mathsf{E} = \mathrm{S}_1\mathrm{S}_2\ldots\mathrm{S}_6,$$

or

$$\mathsf{E}^{-1}\mathrm{e}^{-2i\pi\alpha\sigma_3}\mathsf{E}\cdot(\mathrm{S}_1\mathrm{S}_2\ldots\mathrm{S}_6) = I.$$

This is precisely the equation needed to guarantee the cyclic product condition near A_6:

$$\left(\lambda\mathrm{e}^{2i\pi}\right)^{-\alpha\sigma_3}\mathsf{E}\cdot\mathrm{S}_1\mathrm{S}_2\ldots\mathrm{S}_6\cdot\mathsf{E}^{-1}\lambda^{\alpha\sigma_3} = I.$$

The above cyclic product conditions represent the "smoothness" of the global jump matrix $\mathrm{e}^{\theta}V\mathrm{e}^{-\theta}$ at the intersection points $A_1, \ldots, A_6$. Since the jump matrix is globally smooth, and since it decays for large λ, it is possible to apply to the above Riemann-Hilbert problem the rigorous theory developed in Chapter 3.

(b) We define $\Psi(\lambda, x)$ in terms of $Y(\lambda, x)$ by

$$\Psi = Y\mathrm{e}^{\theta}\lambda^{\alpha\sigma_3}\lambda^{\mathsf{N}}q^{\sigma_3}$$

for λ inside the circle, and by

$$\Psi = Y\mathrm{e}^{\theta}$$

for λ outside the circle. We will show that if Y solves the above RH problem, then Ψ solves equations (5.0.3) and (5.0.4).

We first observe that the symmetry condition $S_{k+3} = \sigma_1 S_k \sigma_1$, $k = 1,2,3$, implies the following symmetry condition for $Y(\lambda, x)$:

$$Y(\lambda e^{i\pi}, x) = \sigma_1 Y(\lambda, x)\sigma_1, \ \lambda \text{ outside the circle.} \tag{5.0.33}$$

This symmetry condition plays a crucial role in what follows.

DERIVATION OF $\Psi_x = \mathsf{U}\Psi$.

(i) We denote by Ψ and Y the functions piecewise defined by Ψ_k, Ψ_0 and Y_k, Y_0. Let U be defined by

$$\mathsf{U} = \Psi_x \Psi^{-1} = (Y_x + Y(-i\lambda\sigma_3))Y^{-1}. \tag{5.0.34}$$

Since the holomorphic functions Y_x and Y have the same jumps on the contour L, it follows that U is holomorphic in the finite complex λ-plane and it has a polynomial dependence on λ at $\lambda = \infty$. Evaluating equation (5.0.34) at $\lambda = \infty$ it follows that

$$\mathsf{U} = U_0(x) + \lambda U_1(x).$$

Hence, equation (5.0.34) becomes

$$Y_x - i\lambda Y\sigma_3 = (U_0(x) + \lambda U_1(x))Y. \tag{5.0.35}$$

Substituting

$$Y = I + \frac{Y_{-1}(x)}{\lambda} + O\left(\frac{1}{\lambda^2}\right), \quad \lambda \to \infty, \tag{5.0.36}$$

into equation (5.0.35) we find

$$U_1 = -i\sigma_3, \quad U_0 = i[\sigma_3, Y_{-1}]. \tag{5.0.37}$$

However, we need to establish that actually $U_0 = u(x)\sigma_1$, where $u(x)$ is some scalar function of x. This follows from equation (5.0.37b) and from the symmetry condition (5.0.33). Indeed, substituting (5.0.36) into (5.0.33) we find

$$Y_{-1}(x) = -\sigma_1 Y_{-1}(x)\sigma_1. \tag{5.0.38}$$

Expanding Y_{-1} in the form $Y_{-1} = a\sigma_1 + b\sigma_2 + c\sigma_3 + dI$ (a, b, c, d scalars), substituting this expression into (5.0.38), and using $\sigma_1\sigma_2 = -\sigma_2\sigma_1$, $\sigma_1\sigma_3 = -\sigma_3\sigma_1$, it follows that $a = d = 0$, i.e.,

$$Y_{-1} = \frac{u(x)}{2}\sigma_2 + c(x)\sigma_3, \tag{5.0.39}$$

where $u(x)$ and $c(x)$ are some scalar functions of x. Substituting equation (5.0.39) into (5.0.37b), and using $[\sigma_3, \sigma_2] = -2i\sigma_1$, we find $U_0 = u(x)\sigma_1$. Hence,

$$\mathsf{U} = U_0 + \lambda U_1 = -i\sigma_3 + u(x)\sigma_1,$$

i.e., $\Psi(\lambda, x)$ solves equation (5.0.4).

DERIVATION OF $\Psi_\lambda = \mathsf{A}\Psi$.

Define A by $\mathsf{A} = \Psi_\lambda \Psi^{-1}$. Then for λ inside the circle

$$\mathsf{A} = \left(Y_\lambda - i(4\lambda^2 + x)Y\sigma_3 - \frac{\alpha}{\lambda}Y\sigma_3\right)Y^{-1}, \tag{5.0.40}$$

while for λ outside the circle

$$\mathsf{A} = (Y_\lambda - i(4\lambda^2 + x)Y\sigma_3)Y^{-1}. \tag{5.0.41}$$

Since the holomorphic functions Y_λ and Y have the same jumps on the contour L, it follows that A is a holomorphic function in $\mathbb{C}\backslash\{0\}$, it has a pole at $\lambda = 0$, and it has a polynomial dependence at $\lambda = \infty$; evaluating equation (5.0.41) at $\lambda = \infty$, it follows that this polynomial dependence involves terms up to and including $O(\lambda^2)$. Thus,

$$A = \frac{1}{\lambda}A_{-1}(x) + A_0(x) + \lambda A_1(x) + \lambda^2 A_2(x). \tag{5.0.42}$$

Equations (5.0.40) and (5.0.42) imply

$$A_{-1}(x) = -\alpha Y(0,x)\sigma_3 Y^{-1}(0,x).$$

However, we must establish that $A_{-1} = \alpha\sigma_1$. This follows from the fact that for λ inside the circle, the solution of the RH problem (5.0.29)–(5.0.31) satisfies

$$Y(0,x)\sigma_3 = -\sigma_1 Y(0,x).$$

In order to determine A_0, A_1, A_2 we use equations (5.0.41) and (5.0.42). These equations imply

$$Y_\lambda - i(4\lambda^2 + x)Y\sigma_3 = \left(\frac{1}{\lambda}A_{-1}(x) + A_0(x) + \lambda A_1(x) + \lambda^2 A_2(x)\right)Y, \tag{5.0.43}$$

where λ is outside the circle. Substituting

$$Y = I + \frac{Y_{-1}(x)}{\lambda} + \frac{Y_{-2}(x)}{\lambda^2} + O\left(\frac{1}{\lambda^2}\right), \quad \lambda \to \infty, \tag{5.0.44}$$

into equation (5.0.43) we find

$$A_2 = -4i\sigma_3, \quad A = 4i[\sigma_3, Y_{-1}], \quad A_0 = -ix\sigma_3 - A_1 Y_{-1} + 4i[\sigma_3, Y_{-2}].$$

Using the expression for Y_{-1} given by (5.0.39), it follows that $A_1 = 4u(x)\sigma_1$, and

$$A_0 = -ix\sigma_3 - 4u(x)\sigma_1 Y_{-1} + 4i[\sigma_3, Y_{-2}]. \tag{5.0.45}$$

Substituting equation (5.0.44) into

$$Y_x + i\lambda[\sigma_3, Y] = u(x)\sigma_1 Y,$$

it follows that

$$(Y_{-1})_x = -i[\sigma_3, Y_{-2}] + u(x)\sigma_1 Y_{-1}. \tag{5.0.46}$$

Thus, equation (5.0.45) becomes

$$A_0 = -ix\sigma_3 - 4(Y_{-1})_x. \tag{5.0.47}$$

Using the expression of Y_{-1} given by (5.0.39), equation (5.0.46) implies

$$\operatorname{diag}(Y_{-1})_x = u(x)\operatorname{diag}(\sigma_1 Y_{-1}) = u(x)\operatorname{diag}(\frac{1}{2}u(x)\sigma_1\sigma_2) = \frac{i}{2}u(x)^2\sigma_3.$$

Also,

$$\text{off diag}(Y_{-1})_x = \frac{1}{2}\frac{du}{dx}\sigma_2.$$

Thus,

$$(Y_{-1})_x = \frac{1}{2}\left(iu^2\sigma_3 + \frac{du}{dx}\sigma_2\right)$$

and (5.0.47) becomes $A_0 = ix\sigma_3 - 2iu^2\sigma_3 - 2\frac{du}{dx}\sigma_2$.

In summary, Ψ solves $\Psi_\lambda = A\Psi$, where A is given by (5.0.42) and

$$A_{-1} = \alpha\sigma_1, \quad A_2 = -4i\sigma_3, \quad A_1 = 4u\sigma_1,$$
$$A_0 = -i(x + 2u^2)\sigma_3 - 2\frac{du}{dx}\sigma_2.$$

□

REMARK 5.3. By virtue of Corollary 3.1 of Chapter 3, in order to prove that the solution of the above RH problem exists and is meromorphic with respect to x, it is sufficient to establish its solvability for at least one value of x. This allows us to use Theorem 5.1 to solve the initial value problem for the Painlevé II equation. Indeed, in this case we can take the point of (5.0.28) corresponding, via the direct monodromy transform, to the initial data. Then the solution of the Riemann-Hilbert problem *a priori* exists at the value x which serves as the initial point.

The proof of the meromorphic solvability for an arbitrary point of the algebraic manifold (5.0.28), can also be established by the alternative "holomorphic bundle" approach which is used in the proof of Theorem 3.4 of Chapter 3 and which is presented in Appendix A. We also note that in Section 5 of this chapter, we will show how one can exploit the "vanishing lemma", i.e., Corollary 3.2 of Chapter 3, to obtain the global solvability results for points of (5.0.28) subject to certain algebraic restrictions.

Theorem 5.1 shows that the Lax pair (5.0.3), (5.0.4) can be used to characterize the solution $u(x)$ of Painlevé II through a RH problem uniquely specified by the monodromy data $\{s_1, s_2, s_3\}$ defined on the algebraic variety (5.0.28). For the asymptotic analysis of $u(x)$, we will use a Lax pair which is related to the above Lax pair by the simple gauge transformation $\Psi = \mathrm{e}^{-\frac{i\pi}{4}\sigma_3}\tilde{\Psi}\mathrm{e}^{\frac{i\pi}{4}\sigma_3}$. Redefining all the canonical solutions Ψ_k, $\Psi^{(0)}$ according to this gauge transformation, the monodromy data are related by the following equations (the superscript FN denotes the objects described in the theory of the Flaschka-Newell Lax pair, i.e., the one we have used so far):

$$s_{2k-1}^{FN} = is_{2k-1}, \quad s_{2k}^{FN} = -is_{2k}, \tag{5.0.48}$$

$$-\mathrm{e}^{i\frac{\pi}{4}\sigma_3}M^{FN}\sigma_1\mathrm{e}^{-i\frac{\pi}{4}\sigma_3} = M\sigma_2 = \begin{pmatrix} -\mathrm{e}^{i\pi\alpha} & J_+ \\ J_- & \mathrm{e}^{-i\pi\alpha} \end{pmatrix}, \tag{5.0.49}$$

$$J_\pm = \mp iJ_\pm^{FN} = \mp i\pi j_\pm^{FN}\mathrm{e}^{\pm i\pi\alpha}q^{\mp 2}, \quad j_\pm = \pm ij_\pm^{FN}, \quad J_+J_- = 0,$$

$$\mathrm{e}^{i\frac{\pi}{4}\sigma_3}\mathsf{E}^{FN}\mathrm{e}^{-i\frac{\pi}{4}\sigma_3} = \mathsf{E}. \tag{5.0.50}$$

We now summarize the results of the direct problem associated with the Lax pair satisfied by $\tilde{\Psi}$ (omitting the tilde), namely

$$\frac{\partial\Psi}{\partial\lambda} = \mathsf{A}\Psi, \tag{5.0.51}$$
$$\mathsf{A}(\lambda, x) = -i(4\lambda^2 + x + 2u^2)\sigma_3 - \left(4\lambda u + \tfrac{\alpha}{\lambda}\right)\sigma_2 - 2v\sigma_1,$$

$$\frac{\partial\Psi}{\partial x} = \mathsf{U}\Psi, \quad \mathsf{U}(\lambda, x) = -i\lambda\sigma_3 - u\sigma_2.$$

This Lax pair is the one which has already appeared in Chapter 4; see (4.2.36).

PROPOSITION 5.2. (The direct problem for another Lax pair). *Let u, v solve the system (5.0.22). Let $\Psi_k(\lambda, x)$, $k = 1, \ldots, 6$, be the unimodular solutions of the linear ODE (5.0.51), which are uniquely specified by the asymptotic condition*

$$\Psi_k(\lambda, x) = \Big(I + O(\lambda^{-1})\Big)e^{-i(\frac{4}{3}\lambda^3 + x\lambda)\sigma_3}, \quad \lambda \to \infty, \quad \lambda \in \Omega_k,$$

where the sectors Ω_k, $k = 1, \ldots, 6$, are defined in (5.0.23). Then the functions $\Psi_k(\lambda, x)$ are related by the equations

$$\Psi_{k+1} = \Psi_k \mathsf{S}_k; \quad \Psi_7 = \Psi_1,$$

where

$$\mathsf{S}_{2k-1} = \begin{pmatrix} 1 & 0 \\ s_{2k-1} & 1 \end{pmatrix}, \quad \mathsf{S}_{2k} = \begin{pmatrix} 1 & s_{2k} \\ 0 & 1 \end{pmatrix}, \quad s_{k+3} = -s_k,$$

and the Stokes multipliers $\{s_1, s_2, s_3\}$ satisfy

$$s_1 - s_2 + s_3 + s_1 s_2 s_3 = -2\sin\alpha\pi. \tag{5.0.52}$$

This equation describes a 2-dimensional complex surface in $\mathbb{C}^3$ with the coordinates (s_1, s_2, s_3). For $\alpha = \frac{1}{2} + n$, $n \in \mathbb{Z}$, this surface has a unique special point, $s_1 = -s_2 = s_3 = (-1)^{n+1}$. The complete set of the monodromy data consists of the union of the surface (5.0.52) and the space $\mathbb{CP}^1$ associated to the special point of the surface.

Given α, and (s_1, s_2, s_3) belonging to the above surface, let the matrix E be defined by the equations (5.0.50) and (5.0.17)–(5.0.21). Then the unimodular matrix $\Psi^{(0)}(\lambda, x)$ defined by

$$\Psi_1(\lambda, x) = \Psi^{(0)}(\lambda, x)\mathsf{E},$$

has the asymptotics

$$\Psi^{(0)}(\lambda, x) = \frac{1}{\sqrt{2}}(I - i\sigma_1)\hat{\Psi}^{(0)}(\lambda, x)\lambda^{\alpha\sigma_3}\lambda^{\mathsf{N}}q^{\sigma_3},$$

where $\hat{\Psi}^{(0)}(\lambda, x)$ is holomorphic at $\lambda = 0$.

PROOF. This proposition follows immediately from Proposition 5.1 using the above gauge transformation. □

The symmetry with respect to $\lambda \mapsto -\lambda$ of the Lax pairs (5.0.3)–(5.0.4) and (5.0.51), implies the existence of a Lax pair on the "folded" complex λ-plane. Indeed, the change of the variable $\lambda^2 = \xi$ supplemented by the gauge transformation

$$\Psi(\xi, x) = \lambda^{\sigma_3/2}\frac{1}{\sqrt{2}}(I + i\sigma_2)\Psi^{FN}(\lambda, x), \quad \xi = \lambda^2, \tag{5.0.53}$$

maps the Flaschka-Newell Lax pair (5.0.3)–(5.0.4), into the following Lax pair,

$$\frac{\partial \Psi}{\partial \xi} = \mathsf{A}\Psi, \quad \frac{\partial \Psi}{\partial x} = \mathsf{U}\Psi,$$

$$\mathsf{A} = \begin{pmatrix} 2u + \dfrac{\alpha + \frac{1}{2}}{2\xi} & 2i\xi + iq \\ 2i + \dfrac{ir}{\xi} & -2u - \dfrac{\alpha + \frac{1}{2}}{2\xi} \end{pmatrix}, \quad \mathsf{U} = \begin{pmatrix} u & i\xi \\ i & -u \end{pmatrix}, \tag{5.0.54}$$

$$q = v + u^2 + \frac{x}{2}, \quad r = -v + u^2 + \frac{x}{2}. \tag{5.0.55}$$

The compatibility condition of (5.0.54) has the form

$$
\begin{cases} u_x = \frac{1}{2}(q-r), \\ q_x = 2uq - \alpha + \frac{1}{2}, \\ r_x = -2ur + \alpha + \frac{1}{2}. \end{cases} \tag{5.0.56}
$$

This system implies $q + r - 2u^2 - x = const$, which agrees with the definitions of q and r (5.0.55), and which shows that the constant is zero. Eliminating q and r from (5.0.56) yields equation PII, $u_{xx} = 2u^3 + xu - \alpha$. Eliminating u and q or u and r from this system, we obtain (within some scaling) the equation number XXXIV in the list of Painlevé and Gambier [**In**]:

$$
r_{xx} = \frac{r_x^2}{2r} + 2r^2 - xr - \frac{(\alpha+\frac{1}{2})^2}{2r}, \qquad u = -\frac{r_x}{2r} + \frac{\alpha+\frac{1}{2}}{2r}, \tag{5.0.57}
$$

or

$$
q_{xx} = \frac{q_x^2}{2q} + 2q^2 - xq - \frac{(\alpha-\frac{1}{2})^2}{2r}, \qquad u = \frac{q_x}{2q} + \frac{\alpha-\frac{1}{2}}{2q}. \tag{5.0.58}
$$

It is interesting that there exists another Lax pair for PII, which was found by M. Jimbo and T. Miwa [**JM**] (although, in an equivalent form it had already been known to Garnier) and whose set of the monodromy data is not isomorphic to the monodromy set of (5.0.3)–(5.0.4). Therefore, these two different Lax pairs cannot be related by an algebraic gauge transformation. This Jimbo-Miwa-Garnier Lax pair is

$$
\begin{gathered} \frac{\partial \Psi}{\partial \lambda} = \mathsf{A}\Psi; \\ \mathsf{A}(\lambda, x) = (\lambda^2 + \tfrac{x}{2} + z)\sigma_3 + y(\lambda - u)\sigma_+ - \tfrac{2}{y}(\lambda z + uz + \alpha + \tfrac{1}{2})\sigma_-, \end{gathered} \tag{5.0.59}
$$

$$
\frac{\partial \Psi}{\partial x} = \mathsf{U}\Psi; \quad \mathsf{U}(\lambda, x) = \frac{\lambda}{2}\sigma_3 + \frac{y}{2}\sigma_+ - \frac{z}{y}\sigma_-, \tag{5.0.60}
$$

where

$$
\sigma_+ = \begin{pmatrix} 0 & 1 \\ 0 & 0 \end{pmatrix}, \qquad \sigma_- = \begin{pmatrix} 0 & 0 \\ 1 & 0 \end{pmatrix}. \tag{5.0.61}
$$

The compatibility condition of equations (5.0.59), (5.0.60) is equivalent to the following system of three first order ODEs,

$$
\begin{cases} u_x = u^2 + z + \frac{x}{2}, \\ z_x = -2uz - \alpha - \frac{1}{2}, \\ y_x = -uy. \end{cases} \tag{5.0.62}
$$

The third of these equations yields

$$
y = \exp\Big\{-\int^x u\,dx\Big\}, \tag{5.0.63}
$$

which gives y up to an arbitrary constant multiplier. This reflects the invariance of the Painlevé functions with respect to a constant diagonal gauge transformation of the Lax pair (5.0.59)–(5.0.60), $\Psi \mapsto c^{\sigma_3}\Psi c^{-\sigma_3}$.

Eliminating z from the first two equations of (5.0.62) yields PII,

$$z = u_x - u^2 - \frac{x}{2}, \quad u_{xx} = 2u^3 + xu - \alpha, \tag{5.0.64}$$

while eliminating from the same equations the variable u yields (to within scaling) equation XXXIV in the list of Painlevé-Gambier [**In**],

$$u = -\frac{z_x}{2z} - \frac{\alpha + \frac{1}{2}}{2z}, \quad z_{xx} = \frac{z_x^2}{2z} - 2z^2 - xz - \frac{(\alpha + \frac{1}{2})^2}{2z}. \tag{5.0.65}$$

The first of the equations (5.0.64) and (5.0.65) constitute a Bäcklund transformation and its inverse relating equations PII and PXXXIV.

In the following proposition, we formulate the monodromy properties of the Garnier system.

PROPOSITION 5.3 (The direct problem for the Garnier Lax pair). *Let $u(x)$, $z(x)$ and $y(x)$ satisfy the system of the first order ODEs (5.0.62). Let $\Psi_k(\lambda, x)$, $k = 1, \dots, 6$, be the unimodular solutions of the linear ODE (5.0.59), which are uniquely specified by the asymptotic conditions*

$$\Psi_k(\lambda, x) = \left(I + O\left(\frac{1}{\lambda}\right)\right) e^{(\frac{1}{3}\lambda^3 + \frac{x}{2}\lambda)\sigma_3} \left(\frac{1}{\lambda}\right)^{(\alpha+\frac{1}{2})\sigma_3}, \tag{5.0.66}$$
$$\lambda \to \infty, \quad \lambda \in \Omega_k,$$

where the open sectors Ω_k are defined by

$$\Omega_k = \left\{\lambda \in \mathbb{C}: \ \arg\lambda \in \left(-\frac{\pi}{2} + k\frac{\pi}{3}, \frac{\pi}{6} + k\frac{\pi}{3}\right)\right\}, \quad k = 1, \dots, 6.$$

Then entire functions Ψ_k are related by the equations

$$\Psi_{k+1}(\lambda, x) = \Psi_k(\lambda, x) S_k, \tag{5.0.67}$$
$$S_{2k-1} = \begin{pmatrix} 1 & s_{2k-1} \\ 0 & 1 \end{pmatrix}, \quad S_{2k} = \begin{pmatrix} 1 & 0 \\ s_{2k} & 1 \end{pmatrix},$$

where

$$S_{k+6} = e^{2\pi i\alpha\sigma_3} S_k e^{-2\pi i\alpha\sigma_3}, \tag{5.0.68}$$
$$S_1 S_2 S_3 S_4 S_5 S_6 = e^{-2\pi i(\alpha+\frac{1}{2})\sigma_3}, \tag{5.0.69}$$

or, in terms of the Stokes multipliers s_k,

$$s_{k+6} = s_k e^{(-1)^{k+1} 4\pi i\alpha}, \tag{5.0.70}$$

$$1 + s_1 s_2 + e^{-2\pi i\alpha}(1 + s_4 s_5) = 0,$$
$$1 + s_2 s_3 + e^{2\pi i\alpha}(1 + s_5 s_6) = 0,$$
$$s_1 + s_3 + s_1 s_2 s_3 = s_5 e^{-2\pi i\alpha},$$
$$s_4 + s_6 + s_4 s_5 s_6 = s_2 e^{-2\pi i\alpha}. \tag{5.0.71}$$

Three among equations (5.0.71) are independent, these three equations describe a smooth 3-dimensional surface in the space $\mathbb{C}^6$ with the coordinates $(s_1, \dots, s_6)$. For a half-integer α, i.e. $\alpha = \frac{1}{2} + n$, $n \in \mathbb{Z}$, the surface (5.0.71) has the unique special point $s_1 = \dots = s_6 = 0$. This is the intersection point of the two linear subvarieties $s_1 = s_3 = s_5 = 0$, $s_2 + s_4 + s_6 = 0$, and $s_2 = s_4 = s_6 = 0$, $s_1 + s_3 + s_5 = 0$.

The diagonal gauge transformation $\Psi \mapsto \tilde{\Psi} = c^{\sigma_3}\Psi c^{-\sigma_3}$, $c = const$, leaves the Painlevé functions u and z invariant, maps the function y by $c^2 y$, and maps the Stokes matrices to their conjugate, $S \mapsto c^{\sigma_3} S c^{-\sigma_3}$, i.e. $s_{2k-1} \mapsto c^2 s_{2k-1}$ and $s_{2k} \mapsto c^{-2} s_{2k-1}$.

Three independent Stokes multipliers s_k, $k \in \mathbb{Z}$, form a complete set of first integrals to the nonlinear system of the three ODEs (5.0.62). The orbits of the diagonal gauge transformation can be parametrized by two of the products $s_{2k-1}s_{2l}$ or by any pair of the ratios s_{2k}/s_{2l}, s_{2k-1}/s_{2l-1}, $k, l \in \mathbb{Z}$. Thus these combinations, form a complete set of first integrals for PII and for Painlevé XXXIV.

PROOF. Equations (5.0.68) and (5.0.70) follow from the equation

$$\Psi_{k+6}\big(e^{2\pi i}\lambda, x\big) = \Psi_k(\lambda, x)e^{-i\pi(2\alpha+1)\sigma_3}. \tag{5.0.72}$$

The cyclic relation (5.0.69) follows from the fact that all solutions of (5.0.59) are entire functions,

$$\Psi_7(e^{2i\pi}\lambda, x) = \Psi_7(\lambda, x) = \Psi_1(\lambda, x)S_1S_2S_3S_4S_5S_6,$$

as well as from the equation $\Psi_7(e^{2i\pi}\lambda, x) = \Psi_1(\lambda, x)e^{-2\pi i(\alpha+\frac{1}{2})}$.

The geometrical properties of the monodromy surface follow from the elementary algebraic properties of the system (5.0.71).

All the remaining assertions are obvious. □

This proposition can be used to formulate a RH problem for a sectionally holomorphic function $Y(\lambda, x)$:

$$Y(\lambda, x)\Big|_{\lambda\in\omega_k} = Y_k(\lambda, x) = \Psi_k(\lambda, x)e^{-\theta\sigma_3}, \quad \theta = \tfrac{1}{3}\lambda^3 + \tfrac{x}{2}\lambda - (\alpha + \tfrac{1}{2})\ln\lambda, \tag{5.0.73}$$

where

$$\omega_k = \Big\{\lambda \in \mathbb{C}: \quad \arg\lambda \in \big(\tfrac{\pi}{3}(k-1), \tfrac{\pi}{3}k\big)\Big\}, \quad k = 1, \dots, 6,$$

and the main branch of the logarithm is chosen.

Across the rays $\gamma_k = \big\{\lambda \in \mathbb{C}: \quad \arg\lambda = \frac{\pi}{3}k\big\}$, $k = 1, \dots, 6$, oriented toward infinity, the following jump conditions are valid:

$$Y_+(\lambda, x) = Y_-(\lambda, x)e^{\theta\sigma_3}S_k e^{-\theta\sigma_3}, \quad \lambda \in \gamma_k, \quad k = 1, \dots, 6, \tag{5.0.74}$$

where the constant matrices S_k satisfy the condition (5.0.69). In this formula, for the jump across γ_6, ($\arg\lambda = 2\pi$), we have used the following:

$$\Psi_1(e^{-2\pi i}\lambda, x) = \Psi_7(\lambda, x)e^{2\pi i(\alpha+\frac{1}{2})\sigma_3} = \Psi_6(\lambda, x)S_6e^{2\pi i(\alpha+\frac{1}{2})\sigma_3};$$

then using $\Psi_6(\lambda, x) = Y_6(\lambda, x)e^{\theta(\lambda,x)\sigma_3}$ and

$$\Psi_1(e^{-2\pi i}\lambda, x) = Y_1(e^{-2\pi i}\lambda, x)e^{\theta(e^{-2\pi i}\lambda,x)\sigma_3} = Y_1(e^{-2\pi i}\lambda, x)e^{\theta(\lambda,x)\sigma_3}e^{2\pi i(\alpha+\frac{1}{2})\sigma_3},$$

we obtain (5.0.74) for $k = 6$. Observe that, due to the triangular structure of the Stokes matrices S_k, $k = 1, \dots, 6$, the jump matrices in (5.0.74), approach the unit matrix exponentially fast. Furthermore, using the cyclic relation (5.0.69), the product of all the jump matrices approaches the unit matrix as $\lambda \to 0$; this constitutes the continuity of the RH problem at the node point. The above conditions together with the asymptotic condition at infinity,

$$Y(\lambda, x) \to I, \quad \lambda \to \infty,$$

define a well-posed continuous RH problem. Then the usual arguments imply a unique meromorphic in x solution. In fact, within a simple scaling, this RH problem is equivalent to the RH problem we discussed in Chapter 3; see Theorem 3.4.

We conclude this paragraph, by explaining how the relevant Painlevé functions can be obtained from the solution of this RH problem. Using the procedure described in Chapter 1 (see (1.1.43)–(1.1.44)), we find the following asymptotics of $Y(\lambda, x)$ as $\lambda \to \infty$,

$$Y(\lambda, x) = I + \lambda^{-1} Y_1(x) + \lambda^{-2} Y_2(x) + O(\lambda^{-3}), \tag{5.0.75}$$

where

$$Y_1(x) = \begin{pmatrix} -H & -\frac{y}{2} \\ -\frac{z}{y} & H \end{pmatrix}, \tag{5.0.76}$$

$$Y_2(x) = \begin{pmatrix} \frac{1}{2}H^2 + \frac{z}{4} - \frac{x}{4}(\alpha + \frac{1}{2}) & \frac{y}{2}(u - H) \\ -\frac{z}{y}(u + \frac{\alpha + \frac{1}{2}}{z} - H) & \frac{1}{2}H^2 + \frac{z}{4} + \frac{x}{4}(\alpha + \frac{1}{2}) \end{pmatrix},$$

with

$$H = \tfrac{1}{2}z(z + x) + zu^2 + u\big(\alpha + \tfrac{1}{2}\big).$$

Thus, the "residue" $Y_1(x)$, of $Y(\lambda, x)$ at infinity yields:
(a) the integral of the second Painlevé function $\exp\{-\int^x u\, dx\}$,

$$y = -2(Y_1)_{12};$$

(b) the solution z of Painlevé-Gambier function XXXIV,

$$z = 2(Y_1)_{12}(Y_1)_{21},$$

(c) the function H, related to the Hamiltonian of PII (see [**Ok2**]),

$$H = -(Y_1)_{11}.$$

The function u follows by evaluating the logarithmic derivative of y. Alternatively, u is given by

$$u = -\frac{(Y_2)_{12}}{(Y_1)_{12}} - (Y_1)_{11},$$

where $Y_2(x)$ is the next term in the asymptotic expansion of $Y(\lambda, x)$.

Summary of the Isomonodromy Method

Let the unknown complex-valued matrix function $U(x)$, $x \in \mathbb{C}$, satisfy a given system of nonlinear ODEs. Assume that this system can be written as the compatibility condition of two linear equations of the $N \times N$ complex-valued matrix function $\Psi(x, \lambda)$, $\lambda \in \mathbb{C}$,

$$\frac{\partial \Psi(\lambda, x)}{\partial \lambda} = \mathsf{A}(\lambda, x)\Psi(\lambda, x), \tag{5.0.77}$$

$$\frac{\partial \Psi(\lambda, x)}{\partial x} = \mathsf{U}(\lambda, x)\Psi(\lambda, x), \tag{5.0.78}$$

where $\mathsf{A}(\lambda, x)$, $\mathsf{U}(\lambda, x)$ are given $N \times N$ complex-valued matrices.

Assume that the linear ODE (5.0.77) possesses:

(a) n regular singular points $\lambda_1, \ldots, \lambda_n$.
(b) m irregular singular points $\lambda^1, \ldots, \lambda^m$; let $\Omega_1^k, \ldots, \Omega_{M_k}^k$, be the Stokes sectors associated with the point λ^k, $k = 1, \ldots, m$.

The isomonodromy method involves the following steps.

1. Define certain sets of canonical solutions of equation (5.0.77): (a) Ψ_k, $k = 1, \ldots, n$, is a unimodular solution valid in the neighborhood of λ_k. (b) For each $k, k = 1, \ldots, m$, the solutions $\Psi_1^k, \ldots, \Psi_{M_k}^k$ are the canonical solutions associated with λ^k, i.e., these solutions are the unimodular solutions uniquely defined by their asymptotic behavior as $\lambda \to \lambda^k$ in $\Omega_1^k, \ldots, \Omega_{M_k}^k$.

2. Relate the above solutions: (a) For each k, $k = 1, \ldots, m$, the solutions $\Psi_1^k, \ldots, \Psi_{M_k}^k$ are related by appropriate $N \times N$ Stokes matrices $S_1^k, \ldots, S_{M_k}^k$. (b) The solutions $\Psi_1, \ldots, \Psi_n$ and $\Psi_1^1, \ldots, \Psi_1^m$ are related by a set of $N \times N$ unimodular matrices which will be denoted by E^{ij}.

3. Introduce the monodromy data: (a) Use all possible symmetries of equation (5.0.77) to express the entries of the sets of the matrices $\{\mathsf{E}\}$ and $\{S\}$ in terms of a minimal set of complex scalars. (b) Use the fact that equation (5.0.78) defines an isomonodromic deformation of equation (5.0.77), to show that the above complex scalars are actually constants (i.e., x-independent). This set of constant complex numbers will be referred to as the set of monodromy data.

The implementation of the above steps will be referred to as the investigation of the *direct problem.* In summary, the direct problems established that if $U(x)$ satisfies the given system of nonlinear ODEs then there exist canonical solutions of equation (5.0.77) which are: (a) holomorphic in the universal covering of the entire complex λ-plane punctured at the singular points; (b) have a well-defined behavior as $\lambda \to \lambda_k$, $k = 1, \ldots, n$, and as $\lambda \to \lambda^k$, $k = 1, \ldots, m$; (c) are related by x-independent matrices, whose entries can be expressed in terms of the monodromy data.

The above properties can be used to formulate a RH problem for the sectionally holomorphic (on the complex λ-plane) function Y. This function, which has different representations in different regions of the complex λ-plane, is related to Ψ_k, $k = 1, \ldots, n$, and to $\Psi_1^k, \ldots, \Psi_{M_k}^k$, $k = 1, \ldots, m$, through simple explicit transformations (similar to those appearing in equation (5.0.25)). The remaining steps in the implementation of the isomonodromy method involve *postulating* and investigating the above RH problem:

4. *Define* $Y(\lambda, x)$ as the solution of the RH problem suggested by the investigation of the direct problem. Show that it is possible to choose an appropriate contour L such that on this contour, the jump matrix decays for large λ and it is globally smooth. This implies that $Y(\lambda, x)$ is meromorphic in x, if it exists for at least one value of x, and that it can be obtained through a linear integral equation uniquely defined by the monodromy data. At this stage, for the analysis of the meromorphic solvability of the Riemann-Hilbert problem one can use either the techniques of the Fredholm theory related to the "vanishing lemmas", or an alternative approach based on the theory of holomorphic vector bundles. (For general systems, the Fredholm method and the holomorphic bundles approach have been developed in [**BS**] and [**Mal**], respectively.) We note that in the general case, the "meromorphicity in x" means the meromorphicity on the universal covering of a punctured x-plane.

5. *Define* $U(x)$ in terms of $Y(\lambda, x)$ in such a way that $Y(\lambda, x)$ satisfies a system of linear equations similar to (5.0.77), (5.0.78). Show that the compatibility condition of these equations implies that $U(x)$ satisfies the given system of nonlinear ODEs. This implies that $U(x)$ is meromorphic in x (more precisely, in the universal

covering of a punctured x-plane), and that it can be parametrized (through the solution of a linear integral equation) by the monodromy data.

1. The fourth Painlevé transcendent

Let $u(x)$ satisfy the Painlevé IV equation

$$\frac{d^2u}{dx^2} = \frac{1}{2u}\left(\frac{du}{dx}\right)^2 + \frac{3}{2}u^3 + 4xu^2 + (2 + 2x^2 - 4\Theta_\infty)u - \frac{8\Theta^2}{u}, \tag{5.1.1}$$

where Θ and Θ_∞ are constants. Let $y(x)$ satisfy

$$\frac{1}{y}\frac{dy}{dx} = -u - 2x. \tag{5.1.2}$$

Equations (5.1.1) and (5.1.2) possess the following Lax pair,

$$\begin{aligned}\frac{\partial \Psi}{\partial \lambda} = \mathsf{A}\Psi;\quad \mathsf{A} = \left(\lambda + x + \frac{1}{\lambda}(\Theta - z)\right)\sigma_3 \\ +y\left(1 - \frac{u}{2\lambda}\right)\sigma_+ + \frac{2}{y}\left(z - \Theta - \Theta_\infty + \frac{z}{\lambda u}(z - 2\Theta)\right)\sigma_-,\end{aligned} \tag{5.1.3}$$

$$\frac{\partial \Psi}{\partial x} = \mathsf{U}\Psi;\quad \mathsf{U} = \lambda\sigma_3 + y\sigma_+ + \frac{2}{y}(z - \Theta - \Theta_\infty)\sigma_-, \tag{5.1.4}$$

where $z(x)$ is uniquely defined in terms of u by

$$4z(x) = -\frac{du}{dx} + u^2 + 2xu + 4\Theta, \tag{5.1.5}$$

and σ_+, σ_- are defined in equations (5.0.61).

Indeed, the compatibility condition of equations (5.1.3) and (5.1.4) yields equation (5.1.2) and

$$\frac{dz}{dx} = -u(z - \Theta - \Theta_\infty) - \frac{2z}{u}(z - 2\Theta).$$

Substituting in this equation the expression for z from equation (5.1.5), we find PIV.

PROPOSITION 5.4 (The direct problem for the Garnier-Jimbo-Miwa Lax pair). *Let $u(x)$ solve the PIV equation (5.1.1).*

Let $z(x)$ be defined in terms of $u(x)$ by $4z(x) = -u_x + u^2 + 2xu + 4\Theta$, and let $y(x)$ satisfy $y_x/y = -u - 2x$. Assume that $\Theta \neq n/2$, $n \in \mathbb{Z}$. Let $\Psi_k^{(\infty)}(\lambda, x)$, $k = 1, \dots, 4$, be the unimodular solutions of the linear ODE (5.1.3), which are uniquely specified by the asymptotic conditions

$$\Psi_k^{(\infty)}(\lambda, x) = \left(I + O\left(\frac{1}{\lambda}\right)\right)e^{(\frac{\lambda^2}{2} + x\lambda)\sigma_3}\left(\frac{1}{\lambda}\right)^{\Theta_\infty\sigma_3}, \quad \lambda \to \infty, \quad \lambda \in \Omega_k,$$

where the open sectors $\Omega_1, \dots, \Omega_4$ are defined by $\arg\lambda$ in

$$\left(-\frac{3\pi}{4}, \frac{\pi}{4}\right), \quad \left(-\frac{\pi}{4}, \frac{3\pi}{4}\right), \quad \left(\frac{\pi}{4}, \frac{5\pi}{4}\right), \quad \left(\frac{3\pi}{4}, \frac{7\pi}{4}\right), \tag{5.1.6}$$

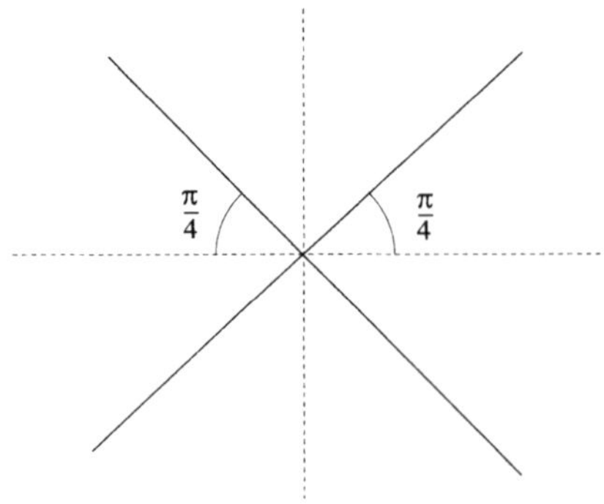

FIGURE 5.5. Stokes rays for PIV

respectively. The functions $\Psi_k^{(\infty)}(\lambda, x)$ *are related by the equations*

$$\Psi_{k+1}^{(\infty)}(\lambda, x) = \Psi_k^{(\infty)}(\lambda, x)\mathsf{S}_k, \quad k = 1, 2, 3;$$

(5.1.7)

$$\Psi_1^{(\infty)}(\lambda, x) = \Psi_4^{(\infty)}(e^{2i\pi}\lambda, x)\mathsf{S}_4 e^{2i\pi\Theta_\infty\sigma_3},$$

where

$$\text{(5.1.8)} \quad \mathsf{S}_1 = \begin{pmatrix} 1 & 0 \\ s_1 & 1 \end{pmatrix}, \quad \mathsf{S}_2 = \begin{pmatrix} 1 & s_2 \\ 0 & 1 \end{pmatrix}, \quad \mathsf{S}_3 = \begin{pmatrix} 1 & 0 \\ s_3 & 1 \end{pmatrix}, \quad \mathsf{S}_4 = \begin{pmatrix} 1 & s_4 \\ 0 & 1 \end{pmatrix},$$

and the constants s_k, $k = 1, \ldots, 4$, *satisfy*

$$\text{(5.1.9)} \quad (1 + s_2 s_3)e^{2i\pi\Theta_\infty} + [s_1 s_4 + (1 + s_3 s_4)(1 + s_1 s_2)]e^{-2i\pi\Theta_\infty} = 2\cos 2\pi\Theta.$$

Let $\Psi^{(0)}(\lambda, x)$ *be defined by*

$$\text{(5.1.10)} \qquad \Psi_1^{(\infty)}(\lambda, x) = \Psi^{(0)}(\lambda, x)\mathsf{E},$$

where E *is any unimodular eigenmatrix of the matrix* $\mathsf{S}_1\mathsf{S}_2\mathsf{S}_3\mathsf{S}_4 e^{2i\pi\Theta_\infty}$, *i.e.,* E *satisfies*

$$\text{(5.1.11)} \qquad \mathsf{S}_1\mathsf{S}_2\mathsf{S}_3\mathsf{S}_4 e^{2i\pi\Theta_\infty} = \mathsf{E}^{-1}e^{-2i\pi\Theta}\mathsf{E}.$$

Then $\Psi^{(0)}$ *satisfies*

$$\Psi^{(0)}(\lambda, x) = \hat{\Psi}^{(0)}(\lambda, x)\lambda^{\Theta\sigma_3},$$

where $\hat{\Psi}^{(0)}(\lambda, x)$ *is analytic at* $\lambda = 0$. *Changing* y *to* y/c^2, $c \neq 0$, *constant, leaves PIV invariant, changes* Ψ *to* $C^{-1}\Psi C$, $C = \mathrm{diag}(c, c^{-1})$, *and maps* $\{s_1, s_2, s_3, s_4\}$ *to* $\{c^2 s_1, c^{-2} s_2, c^2 s_3, c^{-2} s_4\}$; *equation (5.1.9) is invariant under this transformation.*

PROOF. The dominant behavior of the solution $\Psi(\lambda, x)$ of equation (5.1.3) in the neighborhood of $\lambda = \infty$ is given by $\exp[(\frac{\lambda^2}{2} + x\lambda)\sigma_3]$. This implies that the Stokes rays asymptote to the lines $\arg \lambda^2 = \pi/2 + k\pi$, $k \in \mathbb{Z}$. Thus the Stokes rays are given by $\arg \lambda = \frac{\pi}{4}, \frac{3\pi}{4}, \frac{5\pi}{4}, \frac{7\pi}{4}$; see Figure 5.5. The associated Stokes sectors are defined in (5.1.6). Equations (5.1.7) imply

$$\Psi_1^{(\infty)}(\lambda,x)=\Psi_4^{(\infty)}(\mathrm{e}^{2i\pi}\lambda,x)\mathrm{S}_4\mathrm{e}^{2i\pi\Theta_\infty\sigma_3}=\Psi_3^{(\infty)}(\mathrm{e}^{2i\pi}\lambda,x)\mathrm{S}_3\mathrm{S}_4\mathrm{e}^{2i\pi\Theta_\infty\sigma_3}$$
$$=\cdots=\Psi_1^{(\infty)}(\mathrm{e}^{2i\pi}\lambda,x)\mathrm{S}_1\ldots\mathrm{S}_4\mathrm{e}^{2i\pi\Theta_\infty\sigma_3}.$$

This equation together with equations (5.1.8) and (5.1.10) yield (5.1.11). This equation defines E to within a multiplicative diagonal matrix, $\mathrm{diag}(d,d^{-1})$, $d\neq 0$, constant. Again the different choices of E do not affect $u(x)$ and $u(x)$. The trace of equation (5.1.11) yields (5.1.9). It is straightforward to show that $\{s_1,\ldots,s_4\}$ are x-independent. □

THEOREM 5.2. (The solution of equation Painlevé IV). *Let four complex constants $\{s_1,s_2,s_3,s_4\}$ belong to the equivalent class induced by the transformation*

$$\{s_1,s_2,s_3,s_4\}\to\{c^2s_1,c^{-2}s_2,c^2s_3,c^{-2}s_4\}$$

on the algebraic equation

$$(1+s_2s_3)\mathrm{e}^{2i\pi\Theta_\infty}+[s_1s_4+(1+s_3s_4)(1+s_1s_2)]\mathrm{e}^{-2i\pi\Theta_\infty}=2\cos 2\pi\Theta.$$

Let the 2×2 complex-valued matrix function $Y(\lambda,x)$, $\lambda\in\mathbb{C}$, $x\in\mathbb{C}$, have the following representation

$$Y(\lambda,x)=\begin{cases}Y_0(\lambda,x), & \textit{for}\quad \lambda\quad \textit{in the finite complex plane,}\\[2ex] Y_k(\lambda,x), & \textit{for}\quad \lambda\neq 0\quad \textit{and}\lambda \textit{in}\ \Omega_k,\quad k=1,\ldots,4,\end{cases}$$

where the sectors Ω_k are defined in equation (5.1.6).

Let $Y(\lambda,x)$ satisfy the following RH problem

$$Y_{k+1}(\lambda,x)=Y_k(\lambda,x)\mathrm{e}^{\theta(\lambda,x)}\left(\frac{1}{\lambda}\right)^{\Theta_\infty\sigma_3}\mathrm{S}_k\left(\frac{1}{\lambda}\right)^{-\Theta_\infty\sigma_3}\mathrm{e}^{-\theta(\lambda,x)},\quad k=1,2,3,$$

$$Y_1(\lambda,x)=Y_4(\mathrm{e}^{2i\pi}\lambda,x)\mathrm{e}^{\theta(\lambda,x)}\left(\frac{1}{\mathrm{e}^{2i\pi}\lambda}\right)^{\Theta_\infty\sigma_3}\mathrm{S}_4\left(\frac{1}{\mathrm{e}^{2i\pi}\lambda}\right)^{-\Theta_\infty\sigma_3}\mathrm{e}^{-\theta(\lambda,x)},\tag{5.1.12}$$

$$Y_1(\lambda,x)=Y_0(\lambda,x)\mathrm{e}^{\theta(\lambda,x)}\lambda^{\Theta\sigma_3}\mathrm{E}\left(\frac{1}{\lambda}\right)^{-\Theta_\infty\sigma_3}\mathrm{e}^{-\theta(\lambda,x)},$$

$$Y_1(\lambda,x)=I+O\left(\frac{1}{\lambda}\right),\lambda\to\infty,$$

where

$$\theta(\lambda,x)=\left(\frac{\lambda^2}{2}+x\lambda\right)\sigma_3,$$

and the constant matrices S_k, $k=1,\ldots,4$, E are defined in terms of $\{s_1,\ldots,s_4\}$ as follows: The matrices S_k are given by equations (5.1.8); E is any unimodular eigenmatrix of the matrix $\mathrm{S}_1\mathrm{S}_2\mathrm{S}_3\mathrm{S}_4\mathrm{e}^{2i\pi\Theta_\infty}$ (see equation (5.1.11) which defines E to within a right multiplicative diagonal matrix).

Let $u(x)$ be defined by

$$u(x)=-2x-\lim_{\lambda\to\infty}\lambda\partial_x\ln(Y_1(\lambda,x))_{12},\tag{5.1.13}$$

where the subscript 12 denotes the $(1,2)$ entry of the matrix Y_1. Then:

(a) The above RH problem defines a unique 2×2 matrix $Y(\lambda,x)$ which is a meromorphic function of x.

(b) The function $u(x)$ is a meromorphic function of x and it satisfies the Painlevé IV equation (5.1.1) with $\Theta \neq n/2$, $n \in \mathbb{Z}$.

PROOF. Let $\Psi_0(\lambda, x)$ and $\Psi_k(\lambda, x)$, $k = 1, \ldots, 4$, be defined in terms of Y by

$$\Psi_k = Y_k e^{\theta} \left(\frac{1}{\lambda}\right)^{\Theta_\infty \sigma_3}, \quad k = 1, \ldots, 4; \quad \Psi_0 = Y_0 e^{\theta} \lambda^{\Theta \sigma_3}.$$

Equations (5.1.12) imply that the functions Ψ_k, Ψ_0 satisfy the relations (5.1.7).

The proof of the theorem consists of two parts: (a) We will show that it is possible to choose a contour in the complex λ-plane such that on this contour the global jump matrix of the above RH problem is smooth and it decays for large λ. (b) We will show that if $u(x)$ is defined by (5.1.13) and if $y(x)$ is defined by

$$y(x) = -2 \lim_{\lambda \to \infty} \lambda (Y_1(\lambda, x))_{12},$$

then Ψ_0 and Ψ_k satisfy equations (5.1.3) and (5.1.4).

(a) Let L consist of the union of the unit circle and of the anti-Stokes curves extending from the unit circle to infinity; see Figure 5.6. The anti-Stokes curves asymptote to the rays $\arg \lambda^2 = k\pi$, $k \in \mathbb{Z}$, i.e., $\arg \lambda = \frac{k\pi}{2}$.

Equations (5.1.12) imply

$$\begin{aligned}
&Y_0 = Y_1 e^{\theta} f_1 e^{-\theta}, && f_1^{-1} = \lambda^{\Theta \sigma_3} \mathsf{E} \left(\tfrac{1}{\lambda}\right)^{-\Theta_\infty \sigma_3};\\
&Y_2 = Y_0 e^{\theta} f_2 e^{-\theta}, && f_2 = \lambda^{\Theta \sigma_3} \mathsf{E} \mathsf{S}_1 \left(\tfrac{1}{\lambda}\right)^{-\Theta_\infty \sigma_3};\\
&Y_0 = Y_3 e^{\theta} f_3 e^{-\theta}; && f_3^{-1} = \lambda^{\Theta \sigma_3} \mathsf{E} \mathsf{S}_1 \mathsf{S}_2 \left(\tfrac{1}{\lambda}\right)^{-\Theta_\infty \sigma_3};\\
&Y_4 = Y_0 e^{\theta} f_4 e^{-\theta}, && f_4 = \lambda^{\Theta \sigma_3} \mathsf{E} \mathsf{S}_1 \mathsf{S}_2 \mathsf{S}_3 \left(\tfrac{1}{\lambda}\right)^{-\Theta_\infty \sigma_3};\\
&Y_2 = Y_1 e^{\theta} g_1 e^{-\theta}, && g_1 = \left(\tfrac{1}{\lambda}\right)^{\Theta_\infty \sigma_3} \mathsf{S}_1 \left(\tfrac{1}{\lambda}\right)^{-\Theta_\infty \sigma_3};\\
&Y_2 = Y_3 e^{\theta} g_2 e^{-\theta}, && g_2 = \left(\tfrac{1}{\lambda}\right)^{\Theta_\infty \sigma_3} \mathsf{S}_2^{-1} \left(\tfrac{1}{\lambda}\right)^{-\Theta_\infty \sigma_3};\\
&Y_4 = Y_3 e^{\theta} g_3 e^{-\theta}, && g_3 = \left(\tfrac{1}{\lambda}\right)^{\Theta_\infty \sigma_3} \mathsf{S}_3 \left(\tfrac{1}{\lambda}\right)^{-\Theta_\infty \sigma_3};\\
&Y_4 = Y_1 e^{\theta} g_4 e^{-\theta}, && g_4 = \left(\tfrac{1}{\lambda}\right)_+^{\Theta_\infty \sigma_3} \mathsf{S}_4^{-1} \left(\tfrac{1}{\lambda}\right)_+^{-\Theta_\infty \sigma_3},
\end{aligned} \tag{5.1.14}$$

where $(\frac{1}{\lambda})_+^{\Theta_\infty \sigma_3}$ denotes the boundary value of $(\frac{1}{\lambda})^{\Theta_\infty \sigma_3}$ from the + region, i.e., from the left of the branch cut.

The above equations define the following "jump" condition for the matrix $Y(\lambda, x)$,

$$Y_+(\lambda, x) = Y_-(\lambda, x) e^{\theta(\lambda, x)} V(\lambda) e^{-\theta(\lambda, x)}, \quad \lambda \in L,$$

where $V(\lambda) = g_1$ on $\{\arg \lambda = 0\} \cap \{|\lambda| > 1\}$; $V(\lambda) = g_2$ on $\{\arg \lambda = \frac{\pi}{2}\} \cap \{|\lambda| > 1\}$, etc. The jump matrix $e^{\theta} V(\lambda) e^{-\theta}$ decays exponentially for large λ. For example, the jump on $\arg \lambda = 0$ involves the matrix

$$\begin{pmatrix} 1 & 0 \\ s_1 e^{-\lambda^2 - 2\lambda} & 1 \end{pmatrix}.$$

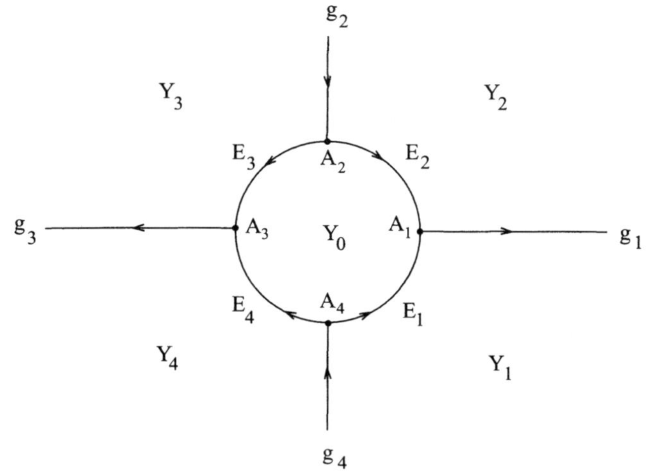

FIGURE 5.6. The RH problem for PIV

Also at $A_1, \ldots, A_4$, appropriate cyclic products of the relevant jump matrices equal the unit matrix. Indeed,

$$g_1 f_2^{-1} f_1^{-1} = I \quad \text{near} \quad A_1, \quad f_2 g_2^{-1} f_3 = I \quad \text{near} \quad A_2,$$

$$f_3^{-1} g_3 f_4^{-1} = I \quad \text{near} \quad A_3, \quad f_4 g_4^{-1} f_1 = I \quad \text{near} \quad A_4.$$

The conditions near A_1, A_2, A_3 follow from the definitions (5.1.14). The condition near A_4 follows from (5.1.14) and the definition of E:

$$f_4(e^{2i\pi}\lambda) g_4^{-1}(e^{2i\pi}\lambda) f_1(\lambda) = I, \quad \text{or} \quad e^{2i\pi\Theta\sigma_3} \mathsf{E} \mathsf{S}_1 \ldots \mathsf{S}_4 e^{2i\pi\Theta_\infty\sigma_3} \mathsf{E}^{-1} = I.$$

(b) We denote by Ψ and Y the solutions piecewise defined by Ψ_k, Ψ_0 and Y_k, Y_0. Let A be defined by $A = \Psi_\lambda \Psi^{-1}$. Then for λ outside the unit circle

$$A = \left(Y_\lambda + (\lambda + x) Y \sigma_3 - \Theta_\infty \lambda^{-1} Y \sigma_3\right) Y^{-1}.$$

Since Ψ_λ and Ψ have the same jumps across L, it follows that

$$A = \lambda^{-1} A_{-1} + A_0 + \lambda A_1. \tag{5.1.15}$$

Thus for λ outside the unit circle

$$Y_\lambda + (\lambda + x) Y \sigma_3 - \Theta_\infty \lambda^{-1} Y \sigma_3 = (\lambda^{-1} A_{-1} + A_0 + \lambda A_1) Y. \tag{5.1.16}$$

Similarly for λ inside the circle,

$$Y_\lambda + (\lambda + x) Y \sigma_3 + \Theta \lambda^{-1} Y \sigma_3 = (\lambda^{-1} A_{-1} + A_0 + \lambda A_1) Y. \tag{5.1.17}$$

Letting

$$Y = I + \frac{Y_{-1}}{\lambda} + \frac{Y_{-2}}{\lambda^2} + O\left(\frac{1}{\lambda^3}\right), \quad \lambda \to \infty,$$

equation (5.1.16) yields

$$\begin{aligned} &A_1 = \sigma_3, \\ (5.1.18) \qquad &A_0 = x\sigma_3 + [Y_{-1}, \sigma_3], \\ &A_{-1} = -\Theta_\infty \sigma_3 + [Y_{-2} + xY_{-1}, \sigma_3] + [\sigma_3, Y_{-1}]Y_{-1}. \end{aligned}$$

Equation (5.1.17) as $\lambda \to 0$ yields

$$A_{-1} = \Theta Y(0,x)\sigma_3 Y^{-1}(0,x), \quad \text{thus} \quad \det A_{-1} = -\Theta^2.$$

We define $y(x)$ and $z(x)$ by

$$y = -2(Y_{-1})_{12}, \qquad z = \Theta + \Theta_\infty - 2(Y_{-1})_{12}(Y_{-1})_{21}.$$

Writing $(Y_{-1})_{12}$ and $(Y_{-1})_{21}$ in terms of y and z, the equation for A_0 (equation (5.1.18b)) becomes

$$A_0 = x\sigma_3 + \begin{pmatrix} 0 & y \\ \frac{2}{y}(z - \Theta - \Theta_\infty) & 0 \end{pmatrix}.$$

Taking the diagonal of the equation for A_{-1} (equation (5.1.18c)) we find

$$(A_{-1})_{\text{diag}} = (2(Y_{-1})_{12}(Y_{-1})_{21} - \Theta_\infty)\,\sigma_3 = (\Theta - z)\sigma_3.$$

We define $(A_{-1})_{12}$ by $(A_{-1})_{12} = -yu/2$. This equation, together with $\det A_{-1} = -\Theta^2$, and $(A_{-1})_{\text{diag}} = (\Theta - z)\sigma_3$, imply $(A_{-1})_{21} = 2z(z - 2\Theta)/yu$. In summary,

$$\begin{aligned} &A_1 = \sigma_3, \quad A_0 = \begin{pmatrix} x & y \\ \frac{2}{y}(z - \Theta - \Theta_\infty) & -x \end{pmatrix}, \\ (5.1.19) \qquad &A_{-1} = \begin{pmatrix} \Theta - z & -\frac{yu}{2} \\ \frac{2z}{yu}(z - 2\Theta) & -\Theta + z \end{pmatrix}. \end{aligned}$$

Equation $\Psi_\lambda = A\Psi$, where A is defined by equations (5.1.15) and (5.1.19) is equation (5.1.3). In a similar way it can be shown that Ψ also solves equation (5.1.4). Hence $u(x)$ and $y(x)$ solve equations (5.1.1) and (5.1.2).

The proof of existence and of the meromorphic dependence on x of the solution of the above RH problem, is similar to the analogous proof for PII. □

A symmetric Lax pair, which is gauge equivalent to the above Lax pair [**KH**] is slightly more convenient for the asymptotic study of equation PIV, [**IK2, K6, K8**]. This symmetric Lax pair is given by the following system of equations,

$$\begin{aligned} &\frac{\partial \Psi}{\partial \lambda} = A\Psi; \\ (5.1.20) \qquad &A = \left(\frac{1}{2}\lambda^3 + \lambda(x + yv) + \frac{\alpha}{\lambda}\right)\sigma_3 + i\left(\lambda^2 y + 2xy + \frac{dy}{dx}\right)\sigma_+ \\ &\quad + i\left(\lambda^2 v + 2xv - \tfrac{dv}{dx}\right)\sigma_-, \end{aligned}$$

$$\frac{\partial \Psi}{\partial x} = \mathrm{U}\Psi;$$
(5.1.21)
$$\mathrm{U} = \left(\frac{1}{2}\lambda^2 + yv\right)\sigma_3 + i\lambda y\sigma_+ + i\lambda v\sigma_-,$$

where σ_+ and σ_- are defined in (5.0.61).

The compatibility of equations (5.1.20) and (5.1.21) yields

(5.1.22) $$\frac{d^2y}{dx^2} = -(1+2\alpha)y - 2x\frac{dy}{dx} + 4xy^2v + 2yv\frac{dy}{dx},$$

(5.1.23) $$\frac{d^2v}{dx^2} = (1-2\alpha)v + 2x\frac{dv}{dx} + 4xyv^2 - 2yv\frac{dv}{dx},$$

Let u be defined by $u = yv$. The two equations above are equivalent to

(5.1.24) $$v\frac{dy}{dx} - y\frac{dv}{dx} = \beta - 2xu + u^2, \quad \beta = \text{const},$$

where u solves PIV. Indeed, equations (5.1.22) and (5.1.23) imply

$$v\frac{d^2y}{dx^2} - y\frac{d^2v}{dx^2} = -2yv - 2x\frac{d}{dx}(yv) + 2yv\frac{d}{dx}(yv),$$

which upon integration becomes equation (5.1.24). Equation (5.1.24) and the equation $\frac{du}{dx} = y\frac{dv}{dx} + v\frac{dy}{dx}$ imply that

$$\frac{dy}{dx} = \frac{1}{2}\frac{du}{dx} + \frac{1}{2}(\beta - 2xu + u^2), \quad y\frac{dv}{dx} = \frac{1}{2}\frac{du}{dx} - \frac{1}{2}(\beta - 2xu + u^2).$$

Using these two equations in (5.1.22) and (5.1.23), it follows that u solves the PIV.

PROPOSITION 5.5 (The direct problem for the symmetric Lax pair of PIV). *Let $u(x)$ satisfy the PIV equation*

(5.1.25)
$$\frac{d^2u}{dx^2} = \frac{1}{2u}\left(\frac{du}{dx}\right)^2 + \frac{3}{2}u^3 + 4xu^2 + 2\left(x^2 + \frac{\beta}{2} - 2\alpha\right)u - \frac{\beta^2}{2u},$$
$$\alpha, \beta = \text{const},$$

and assume that $\alpha \pm \frac{1}{2} \notin \mathbb{Z}$. Let $y(x)$ satisfy the equation

(5.1.26) $$2\frac{u}{y}\frac{dy}{dx} - \frac{du}{dx} + 2xu - u^2 - \beta = 0.$$

Let $v(x)$ be defined by $v = u/y$. Let $\Psi_k^{(\infty)}(\lambda, x)$, $k = 1, \ldots, 8$, be the unimodular solutions of the linear ODE (5.1.20), which are uniquely specified by the asymptotic conditions

$$\Psi_k^{(\infty)}(\lambda, x) = \left(I + O(\tfrac{1}{\lambda})\right) e^{\left(\frac{\lambda^4}{8} + \frac{x\lambda^2}{2}\right)\sigma_3}$$
$$\left(\tfrac{1}{\lambda}\right)^{(\beta-\alpha)\sigma_3}, \quad \lambda \to \infty, \quad \lambda \in \Omega_k, \quad k = 1, \ldots, 8,$$

where the open sectors Ω_k are defined by

$$\Omega_k = \left\{\lambda \in \mathbb{C}\colon\ \arg\lambda \in \left(-\frac{3\pi}{8} + k\frac{\pi}{4}, \frac{\pi}{8} + k\frac{\pi}{4}\right)\right\}, \quad k = 1, \ldots, 8.$$

These functions are related by the equations

$$
\Psi^{(\infty)}_{k+1}(\lambda, x) = \Psi^{(\infty)}_{k}(\lambda, x)\mathsf{S}_k,
$$

(5.1.27)
$$
\mathsf{S}_{2k-1} = \begin{pmatrix} 1 & s_{2k-1} \\ 0 & 1 \end{pmatrix}, \quad \mathsf{S}_{2k} = \begin{pmatrix} 1 & 0 \\ s_{2k} & 1 \end{pmatrix},
$$

where the entries of the Stokes matrices satisfy

(5.1.28)
$$
s_{k+4} = -s_k \mathrm{e}^{2\pi i(-1)^k(\alpha-\beta)},
$$

(5.1.29)
$$
((1+s_1s_2)(1+s_3s_4)+s_1s_4)\,\mathrm{e}^{i\pi(\beta-\alpha)} - (1+s_2s_3)\mathrm{e}^{i\pi(\alpha-\beta)} = -2i\sin\pi\alpha.
$$

Let the unimodular matrix $\Psi^{(0)}(\lambda, x)$ *be defined by*

$$
\Psi^{(\infty)}_1(\lambda, x) = \Psi^{(0)}(\lambda, x)\mathsf{E},
$$

where E *is any unimodular eigenmatrix of* $\mathsf{S}_1\mathsf{S}_2\mathsf{S}_3\mathsf{S}_4\mathrm{e}^{i\pi(\beta-\alpha)\sigma_3}\sigma_3$, *i.e.,* E *satisfies*

(5.1.30)
$$
\mathsf{S}_1\mathsf{S}_2\mathsf{S}_3\mathsf{S}_4\mathrm{e}^{i\pi(\beta-\alpha)\sigma_3}\sigma_3 = \mathsf{E}^{-1}\mathrm{e}^{-i\pi\alpha\sigma_3}\sigma_3\mathsf{E}.
$$

Then

(5.1.31)
$$
\Psi^{(0)}(\lambda, x) = \hat{\Psi}^{(0)}(\lambda, x)\lambda^{\alpha\sigma_3},
$$

where $\hat{\Psi}^{(0)}(\lambda, x)$ *is analytic at* $\lambda = 0$. *The transformation*

$$
\{s_1, s_2, s_3, s_4\} \mapsto \left\{c^2 s_1, c^{-2}s_2, c^2 s_3, c^{-2}s_4\right\}, \quad c \neq 0
$$

*leaves PIV, as well as the equation (*5.1.29*), invariant.*

The proof is similar to that of Proposition 5.4. Equations (5.1.28) follow from

$$
\mathsf{S}_{k+4} = \mathrm{e}^{i\pi(\beta-\alpha)\sigma_3}\sigma_3\mathsf{S}_k\mathrm{e}^{i\pi(\alpha-\beta)\sigma_3},
$$

which are a consequence of the symmetry relations

$$
\sigma_3\Psi^{(\infty)}_{k+4}(\mathrm{e}^{i\pi}\lambda, x)\sigma_3 = \Psi^{(\infty)}_k(\lambda, x)\mathrm{e}^{i\pi(\alpha-\beta)\sigma_3}.
$$

The above relation, together with equations (5.1.27), imply equation (5.1.30). The trace of equation (5.1.30) yields (5.1.29). □

We note that equations (5.1.28) and (5.1.29) imply that there exist three independent monodromy data. This is reflection of the fact that the associated nonlinear equations (5.1.25) and (5.1.26) are a second order and a first order ODEs, respectively.

2. The first Painlevé transcendent

Let $u(x)$ satisfy the Painlevé I equation

(5.2.1)
$$
\frac{d^2u}{dx^2} = 6u^2 + x.
$$

Let $v(x)$ satisfy the equation

(5.2.2)
$$
\frac{dv}{dx} + 2iv^2 + iu = 0.
$$

Equations (5.2.1) and (5.2.2) possess the following Lax pair,

$$\frac{\partial\Psi}{\partial\lambda} = \mathsf{A}\Psi; \quad \mathsf{A}(\lambda,x) = \sum_{j=-1}^{4} A_j(x)\lambda^j, \tag{5.2.3}$$

$$\frac{\partial\Psi}{\partial x} = \mathsf{U}\Psi; \quad \mathsf{U}(\lambda,x) = -i\lambda\sigma_3 - 2iv\sigma_1, \tag{5.2.4}$$

where

$$A_4 = -4i\sigma_3, \quad A_3 = -8iv\sigma_1, \quad A_2 = 4u\sigma_2 + 8iv^2(\sigma_3 - i\sigma_2),$$

$$A_1 = \left(2\frac{du}{dx} - 8ivu\right)\sigma_1, \quad A_0 = \left(8iv^2u - 4v\frac{du}{dx} - i(2u^2+x)\right)(\sigma_3 - i\sigma_2). \tag{5.2.5}$$

PROPOSITION 5.6 (The direct problem). *Let $u(x)$ solve the PI equation (5.2.1) and let $v(x)$ solve $v_x + 2iv^2 = -iu$. Let $\Psi_k^{(\infty)}(\lambda,x)$, $k = 1,\dots,6$, be the unimodular solutions of the linear ODE (5.2.3), which are uniquely specified by the asymptotic conditions*

$$\Psi_k^{(\infty)}(\lambda,x) = \left(I + O(\frac{1}{\lambda})\right) \mathrm{e}^{-i(\frac{4}{5}\lambda^5 + x\lambda)\sigma_3}, \quad \lambda \to \infty, \quad \lambda \in \Omega_k,$$

where the open sectors Ω_k are defined by

$$\Omega_k = \left\{\lambda \in \mathbb{C}\colon \ \arg\lambda \in \left((k-2)\frac{\pi}{5}, \frac{k\pi}{5}\right)\right\}, \quad k = 1,\dots,10. \tag{5.2.6}$$

The functions $\Psi_k^{(\infty)}$ are related by

$$\begin{aligned} &\Psi_{k+1}^{(\infty)}(\lambda,x) = \Psi_k^{(\infty)}(\lambda,x)\mathsf{S}_k, \quad k = 1,\dots,9; \\ &\Psi_1^{(\infty)}(\lambda,x) = \Psi_{10}^{(\infty)}(\mathrm{e}^{2i\pi}\lambda,x)\mathsf{S}_{10}, \end{aligned} \tag{5.2.7}$$

where

$$\mathsf{S}_{2k-1} = \begin{pmatrix} 1 & 0 \\ s_{2k-1} & 1 \end{pmatrix}, \quad \mathsf{S}_{2k} = \begin{pmatrix} 1 & s_{2k} \\ 0 & 1 \end{pmatrix}, \quad k = 1,\dots,5. \tag{5.2.8}$$

These matrices are constant and they satisfy the relations

$$\mathsf{S}_{k+5} = \sigma_1\mathsf{S}_k\sigma_1; \quad \mathsf{S}_1\dots\mathsf{S}_5 = i\sigma_1. \tag{5.2.9}$$

Let $\Psi^{(0)}(\lambda,x)$ be the solution of equation (5.2.3) which in the neighborhood of $\lambda = 0$ is normalized as follows:

$$\Psi^{(0)}(\lambda,x) = \left(\begin{pmatrix} \alpha & \beta \\ -\alpha & \beta \end{pmatrix} + O(\lambda)\right)\lambda^{\sigma_3/2}, \quad |\lambda| < \infty. \tag{5.2.10}$$

The functions $\Psi_1^{(\infty)}(\lambda,x)$ and $\Psi^{(0)}(\lambda,x)$ are related by the unimodular constant matrix E,

$$\Psi_1^{(\infty)}(\lambda,x) = \Psi^{(0)}(\lambda,x)\mathsf{E}, \quad \det\mathsf{E} = 1. \tag{5.2.11}$$

Equations (5.2.9) imply that among the entries of the Stokes matrices, only two are independent. Equations (5.2.10) and (5.2.11) imply that E *involves only one independent parameter. Thus equations (5.2.1) and (5.2.2) are associated with three constant monodromy data.*

PROOF. The dominant behavior of the solution $\Psi(\lambda, x)$ of equation (5.2.3) in the neighborhood of $\lambda = \infty$ is given by $\exp[-i(\frac{4}{5}\lambda^5 + x\lambda)\sigma_3]$. This implies that the Stokes curves asymptote to the rays $\arg\lambda = k\pi/5$, $k \in \mathbb{Z}$. The associated Stokes sectors are defined in (5.2.6).

The matrices A and B defined in (5.2.3)–(5.2.5) satisfy

$$\sigma_1 \mathsf{A}(e^{i\pi}\lambda, x)\sigma_1 = -\mathsf{A}(\lambda, x), \quad \sigma_1 \mathsf{U}(e^{i\pi}\lambda, x)\sigma_1 = \mathsf{U}(\lambda, x).$$

Thus, if $\Psi(\lambda, x)$ is a solution of equations (5.2.3) and (5.2.4), then $\sigma_1\Psi(e^{i\pi}\lambda, x)\sigma_1$ is also a solution. Also if $e^{-i\pi}\lambda \in \Omega_k$, then $\lambda \in \Omega_{k+5}$. These facts imply the important symmetry relationships

$$\Psi_{k+5}^{(\infty)}(\lambda, x) = \sigma_1 \Psi_k^{(\infty)}(e^{-i\pi}\lambda, x)\sigma_1, \quad k \in \mathbb{Z}.$$

These equations, together with (5.2.8), yield (5.2.9a).

The dominant term near $\lambda = 0$ is $-\sigma_1/2\lambda$; since trace $\sigma_1 = 0$ and $\det\sigma_1 = -1$, σ_1 diagonalized to σ_3. Thus, a solution in the neighborhood of $\lambda = 0$ can be obtained as a convergent power series, with the first term given by equation (5.2.10). This equation implies

$$\sigma_1\Psi^{(0)}(e^{i\pi}\lambda, x) = \left(\sigma_1\begin{pmatrix}\alpha & \beta \\ -\alpha & \beta\end{pmatrix} + O(\lambda)\right) i\sigma_3\lambda^{\frac{\sigma_3}{2}} = -i\Psi^{(0)}(\lambda, x).$$

To derive equation (5.2.9b) we use

$$\sigma_1\Psi_1^{(\infty)}(\lambda, x)\sigma_1 = \Psi_6^{(\infty)}(e^{i\pi}\lambda, x) = \Psi_1^{(\infty)}(e^{i\pi}\lambda, x)\mathsf{S}_1\ldots\mathsf{S}_5,$$

i.e.,

$$\sigma_1\Psi^{(0)}(\lambda, x)\mathsf{E}\sigma_1 = \Psi^{(0)}(e^{i\pi}\lambda, x)\mathsf{E}\mathsf{S}_1\ldots\mathsf{S}_5.$$

This equation together with $\Psi^{(0)}(e^{i\pi}\lambda, x) = -i\sigma_1\Psi^{(0)}(\lambda, x)$, yield (5.2.9b). Equations (5.2.9) imply that only two of the entries of the Stokes matrices are independent. We will now show that E essentially involves only one independent parameter. Let $\Psi^{(0)}$ and $\Psi^{0'}$ be two different solutions related to the same solution $\Psi_1^{(\infty)}$ through the two different connection matrices E and E'. Letting $\Psi^{(0)} = \hat{\Psi}^{(0)}\lambda^{\frac{\sigma_3}{2}}$, we find

$$\hat{\Psi}^{(0)} = \hat{\Psi}^{0'}\lambda^{\frac{\sigma_3}{2}}F\lambda^{-\frac{\sigma_3}{2}}, \quad F = \mathsf{E}'\mathsf{E}^{-1}.$$

Since $\hat{\Psi}^{(0)}$ and $\hat{\Psi}^{0'}$ are holomorphic at $\lambda = 0$, and since $\det\hat{\Psi}^{(0)} = \det\hat{\Psi}^{0'} = 1$, it follows that

$$F = \begin{pmatrix} F_1 & F_{12} \\ 0 & F_1^{-1}\end{pmatrix}, \quad F_1 \neq 0.$$

Changing E to $F\mathsf{E}$ changes α of (5.2.10) to α/F_1.

It is straightforward to establish that the Stokes matrices and the connection matrix E are x-independent. □

THEOREM 5.3 (The solution of Painlevé I). *Let S_{2k-1}, S_{2k}, $k = 1, \ldots, 5$, be the constant 2×2 matrices (5.2.8) satisfying the relations (5.2.9). Let E be the 2×2 constant matrix satisfying the relations*

$$\det \mathsf{E} = 1, \qquad F\mathsf{E} \sim \mathsf{E},$$

where F is an upper triangular unimodular matrix and $\sim$ denotes an equivalent class. Let the unimodular 2×2 complex-valued matrix $Y(\lambda, x)$, $\lambda \in \mathbb{C}$, $x \in \mathbb{C}$, have the following representation,

$$Y(\lambda, x) = \begin{cases} Y_0(\lambda, x), & \textit{for } \lambda \textit{ in the finite complex plane} \\ Y_k(\lambda, x), & \textit{for } \lambda \neq 0 \textit{ and } \lambda \textit{ in } \Omega_k, \quad k = 1, \ldots, 10, \end{cases}$$

where the sectors Ω_k are defined in equations (5.2.6). Let Y satisfy the RH problem

$$Y_{k+1}(\lambda, x) = Y_k(\lambda, x) e^{\theta(\lambda,x)} S_k e^{-\theta(\lambda,x)}, \quad k = 1, \ldots, 10; \quad Y_{11} = Y_1.$$

$$Y_1(\lambda, x) = Y_0(\lambda, x) e^{\theta(\lambda,x)} \lambda^{\frac{\sigma_3}{2}} \mathsf{E} e^{-\theta(\lambda,x)}, \tag{5.2.12}$$

$$Y_1(\lambda, x) = I + O\left(\tfrac{1}{\lambda}\right), \lambda \to \infty,$$

where

$$\theta(\lambda, x) = -i\left(\frac{4}{5}\lambda^5 + x\lambda\right)\sigma_3.$$

Then this RH problem defines a unique matrix $Y(\lambda, x)$ which is a meromorphic function of x. Let $v(x)$ be defined by

$$v(x) = -i \lim_{\lambda \to \infty} \lambda (Y_1(\lambda, x))_{12}, \tag{5.2.13}$$

where the subscript 12 denotes the $(1, 2)$ entry of the matrix Y_1. Let $u(x)$ be defined by

$$u(x) = i\frac{dv}{dx} - 2v^2. \tag{5.2.14}$$

Then $u(x)$ is a meromorphic function of x, and it satisfies the PI equation (5.2.1).

PROOF. Let $\Psi_0(\lambda, x)$ and $\Psi_k(\lambda, x)$, $k = 1, \ldots, 10$, be defined in terms of Y by

$$\Psi_k = Y_k e^{\theta}, \quad k = 1, \ldots, 10; \quad \Psi_0 = Y_0 e^{\theta} \lambda^{\frac{\sigma_3}{2}}.$$

Equations (5.2.12) imply that Ψ_k and Ψ_0 satisfy the relations (5.2.8) and (5.2.11). It is possible to choose a contour L in the complex λ-plane such that on this contour the global jump matrix of the above RH problem is smooth and it decays for large λ. For convenience we choose L as the union of the unit circle and of the anti-Stokes curves extending from the unit circle to infinity; see Figure 5.7. The anti-Stokes curves asymptote to $\arg\lambda = (2k+1)\pi/10$, $k = 0, 1, \ldots, 9$. Equations (5.2.21), (5.2.8), (5.2.11) yield

$$Y_+(\lambda, x) = Y_-(\lambda, x) e^{\theta(\lambda,x)} V(\lambda) e^{-\theta(\lambda,x)},$$

where the jumps are $V = (S_1, S_2^{-1}, S_3, S_4^{-1}, \ldots, S_9, S_{10}^{-1})$ on the rays outside the circle and $V = (f_1^{-1}, f_2, f_3^{-1}, f_4, \ldots, f_9^{-1}, f_{10})$ on the circle (see Figure 5.7), where

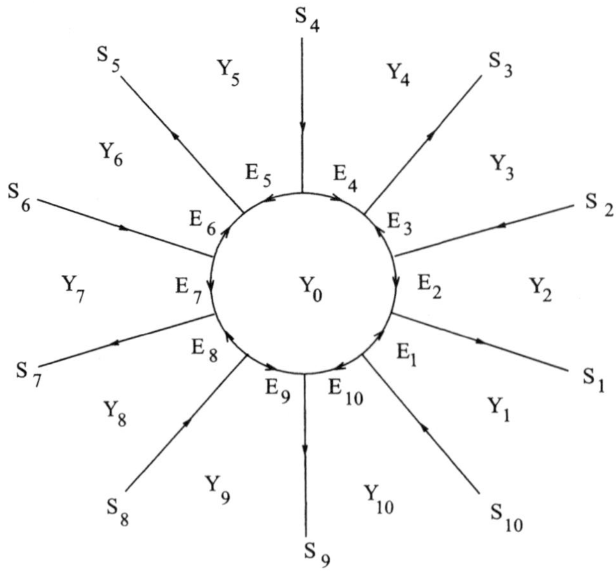

FIGURE 5.7. Stokes rays for equation PI

the f's are defined by

$$f_k = \lambda^{\frac{\sigma_3}{2}} \mathsf{E} \mathsf{S}_1 \mathsf{S}_2 \dots \mathsf{S}_{k-1}, \quad k = 1, \dots, 5;$$

$$f_6 = i\lambda^{\frac{\sigma_3}{2}} \mathsf{E}\sigma_1; \quad f_{6+k} = i\lambda^{\frac{\sigma_3}{2}} \mathsf{E} \mathsf{S}_1 \mathsf{S}_2 \dots \mathsf{S}_k \sigma_1, k = 1, 2, 3, 4.$$

As with Theorems 5.1 and 5.2 it is straightforward to verify that V is smooth and that $\mathrm{e}^{\theta} V \mathrm{e}^{-\theta}$ decays exponentially for large λ. We will now show that if v and u are defined by equations (5.2.13) and (5.2.14), and Ψ is defined by

$$\Psi_k = Y_k \mathrm{e}^{\theta}, \quad \lambda \neq 0, \quad \lambda \in \Omega_k; \quad \Psi_0 = Y_0 \mathrm{e}^{\theta} \lambda^{\frac{1}{2}} \mathsf{E}, \quad |\lambda| < \infty,$$

then Ψ solves equations (5.2.3) and (5.2.4).

We first note that the symmetry condition (5.2.9) implies

$$\Psi_{k+5}(\lambda, x) = \sigma_1 \Psi_k(\mathrm{e}^{-i\pi}\lambda, x)\sigma_1, \quad k \in \mathbb{Z}. \tag{5.2.15}$$

Also,

$$\sigma_1 \Psi_1(\lambda, x)\sigma_1 = \Psi_6(\mathrm{e}^{i\pi}\lambda, x) = \Psi_1(\mathrm{e}^{i\pi}\lambda, x)\mathsf{S}_1 \dots \mathsf{S}_5,$$

or

$$\sigma_1 \Psi_0(\lambda, x)\mathsf{E}\sigma_1 = i\Psi_0(\mathrm{e}^{i\pi}\lambda, x)\mathsf{E}\sigma_1,$$

or

$$\sigma_1 \Psi_0(\mathrm{e}^{i\pi}\lambda, x) = -i\Psi_0(\lambda, x). \tag{5.2.16}$$

Equations (5.2.15) and (5.2.16) will play an important role in proving that the inverse problem solves the direct one.

For large λ, Y has the asymptotic expansion

$$Y \sim I + \sum_{j=1}^{\infty} \frac{Y_{-j}}{\lambda^j}, \quad \lambda \to \infty. \tag{5.2.17}$$

Using (5.2.15), it follows that

$$(5.2.18)\qquad \begin{aligned} Y_{-2j} &= \alpha_{2j} I + \beta_{2j}\sigma_1, \qquad I = \mathrm{diag}(1,1),\\ Y_{-(2j+1)} &= \alpha_{2j+1}\sigma_3 + i\beta_{2j+1}\sigma_2, \quad j = 0,1,2,\ldots, \end{aligned}$$

where α and β are scalar functions of x.

(a) **Derivation of $\Psi_x = \mathrm{U}\Psi$.** Ψ_x and Ψ have the same jumps, thus $\Psi_x\Psi^{-1}$ is holomorphic; also $\Psi_x\Psi^{-1}$ tends to $-i\sigma_3\lambda$ as $\lambda \to \infty$, thus

$$(5.2.19)\qquad \Psi_x\Psi^{-1} = -i\sigma_3\lambda + U_0,$$

or

$$(5.2.20)\qquad Y_x - i\lambda Y\sigma_3 = (-i\lambda\sigma_3 + U_0)Y.$$

Substituting (5.2.17) into (5.2.20) we find

$$O(1):\ U_0 = i[\sigma_3, Y_{-1}].$$

$$(5.2.21)\qquad O\left(\frac{1}{\lambda^j}\right):\ (Y_{-j})_x = -i[\sigma_3, Y_{-(j+1)}] + U_0 Y_{-j}.$$

Using

$$Y_{-1} = \alpha_1\sigma_3 + i\beta_1\sigma_2,$$

we find

$$U_0 = 2i\beta_1\sigma_1.$$

Equation (5.2.19) implies that

$$\Psi_x = (-i\sigma_3\lambda + 2i\beta_1\sigma_1)\Psi,$$

which is equation (5.2.3) provided that we define v by $-\beta_1$, i.e., $v = -(Y_{-1})_{12}$.

(b) **Derivation of $\Psi_\lambda = \mathrm{A}\Psi$.** Equations (5.2.21) with $j = 1,\ldots,4$ yield

$$(5.2.22)\qquad \alpha_{1_x} = -2iv^2, \quad \beta_2 = -iv_x + \alpha_1 v,$$

$$(5.2.23)\qquad \alpha_{2_x} = -\frac{1}{4}(v^2)_x - 2i\alpha_1 v^2, \quad \beta_3 = -v\alpha_2 + \frac{1}{4}v_{xx} + \frac{i}{2}(\alpha_1 v)_x,$$

$$(5.2.24)\qquad \alpha_{3_x} = 2iv\beta_3, \quad \beta_4 = v\alpha_3 + \frac{i}{2}(\beta_3)_x,$$

$$(5.2.25)\qquad \alpha_{4_x} = -2iv\beta_4, \quad \beta_5 = \frac{i}{2}(\beta_4)_x - \alpha_4.$$

These equations will be used below for the derivation of $\Psi_\lambda = \mathrm{A}\Psi$. Similar considerations as before imply

$$Y_\lambda - iY(4\lambda^2 + x)\sigma_3 = \left(\sum_{k=-1}^{4} A_k\lambda^k\right)Y.$$

Substituting (5.2.17) into this equation we find

$$O(\lambda^3):\ A_3 = 4i[\sigma_3, Y_{-1}], \quad O(\lambda^2):\ A_2 = 4i[\sigma_3, Y_{-2}] - A_3Y_{-1},$$

$$O(\lambda):\ A_1 = 4i[\sigma_3, Y_{-3}] - A_3Y_{-2} - A_2Y_{-1},$$

$$O(1):\ A_0 = 4i[\sigma_3, Y_{-4}] - A_3Y_{-3} - A_2Y_{-2} - A_1Y_{-1} - ix\sigma_3.$$

Using (5.2.18) to express the A_j in terms of α_j and β_j and then using (5.2.22)–(5.2.25), the above equations yield

$$A_4 = -4i\sigma_3, \quad A_3 = -8iv\sigma_1,$$
(5.2.26)
$$A_2 = 8iv^2\sigma_3 + 4iv_x\sigma_2, \quad A_1 = i(2v_{xx} + 16v^3)\sigma_1,$$

$$A_0 = -i(v_{xx} + 24v^2 v_x)\sigma_2 + i(2v_x^2 - 4vv_{xx} - 24v^4 - x)\sigma_3.$$

We define u by

$$u = i(v_x + 2iv^2).$$

Then equations (5.2.26) reduce to the expressions for A_j, $j = 1, \dots, 4$, given by (5.2.5), whereas A_0 becomes

$$A_0 = -i\left[8iv^2u - 4v\frac{du}{dx} - i\left(\frac{d^2u}{dx^2} - 4u^2\right)\right]\sigma_2 \tag{5.2.27}$$
$$+ \left[8iv^2u - 4v\frac{du}{dx} - i(2u^2 + x)\right]\sigma_3.$$

We will now prove that

$$A_{-1} = -\frac{\sigma_1}{2}, \quad A_0 = a(\sigma_3 - i\sigma_2), \tag{5.2.28}$$

where a is some scalar function of x. Then the second equation above together with (5.2.28) imply that $u(x)$ solves

$$\frac{d^2u}{dx^2} - 4u^2 = 2u^2 + x,$$

i.e., PI.

To prove equations (5.2.28) we analyze Ψ near $\lambda = 0$. Let

$$\Psi_0 = \left(\varphi_0(x) + \lambda\varphi_1(x) + O(\lambda^2)\right)\lambda^{\frac{\sigma_3}{2}}.$$

The symmetry relationship (5.2.16) implies

$$\sigma_1\varphi_0\sigma_3 = -\varphi_0, \quad \sigma_1\varphi_1\sigma_3 = \varphi_1. \tag{5.2.29}$$

Substituting the expression for Ψ_0 into $\Psi_{0_\lambda} = A\Psi_0$ we find

$$O\left(\frac{1}{\lambda}\right): \ \varphi_0\sigma_3 = 2A_{-1}\varphi_0, \quad \text{or} \quad A_{-1} = \frac{1}{2}\varphi_0\sigma_3\varphi_0^{-1} = -\frac{\sigma_1}{2}.$$

$$O(1): \ A_0\varphi_0 = \varphi_1 + \frac{\varphi_1\sigma_3}{2} - A_{-1}\varphi_1, \quad \text{or} \quad A_0 = (I + \sigma_1)\varphi_1\varphi_0^{-1}.$$

But equations (5.2.29) imply that $\varphi_1\varphi_0^{-1}$ satisfies

$$\sigma_1\varphi_1\varphi_0^{-1}\sigma_1 = -\varphi_1\varphi_0^{-1}, \quad \text{or} \quad \varphi_1\varphi_0^{-1} = \alpha\sigma_3 + i\beta\sigma_1,$$

where α, β are scalar functions of x. Thus $A_0 = (\alpha - \beta)(\sigma_3 - i\sigma_2)$.

The proof of existence and of the meromorphic dependence on x of the solution of the above RH problem, is similar to the analogous proof for PII. □

REMARK 5.4. The symmetry condition (5.2.16) plays a crucial role in proving the above result. If one tries to bypass this equation by using the compatibility condition of the Lax pair, one finds instead of PI, the equation

$$P_x = 4ivP, \quad P = \frac{d^2u}{dx^2} - 6u^2 - x.$$

For the investigation of the asymptotic behavior of $u(x)$ it is convenient to introduce the following Lax pair

$$\begin{aligned}(5.2.30)\quad &\frac{\partial\tilde{\Psi}}{\partial\lambda}=\tilde{\mathrm{A}}\tilde{\Psi};\quad \tilde{A}=(4\tilde{\lambda}^4+2u^2+x)\sigma_3-i(4u\tilde{\lambda}^2+2u^2+x)\sigma_2\\ &-\left(2\tilde{\lambda}\frac{du}{dx}+\frac{1}{2\tilde{\lambda}}\right)\sigma_1,\\ &\frac{\partial\tilde{\Psi}}{\partial x}=\tilde{\mathrm{U}}\tilde{\Psi};\quad \tilde{B}=\left(\tilde{\lambda}+\frac{u}{\tilde{\lambda}}\right)\sigma_3-\frac{iu}{\tilde{\lambda}}\sigma_2.\end{aligned}$$

This Lax pair is related to the Lax pair defined by equations (5.2.3) and (5.2.4), by the equations

$$\lambda=i\tilde{\lambda},\quad \Psi=\left(I+\frac{v}{\lambda}(\sigma_3-i\sigma_2)\right)\tilde{\Psi}. \tag{5.2.31}$$

The connection matrix $\tilde{\mathrm{E}}$ associated with the above Lax pair is x-dependent, however the entries of the Stokes matrices associated with $\tilde{\Psi}$ are still x-independent.

PROPOSITION 5.7 (The direct problem for another Lax pair). *Let $u(x)$ solve the PI equation (5.2.1). Let $\Psi_k^{(\infty)}(\lambda,x)$, $k\in\mathbb{Z}$, be the unimodular solutions of the linear ODE (5.2.31) which are uniquely specified by the asymptotic conditions*

$$\Psi_k^{(\infty)}(\lambda,x)=\left(I+O\left(\frac{1}{\lambda}\right)\right)\mathrm{e}^{(\frac{4}{5}\lambda^5+x\lambda)\sigma_3},\quad \lambda\to\infty,\quad \lambda\in\Omega_k,$$

where the open sectors Ω_k are defined by

$$\Omega_k=\left\{\lambda\in\mathbb{C}:\ \arg\lambda\in\left(\frac{\pi}{10}+(k-2)\frac{\pi}{5},\frac{\pi}{10}+\frac{k\pi}{5}\right)\right\},\quad k\in\mathbb{Z}.$$

These functions are related by the equations

$$\Psi_{k+1}^{(\infty)}(\lambda,x)=\Psi_k^{(\infty)}(\lambda,x)\mathrm{S}_k,\quad \mathrm{S}_{2k-1}=\begin{pmatrix}1 & s_{2k-1}\\ 0 & 1\end{pmatrix},\quad \mathrm{S}_{2k}=\begin{pmatrix}1 & 0\\ s_{2k} & 1\end{pmatrix},$$

where the entries of the Stokes matrices satisfy

$$s_{k+5}=s_5,\quad s_k=i(1+s_{k+2}s_{k+3}),\quad k\in\mathbb{Z}. \tag{5.2.32}$$

PROOF. Equations (5.2.32) are the consequence of the equations (5.2.9). These equations follow from the fact that Ψ and $\tilde{\Psi}$ (see (5.2.31)) satisfy similar symmetry conditions. □

3. The third Painlevé transcendent

Let $u(x)$ satisfy the third Painlevé transcendent,

$$\frac{d^2u}{dx^2}=\frac{1}{u}\left(\frac{du}{dx}\right)^2-\frac{1}{x}\frac{du}{dx}+\frac{4}{x}(\Theta_0u^2+1-\Theta_\infty)+4u^3-\frac{4}{u}. \tag{5.3.1}$$

Let $y(x)$ and $U(x)$ satisfy the first order ODEs,

$$\frac{dy}{dx}=\frac{\Theta_\infty}{x}y+\frac{2y}{u}, \tag{5.3.2}$$

and

$$x\frac{dU}{dx} = -4uU^2 + (4ux + 1 - 2\Theta_\infty)U + (\Theta_0 + \Theta_\infty)x. \tag{5.3.3}$$

Let $w(x)$ and $v(x)$ satisfy the algebraic equations

$$w = -\frac{y}{uU}, \quad \frac{yv}{xu} + \frac{\Theta_0}{2} = \frac{U}{x}\left[\frac{\Theta_\infty}{2}(2 - \frac{x}{U}) + (U - x)u\right]. \tag{5.3.4}$$

The above equations admit the following Lax pair

$$\frac{\partial \Psi}{\partial \lambda} = \mathsf{A}\Psi; \quad A = A_0 + \frac{A_{-1}}{\lambda} + \frac{A_{-2}}{\lambda^2}, \tag{5.3.5}$$

$$\frac{\partial \Psi}{\partial x} = \mathsf{U}\Psi; \quad \mathsf{U} = \lambda B_1 + B_0 + \frac{B_{-1}}{\lambda}, \tag{5.3.6}$$

where
(5.3.7)

$$A_0 = \frac{ix}{2}\sigma_3, \quad A_{-1} = \begin{pmatrix} -\frac{\Theta_\infty}{2} & y \\ v & \frac{\Theta_\infty}{2} \end{pmatrix}, \quad A_{-2} = -i\begin{pmatrix} U - \frac{x}{2} & -wU \\ \frac{U-x}{w} & -U + \frac{x}{2} \end{pmatrix},$$

and

$$B_1 = \frac{i}{2}\sigma_3, \quad B_0 = \frac{1}{x}\begin{pmatrix} 0 & y \\ v & 0 \end{pmatrix}, \quad B_{-1} = -\frac{A_{-2}}{x}. \tag{5.3.8}$$

Indeed the compatibility of equations (5.3.5) and (5.3.6) yields

$$x\frac{dw}{dx} = w\left[-\frac{x}{U}(\Theta_0 + \Theta_\infty) + \frac{2xy}{wU} + \Theta_\infty\right], \tag{5.3.9}$$

$$\frac{dy}{dx} = \frac{\Theta_\infty}{x}y - 2wU, \tag{5.3.10}$$

$$\frac{dv}{dx} = -\frac{\Theta_\infty}{x}v + \frac{2}{w}(x - U), \tag{5.3.11}$$

$$x\frac{dU}{dx} = \frac{4y}{w}U + \left(-\frac{4xy}{wU} - 2\Theta_\infty + 1\right)U + (\Theta_0 + \Theta_\infty)x. \tag{5.3.12}$$

Letting

$$u = -\frac{y}{wU},$$

equations (5.3.10) and (5.3.12) become equations (5.3.2) and (5.3.3). Also u satisfies

$$\frac{du}{dx} = 4Uu^2 - 2xu^2 + (2\Theta_\infty - 1)u + 2x, \tag{5.3.13}$$

and

$$\frac{\Theta_0}{2} = \frac{U}{x}\left[\frac{\Theta_\infty}{2}(2 - \frac{x}{U}) + (U - x)u\right] - \frac{vy}{xu}. \tag{5.3.14}$$

Equation (5.3.14) is equation (5.3.4); differentiating equation (5.3.13) we find (5.3.1).

PROPOSITION 5.8 (The direct problem of PIII). *Let $u(x)$ satisfy the PIII equation (5.3.1). Let $y(x)$, $U(x)$, $w(x)$, $v(x)$ satisfy equations (5.3.2)–(5.3.4). Let $\Psi_k^{(\infty)}(\lambda, x)$, $k = 1, 2$, be the unimodular solutions of the linear ODE (5.3.5) which are uniquely specified by the conditions*

$$\Psi_k^{(\infty)}(\lambda, x) = \left(I + O\left(\frac{1}{\lambda}\right)\right) \mathrm{e}^{\frac{ix\lambda}{2}\sigma_3} \left(\frac{1}{\lambda}\right)^{\frac{\Theta_\infty \sigma_3}{2}}, \quad \lambda \to \infty, \quad \lambda \in \Omega_k^\infty,$$

where

(5.3.15) Ω_1^∞: $-\arg x < \arg \lambda < 2\pi - \arg x$; Ω_2^∞: $-\pi - \arg x < \arg \lambda < \pi - \arg x$.

The functions $\Psi_k^{(\infty)}(\lambda, x)$ are related by

$$\Psi_2^{(\infty)}(\lambda, x) = \Psi_1^{(\infty)}(\lambda, x) \mathsf{S}_1^{(\infty)},$$

(5.3.16)

$$\Psi_1^{(\infty)}(\lambda, x) = \Psi_2^{(\infty)}(\mathrm{e}^{2i\pi}\lambda, x) \mathsf{S}_2^{(\infty)} \mathrm{e}^{i\pi\Theta_\infty\sigma_3},$$

where

$$\mathsf{S}_1^{(\infty)} = \begin{pmatrix} 1 & s_1^\infty \\ 0 & 1 \end{pmatrix}, \quad \mathsf{S}_2^{(\infty)} = \begin{pmatrix} 1 & 0 \\ s_2^\infty & 1 \end{pmatrix}.$$

Let $\Psi_k^{(0)}(\lambda, x)$, $k = 1, 2$, be the unimodular solution of the linear ODE (5.3.5) which is specified by the condition

(5.3.17)

$$\Psi_k^{(0)}(\lambda, x) = \begin{pmatrix} wa & b \\ a & \frac{b}{wU}(U - x) \end{pmatrix} (I + O(\lambda)) \mathrm{e}^{-\frac{ix}{2\lambda}\sigma_3} \lambda^{\frac{\Theta_0 \sigma_3}{2}}, \quad \lambda \to 0, \quad \lambda \in \Omega_k^0,$$

where $a(x)$ and $b(x)$ are certain functions of x satisfying $ab = -U/x$, and the sectors Ω_k^0 are defined by

(5.3.18) Ω_1^0: $\arg x < \arg \lambda < 2\pi + \arg x$; Ω_2^0: $-\pi + \arg x < \arg \lambda < \pi + \arg x$.

The functions $\Psi_k^{(0)}$ are related by

(5.3.19) $\Psi_2^{(0)}(\lambda, x) = \Psi_1^{(0)}(\lambda, x) \mathsf{S}_1^{(0)}$, $\Psi_1^{(0)}(\lambda, x) = \Psi_2^{(0)}(\mathrm{e}^{2i\pi}\lambda, x) \mathsf{S}_2^{(0)} \mathrm{e}^{-i\pi\Theta_0\sigma_3}$,

where

$$\mathsf{S}_1^{(0)} = \begin{pmatrix} 1 & s_1^0 \\ 0 & 1 \end{pmatrix}, \quad \mathsf{S}_2^{(0)} = \begin{pmatrix} 1 & 0 \\ s_2^0 & 1 \end{pmatrix}.$$

Let E be the 2×2 unimodular matrix defined by

(5.3.20) $$\Psi_1^{(\infty)} = \Psi_1^{(0)} \mathsf{E}.$$

The entries of the Stokes matrices $\mathsf{S}_k^{(\infty)}$ and $\mathsf{S}_k^{(0)}$ satisfy the algebraic equation

(5.3.21) $$2\cos \pi\Theta_0 + s_1^0 s_2^0 \mathrm{e}^{-i\pi\Theta_0} = 2\cos \pi\Theta_\infty + s_1^\infty s_2^\infty \mathrm{e}^{i\pi\Theta_\infty}.$$

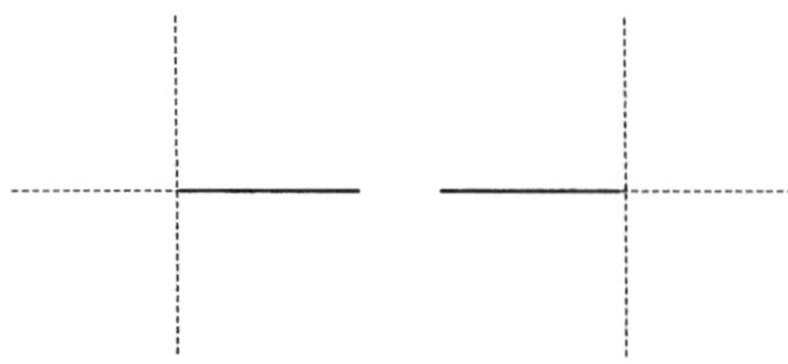

FIGURE 5.8. Stokes lines near $\lambda = \infty$.

The transformation from Ψ to $D^{-1}\Psi D$, where $D = \operatorname{diag}(d, d^{-1})$, $d \neq 0$, constant, leaves the Lax pair (equations (5.3.5) and (5.3.6)) invariant, leaves the algebraic equation (5.3.21) invariant, leaves PIII invariant, and changes y, v, w to y/d^2, vd^2, w/d^2.

The monodromy data $\{s_1^0, s_2^0, s_1^\infty, s_2^\infty\}$ determine E to within a right-multiplicative diagonal matrix.

PROOF. The dominant behavior of the solution $\Psi(\lambda, x)$ of equation (5.3.5) near $\lambda = \infty$ involves $\exp[\frac{ix\lambda}{2}\sigma_3]$. This implies that the Stokes curves asymptote to the ray $\arg(x\lambda) = 0$, π; see Figure 5.8. Thus,

$$\Omega_1^\infty: \quad 0 < \arg x\lambda < 2\pi,$$
$$\Omega_2^\infty: \quad -\pi < \arg x\lambda < \pi.$$

The dominant behavior of the solution $\Psi(\lambda, x)$ of equation (5.3.5) near $\lambda = 0$ involves $\exp[-\frac{ix}{2\lambda}\sigma_3]$. This implies that the Stokes curves asymptote to the lines $\arg(\frac{\lambda}{x}) = 0$, π; see Figure 5.8. Thus,

$$\Omega_1^0: \quad 0 < \arg\frac{\lambda}{x} < 2\pi,$$
$$\Omega_2^0: \quad -\pi < \arg\frac{\lambda}{x} < \pi.$$

The dominant behavior of the solution $\Psi_k^{(0)}$ near zero involves A_{-2}. Since trace $A_{-2} = 0$ and $\det A_{-2} = x^2/4$, it follows that

$$\Psi^{(0)} = \hat{\Psi}^{(0)} e^{-\frac{ix}{2\lambda}\sigma_3} \lambda^{\frac{\Theta_0\sigma_3}{2}},$$

where $\hat{\Psi}^{(0)}$ satisfies

$$\frac{\partial\hat{\Psi}^{(0)}}{\partial\lambda} + \hat{\Psi}^{(0)}\left(\frac{ix}{2\lambda^2}\sigma_3 + \frac{\Theta_0}{2}\frac{\sigma_3}{\lambda}\right) = \left(A_0 + \frac{A_{-1}}{\lambda} + \frac{A_{-2}}{\lambda^2}\right)\hat{\Psi}^{(0)}.$$

Thus, $\hat{\Psi}^{(0)}(0, x)$ satisfies

$$\frac{1}{2}ix\hat{\Psi}^{(0)}(0, x)\sigma_3(\hat{\Psi}^{(0)}(0, x))^{-1} = A_{-2}.$$

Taking $\hat{\Psi}^{(0)}(0, x)$ in the form appearing in (5.3.17) it follows that

$$\det\hat{\Psi}^{(0)}(0, x) = -\frac{abx}{s} = 1.$$

Using (5.3.17), (5.3.19) and (5.3.20) we find the consistency condition

$$\mathsf{S}_1^{(\infty)}\mathsf{S}_2^{(\infty)} e^{i\pi\Theta_\infty\sigma_3} = \mathsf{E}^{-1}\mathsf{S}_1^{(0)}\mathsf{S}_2^{(0)} e^{-i\pi\Theta_0\sigma_3}\mathsf{E}. \tag{5.3.22}$$

The trace of this equation implies the algebraic equation (5.3.21).

We will now show that the monodromy data $\{S_1^{(0)}, S_2^{(0)}, s_1^\infty, s_2^\infty\}$ determine E to within a right-multiplicative diagonal matrix. To establish this fact we let

$$E^{-1}S_1^{(0)}S_2^{(0)}e^{-i\pi\Theta_0\sigma_3}E = \begin{pmatrix} F & \tau \\ \frac{FG-F^2-1}{\tau} & G-F \end{pmatrix}. \tag{5.3.23}$$

The trace of this equation yields

$$G = 2\cos\pi\Theta_0 + S_1^{(0)}S_2^{(0)}e^{-i\pi\Theta_0}. \tag{5.3.24}$$

Equations (5.3.22) and (5.3.23) imply

$$S_1^{(\infty)}S_2^{(\infty)} = \begin{pmatrix} F & \tau \\ \dfrac{FG-F^2-1}{\tau} & G-F \end{pmatrix} e^{-i\pi\Theta_\infty\sigma_3}.$$

The trace of this equation yields

$$2iF\sin\pi\Theta_\infty = Ge^{i\pi\Theta_\infty} - 1 - s_1^\infty s_2^\infty. \tag{5.3.25}$$

Thus, if $\Theta_\infty \notin \mathbb{Z}$, equations (5.3.24) and (5.3.25) imply that both F and G are determined in terms of $s_1^0 s_2^0$ and $s_1^\infty s_2^\infty$. The only freedom in (5.3.23) is the parameter τ which corresponds to a right diagonal matrix, changing E to $E\,\mathrm{diag}\,(d, d^{-1})$ changes τ to $d^2\tau$. Using $\Psi_1^{(\infty)} = \Psi_1^{(0)}E^{-1}$, this transformation has the effect of changing a and b to a/d and bd respectively, which leaves $ab = -s/x$ invariant.

It is straightforward to prove that if Ψ evolves according to (5.3.6), then $\{s_1^0, S_2^{(0)}, s_1^\infty, s_2^\infty\}$ are x-independent. For E to be x-independent, b must satisfy

$$\frac{\partial b}{\partial x} = \frac{yb}{wUx}(U-x) + \frac{w}{x}ab^2\left[\frac{x-U}{wU}\left(\Theta_\infty - \frac{y}{wU}(U-x) - v\right)\right].$$

This equation specifies b to within an integration constant, which is the reflection of the freedom of choosing an arbitrary constant d. □

THEOREM 5.4. (The solution of PIII). *Let the complex constants* $\{s_1^0, s_2^0, s_1^\infty, s_2^\infty\}$ *belong to the equivalent class induced by the transformation*

$$\{s_1^0, s_2^0, s_1^\infty, s_2^\infty\} \to \{s_1^0/c^2, s_2^0c^2, s_1^\infty/c^2, s_2^\infty c^2\}, \quad c \neq 0,$$

on the algebraic equation

$$2\cos\pi\Theta_0 + s_1^0s_2^0e^{-i\pi\Theta_0} = 2\cos\pi\Theta_\infty + s_1^\infty s_2^\infty e^{-i\pi\Theta_\infty}.$$

Let the unimodular 2×2 *complex-valued matrix function* $Y(\lambda, x)$, $\lambda \in \mathbb{C}$, $x \in \mathbb{C}$, *have the following representation*

$$Y(\lambda, x) = \begin{cases} Y_k^0 & \text{for} \quad \lambda \in \Omega_k^0, \quad k = 1,2, \\ Y_k^\infty & \text{for} \quad \lambda \in \Omega_k^\infty, \quad k = 1,2, \end{cases}$$

where the sectors $\Omega_k^0, \Omega_k^\infty$ *are defined in* (5.3.15), (5.3.18). *Let* $Y(\lambda, x)$ *satisfy the following RH problem*

$$Y_2^\infty(\lambda,x) = Y_1^\infty(\lambda,x)\mathrm{e}^{\theta(\lambda,x)}\left(\frac{1}{\lambda}\right)^{\frac{1}{2}\Theta_\infty\sigma_3} \mathsf{S}_1^{(\infty)}\left(\frac{1}{\lambda}\right)^{-\frac{1}{2}\Theta_\infty\sigma_3}\mathrm{e}^{-\theta(\lambda,x)},$$

$$Y_1^\infty(\lambda,x) = Y_2^\infty\left(\mathrm{e}^{2i\pi}\lambda,x\right)\mathrm{e}^{\theta(\lambda,x)}\left(\frac{1}{\mathrm{e}^{2i\pi}\lambda}\right)^{\frac{1}{2}\Theta_\infty\sigma_3} \mathsf{S}_2^{(\infty)}\left(\frac{1}{\mathrm{e}^{2i\pi}\lambda}\right)^{-\frac{1}{2}\Theta_\infty\sigma_3}\mathrm{e}^{-\theta(\lambda,x)},$$

$$Y_2^0(\lambda,x) = Y_1^0(\lambda,x)\mathrm{e}^{\theta(\lambda,x)}\lambda^{\frac{1}{2}\Theta_0\sigma_3}\mathsf{S}_1^{(0)}\lambda^{-\frac{1}{2}\Theta_0\sigma_3}\mathrm{e}^{-\theta(\lambda,x)},$$

$$Y_1^0(\lambda,x) = Y_2^0\left(\mathrm{e}^{2i\pi\Theta}\lambda,x\right)\mathrm{e}^{\theta(\lambda,x)}\left(\mathrm{e}^{2i\pi}\lambda\right)^{\frac{1}{2}\Theta_0\sigma_3}\mathsf{S}_2^{(0)}\left(\mathrm{e}^{2i\pi}\lambda\right)^{-\frac{1}{2}\Theta_0\sigma_3}\mathrm{e}^{-\theta(\lambda,x)},$$

$$Y_1^\infty(\lambda,x) = Y_1^0(\lambda,x)\mathrm{e}^{\theta(\lambda,x)}\lambda^{\frac{1}{2}\Theta_0\sigma_3}\mathsf{E}\left(\frac{1}{\lambda}\right)^{-\frac{1}{2}\Theta_\infty\sigma_3}\mathrm{e}^{-\theta(\lambda,x)},$$

$$Y_1^\infty(\lambda,x) = I + O\left(\frac{1}{\lambda}\right), \quad \lambda\to\infty, \tag{5.3.26}$$

where

$$\theta(\lambda,x) = \frac{ix}{2}\left(\lambda - \frac{1}{\lambda}\right)\sigma_3, \tag{5.3.27}$$

and the 2×2 *matrices* $\mathsf{S}_1^{(0)}, \mathsf{S}_2^{(0)}, s_1^{(\infty)}, \mathsf{S}_2^{(\infty)}, \mathsf{E}$, *are defined in terms of* $\{s_1^0, s_2^0, s_1^\infty, s_2^\infty\}$ *as follows:*

$$\mathsf{S}_1^{(0)} = \begin{pmatrix} 1 & s_1^0 \\ 0 & 1\end{pmatrix}, \quad \mathsf{S}_2^{(0)} = \begin{pmatrix} 1 & 0 \\ s_2^0 & 1\end{pmatrix}, \quad \mathsf{S}_1^{(\infty)} = \begin{pmatrix} 1 & s_1^\infty \\ 0 & 1\end{pmatrix}, \quad \mathsf{S}_2^{(\infty)} = \begin{pmatrix} 1 & 0 \\ s_2^\infty & 1\end{pmatrix};$$

E *is a unimodular* 2×2 *constant matrix satisfying* (5.3.22) *(*E *is specified to within a right multiplicative diagonal matrix). Let* $u(x)$ *be defined by*

$$u(x) = -ix\lim_{\lambda\to\infty}\lambda(Y_1^\infty)_{12},$$

where the subscript 12 *denotes the* (1, 2) *entry of* Y_1^∞. *Then: (a) The above RH problem defines a unique* 2×2 *matrix* $Y(\lambda,x)$ *which is a meromorphic function of* x. *(b) The function* $u(x)$ *is a meromorphic function of* x *and it satisfies the Painlevé III equation* (5.3.1)*. The "meromorphicity of* x*" means meromorphicity on the universal covering of the complex* x*-plane punctured at* $x = 0$.

PROOF. Let $\Psi_k^{(0)}(\lambda,x)$ and $\Psi_k^{(\infty)}(\lambda,x)$ be defined in terms of $Y(\lambda,x)$ by

$$\Psi_k^{(0)} = Y_k^0\mathrm{e}^\theta\lambda^{\frac{1}{2}\Theta_0\sigma_3}, \quad \Psi_k^{(\infty)} = Y_k^\infty\mathrm{e}^\theta\left(\frac{1}{\lambda}\right)^{\frac{1}{2}\Theta_\infty\sigma_3}. \tag{5.3.28}$$

The anti-Stokes lines near $\lambda=\infty$ are given by $\arg x\lambda = \frac{\pi}{2}, \frac{3\pi}{2}$, i.e., $\arg\lambda = \frac{\pi}{2} - \arg x$, $\arg\lambda = \frac{3\pi}{2} - \arg x$. Similarly the anti-Stokes lines near $\lambda = 0$ are given by $\arg\lambda = \frac{\pi}{2} + \arg x$, $\arg\lambda = \frac{3\pi}{2} + \arg x$. Let L be the contour depicted in Figure 5.9. It consists of part of the above anti-Stokes lines as well as of the unit circle at $\lambda = 0$. Equations (5.3.26) imply

$$Y_+(\lambda,x) = Y_-(\lambda,x)\mathrm{e}^{\theta(\lambda,x)}V(\lambda)\mathrm{e}^{-\theta(\lambda,x)}, \quad \lambda\in L,$$

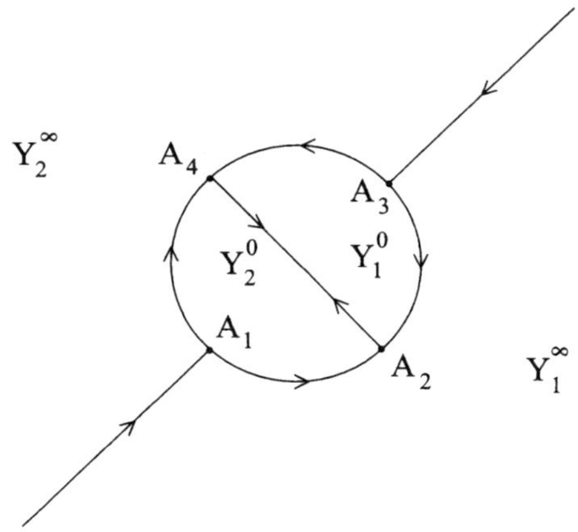

FIGURE 5.9. The RH problem for PIII

where the jumps V are given by

$$
\begin{aligned}
V_{A_3A_2} &= \lambda^{\frac{1}{2}\Theta_0\sigma_3}\mathsf{E}\left(\frac{1}{\lambda}\right)^{-\frac{1}{2}\Theta_\infty\sigma_3},\\
V_{\infty A_3} &= \left(\frac{1}{\lambda}\right)^{\frac{1}{2}\Theta_\infty\sigma_3}(\mathsf{S}_1^{(\infty)})^{-1}\left(\frac{1}{\lambda}\right)^{-\frac{1}{2}\Theta_\infty\sigma_3},\\
V_{OA_4} &= \lambda^{\frac{1}{2}\Theta_0\sigma_3}(\mathsf{S}_1^{(0)})^{-1}\lambda^{-\frac{1}{2}\Theta_0\sigma_3},\\
V_{A_1A_4} &= \lambda^{\frac{1}{2}\Theta_0\sigma_3}(\mathsf{S}_1^{(0)})^{-1}\mathsf{E}\mathsf{S}_1^{(\infty)}\left(\frac{1}{\lambda}\right)^{-\frac{1}{2}\Theta_\infty\sigma_3},\\
V_{A_2O} &= \left(\lambda^{\frac{1}{2}\Theta_0\sigma_3}\right)_+(\mathsf{S}_2^{(0)})^{-1}\left(\lambda^{-\frac{1}{2}\Theta_0\sigma_3}\right)_+,\\
V_{A_1A_2} &= \left(\frac{1}{\lambda}\right)^{\frac{1}{2}\Theta_\infty\sigma_3}\mathsf{E}^{-1}e^{i\pi\Theta_0\sigma_3}(\mathsf{S}_2^{(0)})^{-1}\lambda^{-\frac{1}{2}\Theta_0\sigma_3},\\
V_{A_3A_4} &= \left(\frac{1}{\lambda}\right)^{\frac{1}{2}\Theta_\infty\sigma_3}(\mathsf{E}\mathsf{S}_1^{(\infty)})^{-1}\lambda^{-\frac{1}{2}\Theta_0\sigma_3},\\
V_{\infty A_1} &= \left(\frac{1}{\lambda}\right)_+^{\frac{1}{2}\Theta_\infty\sigma_3}(\mathsf{S}_2^{(\infty)})^{-1}\left(\frac{1}{\lambda}\right)_+^{-\frac{1}{2}\Theta_\infty\sigma_3}.
\end{aligned}
$$

We have associated the branch cuts ∞A_1 and A_2O with $(1/\lambda)^{\frac{1}{2}\Theta_\infty}$ and $\lambda^{\frac{1}{2}\Theta_0}$ respectively. The subscripts $+$ appearing in the definition of $V_{\infty A_1}$ and V_{A_2O} indicate that we consider the relevant boundary values from the $+$ region, i.e., $(\lambda)_+ = |\lambda|e^{2i\pi}$.

By construction, the function Y is continuous at the intersection points.

This of course can be verified directly by checking that at the intersection points the product of jumps is unity. The product conditions at A_2, A_3 and A_4 are satisfied identically; the product condition at $A_1(V_{\infty A_1})^{-1}V_{A_1A_2}(V_{A_1A_4})_+ = I$, is satisfied because of the consistency condition (5.3.22) ($(V_{A_1A_4})_+$ indicates that the $1/\lambda$ term appearing in $V_{A_1A_4}$ must be evaluated as $(1/\lambda)_+$).

Derivation of $\Psi_\lambda = A\Psi$. We define A by $A = \Psi_\lambda \Psi^{-1}$, where Y is defined in terms of Ψ by equations (5.3.28). Since Ψ_λ and Ψ admit the same jumps it follows that A is holomorphic in $\mathbb{C}/\{0\}$. Also,

$$\Psi \sim e^{\left[\frac{ix}{2}(\lambda - \frac{1}{\lambda}) - \frac{1}{2}\Theta_\infty \ln\lambda\right]\sigma_3}, \quad \lambda \to \infty,$$

thus

$$A = A_0 + \frac{A_{-1}}{\lambda} + \frac{A_{-2}}{\lambda^2}.$$

Using (5.3.28) it follows that Y solves

$$Y_\lambda + Y\left(\frac{ix}{2} - \frac{\Theta_\infty}{2\lambda} + \frac{ix}{2\lambda^2}\right)\sigma_3 = \left(A_0 + \frac{A_{-1}}{\lambda} + \frac{A_{-2}}{\lambda^2}\right)Y, \quad \text{near } \infty \tag{5.3.29}$$

$$Y_\lambda + Y\left(\frac{ix}{2} + \frac{\Theta_0}{2\lambda} + \frac{ix}{2\lambda^2}\right)\sigma_3 = \left(A_0 + \frac{A_{-1}}{\lambda} + \frac{A_{-2}}{\lambda^2}\right)Y, \quad \text{near } 0. \tag{5.3.30}$$

For large λ, Y admits the expansion

$$Y = I + \frac{Y_{-1}}{\lambda} + \frac{Y_{-2}}{\lambda^2} + O\left(\frac{1}{\lambda^3}\right), \quad \lambda \to \infty. \tag{5.3.31}$$

Substituting (5.3.31) into (5.3.29) we find

$$O(1)\colon\ A_0 = \frac{ix}{2}\sigma_3; \quad O\left(\frac{1}{\lambda}\right): \ A_{-1} = -\frac{\Theta_\infty}{2}\sigma_3 + \frac{ix}{2}[Y_{-1}, \sigma_3].$$

$$O\left(\frac{1}{\lambda^2}\right): \ A_{-2} = \frac{ix}{2}\sigma_3 - \left(I + \frac{\Theta_\infty}{2}\sigma_3 + A_{-1}\right)Y_{-1} + \frac{ix}{2}\left[Y_{-2}, \sigma_3\right].$$

We define

$$(Y_{-1})_{12} = -\frac{y(x)}{ix}, \quad (Y_{-1})_{21} = \frac{v(x)}{ix}. \tag{5.3.32}$$

Then A_{-1} becomes

$$A_{-1} = \begin{pmatrix} -\dfrac{\Theta_\infty}{2} & y \\ v & \dfrac{\Theta_\infty}{2} \end{pmatrix}.$$

We define

$$(Y_{-1})_{11} = is(x) - \frac{yv}{ix}.$$

Then

$$(A_{-2})_{11} = \frac{ix}{2} - is.$$

Equation (5.3.30) implies

$$\frac{ix}{2}Y(0,x)\sigma_3 = A_{-2}Y(0,x).$$

Thus,

$$\det A_{-2} = \frac{x^2}{2}, \quad \operatorname{Tr} A_{-2} = 0.$$

These equations together with $(A_{-2})_{11} = \frac{ix}{2} - is$, yield the expression for A_{-2} given in (5.3.7).

Derivation of $\Psi_x = \mathrm{U}\Psi$. Similar considerations as above imply that if $B = \Psi_x\Psi^{-1}$, then

$$B = \lambda B_1 + B_0 + \frac{B_{-1}}{\lambda}.$$

Also, Y solves

$$Y_x + \frac{i}{2}Y\left(\lambda - \frac{1}{\lambda}\right)\sigma_3 = \left(\lambda B_1 + B_0 + \frac{B_{-1}}{\lambda}\right)Y. \tag{5.3.33}$$

Substituting (5.3.31) into this equation we find

$$O(\lambda):\ B_1 = \frac{i}{2}\sigma_3.$$

$$O(1):\ B_0 = \frac{i}{2}[Y_{-1}, \sigma_3].$$

$$O\left(\frac{1}{\lambda}\right):\ B_{-1} = \frac{\partial Y_{-1}}{\partial x} + \frac{i}{2}[Y_{-2}, \sigma_3] - \frac{i}{2}\sigma_3 - B_0Y_{-1}. \tag{5.3.34}$$

Using the definitions of $(Y_{-1})_{12}$, $(Y_{-1})_{21}$ given by equations (5.3.32), it follows that B_0 is given by the expression appearing in (5.3.8). Also equation (5.3.34) implies

$$(B_{-1})_{11} = \frac{\partial}{\partial x}(Y_{-1})_{11} - \frac{i}{2} - \frac{yv}{ix^2} = \frac{\partial}{\partial x}\left(iU - \frac{yv}{ix}\right) - \frac{i}{2} - \frac{yv}{ix^2}.$$

Thus,

$$(B_{-1})_{11} = \frac{i}{x}\left(U - \frac{x}{2}\right),$$

iff

$$\frac{\partial}{\partial x}\left(U + \frac{yv}{x}\right) = \frac{1}{x}\left(U - \frac{yv}{x}\right). \tag{5.3.35}$$

Letting $\lambda \to 0$ in equation (5.3.33) we find

$$-\frac{i}{2}Y(0,x)\sigma_3 = B_{-1}Y(0,x).$$

Thus,

$$\det B_{-1} = \frac{1}{4},\quad \operatorname{Tr} B_{-1} = 0.$$

These equations together with $(B_{-1})_{11} = i\left(U - \frac{x}{2}\right)/x$ imply the expression for B_{-1} appearing in (5.3.8). Equation (5.3.35) is consistent with equations (5.3.1)–(5.3.4).

The meromorphic with respect to x solvability of the above RH problem, follows as with the RH problem of PII. □

4. The fifth Painlevé transcendent

Let $u(x)$, $v(x)$ and $y(x)$ satisfy the nonlinear ODEs

$$x\frac{du}{dx} = xu - 2v(u-1)^2 - \frac{1}{2}(u-1)[(\Theta_0 - \Theta_1 + \Theta_\infty)u - (3\Theta_0 + \Theta_1 + \Theta_\infty)], \tag{5.4.1}$$

$$x\frac{dv}{dx} = uv\left[v + \frac{1}{2}(\Theta_0 - \Theta_1 + \Theta_\infty)\right] - \frac{1}{u}(v+\Theta_0)\left[v + \frac{1}{2}(\Theta_0 + \Theta_1 + \Theta_\infty)\right], \tag{5.4.2}$$

(5.4.3)

$$x\frac{dy}{dx} = y\left\{-2v - \Theta_0 + u\left[v + \frac{1}{2}(\Theta_0 - \Theta_1 + \Theta_\infty)\right] + \frac{1}{u}\left[v + \frac{1}{2}(\Theta_0 + \Theta_1 + \Theta_\infty)\right]\right\}.$$

These equations are the compatibility condition of the following Lax pair for the 2×2 matrix-valued function $Y(\lambda, x)$:

$$\frac{\partial \Psi}{\partial \lambda} = \mathsf{A}\Psi; \quad \mathsf{A}(\lambda, x) = \frac{x}{2}\sigma_3 + \frac{A_1(x)}{\lambda} + \frac{A_2(x)}{\lambda - 1}; \tag{5.4.4}$$

$$\frac{\partial \Psi}{\partial x} = \mathsf{U}\Psi; \quad \mathsf{U}(\lambda, x) = \frac{\lambda}{2}\sigma_3 + B_0(x); \tag{5.4.5}$$

where

$$A_1(x) = \begin{pmatrix} v + \frac{\Theta_0}{2} & -y(v + \Theta_0) \\ \frac{v}{y} & -(v + \frac{\Theta_0}{2}) \end{pmatrix}, \tag{5.4.6}$$

$$A_2(x) = \begin{pmatrix} -v - \frac{1}{2}(\Theta_0 + \Theta_\infty) & yu(v + \frac{1}{2}(\Theta_0 + \Theta_\infty - \Theta_1)) \\ -\frac{1}{yu}(v + \frac{1}{2}(\Theta_0 + \Theta_\infty + \Theta_1)) & v + \frac{1}{2}(\Theta_0 + \Theta_\infty) \end{pmatrix}, \tag{5.4.7}$$

(5.4.8)

$$B_0(x) =$$

$$\frac{1}{x}\begin{pmatrix} 0 & -y[v + \Theta_0 - u(v + \frac{1}{2}(\Theta_0 + \Theta_\infty - \Theta_1))] \\ \frac{1}{y}[v - \frac{1}{u}(v + \frac{1}{2}(\Theta_0 + \Theta_\infty + \Theta_1))] & 0 \end{pmatrix}.$$

Using equation (5.4.2) to eliminate v, equation (5.4.1) implies that $u(x)$ solves the PV equation,

$$\begin{aligned} \frac{d^2u}{dx^2} &= \left(\frac{1}{2u} \pm \frac{1}{u-1}\right)\left(\frac{du}{dx}\right)^2 \\ &-\frac{1}{x}\frac{du}{dx} + \frac{(u-1)^2}{x^2}\left(\alpha u + \frac{\beta}{u}\right) + \frac{\gamma u}{x} + \frac{\delta u(u+1)}{u-1}, \end{aligned} \tag{5.4.9}$$

where the constants $\alpha, \beta, \gamma, \delta$ are given by

$$\begin{aligned} \alpha &= \frac{1}{8}(\Theta_0 - \Theta_1 + \Theta_\infty)^2, \\ \beta &= -\frac{1}{8}(\Theta_0 - \Theta_1 - \Theta_\infty)^2, \\ \gamma &= 1 - \Theta_0 - \Theta_1, \delta = -\frac{1}{2}. \end{aligned} \tag{5.4.10}$$

We note that the general case of $\delta \neq 0$ can be reduced to the case of $\delta = -1/2$ by a simple scaling; the case $\delta = 0$ is equivalent to PIII [**Gro3**].

PROPOSITION 5.9 (The direct problem of PV). *Let $u(x)$ satisfy the PV equation (5.4.10) with the parameters $\alpha, \beta, \gamma, \delta$ given by equations (5.4.11) and $\Theta_0 \notin \mathbb{Z}$, $\Theta_1 \notin \mathbb{Z}$. Let $v(x)$ be defined in terms of $u(x)$ by equation (5.4.1). Let $y(x)$ satisfy* *the first order ODE (5.4.3). Let* $\Psi_k^{(\infty)}(\lambda, x)$, $k = 1, 2$, *be the unimodular solutions of the linear ODE (5.4.4), which are uniquely specified by the asymptotic conditions*

$$\Psi_k^{(\infty)}(\lambda, x) = \left(I + O\left(\frac{1}{\lambda}\right)\right) e^{\frac{1}{2}\lambda x \sigma_3} \left(\frac{1}{\lambda}\right)^{\frac{1}{2}\Theta_\infty \sigma_3}, \quad \lambda \to \infty \quad \text{in } \Omega_k,$$

where the open sectors Ω_k, $k = 1, 2$, *are defined by*

$$\Omega_1: \quad -\arg x - \frac{3\pi}{2} < \arg \lambda < \frac{\pi}{2} - \arg x, \quad |\lambda| > \lambda_\infty,$$

$$\Omega_2: \quad -\arg x - \frac{\pi}{2} < \arg \lambda < \frac{3\pi}{2} - \arg x, \quad |\lambda| > \lambda_\infty, \tag{5.4.11}$$

and λ_∞ *is an arbitrary finite positive constraint. The functions* $\Psi_1^{(\infty)}$ *and* $\Psi_2^{(\infty)}$ *are related by the equations*

$$\Psi_2^{(0)}(\lambda, x) = \Psi_1^{(\infty)}(\lambda, x) \mathsf{S}_1, \quad \Psi_1^{(\infty)}(\lambda, x) = \Psi_2^{(\infty)}(e^{2i\pi}\lambda, x) \mathsf{S}_2 e^{i\pi\Theta_\infty \sigma_3}, \tag{5.4.12}$$

where

$$\mathsf{S}_1 = \begin{pmatrix} 1 & 0 \\ s_1 & 0 \end{pmatrix}, \quad \mathsf{S}_2 = \begin{pmatrix} 1 & s_2 \\ 0 & 1 \end{pmatrix},$$

and s_1, s_2 *are arbitrary complex constants.*

Let E_0 *and* E_1 *be* 2×2 *unimodular matrices satisfying*

$$\mathsf{E}_0^{-1} e^{-i\pi\Theta_0\sigma_3} \mathsf{E}_0 = \mathsf{S}_1 \mathsf{S}_2 e^{i\pi\Theta_\infty\sigma_3} \mathsf{E}_1^{-1} e^{i\pi\Theta_1\sigma_3} \mathsf{E}_1. \tag{5.4.13}$$

This equation specifies E_0 *and* E_1 *to within left-multiplicative diagonal matrices* $diag(d_0, d_0^{-1})$, *and* $diag(d_1, d_1^{-1})$, *respectively.*

Let $\Psi^{(0)}(\lambda, x)$ *and* $\Psi^1(\lambda, x)$ *be* 2×2 *unimodular matrices defined in terms of* $\Psi_1^{(\infty)}(\lambda, x)$, E_0 *and* E_1 *by*

$$\Psi_1^{(\infty)}(\lambda, x) = \Psi^{(0)}(\lambda, x) \mathsf{E}_0, \quad \Psi_1^{(\infty)}(\lambda, x) = \Psi^1(\lambda, x) \mathsf{E}_1.$$

Then

$$\Psi^{(0)}(\lambda, x) = \hat{\Psi}^{(0)}(\lambda, x) \lambda^{\frac{1}{2}\Theta_0 \sigma_3}, \quad |\lambda| < 1,$$

and

$$\Psi^1(\lambda, x) = \hat{\Psi}^1(\lambda, x)(\lambda - 1)^{\frac{1}{2}\Theta_1\sigma_3}, \quad |\lambda - 1| < 1,$$

where $\hat{\Psi}^{(0)}(\lambda, x)$, $\hat{\Psi}(\lambda, x)$ *are holomorphic at* $\lambda = 0$, $\lambda = 1$, *respectively.*

We associate with λ^∞ *and* $(\frac{1}{\lambda})^{\Theta_\infty}$ *the branch cut indicated in Figure 5.10; with* $(\lambda - 1)^{\Theta_1}$ *we associate the branch cut from 0 to 1.*

PROOF. The dominant behavior near $\lambda = 0$ is characterized by $\partial\Psi/\partial\lambda \sim A_1/\lambda$; since trace $A_1 = 0$ and $\det A_1 = -\Theta_0^2/4$, it follows that $\Psi = \hat{\Psi}^{(0)} \lambda^{\frac{1}{2}\Theta_0\sigma_3}$. Furthermore, it is possible to choose $\hat{\Psi}^{(0)}$ so that $\det \hat{\Psi}^{(0)} = 1$. Similarly, the dominant behavior near $\lambda = 1$ is characterized by $\partial\Psi/\partial\lambda \sim A_2/\lambda$, and trace $A_2 = 0$, $\det A_2 = -\Theta_1^2/4$.

The dominant behavior of the solution $\Psi(\lambda, x)$ of equation (5.4.4) in the neighborhood of $\lambda = \infty$ is given by $\exp[(\lambda x - \Theta_\infty \ln \lambda)\sigma_3]$. This implies that the Stokes curves asymptote to $\arg(x\lambda) = \frac{\pi}{2}, \frac{3\pi}{2}$; see Figure 5.10. Thus,

FIGURE 5.10. Stokes rays for PV

$$\Omega_1: \quad \frac{-3\pi}{2} < \arg(x\lambda) < \frac{\pi}{2},$$
$$\Omega_2: \quad \frac{-\pi}{2} < \arg(x\lambda) < \frac{3\pi}{2}.$$

We will now show that equation (5.4.13) specifies E_0 and E_1 within left-multiplicative diagonal matrices. We let

$$\mathsf{E}_1\mathsf{S}_1\mathsf{S}_2 e^{i\pi\Theta_\infty\sigma_3}\mathsf{E}_1^{-1} = \begin{pmatrix} F & \tau \\ (FG-F^2-1)/\tau & G-F \end{pmatrix}. \tag{5.4.14}$$

The trace of this equation implies

$$G = 2\cos\pi\Theta_\infty + s_1s_2 e^{-i\pi\Theta_\infty}. \tag{5.4.15}$$

Substituting (5.4.14) into (5.4.13) it follows that

$$\mathsf{E}_1\mathsf{E}_0^{-1}e^{-i\pi\Theta_\infty\sigma_3}\mathsf{E}_0\mathsf{E}_1^{-1} = \begin{pmatrix} F & \tau \\ (FG-F^2-1)/\tau & G-A \end{pmatrix} e^{i\pi\Theta_1\sigma_3}. \tag{5.4.16}$$

The trace of this equation implies

$$2iF\sin\pi\Theta_1 = 2\cos\pi\Theta_0 - Ge^{-i\pi\Theta_1}. \tag{5.4.17}$$

Thus, if $\Theta_1 \notin \mathbb{Z}$, equations (5.4.15) and (5.4.16) yield F and G in terms of s_1s_2. The only freedom in (5.4.14) is the parameter τ. Changing τ to $d_1^2\tau$ changes E_1 to $\operatorname{diag}(d_1, d_1^{-1})\mathsf{E}_1$. Similarly, (5.4.17) yields E_0 to within a left diagonal multiplicative constant matrix $\operatorname{diag}(d_0, d_0^{-1})$.

The above freedom in choosing E_0 and E_1 has the following effect on the normalization of $\hat{\Psi}^{(0)}(0,x)$ and $\hat{\Psi}^1(0,x)$:

$$\hat{\Psi}^{(0)}(0,x) = \begin{pmatrix} a_0\frac{y}{v}(v+\Theta_0) & b_0 y \\ a_0 & b_0 \end{pmatrix},$$

$$\hat{\Psi}^1(0,x) = \begin{pmatrix} a_1 & b_1 yu \\ \frac{a_1}{yu}\frac{2v+\Theta_0+\Theta_\infty+\Theta_1}{2v+\Theta_0+\Theta_\infty-\Theta_1} & b_1 \end{pmatrix},$$

where

$$a_0b_0 = \frac{v}{y\Theta_0}, a_1b_1 = \Theta_1 - \frac{v+\Theta_0+\Theta_\infty}{\Theta_1}.$$

Replacing E_0 and E_1 by

$$\operatorname{diag}(d_0, d_0^{-1})\mathsf{E}_0, \operatorname{diag}(d_1, d_1^{-1})\mathsf{E}_1$$

has the effect of replacing a_0, b_0, a_1, b_1 by

$$d_0a_0, \frac{b_0}{d_0}, d_1a_1, \frac{b_1}{d_1},$$

thus a_0b_0 and a_1b_1 remain invariant.

If $\Psi(\lambda, x)$ evolves in x according to equation (5.4.5), then s_1, s_2 are x-independent. E_0 and E_1 are also x-independent, iff b_0 and b_1 satisfy

$$b_0 = \mathrm{e}^{p_0}, \quad x\frac{dp_0}{dx} = v - \frac{1}{u}\left(v + \frac{\Theta_0+\Theta_\infty+\Theta_1}{2}\right),$$

$$b_1 = \mathrm{e}^{p_1}, \quad x\frac{dp_1}{dx} = uv - \left(v + \frac{\Theta_0+\Theta_\infty+\Theta_1}{2}\right) - \frac{x}{2}.$$

These equations specify b_0 and b_1 within two constants of integration, which reflect the freedom in choosing d_0 and d_1. □

THEOREM 5.5 (The solution of Painlevé V). *Let s_1, s_2 be complex constants. Let the unimodular 2×2 complex-valued matrix $Y(\lambda, x)$, $\lambda\in\mathbb{C}$, $x\in\mathbb{C}$, have the following representation*

$$Y(\lambda, x) = \begin{cases} Y^0(\lambda, x), & |\lambda| < 1; \\ Y^1(\lambda, x), & |\lambda - 1| < 1; \\ Y_k(\lambda, x) & \textit{for} \quad \lambda \neq 0, \quad \lambda \neq 1, \lambda \ \textit{textin} \ \Omega_k, \quad k = 1, 2, \end{cases}$$

where the open sectors Ω_k are defined in equations (5.4.11). Let $Y(\lambda, x)$ satisfy the Riemann-Hilbert problem

$$Y_2(\lambda, x) = Y_1(\lambda, x)\mathrm{e}^{\theta(\lambda,x)}\left(\frac{1}{\lambda}\right)^{\frac{1}{2}\Theta_\infty\sigma_3}\mathsf{S}_1\left(\frac{1}{\lambda}\right)^{-\frac{1}{2}\Theta_\infty\sigma_3}\mathrm{e}^{-\theta(\lambda,x)},$$

$$Y_1(\lambda, x) = Y_2\left(\mathrm{e}^{2i\pi\lambda}, x\right)\mathrm{e}^{\theta(\lambda,x)}\left(\frac{1}{\mathrm{e}^{2i\pi}\lambda}\right)^{\frac{1}{2}\Theta_\infty\sigma_3}\mathsf{S}_2\left(\frac{1}{\mathrm{e}^{2i\pi}\lambda}\right)^{-\frac{1}{2}\Theta_\infty\sigma_3}\mathrm{e}^{-\theta(\lambda,x)},$$

$$Y_1(\lambda, x) = Y^0(\lambda, x)\mathrm{e}^{\theta(\lambda,x)}\lambda^{\frac{1}{2}\Theta_0\sigma_3}\mathsf{E}_0\left(\frac{1}{\lambda}\right)^{-\frac{1}{2}\Theta_\infty\sigma_3}\mathrm{e}^{-\theta(\lambda,x)},$$

$$Y_1(\lambda, x) = Y^1(\lambda, x)\mathrm{e}^{\theta(\lambda,x)}(\lambda-1)^{\frac{1}{2}\Theta_1\sigma_3}\mathsf{E}_1\left(\frac{1}{\lambda}\right)^{-\frac{1}{2}\Theta_\infty\sigma_3}\mathrm{e}^{-\theta(\lambda,x)},$$

(5.4.18)
$$Y_1(\lambda, x) = I + O\left(\frac{1}{\lambda}\right), \quad \lambda\to\infty,$$

where

(5.4.19)
$$\theta(\lambda, x) = \frac{1}{2}\lambda x\sigma_3,$$

and the 2×2 constant matrices S_1, S_2, E_0, E_1, are defined in terms of s_1, s_2 as follows:

$$\mathsf{S}_1 = \begin{pmatrix} 1 & 0 \\ s_1 & 1 \end{pmatrix}, \quad \mathsf{S}_2 = \begin{pmatrix} 1 & s_2 \\ 0 & 1 \end{pmatrix},$$

$$\operatorname{diag}(d_0, d_0^{-1})\mathsf{E}_0 \sim \mathsf{E}_0, \quad \operatorname{diag}(d_1, d_1^{-1})\mathsf{E}_1 \sim \mathsf{E}_1,$$

where d_0 and d_1 are nonzero constants, $\sim$ denotes an equivalent class, and $\mathsf{E}_0, \mathsf{E}_1$ are unimodular matrices satisfying

$$\mathsf{E}_0^{-1}\mathrm{e}^{-i\pi\Theta_0\sigma_3}\mathsf{E}_0 = \mathsf{S}_1\mathsf{S}_2\mathrm{e}^{i\pi\Theta_\infty\sigma_3}\mathsf{E}_1^{-1}\mathrm{e}^{i\pi\Theta_1\sigma_3}\mathsf{E}_1. \tag{5.4.20}$$

Let $u(x)$ be defined by the quadratic algebraic equation

$$\begin{aligned} x^2 (Y_{-1})_{12}(Y_{-1})_{21} &= \left[v + \Theta_0 - u\left(v + \tfrac{\Theta_0+\Theta_\infty-\Theta_1}{2}\right)\right] \\ &\times \left[v - \tfrac{1}{u}\left(v + \tfrac{\Theta_0+\Theta_\infty+\Theta_1}{2}\right)\right], \end{aligned}$$

where

$$Y_{-1}(x) = \lim_{\lambda\to\infty} (Y_1(\lambda, x) - I),$$

and

$$v(x) = -(Y_{-1})_{11} + x(Y_{-1})_{12}(Y_{-1})_{21} - \frac{\Theta_0 + \Theta_\infty + \Theta_1}{2}.$$

Then: (a) The above RH problem defines a unique 2×2 matrix $Y(\lambda, x)$ which is a meromorphic function of x. (b) The function $u(x)$ is a meromorphic function of x and it satisfies the Painlevé V equation (5.4.10). The "meromorphicity of x" means meromorphicity on the universal covering of the complex x-plane punctured at $x = 0$.

PROOF. Let $\Psi_k(\lambda, x)$, $k = 1, 2$, $\Psi^{(0)}(\lambda, x)$, $\Psi^1(\lambda, x)$ be defined in terms of $Y(\lambda, x)$ by

$$\Psi_k(\lambda, x) = Y_k(\lambda, x)\mathrm{e}^{\theta(\lambda,x)}\left(\frac{1}{\lambda}\right)^{\frac{1}{2}\Theta_\infty\sigma_3}, k = 1, 2,$$

$$\Psi^1(\lambda, x) = Y^1(\lambda, x)\mathrm{e}^{\theta(\lambda,x)}(\lambda - 1)^{\frac{1}{2}\Theta_1\sigma_3},$$

$$\Psi^{(0)}(\lambda, x) = Y^0(\lambda, x)\mathrm{e}^{\theta(\lambda,x)}\lambda^{\frac{1}{2}\Theta_0\sigma_3}. \tag{5.4.21}$$

The anti-Stokes lines are defined by $\arg(x\lambda) = 0, \pi$, i.e., $\arg\lambda = -\arg x$, $\arg\lambda = -\arg x + \pi$. Let L be the contour depicted in Figure 5.11. This contour contains part of the anti-Stokes lines, part of the real axis and the circles with centers at 0 and at 1 and of radius $\frac{1}{4}$. Equations (5.4.18) imply

$$Y_+(\lambda, x) = Y_-(\lambda, x)\mathrm{e}^{\theta(\lambda,x)}V(\lambda)\mathrm{e}^{-\theta(\lambda,x)},$$

$$Y = I + O\left(\frac{1}{\lambda}\right), \lambda \to \infty,$$

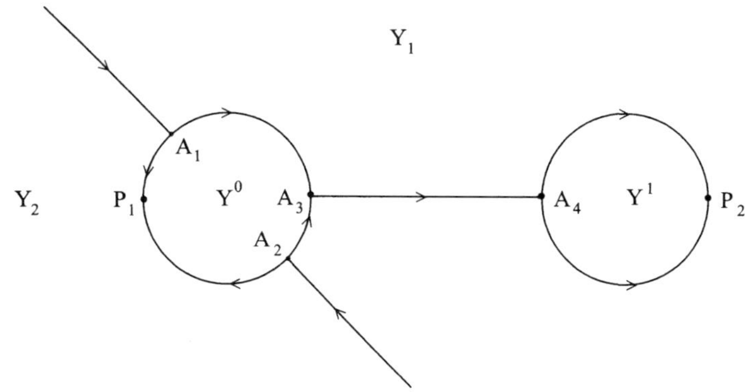

FIGURE 5.11. The RH problem for PV

where the jumps $V(\lambda)$ are given by

$$V_{A_1\infty} = \left(\frac{1}{\lambda}\right)^{\frac{1}{2}\Theta_\infty\sigma_3} \mathrm{S}_1^{-1} \left(\frac{1}{\lambda}\right)^{-\frac{1}{2}\Theta_\infty\sigma_3},$$

$$V_{A_2\infty} : Y_2 = Y_1 \left(\frac{1}{\lambda}\right)_+^{\frac{1}{2}\Theta_\infty\sigma_3} \mathrm{S}_2^{-1} \left(\frac{1}{\lambda}\right)_+^{-\frac{1}{2}\Theta_\infty\sigma_3}, V_{A_1A_3} = \lambda^{\frac{1}{2}\Theta_0\sigma_3}\mathrm{E}_0 \left(\frac{1}{\lambda}\right)_+^{-\frac{1}{2}\Theta_\infty\sigma_3},$$

$$V_{A_1P_1} : \left(\frac{1}{\lambda}\right)^{\frac{1}{2}\Theta_\infty\sigma_3} (\mathrm{E}_0\mathrm{S}_1)^{-1}\lambda^{-\frac{1}{2}\Theta_0\sigma_3}, V_{P_1A_2} = \lambda^{\frac{1}{2}\Theta_0\sigma_3}\mathrm{E}_0\mathrm{S}_1 \left(\frac{1}{\lambda}\right)^{-\frac{1}{2}\Theta_\infty\sigma_3},$$

$$V_{A_3A_4} : \left(\frac{1}{\lambda}\right)^{\frac{1}{2}\Theta_\infty\sigma_3} \mathrm{E}_1^{-1}\mathrm{e}^{i\pi\Theta_1\sigma_3}\mathrm{E}_1 \left(\frac{1}{\lambda}\right)^{-\frac{1}{2}\Theta_\infty\sigma_3},$$

$$V_{\overrightarrow{A_4P_2}} : (\lambda-1)^{\frac{1}{2}\Theta_1\sigma_3}\mathrm{E}_1 \left(\frac{1}{\lambda}\right)^{-\frac{1}{2}\Theta_\infty\sigma_3}, V_{A_4\underrightarrow{P_2}} : \left(\frac{1}{\lambda}\right)^{\frac{1}{2}\Theta_\infty\sigma_3} \mathrm{E}_1^{-1}(\lambda-1)^{-\frac{1}{2}\Theta_1\sigma_3}$$

$$V_{A_2A_3} = \left(\frac{1}{\lambda}\right)^{\frac{1}{2}\Theta_\infty\sigma_3} (\mathrm{e}^{i\pi\Theta_0\sigma_3}\mathrm{E}_0\mathrm{S}_1\mathrm{S}_2\mathrm{e}^{i\pi\Theta_\infty\sigma_3})^{-1}\lambda^{-\frac{1}{2}\Theta_0\sigma_3}.$$

The + subscripts in $V_{A_2\infty}$ indicate that the boundary value is computed from the + region, i.e., $(\lambda)_+ = |\lambda|\mathrm{e}^{2i\pi}$. The product conditions at A_1, A_2, A_4 are identically satisfied:

$$A_1\colon\ (V_{A_1\infty})^{-1}V_{A_1P_1}V_{A_1A_3} = I,$$

$$A_2\colon\ (V_{A_2\infty})^{-1}V_{A_2A_3}(V_{P_1A_2})_+ = I,$$

$$A_4\colon\ (V_{A_3A_4})^{-1}V_{\overrightarrow{A_4P_2}}V_{A_4\underrightarrow{P_2}} = I.$$

The product condition at A_3 is satisfied because of the consistency condition (5.4.20),

$$A_3\colon\ (V_{A_3A_4})^{-1}V_{A_2A_3}V_{A_1A_3} = I.$$

Derivation of $\Psi_\lambda = \mathrm{A}\Psi$. We define A by $A = \Psi_\lambda\Psi^{-1}$. Since Ψ_λ and Ψ admit the same jumps, it follows that A is holomorphic in $\mathbb{C}/(\{0\}+\{1\})$. Furthermore,

$$\Psi \sim \mathrm{e}^{(\frac{1}{2}x\lambda - \frac{1}{2}\Theta_\infty \ln\lambda)\sigma_3} \quad \text{as} \quad \lambda \to \infty,$$

thus

$$A = A_0 + \frac{A_1}{\lambda} + \frac{A_2}{\lambda - 1}.$$

Substituting (5.4.21) into $\Psi_\lambda = A\Psi$ we find

$$Y_\lambda + \frac{x}{2}Y\sigma_3 - \frac{\Theta_\infty}{2\lambda}Y\sigma_3 = \left(A_0 + \frac{A_1}{\lambda} + \frac{A_2}{\lambda-1}\right)Y \qquad \text{near } \lambda = \infty, \tag{5.4.22}$$

$$Y_\lambda + \frac{x}{2}Y\sigma_3 - \frac{\Theta_\infty}{2\lambda}Y\sigma_3 = \left(A_0 + \frac{A_1}{\lambda} + \frac{A_2}{\lambda-1}\right)Y \qquad \text{near } \lambda = 0, \tag{5.4.23}$$

$$Y_\lambda + \frac{x}{2}Y\sigma_3 - \frac{\Theta_1}{2(\lambda-1)}Y\sigma_3 = \left(A_0 + \frac{A_1}{\lambda} + \frac{A_2}{\lambda-1}\right)Y \qquad \text{near } \lambda = 1. \tag{5.4.24}$$

For large λ, Y has the expansion

$$Y = I + \frac{Y_{-1}}{\lambda} + \frac{Y_{-2}}{\lambda^2} + O\left(\frac{1}{\lambda^3}\right), \qquad \lambda \to \infty. \tag{5.4.25}$$

Substituting (5.4.25) into (5.4.22) we find:

$$O(1): \qquad A_0 = \tfrac{x}{2}\sigma_3,$$

$$O\left(\tfrac{1}{\lambda}\right): \qquad A_1 + A_2 = -\tfrac{\Theta_\infty}{2}\sigma_3 + \tfrac{x}{2}[Y_{-1}, \sigma_3],$$

$$O\left(\tfrac{1}{\lambda^2}\right): \qquad \tfrac{x}{2}[Y_{-2}, \sigma_3] = Y_{-1}\left(I + \tfrac{\Theta_\infty}{2}\sigma_3\right) + (A_1 + A_2)Y_{-1} + A_2.$$

Thus,

$$A_1 + A_2 = \begin{pmatrix} -\frac{\Theta_\infty}{2} & -x(Y_{-1})_{12} \\ x(Y_{-1})_{21} & \frac{\Theta_\infty}{2} \end{pmatrix}, \tag{5.4.26}$$

$$(A_2)_{11} + (Y_{-1})_{11} - x(Y_{-1})_{12}(Y_{-1})_{21} = 0. \tag{5.4.27}$$

As $\lambda \to 0$, equation (5.4.23) implies $A_1 = -\frac{\Theta_0}{2}Y^0(0,x)\sigma_3[Y^0(0,x)]^{-1}$. Hence

$$\det A_1 = -\frac{\Theta_0^2}{2}, \quad \operatorname{Tr} A_1 = 0. \tag{5.4.28}$$

As $\lambda \to 1$, equation (5.4.24) implies $A_2 = -\frac{\Theta_1}{2}Y^1(0,x)\sigma_3[Y^1(0,x)]^{-1}$. Hence

$$\det A_2 = -\frac{\Theta_1^2}{2}, \quad \operatorname{Tr} A_2 = 0. \tag{5.4.29}$$

Equations (5.4.28), (5.4.29) imply that A_1 and A_2 can be taken in the form appearing in equations (5.4.6) and (5.4.7). Then equations (5.4.26), (5.4.27) yield

$$x(Y_{-1})_{12} = y(v + \Theta_0) - yu\left(v + \frac{\Theta_0 + \Theta_\infty - \Theta_1}{2}\right),$$

$$x(Y_{-1})_{21} = \frac{v}{y} - \frac{1}{yu}\left(v + \frac{\Theta_0 + \Theta_\infty + \Theta_1}{2}\right),$$

$$-(Y_{-1})_{11} + x(Y_{-1})_{12}(Y_{-1})_{21} = v + \frac{\Theta_0 + \Theta_\infty + \Theta_1}{2}. \tag{5.4.30}$$

Equations (5.4.30) imply y, v, u in terms of $(Y_{-1})_{11}$, $(Y_{-1})_{12}$, $(Y_{-1})_{21}$. In particular, equation (5.4.30c) defines v, while equations (5.4.30a), (5.4.30b) yield

$$x^2(Y_{-1})_{12}(Y_{-1})_{21} = \left[v + \Theta_0 - u\left(v + \tfrac{\Theta_0+\Theta_\infty-\Theta_1}{2}\right)\right]$$
$$\times \left[v - \tfrac{1}{u}\left(v + \tfrac{\Theta_0+\Theta_\infty+\Theta_1}{2}\right)\right].$$

Derivation of $\Psi_x = U\Psi$. We define B by $B = \Psi_x\Psi^{-1}$. Then Y solves

$$Y_x + \frac{\lambda}{2}Y\sigma_3 = (B_0 + \lambda B_1)Y.$$

Substituting (5.4.25) into this equation we find:

$$O(\lambda): \quad B_1 = \tfrac{\sigma_3}{2},$$

$$O(1): \quad B_0 = \tfrac{1}{2}[Y_{-1}, \sigma_3] = \begin{pmatrix} 0 & -(Y_{-1})_{12} \\ (Y_{-1})_{21} & 0 \end{pmatrix}.$$

Using (5.4.30a) and (5.4.30b) to express $(Y_{-1})_{12}$, $(Y_{-1})_{21}$ in terms of y, v, u, we find that B_0 is given by equation (5.4.8).

The meromorphic with respect to x solvability of the above RH problem, follows as with the RH problem of PII. □

5. Global solutions of the Painlevé equations

We will show that if the monodromy data satisfy certain restrictions, and if x is constrained to be on certain rays of the complex x-plane, then the solution of some of the Painlevé equations is global, i.e., it does *not* have poles along these rays.

5.1. The Painlevé II equation.

THEOREM 5.6. *Let $u(x)$ be the solution of Painlevé II equation obtained through the construction of Theorem* 5.1. *Recall that this solution is parametrized through the monodromy data $\{s_1, s_2, s_3\}$, satisfying the equation*

$$s_1 + s_2 + s_3 + s_1s_2s_3 = -2i\sin\alpha\pi, \quad \alpha \neq \frac{2n+1}{2}, \quad n \in \mathbb{Z}.$$

Assume that

(5.5.1) $$|\operatorname{Re}\alpha| < \frac{1}{2}, \quad s_2 = \overline{s}_2, \quad |s_1 - \overline{s}_2| < 2, \quad x \in \mathbb{R},$$

or that

(5.5.2) $$|\operatorname{Re}\alpha| < \frac{1}{2}, \quad s_3 = \overline{s}_3, \quad |s_2 - \overline{s}_1| < 2, \quad x \in e^{\frac{2\pi i}{3}}\mathbb{R},$$

or that

(5.5.3) $$|\operatorname{Re}\alpha| < \frac{1}{2}, \quad s_1 = \overline{s}_1, \quad |s_3 - \overline{s}_2| < 2, \quad x \in e^{\frac{4\pi i}{3}}\mathbb{R}.$$

Then $u(x)$ is a global solution of Painlevé II, i.e., it is free of poles.

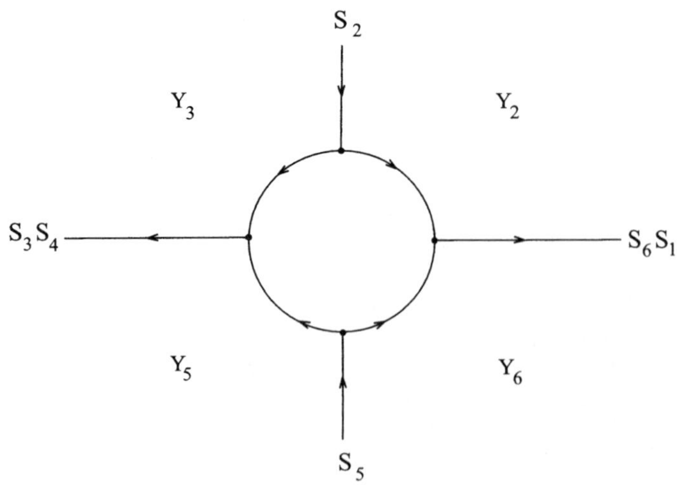

FIGURE 5.12. The Riemann-Hilbert problem for the vanishing lemma of PII.

PROOF. For the derivation of this result is more convenient to choose the contour consisting of a circle of radius ε, and of the real and imaginary axis extending from this circle to infinity (see Figure 5.12). We now formulate a RH problem on this contour instead on the contour of Figure 5.3. This new RH problem has the jumps indicated in Figure 5.12. Recall that $Y_2 \to I$ as $\lambda \to \infty$ for $0 < \arg\lambda < \frac{2\pi}{3}$. Also, $Y_2 = Y_1 e^{\theta} S_1 e^{-\theta}$, $Y_1 = Y_6 e^{\theta} S_6 e^{-\theta}$, thus $Y_2 = Y_6 e^{\theta} S_1 S_6 e^{-\theta}$. Similarly, $Y_5 \to I$ as $\lambda \to \infty$ for $\pi < \arg\lambda < \frac{4\pi}{3}$ and

$$Y_5 = Y_4 e^{\theta} S_4 e^{-\theta} = Y_3 e^{\theta} S_3 S_4 e^{-\theta}.$$

We will now show that if equations (5.5.1) are valid, then the RH problem defined by

$$Y_2 = Y_6 e^{\theta} S_6 S_1 e^{-\theta}, \quad \lambda \in [\varepsilon, \infty); \quad Y_3 = Y_2 e^{\theta} S_2 e^{-\theta}, \quad \lambda \in [i\varepsilon, i\infty);$$
$$Y_5 = Y_3 e^{\theta} S_3 S_4 e^{-\theta}, \quad \lambda \in (-\varepsilon, -\infty); \quad Y_6 = Y_5 e^{\theta} S_5 e^{-\theta}, \quad \lambda \in [-i\varepsilon, -i\infty),$$
$$Y_2 = O\left(\frac{1}{\lambda}\right), \quad \lambda \to \infty \text{ in } 0 < \arg\lambda \leq \frac{\pi}{2},$$

has only the zero solution.

We will use the following notation:

(a) χ will denote Schwartz reflection, i.e. $Y(\lambda)^{\chi} = \overline{Y(\overline{\lambda})}^{T}$, where T denotes transposition.

(b) $*$ will denote complex conjugation and transposition, i.e., $A^* = \overline{A}^T$.

Let $H(\lambda, x)$ be defined by

$$H(\lambda, x) = Y(\lambda, x) Y(\lambda, x)^{\chi}.$$

If $x \in \mathbb{R}$, and if $S_2 = S_3^*$, i.e., $s_2 = \overline{s}_2$, then H is continuous across $[i\varepsilon, i\infty)$. Indeed,

$$H_3 = Y_3 Y_5^{\chi} = Y_2 e^{\theta} S_2 e^{-\theta} Y_5^{\chi},$$
$$H_2 = Y_2 Y_6^{\chi} = Y_2 (Y_5 e^{\theta} S_5 e^{-\theta})^{\chi} = Y_2 e^{\theta} S_5^* e^{-\theta} Y_5^{\chi},$$

where we have used $\theta^{\chi} = -\theta$ (since $x \in \mathbb{R}$).

The function $H(\lambda, x)$ is analytic in the upper-half complex λ-plane and it is of order $(\frac{1}{\lambda})$ as $\lambda \to \infty$. Thus the sum of the integrals of $H(\lambda, x)$ along $(-\infty, -\varepsilon]$, along $[\varepsilon, \infty)$, and along the semi-circle $\{|\lambda| = \varepsilon, \text{ Im } \lambda > 0\}$, vanish. If $|\text{Re } \alpha| < \frac{1}{2}$, then H has an integrable singularity at $\lambda = 0$, thus letting $\varepsilon \to 0$, we find

$$\int_{-\infty}^{0} H(\lambda, x) d\lambda + \oint H(\lambda, x) d\lambda = 0.$$

Hence

$$\int_{-\infty}^{0} Y_3 Y_5^* d\lambda + \oint Y_2 Y_6^* d\lambda = 0,$$

or

$$\int_{-\infty}^{0} Y_3 \left(Y_3 \mathrm{e}^{\theta} \mathrm{S}_3 \mathrm{S}_4 \mathrm{e}^{-\theta}\right)^* d\lambda + \oint Y_6 \mathrm{e}^{\theta} \mathrm{S}_6 \mathrm{S}_1 \mathrm{e}^{-\theta} Y_6^* d\lambda = 0.$$

Letting $Y_3 \mathrm{e}^{\theta} = Y_3$, $Y_6 \mathrm{e}^{\theta} = Y_6$, this equation becomes

$$\int_{-\infty}^{0} Y_3 (\mathrm{S}_3 \mathrm{S}_4)^* Y_3^* d\lambda + \oint Y_6 \mathrm{S}_6 \mathrm{S}_1 Y_6^* d\lambda = 0. \tag{5.5.4}$$

Taking the complex conjugate and the transpose of equation (5.5.4) and adding the resulting equation to equation (5.5.4), we find

$$\int_{-\infty}^{0} Y_3 \left((\mathrm{S}_3 \mathrm{S}_4) + (\mathrm{S}_3 \mathrm{S}_4)^*\right) Y_3^* d\lambda + \oint Y_6 \left(\mathrm{S}_6 \mathrm{S}_1 + (\mathrm{S}_6 \mathrm{S}_1)^*\right) Y_6^* d\lambda = 0.$$

Thus if the matrices $\mathrm{S}_3 \mathrm{S}_4 + (\mathrm{S}_3 \mathrm{S}_4)^*$ and $\mathrm{S}_6 \mathrm{S}_1 + (\mathrm{S}_6 \mathrm{S}_1)^*$ are both positive definite or are both negative definite, then $Y_3 = Y_6 = 0$. These two matrices have the same real eigenvalues. Furthermore, their determinant equals $4 - |s_1 - \overline{s}_3|^2$. If $|s_1 - \overline{s}_3| < 2$, the determinant is positive and hence both eigenvalues have the same sign.

Performing suitable rotations of the λ-complex plane it follows that $u(x)$ is also free of poles provided that either of the sets of conditions given by equations (5.5.2) and (5.5.3) are valid. □

5.2. The RH problem of the Painlevé IV equation.

THEOREM 5.7. *Let $Y(\lambda, x)$ be the solution of the Riemann-Hilbert problem defined in Theorem 5.2. Recall that this solution is parametrized through the monodromy data $\{s_1, s_2, s_3, s_4\}$. These constants belong to the equivalent class induced by the transformation*

$$\{s_1, s_2, s_3, s_4\} \to \{c^2 s_1, c^{-2} s_2, \quad c^2 s_3, c^{-2} s_4\}, \quad c \text{ constant},$$

on the algebraic equation

$$(1 + s_2 s_3) \mathrm{e}^{2i\pi\Theta_\infty} + [s_1 s_4 + (1 + s_3 s_4)(1 + s_1 s_2)] \mathrm{e}^{-2i\pi\Theta_\infty}$$

$$= 2\cos 2\pi\Theta, \quad \Theta \notin \mathbb{Z}/2.$$

Assume that

$$|Re\,\Theta| < \frac{1}{2}, \quad |Re\,\Theta_\infty| < \frac{1}{4}, \tag{5.5.5}$$

and that either of the following equations are valid:

$$|s_2 - \overline{s}_3| < 2, \quad \mathrm{e}^{2\pi \,\mathrm{Im}\, \Theta_\infty} |s_1 - \overline{s}_4| < 2|\cos 2\pi \,\mathrm{Re}\, \Theta_\infty|, \quad x \in \mathrm{e}^{\frac{-\pi i}{4}} \mathbb{R}, \tag{5.5.6}$$

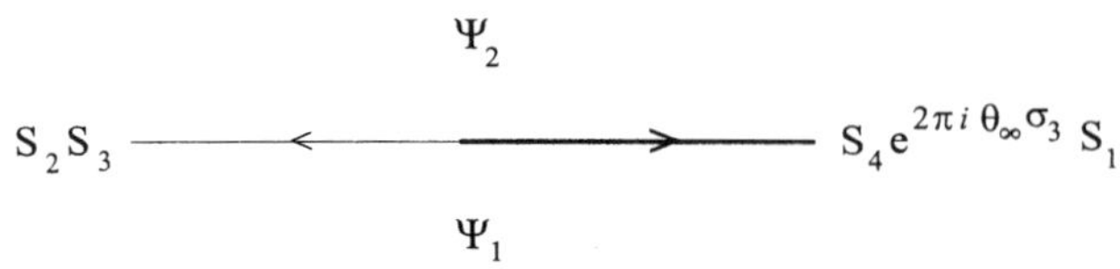

FIGURE 5.13. The RH problem for the vanishing lemma of PIV

or

$$|s_3 - \overline{s}_4| < 2, \quad e^{-2\pi Im\Theta_\infty}|s_1 - \overline{s}_2| < 2|\cos 2\pi \operatorname{Re} \Theta_\infty|, \quad x \in e^{\frac{\pi i}{4}}\mathbb{R}.$$

Then $Y(\lambda, x)$ is a global function of x, i.e., it is free of poles.

PROOF. We first consider $x \in e^{-\frac{\pi}{4}}\mathbb{R}$. For the derivation of this result we need $\theta^\chi = -\theta$, where $\theta = \left(\frac{\lambda^2}{2} + x\lambda\right)\sigma_3$. So we perform the change of variables $\lambda = e^{-\frac{i\pi}{4}}\lambda'$, to obtain $\theta = -i\left(\frac{\lambda'^2}{2} + xe^{\frac{i\pi}{4}}\lambda'\right)\sigma_3$. Thus if $xe^{\frac{i\pi}{4}} \in \mathbb{R}$, $\theta^\chi = -\theta$. It is more convenient to consider the RH problem defined on the real axis instead of the contour of Figure 5.13. On this contour, the jumps are given in Figure 5.13. In order to prove a vanishing lemma for this RH problem we need to investigate the matrices

$$F = S_2S_3 + (S_2S_3)^*, \quad M = S_4e^{2i\pi\Theta_\infty\sigma_3}S_1 + \left(S_4e^{2i\pi\Theta_\infty\sigma_3}S_1\right)^*. \tag{5.5.7}$$

The determinant and the trace of F are given by

$$\det F = 4 - |s_2 - \overline{s}_3|^2, \quad \operatorname{Tr} F = 4 + s_2s_3 + \overline{s}_2\overline{s}_3.$$

The matrix F is definite provided that $\det F > 0$, i.e. $|s_2 - \overline{s}_3| < 2$. Also, noting that $\operatorname{Tr} F = \det F + |s_2|^2 + |s_3|^2$, it follows that if $|s_2 - \overline{s}_3| < 2$, then F is actually positive definite.

Using equation (5.5.7b) to compute $\det M$, we find

$$\det M = 2 + 2\cos 4\pi Re\Theta_\infty - e^{2i\pi(\overline{\Theta}_\infty - \Theta_\infty)}|s_1 - \overline{s}_4|^2,$$

or

$$\det M = 4\cos^2 2\pi Re\Theta_\infty - e^{4\pi \operatorname{Im} \Theta_\infty}|s_1 - \overline{s}_4|^2.$$

Hence $\det M > 0$ iff equation (5.5.6b) is valid. The matrix M has the form

$$M = \begin{pmatrix} A_R & B \\ \overline{B} & B_R \end{pmatrix}, \tag{5.5.8}$$

where A_R and B_R are real and

$$B_R = e^{2i\pi\overline{\Theta}_\infty} + e^{-2i\pi\Theta_\infty} = 2e^{2\pi Im\Theta_\infty}\cos 2\pi Re\Theta_\infty.$$

Since $|\operatorname{Re}\Theta_\infty| < \frac{1}{4}$, it follows that $B_R > 0$. Also since $\det M > 0$, it follows that $A_RB_R > |B|^2$, thus $A_R > 0$. Hence $\operatorname{Tr} M > 0$, which implies that M is positive definite.

The conditions (5.5.5) guarantee that the singularities of $\Psi\Psi^*$ at $\lambda = 0$ and $\lambda = \infty$ are integrable.

For the case of $x \in e^{\frac{i\pi}{4}}\mathbb{R}$, we first make the change of variables $\lambda = e^{-\frac{3i\pi}{4}}\lambda'$, and then consider the RH problem depicted in Figure 5.14.

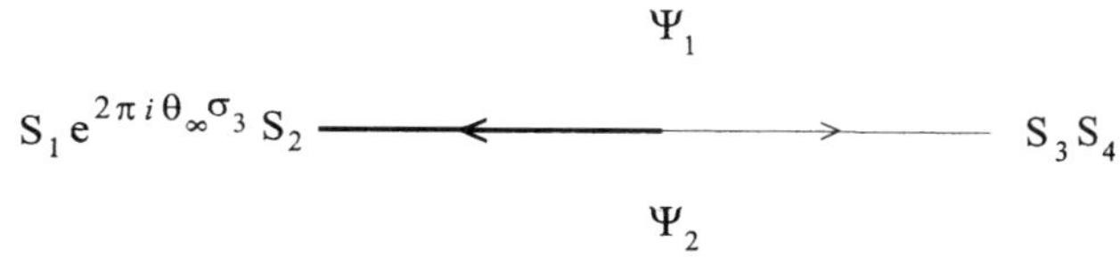

FIGURE 5.14

REMARK 5.5. The above result cannot be used to prove global existence of $u(x)$, since u will, in general, have a pole where $(Y_{-1})_{12}$ has a zero.

5.3. The Painlevé III Equation.

THEOREM 5.8. *Let $u(x)$ be the solution of Painlevé III equation obtained through the construction of Theorem 5.4. Recall that this solution is parametrized through the monodromy data $\{s_1^0, s_1^0, s_1^\infty, s_2^\infty\}$ and through the 2×2 unimodular matrix E. The monodromy data belongs to the equivalent class induced by the transformation*

$$\{s_1^0, s_2^0, s_1^\infty, s_2^\infty\} \to \{s_1^0/c^2, s_2^0c^2, s_1^\infty/c^2, s_2^\infty c^2\}$$

on the algebraic equation

$$2\cos\pi\Theta_0 + s_1^0 s_2^0 \mathrm{e}^{-i\pi\Theta_0} = 2\cos\pi\Theta_\infty + s_1^\infty s_2^\infty \mathrm{e}^{-i\pi\Theta_0}. \tag{5.5.9}$$

The matrix E satisfies

$$\begin{pmatrix} 1 & s_1^\infty \\ 0 & 1 \end{pmatrix}\begin{pmatrix} 1 & 0 \\ s_2^\infty & 1 \end{pmatrix}\mathrm{e}^{i\pi\Theta_\infty\sigma_3} = \mathsf{E}^{-1}\begin{pmatrix} 1 & s_1^0 \\ 0 & 1 \end{pmatrix}\begin{pmatrix} 1 & 0 \\ s_2^0 & 1 \end{pmatrix}\mathrm{e}^{-i\pi\Theta_0\sigma_3}\mathsf{E}, \tag{5.5.10}$$

$$\operatorname{diag}(d, d^{-1})\mathsf{E} \sim \mathsf{E},$$

where $\sim$ denotes an equivalent class.

Assume that

$$|\operatorname{Re}\,\Theta_\infty| < \frac{1}{2}, \quad |\operatorname{Re}\,\Theta_0| < \frac{1}{2}, \quad x \in \mathbb{R}, \tag{5.5.11}$$

$$\begin{aligned} &\left|\overline{s}_1^\infty \mathrm{e}^{-i\pi\Theta_\infty} - s_2^\infty \mathrm{e}^{i\pi\Theta_\infty}\right| < 2\cos\pi\,\operatorname{Re}\,\Theta_\infty, \\ &\left|\overline{s}_1^0 \mathrm{e}^{i\pi\Theta_0} - s_2^0 \mathrm{e}^{-i\pi\Theta_0}\right| < 2\cos\pi\,\operatorname{Re}\,\Theta_0. \end{aligned} \tag{5.5.12}$$

Then $u(x)$ is a global solution of Painlevé III, i.e., it is free of poles.

PROOF. If $x \in \mathbb{R}$ and $\theta = \frac{ix}{2}(\lambda - \frac{1}{\lambda})\sigma_3$ (see (5.3.27)), then $\theta = \theta^\chi$. Thus we look at the jumps satisfied by $\Psi(\lambda, x)$ instead of $Y(\lambda, x)$. It is convenient to formulate these jumps on the Stokes lines, which since $\arg x = 0$, are given by $\arg\lambda = 0$ and $\arg\lambda = \pi$. Using

$$\Psi_2^{(\infty)}(\lambda, x) = \Psi_1^{(\infty)}(\lambda, x)\mathsf{S}_1^{(\infty)}, \quad \Psi_1^{(\infty)}(\lambda, x) = \Psi_2^{(\infty)}(\mathrm{e}^{2i\pi}\lambda, x)\mathsf{S}_2^{(\infty)}\mathrm{e}^{i\pi\Theta_\infty\sigma_3},$$

we find

$$\Psi_1^{(\infty)}(\lambda, x) = \Psi_1^{(\infty)}(\mathrm{e}^{2i\pi}\lambda, x)\mathsf{S}_1^{(\infty)}\mathsf{S}_2^{(\infty)}\mathrm{e}^{i\pi\Theta_\infty\sigma_3}. \tag{5.5.13}$$

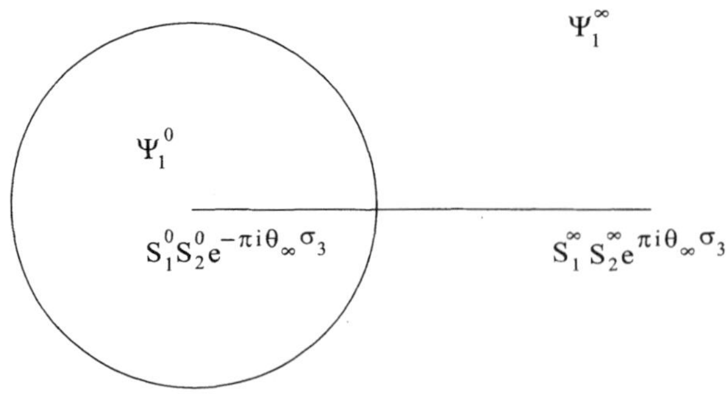

FIGURE 5.15. The RH problem for PIII

Similarly, using

$$\Psi_2^{(0)}(\lambda,x) = \Psi_1^{(0)}(\lambda,x)S_1^{(0)},\, \Psi_1^{(0)}(\lambda,x) = \Psi_2^{(0)}(e^{2i\pi}\lambda,x)S_1^{(0)}e^{-i\pi\Theta_0\sigma_3},$$

we find

$$\Psi_1^{(0)}(\lambda,x) = \Psi_1^{(0)}(e^{2i\pi}\lambda,x)S_1^{(0)}S_2^{(0)}e^{-i\pi\Theta_0\sigma_3}. \tag{5.5.14}$$

Also,

$$\Psi_1^{(\infty)}(\lambda,x) = \Psi_1^{(0)}(\lambda,x)E. \tag{5.5.15}$$

The jumps of the functions $\Psi_1^{(0)}$ and $\Psi_1^{(\infty)}$ are indicated in Figure 5.15. The circle around $\lambda = 0$ is of radius ε. By considering the limit of $\varepsilon \to 0$ it follows that there exists a vanishing lemma provided that the singularity at $\lambda = 0$ and $\lambda = \infty$ are integrable and that the following conditions are satisfied: The matrices

$$\begin{aligned} F &= S_1^{(0)}S_2^{(0)}e^{-i\pi\Theta_0\sigma_3} + (S_1^{(0)}S_2^{(0)}e^{-i\pi\Theta_0\sigma_3^*}, \\ M &= S_1^{(\infty)}S_2^{(\infty)}e^{i\pi\Theta_\infty\sigma_3} + (S_1^{\infty}S_2^{(\infty)}e^{i\pi\Theta_\infty\sigma_3})^* \end{aligned} \tag{5.5.16}$$

are either both positive definite or are both negative definite. Using the definition of F we find

$$\det F = 4\cos^2\pi Re\Theta_\infty - \left|\overline{s_1^\infty}e^{-i\pi\Theta_\infty} - s_2^\infty e^{i\pi\Theta_\infty}\right|^2.$$

Thus, $\det F > 0$ iff equation (5.5.12a) is valid. The matrix F has the form (5.5.8), where

$$B_R = e^{-i\pi\Theta_\infty} + e^{i\pi\overline{\Theta}_\infty} = 2e^{\pi\,\mathrm{Im}\,\Theta_\infty}\cos\pi\,\mathrm{Re}\,\Theta_\infty.$$

Since $|\,\mathrm{Re}\,\Theta_\infty| < \frac{1}{2}$, it follows that $B_R > 0$. Also, since $\det F > 0$, it follows that $A_R B_R > |B|^2$, thus $A_R > 0$. Hence Tr $F > 0$, which implies that F is positive definite.

The matrix M defined in (5.5.16) can be obtained from the matrix F by replacing $s_1^\infty, s_2^\infty, \Theta_\infty$ by s_1^0, s_2^0, Θ_0. Thus the condition (5.5.12a) yields the condition (5.5.12b). The conditions $|\,\mathrm{Re}\,\Theta_\infty| < \frac{1}{2}$ and $|\,\mathrm{Re}\,\Theta_0| < \frac{1}{2}$, imply that the singularities at $\lambda = \infty$ and $\lambda = 0$ are integrable.

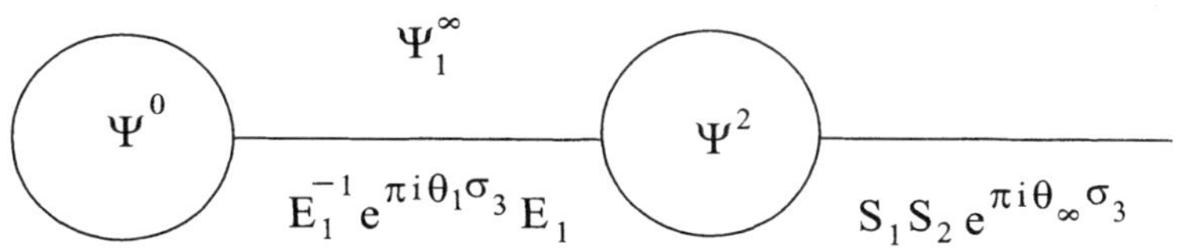

FIGURE 5.16. The RH problem for PV

5.4. The RH problem of the Painlevé V equation.

THEOREM 5.9. *Let $Y(\lambda, x)$ be the solution of the Riemann-Hilbert problem defined in Theorem 5.5. Recall that this solution is parametrized through the monodromy data s_1, s_2 and through the 2×2 unimodular matrices E_0 and E_1 satisfying*

$$\mathsf{E}_0^{-1} e^{-i\pi\Theta_0\sigma_3}\mathsf{E}_0 = \begin{pmatrix} 1 & 0 \\ s_1 & 1 \end{pmatrix}\begin{pmatrix} 1 & s_2 \\ 0 & 1 \end{pmatrix} e^{i\pi\Theta_\infty\sigma_3}\mathsf{E}_1^{-1}e^{i\pi\Theta_1\sigma_3}\mathsf{E}_1,$$

$$\operatorname{diag}(d_1, d_1^{-1})\mathsf{E}_1 \sim \mathsf{E}_1, \quad \operatorname{diag}(d_0, d_0^{-1})\mathsf{E}_0 \sim \mathsf{E}_0$$

where $\sim$ denotes an equivalent class.

Assume that

$$x \in i\mathbb{R}, \quad |\operatorname{Re}\,\Theta_\infty| < \tfrac{1}{2},$$
$$(5.5.17)\qquad |\operatorname{Re}\,\Theta_1| < \tfrac{1}{2}, \quad 2\cos\pi Re\Theta_\infty > e^{\pi Im\Theta_\infty}|\bar{s}_1 - s_1|,$$

and that the matrix

$$\mathsf{E}_1^{-1}e^{i\pi\Theta_1\sigma_3}\mathsf{E}_1 + \left(\mathsf{E}_1^{-1}e^{i\pi\Theta_1\sigma_3}\mathsf{E}_1\right)^*,$$

is positive definite (denotes complex conjugation and transposition). Then $Y(\lambda, x)$ is a global function of x, i.e., it is free of poles.*

PROOF. If $x \in i\mathbb{R}$ and $\theta = \frac{1}{2}x\lambda\sigma_3$ (see (5.4.19)), then $\theta = \theta^\chi$. Thus we look at the jumps satisfied by $\Psi(\lambda, x)$ instead of $Y(\lambda, x)$. It is convenient to look at these jumps on the Stokes lines, which since $\arg x = \pi/2$, are given by $\arg\lambda = 0$ and $\arg\lambda = \pi$. Using equations (5.4.12) we find

$$\Psi_1^{(\infty)}(\lambda, x) = \Psi_1^{(\infty)}(e^{2i\pi}\lambda, x)\mathsf{S}_1\mathsf{S}_2 e^{i\pi\Theta_\infty\sigma_3}.$$

Also,

$$\Psi_1^{(\infty)}(|\lambda - 1|, x) = \Psi^1(|\lambda - 1|, x)\mathsf{E} = \Psi^1(e^{-2i\pi}|\lambda - 1|, x)e^{i\pi\Theta_1\sigma_3}\mathsf{E}_1.$$

Thus,

$$\Psi_1^{(\infty)}(|\lambda - 1|, x) = \Psi_1^{(\infty)}(e^{-2i\pi}|\lambda - 1|, x)\mathsf{E}_1^{-1}e^{i\pi\Theta_1\sigma_3}\mathsf{E}_1.$$

The jumps of the RH problem satisfied by $\Psi_1^{(\infty)}$ occur on part of the real axis; see Figure 5.16. The circles around $\lambda = 0$ and $\lambda = 1$ are of radius ε. By considering the limit of $\varepsilon \to 0$ it follows that there exists a vanishing lemma provided that the singularities at $\lambda = 0$, $\lambda = 1$ are integrable and that the following conditions are satisfied: The matrices

$$(5.5.18)\qquad \mathsf{S}_1\mathsf{S}_2 e^{i\pi\Theta_\infty\sigma_3} + \left(\mathsf{S}_1\mathsf{S}_2 e^{i\pi\Theta_\infty\sigma_3}\right)^*, \quad \mathsf{E}_1^{-1}e^{i\pi\Theta_1\sigma_3}\mathsf{E}_1 + \left(\mathsf{E}_1^{-1}e^{i\pi\Theta_1\sigma_3}\mathsf{E}_1\right)^*$$

are both are either positive definite or negative definite. The first matrix in (5.5.18) is similar to the one appearing in the vanishing lemma of PIII; its determinant is given by

$$4\cos^2 \pi Re\Theta_\infty - \left|\overline{s}_1 \mathrm{e}^{i\pi\Theta_\infty} - s_2 \mathrm{e}^{-i\pi\Theta_\infty}\right|^2 .$$

Furthermore, $|\Theta_\infty| < \frac{1}{2}$, thus similar arguments like those used for the vanishing lemma of PIII yield that if (5.5.17d) is satisfied, the first matrix is positive definite. $\square$

CHAPTER 6

Bäcklund Transformations

0. Introduction

Let $u = u(x|\nu)$ solve a Painlevé equation, or, more precisely, let a vector u solve an equivalent Painlevé system with the vector of parameters ν. If there exists a transformation

$$\tilde{u} = B(x, u, \nu), \quad \tilde{\nu} = C(\nu), \quad \tilde{x} = \varphi(x),$$

such that $\tilde{u}(\tilde{x})$ also solves a Painlevé system with parameters $\tilde{\nu}$, then we call B a *Bäcklund transformation.* If both vector functions u and $\tilde{u}$ solve the same equation (but with a different set of parameters), we call B an *auto-Bäcklund transformation.* For instance, the second Painlevé system, $u_x = v$, $v_x = 2u^3 + xu - \alpha$, admits the auto-Bäcklund transformations,

$$\tilde{\alpha} = -\alpha - \sigma, \quad \tilde{u} = u + \frac{\frac{1}{2} + \sigma\alpha}{v - \sigma(u^2 + \frac{x}{2})}, \quad \tilde{v} = v + \sigma(\tilde{u}^2 - u^2), \quad \sigma = \pm 1. \tag{6.0.1}$$

These equations determine the new Painlevé function $\tilde{u}$ in terms of u and its derivative $u_x = v$, and thus they describe a *contact* transformation of the Painlevé function. We do not consider Bäcklund transformations the point transformations, such as the reflection $u(x|\alpha) \mapsto -u(x| - \alpha)$, or the transformations given by equations (7.2.10) and (7.2.11) of Chapter 7.

Historically the first Bäcklund and auto-Bäcklund transformations were found by Painlevé and Gambier during the classification of second order ODEs with the Painlevé properties [**Gam, In**]. Interest in the transformations of the classical Painlevé equations has been renewed since the 1960s; see [**Lu1, Lu2, Lu3, Gro3, Gro4, Ai, FoY, FoA2**]. The transformations obtained in this period, were mainly found by manipulating the various representations of the Painlevé equations as nonlinear systems of ODEs (for more details see the monographs [**GroL2, GroLS**]). Recently, a more systematic approach to derivating the auto-Bäcklund transformations was developed in [**GJP, ConM**], the so-called a *truncation method*; however, apparently no new transformations were found in this way.

One of the first applications of the auto-Bäcklund transformations was the description and the classification of families of particular solutions to Painlevé equations, namely families of rational, algebraic and classical solutions, i.e., solutions expressible in terms of classical transcendental functions; see [**Ai, Bor1, Bor2, BCH, Gro3, Gro5, GroL1, MCB, Mur**]. We note that the classification of the algebraic solutions to PVI has not been completed yet; the most effective approach to this problem is presented in [**DM**] (see also [**Bo4**]) and is based on the classification of monodromy groups admitting a finite braid group orbit. Lists of the exact solutions obtained via Bäcklund transformations are presented in [**GroLS, BCH**].

It is also understood that the auto-Bäcklund transformations can be thought of as a class of discrete integrable systems [**JMU, JM**], or more precisely as discrete Painlevé equations [**RGH, GroT, NiRGO, NiW**]. For instance, letting in (6.0.1) $\sigma \mapsto \sigma_n$, $\alpha \mapsto \alpha_n$, $u \mapsto u_n$, $v \mapsto v_n$, $\tilde{u} \mapsto u_{n+1}$, $\tilde{v} \mapsto v_{n+1}$, putting $\sigma_n = (-1)^n\sigma$, $\sigma = \pm 1$, iterating the transformation (6.0.1) twice and eliminating from the resulting system all v-variables, we find the second order difference equation

$$\sigma_n\big(2u_n^2 + x\big) + \frac{\frac{1}{2} + \sigma_n\alpha_n}{u_{n+1} - u_n} + \frac{\frac{1}{2} - \sigma_n\alpha_n}{u_n - u_{n-1}} = 0, \quad \alpha_n = (-1)^n\big(\alpha_0 + n\sigma\big),$$

often referred to as the dPI.

In the modern theory of isomonodromic deformations the auto-Bäcklund transformations can be interpreted as the compatibility conditions of the so-called Schlesinger transformations and the equations of the relevant Lax pair; see e.g. [**FoMA**]. In the language of integrable PDEs, a Bäcklund transformation is induced by a "rational dressing" of the Ψ-function. In particular, $B\colon\ u \mapsto \tilde{u},\ \nu \mapsto \tilde{\nu}$, is generated by the left-hand side multiplication of the Ψ-function with a rational in λ matrix R,

$$\Psi(\lambda) \mapsto R(\lambda)\Psi(\lambda).$$

The matrix R is chosen in such a way, that the new Ψ-function is again a solution of the λ-equation (with some new parameters associated with $\tilde{u}(x|\tilde{\nu})$).

In general, a left multiplication by a rational matrix does not change the local monodromy of Ψ around a pole of R. However, this multiplication does change the asymptotic behavior of Ψ near the pole which induces a conjugation of some of the monodromy data; see the next section. Thus, we call a rational gauge transformation a *Schlesinger transformation* if it preserves the transcendental part of the monodromy data, such as the Stokes multipliers and the invariants of the monodromy group including the traces of the monodromy matrix products $\operatorname{Tr}(M_iM_j)$, but shifts the (formal) monodromy exponents by integers.

Recently, the theory of Schlesinger transformations and of the induced Bäcklund transformations has been extended to the case of algebraic Schlesinger transformations which shift the formal monodromy exponents by half-integers [**Ki8**], as well as to the case of isomonodromic deformations on Riemann surfaces of genus 1 [**KoMS, Ko1**].

Also, the group-theoretic interpretation of Bäcklund transformations has been significantly developed. Namely, it was pointed out in [**Mq**] that the Painlevé equations have equivalent canonical forms with polynomial Hamiltonians, and furthermore it was shown in [**Ok4**] that the Hamiltonian structures of all Painlevé equations are implied by the isomonodromic deformations. On this basis, a systematic theory of birational canonical transformations (Bäcklund transformations) was developed for the Painlevé systems [**Ok5**].

The group of Bäcklund transformations for each of the Painlevé equations from PII through PVI provides a representation of one of the affine Weyl groups, or more precisely it provides a lift of the relevant Weyl symmetry action from the parameter space into a polynomial ring $K[p,q]$. The parameter space to each of the Painlevé equations can be identified with the Cartan subalgebra of a semisimple Lie algebra, and along each of the associated hyperplanes the Painlevé equation has a classical solution of hypergeometric type. Actually, the classical solutions to Painlevé equations are defined by certain invariant principal ideals of the polynomial ring $K[p,q]$. This fact yields a complete theory of classical solutions of Painlevé equations, and gives rise to certain important irreducibility theorems; see [**Um1, NoO, UmW, Wa1, Wa2**].

Another important development in the group-theoretic interpretation of the Painlevé equations, is the so-called symmetric form of Painlevé equations introduced in [**Adl, NoY1**]. In this form of PII, PIII, PIV, PV and PVI, the Bäcklund transformations can be described in a unified manner, not only for the root systems $A_1^{(1)}$, $A_1^{(1)} \oplus A_1^{(1)}$, $A_2^{(1)}$, $A_3^{(1)}$ and $D_4^{(1)}$, respectively, but also for a general root system $A_n^{(1)}$; for more details see the monograph [**No2**]. Furthermore, the so-called q-difference and ultra-difference analogs of Painlevé equations can be considered in a similar manner; see [**No1**] and references therein.

The group-theoretic interpretation of Painlevé equations has important geometrical implications. Recently [**S**], all rational surfaces were classified with respect to those admissible birational transformations, which provide representations of certain affine Weyl groups treated as symmetries for these surfaces. The translation part of these symmetries determines difference systems. Thus, a classification of discrete Painlevé equations appears as well. Some (but not all) of the rational surfaces arising from this classification, coincide with the spaces of initial conditions for the Painlevé differential equations introduced by Okamoto [**Ok1**]. Moreover, depending on the type of surface, all discrete Painlevé equations can be divided into three classes called elliptic-difference, q-difference and usual difference equations; some of the discrete Painlevé equations constructed in this way were not known before (see [**Tk, MuSY**] and papers mentioned there).

Another advantage of the symmetry approach to Painlevé equations, is the possibility of constructing a Lax representation whose Schlesinger transformation group represents the symmetry group of a given Painlevé equation [**No2**]. For instance, it was known for a long time that one of the Okamoto's transformations for PVI does not have an isomonodromic interpretation in the framework of the classical Fuchsian 2×2 Lax pair; however, using the above construction it was shown that the transformation is related with an appropriate 8×8 Lax pair [**NoY2**].

One of the most interesting problems in the theory of Bäcklund transformations of Painlevé equations, is the problem of describing their iterated actions. This is particularly important for the description of the families of exact solutions. In some important cases, this action can be described by the use of determinant formulae. The very first determinant formulae were presented in [**Ai**] for the family of the classical solutions of PII, and in [**FN1**] for the family of the rational solutions of PII. In [**KO1**], two inequivalent determinant formulae for the rational solutions of PII were derived, using the bilinear Hirota formalism of the determinant solutions of the KP hierarchy and of the Toda lattice. In [**KM**] (see also [**KMNOY**]) a determinant formula suitable for transcendent solutions of PII was found using the Hirota formalism. A similar approach was also applied to PIV in [**KO2**], and to PV in [**MOK**]. Generalization of the determinant formulae for higher root systems can be found in [**No2, Y**]. The asymptotics of the second Painlevé transcendents resulting from the action of a long chain of iterated Bäcklund transformations is described in [**KK1, K4**]. For other ν-large asymptotic results; see [**K8, K9**].

We note that the theory of Bäcklund transformations is still not complete, since there exist non-Weyl transformations of PII [**Gam**], PIII [**Wt**] and PIV [**Ok5**], which do not have an isomonodromic interpretation. For instance, if $u(x)$ solves equation PII with $\alpha = 0$, then the function $y(t)$,

$$y(t) = -2^{-1/3}\frac{u'(-2^{-1/3}t)}{u(-2^{-1/3}t)}, \tag{6.0.2}$$

(where the prime denotes derivative) satisfies equation PII with $(\alpha = \frac{1}{2})$.

The purpose of the present chapter is rather modest: We will derive Schlesinger transformations and the induced Bäcklund transformations for PII, PIII and PIV in two different ways. The first way is described in [**JM**] and consists of the construction of a rational gauge matrix R, which preserves the singularity structure and the essential monodromy property of the Ψ-function. The other approach is presented in [**FoMA**], and it constructs the gauge matrix R in such a way that the transformed matrix $\tilde{A} = R_\lambda R^{-1} + RAR^{-1}$ inherits the singularity structure of the matrix A.

1. PII equation

The second Painlevé equation PII,

$$u_{xx} = xu + 2u^3 - \alpha, \tag{6.1.1}$$

is the compatibility condition to the Flaschka-Newell Lax pair (5.0.3)–(5.0.4). We look for a rational gauge transformation,

$$\tilde{\Psi}(\lambda, x) = R(\lambda, x)\Psi(\lambda, x), \tag{6.1.2}$$

which preserves the canonical asymptotic structure of the Ψ-function. First, since $\Psi(\lambda, x)$ is unimodular and has two singular points at $\lambda = 0$ and $\lambda = \infty$, the rational matrix $R(\lambda, x)$ is also unimodular, with possible poles at $\lambda = 0$ and at $\lambda = \infty$. Using the canonical asymptotics at infinity (5.0.6), i.e., $\Psi_k(\lambda, x) = \big(I + O(\lambda^{-1})\big)\exp\big\{-i\big(\frac{4}{3}\lambda^3 + x\lambda\big)\sigma_3\big\}$, and the identity

$$I = R\Psi\tilde{\Psi}^{-1}, \tag{6.1.3}$$

we conclude that $R(\lambda, x) = I + O(\lambda^{-1})$. Indeed, the absence of the formal monodromy at infinity implies that $R(\lambda, x)$ is constant at infinity, i.e., $R = const + O(\lambda^{-1})$. Actually, the exponential behavior of $\Psi(\lambda, x)$ as $\lambda \to \infty$, is consistent with a diagonal constant term in the expansion of R. However, the nonunit diagonal constant term leads to a conjugation of the Stokes matrices.

Thus the rational matrix $R(\lambda, x)$ can have a pole only at $\lambda = 0$. The simplest matrix of such type is

$$R(\lambda, x) = I + \frac{1}{\lambda}R_0(x). \tag{6.1.4}$$

Using two terms in the canonical asymptotics of Ψ_k (see (5.0.7))

$$\Psi_k(\lambda, x) = \Big(I + \lambda^{-1}\big(-\tfrac{i}{2}D\sigma_3 + \tfrac{1}{2}u\sigma_2\big) + O(\lambda^{-2})\Big)e^{-i(\frac{4}{3}\lambda^3 + x\lambda)\sigma_3}, \tag{6.1.5}$$

it follows that equation (6.1.3) yields

$$R_0(x) = -\frac{i}{2}(\tilde{D} - D)\sigma_3 + \frac{1}{2}(\tilde{u} - u)\sigma_2. \tag{6.1.6}$$

Assuming for simplicity that $\alpha - \frac{1}{2} \notin \mathbb{Z}$, we obtain the following two term refinement of the expansion of the Ψ-function at the origin (a refinement of (5.0.8)),

(6.1.7)
$$\Psi^{(0)}(\lambda, x) = \frac{1}{\sqrt{2}}(I - i\sigma_2)\hat{\Psi}^{(0)}(\lambda, x)\lambda^{\alpha\sigma_3}q^{\sigma_3},$$
$$\hat{\Psi}^{(0)}(\lambda, x) = I + \lambda\psi_1^0(x) + O(\lambda^2),$$
$$\psi_1^0(x) = \begin{pmatrix} 0 & 2i\dfrac{v + u^2 + \frac{x}{2}}{1 - 2\alpha} \\ -2i\dfrac{v - u^2 - \frac{x}{2}}{1 + 2\alpha} & 0 \end{pmatrix}.$$

The equation of the monodromy surface (5.0.24), $s_1 + s_2 + s_3 + s_1 s_2 s_3 = -2i\sin\pi\alpha$, allows the monodromy exponent α to be replaced by $\tilde{\alpha} = \alpha + 2n$ or $\tilde{\alpha} = -\alpha - 1 - 2n$, $n \in \mathbb{Z}$. Thus, a suitable transformation is $\alpha \mapsto \tilde{\alpha} = -\alpha - \sigma$, $\sigma = \pm 1$. Therefore, an elementary Schlesinger transformation leads to a shift of the monodromy exponent and to a right multiplication of the $\Psi^{(0)}$ by the quasi-permutation matrix (the latter yields a left multiplication of the connection matrix E in the quasi-permutation matrix, as can be verified by a straightforward computation using (5.0.17)). Letting

$$\tilde{\Psi}^{(0)}(\lambda, x) = \frac{1}{\sqrt{2}}(I - i\sigma_2)\big(I + O(\lambda)\big)\lambda^{\tilde{\alpha}\sigma_3}\tilde{q}^{\sigma_3}\sigma_1,$$

and using elementary manipulations, the identity (6.1.3) yields

(6.1.8)
$$\big(\lambda^{-1}\hat{R}_0 + I + \hat{R}_0\psi_1^0 + O(\lambda)\big)\lambda^{(\alpha+\tilde{\alpha})\sigma_3} = \big(I + O(\lambda)\big)\tilde{q}^{\sigma_3}\sigma_1 q^{-\sigma_3},$$

where $\alpha + \tilde{\alpha} = -\sigma = \pm 1$, and

$$\hat{R}_0 = \frac{1}{\sqrt{2}}(I + i\sigma_2)R_0\frac{1}{\sqrt{2}}(I - i\sigma_2)$$
$$= \begin{pmatrix} 0 & -\frac{i}{2}\big((\tilde{u} - u) - (\tilde{D} - D)\big) \\ \frac{i}{2}\big((\tilde{u} - u) + (\tilde{D} - D)\big) & 0 \end{pmatrix}.$$

The right-hand side of equation (6.1.8) is holomorphic at $\lambda = 0$. The condition of the nonsingularity of $\lambda^{-1}\hat{R}_0\lambda^{-\sigma\sigma_3}$, means that if $\sigma = -1$, then the second column of $\hat{R}_0$ vanishes, while if $\sigma = +1$, then the first column of $\hat{R}_0$ vanishes. Hence,

$$\sigma = -1: \quad \tilde{D} - D = \tilde{u} - u, \quad \hat{R}_0 = i(\tilde{u} - u)\sigma_-,$$
$$\sigma = +1: \quad \tilde{D} - D = -(\tilde{u} - u), \quad \hat{R}_0 = -i(\tilde{u} - u)\sigma_+.$$

The nonsingularity of the term $\big(I + \hat{R}_0\psi_1^0\big)\lambda^{-\sigma\sigma_3}$ for $\sigma = -1$, implies that the second column of the sum $I + \hat{R}_0\psi_1^0$ vanishes. Similarly, its nonsingularity for $\sigma = +1$, implies that the first column of the sum $I + \hat{R}_0\psi_1^0$ vanishes. Thus, taking into account the explicit form of ψ_1^0 (see (6.1.7)) we find

(6.1.9)
$$\tilde{\alpha} = -\alpha - \sigma, \quad \tilde{u} = u + \frac{\frac{1}{2} + \sigma\alpha}{v - \sigma(u^2 + \frac{x}{2})}, \quad R = I + \frac{i\sigma(\tilde{u} - u)}{2\lambda}(\sigma_3 - i\sigma\sigma_2), \quad \sigma = \pm 1.$$

Let us now describe a Schlesinger transformation for the Garnier Lax pair (5.0.59)–(5.0.60). In this case, the Ψ-function has only one singularity at $\lambda = \infty$, thus the Schlesinger matrix $R(\lambda, x)$ is a polynomial. The equations determining the monodromy surface (5.0.69) or (5.0.71) admit a shift of the formal monodromy exponent $\alpha + \frac{1}{2}$ by an integer. In the simplest case of shift by one, the unimodular matrix $R(\lambda, x)$ is polynomial of degree one, $R(\lambda, x) = R_0 + \lambda R_1$, with a nilpotent matrix R_1. Observing that the Schlesinger transformation does not affect the Stokes multipliers, and therefore does not change the canonical asymptotics (5.0.66) except for the formal monodromy exponent, $\alpha \mapsto \tilde{\alpha} = \alpha + \sigma$, $\sigma = \pm 1$, we can use in (6.1.3) the extended asymptotics (5.0.75) multiplied in the canonical exponential term. In this way, we find the equation,

$$\lambda R_1 \lambda^{\sigma\sigma_3} + \big(R_0 + R_1 Y_1\big)\lambda^{\sigma\sigma_3} + \lambda^{-1}\big(R_0 Y_1 + R_1 Y_2\big)\lambda^{\sigma\sigma_3} = I + O(\lambda^{-1}). \tag{6.1.10}$$

This equation immediately yields

$$R(\lambda, x) = \lambda R_1 + R_0, \tag{6.1.11}$$

where the matrices R_0 and R_1 are given either by $(\sigma = +1)$

$$R_1 = \begin{pmatrix} 0 & 0 \\ 0 & 1 \end{pmatrix}, \quad R_0 = \begin{pmatrix} 0 & -\frac{y}{z} \\ \frac{z}{y} & -u - \frac{\alpha + \frac{1}{2}}{z} \end{pmatrix}, \tag{6.1.12}$$

which imply the Bäcklund transformation

$$\tilde{\alpha} = \alpha + 1, \quad \tilde{z} = -z - x - 2\Big(u + \frac{\alpha + \frac{1}{2}}{z}\Big)^2, \quad \tilde{y} = 2\frac{y}{z}, \quad \tilde{u} = -u - \frac{\alpha + \frac{1}{2}}{z}, \tag{6.1.13}$$

or by $(\sigma = -1)$

$$R_1 = \begin{pmatrix} 1 & 0 \\ 0 & 0 \end{pmatrix}, \quad R_0 = \begin{pmatrix} u & \frac{y}{2} \\ -\frac{2}{y} & 0 \end{pmatrix}, \tag{6.1.14}$$

which imply the Bäcklund transformation,

$$\tilde{\alpha} = \alpha - 1, \quad \tilde{z} = -z - x - 2u^2, \quad \tilde{y} = \frac{1}{2} y \tilde{z}, \quad \tilde{u} = -u - \frac{\alpha - \frac{1}{2}}{\tilde{z}}. \tag{6.1.15}$$

2. PIII equation

The third Painlevé equation,

$$u_{xx} = \frac{(u_x)^2}{u} - \frac{u_x}{x} + \frac{\alpha u^2 + \beta}{x} + \gamma u^3 + \frac{\delta}{u}, \tag{6.2.1}$$

is the compatibility condition of the linear systems of equations (see [**JU**], [**FoMA**])

$$\Psi_\lambda(\lambda) = \mathsf{A}(\lambda)\Psi(\lambda), \quad \Psi_x(\lambda) = \mathsf{U}(\lambda)\Psi(\lambda),$$

where

$$\begin{aligned} \mathsf{A}(\lambda) &= \frac{x}{2}\begin{pmatrix} 1 & 0 \\ 0 & -1 \end{pmatrix} + \frac{1}{\lambda}\begin{pmatrix} -\mu & y \\ z & \mu \end{pmatrix} + \frac{1}{\lambda^2}\begin{pmatrix} v - \frac{x}{2} & -wv \\ \frac{v - x}{w} & -v + \frac{x}{2} \end{pmatrix}, \\ \mathsf{U}(\lambda) &= \frac{\lambda}{2}\begin{pmatrix} 1 & 0 \\ 0 & -1 \end{pmatrix} + \frac{1}{x}\begin{pmatrix} 0 & y \\ z & 0 \end{pmatrix} - \frac{1}{x\lambda}\begin{pmatrix} v - \frac{x}{2} & -wv \\ \frac{v - x}{w} & -v + \frac{x}{2} \end{pmatrix}. \end{aligned} \tag{6.2.2}$$

The entries of the matrices (6.2.2) satisfy the following system of equations:

$$
\begin{aligned}
y_x &= \frac{\mu}{x}y - 2wv, \quad v_x = -\frac{\mu}{x}z + \frac{2}{w}(x - v),\\
xv_x &= -4uv^2 + (4xu - 4\mu + 1)v + (\nu + 2\mu)x,\\
xw_x &= w\Big(\frac{x}{v}(\nu + 2\mu) - 2xu + 2\mu\Big),\\
xu_x &= 4vu^2 - 2xu^2 + (4\mu - 1)u + 2x,
\end{aligned} \tag{6.2.3}
$$

where

$$u = -\frac{y}{vw}$$

and

$$\frac{\nu}{2} = -\frac{v - x}{xw}(y - \mu w) + \frac{v}{x}(wz + \mu). \tag{6.2.4}$$

Thus, u satisfies PIII equation (6.2.1) with the parameters

$$\alpha = 4\nu, \quad \beta = 4(1 - 2\mu), \quad \gamma = 4, \quad \delta = -4.$$

We look for a Schlesinger transformation in the form

$$R(\lambda) = \lambda^n(R_1\lambda + R_0),$$

and solve the equation

$$R_\lambda = \hat{A}R - RA, \tag{6.2.5}$$

where

$$
\begin{aligned}
A &= a + b\lambda^{-1} + c\lambda^{-2},\\
\hat{A} &= \hat{a} + \hat{b}\lambda^{-1} + \hat{c}\lambda^{-2},
\end{aligned}
$$

with the constant matrices a, b and c defined in (6.2.2).

The equation (6.2.5) splits into four equations,

$$
\begin{aligned}
\lambda:&\quad [a, R_1] = 0,\\
\lambda^0:&\quad (n+1)R_1 = [a, R_0] + \hat{b}R_1 - R_1 b,\\
\lambda^{-1}:&\quad nR_0 = \hat{b}R_0 - R_0 b + \hat{c}R_1 - R_1 c,\\
\lambda^{-2}:&\quad 0 = \hat{c}R_0 - R_0 c.
\end{aligned}
$$

The first equation yields $R_1 = \operatorname{diag}(r_1, r_2)$, and the second equation reads

$$(n+1)r_1 = (\hat{\mu} - \mu)r_1, \quad (n+1)r_2 = -(\hat{\mu} - \mu)r_2. \tag{6.2.6}$$

$$(R_0)_{12} = \frac{r_1 y}{x}, \quad (R_0)_{21} = \frac{r_1\hat{z}}{x}.$$

Since $\det R = \text{const}$, independent of λ, we find

$$n = -\frac{1}{2}, \quad \hat{\mu} = \mu + \frac{1}{2}, \quad \det R_0 = 0. \tag{6.2.7}$$

These equations and (6.2.6), imply that we can choose the matrices R_1 and R_0 in the form

$$R_1 = \begin{pmatrix} 1 & 0\\ 0 & 0 \end{pmatrix}, \quad R_2 = \begin{pmatrix} \dfrac{\hat{z}y}{x^2} & \dfrac{y}{x}\\ \dfrac{\hat{z}}{x} & 1 \end{pmatrix}. \tag{6.2.8}$$

Now the two equations at λ^{-1} and λ^{-2} imply

$$\hat{z} = -\frac{x}{w}, \quad \hat{y} = \frac{y}{x}\Big(2\hat{\mu} + uv + \frac{x}{u}\Big),$$

$$\hat{v} = \frac{yz}{x} - 2\mu\frac{uv}{x} - \frac{u^2v^2}{x}, \quad \hat{w} = \frac{y}{x}\Big(1 - \frac{x}{\hat{v}}\Big),$$

$$\hat{v}\hat{w} = -\frac{y}{x}\big(x + (\nu+\mu)u + vu^2\big), \tag{6.2.9}$$

where in order to derive the last equation we have used the conservation law (6.2.4). Using the definition of u in terms of y, v and w, we find

$$\begin{aligned} \hat{y} &= -\frac{\hat{y}}{\hat{v}\hat{w}} = \frac{2\hat{\mu} + uv + x/u}{x + (\mu+\nu) + vu^2} \\ &= \frac{1}{u} + \frac{2\hat{\mu} - \mu - \nu}{x + (\mu+\nu) + vu^2}. \end{aligned} \tag{6.2.10}$$

Expressing v in terms of u and u_x via the last equation of (6.2.3),

$$vu^2 = \frac{1}{4}\big(xu_x + 2xu^2 + 2x + (4\mu+1)u\big),$$

we find the Bäcklund transformation

$$\begin{aligned} \hat{u} &= \frac{1}{u} + \frac{4\mu - 4\nu - 4}{xu_x + 2xu^2 + 2x + (4\mu+1)u} \\ &= \frac{1}{u} - \frac{4 - \alpha - \beta}{2xu_x + 4xu^2 + 4x + (\alpha+2)u}. \end{aligned} \tag{6.2.11}$$

The function $\hat{u}$ satisfies the PIII equation (6.2.1) with parameters

$$\hat{\alpha} = \alpha + 4, \quad \hat{\beta} = \beta + 4, \quad \gamma = -\delta = 4.$$

The associated Schlesinger transformation has the form

$$R(\lambda) = \begin{pmatrix} 1 & 0 \\ 0 & 0 \end{pmatrix}\sqrt{\lambda} + \begin{pmatrix} -\dfrac{y}{xw} & \dfrac{y}{x} \\ -\dfrac{1}{w} & 1 \end{pmatrix}\frac{1}{\sqrt{\lambda}}.$$

3. PIV equation

The fourth Painlevé equation

$$u_{xx} = \frac{(u')^2}{2u} + \frac{3}{2}u^3 + 4xu^2 + (-4\alpha + \beta + 2x^2)u - \frac{\beta^2}{2u} \tag{6.3.1}$$

is the compatibility condition of the two linear systems

$$\Psi_\lambda(\lambda) = \mathsf{A}(\lambda)\Psi(\lambda), \quad \Psi_x(\lambda) = \mathsf{U}(\lambda)\Psi(\lambda),$$

with

$$
\mathsf{A}(\lambda) = \frac{1}{2}\begin{pmatrix} 1 & 0 \\ 0 & -1 \end{pmatrix}\lambda^3 + \begin{pmatrix} 0 & iw \\ iv & 0 \end{pmatrix}\lambda^2 + \begin{pmatrix} x+wv & 0 \\ 0 & -x-wv \end{pmatrix}\lambda
$$

$$
+ \begin{pmatrix} 0 & 2xw + w_x \\ 2xv - v_x & 0 \end{pmatrix}\lambda + \begin{pmatrix} \alpha & 0 \\ 0 & -\alpha \end{pmatrix}\frac{1}{\lambda}, \tag{6.3.2}
$$

$$
\mathsf{U}(\lambda) = \frac{\lambda}{2}\begin{pmatrix} 1 & 0 \\ 0 & -1 \end{pmatrix}\lambda^2 + \begin{pmatrix} 0 & iw \\ iv & 0 \end{pmatrix}\lambda + \begin{pmatrix} wv & 0 \\ 0 & -wv \end{pmatrix}.
$$

Here the functions w and $v4$ satisfy the constraint

$$
\beta = w_x v - w v_x + 2xwv - w^2 v^2 \equiv \text{const}, \tag{6.3.3}
$$

and their product

$$
u = vw
$$

is a solution of PIV equation (6.3.1).

We look for a Schlesinger transformation in a form

$$
R(\lambda) = \lambda^n (R_1 \lambda + R_0),
$$

with

$$
\begin{aligned}
A &= a\lambda^3 + b\lambda^2 + c\lambda + d + e\lambda^{-1}, \\
\hat{A} &= \hat{a}\lambda^3 + \hat{b}\lambda^2 + \hat{c}\lambda + \hat{d} + \hat{e}\lambda^{-1}.
\end{aligned}
$$

The compatibility condition yields

$$
\begin{aligned}
\lambda^3 &: \quad [a, R_1] = 0, \\
\lambda^2 &: \quad \hat{b}R_1 - R_1 b + [a, R_0] = 0, \\
\lambda^1 &: \quad \hat{c}R_1 - R_1 c + \hat{b}R_1 - R_1 b = 0, \\
\lambda^0 &: \quad \hat{d}R_1 - R_1 d + \hat{c}R_1 - R_1 c = 0, \\
\lambda^{-1} &: \quad \hat{e}R_1 - R_1 e + \hat{d}R_1 - R_1 d = (n+1)R_1, \\
\lambda^{-2} &: \quad \hat{e}R_0 - R_0 e = nR_0.
\end{aligned}
$$

The first three equations are solved as follows:

$$
\lambda^3 : \quad R_1 = \begin{pmatrix} r_1 & 0 \\ 0 & r_2 \end{pmatrix},
$$

$$
\lambda^2 : \quad \begin{cases} (R_0)_{12} = i(wr_1 - \hat{w}r_2), \\ (R_0)_{21} = i(\hat{v}r_1 - vr_2), \end{cases}
$$

$$
\lambda^1 : \quad \begin{cases} i\hat{w}(R_0)_{22} = iw(R_0)_{11}, \\ i\hat{v}(R_0)_{11} = iv(R_0)_{22}. \end{cases}
$$

Thus, $(R_0)_{11} = (R_0)_{22} = 0$, and the last equation at λ^{-2} gives

$$
(\hat{\alpha} + \alpha)(R_0)_{12} = n(R_0)_{12}, \quad -(\hat{\alpha} + \alpha)(R_0)_{21} = n(R_0)_{21}.
$$

This leads to three possibilities

i) $(R_0)_{12} = 0, \quad (R_0)_{21} \neq 0, \quad \hat{\alpha} + \alpha = -n,$
ii) $(R_0)_{12} \neq 0, \quad (R_0)_{21} = 0, \quad \hat{\alpha} + \alpha = n,$
iii) $(R_0)_{12} \neq 0, \quad (R_0)_{12} \neq 0, \quad \hat{\alpha} = -\alpha, \quad n = 0.$

Consider first the case i), and put for simplicity $r_1 = r_2 = 1$. The equations at λ^0 and λ^{-1} now read

$$\begin{aligned}
&\hat{w} = w,\\
&2v\hat{v} - \hat{v}_x - 2xv + v_x - (\hat{v} - v)(2x + \hat{w}\hat{v} + wv) = 0,\\
&\hat{\alpha} - \alpha - (2xw + w_x)(\hat{v} - v) = n + 1,\\
&\alpha - \hat{\alpha} + (2xw + w_x)(\hat{v} - v) = n + 1,
\end{aligned}$$

which yields

$$\hat{\alpha} = -\alpha + 1, \quad n = -1,$$

$$\hat{v} = v + \frac{1 - 2\alpha}{2xw + w_x}. \tag{6.3.4}$$

Since $\hat{w} = w$ and $\hat{u} = \hat{w}\hat{v}$, the equation (6.3.4) produces the Bäcklund transformation

$$\begin{aligned}
\hat{u} &= u + \frac{1 - 2\alpha}{2x + w_x/w} = u + \frac{1 - 2\alpha}{2x + \left(-x + \frac{u_x}{2u} + \frac{u}{2} + \frac{\beta}{2u}\right)}\\
&= u + 2\frac{(1 - 2\alpha)u}{u_x + 2xu + u^2 + \beta},
\end{aligned} \tag{6.3.5}$$

where we have used the conservation law (6.3.3),

$$\frac{w_x}{w} = \frac{v_x}{v} - 2x + u - \frac{\beta}{u},$$

together with the obvious constraint $u_x = w_x v + w v_x$.

The Schlesinger transformation associated with (6.3.5) has the form

$$R(\lambda) = \begin{pmatrix} 1 & 0 \\ \dfrac{i}{\lambda}\dfrac{1 - 2\alpha}{2xw + w_x} & 1 \end{pmatrix}.$$

The case ii) is treated in a similar way and is left as an exercise.

Proceeding with the case iii), we note that the condition $\det R(\lambda) = \text{const}$ leads to $r_1 r_2 = 0$. Letting, for example, $r_1 = 1$, $r_2 = 0$, we find

$$R(\lambda) = \begin{pmatrix} \lambda & iw \\ i\hat{v} & 0 \end{pmatrix}.$$

The equations at λ^0 and λ^{-1} are

$$w_x = w(\hat{w}\hat{v} + wv), \quad v_x = -v(\hat{w}\hat{v} + wv). \tag{6.3.6}$$

They have the simple conserved quantity $w\hat{v} = \text{const}$. Choosing

$$\hat{v} = \frac{1}{w},$$

the first equation in (6.3.6) reads

$$\hat{w}_x = w_x - w^2 v. \tag{6.3.7}$$

Finally, applying the conservation law (6.3.3), we obtain the Bäcklund transformation

$$\hat{u} = \frac{u_x}{2u} + \frac{1}{2}u - x + \frac{\beta}{2u}.$$

Part 2

Asymptotics of the Painlevé II Transcendent. A Case Study

As it was mentioned earlier, the RH representations for the Painlevé equations, play the same role that the classical integral representations play for the "linear" special functions. These RH representations provide a powerful tool for deriving asymptotic expansions as well as asymptotic connection formulae at critical points of the Painlevé functions in the complex plane.

We will use the PII equation,

$$u_{xx} = xu + 2u^3 - \alpha,$$

to illustrate how these asymptotic results can be obtained. The reason for this choice is the simplicity of both PII and the related RH problem. Another simplification of PII is that it contains only one parameter. One might argue that PI equation has no parameters at all, but its RH problem is more complicated (10 sectors in the λ-plane). The simplicity of PII makes it possible to present complete proofs of almost all major asymptotic results.

We start with the asymptotic behavior of the generic two-parameter PII transcendent in the complex plane. We consider two basic approaches. First, the WKB study of the direct monodromy problem is presented in Chapter 7. Second, the Deift-Zhou method of the "asymptotic undressing" of the inverse monodromy problem is studied in Chapter 8. The results of Chapters 8 and 9 are true as $x \in \mathbb{C}$, except for x on the rays $\arg x = \frac{\pi n}{3}, \quad n = 1, 2, \ldots, 6$. In Chapters 9 and 10 we concentrate on the asymptotic behavior of the PII transcendent on these rays and present connection formulas, which are useful for many applications of PII.

Throughout Chapters 7–10, for simplicity, we take the parameter α to be zero,

$$u_{xx} = xu + 2u^3.$$

In Chapter 11 we return to the general PII equation and apply the Deift-Zhou method in order to specify exponentially small corrections to PII asymptotics. This reveals another feature of Painlevé transcendents, which is similar to the "linear" special functions — the Stokes phenomenon.

CHAPTER 7

Asymptotic Solutions of the Second Painlevé Equation in the Complex Plane. Direct Monodromy Problem Approach

0. Introduction

The asymptotic behavior of the solutions of the first and second Painlevé equations were first studied in the classical work of P. Boutroux [**Bou**]. In particular, Boutroux expressed certain asymptotic formulae in terms of elliptic functions. The Boutroux elliptic ansatz contains the phase shift as a free parameter.

N. Joshi and M. Kruskal revised the result of Boutroux in [**JoKr1**], using the multiple-scale expansion method, and found the form of the phase shift in the Boutroux elliptic ansatz. Specifically, Joshi and Kruskal showed that for the second Painlevé equation,

$$\frac{d^2u}{dx^2} = xu + 2u^3, \tag{7.0.1}$$

the large $|x|$ asymptotics of its generic solution $u(x)$ is given by

$$u(x) \sim c_1\sqrt{x}\ \mathrm{sn}\left(\frac{2}{3}c_2x^{3/2} + \mu\omega_1 + \nu\omega_2|D\right), \quad x \in \mathbb{C}. \tag{7.0.2}$$

Here the periods ω_j, $j = 1, 2$, of the elliptic sine, the multipliers $c_{1,2}$, and the module D are (transcendental) functions of $\arg x$, while the parameters μ, ν are constant in the interior of any sector $\arg x \in \left(\frac{\pi}{3}k, \frac{\pi}{3}(k+1)\right)$, $k = 0, 1, \ldots, 5$.

In what follows we will prove the Boutroux elliptic ansatz (7.0.2) for PII transcendent with the help of the RH approach discussed in Chapter 4. This will enable us not only to prove the existence theorem rigorously, but also to derive explicit formulas linking the periods and the phase shift of the elliptic function appearing in (7.0.2) with the monodromy data parametrizing PII transcendent.

The outline of this chapter is: In Section 1, we give a short exposition of Boutroux results and discuss the analytic structure of the PII transcendent in the complex plane. In Section 2, we recall the setting of the direct monodromy problem related to equation (7.0.1) and formulate the main theorem (Theorem 7.1). In Section 3, we assume the Boutroux ansatz for $u(x; s)$ and present the WKB formalism, starting with the "exterior" asymptotic expansion of the matrix Ψ-function outside the turning points. Then we study the "interior" expansion by integrating a model equation for the Ψ-function in the neighborhood of the turning points. The model equation is solved in terms of the Airy functions, and this yields explicit asymptotic formulae, which match the "external" expansion. In Section 5, we combine "interior" and "exterior" functions and construct an approximate solution of the direct monodromy problem which in turn yields a global function which approximates the solution of the RH problem uniformly in λ. Finally, in Section 6 we present a proper uniformization of the WKB solution in terms of theta-functions, which completes the proof of the main theorem. The justification of the Boutroux ansatz is based on Kitaev's method [**Ki1**] and it is presented in Section 6.1.

1. Preliminary remarks. The Boutroux ansatz

Let $u(x)$ be a solution of equation (7.0.1). Under the scaling

$$t = \tfrac{2}{3}x^{3/2}, \quad v(t) = x^{-1/2}u(x), \tag{7.1.1}$$

the equation (7.0.1) takes the form

$$v_{tt} - v - 2v^3 = -\frac{v_t}{t} + \frac{v}{9t^2}.$$

Fix any initial point t_0 and choose Cauchy data

$$v_0 = v(t_0), \quad v_1 = v'(t_0),$$

in such a way that the value of D_0 defined by

$$D_0 := v_1^2 - v_0^4 - v_0^2$$

satisfies the inequalities

$$D_0 \neq 0, \tfrac{1}{4}.$$

Construct an elliptic function $V(t, D_0)$ solving the Cauchy problem

$$\begin{aligned} &V_t^2 = V^4 + V^2 + D_0, \\ &V(t_0) = v_0, \quad V_t(t_0) = v_1. \end{aligned}$$

The solution V of (7.1.2) can be expressed as the Jacobi sine function

$$V(t) = \frac{\varkappa}{i\sqrt{1+\varkappa^2}} \operatorname{sn}\left(\frac{t-\hat{t_0}}{i\sqrt{1+\varkappa^2}}\Big|\varkappa\right), \quad D_0 = \frac{\varkappa^2}{(1+\varkappa^2)^2}.$$

It is clear that

$$\Big|v(t) - V(t-t_0, D_0)\Big| = O(t_0^{-1}) \tag{7.1.2}$$

for any t running within a finite-radius circle and avoiding the lattice of poles of V.

Choose a cycle $\mathcal{L} = ka + lb$ on the elliptic curve $W^2 = V^4 + V^2 + D_0$, where a, b are basic cycles and $k, l \in \mathbb{Z}$ are arbitrary integers. Construct sequences of points $\{t_n\}$, modular parameters $\{D_n\}$ and the curves $W^2 = V^4 + V^2 + D_n$, $n = 0, 1, \ldots$, by the recurrent formula

$$\omega_n = \oint_{\mathcal{L}} \frac{dV}{\sqrt{V^4 + V^2 + D_n}},$$

$$t_{n+1} = t_n + \omega_n,$$

$$D_{n+1} = V_t^2(t_{n+1}|D_n) - V^4(t_{n+1}|D_n) - V^2(t_{n+1}|D_n).$$

If the sequence of the points $\{t_n\}$, called by Boutroux a "line of periods", admits an infinite extension, then the sequence of the modular parameters $\{D_n\}$ tends to a finite limit:

$$\omega_n = \omega_\infty + O(t_n^{-1}),$$

$$D_n = D_\infty + O(t_n^{-1}),$$

$$\omega_\infty = \oint_{\mathcal{L}} \frac{dV}{\sqrt{V^4 + V^2 + D_\infty}}, \tag{7.1.3}$$

where D_∞ satisfies the transcendental equation

$$\oint_{\mathcal{L}} \sqrt{V^4 + V^2 + D_\infty}\, dV = 0. \tag{7.1.4}$$

This result expands the estimate (7.1.2) beyond the circle $|t - t_0| \leq const$ to a large domain with ranges of order $O(t)$, $t \gg 1$. Now the elliptic function (7.1.2) becomes "modulated" with respect to $\phi = \arg t$ which varies by $o(t_{n+1} - t_n)$ as $t_{n+1} - t_n = O(1)$:

$$V(t) = V\big(t - t_\infty(\phi)|D_\infty(\phi)\big). \tag{7.1.5}$$

Here for the modulus D_∞ of the elliptic function the equation (7.1.4) takes the form

$$\arg \oint_a \sqrt{V^4 + V^2 + D_\infty}\, dV = \arg \oint_b \sqrt{V^4 + V^2 + D_\infty}\, dV = -\arg t - \frac{\pi}{2}. \tag{7.1.6}$$

Equations (7.1.6) are called the *Boutroux equations.* They can be considered as modulation equations on the modulus $\varkappa = \varkappa(\phi)$ of Jacobi sine function (7.1.5):

$$\begin{cases} \operatorname{Re}\Big[\Big(\dfrac{e^{i\phi}}{1+\varkappa^2}\Big)^{3/2}(-2\varkappa^2 K' + (1+\varkappa^2)E')\Big] = 0, \\ \operatorname{Im}\Big[\Big(\dfrac{e^{i\phi}}{1+\varkappa^2}\Big)^{3/2}((\varkappa^2 - 1)K + (1+\varkappa^2)E)\Big] = 0, \end{cases} \tag{7.1.7}$$

where

$$D_\infty = \frac{\varkappa^2}{(1+\varkappa^2)^2}$$

and $E(\varkappa)$ and $K(\varkappa)$ are standard complete elliptic integrals:

$$K = K(\varkappa) = \int_0^1 \frac{dz}{\sqrt{(1-z^2)(1-\varkappa^2 z^2)}}, \quad K' = K(\varkappa'),$$

$$E = E(\varkappa) = \int_0^1 \sqrt{\frac{1-\varkappa^2 z^2}{1-z^2}}\, dz, \quad E' = E(\varkappa'),$$

$$\varkappa' = \sqrt{1-\varkappa^2}, \quad 0 < \arg \varkappa' < \frac{\pi}{2}, \quad -\frac{\pi}{2} < \arg(1+\varkappa^2) < 0.$$

In the framework of the RH problem approach to the Painlevé equations, Boutroux equations (7.1.7) come from the leading order isomonodromy condition, i.e., from independence of generic Stokes multipliers from x. More discussion of the Boutroux equations including their isomonodromic origin and their solvability can be found in [**K9**].

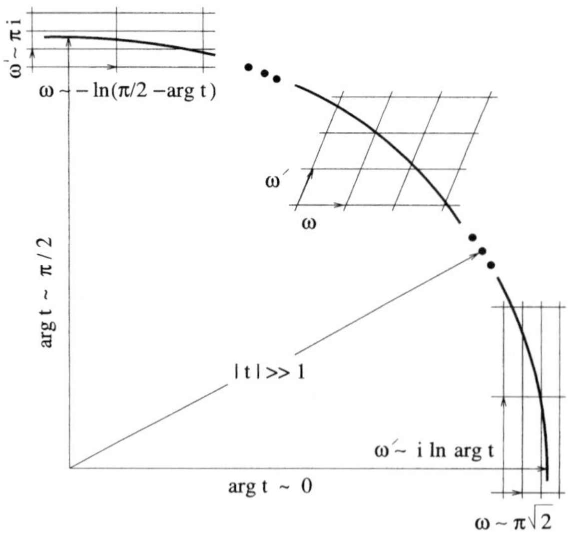

FIGURE 7.1. The deformation of the period lattice for the elliptic function V in the first quadrant $0 < \arg t < \pi/2$.

One can show that system (7.1.7) has a unique differentiable w.r.t. φ solution normalized by the conditions,

$$\varphi \to \frac{2\pi}{3} \Rightarrow \varkappa \to 1, \quad \varphi \to \pi \Rightarrow \varkappa \to 0, \quad -\frac{\pi}{2} < \arg \varkappa < 0.$$

The lattice of periods for the approximate solution V is shown on Figure 7.1. This can be seen by decomposition of the integral (7.1.4) over the basic cycles a and b,

$$\mathcal{L} = ka + lb$$

for the Abelian differential $\sqrt{V^4 + V^2 + D_\infty}dV$. Introducing basic periods

$$\omega = \oint_a \frac{dV}{\sqrt{V^4 + V^2 + D_\infty}}, \quad \omega' = \oint_b \frac{dV}{\sqrt{V^4 + V^2 + D_\infty}},$$

alongside with the integrals

$$J = \oint_a \sqrt{V^4 + V^2 + D_\infty}dV, \quad J' = \oint_b \sqrt{V^4 + V^2 + D_\infty}dV$$

one can easily rewrite the Boutroux equation (7.1.4) in a form of the well-known Legendre constraint, appearing in the theory of Weierstrass $\wp$-function

$$J'\overline{\omega} - J\overline{\omega}' = \frac{2\pi i}{3} e^{3i\phi/2}, \tag{7.1.8}$$

where $\phi = \arg x$ and the bar denotes the complex conjugate.

Boutroux had also investigated degenerate cases of elliptic solution (7.1.2). Namely, it goes trigonometric

$$V(t) = \pm\frac{i}{\sqrt{2}} \cot\frac{t - t_k}{\sqrt{2}}, \quad D_\infty = \frac{1}{4}, \quad \omega = \pi\sqrt{2}$$

on the rays

$$\arg t = \pi k, \quad k \in \mathbb{Z},$$

and it goes hyperbolic

$$V(t) = \frac{i}{\sinh(t - t_k)}, \quad D_\infty = 0, \quad \omega' = 2\pi i,$$

on the rays

$$\arg t = \pi k, \quad k \in \mathbb{Z}.$$

The correct asymptotics of PII transcendent on canonical six-rays $\arg x = \frac{\pi k}{3}$ will be discussed below in Chapter 10. Note that it is far more complicated than direct degeneration of the Boutroux ansatz (7.1.5).

2. Direct monodromy problem. Formulation of the main theorem

Recall the λ-equation corresponding to the second Painlevé equation PII with $\alpha = 0$ (see (4.2.36))

$$\frac{d\Psi}{d\lambda} = \left[-(4i\lambda^2 + ix + 2iu^2)\sigma_3 - 4u\lambda\sigma_2 - 2w\sigma_1\right]\Psi, \tag{7.2.1}$$

where σ_1, σ_2, σ_3 are the Pauli matrices

$$\sigma_3 = \begin{pmatrix} 1 & 0 \\ 0 & -1 \end{pmatrix}, \quad \sigma_2 = \begin{pmatrix} 0 & -i \\ i & 0 \end{pmatrix}, \quad \sigma_1 = \begin{pmatrix} 0 & 1 \\ 1 & 0 \end{pmatrix},$$

x, u, w are complex parameters. The set of the monodromy data of this equation consists of the six Stokes matrices,

$$S_{2k} = \begin{pmatrix} 1 & s_{2k} \\ 0 & 1 \end{pmatrix}, \quad S_{2k+1} = \begin{pmatrix} 1 & 0 \\ s_{2k+1} & 1 \end{pmatrix}, \quad k = 1, 2, 3$$

defined through the equations

$$S_k = \Psi_k^{-1}\Psi_{k+1}, \quad k = 1, \ldots, 6, \tag{7.2.2}$$

where $\Psi_1(\lambda), \ldots, \Psi_7(\lambda)$ are the canonical solutions of the equation (7.2.1). These solutions are fixed by the asymptotic condition (Chapter 2, (2.1.88))

$$\Psi_k(\lambda) = (I + O(\lambda^{-1}))\exp\left\{-\frac{4}{3}i\lambda^3\sigma_3 - ix\lambda\sigma_3\right\}, \tag{7.2.3}$$

$$\lambda \to \infty, \quad \frac{\pi}{3}(k-2) < \arg\lambda < \frac{\pi}{3}k, \quad k = 1, \ldots, 7.$$

Note that $\Psi_7 = \Psi_1$. The following general constraints take place:

$$S_{k+3} = \sigma_2 S_k \sigma_2, \quad k = 1, 2, 3,$$

$$S_1 S_2 \ldots S_6 = I,$$

or, in terms of the Stokes multipliers s_k,

$$s_{k+3} = -s_k, \quad k = 1, 2, 3, \quad s_1 - s_2 + s_3 + s_1 s_2 s_3 = 0. \tag{7.2.4}$$

The Painlevé function $u(x)$ admits the following representation in terms of the solution $\Psi(\lambda)$ of the RH problem (7.2.2),

$$u(x) = 2 \lim_{\lambda\to\infty} \left(\lambda\Psi_{12}(\lambda)e^{-\theta(\lambda,x)}\right).$$

Moreover, the map,

$$\{s\} \equiv \{s_k: \quad s_{k+3} = -s_k, \quad s_1 - s_2 + s_3 + s_1 s_2 s_3 = 0\}$$

$$\Downarrow \tag{7.2.5}$$

$$\text{solutions } u(x) \text{ of PII equation (7.0.1)}$$

is a bijection (see Chapter 4, Corollary 4.4). Hence we adopt the notation

$$u(x) \equiv u(x; s)$$

for the second Painlevé transcendent which is solution of equation (7.0.1).

The main result concerning the asymptotics of $u(x; s)$ for complex x can be formulated as the following theorem.

THEOREM 7.1. *Let*

$$s_3 \neq 0, \quad 1 - s_1 s_3 \neq 0. \tag{7.2.6}$$

Then

$$u(x; s) = -\frac{i\varkappa}{\sqrt{1+\varkappa^2}} x^{1/2} \operatorname{sn}\left(\frac{2i}{3}\frac{x^{3/2}}{\sqrt{1+\varkappa^2}} - \frac{2iK}{\pi}\ln(s_3) + \frac{K'}{\pi}\ln(1 - s_1 s_3) + O\big(x^{-3/2}\big)\big|\varkappa\right), \tag{7.2.7}$$

$$x \to \infty, \quad \arg x = \varphi + O\big(x^{-3/2}\big), \quad \varphi \in \left(\frac{2\pi}{3}, \pi\right), \tag{7.2.8}$$

where $\varkappa = \varkappa(\varphi)$ satisfies Boutroux equations (7.1.7) and K, K' are the complete elliptic integrals. This asymptotics is uniform on the cheese-type domain

$$\mathcal{D} = \left\{x \in \mathbb{C}: \quad |\arg x - \varphi| < C|x|^{-3/2}\right\} \cap D_\varepsilon, \tag{7.2.9}$$

where D_ε is the complement to the union of the ε-neighborhoods of all the poles of the above elliptic function.

We complete this section by noting that Theorem 7.1 yields, in fact, the asymptotic description for the rest of the sectors. This is due to the following important symmetry properties of the Painlevé functions.

PROPOSITION 7.1. *The solution $u(x; s)$ of the second Painlevé equation (7.0.1) corresponding to the monodromy data s satisfies the symmetry relations,*

$$u(x; s) = e^{i\frac{2\pi}{3}l} u(xe^{i\frac{2\pi}{3}l}; s^{2l}), \tag{7.2.10}$$

$$u(x; s) = \overline{u}(\overline{x}; s^*), \tag{7.2.11}$$

where $s \mapsto s^d$, $d \in \mathbb{Z}$ and $s \mapsto s^*$ denote the automorphisms of the set of monodromy data $\{s\}$ defined by the relations

$$(s^d)_j := s_{j+d}, \quad mod\ (s_{k+3} = -s_k)$$

and

$$(s^*)_j := \overline{s}_{4-j}, \quad mod\ (s_{k+3} = -s_k).$$

PROOF. To prove, say, equation (7.2.10), we observe the following obvious fact. If $\Psi_k(\lambda, x)$ is the canonical solution corresponding to the sector $\Omega_k := \{\lambda : \frac{\pi}{3}(k-2) < \arg\lambda < \frac{\pi}{3}k\}$ in the λ-plane, then the function

$$\tilde{\Psi}(\lambda, x) := \Psi_k(e^{-i\frac{2\pi}{3}l}\lambda, e^{i\frac{2\pi}{3}l}x) \tag{7.2.12}$$

satisfies the canonical asymptotics (7.2.3) in the sector Ω_{k+2l}. Hence, the collection

$$\tilde{\Psi}_k(\lambda, x) := \Psi_{k-2l}(e^{-i\frac{2\pi}{3}l}\lambda, e^{i\frac{2\pi}{3}l}x),$$

defined for $k = 1, \ldots, 7$, modulo the relation $\tilde{\Psi}_{k+3}(\lambda, x) = \sigma_2\tilde{\Psi}_k(-\lambda, x)\sigma_2$, form the new set of matrix-valued functions satisfying the canonical asymptotics (7.2.3) in the respective sectors Ω_k. Furthermore,

$$\tilde{\Psi}_k^{-1}(\lambda, x)\tilde{\Psi}_{k+1}(\lambda, x)$$
$$= \Psi_{k-2l}^{-1}(e^{-i\frac{2\pi}{3}l}\lambda, e^{i\frac{2\pi}{3}l}x)\Psi_{k+1-2l}(e^{-i\frac{2\pi}{3}l}\lambda, e^{i\frac{2\pi}{3}l}x) = \mathrm{S}_{k-2l}.$$

In other words, the piecewise analytic function $\tilde{\Psi}(\lambda, x)$ defined by the collection $\{\tilde{\Psi}_k(\lambda, x)\}$ solves the Riemann-Hilbert problem (7.2.2) defined by the Stokes matrices,

$$\tilde{\mathrm{S}}_k := \mathrm{S}_{k-2l} \mod (\mathrm{S}_{k+3} = \sigma_2\mathrm{S}_k\sigma_2). \tag{7.2.13}$$

Let $\tilde{u}(x) \equiv u(x; \tilde{s})$ be the Painlevé function corresponding to the monodromy data $\tilde{\mathrm{S}}_k$. Then, equation (7.2.12) tells us that

$$u(x; \tilde{s}) = 2\lim_{\lambda\to\infty}\left(\lambda\tilde{\Psi}_{12}(\lambda)e^{-\theta(\lambda,x)}\right)$$
$$= 2\lim_{\lambda\to\infty}\left(\lambda\Psi_{12}(e^{-i\frac{2\pi}{3}l}\lambda, e^{i\frac{2\pi}{3}l}x)e^{-\theta(\lambda,x)}\right)$$
$$= e^{i\frac{2\pi}{3}l}u(e^{i\frac{2\pi}{3}l}x; s).$$

Simultaneously, equation (7.2.13) indicates that

$$\tilde{s}_k = s_{k-2l}.$$

Relation (7.2.10) is proven. Symmetry (7.2.11) can be dealt with in a similar way. □

COROLLARY 7.1. *If $s_{2+2l} \neq 0$ and $1 + s_{2+2l}s_{3+2l} \neq 0$ for $l = 0, \pm 1$, then the leading asymptotic term of the Painlevé function in the interior of the sector $\arg x \in \left(-\frac{2\pi}{3}l; \frac{\pi}{3} - \frac{2\pi}{3}l\right)$ is elliptic.*

If $s_{2+2l} \neq 0$ and $1 + s_{1+2l}s_{2+2l} \neq 0$ for $l = 0, \pm 1$, then the leading asymptotic term of the Painlevé function in the interior of the sector $\arg x \in \left(-\frac{\pi}{3} - \frac{2\pi}{3}l; -\frac{2\pi}{3}l\right)$ is elliptic.

The phase shift of the elliptic asymptotics in each of the sectors is given by the expression,

$$\frac{2iK}{\pi}\ln s_K \pm \frac{K'}{\pi}\ln s_{K'},$$

where the choice of the monodromy parameters s_K and $s_{K'}$ is indicated in Figure 7.2. Together with the monodromy equations (7.2.4) this constitutes the connection formulae for the Painlevé function $u(x; s)$ in the complex domain.

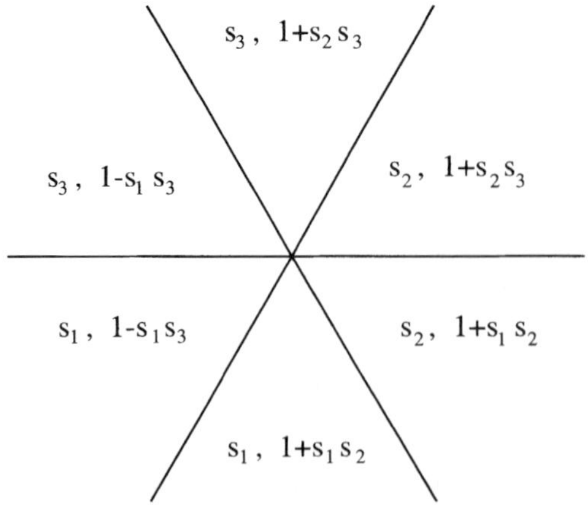

FIGURE 7.2. Combinations of the Stokes multipliers, whose non-triviality yields the elliptic asymptotic behavior of the Painlevé function in the corresponding sector

REMARK 7.1. A complete description of the asymptotic behavior of $u(x;s)$, which includes the "Stokes rays" $\arg x = \frac{\pi}{3}k$, is given in [**K1**]. In fact, the complete description of the asymptotic behavior of the solutions of the second Painlevé equation obtained in [**K1**] covers the general case, i.e., the equation

$$\frac{d^2u}{dx^2} = 2u^3 + xu - \alpha, \quad \alpha \in \mathbb{C}. \tag{7.2.14}$$

This result implies that the leading order asymptotic term (7.2.7) does not depend on the parameter α.

3. WKB-analysis for the Ψ-function.

To solve the direct monodromy problem one needs to calculate two independent components of Stokes matrices, say s_1 and s_3. The most convenient way to achieve this is to consider the product

$$\mathrm{S}_0\mathrm{S}_1 = \sigma_2\mathrm{S}_3\sigma_2\mathrm{S}_1 = \begin{pmatrix} 1 - s_1s_3 & -s_3 \\ s_1 & 1 \end{pmatrix} = \Psi_0^{-1}\Psi_2, \tag{7.3.1}$$

where Ψ_0, Ψ_2 are solutions of the system (7.2.1). This product will be calculated asymptotically as $x \to \infty$ with the help of the WKB method (see [**He**]).

We begin with "equation in λ", which we rewrite in the scaled variables

$$u = (\mathrm{e}^{-i\phi}x)^{1/2}v, \quad u_x = \mathrm{e}^{-2i\phi}x\Big(v' + \frac{v}{3t}\Big), \quad t = \frac{2}{3}(\mathrm{e}^{-i\phi}x)^{2/3}, \tag{7.3.2}$$

$$\mathrm{e}^{-2i\phi}\Big(v' + \frac{v}{3t}\Big)^2 - v^4 - \mathrm{e}^{i\phi}v^2 = A(\phi) + \frac{1}{t}B(\phi,t). \tag{7.3.3}$$

The functions v and v' are assumed to be bounded by constants independent of t:

$$|v| < C_1, \quad |v'| < C_2, \quad |B(\phi,t)| < C_3. \tag{7.3.4}$$

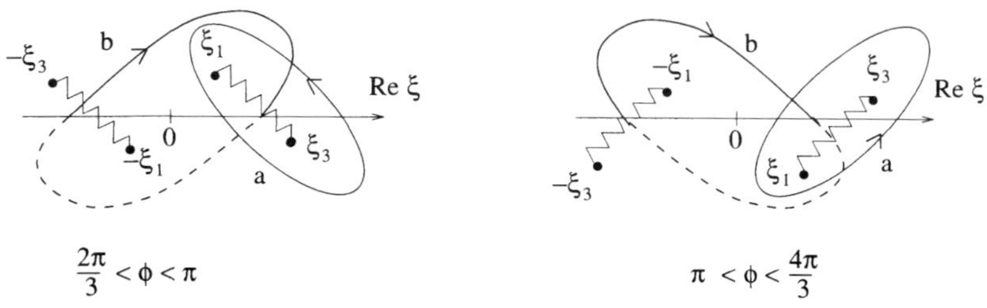

FIGURE 7.3. The Riemann surface $\Gamma(t)$ and the associated basic cycles a and b.

The decomposition (7.3.3) together with the estimates (7.3.4) are based in fact on the Boutroux ansatz, formulated in §1 for the v function.

We rescale λ and we rewrite equation (7.2.1) in a form suitable for the WKB analysis:

$$\frac{d\Psi}{dz} = \frac{3t}{2}\big(b_3\sigma_3 + b_2\sigma_2 + b_1\sigma_1\big)\Psi, \tag{7.3.5}$$

$$\begin{aligned} \lambda &= \big(\mathrm{e}^{-i\phi}x\big)^{1/2}z,\\ b_1 &= -2\mathrm{e}^{-i\phi}\Big(v' + \frac{v}{3t}\Big),\\ b_2 &= -4vz,\\ b_3 &= -i\big(4z^2 + \mathrm{e}^{i\phi} + 2v^2\big). \end{aligned} \tag{7.3.6}$$

The squares of the eigenvalues of $\sum b_k\sigma_k$ have the form

$$\mu^2 = -16z^4 - 8\mathrm{e}^{i\phi}z^2 - \mathrm{e}^{2i\phi} + 4A(\phi).$$

This defines the elliptic Riemann surface

$$\Gamma(t):\ w = \sqrt{(z^2 - z_1^2)(z^2 - z_3^2)} \equiv \frac{\mu}{4i} = \frac{1}{4i}\sqrt{b_1^2 + b_2^2 + b_3^2}. \tag{7.3.7}$$

The branch points $\pm z_1$, $\pm z_3$ satisfy a system of algebraic equations, which follows from the definitions (7.3.6), (7.3.7):

$$\begin{cases} z_1^2 + z_3^2 = -\frac{1}{2}\mathrm{e}^{i\phi}, \\ (z_1^2 - z_3^2)^2 = A(\phi) + \frac{1}{t}B(\phi, t), \end{cases} \qquad \mathrm{Re}\ z_1 > 0, \quad \frac{2\pi}{3} \le \phi < \frac{4\pi}{3}. \tag{7.3.8}$$

By using the simple reductions

$$A(-\phi) = \overline{A}(\phi), \quad A\Big(\phi - \frac{2\pi}{3}\Big) = \mathrm{e}^{2\pi i/3}A(\phi), \quad A\Big(\frac{2\pi}{3}\Big) = \frac{1}{4}\mathrm{e}^{4\pi i n/3}, \quad n \in \mathbb{Z},$$

we find the correct definitions of z_1 and z_3 in the remaining sectors of the z-plane

$$z_k\Big(\phi + \frac{2\pi}{3}n\Big) = z_k(\phi)\mathrm{e}^{i\pi n/3}, \quad k = 1, 2, \quad n \in \mathbb{Z}. \tag{7.3.9}$$

The upper Γ^+ and lower Γ^- sheets of the surface Γ are glued over the cuts along $[z_1,\, z_3]$ and $[-z_1,\, -z_3]$, as shown on Figure 7.3. The sheets are distinguished by the sign of the asymptotics $w \to \pm z^2$, $z \in \Gamma^{\pm}$. The basic cycles a and b on the surface pass over the segments $[z_1,\, z_3]$ and $[-z_1,\, z_]$, respectively. Define the complete elliptic integrals over the cycles a and b,

$$\omega_{a,b} = 4i \oint\limits_{a,b} \frac{dz}{\mu(z)}, \qquad \mathcal{J}_{a,b} = \frac{1}{4i} \oint\limits_{a,b} \mu(z)\, dz. \tag{7.3.10}$$

Consider the Stokes graph on the upper sheet $\Gamma^+(t) \bigcup \infty^+$ consisting of the vertices z_k and ∞^+ connected by the Stokes lines

$$\mathrm{Re} \int\limits_{z_k}^{z} \mu(z)\, dz = 0.$$

The six canonical domains $\tilde{\Omega}_j$, bounded by the Stokes lines, correspond to the six original sectors for the Ψ-function. More precisely, for any point $z \in \tilde{\Omega}_j$ there exist contours $\mathcal{C}_1(z)$ and $\mathcal{C}_2(z)$, ending at z, such that

$$\mathrm{Re} \int\limits_{\tau_1}^{z} \mu(z) dz \to -\infty, \quad \tau_1 \to \infty, \quad \tau_1 \in \mathcal{C}_1(z),$$

$$\mathrm{Re} \int\limits_{\tau_2}^{z} \mu(z) dz \to +\infty, \quad \tau_2 \to \infty, \quad \tau_2 \in \mathcal{C}_2(z). \tag{7.3.11}$$

As a rule, the contours $\mathcal{C}_{1,2}(z)$ can be chosen as parts of anti-Stokes lines.

For $\pi < \varphi < \frac{4\pi}{3}$, the Stokes graph together with the canonical domains is shown on Figure 7.4.

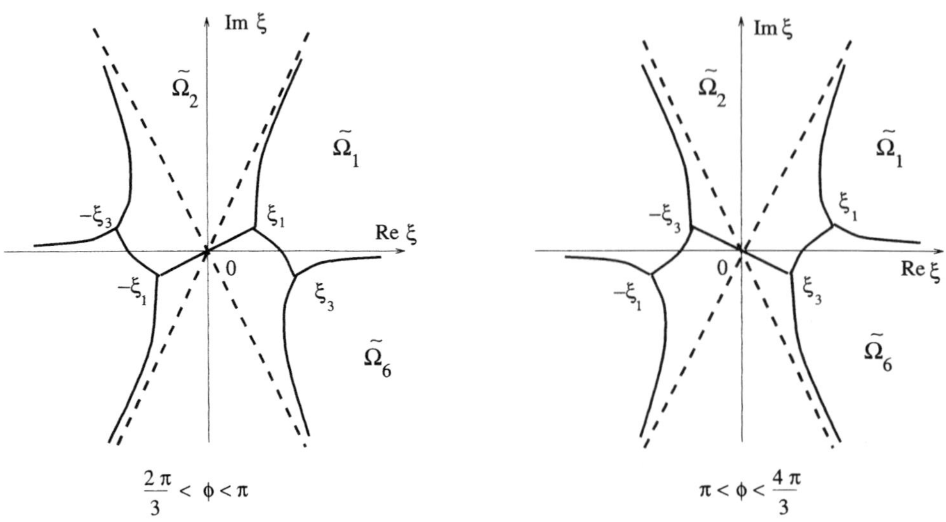

FIGURE 7.4. Stokes graph on $\Gamma^+(t)$

In each of the canonical domains $\tilde{\Omega}_j$ there exists a WKB-solution of the system (7.3.5) having the form

$$\Psi_{WKB}(z_0, z) = T(z)\chi(z,t) \exp\Big\{ \int\limits_{z_0}^{z} \Lambda(\zeta)\, dz \Big\}, \tag{7.3.12}$$

where z_0 is a normalization point different from the turning points and the zeros of $b_1 \pm ib_2$,

$$T(z) = \begin{pmatrix} 1 & (b_3+\mu)(ib_2+b_1)^{-1} \\ (b_3+\mu)(ib_2-b_1)^{-1} & 1 \end{pmatrix},$$

$$\Lambda(z) = -\tfrac{3}{2}t\mu(z)\sigma_3 - \operatorname{diag}\left(T^{-1}(z)T_z(z)\right).$$

The leading term of the WKB solution (7.3.12) is defined by diagonalizing the matrix of coefficients in equation (7.3.5),

$$b_3\sigma_3 + b_2\sigma_2 + b_1\sigma_1 = T(z)\begin{pmatrix} -\mu(z) & 0 \\ 0 & \mu(z) \end{pmatrix} T^{-1}(z). \tag{7.3.13}$$

We need a uniform estimate for χ as $|t| \to \infty$ and $z \in \Omega_j \setminus \varepsilon(z_k)$, i.e., outside a small neighborhood of the turning points z_k, $k = 1,3$, $\varepsilon(z_k) = \{z \in \mathbb{C}\colon\ |z - z_k| \le |t|^{-\frac{2}{3}+\varepsilon}\}$, $\varepsilon > 0$. This estimate is given by the following theorem.

THEOREM 7.2. *Let the PII functions v and v' in (7.3.5) are such that the following estimates hold:*

$$\begin{aligned} &|v| < C_1, \quad |v'| < C_2, \\ &|ib_2 \pm b_1| = \left|4ivz \pm 2\mathrm{e}^{-i\varphi}(v' + v/3t)\right| > C_3, \\ &|z - z_k| > C_4|t|^{-\frac{2}{3}+\varepsilon}, \quad \varepsilon > 0, \quad k = 1,3, \end{aligned} \tag{7.3.14}$$

with C_j, $j = 1,2,3,4$, independent of t.

Then the correction function χ satisfying the integral equation (7.3.19), satisfies

$$\|\chi(z,t) - I\| < Ct^{-1}|z - z_k|^{-3/2}, \tag{7.3.15}$$

where the constant C depends only of C_j, $j = 1,2,3,4$.

PROOF. Introduce the difference

$$R(z,t) = T^{-1}T_z - \operatorname{diag}\left(T^{-1}T_z\right) = \begin{pmatrix} 0 & r_1(z,t) \\ r_2(z,t) & 0 \end{pmatrix}, \tag{7.3.16}$$

where the matrix elements r_1, r_2 are given by

$$r_1 = \frac{\mu - b_3}{2\mu}\frac{\partial}{\partial z}\left(\frac{\mu + b_3}{ib_2 + b_1}\right), \quad r_2 = \frac{\mu - b_3}{2\mu}\frac{\partial}{\partial z}\left(\frac{\mu + b_3}{ib_2 - b_1}\right). \tag{7.3.17}$$

We will look for the correction function $\chi(z,t)$ in the form of the product

$$\chi(z,t) = \left(I + T_1(z,t)\right)\left(I + \chi_1(z,t)\right), \tag{7.3.18}$$

where

$$T_1(z,t) = \frac{1}{3t\mu}\begin{pmatrix} 0 & -r_1 \\ r_2 & 0 \end{pmatrix};$$

the function $\chi_1(z,t)$ satisfies the integral equation,

$$\chi_1(z,t) = \int\limits_{\mathcal{C}(z)} e^{\int\limits_{\zeta}^{z} \Lambda(\nu)\,d\nu} \big(I + T_1(\zeta)\big)^{-1} \rho(\zeta,t)\Big(I + \chi_1(\zeta,t)\Big) e^{-\int\limits_{\zeta}^{z} \Lambda(\nu)\,d\nu}\, d\zeta,$$

where

$$\rho(z,t) = T_1 \operatorname{diag}\left(T^{-1}T_z\right) - T^{-1}T_z T_1 - (T_1)_z. \tag{7.3.19}$$

Possible singularities of the kernel of the Volterra integral equation (7.3.19) can be located at the turning points z_k, at the zeros of $b_1 \pm ib_2$ and at infinity. By the assumption (7.3.14), we have excluded the finite neighborhoods of the zeros of $b_1 \pm ib_2$ from consideration. In the neighborhoods of the other possible singularities, we find

$$\begin{aligned} &\operatorname{diag}\left(T^{-1}T_z\right) = O(z^{-3}), \quad R(z) = O(z^{-2}), \quad T_1 = O(t^{-1}z^{-4}), \\ &\rho(z,t) = O(t^{-1}z^{-5}), \quad z \to \infty, \end{aligned} \tag{7.3.20}$$

and

$$\begin{aligned} &\operatorname{diag}\left(T^{-1}T_z\right) = \frac{\mu_z}{2\mu} I + O(\mu^{-1}), \quad R(z) = O(\mu^{-2}), \quad T_1 = O(t^{-1}\mu^{-3}), \\ &\rho(z,t) = O(t^{-1}\mu^{-5}), \quad z \to z_k, \quad k = 1,3. \end{aligned} \tag{7.3.21}$$

Due to (7.3.11), we have

$$\begin{aligned} &\operatorname{Re} \int_{\zeta}^{z} \mu(\nu)\,d\nu \le 0, \quad z \in \Omega_j, \quad \zeta \in \mathcal{C}_1(z), \\ &\operatorname{Re} \int_{\zeta}^{z} \mu(\nu)\,d\nu \ge 0, \quad z \in \Omega_j, \quad \zeta \in \mathcal{C}_2(z). \end{aligned} \tag{7.3.22}$$

By definition, $\Lambda = -\frac{3}{2}t\mu\sigma_3 - \operatorname{diag}(T^{-1}T_z)$. Thus, taking into account that the scalar components of the integrals of Λ cancels each other, that the σ_3 component is integrable, and that $I + T_1(z,t)$ is uniformly bounded in the asymptotic domain (7.3.14), the integral equation (7.3.19) yields the estimate,

$$\|\chi_1(z,t)\| \le \int_{\mathcal{C}_1(z)\cup\mathcal{C}_2(z)} \|\rho(\zeta,t)\| \cdot \|I + \chi_1(\zeta,t)\|\, |d\zeta|,$$

Therefore, if

$$M(z,t) \equiv \int\limits_{\mathcal{C}_1(z)\cup\mathcal{C}_2(z)} \|\rho(\zeta,t)\|\, |d\zeta| < 1$$

for $z \in \Omega_j/\varepsilon(z_k)$, the integral operator (7.3.19) is contracting. This guarantees the existence of χ_1, as well as the estimate

$$\|\chi_1(z,t)\| \le e^{M(z,t)} - 1. \tag{7.3.23}$$

The estimates for $\rho(z,t)$ in (7.3.20) and (7.3.21) provide the estimate for M,

$$M(z,t) = O(t^{-1}\mu^{-3}).$$

This estimate, equation (7.3.23), and the estimates for $T_1(z,t)$ in (7.3.20), (7.3.21), yield the estimate (7.3.15) for $\chi(z,t)$. □

4. Local solutions near turning points

Consider a small neighborhood of each of the two turning points z_1 and z_3 and introduce the local variables

$$\xi = z - z_1, \quad \tau = z - z_3.$$

It is natural to expect, that in terms of a "stretched" variable ζ, the system (7.3.5) reduces locally to a model problem. Since the turning points are simple and isolated, the general theory of WKB-approximations (see, for example, [**W1**]) implies that this model problem should be solvable in terms of Airy functions.

We will present the relevant asymptotic analysis only for one point, say z_1. In the generic case the function $b_3(z)$ is nonzero at this point. For the solution $\Psi(z)$ we perform the gauge transformation

$$\Psi(z_1 + \xi) = \mathcal{V}(\xi)\tilde{\Phi}(\xi), \tag{7.4.1}$$

where

$$\mathcal{V}(\xi) = V_0 \cdot \big(I + V_1(\xi)\big), \quad V_0 = \frac{1}{\sqrt{2}} \begin{pmatrix} 1 & -\dfrac{c_3}{ic_2 + c_1} \\ \dfrac{c_3}{ic_2 - c_1} & 1 \end{pmatrix}, \tag{7.4.2}$$

$$V_1(\xi) = \begin{pmatrix} \dfrac{c_3(b_3 - c_3) - c_2(b_2 - c_2)}{4c_3^2} & \dfrac{(c_2'c_c - c_3'c_3)^2\xi}{8ic_1c_2'(ic_2 + c_1)} \\ -\dfrac{ic_2(b_2 - c_2)}{2c_3(ic_2 - c_1)} & -\dfrac{c_3(b_3 - c_3) - c_2(b_3 - c_2)}{4c_3^2} \end{pmatrix},$$

$$c_k := b_k(z_1), \quad c_k' := \frac{\partial b_k}{\partial z}\bigg|_{z_1}. \tag{7.4.3}$$

The matrix function $\tilde{\Phi}(\xi)$ satisfies the equation

$$\frac{\partial \tilde{\Phi}}{\partial \xi} = \frac{3t}{2}\tilde{\mathsf{B}}\tilde{\Phi},$$

where

$$\tilde{\mathsf{B}} = \mathcal{V}^{-1}\big(b_3\sigma_3 + b_2\sigma_2 + b_1\sigma_1\big)\mathcal{V} - \frac{2}{3t} \cdot \mathcal{V}^{-1}\frac{\partial \mathcal{V}}{\partial \xi}.$$

Ignoring the terms of order $O(\xi^2)$, we find the model equation

$$\frac{\partial \tilde{\Phi}}{\partial \xi} = \frac{3t}{2} \begin{pmatrix} \tilde{p} & \tilde{q} \\ \tilde{r} + \tilde{s}\zeta & -\tilde{p} \end{pmatrix} \tilde{\Phi} = \frac{3t}{2}\tilde{\mathsf{B}}\tilde{\Phi}, \tag{7.4.4}$$

where

$$\tilde{p} = \frac{2}{3t} \cdot \frac{1}{4c_3^2}(c_2' - c_3' c_3),$$

$$\tilde{q} = -2(ic_2 - c_1) - \frac{2}{3t} \cdot \frac{(c_2' c_2 - c_3' c_3)^2}{8ic_1 c_2'(ic_2 + c_1)},$$

$$\tilde{r} = \frac{2}{3t} \cdot \frac{ic_1 c_2'}{2c_3(ic_2 - c_1)}, \quad s = -\frac{c_3' c_3 + c_2' c_2}{ic_2 - c_1}. \tag{7.4.5}$$

After the rescaling

$$\xi = \beta\zeta, \quad \beta = \left(\frac{3t}{2}\right)^{-2/3} (\tilde{q}s)^{-1/3} \tag{7.4.6}$$

equation (7.4.4) takes standard form

$$\frac{\partial \Phi}{\partial \zeta} = \begin{pmatrix} p & q \\ r + \frac{\zeta}{q} & -p \end{pmatrix} \Phi, \tag{7.4.7}$$

where

$$p = \frac{3t}{2}\beta p, \quad q = \frac{3t}{2}\beta q, \quad r = \frac{3t}{2}\beta r.$$

Equation (7.4.7) can be solved in terms of Airy functions

$$\Phi(\zeta) = \begin{pmatrix} 1 & 0 \\ \frac{p}{q} & \frac{1}{q} \end{pmatrix} \cdot \begin{pmatrix} \mathrm{Ai}(\zeta - \zeta_0) & \mathrm{Bi}(\zeta - \zeta_0) \\ \mathrm{Ai}'(\zeta - \zeta_0) & \mathrm{Bi}'(\zeta - \zeta_0) \end{pmatrix}, \tag{7.4.8}$$

where

$$\zeta_0 = -rq - p^2.$$

The well-known asymptotic expansions for the Airy functions (see [**BE**]) lead to normalization conditions for the local solutions at infinity

$$\Phi^{(k)}(\zeta) = \begin{pmatrix} 1 & 0 \\ -\frac{p}{q} & \frac{1}{q} \end{pmatrix} \exp\Big\{-\frac{1}{4}\log\zeta \cdot \sigma_3\Big\}(\sigma_3 + \sigma_1)$$

$$\times (I + O(\zeta^{-3/2})) \exp\Big\{\frac{2}{3}\zeta^{3/2} \cdot \sigma_3\Big\}, \quad k \in \mathbb{Z}, \tag{7.4.9}$$

$$\zeta \to \infty, \quad z \in \omega_k = \Big\{\zeta \in \mathbb{C} : \arg\zeta \in \Big(\frac{\pi}{3}(2k-3), \frac{\pi}{3}(2k+1)\Big)\Big\}.$$

The above canonical solutions $\Phi^{(k)}$ are related by the following normalized Stokes matrices

$$\Phi^{(k+1)}(\zeta) = \Phi^{(k)}(\zeta)\mathsf{G}^{(k)}, \quad \mathsf{G}^{(k+2)} = \mathsf{G}^{(k)},$$

$$\mathsf{G}^{(1)} = \begin{pmatrix} 1 & -i \\ 0 & 1 \end{pmatrix}, \quad \mathsf{G}^{(2)} = \begin{pmatrix} 1 & 0 \\ -i & 1 \end{pmatrix}.$$

In order to estimate the terms which were discarded in the above derivation we define the correction function $\chi(z)$,

$$\tilde{\Phi} = \chi(\zeta)\Phi(\zeta).$$

THEOREM 7.3. *The following estimate is true for the correction function* $\chi(z)$,

$$\|\chi(\zeta) - I\| < Ct^{-1/3}|\zeta|^2, \quad |\zeta| < t^{1/6-\delta}, \tag{7.4.10}$$

where $0 < \delta < 1/6$ *and the constant* C *depends only on* c_k, c_k', $k = 1, 2, 3$, *in (7.4.2).*

Proof We consider the integral equation (see (7.3.19))

$$\chi(\zeta) = I + \int_0^\zeta \Phi(\zeta)\Phi^{-1}(\nu)\mathsf{R}(\nu)\chi(\nu)\Phi(\nu)\Phi^{-1}(\zeta)d\nu,$$

where the kernel is the difference between the exact and the approximate matrix coefficients,

$$\mathsf{R}(\zeta) = \frac{3t}{2}\big(\tilde{B}(z) - B(z)\big) = O(t\zeta^2) = O(t^{-1/3}z^2).$$

Using the exact form of $\Phi(\zeta)$ given by (7.4.8), one can prove that in the class of functions with power-like growth, the norm of the kernel does not exceed $O(t^{-1/3}z^2)$. This implies that the integral operator is contracting for $|\zeta| < t^{1/6-\delta}$, so the estimate (7.4.10) follows.

5. Approximation of the monodromy data

We will now obtain the asymptotics of the Stokes matrices of the system (7.3.5), by constructing matched asymptotic expansions for appropriate global solutions. The WKB-approximation (7.3.10) is valid everywhere in the canonical domains except at the neighborhood of the branch points z_k. It can be matched at the boundaries of these neighborhoods with the local solutions (7.4.8), (7.4.9), expressed in terms of the Airy functions. Note that all z_k are *simple turning points* since we restrict ourselves to the nondegenerate case of $\phi \neq \frac{\pi n}{3}$, and the assumptions (7.3.3), (7.3.9) hold.

It will be shown below, that all Stokes multipliers entering S_j, are proportional to

$$\exp\Big\{\frac{3t}{2}\oint_a \mu(z)\,dz\Big\}, \quad \exp\Big\{\frac{3t}{2}\oint_b \mu(z)\,dz\Big\},$$

where a and b are the basic cycles shown in Figure 7.3. Since the Stokes multipliers are independent of t, one can pose the *isomonodromy condition*, which to the leading

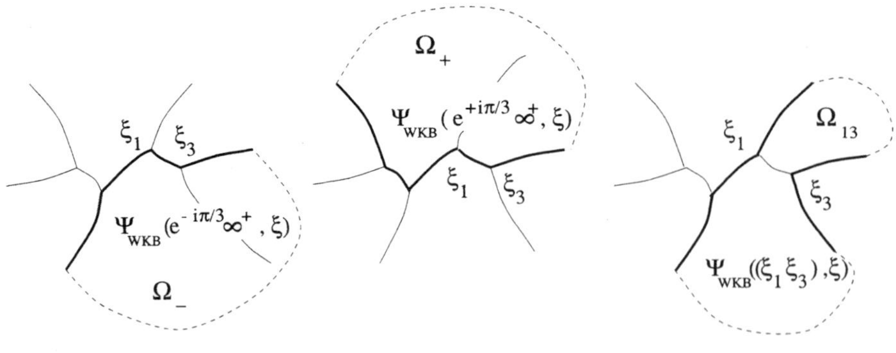

FIGURE 7.5. The domains for the WKB - solutions.

order is equivalent to the following:

$$\operatorname{Im} \mathcal{J}_a = \operatorname{Im} \oint_a \sqrt{(z^2 - z_1^2)(z^2 - z_3^2)}\, dz = 0,$$

$$\operatorname{Im} \mathcal{J}_b = \operatorname{Im} \oint_b \sqrt{(z^2 - z_1^2)(z^2 - z_3^2)}\, dz = 0. \tag{7.5.1}$$

These constraints and equation (7.3.8) can be treated as equations for $A(\phi)$; the existence of a solution follows from the invertibility of the Abelian map (see [**N2**])

$$(z_1, z_3) \mapsto (\mathcal{J}_a, \mathcal{J}_b).$$

In fact, equations (7.5.1), under the simple transformations

$$\mathcal{J}_a = J', \quad \mathcal{J}_b = -\frac{1}{8} J, \quad A(\phi) = \frac{1}{4} - D_\infty, \tag{7.5.2}$$

become Boutroux equation (7.3.5).

Due to the symmetry (7.2.10) it is sufficient to consider only the two cases of $\frac{2\pi}{3} < \phi < \pi$ and $\pi < \phi < \frac{4\pi}{3}$. Furthermore, using the symmetry (7.2.11), it follows that it is sufficient to consider only one of these sectors.

Equation (7.3.1) contains the two independent parameters s_1 and s_3, thus it is convenient to calculate the product

$$\mathrm{S}_0 \mathrm{S}_1 = \Psi_0^{-1}(z) \Psi_2(z), \tag{7.5.3}$$

and to match the approximate solutions near the turning points z_1 and z_3. The relevant local solutions are $\Phi_1(\zeta)$ and $\Phi_3(\zeta)$, and the WKB-approximations are $\Psi_{WKB}(\mathrm{e}^{\pm \pi i/3}\infty^+, z)$ and $\Psi_{WKB}(z_1, z)$, valid in the domains shown on Figure 7.5. For the sector $\frac{2\pi}{3} < \phi < \pi$, one can decompose the product (7.5.3) into the following factors:

$$
\begin{aligned}
\mathrm{S}_0\mathrm{S}_1 &= \begin{pmatrix} 1-s_1s_3 & -s_3 \\ s_1 & 1 \end{pmatrix} \simeq \Big(\Psi_0^{-1}(z)\Psi_{WKB}\big(e^{-\pi i/3}\infty^+, z_3\big)\Big) \\
&\times\Big(\Psi_{WKB}^{-1}\big(e^{\pm\pi i/3}\infty^+, z_3\big)\Phi\big(e^{-\pi i/3}\infty^+, z_3\big)\Big) \\
&\times\Big(\Phi^{-1}\big(e^{-\pi i/3}\infty^+, z_3\big)\Phi\big(\infty^+, z_3\big)\Big) \\
&\times\Big(\Phi^{-1}\big(\infty^+, z_3\big)\Phi\big(z_3, z_1\big)\Big)\Big(\Phi^{-1}\big(z_3, z_1\big)\Psi_{WKB}\big(z_1, z_3\big)\Big) \\
&\times\Big(\Psi_{WKB}^{-1}\big(z_1, z_3\big)\Phi\big(z_1, z_3\big)\Big)\Big(\Phi_1^{-1}\big(z_1, z_3\big)\Phi\big(e^{\pi i/3}\infty^+, z_1\big)\Big) \\
&\times\Big(\Phi^{-1}\big(e^{\pi i/3}\infty^+, z_1\big)\Psi_{WKB}\big(e^{\pi i/3}\infty^+, z_1\big)\Big) \\
&\times\Big(\Psi_{WKB}^{-1}\big(e^{\pi i/3}\infty^+, z_1\big)\Psi_2(z)\Big).
\end{aligned} \tag{7.5.4}
$$

Here we have used the notations

$$
\begin{aligned}
\Phi\big(z_1, z_3\big) &:= \mathcal{V}_1^{-1}(\zeta)\Big|_{z\to z_1,\, z\in\Omega_+} \cdot \mathcal{V}_1(\zeta)\Big|_{z\to z_1,\, z\in\Omega_{13}}, \\
\Phi\big(z_3, z_1\big) &:= \mathcal{V}_3^{-1}(\tau)\Big|_{z\to z_3,\, z\in\Omega_-} \cdot \mathcal{V}_3(\tau)\Big|_{z\to z_3,\, z\in\Omega_{13}}, \\
\Phi\big(e^{\pi i/3}\infty^+, z_1\big) &:= \mathcal{V}_1(\zeta)\Phi_1(\zeta)\Big|_{z\to z_1,\, z\in\Omega_+}, \\
\Phi\big(e^{\pi i/3}\infty^+, z_3\big) &:= \mathcal{V}_3(\tau)\Phi_3(\tau)\Big|_{z\to z_3,\, z\in\Omega_-}.
\end{aligned}
$$

Figure 7.6 illustrates the domains of the right-hand side in equation (7.5.4). It shows a fragment of the Stokes graph near the cut $[z_1, z_3]$. On the edges outside the vertices the fundamental solutions of (7.3.12), $\Psi_{WKB}(z_i, z_j)$, are marked, which are uniquely normalized on the corresponding interval

$$(z_i, z_j), \quad i, j = 1, 3, \infty, \quad z_\infty = e^{i\pi k/3}\infty^+, \ k = 0, \pm 1.$$

On the edges near the vertices the local fundamental solutions $\Phi(z_i, z_j)$ are marked, which are normalized in terms of the Airy functions (see (7.4.9)). The filled pieces show overlapping domains where both interior (Φ) and exterior (Ψ) approximations are valid. Substituting the expressions (7.3.12) and (7.4.8) into equation (7.5.4), and taking into account the following choice for the branch of the radicals in (7.4.6)

$$\frac{3t}{2}\beta_{1,3}\mu(\zeta) = -z_{1,3}^{-1/2}, \quad \zeta = z - z_1 \to 0, \quad \tau = z - z_3 \to 0,$$

we find the following:

$$S_0S_1 \simeq \exp\Big\{ \int_{z_3+\tau}^{z} \Lambda(z)dz + i\Big(\frac{4}{3}\lambda^3 + x\lambda\Big)\sigma_3$$

$$+\frac{2}{3}\tilde{z}_3^{3/2}\cdot\sigma_3 - \frac{1}{4}\log\tilde{z}_3\cdot I\Big\}\Big|_{z\to\infty,\ \tau\in\Omega_-}\cdot\begin{pmatrix} 0 & \frac{id_2-d_1}{id_3} \\ -\frac{id_3}{id_2-d_1} & 1\end{pmatrix}$$

$$\times\exp\Big\{ \int_{z_1+\zeta}^{z_3+\tau} \Lambda(z)dz - \frac{2}{3}z_3^{3/2}\cdot\sigma_3 + \frac{1}{4}\log z_3\cdot I \tag{7.5.5}$$

$$+\frac{2}{3}z_1^{3/2}\cdot\sigma_3 - \frac{1}{4}\log z_1\cdot I\Big\}\Big|_{\zeta,\tau\in\Omega_{13}}\begin{pmatrix} 1 & 0 \\ -\frac{ic_3}{ic_2-c_1} & 1\end{pmatrix}$$

$$\times\exp\Big\{ \int_{z_1+\tilde\zeta}^{z} \Lambda(z)dz - i\Big(\frac{4}{3}\lambda^3 + x\lambda\Big)\sigma_3 - \frac{2}{3}\tilde{z}_1^{3/2}\cdot\sigma_3$$

$$-\frac{1}{4}\log\tilde{z}_1\cdot I\Big\}\Big|_{z\to\infty,\ \tilde\zeta\in\Omega_-},$$

where

$$c_k = b_k(z_1),\quad d_k = b_k(z_3),\quad k = 1, 2, 3, \tag{7.5.6}$$
$$\beta_1 = \beta(t, c_k, c_k'),\quad \beta_3 = \beta(t, d_k, d_k').$$

Consider now the entries of the exponent in (7.5.5). First note that we have only diagonal matrices proportional to σ_3, since the scalar components appear under an integral over a contour which is homologous to zero. Second, the σ_3-component satisfies the obvious constraint

$$\int_{-z_i}^{-z_j} \Lambda(z)dz = -\int_{-z_i}^{-z_j} \Lambda(z)dz.$$

Denote

$$\Big(\int_{z_1+\zeta}^{z} \Lambda(z)dz + i\Big(\frac{4}{3}\lambda^3 + x\lambda\Big) + \frac{2}{3}z_3^{3/2}\Big)\Big|_{z\to\infty,\ \tilde\zeta\to 0} = \frac{1}{4}J_2,$$

$$\Big(\int_{z_1+\zeta}^{z+\tau} \Lambda(z)dz - \frac{2}{3}z_3^{3/2} + \frac{2}{3}z_1^{3/2}\Big)\Big|_{z\to\infty,\ \tilde\zeta\to 0} = -\frac{1}{2}J_1.$$

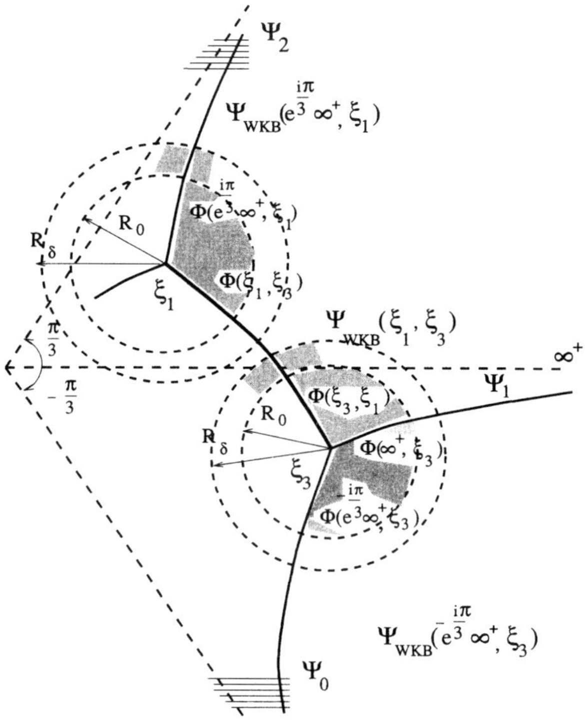

FIGURE 7.6. The Stokes lines are shown in solid, the thick line denote the cut, and the dotted rays denote the Stokes lines asymptotes. The discs enclosed by the inner dotted circles of radii $R_0 = O(t^{-2/3})$, are the domains where the WKB asymptotics fails, whereas the annular regions denote the regions of asymptotic matching; $R_\delta = R_0 t^\delta$, $\delta > 0$.

Finally, taking into account the second order correction $\frac{1}{t}B(\phi, t)$, we find in the sector $\frac{2}{3}\pi < \phi < \pi$, the following leading terms for the Stokes multipliers,

$$
\begin{aligned}
1 - s_1 s_3 &= -\frac{c_3}{ic_2 - c_1} \cdot \frac{id_2 - d_1}{d_3} \exp(J_1), \\
s_3 &= -\frac{id_2 - d_1}{id_3} \exp\Big(\frac{1}{2}J_2\Big), \\
s_1 &= -\Big(\frac{id_3}{id_2 - d_1} + \frac{ic_2}{ic_2 - c_1} \exp J_1\Big) \exp\Big(-\frac{1}{2}J_2\Big).
\end{aligned} \tag{7.5.7}
$$

Similarly, in the sector $\pi < \phi < \frac{4}{3}\pi$,

$$
\begin{aligned}
1 - s_1 s_3 &= -\frac{d_3}{id_2 - d_1} \cdot \frac{ic_2 - c_1}{c_3} \exp(J_1), \\
s_1 &= -\frac{d_3}{id_2 - d_1} \exp\Big(-\frac{1}{2}J_2\Big), \\
s_1 &= -\Big(\frac{id_2 - d_1}{id_3} + \frac{ic_2 - c_1}{ic_2} \exp J_1\Big) \exp\Big(\frac{1}{2}J_2\Big).
\end{aligned} \tag{7.5.8}
$$

6. Uniformization by theta-functions. Justification of the Boutroux ansatz

The elliptic integrals of the phase function

$$\Lambda(z) = -\frac{3t}{2}\mu - \frac{1}{2}\frac{\partial b_2}{\partial z}\Big(\mu(\mu - b_3)\Big)^{-1},$$

where

$$\mu(z) = 4i\sqrt{(z^2 - z_1^2)(z^2 - z_3^2)} = 4iw(z)$$

can be expressed through the standard elliptic integrals

$$I_k(z, z_0) = \int \frac{(z - z_0)^k}{w(z)} dz.$$

The decomposition here goes via integration by parts

$$\begin{aligned}\int \Lambda(z)dz &= -2it\Big(z - \frac{z_1^2 + z_3^2}{z}\Big)w(z)\\ &+\Big(\frac{z_0}{2} - 4itz_1^2 z_3^2\Big)I_0(z) - \frac{1}{4}\log\frac{z - z_0}{z + z_0}\\ &+\frac{1}{4}w(z_0^+)\big(I_{-1}(z, z_0) - I_{-1}(z, -z_0)\big)\\ &+2it(z_1^2 + z_3^2)z_1^2 z_3^2 I_{-2}(z, 0),\end{aligned} \tag{7.6.1}$$

where

$$z_0 \simeq \frac{i}{2}e^{-i\phi}\frac{v'}{v},$$

and z_0^+ is the lift of z_0 on the upper sheet of $\Gamma(t)$.

Parametrization via Jacobi theta-functions

As usual the Jacobi theta-functions $\Theta(z|\tau)$ are a convenient tool for uniformizing analytic objects over an elliptic curve. We decompose the following differentials which are of single-valued on Γ,

$$\frac{dz}{(z - z_0)w(z)}, \quad d\log\frac{\Theta(z_+ + \nu|\tau)}{\Theta(z_- + \nu|\tau)},$$

where

$$z_\pm = \frac{1}{\omega_a}\int_{z_0^\pm}^{z} \frac{dz}{w(z)}, \quad \omega_{a,b} = \oint_{a,b} \frac{dz}{w(z)},$$

$$\tau = \frac{\omega_b}{\omega_a}, \quad \nu = \frac{1 + \tau}{2},$$

into Laurent series near one of their two simple poles

$$z_0^+ = \big(z_0, w(z_0^+)\big) \quad \text{or} \quad z_0^- = \big(z_0, w(z_0^-)\big).$$

This yields

$$\frac{dz}{(z-z_0)w(z)} = \frac{1}{w(z_0^+)}d\log\frac{\Theta(z_+ + \nu|\tau)}{\Theta(z_- + \nu|\tau)} \tag{7.6.2}$$

$$-\frac{1}{w(z_0^+)}\left\{\frac{1}{2}w'(z_0^+) + \frac{1}{\omega_a}\left(\frac{\Theta''(\nu|\tau)}{2\Theta'(\nu|\tau)} - \frac{\Theta'(z_-^+ + \nu|\tau)}{\Theta(z_-^+ + \nu|\tau)}\right)\right\}du,$$

where

$$z_-^+ = z_- - z_+, \quad du = \frac{dz}{w(z)}.$$

Note that all differentials in (7.6.2) are single-valued, i.e., they are Abelian differentials of the third kind. Moreover, the value of the right-hand side in (7.6.2) does not depend on the choice of the contours $(z_0^\pm, z)$, defined so far up to the additions of $a-$ and $b-$ cycles. In fact, since theta-functions are 1-periodic, it is sufficient to consider (7.6.2) under other shifts of τ. This yields

$$d\log\frac{\Theta(z_+ + \nu + m_+\tau|\tau)}{\Theta(z_- + \nu + m_-\tau|\tau)} + \frac{1}{\omega_a}\frac{\Theta'(z_-^+ + \nu + (m_- - m_+)\tau|\tau)}{\Theta'(z_-^+ + \nu + (m_- - m_+)\tau|\tau)}du$$

$$= d\log\frac{\Theta(z_+ + \nu|\tau)}{\Theta(z_- + \nu|\tau)} + 2\pi i(m_-dz_- - m_+dz_+)$$

$$+\frac{1}{\omega_a}\frac{\Theta'(z_-^+ + \nu|\tau)}{\Theta'(z_-^+ + \nu|\tau)}du - \frac{2\pi i}{\omega_a}(m_- - m_+)\,du, \quad dz_\pm = \frac{1}{\omega_a}du,$$

which proves the single-valuedness of (7.6.2).

However, the integral of (7.6.2) over an arbitrary contour is no longer a single-valued function on Γ; it has logarithmic singularities at the points $z_0^\pm$. Taking into account that

$$w(-z_0^+) = w(z_0^+) = -\frac{1}{4i}b_3(z_0), \quad w'(z_0^+) = -w'(-z_0^+),$$

$$\frac{\Theta'(-z+\nu|\tau)}{\Theta(-z+\nu|\tau)} = -2\pi i - \frac{\Theta'(z+\nu|\tau)}{\Theta(z+\nu|\tau)},$$

$$\frac{\Theta''(\nu|\tau)}{\Theta'(\nu|\tau)} = -2\pi i, \quad \nu = \frac{1+\tau}{2}, \tag{7.6.3}$$

we find

$$I_{-1}(z,z_0)\Big|_a = -\frac{w'(z_0^+)}{2w(z_0^+)}\omega_a + \frac{1}{w(z_0^+)}\left(\pi i + \frac{\Theta'(z_-^+ + \nu|\tau)}{\Theta(z_-^+ + \nu|\tau)}\right),$$

$$I_{-1}(z,z_0)\Big|_b = \frac{2\pi i}{w(z_0^+)}z_-^+ + \tau I_{-1}(z,z_0)\Big|_a. \tag{7.6.4}$$

The difference z_-^+ can be made single-valued by defining its contour (z_0^-, z_0^+) on a universal covering $\tilde{\Gamma}$, which emerges by cutting the surface Γ over $a-$ and $b-$ cycles. For this construction one must assign to z_-^+, at the point $z_0 = ie^{-\phi}v'/2v$ when it crosses the cuts, the value $z_-^+ + 2k + 2\tau l$, where the integers k and l characterize a homotopy class of the trajectory of z_0. Similarly, the integrals

$$L(z_0) = I_{-1}(z,z_0)\Big|_a \quad \text{and} \quad M(z_0) = I_{-1}(z,z_0)\Big|_b$$

should be replaced by

$$L(z_0) - \frac{4\pi i}{w(z_0^+)}l \quad \text{and} \quad M(z_0) + \frac{4\pi i}{w(z_0^+)}k,$$

respectively.

Making use of (7.6.4) it is easy to find analogous expressions for the integrals I_{-2},

$$I_{-2}(z, z_0)\Big|_a = \frac{z_1^2 + z_3^2}{2z_1^2 z_3^2}\omega_a + \frac{2}{z_1^2 z_3^2\omega_a}\left\{\frac{\Theta''(\frac{1}{2}|\tau)}{\Theta(\frac{1}{2}|\tau)} - \left(\frac{\Theta'(\frac{1}{2}|\tau)}{\Theta(\frac{1}{2}|\tau)}\right)^2\right\}$$

$$I_{-2}(z, z_0)\Big|_b = \frac{4\pi i}{z_1^2 z_3^2\omega_a} + \tau I_{-2}(z, z_0)\Big|_a. \tag{7.6.5}$$

Taking into account the constraints

$$z_1^2 + z_3^2 = -\frac{1}{2}e^{i\phi}, \quad (z_1^2 - z_3^2)^2 = A(\phi) + \frac{1}{t}B(\phi, t),$$

$$I_{-1}(z, -z_0)\Big|_{a,b} = I_{-1}(z, z_0)\Big|_{a,b},$$

$$I_{-l}(z_3, -z_0) - I_{-1}(z_1, z_0) = -\frac{1}{2}g_k I_{-l}(z, -z_0)\Big|_a,$$

$$I_{-l}(\infty, -z_0) - I_{-1}(z_3, z_0) = -\frac{1}{2}g_k I_{-l}(z, -z_0)\Big|_b, \quad l = 0, 1, 2,$$

where

$$k = 3, 1, \quad g_3 = 1, \quad g_1 = -1, \quad \arg x \in \begin{cases} \left(\frac{2\pi}{3}, \pi\right), & k = 3, \\ \left(\pi, \frac{4\pi}{3}\right), & k = 1, \end{cases}$$

and noting that a and b are simple cycles (see Figure 7.3), we find the following theta-function parametrization of phase integrals in (7.5.7):

$$J_1 = g_k F_1 + \frac{1}{2}\log\left(\frac{z_3 - z_0}{z_3 + z_0} \cdot \frac{z_1 + z_0}{z_1 - z_0}\right),$$

$$J_2 = \pi i z_-^+ + \frac{4\pi e^{i\phi}}{\omega_a}t + \tau F_1 + \log\frac{z_3 - z_0}{z_3 + z_0}, \tag{7.6.6}$$

where

$$F_1 = -6it\mathcal{J}_a + \frac{\Theta'\left(\frac{1}{2}(z_-^+ + 1) + \frac{\tau}{4}|\tau\right)}{\Theta\left(\frac{1}{2}(z_-^+ + 1) + \frac{\tau}{4}|\tau\right)},$$

$$\mathcal{J}_a = \oint_a w(z)dz = -\frac{1}{6}\left(A(\phi) + \frac{1}{t}B(\phi, t)\right)\omega_a$$

$$+\frac{e^{i\phi}}{3\omega_a}\left\{\frac{\Theta''(\frac{1}{2}|\tau)}{\Theta(\frac{1}{2}|\tau)} - \left(\frac{\Theta'(\frac{1}{2}|\tau)}{\Theta(\frac{1}{2}|\tau)}\right)^2\right\}.$$

Here we have made use of the constraint

$$\left(\frac{z_0}{2} - \frac{1}{4}w'(z_0^+)\right)\omega_a + \frac{\Theta'(z_- + \nu|\tau)}{2\Theta(z_-^+ + \nu|\tau)} - \frac{\Theta''(\nu|\tau)}{4\Theta'(\nu|\tau)} = \frac{\Theta'(\frac{1}{2}(z_-^+ + 1) + \frac{\tau}{4}|\tau)}{\Theta(\frac{1}{2}(z_-^+ + 1) + \frac{\tau}{4}|\tau)}.$$

Due to the identities

$$\frac{ic_2 - c_1}{c_3} = \Big(\frac{ic_2 - c_1}{ic_2 + c_1}\Big)^{1/2} = \Big(\frac{z_1 + z_0}{z_1 - z_0}\Big)^{1/2}, \qquad \frac{id_2 - d_1}{d_3} = \Big(\frac{z_3 + z_0}{z_3 - z_0}\Big)^{1/2}$$

the equations (7.5.7) and (7.5.8) take the form

$$\log(s_1 s_3 - 1) = -F_1 + o(1), \tag{7.6.7}$$

$$2g_k \log s_k = \pi i z_-^+ + \frac{4\pi e^{i\phi}}{\omega_a} t + \tau F_1 - i\frac{\pi\tau}{2} + o(1).$$

These equations are in fact asymptotic expressions of the integrals of motion for PII equation as $t \to \infty$,

$$2g_k \log s_k + \tau \log(s_1 s_3 - 1) = \pi i z_-^+ + \frac{4\pi e^{i\phi}}{\omega_a} t - i\frac{\pi\tau}{2} + o(1). \tag{7.6.8}$$

These expressions yield an asymptotic formula for the logarithmic derivative of the PII solution $v(t)$,

$$z_0 = \frac{i}{2} e^{i\phi} \frac{v'}{v} = -z_3 \,\mathrm{sn}\left(2ie^{i\phi} z_1 t + \psi | \varkappa\right), \tag{7.6.9}$$

$$\psi = \frac{\omega_a z_1}{\pi i} g_k \log s_k + \frac{\omega_b z_1}{2\pi i} \log(s_1 s_3 - 1) + \frac{\omega_b z_1}{4} + \psi_1 + o(1),$$

where

$$\varkappa = \frac{z_3}{z_1}, \qquad \mathrm{sn}\left(\psi_1 | \varkappa\right) = \frac{z_1}{z_3}.$$

An exact phase shift of the Boutroux ansatz

First, note that system (7.3.8) yields

$$e^{-2i\phi} (v')^2 = \big(v^2 - (z_1 + z_3)^2\big)\big(v^2 - (z_1 - z_3)^2\big) + O(t^{-1}),$$

so that the elliptic function ansatz for $v(t)$ has the form

$$v(t) = (z_1 - z_3) \,\mathrm{sn}\left(e^{i\phi}(z_1 + z_3)(t - t_0) | \tilde\varkappa\right), \tag{7.6.10}$$

where $\tilde\varkappa = \dfrac{z_1 - z_3}{z_1 + z_3}$ and t_0 is an arbitrary constant. Using the modular transformation (see [**BE**]) $\varkappa \mapsto \tilde\varkappa = (1 - \varkappa)/(1 + \varkappa)$,

$$\lambda = i(1 + \varkappa) z \big((1 - \zeta^2)(1 - \varkappa^2\zeta^2)\big)^{-1/2},$$

we obtain

$$z_-^+ = \frac{2}{\omega_a} \int_{-z_3}^{z_0^+} \frac{dz}{w(z)} = -\frac{2}{\omega_a z_1} \int_{-1}^{-z_0^+/z_3} \big((1 - \zeta^2)(1 - \varkappa^2\zeta^2)\big)^{-1/2} d\zeta$$

$$= \frac{2i}{\omega_a} \frac{1}{z_1 + z_3} \int_{\infty}^{-iz_0(z_1+z_3)/w(z_0^+)} \big((1 - z^2)(1 - \varkappa^2 z^2)\big)^{-1/2} dz.$$

Hence the phase integrand at the point z_0^+ yields

$$\frac{iz_0}{w(z_0^+)} = -\frac{1}{z_1 + z_3} \,\mathrm{sn}\left(2e^{i\phi}(z_1 + z_3)t + 2\chi | \tilde\varkappa\right),$$

where

$$\chi = i\frac{\omega_a}{4}(z_1+z_3)\Big\{\frac{2g_k}{\pi i}\log s_k + \frac{\tau}{\pi i}\log(1-s_1s_3)\Big\} + o(1), \tag{7.6.11}$$

$$\tilde{\varkappa} = \frac{z_1-z_3}{z_1+z_3}.$$

On the other hand, equation (7.6.10) yields

$$\begin{aligned}\frac{w(z_0^+)}{iz_0} &= -\frac{1}{2}e^{-i\phi}\frac{z_0'}{z_0} = -\frac{1}{2}e^{-i\phi}\Big(\frac{v''}{v'}-\frac{v'}{v}\Big)\\ &= -\frac{1}{2}(z_1+z_3)\Big\{\frac{\operatorname{sn}\alpha}{\operatorname{cn}\alpha\cdot\operatorname{dn}\alpha} - \frac{\operatorname{cn}\alpha\cdot\operatorname{dn}\alpha}{\operatorname{sn}\alpha}\Big\} = -\frac{z_1+z_3}{\operatorname{sn}2\alpha},\end{aligned}$$

where

$$\alpha = \pm e^{i\phi}(z_1+z_3)(t-t_0).$$

By comparing the above two formulas for $\dfrac{iz_0}{w(z_0^+)}$, we find

$$v(t) = (z_1-z_3)\operatorname{sn}\Big(e^{i\phi}(z_1+z_3)t+\chi\big|\tilde{\varkappa}\Big), \tag{7.6.12}$$

where the phase shift χ is defined in (7.6.11). Simple manipulations of the relevant elliptic integrals transform equation (7.6.12) into the asymptotic relation (7.2.7).

6.1. Proof of the main theorem. Justification of the asymptotic expansions based on Kitaev's method. The above constructions provide the leading term of the asymptotics of the PII transcendent in the complex plane as $|x|\to\infty$. We will justify this statement by using the scheme suggested by A. Kitaev in [**Ki1**]. The first step in this scheme is establishing the following fact.

LEMMA 7.1. *Let $\vec{s}_{\text{asym}}$ be a set of monodromy data corresponding to the leading term of the asymptotic solution $u(x)$, i.e., the right-hand side of (7.2.7). Then there exist positive constants δ and C, independent of t, such that*

$$\|\vec{s}-\vec{s}_{\text{asym}}\| < Ct^{-\delta}, \tag{7.6.13}$$

where $\|\cdot\|$ is a norm in $\mathbb{C}^3$.

PROOF. We have already calculated the approximate monodromy data in (7.5.7), (7.5.8), by evaluating the right-hand side via the explicit formulas (7.5.6), (7.6.6) and (7.6.8). However, so far these results are not truly asymptotic, since we have the terms $t\cdot 4\pi e^{i\phi}/\omega_a(z_1,z_3)$ and itF_1, which might produce finite-order corrections; furthermore, z_1 and z_3 involve the small correction $\frac{1}{t}B(\phi,t)$ in (7.3.8). We will now prove the asymptotic nature of these results.

The only finite-order in t term that is generated by the phase integral $\omega_a(z_1,z_3)$ in (7.6.8), equals

$$\frac{\pi i}{8}\zeta_-^+ B(\phi,t),$$

where

$$\zeta_-^+ = \frac{1}{\omega_a^{(\infty)}}\int_{z_0^-}^{z_0^+}\frac{dz}{w(z)^3}, \quad \omega_a^{(\infty)} = \lim_{t\to\infty}\omega_a, \quad |B(\phi,t)| < C_1.$$

Since

$$\zeta_-^+ = 2n + 2m\tau + \tilde{z}_-^+,$$

and $\tilde{z}_-^+$ is defined on the universal covering $\tilde{\Gamma}$ and is bounded, it follows that

$$\zeta_-^+ = \frac{1}{\omega_a^\infty}(2nj_a + 2mj_b) + \tilde{\zeta}_-^+,$$

where

$$j_{a,b} = \oint\limits_{a,b} \frac{dz}{w^3(z)}.$$

Similarly, $\tilde{\zeta}_-^+$ runs over the universal covering $\tilde{\Gamma}$ and is bounded in a neighborhood of the branch points $\pm z_1$, $\pm z_3$.

The boundedness of $B(\phi, t)$ in the strip $|\mathrm{Im}t| < C$, and the isomonodromy condition (7.6.8), yield

$$\begin{aligned} it\mathcal{J}_a^\infty &= it \lim_{t\to\infty} \mathcal{J}_a = 2\pi i m + O(1), \\ it\mathcal{J}_b^\infty &= it \lim_{t\to\infty} \mathcal{J}_b = -2\pi i n + O(1). \end{aligned}$$

Hence

$$\begin{aligned} \lim_{t\to\infty} \frac{\zeta_-^+}{t} &= \frac{1}{\pi\omega_a^\infty}(\mathcal{J}_a^\infty j_b - \mathcal{J}_b^\infty j_a) \\ &= -\frac{64}{\pi\omega_a^\infty}\left(\mathcal{J}_a^\infty \frac{\partial^2 \mathcal{J}_b^\infty}{\partial A^2} - \mathcal{J}_b^\infty \frac{\partial^2 \mathcal{J}_a^\infty}{\partial A^2}\right) \\ &= \frac{8}{\pi\omega_a^\infty}\frac{\partial}{\partial A}\left(\mathcal{J}_a^\infty \omega_b^\infty - \mathcal{J}_b^\infty \omega_a^\infty\right) \\ &= \frac{8}{\pi\omega_a^\infty}\frac{\partial}{\partial A}(4\pi i e^{i\phi}) = 0. \end{aligned} \tag{7.6.14}$$

An estimate for the term F_1 is simpler, since its only "dangerous" entry is $\mathcal{J}_a$. However, the derivative of F_1 with respect to A is double-periodic and any small correction of A does not change its limit:

$$F_1 = it\mathcal{J}_a + \frac{\Theta'\left(\frac{1}{2}(z_-^+ + 1) + \frac{\tau}{4}|\tau\right)}{\Theta\left(\frac{1}{2}(z_-^+ + 1) + \frac{\tau}{4}|\tau\right)} + \frac{3i}{4}\omega_a^\infty B(\phi, t) + o(1).$$

Choose now the parameter x defined by (7.2.9), so that the function $u(x)$ and its derivatives are uniformly bounded by some constant. This yields the boundedness of all functions involving u, u' and the relevant elliptic integrals over all finite contours. Due to the analyticity of these functions with respect to the shift $\frac{1}{t}B(\phi, t)$, $|B| < C_1$ it follows that the remainder terms in (7.5.6) are of order $O(t^{-\delta})$, and the lemma is proved.

The rest of the proof of the Theorem 7.1 involves four steps.

1. Define the approximate solution $u_{\rm asym}(x,\vec{s})$ on the set $\mathcal{D}$ by the r.h.s. of (7.2.7) (without the term $O\left(x^{-3/2}\right)$ in the argument of the elliptic sine), i.e., let

$$u_{\rm asym}(x,\vec{s}) := -\frac{i\varkappa}{\sqrt{1+\varkappa^2}}\,x^{1/2}\,\mathrm{sn}\left(\frac{2i}{3}\,\frac{x^{3/2}}{\sqrt{1+\varkappa^2}} - \frac{2iK}{\pi}\ln(s_3) + \frac{K'}{\pi}\ln(1-s_1s_3)\Big|\varkappa\right),$$

where $\vec{s}$ is the given monodromy data (7.2.5). Substitute this function into the system (7.3.5) and solve it for $|x|\to\infty$.

2. Since the conditions (7.3.2)–(7.3.4) are true by definition for u_{asym}, we can apply the WKB-analysis of §2–§4 to calculate the monodromy data $\vec{s}_{\rm asym}(x,\vec{s})$ corresponding to $u_{\rm asym}(x,\vec{s})$. Lemma 7.1 guarantees the estimate (7.6.13) for the approximate monodromy data,

$$\|\vec{s}-\vec{s}_{\rm asym}\| < C|x|^{-\delta}.$$

In fact, the smooth dependence of the elliptic ansatz $u_{\rm asym}(x,\vec{s})$ on the monodromy parameters s_j yields the smooth dependence on the monodromy data of the kernels of all the Volterra integral operators involved in the derivation of the estimate (7.6.13). This in turn implies that estimate (7.6.13) is in fact uniform in the following sense. Suppose $\vec{s}_0$ satisfies the conditions (7.2.6) of Theorem 7.1, and let $D(\vec{s}_0,\epsilon_0)$ be a closed ϵ_0-disc in the space of monodromy parameters centered at $\vec{s}_0$ and such that every point $\vec{s}$ from $D(\vec{s}_0,\epsilon_0)$ satisfies the same conditions (7.2.6). Then, there exist positive constants δ, C, and ρ such that

$$\|\vec{s}-\vec{s}_{\rm asym}(x,\vec{s})\| < C|x|^{-\delta} \tag{7.6.15}$$

for all $x\in\mathcal{D}$, $|x|\geq\rho$, and

$$\vec{s}\in D(\vec{s}_0,\epsilon_0).$$

3. Define the function

$$\vec{r}(x,\vec{s}) := \vec{s}_0 - \vec{s}_{\rm asym}(x,\vec{s}) + \vec{s}.$$

Note that the function $\vec{r}(x,.)$ maps the disc $D(\vec{s}_0,\epsilon_0)$ into itself for all $|x| > \max\{\rho,(C/\epsilon_0)^{1/\delta}\}\equiv\rho_1$. In virtue of the smooth dependence of the canonical solutions of the linear equation (7.3.5) on its coefficients, and hence on the parameter $\vec{s}$, the function $\vec{r}(x,.)$ is continuous on $D(\vec{s}_0,\epsilon_0)$. Thus the Brouwer Fixed Point Theorem implies (cf. [**Ki1**]) that for each $|x|>\rho_1$ there exists at least one solution of the equation

$$\vec{r}(x,\vec{s}) = s,$$

or, equivalently, of the equation

$$\vec{s}_{\rm asym}(x,\vec{s}) = \vec{s}_0. \tag{7.6.16}$$

The important point is that this solution is unique. Indeed, let $\tilde{\vec{s}}$ be another solution. Consider the corresponding function $u_{\rm asym}(x,\tilde{\vec{s}})$. The linear system (7.2.1) generated by this function has exactly the same monodromy data as the system generated by $u_{\rm asym}(x,\vec{s})$. Indeed, because of (7.6.16) it follows that both systems have the set $\vec{s}_0$ as the sets of their monodromy data. From the injectivity of the monodromy map it follows then that

$$u_{\rm asym}(x,\tilde{\vec{s}}) = u_{\rm asym}(x,\vec{s}),\quad \frac{d}{dx}u_{\rm asym}(x,\tilde{\vec{s}}) = \frac{d}{dx}u_{\rm asym}(x,\vec{s}).$$

By simple algebra (involving elliptic functions) we conclude from these relations that $\tilde{\vec{s}} = \vec{s}$.

4. The results of the previous step allow us to introduce the function $\vec{s}(x, \vec{s}_0)$, such that

$$\vec{s}_{asym}\big(x, \vec{s}(x, \vec{s}_0)\big) = \vec{s}_0, \quad x \in \mathcal{D}, \quad |x| > \rho_1. \tag{7.6.17}$$

Moreover, from (7.6.15) it follows that

$$\|\vec{s}(x, \vec{s}_0) - \vec{s}_0\| < C|x|^{-\delta}. \tag{7.6.18}$$

Consider the functions

$$u_0(x) := u_{\mathrm{asym}}\big(x, \vec{s}(x, \vec{s}_0)\big),$$

and

$$w_0(x) := \frac{d}{dx} u_{\mathrm{asym}}(x, \vec{s})|_{\vec{s}=\vec{s}(x,\vec{s}_0)}.$$

The linear system (7.2.1) that these functions generate has a constant monodromy data. In fact, $s_j \equiv s_{0j}$. Due to the bijectivity (see (7.2.5)) of the monodromy map, this means that the function $u_0(x)$ is smooth (in fact, it is meromorphic on the whole x-plane) and it satisfies the second Painlevé equation (7.0.1). Indeed, according to our notational convention,

$$u_0(x) = u_{\mathrm{asym}}\big(x, \vec{s}(x, \vec{s}_0)\big) \equiv u(x; s_0),$$

which, together with the estimate (7.6.18), completes the proof of the theorem. □

CHAPTER 8

Asymptotic Solutions of the Second Painlevé Equation in the Complex Plane. Inverse Monodromy Problem Approach

In this chapter we present an alternative method for the asymptotic analysis of the RH problem of PII equation

$$\frac{d^2u}{dx^2} = 2u^3 + xu. \tag{8.-1.1}$$

The reader who is interested only in final asymptotic results can skip this chapter, since its main result (formula (8.7.3)) rederives the asymptotic formulae of Theorem 7.1. However, the new method discussed below plays an important role in the framework of the RH approach to Painlevé functions. It starts with the Inverse Monodromy Problem of PII (see Chapter 3) and derives its asymptotic solution "from first principles". This means that no *a priori* ansatz of the behavior of the PII function is needed, in contrast with the WKB method of Chapter 7, where the Boutroux ansatz was the starting point of the analysis. In what follows the leading term of the elliptic asymptotics, modulated by the Boutroux equation, follows naturally from a proper "asymptotic undressing" of the RH problem.

0. Preliminary remarks. The Deift-Zhou method

In the beginning of the 1990s, P. Deift and X. Zhou introduced a nonlinear version of the classical steepest descent method for oscillatory RH problems [**DeZ1**]. This led to a new asymptotic approach for solving the Painlevé equations, which is based on the asymptotic solution of the relevant *Inverse* Monodromy Problems, which can be formulated as oscillatory matrix RH problems. This approach, which has the advantage that it does not use any a priori information for the solutions of the Painlevé equations, has been applied to the asymptotic analysis of the Painlevé equations in [**DeZ2, IFoK, IK4, IK5, K10**]. In this chapter, we will follow [**IK4**].

Another advantage of the Deift-Zhou nonlinear steepest descent scheme is that it provides a more straightforward way to justify the asymptotic formulae than the method combining the WKB results of [**IN1**] with the rigorous scheme of Kitaev.

The justification of the asymptotic results obtained by the nonlinear steepest descent method is based upon the following stability type theorem.

THEOREM 8.1. *Consider the Riemann-Hilbert problem determined by the pair (Γ, G), i.e., the problem of finding a matrix-valued function $Y(\lambda)$ such that,*

$$Y(\lambda) \quad \text{is analytic on} \quad \mathbb{C}\setminus\Gamma,$$

$$Y_+(\lambda) = Y_-(\lambda)G(\lambda), \quad \lambda \in \Gamma, \tag{8.0.1}$$

$$Y(\lambda) \to I, \quad \lambda \to \infty.$$

Suppose that the jump contour Γ and the jump matrix $G(\lambda)$ satisfy all the well-posedness conditions discussed in Chapter 3. Furthermore, assume that the matrix $G(\lambda)$ and the corresponding matrix factors $W^{\pm}(\lambda)$ defined in (3.1.14) depend analytically on an extra parameter t, and that the matrix $G(\lambda)$ is close to the unit matrix, as $t \to \infty$. More precisely, assume that

$$\|G - I\|_{L^2(\Gamma)\cap L^\infty(\Gamma)} < \frac{C}{t^\varepsilon}, \quad t \geq t_0, \quad \varepsilon > 0. \tag{8.0.2}$$

Then, for sufficiently large t, the Riemann-Hilbert problem is uniquely solvable, its solution $Y(\lambda) \equiv Y(\lambda, t)$ depends analytically on t, and it satisfies the following uniform estimate:

$$\|Y(\lambda, t) - I\| < \frac{C}{(1+|\lambda|)t^\varepsilon}, \quad \lambda \in \mathbb{K}, \quad t \geq t_1 \geq t_0, \tag{8.0.3}$$

where $\mathbb{K}$ is a closed subset of $\mathbb{C}P^1 \setminus \Gamma$ satisfying,

$$\frac{dist\{\lambda; \Gamma\}}{1+|\lambda|} \geq c(\mathbb{K}), \quad \forall \lambda \in \mathbb{K}.$$

PROOF. Following the general L^2-theory outlined in Chapter 3, we represent Y in a form of Cauchy integral over the contour Γ

$$Y(\lambda, t) = I + \frac{1}{2\pi i}\int_\Gamma \frac{\rho(\mu, t)(G(\mu, t) - I)}{\mu - \lambda} d\mu, \quad \lambda \notin \Gamma. \tag{8.0.4}$$

Then the RH problem is equivalent to the singular integral equation (cf. (3.1.3)),

$$\rho(\lambda, t) = I + \frac{1}{2\pi i}\int_\Gamma \frac{\rho(\mu, t)(G(\mu, t) - I)}{\mu - \lambda_-} d\mu, \quad \lambda \notin \Gamma.$$

The later equation can be rewritten as the following equation in $L^2(\Gamma)$.

$$\rho_0(\lambda, t) = F(\lambda, t) + \mathcal{K}[\rho_0](\lambda, t), \quad \lambda \in \Gamma, \tag{8.0.5}$$

where

$$\rho_0(\lambda, t) := \rho(\lambda, t) - I,$$

$$F(\lambda, t) = \frac{1}{2\pi i}\int_\Gamma \frac{G(\mu, t) - I}{\mu - \lambda_-} d\mu \equiv C_-[G - I](\lambda, t),$$

$$\mathcal{K}[\rho_0](\lambda, t) = \frac{1}{2\pi i}\int_\Gamma \frac{\rho_0(\mu, t)(G(\mu, t) - I)}{\mu - \lambda_-} d\mu \equiv C_-[\rho_0(G - I)](\lambda, t).$$

Due to the estimate (8.0.2) on $G - I$ and the *boundedness* of the Cauchy operator C_- in the space $L^2(\Gamma)$, it follows that F and $\mathcal{K}$ satisfy the estimates

$$\|F\|_{L^2} \le \|C_-\|_{L^2} \cdot \|G - I\|_{L^2} < \frac{C}{t^\varepsilon},$$

$$\|\mathcal{K}\|_{L^2} \le \|C_-\|_{L^2} \cdot \|G - I\|_{L^\infty} < \frac{C}{t^\varepsilon}.$$

This proves that the integral operator $\mathcal{K}$ is a contraction, thus the solution ρ_0 of equation (8.0.5) exists, and, moreover, it satisfies

$$\|\rho_0\|_{L^2} \le 2\|F\|_{L^2} \le \frac{C}{t^\varepsilon}, \quad t \ge t_1 \ge t_0.$$

This in turn implies the solvability of the Riemann-Hilbert problem. In addition, using the integral representation (8.0.4) we find

$$\|Y(\lambda, t) - I\| \le \left\| \int_\Gamma \frac{I - G(\mu, t)}{\mu - \lambda} d\mu \right\| + \left\| \int_\Gamma \frac{\rho_0(\mu, t)\big(I - G(\mu, t)\big)}{\mu - \lambda} d\mu \right\|$$

$$\le \|I - G\|_{L^2} \sqrt{\int_\Gamma \frac{d\mu}{|\mu - \lambda|^2}} + \frac{1}{\operatorname{dist}\{\lambda, \Gamma\}} \|I - G\|_{L^2} \|\rho_0\|_{L^2}$$

$$\le \frac{C}{(1 + |\lambda|) t^\varepsilon}, \quad t \ge t_1 \ge t_0,$$

for λ belonging to a subset $\mathbb{K}$. This proves the estimate (8.0.3). Finally, the analyticity of $Y(\lambda, t)$ with respect to t is, a direct consequence of Proposition 3.2 of Chapter 3, of the solvability of the Riemann-Hilbert problem, and of the assumptions made about $G(\lambda)$. The theorem is proven. □

The Deift-Zhou method combines the above Theorem 8.1 with an appropriate procedure of RH "undressing" so that the new jump matrix is close to unity. This procedure is possible due to the local analyticity of the relevant jump matrices. Indeed, the local analyticity together with the smoothness conditions at the node points (see Chapter 3) imply that Y_+, Y_- are analytic in some enlarged domains containing the components of the original jump contour (see also Chapter 5). Thus one can deform (cf. again Chapter 3) the contour Γ, moving it to a desired curve, or making it pass through a desired (stationary, saddle) point. In this way the initial RH problem is decomposed to a number of model problems having explicit exact or asymptotic solutions.

We will illustrate the Deift-Zhou method using as an example the RH problem (3.3.94)–(3.3.99) under the additional assumptions

$$\nu = 0, \quad S_{k+3} = \sigma_2 S_k \sigma_2.$$

As we know from Chapters 4 and 5, this RH problem linearizes the PII equation (8.-1.1). We will outline the asymptotic analysis of this problem as $x \to -\infty$. The details of this analysis will be presented in Section 4 of Chapter 9.

- Assuming $x < 0$ we scale the variables of the original RH problem (3.3.94)–(3.3.99)

$$\lambda = (-x)^{1/2} z, \quad (-x)^{3/2} = t.$$

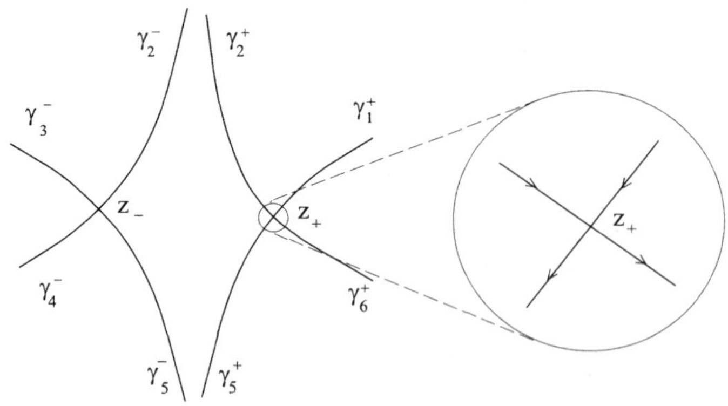

FIGURE 8.1. The oscillatory RH undressing by the Deift-Zhou method. The stationary phase points $z_\pm$ lead to a model RH problem on a cross, which can be solved in terms of Weber-Hermite functions

Then, given the set of monodromy data $s = \{S_k\}$, $k = 1, 2, \dots, 6$, we consider the RH problem for Y_k in the canonical domains Ω_k^0,

$$
\begin{aligned}
&Y_{k+1} = Y_k e^{-it\theta(z)\sigma_3} S_k e^{it\theta(z)\sigma_3}, \quad z \in \gamma_k,\\
&Y_k \to I, \quad z \to \infty, \quad z \in \overline{\Omega^0}_k, \qquad (8.0.6)\\
&\theta(z) = \tfrac{4}{3}z^3 - z.
\end{aligned}
$$

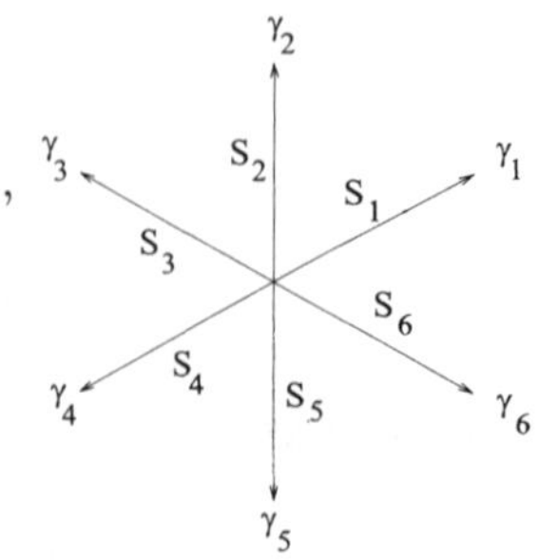

- We begin the asymptotic undressing of this RH problem by deforming the contour $\cup\gamma_k$ to the anti-Stokes lines Re $\theta(z) = 0$, where $\theta(z)$ is the phase function and $z_\pm = \pm\frac{1}{2}$ are the stationary phase points ($\theta'(z_\pm) = 0$).

  The solutions Y_k^{out} of the deformed RH problem (defined on the anti-Stokes graph) are found explicitly. These solutions satisfy the normalization and the jump conditions (8.0.6), up to the exponentially small correction, everywhere except at the stationary points. Moreover, Y_k^{out} have singularities in $z_\pm$, thus at these points Y_k^{out} do not satisfy the analyticity requirement of the original Y_k.
- In order to analyze these singular points, consider a "magnified" neighborhood of the points $z_\pm$. After a proper scaling, the RH problem is reduced to a model RH problem on a cross. This cross is the mapping (using "stretched" variables) of an intersection of γ_k (see Figure 8.1). The model RH problem on the cross is solved explicitly in terms of Weber-Hermite functions and its solutions, say Y_k^{in}, are normalized at infinity so that they to match the "external" solutions Y_k^{out}.

- Finally, combining the "external" and "internal" solutions and using relation,

$$u(x) = 2 \lim_{z\to\infty} z(Y_k)_{12},$$

we find the asymptotics of PII as $x \to -\infty$.

A crucial fact of the above analysis is the boundedness of the exponent $e^{it\theta(z)\sigma_3}$ at the stationary points $z_\pm$. This implies that the model RH problem provides an approximation which is uniformly valid in the entire discs around $z_\pm$. However, this is not the case if x is complex, since in this case

$$\theta(z) = \frac{4}{3}z^3 + e^{i\arg x} z,$$

and the matrix exponent $e^{it\theta(z)\sigma_3}$ is no longer bounded at the stationary points $z_\pm = \pm\frac{1}{2}e^{i\arg x}$. This is not simply a technical obstacle of the method, but a new phenomenon in the asymptotic analysis of the RH problem. The proper modification of the Deift-Zhou method needed in order to overcome this new difficulty will be presented in this chapter following [**IK4**].

The outline of the chapter is: In Section 1 we recall the setting of the RH problem associated to equation (8.-1.1). In Section 2 we map this RH problem to one involving a certain function $g(z)$ and formulated on the graph consisting of the anti-Stokes lines of the function $g(z)$ emanating from its critical points and connected by three finite Stokes lines. In Section 3 we construct explicitly the function $g(z)$[1]. In Section 4 we construct the solution of the reduced RH problem in terms of the Jacobi theta functions. In Section 5 we construct the approximate solutions of the RH problem near the critical points of $g(z)$, in terms of the Airy functions. In Section 6 we combine all the relevant functions and prove that this combined function approximates the solution of the original RH problem. In Section 7 we use the approximate solution of the RH problem to construct the approximate solution of the second Painlevé equation.

1. RH parametrization of PII in the complex plane

As shown in Chapter 3, the RH problem corresponding to the second Painlevé equation (8.-1.1) has a (unique) solution $\Psi(\lambda, x)$, which is meromorphic along $\Omega_k^0 \times X_s$ and holomorphic on $\Omega_k^0 \times (\mathbb{C}/X_s)$. The set X_s is the set of the poles of the function $\Psi(\lambda, x)$ and of the corresponding Painlevé function on the complex x-plane. The set X_s depends on the monodromy data $s = \{S_k\}$. Let us recall the construction of the Ψ-function.

The RH problem associated with the second Painlevé equation (8.-1.1) consists of finding the piecewise holomorphic 2×2 matrix-valued function $\Psi(\lambda)$ of the complex variable λ such that:

1) $\Psi(\lambda)$ is holomorphic for $\lambda \in \mathbb{C}\setminus\cup\{\gamma_k\}$, where γ_k are the rays $\gamma_k = \{\lambda \in \mathbb{C}\colon \arg\lambda = \frac{\pi}{6} + \frac{\pi}{3}(k-1)\}$, $k = 1,\dots,6$, oriented from zero to infinity.
2) On the rays γ_k the following jump conditions hold:

$$\Psi_+(\lambda) = \Psi_-(\lambda)S_k, \quad \lambda \in \gamma_k, \tag{8.1.1}$$

[1] In recent applications of the nonlinear steepest descent method to the orthogonal polynomials (see e.g. [**De1**]), this step corresponds to the construction of the so-called equilibrium measure.

where

$$S_{2k-1} = \begin{pmatrix} 1 & 0 \\ s_{2k-1} & 1 \end{pmatrix}, \quad S_{2k} = \begin{pmatrix} 1 & s_{2k} \\ 0 & 1 \end{pmatrix}, \tag{8.1.2}$$

and the parameters s_k, which do not depend on x and on λ, satisfy the constraints

$$s_{k+3} = -s_k, \quad s_1 - s_2 + s_3 + s_1 s_2 s_3 = 0. \tag{8.1.3}$$

Here, $\Psi_+(\lambda)$ and $\Psi_-(\lambda)$ are the limits of $\Psi(\lambda)$ to the left and to the right of γ, respectively.

3) The behavior of $\Psi(\lambda)$ at $\lambda = \infty$ is described by the asymptotic condition

$$\Psi(\lambda)e^{\theta(\lambda,x)\sigma_3} \to I, \quad \lambda \to \infty, \quad \theta(\lambda, x) = i\left(\frac{4}{3}\lambda^3 + x\lambda\right), \tag{8.1.4}$$

$$\sigma_3 = \begin{pmatrix} 1 & 0 \\ 0 & -1 \end{pmatrix}.$$

The above RH problem is depicted in Figure 8.2.

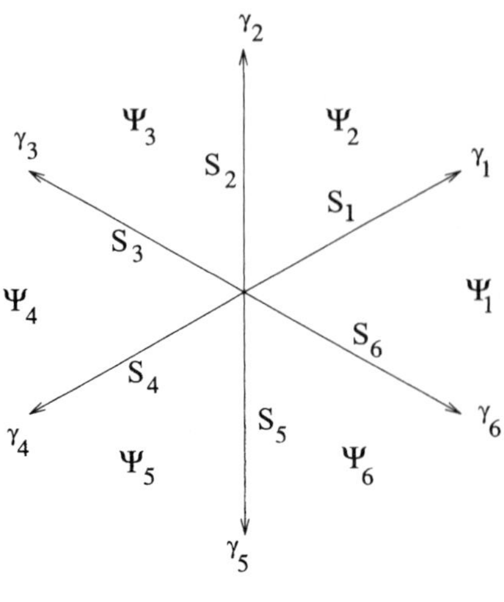

FIGURE 8.2. The RH problem graph γ for the Ψ-function

It is worth noticing that the first of the relations (8.1.3) and the symmetry of the RH problem graph, imply the symmetry equation

$$\Psi(\lambda) = \sigma_2 \Psi(-\lambda)\sigma_2, \quad \sigma_2 = \begin{pmatrix} 0 & -i \\ i & 0 \end{pmatrix}. \tag{8.1.5}$$

Also, the special triangular structure of the matrices S_k ensures that the product $e^{-\theta\sigma_3} S_k e^{\theta\sigma_3} \to I$ as $\lambda \to \infty$, i.e., it insures the well-posedness of the RH problem.

The piecewise holomorphic function $\Psi(\lambda)$ is in fact a collection of the canonical fundamental solutions of the linear 2×2 matrix ODE,

$$\frac{d\Psi}{d\lambda} = \left\{-i(4\lambda^2 + x + 2u^2)\sigma_3 - 4u\lambda\sigma_2 - 2v\sigma_1\right\}\Psi, \tag{8.1.6}$$

$$\sigma_1 = \begin{pmatrix} 0 & 1 \\ 1 & 0 \end{pmatrix},$$

where v and u are (meromorphic) functions of x such that $v = u_x$ and $u(x)$ satisfies the second Painlevé equation (8.-1.1). The jump matrices S_k of the RH problem have the meaning of the Stokes matrices for (8.1.6) and, by construction, do not depend on x. Therefore, as it has already been stated several times, the second Painlevé equation (8.-1.1) describes the isomonodromy deformations of the system (8.1.6).

The Painlevé function $u(x)$ admits the following representation in terms of the solution $\Psi(\lambda) \equiv \Psi(\lambda, x)$ of the RH problem,

$$u(x) = 2 \lim_{\lambda\to\infty} \left(\lambda\Psi_{12}(\lambda)e^{-\theta(\lambda,x)}\right). \tag{8.1.7}$$

Moreover, (cf. Chapter 7) the respective map,

$$\{s\} \equiv \{s_k\colon\ s_{k+3} = -s_k,\ s_1 - s_2 + s_3 + s_1s_2s_3 = 0\} \quad \mapsto \quad \{\text{solutions of (8.-1.1)}\},$$

is a bijection (see Chapter 4, Corollary 4.4). Hence, as in Chapters 4, 5, and 7 we adopt the notation

$$u(x) \equiv u(x; s)$$

for the second Painlevé transcendent.

The following important symmetry properties are valid,

$$u(x; s) = e^{i\frac{2\pi}{3}l}u(xe^{i\frac{2\pi}{3}l}; s^{2l}), \tag{8.1.8}$$

$$u(x; s) = \overline{u}(\overline{x}; s^*), \tag{8.1.9}$$

where $s \mapsto s^d$, $d \in \mathbb{Z}$ and $s \mapsto s^*$ denote the automorphisms of the set of the monodromy data $\{s\}$ defined by the relations

$$(s^d)_j := s_{j+d},$$

and

$$(s^*)_j := \overline{s}_{4-j}$$

(see Proposition 7.1 of Chapter 7). Symmetries (8.1.8) and (8.1.9) show that in order to obtain the complete asymptotic description of the function $u(x; s)$ as $x \to \infty$, $x \in \mathbb{C}$, it is sufficient to find, in terms of s, the asymptotic behavior in only one of the sectors $\left(\frac{\pi}{3}k, \frac{\pi}{3}(k+1)\right)$, say in the sector $\left(\frac{2\pi}{3}, \pi\right)$. The asymptotics of $u(x; s)$ in this sector will be obtained below.

2. The transformation of the RH problem

Let us scale the independent variable λ by

$$\lambda = (e^{-i\varphi}x)^{1/2}z, \quad (e^{-i\varphi}x)^{3/2} = t, \quad \varphi = \text{const} \in (\frac{2\pi}{3}, \pi), \tag{8.2.1}$$
$$\text{Re}\, t \to +\infty, \quad |\text{Im}\, t| < \text{const}.$$

The stationary points of the phase $\theta = it\left(\frac{4}{3}z^3 + e^{i\varphi}z\right)$ are independent of t, and formally, the problem can be treated in precisely the same way as outlined in Section 1, where the case of $\varphi = \pi$ was considered. Indeed, applying the arguments which have been outlined in Section 1 (and which will be presented in detail in Chapter 9; see also [**IFoK**]), one can replace the rays γ_k by the anti-Stokes lines $\gamma_k^\pm = \{\text{Im}\, \theta = \text{Im}\, \theta(z_\pm)\}$ indicated in Figure 8.3.

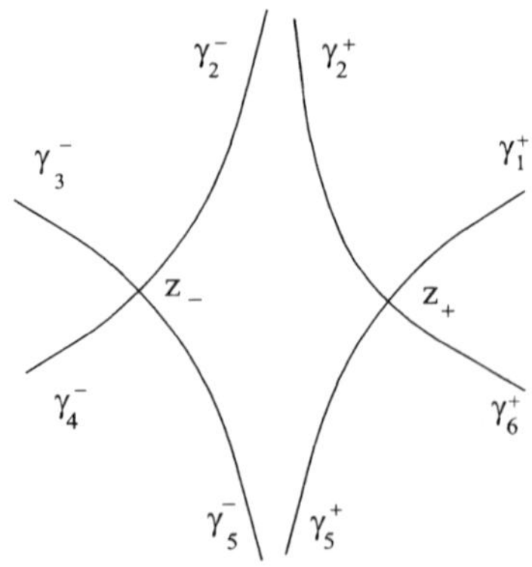

FIGURE 8.3. The anti-Stokes lines for the phase $\theta(z) = it\left(\frac{4}{3}z^3 + e^{i\varphi}z\right)$

However, at the stationary phase points, $z_\pm = \mp\frac{i}{2}e^{i\varphi/2}$ (in contrast to the case $\varphi = \pi$), the exponent $\theta(z)$ has nonzero real parts:

$$\text{Re}\, \theta(z_\pm) \to \pm\frac{t}{3}\cos\frac{3\varphi}{2}, \quad t \to +\infty;$$

since $\varphi \in (\frac{2}{3}\pi, \pi)$, for sufficiently large t,

$$\text{Re}\, \theta(z_+) < 0, \quad \text{Re}\, \theta(z_-) > 0. \tag{8.2.2}$$

Hence, after the normalization of the RH problem (i.e., $\Psi \to \Phi = \Psi e^{\theta(z)\sigma_3}$), only *half* of the jump matrices, namely the matrices corresponding to the lines γ_1^+, γ_2^-, γ_4^-, γ_5^+, approach the identity as $t \to \infty$ *uniformly* for $z \in \gamma_k^\pm$. The other half of the jump matrices, i.e., the matrices associated with the lines γ_2^+, γ_3^-, γ_5^-, γ_6^+, blow up as $t \to \infty$ in the finite neighborhoods of the points $z_\pm$. Therefore, the RH problem under consideration cannot be reduced to the Weber-Hermite model RH problems associated to the stationary phase points $z_\pm$. Thus, the principal idea of the asymptotic analysis in the case $\varphi = \pi$, which has been outlined in Section 1, fails for $\varphi \in (\frac{2\pi}{3}, \pi)$.

An algorithmic procedure that allows us to overcome the above obstacle is based on the following construction. First, we "split" the stationary phase points $z_\pm$ into four new points

$$z_1, \; z_2 = -z_1, \; z_3, \; z_4 = -z_3,$$

where z_1, z_3 lie in a neighborhood of the stationary phase point z_+, while z_2, z_4 are associated in the same way with the stationary point z_-. Secondly, we assume that there exists an analytic function $g(z)$, which satisfies the following conditions:

a) $g(z) = it\left(\frac{4}{3}z^3 + e^{i\varphi}z\right) + O\left(\frac{1}{z}\right)$, as $z \to \infty$;

b) points $\pm z_1$, $\pm z_3$ are the branch points of $g'(z)$ of order 2 and, in their neighborhoods,

$$g'(z) = c^{\mp}_{1,3}(z \pm z_{1,3})^{1/2} + \dots,$$

with $c^{\mp}_{1,3} \neq 0$.

c) $\mathrm{Re}\; g(\pm z_{1,3}) = 0$.

d) $g' \neq 0$ on $\mathbb{C}\backslash\{\pm z_{1,3}\}$.

In fact, conditions a)–d) define a certain well-posed problem in the theory of analytic functions. Simultaneously, they are the defining restrictions on the points $\pm z_{1,3}$. In Section 3, we will solve this problem, using an elliptic integral associated with the elliptic curve $w^2 = (z^2 - z_1^2)(z^2 - z_3^2)$. We will see then that condition c) is just the classical Boutroux equations, which always appear in the asymptotic analysis of the Painlevé equations. It also should be emphasized that the flexibility gained by introducing four parameters, i.e., the points $\pm z_{1,3}$, plays a crucial role for the solvability of the problem a)–d).

Having assumed the existence of the function $g(z)$, we can now replace the original phase $\theta = it\left(\frac{4}{3}z^3 + e^{i\varphi}z\right)$ with the function $g(z)$. Because of the asymptotic equation a), this replacement preserves the normalization condition (8.1.4). Moreover, due to condition c), we have now eliminated the main obstacle, namely inequalities (8.2.2). However, there is a price to be paid. Now, we have to deform the initial RH problem to the RH problem formulated on the anti-Stokes lines $\hat{\gamma}^{(k)}$ corresponding to the function $g(z)$:

$$\mathrm{Im}\; g(z) = \mathrm{Im}\; g(z_k). \tag{8.2.3}$$

The anti-Stokes graph defined in (8.2.3) is more complicated than the θ-graph depicted in Figure 8.2. Properties a), b) and d) lead to the g-anti-Stokes lines $\hat{\gamma}_s^{(k)}$ as shown in Figure 8.4. Our main task for this section, is the transformation of the initial RH graph presented in Figure 8.2 into the one given in Figure 8.4.

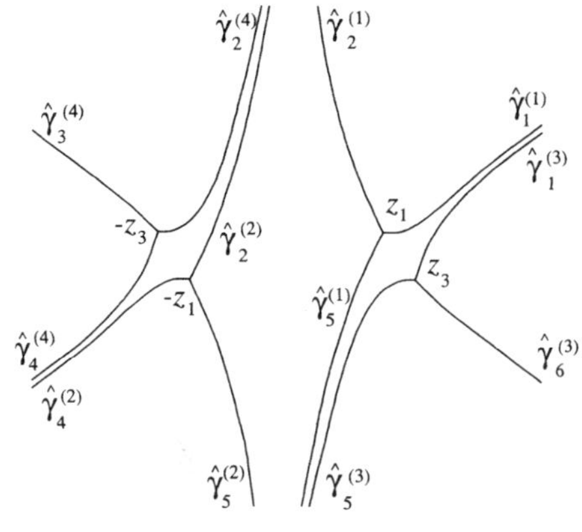

FIGURE 8.4. The anti-Stokes lines $\hat{\gamma}$ for the equation (8.2.3) for $x \to \infty$, $\arg x \in \left(\frac{2\pi}{3}, \pi\right)$

The notation $\hat{\gamma}_s^{(k)}$ for the anti-Stokes line used here refers to its starting point z_k and to the anti-Stokes ray γ_s which is the asymptote for the line $\hat{\gamma}_s^{(k)}$.

Due to the very different topological properties of the graphs presented in Figure 8.2 and Figure 8.4, it is not possible to deform the relevant graphs. In particular, the RH problem graph γ for the Ψ function is connected, while the anti-Stokes graph $\hat{\gamma}$ is disconnected. The best we can do with the original problem by means of trivial deformations, is shown in Figure 8.5.

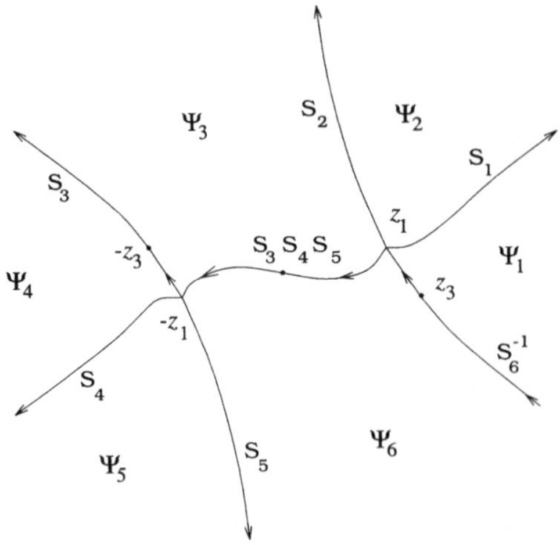

FIGURE 8.5. The continuously deformed RH problem graph

Note that after such deformation, a new boundary curve between the third and the sixth domains arises where the jump is described by the ratio of the sixth and the third canonical solutions:

$$\Psi_6(z) = \Psi_3(z)S_3S_4S_5, \tag{8.2.4}$$

where

$$S_3S_4S_5 = \begin{pmatrix} 1+s_1s_2 & -s_1 \\ -s_1 & 1-s_1s_3 \end{pmatrix}.$$

Let us consider the following function $\delta(z)$, which depends on z_k, $k=1,3$, and some parameters ν_l, $l=1,2,3$,

$$\delta(z) = \left(\frac{z+z_1}{z+z_3}\right)^{\nu_3}\left(\frac{z-z_1}{z+z_1}\right)^{\nu_2}\left(\frac{z-z_3}{z-z_1}\right)^{\nu_1}. \tag{8.2.5}$$

This function is defined on the complex plane cut along the broken line $[z_3,z_1]\cup[z_1,-z_1]\cup[-z_1,-z_3]$, and it is fixed by the condition

$$\delta(z) \underset{z\to\infty}{\longrightarrow} 1. \tag{8.2.6}$$

The function $\delta(z)$ has the following jumps on the relevant cuts

$$\begin{aligned} \delta_-(z) &= \delta_+(z)e^{2\pi i\nu_1}, \quad & z\in(z_3,z_1), \\ \delta_-(z) &= \delta_+(z)e^{2\pi i\nu_2}, \quad & z\in(z_1,-z_1), \\ \delta_-(z) &= \delta_+(z)e^{2\pi i\nu_3}; \quad & z\in(-z_1,-z_3), \end{aligned} \tag{8.2.7}$$

see Figure 8.6.

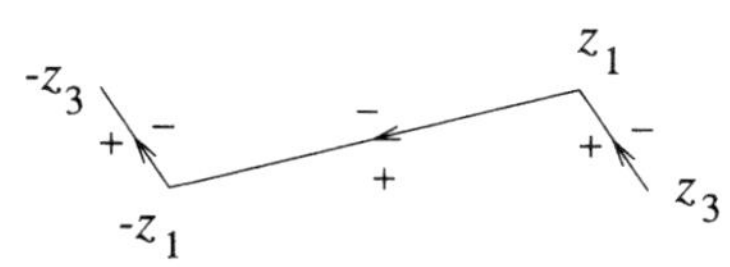

FIGURE 8.6. The graph of the scalar RH problem solved by the function $\delta(z)$

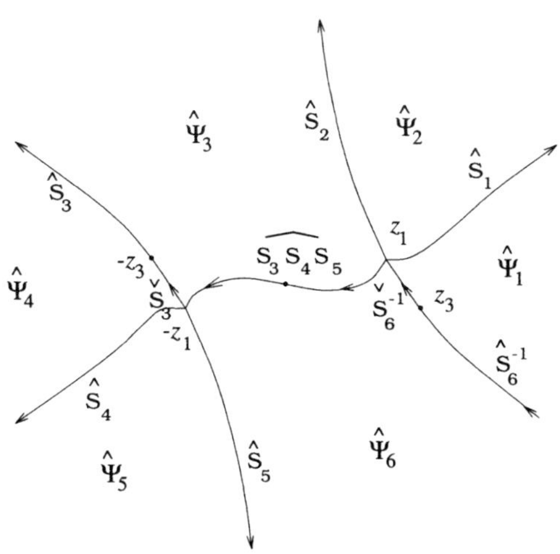

FIGURE 8.7. The RH problem for $\hat{\Psi}$

The next step is of crucial importance for the transformation of the original RH problem. Let us introduce the new function

$$\hat{\Psi} = \Psi \delta^{-\sigma_3}. \tag{8.2.8}$$

This new function does not differ from the original one at infinity, but its jumps are described by the new matrices:

$$\hat{\Psi}_+ = \hat{\Psi}_- \delta_-^{\sigma_3} S \delta_+^{-\sigma_3} = \hat{\Psi}_- \hat{S}. \tag{8.2.9}$$

The new RH graph is shown in Figure 8.7, whereas the new jump matrices $\hat{S}$ are given by the equations,

(8.2.10) $$\hat{S}_{2k-1} = \begin{pmatrix} 1 & 0 \\ s_{2k-1}\delta^{-2} & 1 \end{pmatrix}, \quad \hat{S}_{2k} = \begin{pmatrix} 1 & s_{2k}\delta^2 \\ 0 & 1 \end{pmatrix},$$

$$\check{S}_3 = \begin{pmatrix} e^{2\pi i\nu_3} & 0 \\ s_3\delta_-^{-2}e^{2\pi i\nu_3} & e^{-2\pi i\nu_3} \end{pmatrix}, \quad (\check{S}_6)^{-1} = \begin{pmatrix} e^{2\pi i\nu_1} & s_3\delta_-^2 e^{-2\pi i\nu_1} \\ 0 & e^{-2\pi i\nu_1} \end{pmatrix},$$

$$\widehat{S_3S_4S_5} = \begin{pmatrix} (1+s_1s_2)e^{2\pi i\nu_2} & -s_1\delta_-^2 e^{-2\pi i\nu_2} \\ -s_1\delta_-^{-2}e^{2\pi i\nu_2} & (1-s_1s_3)e^{-2\pi i\nu_2} \end{pmatrix}.$$

The importance of the above transformation is that it yields new jump matrices which can be factorized in such a way, that a simple deformation of the RH graph is now sufficient in order to approach the anti-Stokes lines indicated in Figure 8.4 (cf. [**IFoK**]) and simultaneously preserve the asymptotic behaviour of the function $\hat{\Psi}(z)$ at $z = \infty$. Indeed, let us assume the generic conditions,

(8.2.11) $$s_3 \neq 0, \quad 1 - s_1s_3 \neq 0.$$

If we set

(8.2.12) $$e^{-2\pi i\nu_3} = s_3,$$

it is easy to verify the equation

(8.2.13) $$\check{S}_3 = \hat{S}_2^U\hat{S}_A\hat{S}_4^U,$$

where

(8.2.14) $$\hat{S}_2^U = \begin{pmatrix} 1 & \delta_-^2 s_3^{-1} \\ 0 & 1 \end{pmatrix}, \quad \hat{S}_A = \begin{pmatrix} & -\delta_-^2 \\ \delta_-^{-2} & \end{pmatrix}, \quad \hat{S}_4^U = \begin{pmatrix} 1 & \delta_+^2 s_3^{-1} \\ 0 & 1 \end{pmatrix}.$$

Similarly, letting

(8.2.15) $$e^{2\pi i\nu_1} = s_3, \quad \text{or} \quad \nu_1 = -\nu_3,$$

we find the factorization

(8.2.16) $$\check{S}_6^{-1} = \hat{S}_1^L\hat{S}_B\hat{S}_5^L,$$

where

(8.2.17)
$$\hat{S}_1^L = \begin{pmatrix} 1 & 0 \\ (\delta_-^2 s_3)^{-1} & 1 \end{pmatrix}, \quad \hat{S}_B = \begin{pmatrix} & \delta_-^2 \\ -\delta_-^{-2} & \end{pmatrix}, \quad \hat{S}_5^L = \begin{pmatrix} 1 & 0 \\ (\delta_+^2 s_3)^{-1} & 1 \end{pmatrix}.$$

At the same time, letting

(8.2.18) $$e^{2\pi i\nu_2} = 1 - s_1s_3,$$

we find

(8.2.19) $$\widehat{S_3S_4S_5} = \hat{S}^U\hat{S}^L,$$

with

$$\hat{S}^U = \begin{pmatrix} 1 & -\frac{s_1}{1-s_1s_3}\delta_-^2 \\ 0 & 1 \end{pmatrix}, \quad \hat{S}^L = \begin{pmatrix} 1 & 0 \\ -\frac{s_1}{1-s_1s_3}\delta_+^{-2} & 1 \end{pmatrix}. \tag{8.2.20}$$

The "intermediate" RH graph is shown in Figure 8.8, and the final graph is presented in Figure 8.9.

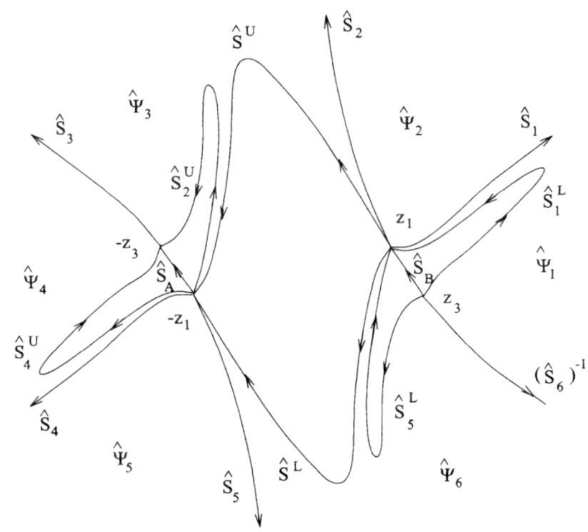

FIGURE 8.8. Deformation of the RH problem graph for $\hat{\Psi}$

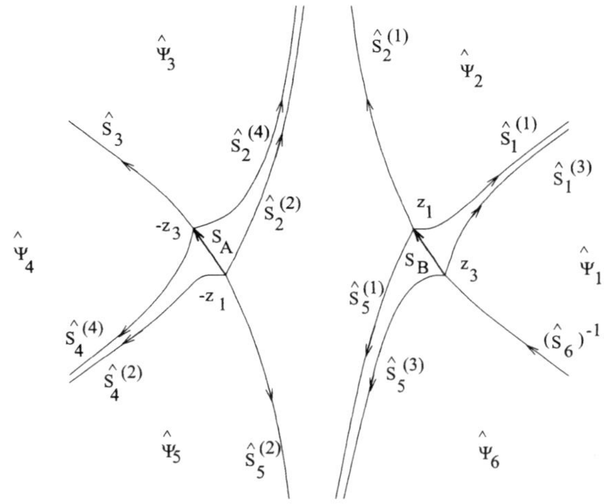

FIGURE 8.9. The final RH problem graph for $\hat{\Psi}$

Here, the jump matrices are

$$\hat{S}_5^{(3)} = \begin{pmatrix} 1 & 0 \\ \frac{1}{\delta^2 s_3} & 1 \end{pmatrix}, \quad \hat{S}_6^{-1} = \begin{pmatrix} 1 & \delta^2 s_3 \\ 0 & 1 \end{pmatrix}, \quad \hat{S}_1^{(3)} = \begin{pmatrix} 1 & 0 \\ \frac{1}{\delta^2 s_3} & 1 \end{pmatrix}, \tag{8.2.21}$$

$$\hat{S}_1^{(1)} = \begin{pmatrix} 1 & 0 \\ -\frac{1-s_1 s_3}{\delta^2 s_3} & 1 \end{pmatrix}, \quad \hat{S}_2^{(1)} = \begin{pmatrix} 1 & \frac{\delta^2 s_3}{1-s_1 s_3} \\ 0 & 1 \end{pmatrix}, \quad \hat{S}_5^{(1)} = \begin{pmatrix} 1 & 0 \\ -\frac{1}{\delta^2 s_3(1-s_1 s_3)} & 1 \end{pmatrix}, \tag{8.2.22}$$

$$\hat{S}_4^{(2)} = \begin{pmatrix} 1 & \delta^2 \frac{1-s_1 s_3}{s_3} \\ 0 & 1 \end{pmatrix}, \quad \hat{S}_5^{(2)} = \begin{pmatrix} 1 & 0 \\ -\frac{s_3}{\delta^2(1-s_1 s_3)} & 1 \end{pmatrix}, \quad \hat{S}_2^{(2)} = \begin{pmatrix} 1 & \frac{\delta^2}{s_3(1-s_1 s_3)} \\ 0 & 1 \end{pmatrix}, \tag{8.2.23}$$

$$\hat{S}_2^{(4)} = \begin{pmatrix} 1 & -\frac{\delta^2}{s_3} \\ 0 & 1 \end{pmatrix}, \quad \hat{S}_3 = \begin{pmatrix} 1 & 0 \\ \frac{s_3}{\delta^2} & 1 \end{pmatrix}, \quad \hat{S}_4^{(4)} = \begin{pmatrix} 1 & -\frac{\delta^2}{s_3} \\ 0 & 1 \end{pmatrix}, \tag{8.2.24}$$

$$\hat{S}_A = \begin{pmatrix} & -\delta_-^2 \\ \frac{1}{\delta_-^2} & \end{pmatrix}, \quad \hat{S}_B = \begin{pmatrix} & \delta_-^2 \\ -\frac{1}{\delta_-^2} & \end{pmatrix}. \tag{8.2.25}$$

We have also taken into account that, by virtue of equations (8.2.12), (8.2.15) and (8.2.18), the jump conditions (8.2.7) for the function $\delta(z)$ (8.2.5) can be rewritten as follows,

$$\begin{aligned} \delta_-(z) &= \delta_+(z) s_3, & z &\in (z_3, z_1), \\ \delta_-(z) &= \delta_+(z)(1 - s_1 s_3), & z &\in (z_1, -z_1), \\ \delta_-(z) &= \delta_+(z) \cdot \tfrac{1}{s_3}, & z &\in (-z_1, -z_3). \end{aligned} \tag{8.2.26}$$

In this form, the RH problem graph for the $\hat{\Psi}$-function coincides with the anti-Stokes graph $\hat{\gamma}$ (see Figure 8.4) except for the additional segments $[z_3, z_1]$ and $[-z_1, -z_3]$.

Now, we do not need the function δ anymore and can return to the original Ψ-function by the use of (8.2.8),

$$\Psi = \hat{\Psi} \delta^{\sigma_3}. \tag{8.2.27}$$

The calculations above show that the solution of the second Painlevé equation can be obtained from the Ψ-function with the jumps on the graph shown in Figure 8.10. The new RH problem, which replaces the problem (8.1.1)–(8.1.4), can be formulated as follows:

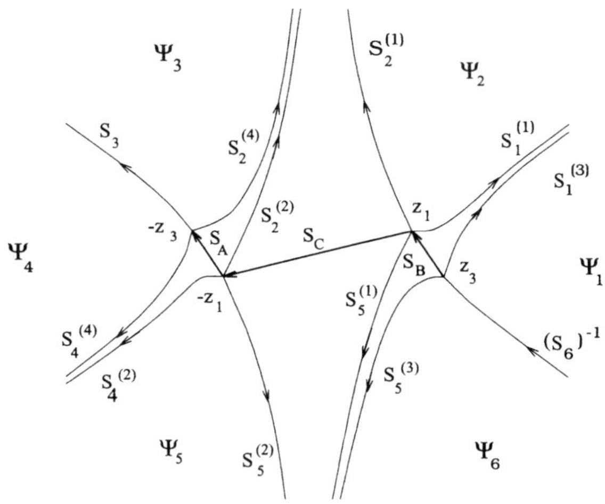

FIGURE 8.10. The ultimate RH problem graph for Ψ

1) Find a piecewise holomorphic matrix-valued 2×2 function Ψ of the complex variable z with an irregular singular point at infinity where it has the exponential asymptotic behavior of the type (8.1.4)

$$\Psi(z) = \big(I + O(z^{-1})\big)e^{-t\theta(z)\sigma_3}, \quad \theta(z) = i\Big(\frac{4}{3}z^3 + e^{i\varphi}z\Big), \quad z \to \infty. \tag{8.2.28}$$

2) On the piecewise smooth oriented graph $\hat{\gamma}$ (see Figure 8.10) the following jump condition is valid:

$$\Psi_+(z) = \Psi_-(z)\mathrm{S}, \quad \lambda \in \hat{\gamma}. \tag{8.2.29}$$

Here, $\Psi_+(z)$ and $\Psi_-(z)$ are the limits of the function $\Psi(z)$ on the contour $\hat{\gamma}$ to the left and to the right, respectively, and the piecewise constant matrix S is described by the following equations:

$$\mathrm{S}_5^{(3)} = \begin{pmatrix} 1 & 0 \\ \frac{1}{s_3} & 1 \end{pmatrix}, \quad \mathrm{S}_6^{-1} = \begin{pmatrix} 1 & s_3 \\ 0 & 1 \end{pmatrix}, \quad \mathrm{S}_1^{(3)} = \begin{pmatrix} 1 & 0 \\ \frac{1}{s_3} & 1 \end{pmatrix}, \tag{8.2.30}$$

$$\mathrm{S}_1^{(1)} = \begin{pmatrix} 1 & 0 \\ -\frac{1-s_1s_3}{s_3} & 1 \end{pmatrix}, \quad \mathrm{S}_2^{(1)} = \begin{pmatrix} 1 & \frac{s_3}{1-s_1s_3} \\ 0 & 1 \end{pmatrix}, \quad \mathrm{S}_5^{(1)} = \begin{pmatrix} 1 & 0 \\ -\frac{1}{s_3(1-s_1s_3)} & 1 \end{pmatrix}, \tag{8.2.31}$$

$$\mathrm{S}_4^{(2)} = \begin{pmatrix} 1 & \frac{1-s_1s_3}{s_3} \\ 0 & 1 \end{pmatrix}, \quad \mathrm{S}_5^{(2)} = \begin{pmatrix} 1 & 0 \\ -\frac{s_3}{1-s_1s_3} & 1 \end{pmatrix}, \quad \mathrm{S}_2^{(2)} = \begin{pmatrix} 1 & \frac{1}{s_3(1-s_1s_3)} \\ 0 & 1 \end{pmatrix}, \tag{8.2.32}$$

$$\text{(8.2.33)} \qquad \mathrm{S}_2^{(4)} = \begin{pmatrix} 1 & -\frac{1}{s_3} \\ 0 & 1 \end{pmatrix}, \quad \mathrm{S}_3 = \begin{pmatrix} 1 & 0 \\ s_3 & 1 \end{pmatrix}, \quad \mathrm{S}_4^{(4)} = \begin{pmatrix} 1 & -\frac{1}{s_3} \\ 0 & 1 \end{pmatrix},$$

$$\text{(8.2.34)} \quad S_A = \begin{pmatrix} & -\frac{1}{s_3} \\ s_3 & \end{pmatrix}, \quad S_B = \begin{pmatrix} & s_3 \\ -\frac{1}{s_3} & \end{pmatrix}, \quad S_C = \begin{pmatrix} \frac{1}{1-s_1 s_3} & \\ & 1 - s_1 s_3 \end{pmatrix}.$$

The parameters s_k satisfy the constraint (8.1.3),

$$s_1 - s_2 + s_3 + s_1 s_2 s_3 = 0, \tag{8.2.35}$$

and do not depend on t and on z.

REMARK 8.1. Given the jump contour $\hat{\gamma}$, the resulting jump conditions (8.2.29)–(8.2.34) can be reconstructed directly from the jump conditions (8.1.1)–(8.1.2) for the contour γ (see Figure 8.2) using, instead of the "ghost" function $\delta(z)$, the monodromy equations (8.1.3).

3. Construction of the function $g(z)$

In this section, we construct the function $g(z)$ described in Section 3. Consider the function

$$w(z) = \sqrt{(z^2 - z_1^2)(z^2 - z_3^2)}, \quad z_1^2 + z_3^2 = -\frac{1}{2}e^{i\varphi}, \tag{8.3.1}$$

defined on the Riemann surface Γ (see Figure 8.11) consisting of two complex planes cut along the intervals $[z_3, z_1]$ and $[-z_1, -z_3]$, whereas the branch of the root on the upper sheet is chosen in such a way that

$$w = z^2 - \frac{1}{2}(z_1^2 + z_3^2) + O(z^{-2}) \quad \text{as} \quad z \to \infty^+.$$

Here ∞^+ denotes the infinite point of the upper sheet. We choose the basis $\{a, b\}$ of the group $H_1(\Gamma)$, as indicated in Figure 8.11.

Introduce the Abelian (elliptic, in fact) integral of the second kind

$$\hat{g}(z) = 4i \int_{0_+}^{z} w(z)\, dz. \tag{8.3.2}$$

We will consider $\hat{g}(z)$ as a single-valued function defined on the complex plane $\mathbb{C}^*$ cut along the sum of the intervals $[z_3, z_1] \cup [z_1, -z_1] \cup [-z_1, -z_3]$. Moreover, the starting point $z = 0_+$ of the integration path lies on the left side of the oriented cut $[z_1, -z_1]$. Introduce the periods of (8.3.2),

$$A = 4i\mathcal{J}_a = 4i \oint_a w(z)\, dz\,, \quad B = 4i\mathcal{J}_b = 4i \oint_b w(z)\, dz\,. \tag{8.3.3}$$

Taking into account the symmetry $w(-z) = w(z)$, we find that

$$\hat{g}(z) + \hat{g}(-z) = 4i\Big(\int_{0_+}^{z} + \int_{0_+}^{z_1} + \oint_a + \int_{z_1}^{0_-} + \int_{0_-}^{-z}\Big) w(z)\, dz = A.$$

Thus the function

$$g(z) = \hat{g}(z) - \frac{1}{2}A, \tag{8.3.4}$$

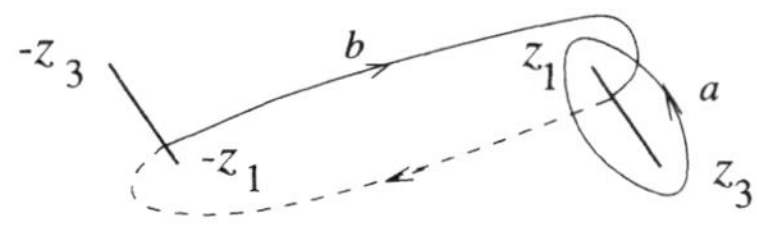

FIGURE 8.11. The Riemann surface of the function $w(z)$ and the respective a and b basic cycles of this surface.

considered on $\mathbb{C}^*$, is odd,

$$g(-z) = -g(z).$$

Using definition (8.3.4), it follows that properties b) and d) of $g(z)$ are satisfied.

In order to find the asymptotics of $g(z)$ as $z \to \infty$, we introduce the regularized integral

$$4i \int_{0_+}^{z} \Big(w(z) - z^2 + \frac{1}{2}(z_1^2 + z_3^2)\Big)\, dz = \hat{g}(z) - i\big(\frac{4}{3}z^3 + e^{i\varphi} z\big).$$

Using the symmetry $w(z) = w(-z)$, this integral equals

$$2i\Big(\int_{-z}^{0_-} + \int_{0_+}^{z}\Big)\Big(w(z) - z^2 + \frac{1}{2}(z_1^2 + z_3^2)\Big)\, dz$$

$$= 2i \oint_a \Big(w(z) - z^2 + \frac{1}{2}(z_1^2 + z_3^2)\Big)\, dz + O(z^{-1}) = \frac{1}{2}A + O(z^{-1}).$$

Thus,

$$g(z) = i\big(\frac{4}{3}z^3 + e^{i\varphi} z\big) + O\big(z^{-1}\big) \quad \text{as} \quad z \to \infty, \tag{8.3.5}$$

which establishes the property a) of the function $g(z)$. The property c) of the function $g(z)$, i.e., Re $g(\pm z_{1,3}) = 0$, is equivalent to the system of equations Re $A =$ Re $B = 0$, and thus takes the integral form,

$$\operatorname{Im} \oint_{a,b} w(z)\, dz = 0. \tag{8.3.6}$$

Equations (8.3.6), known as the *Boutroux equations* (7.1.6), play an important role in our proof below of the solvability of the RH problem.

Denote by $D = z_1^2 z_3^2$ the module corresponding to the elliptic curve. We observe that (8.3.6) determines D as a differentiable function $D(\varphi)$ of the angle parameter $\varphi \in (\frac{2\pi}{3}, \pi)$. Indeed, the l.h.s. of (8.3.6) yields the complete set of the independent first integrals of the first order nonlinear ODE

$$\frac{dD}{d\varphi} = -\frac{1}{4\operatorname{Im}(\omega_a \overline{\omega}_b)}\big(\overline{\omega}_a(e^{i\varphi}\Omega_a - e^{-i\varphi}\overline{\Omega}_a) - \omega_b(e^{i\varphi}\Omega_b - e^{-i\varphi}\overline{\Omega}_b)\big), \tag{8.3.7}$$

where

$$\omega_{a,b} = \oint_{a,b} \frac{dz}{w(z)}, \qquad \Omega_{a,b} = \oint_{a,b} \frac{z^2\, dz}{w(z)}.$$

Provided that the elliptic curve $w(z)$ does not degenerate, the r.h.s. of (8.3.7) is a smooth function of φ, and, given the initial point, equation (8.3.7) is uniquely solvable. To find the initial point, observe that equations (8.3.6) become degenerate at $z_1 = z_3$ for $\varphi = \pi$ and at $z_1 = 0$ for $\varphi = \frac{2\pi}{3}$. At these boundary points, the r.h.s. of (8.3.7) remains continuous, which ensures its solvability. The classical uniqueness theorem for ODEs breaks down at these boundary points, however, only one of the solutions of the differential equation (8.3.7) passing through the initial boundary point satisfies the integral equations (8.3.6).

Define the Stokes line emanating from the point z_k, $k = 1, 2, 3, 4$, as the set of the points z satisfying the equation

$$\operatorname{Im} \int_{z_k}^{z} w(z)\, dz = 0. \tag{8.3.8}$$

Define the Stokes graph as the union of all the Stokes lines. The conditions (8.3.6) mean that the Stokes graph is connected. On the other hand, conditions (8.3.6) allow us to choose the arcs of the Stokes lines as the segments $[z_3, z_1]$, $[z_1, -z_1]$ and $[-z_1, -z_3]$.

For the following, we need to know the jump conditions of the function $g(z)$. These are described by the equations

$$\begin{aligned} z \in (z_3, z_1):& \quad g_+ + g_- = \frac{1}{2}B;\\ z \in (z_1, -z_1):& \quad g_+ - g_- = -A;\\ z \in (-z_1, -z_3):& \quad g_+ + g_- = -\frac{1}{2}B. \end{aligned} \tag{8.3.9}$$

4. The model Baker-Akhiezer RH problem

We recall that the reason for replacing the original RH problem, with the one presented in Figure 8.10, is that

$$e^{-tg(z)\sigma_3} S e^{tg(z)\sigma_3} \to I, \tag{8.4.1}$$

as $|t| \to \infty$ for all z belonging to the infinite branches of the jump contour $\hat{\gamma}$ and $z \neq \pm z_k$. Indeed, equation (8.4.1) means that when we normalize the RH problem, i.e., when we pass from the function Ψ to the function

$$Y(z) = \Psi(z) e^{tg(z)\sigma_3},$$

so that $Y(z) \to I$ as $z \to \infty$, the Y-jump matrices will approach the identity matrix as $|t| \to \infty$ exponentially fast everywhere on the infinite branches of $\hat{\gamma}$. Therefore, one can expect that the main contribution into the asymptotics of the solution of the RH problem will be made by the truncated RH problem posed on the sum of the segments (see Figure 8.10) $(z_3, z_1) \cup (z_1, -z_1) \cup (-z_1, -z_3)$. In other words, we arrive at the model problem consisting of finding the piecewise holomorphic function $\Psi^{(BA)}(z)$ with the following properties:

1) near infinity

$$\Psi^{(BA)}(z) = \big(I + O(z^{-1})\big) e^{-t\theta\sigma_3}, \quad z \to \infty, \tag{8.4.2}$$

where $\theta = i\big(\frac{4}{3}z^3 + e^{i\varphi} z\big)$;

2) on the sum of the oriented contours $(z_3, z_1) \cup (z_1, -z_1) \cup (-z_1, -z_3)$, the following jump conditions are valid

$$z \in (-z_1, -z_3): \quad \Psi_+^{(BA)}(z) = \Psi_-^{(BA)}(z) S_A, \quad S_A = \begin{pmatrix} & -s_3^{-1} \\ s_3 & \end{pmatrix}, \tag{8.4.3}$$

$$z \in (z_3, z_1): \quad \Psi_+^{(BA)}(z) = \Psi_-^{(BA)}(z) S_B, \quad S_B = \begin{pmatrix} & s_3 \\ -s_3^{-1} & \end{pmatrix}, \tag{8.4.4}$$

$$z \in (z_1, -z_1): \quad \Psi_+^{(BA)}(z) = \Psi_-^{(BA)}(z) S_C, \tag{8.4.5}$$

$$S_C = \begin{pmatrix} (1 - s_1 s_3)^{-1} & \\ & 1 - s_1 s_3 \end{pmatrix};$$

3) the function $\Psi^{(BA)}(z)$ has weaker than L^2-integrable singularities at the points $\pm z_1$, $\pm z_3$.

We use the superscript (BA) since we are going to construct $\Psi^{(BA)}(z)$ in terms of the Baker-Akhiezer functions of the Riemann surface Γ. On this Riemann surface, there exists only one, up to an arbitrary constant multiplier, holomorphic differential $\frac{dz}{w(z)}$ with the periods $\omega_{a,b}$. Introduce the canonical Abelian (elliptic) integral of the first kind $U(z)$,

$$U(z) = \frac{1}{\omega_a} \int_{\infty^+}^{z} \frac{dz}{w(z)}, \tag{8.4.6}$$

and the Abelian integral $h(z)$ which differs from $g(z)$ (8.3.4) by some integral of the first kind,

$$h(z) = g(z) + cU(z), \tag{8.4.7}$$

where the constant c will be determined later. Since $U(z) = O(z^{-1})$ as $z \to \infty$, the integral $h(z)$ has the same canonical asymptotics as the integral $g(z)$,

$$h(z) = i\big(\frac{4}{3} z^3 + e^{i\varphi} z\big) + O(z^{-1}). \tag{8.4.8}$$

Denote the corresponding complete integrals by the letters A' and B':

$$\begin{aligned} A' &= 4i \oint_a w(z)\, dz + \frac{c}{\omega_a} \oint_a \frac{dz}{w(z)} \equiv A + c, \\ B' &= 4i \oint_b w(z)\, dz + \frac{c}{\omega_a} \oint_b \frac{dz}{w(z)} \equiv B + c\tau, \quad \tau = \frac{\omega_b}{\omega_a}. \end{aligned} \tag{8.4.9}$$

Let us introduce the "intermediate" $\overset{o}{\Psi}$-function,

$$\overset{o}{\Psi}(z) = \Psi^{(BA)}(z) e^{t_0 h(z) \sigma_3}, \tag{8.4.10}$$

where the constant t_0 will also be determined later. The asymptotics of this function at infinity,

$$\overset{o}{\Psi}(z) = \big(I + O(z^{-1})\big) e^{-(t - t_0)\theta \sigma_3}, \quad z \to \infty, \tag{8.4.11}$$

differs from the asymptotics of canonical functions by the fact that t is replaced with $t - t_0$. The jump properties of $\overset{o}{\Psi}$ are similar to those of the function $\Psi^{(BA)}$ except that they are described in terms of the new parameters t_0, A' and B':

$$z \in (-z_1, -z_3): \qquad \overset{o}{\Psi}_+(z) = \overset{o}{\Psi}_-(z)\hat{G}_A(z), \tag{8.4.12}$$

$$z \in (z_3, z_1): \qquad \overset{o}{\Psi}_+(z) = \overset{o}{\Psi}_-(z)\hat{G}_B(z), \tag{8.4.13}$$

$$z \in (z_1, -z_1): \qquad \overset{o}{\Psi}_+(z) = \overset{o}{\Psi}_-(z)\hat{G}_C(z), \tag{8.4.14}$$

where

$$\hat{G}_A = \begin{pmatrix} & -s_3^{-1}e^{t_0 B'/2} \\ s_3 e^{-t_0 B'/2} & \end{pmatrix}, \quad \hat{G}_B = \begin{pmatrix} & s_3 e^{-t_0 B'/2} \\ -s_3^{-1}e^{t_0 B'/2} & \end{pmatrix},$$

$$\hat{G}_C = \begin{pmatrix} (1 - s_1 s_3)^{-1} e^{-t_0 A'} & \\ & (1 - s_1 s_3) e^{t_0 A'} \end{pmatrix}.$$

Taking into account the conditions (8.2.11), we can choose the constant parameters c and t_0 in such a way that

$$s_3 = e^{t_0 B'/2}, \quad 1 - s_1 s_3 = e^{-t_0 A'}. \tag{8.4.15}$$

The system (8.4.15) yields immediately the parameter values

$$\begin{aligned} t_0 &= -2\frac{\omega_a}{\Delta}\ln s_3 - \frac{\omega_b}{\Delta}\ln(1 - s_1 s_3) \quad \left(\mathrm{mod}\ \frac{4\pi i \omega_a}{\Delta}\right) \quad \left(\mathrm{mod}\ \frac{2\pi i \omega_b}{\Delta}\right), \\ \frac{c}{\omega_a}t_0 &= 2\frac{A}{\Delta}\ln s_3 + \frac{B}{\Delta}\ln(1 - s_1 s_3) \quad \left(\mathrm{mod}\ \frac{4\pi i A}{\Delta}\right) \quad \left(\mathrm{mod}\ \frac{2\pi i B}{\Delta}\right), \end{aligned} \tag{8.4.16}$$

where

$$\Delta = A\omega_b - B\omega_a = 4i\big(\mathcal{J}_a\omega_b - \mathcal{J}_b\omega_a\big) = -\frac{16\pi}{3}\left(z_1^2 + z_3^2\right) = \frac{8\pi}{3}e^{i\varphi}$$

due to the Riemann bilinear identity.

Under the conditions (8.4.15), the RH problem for the function $\overset{o}{\Psi}$ simplifies dramatically:

1) near infinity, it behaves like the exponent

$$\overset{o}{\Psi}(z) = \big(I + O(z^{-1})\big)e^{-(t-t_0)\theta\sigma_3}, \quad z \to \infty; \tag{8.4.17}$$

2) it jumps on the intervals $(-z_1, -z_3)$ and (z_3, z_1):

$$\begin{aligned} z \in (-z_1, -z_3): &\qquad \overset{o}{\Psi}_+(z) = \overset{o}{\Psi}_-(z)\begin{pmatrix} & -1 \\ 1 & \end{pmatrix}, \\ z \in (z_3, z_1): &\qquad \overset{o}{\Psi}_+(z) = \overset{o}{\Psi}_-(z)\begin{pmatrix} & 1 \\ -1 & \end{pmatrix}; \end{aligned} \tag{8.4.18}$$

3) it has weaker than L^2-integrable singularities at $\pm z_1$, $\pm z_3$.

With the exception of the countable set of values of t described below in (8.4.26), this problem can be solved explicitly; cf. [**DeIZ**]. Let us introduce the function

$$\mu = \frac{1}{2}\begin{pmatrix} \beta+\beta^{-1} & -i(\beta-\beta^{-1}) \\ i(\beta-\beta^{-1}) & \beta+\beta^{-1} \end{pmatrix}, \tag{8.4.19}$$

where

$$\beta = \left(\frac{z^2-z_1^2}{z^2-z_3^2}\right)^{1/4} \tag{8.4.20}$$

is defined on the upper sheet of the Riemann surface Γ (see Figure 8.11) in such a way that

$$\beta \to 1 \quad \text{as} \quad z \to \infty,$$

so that

$$\mu \underset{z\to\infty}{\longrightarrow} I. \tag{8.4.21}$$

The discontinuity of the scalar function β is described by the relations

$$\begin{aligned} z \in (-z_1, -z_3):& \qquad \beta_+ = -i\beta_-, \\ z \in (z_3, z_1):& \qquad \beta_+ = i\beta_-, \end{aligned}$$

and therefore, the function μ satisfies the jump conditions

$$\begin{aligned} z \in (-z_1, -z_3):& \qquad \mu_+ = \mu_-\begin{pmatrix} & -1 \\ 1 & \end{pmatrix}, \\ z \in (z_3, z_1):& \qquad \mu_+ = \mu_-\begin{pmatrix} & 1 \\ -1 & \end{pmatrix}. \end{aligned} \tag{8.4.22}$$

The function $\overset{o}{\Psi}$ is constructed in terms of the Baker-Akhiezer functions as follows:

$$\overset{o}{\Psi}(z) = \begin{pmatrix} \mu_{11}\overset{(1)}{\Psi}_{BA} & \mu_{12}\overset{(1)}{\Psi}{}^*_{BA} \\ \mu_{21}\overset{(2)}{\Psi}_{BA} & \mu_{22}\overset{(2)}{\Psi}{}^*_{BA} \end{pmatrix}, \tag{8.4.23}$$

where

$$\begin{aligned} \overset{(1)}{\Psi}_{BA}(z) &= e^{-(t-t_0)\Omega(z)}\frac{\Theta\big(U(z)+V(t-t_0)+\frac{1}{2}\big)\Theta\big(\frac{1}{2}\big)}{\Theta\big(U(z)+\frac{1}{2}\big)\Theta\big(V(t-t_0)+\frac{1}{2}\big)}, \\ \overset{(2)}{\Psi}_{BA} &= e^{-(t-t_0)\left(\Omega(z)-\pi iV\right)}\frac{\Theta\big(U(z)+V(t-t_0)+\frac{1+\tau}{2}\big)\Theta\big(\frac{1}{2}\big)}{\Theta\big(U(z)+\frac{1+\tau}{2}\big)\Theta\big(V(t-t_0)+\frac{1}{2}\big)}, \\ \overset{(1)}{\Psi}{}^*_{BA}(z) &= e^{(t-t_0)\left(\Omega(z)-\pi iV\right)}\frac{\Theta\big(U(z)-V(t-t_0)+\frac{1+\tau}{2}\big)\Theta\big(\frac{1}{2}\big)}{\Theta\big(U(z)+\frac{1+\tau}{2}\big)\Theta\big(V(t-t_0)+\frac{1}{2}\big)}, \\ \overset{(2)}{\Psi}{}^*_{BA} &= e^{(t-t_0)\Omega(z)}\frac{\Theta\big(U(z)-V(t-t_0)+\frac{1}{2}\big)\Theta\big(\frac{1}{2}\big)}{\Theta\big(U(z)+\frac{1}{2}\big)\Theta\big(V(t-t_0)+\frac{1}{2}\big)}, \end{aligned} \tag{8.4.24}$$

and

$$\Omega(z) = g(z) - AU(z), \quad U(z) = \frac{1}{\omega_a}\int_{\infty^+}^{z} \frac{dz}{w(z)}, \qquad \tau = \frac{\omega_b}{\omega_a}, \tag{8.4.25}$$

$$V = \frac{\Delta}{2\pi i \omega_a} = \frac{8i}{3\omega_a}\left(z_1^2 + z_3^2\right) = -\frac{4i}{3\omega_a}e^{i\varphi},$$

$$t \neq t_0 + \frac{\tau}{2V} \quad (\mathrm{mod}\ \frac{1}{V}) \quad (\mathrm{mod}\ \frac{\tau}{V}), \tag{8.4.26}$$

with $\Theta(z)$ denoting the Riemann theta-function of Γ (the third Jacobi theta-function), $\Theta(z) = \sum_{n\in\mathbb{Z}} e^{\pi i n^2\tau + 2\pi i n z}$.

It is worth noting that all $\overset{(k)}{\Psi}_{BA}(z)$, $\overset{(k)}{\Psi}{}^{*}_{BA}(z)$ are single valued on $\mathbb{C}^*$ cut along $[z_3, z_1] \cup [-z_1, -z_3]$ (see Figure 8.11) because $\oint_a d\Omega = 0$. Also, on the Riemann surface parametrized by the pairs $P = (z, w)$ with the involution $*$ defined by $P^* = (z, -w)$, the Baker-Akhiezer functions satisfy

$$\overset{(k)}{\Psi}_{BA}(z) = \overset{(k)}{\Psi}_{BA}(P)$$

and

$$\overset{(k)}{\Psi}{}^{*}_{BA}(z) = \overset{(k)}{\Psi}_{BA}(P^*).$$

Now, let us verify that the function (8.4.23) yields the solution of the RH problem (8.4.17)–(8.4.18). First of all note that $\overset{o}{\Psi}(z)$ has no pole on the upper sheet of Γ. Indeed, $\Theta\big(U(z) + \frac{1}{2}\big) \neq 0\ \forall z \in \mathbb{C}^*$. Taking into account the equation $\Theta\big(\frac{1+\tau}{2}\big) = 0$, the only zero of $\Theta\big(U(z) + \frac{1+\tau}{2}\big)$ lies at ∞^+. Thus $\Theta\big(U(z) + \frac{1+\tau}{2}\big) = O(z^{-1})$, while both μ_{12} and μ_{21} vanish as $O(z^{-2})$.

Next, we verify the asymptotics of $\overset{o}{\Psi}(z)$ near infinity on the upper sheet of Γ:

$$U(z) = -\frac{1}{\omega_a z} + O(z^{-2}), \quad z \to \infty^+,$$

and therefore

$$\big(\overset{o}{\Psi}\big)_{11} = e^{-i(t-t_0)(\frac{4}{3}z^3 + e^{i\varphi}z)}\big(1 + O(z^{-1})\big),$$

$$\big(\overset{o}{\Psi}\big)_{22} = e^{i(t-t_0)(\frac{4}{3}z^3 + e^{i\varphi}z)}\big(1 + O(z^{-1})\big).$$

Similarly, using the asymptotics

$$\mu_{12} = -i\frac{z_3^2 - z_1^2}{4z^2} + O(z^{-4}), \quad \mu_{21} = i\frac{z_3^2 - z_1^2}{4z^2} + O(z^{-4}),$$

$$\Theta\big(U(z) + \frac{1+\tau}{2}\big) = -\frac{1}{\omega_a z}\Theta'\big(\frac{1+\tau}{2}\big) + O(z^{-2}),$$

we obtain

$$\big(\overset{o}{\Psi}\big)_{21} = e^{-i(t-t_0)(\frac{4}{3}z^3 + e^{i\varphi}z)} \cdot O(z^{-1}), \quad \big(\overset{o}{\Psi}\big)_{12} = e^{i(t-t_0)(\frac{4}{3}z^3 + e^{i\varphi}z)} \cdot O(z^{-1}),$$

and the asymptotics (8.4.17) is verified. In order to verify the jump conditions (8.4.18), it is sufficient, due to (8.4.22), to show that

$$\overset{(k)}{\Psi}_{BA}(z)_{\pm} = \overset{(k)}{\Psi}{}^{*}_{BA}(z)_{\mp}$$

on the segments (z_3, z_1) and $(-z_1, -z_3)$. Let us consider the first of them. On the segment (z_3, z_1),

$$U_+ + U_- = \frac{\tau}{2}, \quad \Omega_+ + \Omega_- = \frac{1}{2}(B - A\tau) = -\frac{\Delta}{2\omega_a} = -\pi i V.$$

Therefore,

$$\begin{aligned}\overset{(1)}{\Psi}_{BA}(z)_+ &= e^{-(t-t_0)\Omega_+(z)} \frac{\Theta\big(U_+(z) + V(t-t_0) + \frac{1}{2}\big)\Theta(\frac{1}{2})}{\Theta\big(U_+(z) + \frac{1}{2}\big)\Theta\big(V(t-t_0) + \frac{1}{2}\big)} \\ &= e^{(t-t_0)\left(\Omega_-(z)+\pi i V\right)} \frac{\Theta\big(-U_-(z) + \frac{\tau}{2} + V(t-t_0) + \frac{1}{2}\big)\Theta(\frac{1}{2})}{\Theta\big(-U_-(z) + \frac{\tau}{2} + \frac{1}{2}\big)\Theta\big(V(t-t_0) + \frac{1}{2}\big)} \\ &= e^{(t-t_0)\left(\Omega_-(z)-\pi i V\right)} \frac{\Theta\big(U_-(z) + \frac{1+\tau}{2} - V(t-t_0)\big)\Theta(\frac{1}{2})}{\Theta\big(U_-(z) + \frac{1+\tau}{2}\big)\Theta\big(V(t-t_0) + \frac{1}{2}\big)} = \overset{(1)}{\Psi}{}^*_{BA}(z)_-,\end{aligned}$$

where we have used the properties $\Theta(-z) = \Theta(z)$, $\Theta(z+1) = \Theta(z)$, $\Theta(z \pm \tau) = e^{-\pi i \tau \mp 2\pi i z}\Theta(z)$. The other condition can be checked similarly.

Thus we obtain the following representation for the function $\Psi^{(BA)}$:

$$\Psi^{(BA)} = \overset{o}{\Psi} e^{-t_0 h \sigma_3} \begin{pmatrix} \mu_{11} \overset{(1)}{\Psi}_{BA} & \mu_{12} \overset{(1)}{\Psi}{}^*_{BA} \\ \mu_{21} \overset{(2)}{\Psi}_{BA} & \mu_{22} \overset{(2)}{\Psi}{}^*_{BA} \end{pmatrix} e^{-t_0 h \sigma_3} \tag{8.4.27}$$

$$= \begin{pmatrix} \mu_{11} \overset{(1)}{\Theta} & \mu_{12} e^{-\pi i V(t-t_0)} \overset{(1)}{\Theta}{}^* \\ \mu_{21} e^{\pi i V(t-t_0)} \overset{(2)}{\Theta} & \mu_{22} \overset{(2)}{\Theta}{}^* \end{pmatrix} e^{(tA - t_0 A')U\sigma_3} e^{-tg\sigma_3},$$

where

$$\overset{(1)}{\Theta} = \frac{\Theta\big(U(z) + V(t-t_0) + \frac{1}{2}\big)\Theta(\frac{1}{2})}{\Theta\big(U(z) + \frac{1}{2}\big)\Theta\big(V(t-t_0) + \frac{1}{2}\big)}, \quad \overset{(1)}{\Theta}{}^* = \frac{\Theta\big(U(z) - V(t-t_0) + \frac{1+\tau}{2}\big)\Theta(\frac{1}{2})}{\Theta\big(U(z) + \frac{1+\tau}{2}\big)\Theta\big(V(t-t_0) + \frac{1}{2}\big)},$$

$$\overset{(2)}{\Theta} = \frac{\Theta\big(U(z) + V(t-t_0) + \frac{1+\tau}{2}\big)\Theta(\frac{1}{2})}{\Theta\big(U(z) + \frac{1+\tau}{2}\big)\Theta\big(V(t-t_0) + \frac{1}{2}\big)}, \quad \overset{(2)}{\Theta}{}^* = \frac{\Theta\big(U(z) - V(t-t_0) + \frac{1}{2}\big)\Theta(\frac{1}{2})}{\Theta\big(U(z) + \frac{1}{2}\big)\Theta\big(V(t-t_0) + \frac{1}{2}\big)}.$$

Note finally, that the determinant of $\Psi^{(BA)}(z)$ might have singularities at the branch points $\pm z_1$, $\pm z_3$ of order $O((z - z_k)^{-1/2})$. However, because of the asymptotics $\det \Psi^{(BA)}(z) = 1 + O\big(z^{-1}\big)$ as $z \to \infty$, and the absence of any jump, the determinant is an entire function of z which therefore, due to the Liouville theorem, is constant, $\det \Psi^{(BA)}(z) \equiv 1$.

5. Local RH problems near the branch points

5.1. A RH problem solvable by the Airy functions. The function $\Psi^{(BA)}$ has branch points at $z = z_k$, $k = 1, 2, 3, 4$, and does not approach the solution $\Psi(z)$ of the problem (8.2.28)–(8.2.35). In this section, we will construct such an approximation using a model problem related to the Airy functions. Consider the Wronskian matrix of the independent solutions $y_i(\zeta)$ of the Airy equation $y'' = \zeta y$,

$$\Phi^0(\zeta) = \begin{pmatrix} y_1(\zeta) & y_2(\zeta) \\ y_1'(\zeta) & y_2'(\zeta) \end{pmatrix}.$$

Using the asymptotic formulae presented in [**BE**], one can find the proper constant matrices C_k and introduce the matrix functions $\Phi_k(\zeta) \equiv \Phi^0(\zeta)C_k$ with the standard asymptotic behavior as $\zeta \to \infty$,

$$\Phi_k(\zeta) = \zeta^{-\frac{\sigma_3}{4}} \begin{pmatrix} 1 & 1 \\ 1 & -1 \end{pmatrix} (I + O(\zeta^{-3/2}))e^{\frac{2}{3}\zeta^{3/2}\sigma_3}, \quad \zeta^{3/2} \to \infty, \tag{8.5.1}$$

$$\zeta \in \omega_k = \Big\{\zeta \in \mathbb{C}\colon \ \arg\zeta \in \big(\frac{2\pi}{3}(k - \frac{3}{2}) + \epsilon, \ \frac{2\pi}{3}(k + \frac{1}{2}) - \epsilon\big)\Big\}. \quad \epsilon > 0.$$

These functions are related by the Stokes multipliers

$$\Phi_{k+1}(\zeta) = \Phi_k(\zeta)G_k, \quad G_{2k-1} = \begin{pmatrix} 1 & -i \\ 0 & 1 \end{pmatrix}, \quad G_{2k} = \begin{pmatrix} 1 & 0 \\ -i & 1 \end{pmatrix}. \tag{8.5.2}$$

These matrix functions allow us to construct an exact solution $\Phi(\zeta)$ of the following RH problem (see Figure 8.12):

1) near infinity

$$\Phi = \zeta^{-\frac{\sigma_3}{4}} \begin{pmatrix} 1 & 1 \\ 1 & -1 \end{pmatrix} (I + O(\zeta^{-3/2}))e^{\frac{2}{3}\zeta^{3/2}\sigma_3}, \quad \zeta^{3/2} \to \infty, \tag{8.5.3}$$

where $\zeta^{3/2}$ is defined on the plane cut along the ray $\arg\zeta = -\pi/3$;

2) on the anti-Stokes rays $\arg\zeta = \frac{2\pi}{3}k$, $k = 0, 1, 2$, oriented to infinity, Φ satisfies the jump conditions

$$\Phi_+ = \Phi_- G_k, \quad \arg\zeta = \frac{2\pi}{3}k, \quad k = 0, 1, 2; \tag{8.5.4}$$

3) on the ray $\arg\zeta = -\pi/3$ oriented to infinity, Φ satisfies the jump condition

$$\Phi_+ = \Phi_- G_\infty, \quad G_\infty = i\sigma_1 = \begin{pmatrix} & i \\ i & \end{pmatrix}. \tag{8.5.5}$$

5.2. Parametrix in a neighborhood of the branch point z_3. Let us introduce the change of variables

$$\zeta(z) = (-\frac{3t}{2}g_3(z))^{2/3}, \quad g_3(z) = 4i\int_{z_3}^{z} w(z)\,dz \equiv g(z) - \frac{1}{4}B. \tag{8.5.6}$$

This change is defined on the complex z-plane cut along the intervals $[z_1, -z_1] \cup [-z_1, -z_3] \cup [-z_3, \infty)$; it maps the domain $\mathcal{D}_3$ of the z-plane bounded by the Stokes line chain shown in Figure 8.13, onto the complex ζ-plane cut along the ray $\arg\zeta = -\pi/3$. The function

$$\check{\Phi}(z) = \Phi(\zeta(z)) \tag{8.5.7}$$

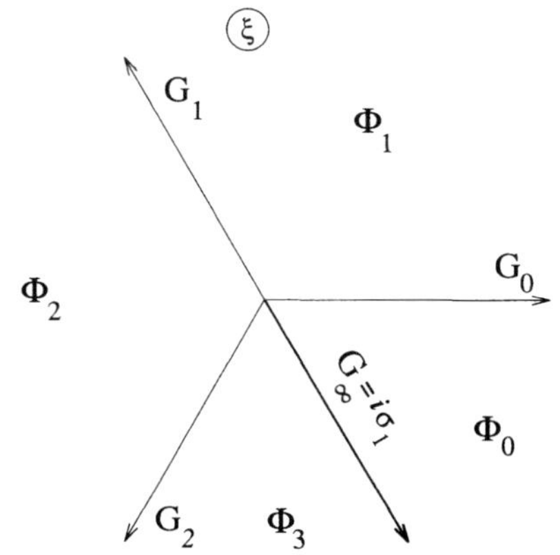

FIGURE 8.12. The RH problem graph for the Wronskian matrix of the Airy functions

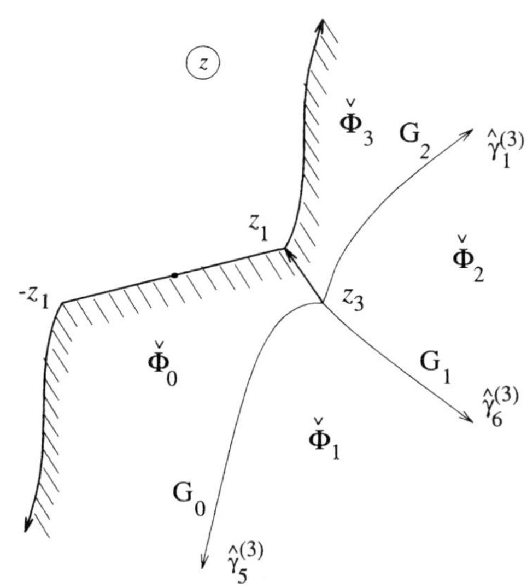

FIGURE 8.13. The jump graph for the fundamental system of the Airy functions after the change of variables $\zeta = \big(-3tg_3(z)/2\big)^{2/3}$

has the following behavior at infinity,
(8.5.8)

$$\check{\Phi}(z) = \big(-\frac{3t}{2}g_3(z)\big)^{-\frac{\sigma_3}{6}} \begin{pmatrix} 1 & 1 \\ 1 & -1 \end{pmatrix} \big(I + O\left(\frac{1}{tg_3(z)}\right)\big)e^{-tg_3(z)\sigma_3}, \quad tg_3(z) \to \infty,$$

and has jumps on the graph shown in Figure 8.13.

The rays $\arg\zeta = 2\pi k/3$, $k = 0, 1, 2$, are mapped into the anti-Stokes lines $\hat{\gamma}_5^{(3)}$, $\hat{\gamma}_6^{(3)}$ and $\hat{\gamma}_1^{(3)}$, respectively (see Figure 8.6); moreover, on these lines the $\check{\Phi}$-jump matrices coincide with the respective Φ-jump matrices. The cut $\arg\zeta = -\pi/3$ is mapped into the segment (z_3, z_1), which splits at the point z_1 into the boundary of the z-domain.

Therefore, the jumps of the function

$$\Phi^{(3)}(z) = \check{\Phi}(z) e^{i\frac{\pi}{4}\sigma_3} s_3^{-\frac{\sigma_3}{2}}, \tag{8.5.9}$$

on the graph shown on Figure 8.13 are given by the matrices (8.2.30), (8.2.34),

$$\begin{gathered} z \in \hat{\gamma}_5^{(3)}: \quad S_5^{(3)} = \begin{pmatrix} 1 & 0 \\ s_3^{-1} & 1 \end{pmatrix}; \qquad z \in \hat{\gamma}_6^{(3)}: \quad S_6 = \begin{pmatrix} 1 & -s_3 \\ 0 & 1 \end{pmatrix}, \\ z \in \hat{\gamma}_1^{(3)}: \quad S_1^{(3)} = \begin{pmatrix} 1 & 0 \\ s_3^{-1} & 1 \end{pmatrix}; \qquad z \in (z_3, z_1): \quad S_B = \begin{pmatrix} & s_3 \\ -s_3^{-1} & \end{pmatrix}. \end{gathered} \tag{8.5.10}$$

Furthermore, $\Phi^{(3)}(z)$ has the following asymptotics near infinity

$$\begin{aligned} \Phi^{(3)}(z) &= \left(-\frac{3t}{2} g_3(z)\right)^{-\frac{\sigma_3}{6}} \begin{pmatrix} 1 & 1 \\ 1 & -1 \end{pmatrix} \left(I + O\left(\frac{1}{t g_3(z)}\right)\right) \\ &\times e^{\frac{1}{4}(tB - t_0 B')\sigma_3} e^{i\frac{\pi}{4}\sigma_3} e^{-t g \sigma_3}, \quad t g_3(z) \to \infty. \end{aligned} \tag{8.5.11}$$

The jumps of the matrix functions $\Psi(z)$ and $\Phi^{(3)}(z)$ coincide on the contours $\hat{\gamma}_5^{(3)}$, $\hat{\gamma}_6^{(3)}$, $\hat{\gamma}_1^{(3)}$ and (z_3, z_1). Thus these functions differ by a left matrix multiplier $M^{(3)}(z)$ holomorphic in some bounded neighborhood of the node point $z = z_3$,

$$\Psi(z) = M^{(3)}(z) \Phi^{(3)}(z).$$

In addition to the Airy fundamental matrix set $\Phi(\zeta)$, we also introduce the function defined by the associated reduced asymptotic form

$$\Phi_{as}(\zeta) = \zeta^{-\frac{\sigma_3}{4}} \begin{pmatrix} 1 & 1 \\ 1 & -1 \end{pmatrix} e^{\frac{2}{3}\zeta^{3/2}\sigma_3}.$$

This function solves the RH problem on the ray $\arg\zeta = -\pi/3$ oriented from zero to infinity, satisfying

$$\Phi_{as_+} = \Phi_{as_-} G_\infty, \quad G_\infty = i\sigma_1 = \begin{pmatrix} & i \\ i & \end{pmatrix}. \tag{8.5.12}$$

Using the transformations (8.5.6) and (8.5.9), the matrix

$$\Phi_{as}^{(3)}(z) = \Phi_{as}(\zeta(z)) e^{i\frac{\pi}{4}\sigma_3} s_3^{-\frac{\sigma_3}{2}}$$

has a jump on the Stokes segment (z_3, z_1), and the relevant jump matrix is (8.2.34)

$$\Phi^{(3)}_{as_+} = \Phi^{(3)}_{as_-} S_B, \quad S_B = \begin{pmatrix} & s_3 \\ -s_3^{-1} & \end{pmatrix}, \qquad z \in (z_3, z_1).$$

Since the jumps of the matrix functions $\Psi^{(BA)}(z)$ and $\Phi^{(3)}_{as}(z)$ coincide on the segment (z_3, z_1), the functions differ by a left matrix multiplier $N^{(3)}(z)$ holomorphic in some bounded neighborhood of the point $z = z_3$,

$$\Psi^{(BA)}(z) = N^{(3)}(z)\Phi^{(3)}_{as}(z). \tag{8.5.13}$$

Let us define

$$\Psi^{(3)}(z) = N^{(3)}(z)\Phi^{(3)}(z), \quad N^{(3)}(z) = \Psi^{(BA)}(z)\big(\Phi^{(3)}_{as}(z)\big)^{-1}. \tag{8.5.14}$$

5.3. Parametrix in a neighborhood of the branch point z_1. To construct the parametrix near $z = z_1$, let us introduce the function $\hat{\Phi}^{(1)}(\zeta)$,

$$\hat{\Phi}^{(1)}(\zeta) = \begin{cases} \Phi_0(\zeta)e^{-i\frac{\pi}{4}\sigma_3}\big(\frac{s_3}{1-s_1s_3}\big)^{-\frac{\sigma_3}{2}} & \text{for} \quad \arg\zeta \in \big(-\frac{\pi}{3}, 0\big), \\ \Phi_1(\zeta)e^{-i\frac{\pi}{4}\sigma_3}\big(\frac{s_3}{1-s_1s_3}\big)^{-\frac{\sigma_3}{2}} & \text{for} \quad \arg\zeta \in \big(0, \frac{2\pi}{3}\big), \\ \Phi_2(\zeta)e^{-i\frac{\pi}{4}\sigma_3}\big(\frac{s_3}{1-s_1s_3}\big)^{-\frac{\sigma_3}{2}} & \text{for} \quad \arg\zeta \in \big(\frac{2\pi}{3}, \pi\big), \\ \Phi_2(\zeta)e^{-i\frac{\pi}{4}\sigma_3}\big(s_3(1-s_1s_3)\big)^{-\frac{\sigma_3}{2}} & \text{for} \quad \arg\zeta \in \big(\pi, \frac{4\pi}{3}\big), \\ \Phi_3(\zeta)e^{-i\frac{\pi}{4}\sigma_3}\big(s_3(1-s_1s_3)\big)^{-\frac{\sigma_3}{2}} & \text{for} \quad \arg\zeta \in \big(\frac{4\pi}{3}, \frac{5\pi}{3}\big). \end{cases} \tag{8.5.15}$$

This matrix function satisfies the following jump conditions on the rays oriented to infinity,

$$\hat{\Phi}^{(1)}_+(\zeta) = \hat{\Phi}^{(1)}_-(\zeta)S^{(1)}_1, \quad S^{(1)}_1 = \begin{pmatrix} 1 & 0 \\ -\frac{1-s_1s_3}{s_3} & 1 \end{pmatrix}, \qquad \arg\zeta = 0, \tag{8.5.16}$$

$$\hat{\Phi}^{(1)}_+(\zeta) = \hat{\Phi}^{(1)}_-(\zeta)S^{(1)}_2, \quad S^{(1)}_2 = \begin{pmatrix} 1 & \frac{s_3}{1-s_1s_3} \\ 0 & 1 \end{pmatrix}, \qquad \arg\zeta = \frac{2\pi}{3},$$

$$\hat{\Phi}^{(1)}_+(\zeta) = \hat{\Phi}^{(1)}_-(\zeta)S_C, \quad S_C = \begin{pmatrix} \frac{1}{1-s_1s_3} & \\ & 1-s_1s_3 \end{pmatrix}, \qquad \arg\zeta = \pi,$$

$$\hat{\Phi}^{(1)}_+(\zeta) = \hat{\Phi}^{(1)}_-(\zeta)S^{(1)}_5, \quad S^{(1)}_5 = \begin{pmatrix} 1 & 0 \\ -\frac{1}{s_3(1-s_1s_3)} & 1 \end{pmatrix}, \qquad \arg\zeta = \frac{4\pi}{3},$$

$$\hat{\Phi}^{(1)}_+(\zeta) = \hat{\Phi}^{(1)}_-(\zeta)S_B^{-1}, \quad S_B^{-1} = \begin{pmatrix} & -s_3 \\ \frac{1}{s_3} & \end{pmatrix}, \qquad \arg\zeta = -\frac{\pi}{3}.$$

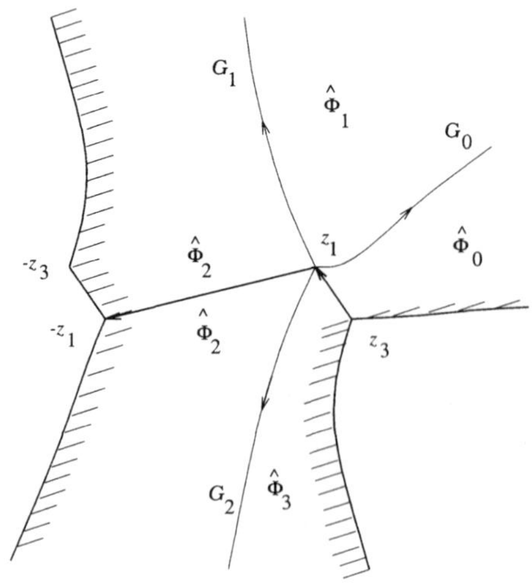

FIGURE 8.14. The jump graph for the Wronskian matrix associated with the Airy functions after the change of variables $\zeta = \left(-3tg_1(z)/2\right)^{2/3}$

Introduce the change of variables

$$\zeta(z) = \left(-\frac{3t}{2}g_1(z)\right)^{2/3}, \tag{8.5.17}$$

where

$$g_1(z) = 4i\int_{z_1}^{z} w(z)\,dz$$

is defined on the z-plane cut along $(\infty, z_3) \cup (z_3, z_1) \cup (\infty, -z_1) \cup (-z_1, -z_3)$. This transformation is holomorphic near the point z_1, and hence it is defined on the z-plane cut along the intervals $(z_3, \infty) \cup (\infty, -z_1) \cup (-z_1, -z_3)$; it maps the domain $\mathcal{D}_1$ of the z-plane with the boundary formed by the Stokes lines shown on Figure 8.14, onto the complex ζ-plane cut along the rays $\arg\zeta = -\pi/3$, $\arg(\zeta - \zeta_2) = \pi$, with some ζ_2 satisfying $\arg\zeta_2 = \pi$.

Thus, in the area to the right of the cut $(z_3, z_1) \cup (z_1, -z_1)$, see Figure 8.14, or, equivalently, for $\arg\left(-\frac{3t}{2}g_1(z)\right) \in \left(-\frac{\pi}{2}, \frac{3\pi}{2}\right)$,

$$g_1(z) = g(z) - \frac{1}{4}B - \frac{1}{2}A; \tag{8.5.18}$$

similarly, in the area to the left of the cut $(z_3, z_1) \cup (z_1, -z_1)$, i.e., for z satisfying the condition $\arg\left(-\frac{3t}{2}g_1(z)\right) \in \left(\frac{3\pi}{2}, \frac{5\pi}{2}\right)$,

$$g_1(z) = g(z) - \frac{1}{4}B + \frac{1}{2}A. \tag{8.5.19}$$

The function

$$\Phi^{(1)}(z) = \hat{\Phi}^{(1)}(\zeta(z)) \tag{8.5.20}$$

has the following asymptotics at infinity

$$\Phi^{(1)}(z) = (-\frac{3t}{2}g_1(z))^{-\frac{\sigma_3}{6}} \begin{pmatrix} 1 & 1 \\ 1 & -1 \end{pmatrix} (I + O(\frac{1}{tg_1(z)}))$$
$$\times e^{-tg\sigma_3} e^{-i\frac{\pi}{4}\sigma_3} e^{\frac{1}{4}(tB - t_0 B')\sigma_3} e^{\frac{1}{2}(tA - t_0 A')\sigma_3},$$
$$tg_1(z) \to \infty, \quad \arg(-tg_1(z)) \in (-\frac{\pi}{2}, \frac{3\pi}{2}),$$
$$\Phi^{(1)}(z) = (-\frac{3t}{2}g_1(z))^{-\frac{\sigma_3}{6}} \begin{pmatrix} 1 & 1 \\ 1 & -1 \end{pmatrix} \left(I + O\left(\frac{1}{tg_1(z)}\right)\right)$$
$$\times e^{-tg\sigma_3} e^{-i\frac{\pi}{4}\sigma_3} e^{\frac{1}{4}(tB - t_0 B')\sigma_3} e^{-\frac{1}{2}(tA - t_0 A')\sigma_3}, \tag{8.5.21}$$
$$tg_1(z) \to \infty, \quad \arg(-tg_1(z)) \in (\frac{3\pi}{2}, \frac{5\pi}{2});$$

the jumps of $\Phi^{(1)}$ are shown on Figure 8.14. Under the transformation (8.5.17), the interval (z_3, z_1) is mapped to a segment of the ray $\arg\zeta = -\frac{\pi}{3}$ (if the orientation of the ray is from infinity to the origin), the curve $\hat{\gamma}_1^{(1)}$ becomes the ray $\arg\zeta = 0$, the curve $\hat{\gamma}_2^{(1)}$ transforms into the ray $\arg\zeta = \frac{2\pi}{3}$, the Stokes line $(z_1, -z_1)$ becomes the ray $\arg\zeta = \pi$ and the curve $\hat{\gamma}_5^{(1)}$ becomes the ray $\arg\zeta = \frac{4\pi}{3}$.

Since the discontinuities of the functions $\Psi(z)$ and $\Phi^{(1)}(z)$ are the same on the contours $\hat{\gamma}_k^{(1)}$ and on the Stokes segments (z_3, z_1) and $(z_1, -z_1)$, two of these functions coincide up to a left matrix multiplier $M^{(1)}(z)$ holomorphic in some bounded neighborhood of the point z_1:

$$\Psi(z) = M^{(1)}(z)\Phi^{(1)}(z).$$

In addition to the model function $\Phi^{(1)}(z)$, we also introduce the function defined by the asymptotics of $\Phi^{(1)}(z)$:

$$\Phi^{(1)}_{as}(z) = (-\frac{3t}{2}g_1(z))^{-\frac{\sigma_3}{6}} \begin{pmatrix} 1 & 1 \\ 1 & -1 \end{pmatrix}$$
$$\times e^{-tg\sigma_3} e^{-i\frac{\pi}{4}\sigma_3} e^{\frac{1}{4}(tB - t_0 B')\sigma_3} e^{\frac{1}{2}(tA - t_0 A')\sigma_3},$$
$$tg_1(z) \to \infty, \quad \arg(-tg_1(z)) \in (-\frac{\pi}{2}, \frac{3\pi}{2}),$$
$$\Phi^{(1)}_{as}(z) = (-\frac{3t}{2}g_1(z))^{-\frac{\sigma_3}{6}} \begin{pmatrix} 1 & 1 \\ 1 & -1 \end{pmatrix}$$
$$\times e^{-tg\sigma_3} e^{-i\frac{\pi}{4}\sigma_3} e^{\frac{1}{4}(tB - t_0 B')\sigma_3} e^{-\frac{1}{2}(tA - t_0 A')\sigma_3}, \tag{8.5.22}$$
$$tg_1(z) \to \infty, \quad \arg(-g_1(z)) \in (\frac{3\pi}{2}, \frac{5\pi}{2}).$$

This function is continuous across the anti-Stokes curves $\hat{\gamma}_k^{(1)}$ and is characterized by the jump conditions

$$\Phi_{as_+}^{(1)}(z) = \Phi_{as_-}^{(1)}(z) S_C, \quad S_C = \begin{pmatrix} (1-s_1s_3)^{-1} & \\ & 1-s_1s_3 \end{pmatrix}, \quad z \in (z_1, -z_1),$$

$$\Phi_{as_+}^{(1)}(z) = \Phi_{as_-}^{(1)}(z) S_B, \quad S_B = \begin{pmatrix} & s_3 \\ -s_3^{-1} & \end{pmatrix}, \quad z \in (z_3, z_1).$$

Note again, that the jumps of the functions $\Psi^{(BA)}(z)$ and $\Phi_{as}^{(1)}(z)$ coincide on the segments (z_3, z_1) and $(z_1, -z_1)$; hence the functions coincide up to a left multiplier $N^{(1)}(z)$ holomorphic in some bounded neighborhood of the point z_1,

$$\Psi^{(BA)}(z) = N^{(1)}(z)\Phi_{as}^{(1)}(z). \tag{8.5.23}$$

Define

$$\Psi^{(1)}(z) = N^{(1)}(z)\Phi^{(1)}(z), \quad N^{(1)}(z) = \Psi^{(BA)}(z)\left(\Phi_{as}^{(1)}(z)\right)^{-1}. \tag{8.5.24}$$

The functions solving the model RH problem associated with the branch points $-z_1$ and $-z_3$ can be constructed in a similar way. Alternatively, they can be defined using the symmetry relations:

$$\Psi^{(2)}(z) = \sigma_2\Psi^{(1)}(-z)\sigma_2, \quad \Psi^{(4)}(z) = \sigma_2\Psi^{(3)}(-z)\sigma_2. \tag{8.5.25}$$

6. Asymptotic solution of the main RH problem

In this section, we will construct the approximate solution $\tilde{\Psi}(z)$ of the RH problem (8.2.28)–(8.2.34) and will also prove the solvability of this problem. To this end, let us introduce four nonoverlapping disks B_k, $k = 1, 2, 3, 4$, centered at the branch points z_k, and four circles $C_k = \partial B_k$. We define the piecewise holomorphic function $\tilde{\Psi}(z)$ by

$$\tilde{\Psi}(z) = \begin{cases} \Psi^{(k)}(z), & z \in B_k, \quad k = 1,2,3,4, \\ \Psi^{(BA)}(z), & z \in \mathbb{C}\setminus \cup_{k=1}^4 B_k. \end{cases} \tag{8.6.1}$$

The function $\tilde{\Psi}$ solves the following RH problem on the contour depicted on Figure 8.15:

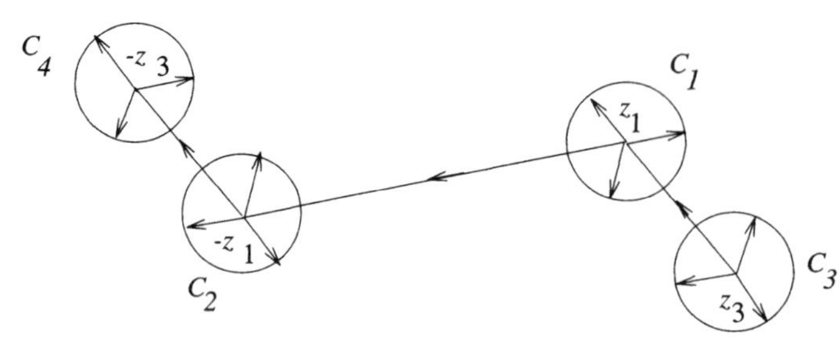

FIGURE 8.15. The model Riemann-Hilbert graph

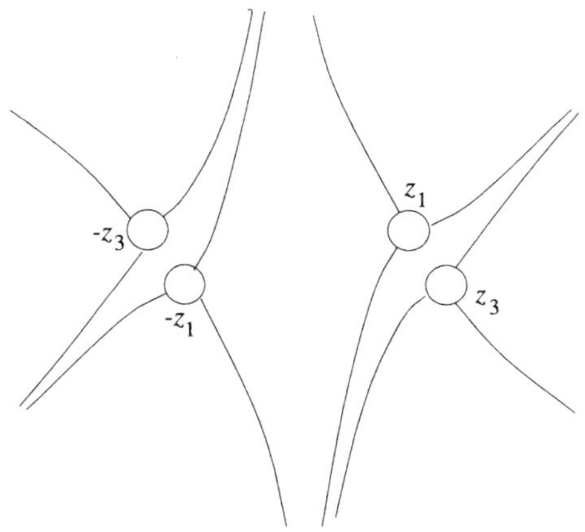

FIGURE 8.16. The RH graph for the correction function $Z(z)$

1) near infinity, due to (8.4.2),

$$\tilde{\Psi}(z) = \Psi^{(BA)}(z) = \big(I + O(z^{-1})\big)e^{-t\theta\sigma_3}, \quad z \to \infty, \tag{8.6.2}$$

where $\theta = i\big(\frac{4}{3}z^3 + e^{i\varphi}z\big)$;

2) on the parts $\hat{\gamma}_s^{(k)} \cap B_k$ of the anti-Stokes lines $\hat{\gamma}_s^{(k)}$ which are in the interior B_k of the circle C_k, and on the segments (z_3, z_1), $(z_1, -z_1)$, $(-z_1, -z_3)$, $\tilde{\Psi}$ satisfies the jump conditions

$$\tilde{\Psi}_+(z) = \tilde{\Psi}_-(z)\mathsf{S}, \tag{8.6.3}$$

where the jump matrices S are defined in (8.2.30)–(8.2.34);

3) on the circles C_k, $k = 1, 2, 3, 4$, oriented counterclockwise, $\tilde{\Psi}$ satisfies the jump conditions

$$\tilde{\Psi}_+(z) = \tilde{\Psi}_-(z)H(z), \tag{8.6.4}$$

where, in accord with (8.5.13), (8.5.14), (8.5.23), (8.5.24), (8.1.5) and (8.5.25),

$$H(z) = (\Psi^{(BA)}(z))^{-1}\Psi^{(k)}(z) = (\Phi_{as}^{(k)}(z))^{-1}\Phi^{(k)}(z), \quad z \in C_k.$$

To describe the difference between the model problem solution $\tilde{\Psi}(z)$ and the solution $\Psi(z)$ of the original problem, let us construct the "ratio"

$$Z(z) = \Psi(z)\tilde{\Psi}^{-1}(z). \tag{8.6.5}$$

The correction function $Z(z)$ satisfies the following RH problem on the jump contour shown on Figure 8.16:

1) $Z(z)$ is normalized at infinity

$$Z \underset{z\to\infty}{\longrightarrow} I; \tag{8.6.6}$$

2) on the parts of the anti-Stokes lines outside the circles C_k and on the circles themselves, Z satisfies the jump conditions

$$Z_+(z) = Z_-(z)\mathsf{G}_s^{(k)}(z), \quad z \in \hat{\gamma}_s^{(k)} \cap (\mathbb{C}\backslash B_k), \tag{8.6.7}$$

where, on the anti-Stokes tails,

$$\mathsf{G}_s^{(k)}(z) = \Psi^{(BA)}(z)\mathsf{S}_s^{(k)}\big(\Psi^{(BA)}(z)\big)^{-1},$$

and, on the circles C_k, $k = 1, 2, 3, 4$,

(8.6.8) $$Z_+(z) = Z_-(z)\mathrm{G}_k(z), \quad z \in C_k,$$

where

$$\mathrm{G}_k(z) = \Psi^{(BA)}(z)\big(\Psi^{(k)}(z)\big)^{-1}$$
$$= N^{(k)}(z)\Phi_{as}^{(k)}(z)\big(\Phi^{(k)}(z)\big)^{-1}\big(N^{(k)}(z)\big)^{-1}.$$

For t different than the points (8.4.26), the jump matrices on the anti-Stokes lines have the asymptotics

$$\mathrm{G}_s^{(k)}(z) = I + \big(e^{-2|iptRe(tg(z))|}\big) = I + \big(e^{-2|iptRe(tg_k(z))|}\big),$$

where we have used the Boutroux equations (8.3.6). Simultaneously, for t different than the points (8.4.26), using the definition of $\Phi_{as}^{(k)}(z)$ and the boundedness of the matrices $N^{(k)}(z)$ and their inverses, we find the estimate

$$\mathrm{G}_k(z) = I + (t^{-1}g_k^{-1}(z)).$$

Now, we will prove the solvability of the RH problem (8.6.6)–(8.6.8).

First we show that the jump matrices satisfy the smoothness condition (see Chapter 3) at the node points of the RH problem graph.

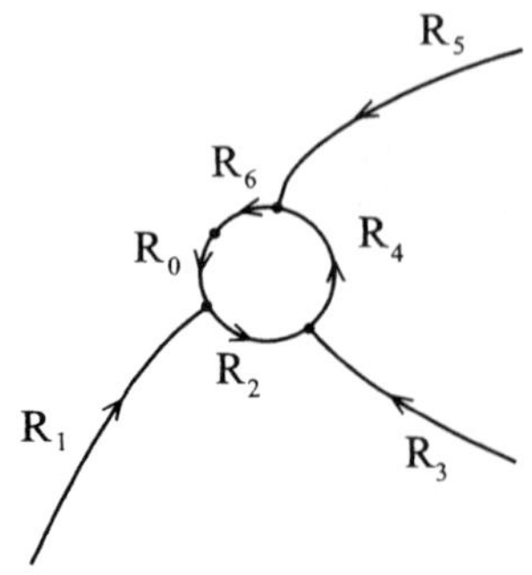

FIGURE 8.17. The connected component of the RH problem graph associated with the point z_3.

Consider the connected component of the graph associated with the point z_3 (see Figure 8.17). For convenience, we orient the anti-Stokes lines $\hat{\gamma}_s^{(3)}$ in the direction from infinity to z_3. Denote the jump matrices for the function $Z(z)$ as $R(z)$, i.e., $Z_+(z) = Z_-(z)R(z)$, moreover;

on the part $C_{25}^{(3)}$ of the circle C_3 between the points of its intersection with the Stokes line (z_3, z_1) and anti-Stokes line $\hat{\gamma}_5^{(3)}$, the jump matrix $R(z) = R_0(z)$ is given by

(8.6.9) $$R_0(z) = N^{(3)}(z)\Phi_{as}^{(3)}(z)(\Phi_0^{(3)}(z))^{-1}(N^{(3)}(z))^{-1} = I + O(t^{-1}g_3^{-1}(z));$$

- on the anti-Stokes line $\hat{\gamma}_5^{(3)}$, the jump matrix $R(z) = R_1(z)$ is given by

(8.6.10) $$R_1(z) = \Psi^{(BA)}(z)(\mathrm{S}_5^{(3)})^{-1}(\Psi^{(BA)}(z))^{-1} = I + O(e^{2tg(z)});$$

- on the part $C_{56}^{(3)}$ of the circle C_3 between the points of its intersection with the anti-Stokes lines $\hat{\gamma}_5^{(3)}$ and $\hat{\gamma}_6^{(3)}$, the jump matrix $R(z) = R_2(z)$ is given by

$$\begin{aligned} R_2(z) &= N^{(3)}(z)\Phi_{as}^{(3)}(z)(\Phi_1^{(3)}(z))^{-1}(N^{(3)}(z))^{-1} \\ &= N^{(3)}(z)\Phi_{as}^{(3)}(z)(S_5^{(3)})^{-1}(\Phi_0^{(3)}(z))^{-1}(N^{(3)}(z))^{-1} \\ &= R_1(z)R_0(z) = I + O(t^{-1}g_3^{-1}(z)); \end{aligned} \tag{8.6.11}$$

- on the anti-Stokes line $\hat{\gamma}_6^{(3)}$, the jump matrix $R(z) = R_3(z)$ is given by

$$R_3(z) = \Psi^{(BA)}(z)(S_6)^{-1}(\Psi^{(BA)}(z))^{-1} = I + O(e^{-2tg(z)}); \tag{8.6.12}$$

- on the part $C_{61}^{(3)}$ of the circle C_3 between the points of its intersection with the anti-Stokes lines $\hat{\gamma}_6^{(3)}$ and $\hat{\gamma}_1^{(3)}$, the jump matrix $R(z) = R_4(z)$ is given by

$$\begin{aligned} R_4(z) &= N^{(3)}(z)\Phi_{as}^{(3)}(z)(\Phi_2^{(3)}(z))^{-1}(N^{(3)}(z))^{-1} \\ &= N^{(3)}(z)\Phi_{as}^{(3)}(z)(S_6)^{-1}(S_5^{(3)})^{-1}(\Phi_0^{(3)}(z))^{-1}(N^{(3)}(z))^{-1} \\ &= R_3(z)R_1(z)R_0(z) = I + O(t^{-1}g_3^{-1}(z)); \end{aligned} \tag{8.6.13}$$

- on the anti-Stokes line $\hat{\gamma}_1^{(3)}$, the jump matrix $R(z) = R_5(z)$ is given by

$$R_5(z) = \Psi^{(BA)}(z)(S_1^{(3)})^{-1}(\Psi^{(BA)}(z))^{-1} = I + O(e^{-2tg(z)}); \tag{8.6.14}$$

- on the part $C_{12}^{(3)}$ of the circle C_3 between the points of its intersection with the anti-Stokes line $\hat{\gamma}_1^{(3)}$ and the Stokes line (z_3, z_1), the jump matrix $R(z) = R_6(z)$ is given by

$$\begin{aligned} R_6(z) &= N^{(3)}(z)\Phi_{as}^{(3)}(z)(\Phi_3^{(3)}(z))^{-1}(N^{(3)}(z))^{-1} \\ &= N^{(3)}(z)\Phi_{as}^{(3)}(z)(S_1^{(3)})^{-1}(S_6)^{-1}(S_5^{(3)})^{-1}(\Phi_0^{(3)}(z))^{-1}(N^{(3)}(z))^{-1} \\ &= R_5(z)R_3(z)R_1(z)R_0(z) = I + O(t^{-1}g_3^{-1}(z)). \end{aligned} \tag{8.6.15}$$

The above obtained relations,

$$R_2 = R_1R_0, \quad R_4 = R_3R_2, \quad R_6 = R_5R_4, \tag{8.6.16}$$

establish the smoothness of the RH problem at the node points $C_3 \cap \hat{\gamma}_s^{(3)}$ (see [**Zh**]; see also Chapter 3). Indeed, this follows from the fact that $R(z)$ admits analytic continuation from the contours C_3, $\hat{\gamma}_s^{(3)}$ and the fact that the graph can be disjointed at the node points in accord with (8.6.16) and can be transformed into the union of the circle C_3 and of three smooth curves with the common endpoint $P = C_3 \cap (z_3, z_1)$. The smoothness of the RH problem at the point P follows from the equation $R_0(z) = R_6(z)$. Indeed, at this point

$$\begin{aligned} R_0(z) &= \Psi_+^{(BA)}(z)(\Phi_0^{(3)}(z))_+^{-1}(N^{(3)}(z))^{-1} \\ &= \Psi_-^{(BA)}(z)S_B(\Phi_3^{(3)}(z)S_B)_+^{-1}(N^{(3)}(z))^{-1} \\ &= \Psi_-^{(BA)}(z)(\Phi_3^{(3)}(z))_+^{-1}(N^{(3)}(z))^{-1} = R_6(z), \end{aligned} \tag{8.6.17}$$

where plus and minus mean the left and right limits on the segment (z_3, z_1). In a similar manner, one can check the smoothness of the RH problem on the other connected components of the graph.

All the jump matrices are uniformly close to identity. More precisely, for t different than the points (8.4.26), the following estimates are valid

$$\|R(z) - I\| < \begin{cases} c_1 e^{-2|iptRe(tg(z))|}, & z \in \hat{\gamma}_s^{(k)}, \\ c_1 |t|^{-1} |g_k(z)|^{-1}, & z \in C_k, \end{cases} \tag{8.6.18}$$

where the concrete value of the constant c_1 is not important for our discussion. The estimates (8.6.18) imply that the differences $R(z) - I$ are square integrable on the RH problem graph $\Sigma = \cup_k C_k \cup_{k,s} \hat{\gamma}_s^{(k)}$, i.e., $R(z) - I \in L^2(\Sigma)$. In fact, we have (cf. Section 1 of this chapter),

$$||I - R||_{L^2(\Sigma)\cap L^\infty(\Sigma)} \leq c_2 |t|^{-1}, \tag{8.6.19}$$

where the precise value of the constant c_2 is not important for our discussion. The RH problem (8.6.6)–(8.6.8), i.e.,

$$Z \underset{z\to\infty}{\longrightarrow} I,$$

$$Z_+(z) = Z_-(z)R(z), \quad z \in \Sigma, \tag{8.6.20}$$

is equivalent to the following system of singular integral equations for the function $\rho(\zeta) \equiv Z_-(\zeta)$,

$$\rho(z) = I + \frac{1}{2\pi i}\int_\Sigma \rho(\zeta)(R(\zeta) - I)\frac{d\zeta}{\zeta - z_-}, \tag{8.6.21}$$

where z_- denotes the right limit of z on the contour of integration. This system can be represented in the form

$$\rho = I + C_-\big[M\rho\big], \tag{8.6.22}$$

where C_- is the Cauchy operator, and M is the operator of the right multiplication in the matrix $R(\zeta) - I \in L^\infty(\Sigma)$. Because the Cauchy operator is bounded in $L^2(\Sigma)$, see [**LiS, BDT, Zh**] (see also Chapter 3), estimate (8.6.19) implies (see the proof of Theorem 8.1 in Section 1 of this Chapter) that

$$\|C_-M\|_{L^2(\Sigma)} < c_3 |t|^{-1}. \tag{8.6.23}$$

Therefore, for sufficiently large values of t different than the points (8.4.26), the operator C_-M is a contraction. Hence (cf. the proof of Theorem 8.1) the equation for the function $\chi = \rho - I$,

$$\chi = C_-[MI] + C_-\big[M\chi\big], \tag{8.6.24}$$

is solvable in the space $L^2(\Sigma)$ by iterations, and the following estimate is valid,

$$\|\rho - I\|_{L^2(\Sigma)} < c|t|^{-1}. \tag{8.6.25}$$

The solution of the RH problem (8.6.20) for the function $Z(z)$ is given by the Cauchy integral

$$Z(z) = I + \frac{1}{2\pi i}\int_\Sigma \rho(\zeta)(R(\zeta) - I)\frac{d\zeta}{\zeta - z},$$

which implies the asymptotic relation

$$\Psi(z) = \big(I + O((1 + |z|)^{-1}t^{-1})\big)\tilde{\Psi}(z), \tag{8.6.26}$$

in any closed subset $\mathbb{K}$ of $\mathbb{C}P^1 \setminus \Sigma$ satisfying,

$$\frac{\text{dist}\,\{z; \Sigma\}}{1 + |z|} \geq c(\mathbb{K}), \quad \forall z \in \mathbb{K}.$$

(cf. Theorem 8.1).

REMARK 8.2. The meromorphic, with respect to t, solvability of the RH problem (8.2.28)–(8.2.34) follows from the general theory presented in Chapter 3; see Theorem 3.4. Estimate (8.6.26) shows that the points (8.4.26) describe the singular set X_s (i.e., the set of all x for which the problem is not solvable) in the neighborhood of $x = \infty$.

7. Asymptotics of the Painlevé function

Now we can finally compute the asymptotics of the Painlevé function. The expansion of $\Psi(\lambda)$ near infinity is given by

$$\begin{aligned}\Psi(\lambda) &= \big(I + \frac{1}{\lambda}(-\frac{i}{2}D\sigma_3 + \frac{1}{2}u\sigma_1) + O(\frac{1}{\lambda^2})\big)e^{-i(\frac{4}{3}\lambda^3 + x\lambda)\sigma_3}\\ &= \big(I + \frac{1}{t^{1/3}z}(-\frac{i}{2}D\sigma_3 + \frac{1}{2}u\sigma_1) + O(\frac{1}{t^{2/3}z^2})\big)e^{-it(\frac{4}{3}z^3 + e^{i\varphi}z)\sigma_3},\end{aligned} \tag{8.7.1}$$

where u is the Painlevé function (cf. (8.1.7)), $D = v^2 - u^4 - xu^2$, and

$$\Psi(z) = \big(I + O(\frac{1}{tz})\big)\tilde{\Psi}(z) = \big(I + O(t^{-1}z^{-1})\big)\,\overset{o}{\Psi}(z)e^{-t_0 h(z)\sigma_3}. \tag{8.7.2}$$

In order to find the asymptotics of u up to terms of order t^{-1}, it is sufficient to expand the Baker-Akhiezer function $\overset{(1)}{\Psi}{}^{*}_{BA}$ or $\overset{(2)}{\Psi}_{BA}$ near infinity:

$$\begin{aligned}\frac{u}{2t^{1/3}} &= \lim_{z\to\infty^+} z e^{\pi i(t-t_0)V}\frac{\Theta(U(z) + V(t-t_0) + \frac{1+\tau}{2})\Theta(\frac{1}{2})}{\Theta(U(z) + \frac{1+\tau}{2})\Theta(V(t-t_0) + \frac{1}{2})}\mu_{21} + O(t^{-1})\\ &= i\frac{\omega_a}{4}(z_1^2 - z_3^2)e^{\pi i(t-t_0)V}\frac{\Theta(V(t-t_0) + \frac{1+\tau}{2})\Theta(\frac{1}{2})}{\Theta(V(t-t_0) + \frac{1}{2})\Theta'(\frac{1+\tau}{2})} + O(t^{-1}).\end{aligned} \tag{8.7.3}$$

Equations (8.3.6), (8.4.16), and (8.7.3) yield the statement of Theorem 7.1 using routine manipulations with the elliptic functions and integrals.

CHAPTER 9

PII Asymptotics on the Canonical Six-rays. The Purely Imaginary Case

0. Introduction

This chapter is devoted to the purely imaginary solutions of PII, i.e., $u \in i\mathbb{R}$ and x is real,

$$u_{xx} = xu + 2u^3, \quad x \in \mathbb{R}, \quad u \in i\mathbb{R}. \tag{9.0.1}$$

The large x asymptotic behavior of these solutions *cannot* be described by the generic formulae discussed in Chapters 7 and 8. The reason is that the six rays, $\arg x = 0,\ e^{\pm\pi i/3},\ e^{\pm 2\pi i/3},\ \pi$, are the asymptotic boundaries as $|x| \to \infty$ of the domains, where the asymptotic formulae of Theorem 7.1 are valid. Actually, the modulated elliptic function

$$u(x) \sim c_1\sqrt{x}\ \mathrm{sn}\left(\frac{2}{3}c_2(x^{3/2} - x_0^{3/2})\Big|\varkappa(\varphi)\right), \quad \varphi = \arg x, \quad |x| \to \infty, \tag{9.0.2}$$

becomes degenerate as $\varphi \to \frac{\pi k}{3}$, i.e., it takes the form of a trigonometric or a hyperbolic function (see Section 1 below). The relevant direct limit is singular, since the phase shift $x_0 = x_0(\varphi)$ tends to infinity as $\varphi \to \frac{\pi k}{3}$. This *double* limit has not been studied for PII, thus a *uniform* asymptotic formulae describing $u(x)$ near the canonical six-rays for large $|x|$, is not yet known.

Due to the symmetry relations (7.2.10) and (7.2.11), it is sufficient to investigate the asymptotic behavior along just two rays, say $\arg x = 0$ and $\arg x = \pi$, i.e., along the real line. Also, the reduction $u \in i\mathbb{R}$ for $x \in \mathbb{R}$, is equivalent to the constraints $s_1 = -\bar{s}_3 = \bar{s}_6$, and it guarantees the absence of poles of $u(x)$ on the real line, which in turn simplifies substantially the asymptotic analysis. The complimentary case, i.e., the case $u \in \mathbb{R}$ for $x \in \mathbb{R}$, will be discussed in the next chapter. As we will see, in both cases, $u(x)$ exhibits a power-like behavior as $x \to \pm\infty$, involving trigonometric oscillations whose parameters are dependent on one complex constant $s = s_1$. The explicit asymptotics of this kind give rise to *connection formulae*, linking together the asymptotic parameters of $x \to +\infty$ with those of $x \to -\infty$.

Following our general methodology, we will again present the two basic schemes of the analysis: one is based on the asymptotic solution of the direct monodromy problem for the λ-equation (7.2.1), and the other is based on the asymptotic solution of the inverse monodromy problem.

We also note that by letting $u \to iu$ in (9.0.1), the results of this chapter are transformed to the asymptotic description of the real solutions of the equation

$$u_{xx} = xu - 2u^3, \quad x \in \mathbb{R},$$

as $x \to \pm\infty$.

1. Formulation of results. Main theorems and discussion

Recall that the λ-equation corresponding to the second Painlevé equation (see Chapter 4) is

$$\frac{d\Psi}{d\lambda} = \left[-(4i\lambda^2 + ix + 2iu^2)\sigma_3 - 4u\lambda\sigma_2 - 2w\sigma_1\right]\Psi, \tag{9.1.1}$$

where σ_1, σ_2, σ_3 are the Pauli matrices

$$\sigma_3 = \begin{pmatrix} 1 & 0 \\ 0 & -1 \end{pmatrix}, \quad \sigma_2 = \begin{pmatrix} 0 & -i \\ i & 0 \end{pmatrix}, \quad \sigma_2 = \begin{pmatrix} 0 & 1 \\ 1 & 0 \end{pmatrix},$$

x is real and u, w are purely imaginary parameters. The set of the monodromy data of this equation consists of the six Stokes matrices,

$$S_{2k} = \begin{pmatrix} 1 & s_{2k} \\ 0 & 1 \end{pmatrix}, \quad S_{2k+1} = \begin{pmatrix} 1 & s_{2k+1} \\ 0 & 1 \end{pmatrix}, \quad k = 1, 2, 3,$$

defined through the equations

$$S_k = \Psi_k^{-1}\Psi_{k+1}, \quad k = 1, \dots, 6, \tag{9.1.2}$$

where $\Psi_1(\lambda), \dots, \Psi_7(\lambda)$ are the canonical solutions of equation (9.1.1). These solutions are fixed by the asymptotic condition (Chapter 5, (5.0.23))

$$\Psi_k(\lambda) = (I + O(\lambda^{-1}))\exp\left\{-\frac{4}{3}i\lambda^3\sigma_3 - ix\lambda\sigma_3\right\}, \tag{9.1.3}$$

$$\lambda \to \infty, \quad \frac{\pi}{3}(k-2) < \arg\lambda < \frac{\pi}{3}k, \quad k = 1, \dots, 7.$$

Note that $\Psi_7 = \Psi_1$. The following constraints are valid,

$$S_{k+3} = \sigma_2 S_k \sigma_2, \quad k = 1, 2, 3, \tag{9.1.4}$$

$$S_1 S_2 \dots S_6 = I.$$

Using equations (9.1.4) together with "purely-imaginary reduction" (see (4.2.49))

$$u \in i\mathbb{R} \quad \Leftrightarrow \quad s_1 = -\overline{s}_3 = \overline{s}_6,$$

it follows that all six Stokes matrices can be parametrized by the single complex parameter $s = s_1$,

$$S_1 = \begin{pmatrix} 1 & 0 \\ s & 1 \end{pmatrix}, \quad S_2 = \begin{pmatrix} 1 & \frac{s-\overline{s}}{1+|s|^2} \\ 0 & 1 \end{pmatrix}, \quad S_3 = \begin{pmatrix} 1 & 0 \\ -\overline{s} & 1 \end{pmatrix},$$

$$S_4 = \begin{pmatrix} 1 & -s \\ 0 & 1 \end{pmatrix}, \quad S_5 = \begin{pmatrix} 1 & 0 \\ \frac{\overline{s}-s}{1+|s|^2} & 1 \end{pmatrix}, \quad S_6 = \begin{pmatrix} 1 & \overline{s} \\ 0 & 1 \end{pmatrix}.$$

The Stokes multiplier s is a function of the coefficients of the system (9.1.1), i.e., $s = s(x, u, w)$. The essence of the Isomonodromy Method is that this complex-valued function is the first integral of the nonlinear equation (9.0.1). More precisely (Flaschka-Newell theorem; see Chapter 4),

$$s \equiv \text{const} \quad \Leftrightarrow \quad w = u_x, \quad u_{xx} - xu - 2u^3 = 0, \tag{9.1.5}$$

and this s can be taken as the proper parametrization of the purely imaginary solutions of PII:

$$u = u(x; s).$$

It is obvious that $u(x; 0) \equiv 0$, therefore the interesting case to consider is when $s \neq 0$. The main results about the asymptotics of the purely-imaginary solution of PII as $x \to \pm\infty$ are described by the following two theorems.

THEOREM 9.1. *Let s be an arbitrary nonzero complex number. Then the purely imaginary solution $u(x) \equiv u(x; s)$, x real, of the PII equation (9.0.1) has the following asymptotics as $x \to -\infty$:*

$$\begin{aligned} u(x) \quad &= id(-x)^{-1/4} \sin\left\{\frac{2}{3}(-x)^{3/2} + \frac{3}{4}d^2 \log(-x) + \varphi\right\} \\ &+0((-x)^{-1}), \quad x \to -\infty, \end{aligned} \tag{9.1.6}$$

where

$$d^2 = \frac{1}{\pi} \log(1 + |s|^2), \quad d > 0, \tag{9.1.7}$$

$$\varphi = \frac{3}{2}d^2 \log 2 - \frac{\pi}{4} - \arg\Gamma\left(i\frac{d^2}{2}\right) - \arg s. \tag{9.1.8}$$

THEOREM 9.2. *Let s be an arbitrary nonzero complex number. Then the purely imaginary solution $u(x) \equiv u(x; s)$, x real, of the PII equation (9.0.1) has the following asymptotics as $x \to +\infty$:*

1) *If* $\operatorname{Im} s \neq 0$, *then*

$$\begin{aligned} u(x) \quad &= i\sigma\sqrt{\frac{x}{2}} + i\sigma(2x)^{-1/4}\rho\cos\left\{\frac{2\sqrt{2}}{3}x^{3/2} - \frac{3}{2}\rho^2 \log x + \theta\right\} \\ &+O(x^{-1}), \quad x \to +\infty, \end{aligned} \tag{9.1.9}$$

where

$$\rho^2 \quad = \frac{1}{\pi} \log \frac{1 + |s|^2}{2|\operatorname{Im} s|}, \quad \rho > 0, \quad \sigma = -\mathit{sign}\operatorname{Im} s, \tag{9.1.10}$$

$$\theta \quad = -\frac{3\pi}{4} - \frac{7}{2}\rho^2 \log 2 + \arg\Gamma(i\rho^2) + \arg(1 + s^2). \tag{9.1.11}$$

2) *If* $\operatorname{Im} s = 0$, *then*

$$u(x) = \frac{is}{2\sqrt{\pi}} x^{-1/4} e^{-\frac{2}{3}x^{3/2}} \left(1 + O(x^{-3/4})\right), \quad x \to +\infty. \tag{9.1.12}$$

Theorem 9.1 and the first statement of Theorem 9.2 (the generic case) will be proven in this chapter. The second statement of Theorem 9.2 (special solution) will be proven in Chapter 11.

Since the expressions (9.1.10) and (9.1.7) involve the *same* parameter s, we can eliminate s and obtain in this way an explicit *connection formulae* between the asymptotic parameters (d, φ) at $-\infty$ and the asymptotic parameters (ρ, θ) at $+\infty$:

$$\rho^2 = d^2 - \frac{1}{\pi}\log 2(e^{\pi d^2} - 1)^{1/2}|\sin\Psi(d,\varphi)|, \tag{9.1.13}$$

$$\sigma = -\text{sign}\,\sin\Psi(d,\varphi),$$

$$\theta = -\frac{3\pi}{4} - \frac{7}{2}\rho^2\log 2 + \arg\Gamma(i\rho^2) + \arg\left(1 + (e^{\pi d^2} - 1)e^{2i\Psi(d,\varphi)}\right),$$

where

$$\Psi(d,\varphi) = \frac{3}{2}d^2\log 2 - \frac{\pi}{4} - \arg\Gamma(id^2/2) - \varphi. \tag{9.1.14}$$

This is the *generic* connection formulae, i.e., we have assumed that

$$\frac{3}{2}d^2\log 2 - \frac{\pi}{4} - \arg\Gamma\left(i\frac{d^2}{2}\right) - \varphi \neq \pi n, \quad n \in \mathbb{Z}. \tag{9.1.15}$$

If, instead, the asymptotic parameters (d, φ) satisfy the constraint,

$$\frac{3}{2}d^2\log 2 - \frac{\pi}{4} - \arg\Gamma\left(i\frac{d^2}{2}\right) - \varphi = \pi n, \quad n \in \mathbb{Z}, \tag{9.1.16}$$

then the asymptotic behavior as $x \to +\infty$ of $u(x)$ is given by equation (9.1.12) and the relevant connection formula becomes

$$s = (-1)^n\left(e^{\pi d^2} - 1\right)^{\frac{1}{2}}. \tag{9.1.17}$$

This is the Ablowitz-Segur [**AS**] one-parameter family of solutions of (9.0.1).

A *formal* derivation of the above asymptotic results can be obtained by using the following theorems, which are of interest in their own right and which constitute the *asymptotic solution of the direct monodromy problem for the λ-equation (9.1.1).*

THEOREM 9.3 ([**I3**]). *Let u and w be purely imaginary functions of x defined on $(-\infty, x_-]$ satisfying the following conditions:*

a)

$$u(x) = O((-x)^{-1/4}), \quad w(x) = O((-x)^{1/4}), \quad x \to -\infty. \tag{9.1.18}$$

b) *There exist constants $c > 0$ and $X_- \le x_-$, such that*

$$|v(x)| \ge c \qquad \forall x \in (-\infty, X_-], \tag{9.1.19}$$

where

$$v = \frac{i}{\sqrt{2}}e^{i\pi/4}[(-x)^{1/4}u + i(-x)^{-1/4}w].$$

Then the associated monodromy data $s = s(x, u, w)$, satisfy the following asymptotic equation

$$s = \frac{i}{v}\frac{\sqrt{2\pi}}{\Gamma(i|v|^2)}\exp\Big\{\frac{2}{3}i(-x)^{3/2} + i\frac{3}{2}|v|^2\log(-x)$$

$$+i|v|^2\log 8 + \frac{\pi}{2}|v|^2\Big\} + o(1), \quad x \to -\infty, \tag{9.1.20}$$

where $\Gamma(z)$ is the Euler gamma-function.

THEOREM 9.4 ([**IK1**]). *Let u, w be purely imaginary functions of x defined on $[x_+, \infty)$ and satisfying the following conditions:*

a)

$$u(x) = i\sqrt{\frac{x}{2}} + u_0(x), \quad u_0 = O(x^{-1/4}), \quad x \to +\infty,$$

$$w(x) = \frac{i}{2}\sqrt{\frac{1}{2x}} + w_0(x), \quad w_0 = O(x^{1/4}), \quad x \to +\infty. \tag{9.1.21}$$

b) *There exist constants $c > 0$ and $X_+ \geq x_+$, such that*

$$|v_0(x)| \geq c, \qquad \forall x \in [X_+, \infty), \tag{9.1.22}$$

where

$$v_0 = e^{i\pi/4}\big[(2x)^{1/4}u_0 + i(2x)^{-1/4}w_0\big].$$

Then the associated monodromy data $s = s(x, u, w)$, satisfy the following asymptotic equations

$$\frac{\overline{s} - s}{1 + |s|^2} = ie^{-\pi|v_0|^2} + o(1), \qquad x \to +\infty,$$

$$\frac{1 + s^2}{1 + |s|^2} = -\frac{i}{v_0}\frac{\sqrt{2\pi}}{\Gamma(i|v_0|^2)}\exp\Big\{-\frac{2i\sqrt{2}}{3}x^{3/2} \tag{9.1.23}$$

$$+i\frac{3}{2}|v_0|^2\log x + i\frac{7}{2}|v_0|^2\log 2 - \frac{\pi}{2}|v_0|^2\Big\} + o(1),$$

$$x \to +\infty.$$

Using the above theorems, it is straightforward to obtain the formal asymptotics (and the connection formulae) for the purely imaginary solutions of the Painlevé equation (9.0.1): Suppose that u and w are not arbitrary functions of x satisfying (9.1.18), (9.1.19) and (9.1.21), (9.1.22) but such that $w = u_x$, and $u(x)$ is a solution of the Painlevé equation (9.0.1). Then, by (9.1.5), the parameter s is a given constant. Solving (9.1.20) and (9.1.23) for $u(x)$, we obtain the leading terms of the asymptotics announced in Theorem 9.1 and Theorem 9.2, respectively.

The above formal derivation of the asymptotics (9.1.6) and (9.1.9), can be made rigorous with the help of Kitaev's method (see [**Ki1**] and Section 6.1 of Chapter 7). This is described in detail in [**IFoK**], so will not be presented here. In the next two sections we will present, however, the detailed proofs of Theorems 9.3 and 9.4 (following [**IN1**]). The proofs of Theorems 9.1 and 9.2 will be presented in the last two sections, using the *asymptotic solution of the inverse monodromy* *problem of the λ-equation (9.1.1).* There, we will present in detail the nonlinear steepest descent analysis of the corresponding oscillatory Riemann-Hilbert problem (this presentation is a slightly modified version of [**DeZ2**]).

There exists one more approach of using Theorems 9.3 and 9.4 to prove Theorems 9.1 and 9.2. This is based on the following *local* asymptotic fact, which can be proven by the conventional ODE techniques.

PROPOSITION 9.1. *Given any $d > 0$ and $\varphi \in \mathbb{R} \pmod{2\pi}$, there exists a solution $u(x)$ of the PII equation (9.0.1), which has the differentiable in x asymptotics*

$$\begin{aligned} u(x) &= id(-x)^{-1/4}\sin\left\{\frac{2}{3}(-x)^{3/2} + \frac{3}{4}d^2\log(-x) + \varphi\right\} \\ &\quad +0((-x)^{-1}), \quad x \to -\infty. \end{aligned} \tag{9.1.24}$$

The functions $u(x)$ and $w(x) := u_x(x)$ satisfy the conditions (9.1.18) and (9.1.19), hence, using Theorem 9.3, the formula (9.1.20) for the monodromy parameter s corresponding to $u(x)$ follows. Since s does not depend on x, we can take the limit of the right-hand side as $x \to -\infty$, and, using (9.1.24), arrive at the relation

$$s = \frac{2\sqrt{\pi}e^{\frac{\pi}{4}d^2}}{\Gamma\left(i\frac{d^2}{2}\right)}e^{-i\frac{\pi}{4}+i\frac{3}{2}\log 2 - i\varphi}.$$

This complex relation, with the help of some simple algebra, can be re-written in the form

$$|s|^2 = e^{\pi d^2} - 1,$$

and

$$\arg s = \frac{3}{2}d^2\log 2 - \frac{\pi}{4} - \arg\Gamma\left(i\frac{d^2}{2}\right) - \varphi.$$

These equations define a *bijection* from the real pairs (d, φ) to the complex numbers $s \in \mathbb{C}$; indeed, the inverse map is given by equations (9.1.7) and (9.1.8). This immediately implies the *global* asymptotic statement of Theorem 9.1.

Proposition 9.1, which was rigorously proven by A.S. Abdullaev in [**Abd**], was used in [**IN1**] to justify equations (9.1.6)–(9.1.8). The same approach is less effective in the case of Theorem 9.2. Even formal derivation of the asymptotic formulae (9.1.9) is much more difficult, due to the fact that this asymptotics is *not* quasi-linear. It was eventually derived by A.S. Abdullaev, but later than the appearance of the complete set of the formulae (9.1.9)–(9.1.11) in the paper [**IK1**] (which was based on the RH approach).

Since the late 1980s, and especially after the introduction of the nonlinear steepest descent method, the rigorous justification of the Riemann-Hilbert scheme has been resolved. Therefore, we will not discuss in this book the techniques based on the local analysis anymore.

It should be mentioned that the first rigorous derivation of the Ablowitz-Segur connection formulae was achieved in [**Su1**] (along the lines just described) and in [**ClM**] (using the Gelfand-Levitan-Marchenko integral equations, an earlier version of the Riemann-Hilbert method).

REMARK 9.1. The issue of the higher corrections to the asymptotic formulae will be briefly addressed in Chapter 11 in relation with some special classes of the Painlevé functions. The construction of the full asymptotic series, which have the expressions (9.1.6) and (9.1.9) as their leading terms, can be obtained with the use of the Riemann-Hilbert problem. Indeed, from the Riemann-Hilbert analysis one can prove the existence of a full asymptotic series, whose concrete forms can be then specified using the differential equation (9.0.1) (cf. [**DeZ3**]). The crucial point is that all the higher coefficients are uniquely determined, via the relevant recurrence relations, from the leading terms (given by equations (9.1.6) and (9.1.9) in the case of the purely imaginary PII solutions).

2. The direct monodromy problem approach. The proof of Theorem 9.3

Main guidelines of the proof

Before presenting the proof of the Theorem 9.3 we first outline its main steps. We begin the asymptotic study of the system (9.1.1) by the change of variable

$$z = (-x)^{-1/2}\lambda.$$

The system (9.1.1) takes a new form

$$\frac{d\Psi}{dz} = \tau A_0(z,\tau)\Psi, \tag{9.2.1}$$

where

$$\tau = (-x)^{3/2},$$

$$A_0(z,\tau) = -\Big(4iz^2 - i - \frac{2iu^2}{x}\Big)\sigma_3 - 4uz(-x)^{-1/2}\sigma_2 + \frac{2w}{x}\sigma_1.$$

In equation (9.2.1) τ is a large parameter and the matrix $A_0(z,\tau)$, according to (9.1.18), satisfies the estimate

$$A_0(z,\tau) = O(1), \quad \tau \to +\infty,$$

where z is fixed. In other words, we arrive at a situation where the WKB-method can be applied. According to this method, it is necessary to determine the turning points of equation (9.2.1) and then to seek an asymptotic solution away from these points in the form

$$\Psi_{WKB} \simeq T(z)\exp\Big\{\tau\int^{\tau}\Lambda(\eta)\,d\eta\Big\} \tag{9.2.2}$$

where $\Lambda(z)$ is the diagonalization of A_0 and T is the corresponding conjugation matrix,

$$\Lambda = T^{-1}A_0T.$$

Because the matrix A_0 is polynomial in z, the WKB-approximation (9.2.2) is double (see [**He**]), i.e., it is also an asymptotic approximation as $z \to \infty$. Thus, we can calculate the leading term in τ of the matrices $C_k(\tau)$ connecting the canonical solutions $\Psi_k(z)$ with WKB-solutions $\Psi^{(k)}_{WKB}(z)$ in the corresponding sectors Ω_k:

$$\Psi_k = \Psi_{WKB}\cdot C_k.$$

Furthermore, due to the double nature of the WKB-approximation, the formula (9.2.2) for the solution $\Psi^{(k)}_{WKB}$ is in some annular domain near the turning points. In the neighborhood of these points the original system (9.1.1) can be simplified and reduced to one which can be solved exactly. It will be shown below that the system (9.2.1) has two multiple turning points[1], $z_\pm = \pm\frac{1}{2}$, in the neighborhood of which the *turning point solutions*, which we denote by $\breve{\Psi}_\pm(z)$, may be matched with the WKB-solutions Ψ^k_{WKB}, $k = 1, 2, \ldots, 6$. This means that we can calculate the matrices N_k,

$$\mathsf{N}_k = \left(\breve{\Psi}_+\right)^{-1} \Psi^k_{WKB}, \tag{9.2.3}$$

to the leading order in τ.

The monodromy data for the system (9.1.1) can be recovered from the Stokes matrix S_1, which in turn can be represented as the following matrix product,

$$\mathsf{S}_1 = \mathsf{C}_1^{-1}\mathsf{N}_1^{-1}\mathsf{N}_2\mathsf{C}_2. \tag{9.2.4}$$

Thus we can compute directly[2] the leading term of asymptotics in τ of the parameter s.

In summary, the proof of Theorem 9.3 involves the following steps:

I. Construction of the WKB-solutions Ψ^k_{WKB}, $k = 1, 2$.
II. Construction of the solution $\breve{\Psi}_+(z)$ in the neighborhood of the turning point.
III. Calculation of the leading asymptotics in τ of the matrices C_k.
IV. Calculation of the leading asymptotics in τ of the matrices N_k.
V. Calculation of s, using equation(9.2.4).

I. The WKB-solutions

The eigenvalues $\mu_{1,2}(z,\tau)$ of the matrix $\mathsf{A}_0(z,\tau)$ are given by the equations

$$\mu_{1,2} = \pm\mu, \tag{9.2.5}$$

$$\mu = 4i\sqrt{\left(z^2 - \frac{1}{4}\right)^2 - \frac{w^2}{4x^2} + \frac{u^2}{4x} + \frac{u^4}{4x^2}}. \tag{9.2.6}$$

The condition $\mu = 0$ characterizes the turning points. A priori, there exist the four turning points

$$z_{1,3} = \sqrt{\frac{1}{4} \pm \sqrt{\frac{w^2}{4x^2} - \frac{u^2}{4x} - \frac{u^4}{4x^2}}},$$

$$z_{2,4} = -z_{1,3} = -\sqrt{\frac{1}{4} \pm \sqrt{\frac{w^2}{4x^2} - \frac{u^2}{4x} - \frac{u^4}{4x^2}}}.$$

However, due to equation (9.1.18), z_1 and z_3 tend to a single point as $\tau \to \infty$; similarly for z_2 and z_4:

$$z_{1,3} \to z_+ = \frac{1}{2},$$

$$z_{2,4} \to z_- = -\frac{1}{2}, \quad \tau \to \infty.$$

[1]It is worth mentioning that these turning points coincide with the stationary phase points in the kernels of the singular integral equations associated with the inverse monodromy problem.

[2]It is possible to obtain the full asymptotic expansion for s by calculation an appropriate number of terms for C_k and N_k. However, for our purpose, i.e., the construction of the asymptotics for PII, it is sufficient to compute only the leading term.

The relevant rate of convergence is of order $O(\tau^{-1/2})$, thus it is natural to apply the WKB-asymptotics (9.2.2) in the domains where

$$\sqrt{\tau}\Big|\Big(z\pm\frac{1}{2}\Big)\Big|\to\infty. \tag{9.2.7}$$

It is easy to show that the matrices $\Lambda(z,\tau)$ and $T(z,\tau)$ for the system (9.2.1) are given by

$$\Lambda=-\mu\sigma_3, \tag{9.2.8}$$

$$T=\frac{1}{8i\big(z^2-\frac14\big)}\left[\Big(4iz^2-i-2i\frac{u^2}{x}+\mu\Big)\sigma_3+\frac{4izu}{(-x)^{1/2}}\sigma_1+\frac{2iw}{x}\sigma_2\right],\quad \sigma_0=I.$$

Equation (9.2.5) and condition a) in Theorem 9.3 imply that $\mu(z,\tau)$ satisfies

$$\mu=4i\Big(z^2-\frac{1}{4}\Big)+O\Big(\tau^{-1}\big(z^2-\frac{1}{4}\big)^{-1}\Big).$$

Thus, in the domain where equation (9.2.7) holds, the following statements are true:

1) $T(z,\tau)=I+O\Big(\tau^{-1/2}z\big(z^2-\frac{1}{4}\big)^{-1}\Big).$

2) $\int\limits_z^\infty R(\eta,\tau)d\eta=O\Big(\tau^{-1/2}z\big(z^2-\frac{1}{4}\big)^{-1}\Big)$, where $R=T^{-1}\dfrac{dT}{dz}$ is the remainder obtained after substituting (9.2.2) into equation (9.2.1).

3) The asymptotics

$$\int^z\mu(\eta,\tau)d\eta\simeq\frac{4}{3}iz^3-iz$$

implies that the Stokes lines

$$\operatorname{Re}\int\limits_{z_{1,2,3,4}}^{z}\mu(\eta,\tau)d\eta=0,$$

as $\tau\to\infty$, tend to the real axis and to the hyperbola

$$\eta^2=3\Big(\xi^2-\frac{1}{4}\Big),\qquad z=\xi+i\eta,$$

(see Figure 9.1).

Applying the standard WKB-technique of integral equations (see, for example [**Ol**]), it is easy to prove the following lemma.

LEMMA 9.1. *Let $\tilde\Omega_k$, $k=1,2,\dots,6$, be nonintersecting open domains bounded by intervals*

$$\Big(-\infty,-\frac{1}{2}\Big],\quad\Big[-\frac{1}{2},\frac{1}{2}\Big],\quad\text{and}\quad\Big[\frac{1}{2},+\infty\Big)$$

and by the hyperbola lines $\eta^2=3(\xi^2-\frac14)$, such that

$$\bigcup_k\bar{\tilde\Omega}_k=\mathbb{C}.$$

Then we have:

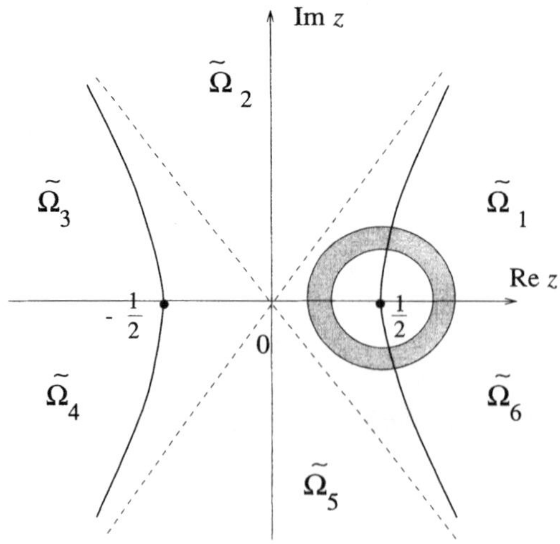

FIGURE 9.1. The Stokes lines and the fundamental domains for equation (9.2.1); the shaded annular domain is the matching region Ω_+^ε

a) *In the domain* $\tilde{\Omega}_k$, *equation* (9.2.1) *has a solution* Ψ^k_{WKB} *with the double asymptotics of the form*

$$\Psi^k_{WKB}(z,\tau) = \left[I + O\left(\frac{z}{\sqrt{\tau}(z^2-\frac{1}{4})}\right) + O\left(\frac{1}{\sqrt{\tau}(z^2-\frac{1}{4})}\right)\right.$$

$$\left. + O\left(\frac{1}{\sqrt{\tau}}\log\frac{z-\frac{1}{2}}{z+\frac{1}{2}}\right)\right]\exp\left\{-\tau\sigma_3\int_{1/2}^{z}\mu(\eta,\tau)d\eta\right\},$$

(9.2.9) $$\left|\sqrt{\tau}\left(z^2-\frac{1}{4}\right)\right| \to \infty.$$

b) *There is a unique extension of the solution* Ψ^k_{WKB} *to the domain* $\tilde{\Omega}_{k+1}$, *or* $\tilde{\Omega}_{k-1}$ *such that the asymptotics* (9.2.9) *is also valid in the extended domain.*

In what follows we will only use the solutions Ψ^k_{WKB} with $k=1$ and 2. We will assume that they are extended to the domains $\tilde{\Omega}_6$ and $\tilde{\Omega}_1$, respectively. The branches of function $\mu(\eta)$ are fixed by the condition

$$-\pi < \arg\eta < \pi.$$

II. THE NEIGHBORHOOD OF THE TURNING POINT z_+

We fix $\varepsilon > 0$ and consider the following domain near the turning point $z_+ = \frac{1}{2}$:

$$\mathcal{D}_+^\varepsilon = \left\{z : \left|z - \frac{1}{2}\right| \leqslant \tau^{-1/2+\varepsilon}\right\}.$$

Assuming $z \in \mathcal{D}_+^\varepsilon$, we substitute in equation (9.2.1) the Taylor series expansion of the matrix $A_0(z,\tau)$ near $z_+ = \frac{1}{2}$, and we cancel the terms quadratic in $z-\frac{1}{2}$ or τ;

this yields, the equation

$$\frac{d\Psi_0}{dz} = \tau \cdot \left[-4i\left(z - \frac{1}{2}\right)\sigma_3 - 2u \cdot (-x)^{-1/2}\sigma_2 - \frac{2w}{x}\sigma_1 \right] \Psi_0 \,. \tag{9.2.10}$$

Introducing new variables

$$\begin{aligned} \zeta &= 2\sqrt{2\tau}\left(z - \frac{1}{2}\right)\mathrm{e}^{-i\pi/4}, \\ v &= \sqrt{2}(-x)^{3/4}\mathrm{e}^{i\pi/4}\left(\frac{i}{2}u(-x)^{-1/2} + \frac{w}{2}x^{-1}\right), \\ v^* &= -\sqrt{2}(-x)^{3/4}\mathrm{e}^{i\pi/4}\left(\frac{i}{2}u(-x)^{-1/2} - \frac{w}{2}x^{-1}\right) \equiv -i\overline{v}, \end{aligned} \tag{9.2.11}$$

equation (9.2.12) takes the form

$$\frac{d\Psi_0}{d\zeta} = \begin{pmatrix} \frac{\zeta}{2} & v \\ v^* & -\frac{\zeta}{2} \end{pmatrix} \Psi_0. \tag{9.2.12}$$

Equation (9.2.12) can be solved in terms of the parabolic cylinder functions. Indeed differentiating with respect to ζ we find

$$\begin{aligned} \frac{d^2\Psi_0}{d\zeta^2} &= \left[\begin{pmatrix} \frac{1}{2} & 0 \\ 0 & -\frac{1}{2} \end{pmatrix} + \begin{pmatrix} \frac{\zeta}{2} & v \\ v^* & -\frac{\zeta}{2} \end{pmatrix}^2 \right] \Psi_0 \\ &= \begin{pmatrix} \frac{\zeta^2}{4} + vv^* + \frac{1}{2} & 0 \\ 0 & \frac{\zeta^2}{4} + vv^* - \frac{1}{2} \end{pmatrix} \Psi_0. \end{aligned}$$

It is clear that the elements of the first row of Ψ_0 can be expressed through Weber-Hermite functions $D_\nu(\zeta)$ or $D_{-1-\nu}(i\zeta)$, with the symbol

$$\nu = -vv^* - 1 = 2i\tau\left(\frac{u^2}{4x} - \frac{w^2}{4x^2}\right) - 1. \tag{9.2.13}$$

The elements of the second line of Ψ_0 can be constructed directly from equation (9.2.12).

Thus, we obtain an exact formula for solutions of equation (9.2.1) valid in the small neighborhood $\mathcal{D}_+^\varepsilon$ of the turning point $z_+ = \frac{1}{2}$,

$$\Psi_0(z) = \begin{pmatrix} D_{-1+\nu}(i\zeta) & D_\nu(\zeta) \\ D^\circ_{-1+\nu}(i\zeta) & D^\circ_\nu(\zeta) \end{pmatrix}, \tag{9.2.14}$$

where "∘" means a transform

$$\phi(z) \mapsto \phi^\circ(\zeta) = \frac{1}{v}\left(f_\zeta - \frac{\zeta}{2}f\right), \tag{9.2.15}$$

and the variables ζ, v, and ν are defined by equations (9.2.11) and (9.2.13).

We now discuss rigorously in what sense the formula (9.2.14) is an asymptotic formula for the solution of equation (9.2.1) in the domain $\mathcal{D}_+^\varepsilon$. Denoting the associated error term in the approximation of $A_0(z,\tau)$ by $R_+(z)$, we have

$$R_+(z) = O\Big(\tau^{-1+2\varepsilon}\Big), \quad \tau \to \infty, \quad z \in \mathcal{D}_+^\varepsilon.$$

The solution of equation (9.2.1) in the domain $\mathcal{D}_+^\varepsilon$ can be represented in the form

$$\Psi(z) = \chi(z)\Psi_0(z).$$

The matrix $\chi(z)$ satisfies an integral equation, which is equivalent to the differential equation (9.2.1), with the initial condition $\Psi(\frac{1}{2}) = \Psi_0(\frac{1}{2})$,

$$\chi(z) = I + \tau \int_{\frac{1}{2}}^{z} \Psi_0(z)\Psi_0^{-1}(\eta)R_+(\eta)\chi(\eta)\Psi_0(\eta)\Psi_0^{-1}(z)d\eta. \tag{9.2.16}$$

This equation is of Volterra type with an analytic kernel. Thus, there exists a unique solution $\chi(z)$ in the domain $\mathcal{D}_+^\varepsilon$, which is analytic function of z. In order to obtain the asymptotic properties of χ, as $\tau \to \infty$, we use the condition b) of Theorem 9.3. Assuming that it holds for the quantity v, then the division by v in equation (9.2.15) is well defined from an asymptotic point of view. Moreover, due to the condition a), the parameter ν in equation (9.2.14) remains bounded as $\tau \to \infty$. The transformation $\Psi_0 \mapsto (\Psi)^{-1}$ does not generate any singularities in equation (9.2.16), because of the well-known formula for the parabolic cylinder functions **[Ol]**

$$\det\Psi_0(z) = i\exp\Big\{-\frac{\nu\pi i}{2}\Big\}\cdot v^{-1}. \tag{9.2.17}$$

It is essential that the domain $\mathcal{D}_+^\varepsilon$ contains an annular subdomain

$$\Omega_+^\varepsilon = \Big\{z \in \mathcal{D}_+^\varepsilon: \quad z - \frac{1}{2} \simeq \tau^{-1/2+\varepsilon}\Big\},$$

where the variable ζ becomes large. The condition c) together with the well-known asymptotics of $D_\nu(\zeta)$ and $D_{1-\nu}(i\zeta)$ (see **[BE]**; see also Chapter 2), imply the following estimate for the function $\Psi_0(z)$ in the entire neighborhood $\mathcal{D}_+^\varepsilon$:

$$\big|\Psi_0(z)\big| \leqslant C\exp\Big\{\big|\operatorname{Re}\zeta^2\big|\Big\}, \quad \tau \to \infty. \tag{9.2.18}$$

The above analysis, together with the estimate (9.2.18) allow us to evaluate the kernel of the integral operator (9.2.16)

$$\begin{aligned}
&\int_{\frac{1}{2}}^{z} \Big|\Psi_0(z)\Psi_0^{-1}(\eta)\Big|\cdot\Big|R_+(\eta)\Big|\cdot\Big|\Psi_0(\eta)\Psi_0^{-1}(z)\Big|\,d\eta \\
&= O\Big(\tau^{-3/2+3\varepsilon}\Big), \quad \tau \to \infty, \quad z \in \mathcal{D}_+^\varepsilon, \\
&\arg\big(z - \frac{1}{2}\big) = \arg\big(\eta - \frac{1}{2}\big) = \frac{\pi k}{2}, \quad k = -1, 0, 1, 2.
\end{aligned} \tag{9.2.19}$$

As a consequence of this estimate, it is possible to evaluate asymptotically the solution $\breve{\Psi}_+(z)$ in the neighborhood of the turning point $z_+ = \frac{1}{2}$.

LEMMA 9.2. *There exits a solution $\breve{\Psi}_+(z)$ of equation (9.2.1), which has the following asymptotic behavior in the domain $\mathcal{D}_+^\varepsilon$, $\varepsilon < \frac{1}{6}$,*

$$\breve{\Psi}_+ = \big(I + o(1)\big)\Psi_0(z), \quad \tau \to \infty,$$

$$z \in \mathcal{D}_+^\varepsilon, \quad \arg\big(z - \frac{1}{2}\big) = \frac{\pi}{2}j, \quad j = -1, 0, 1, 2,$$

where the function $\Psi_0(z)$ is described by equations (9.2.11), (9.2.13)–(9.2.15).

III. CALCULATION OF THE MATRICES C_k

Due to the almost canonical form of the WKB-asymptotics (9.2.9) as well as to our choice of normalization of Ψ^k_{WKB}, $k = 1, 2$, we find that the matrices

$$\mathrm{C}_k = \left[\Psi^k_{WKB}\right]^{-1}\Psi_k,$$

are diagonal, and can be expressed as follows

$$\mathrm{C}_k = \lim_{z\to\infty} \exp\Big\{\tau\int_{\frac{1}{2}}^{z}\mu(\eta,\tau)d\eta\sigma_3 - \tau\Big(\frac{4}{3}iz^3 - iz\Big)\sigma_3\Big\},$$

$$\arg z = \frac{\pi}{3}(k-1), \quad k = 1, 2. \tag{9.2.20}$$

Thus, in order to calculate the leading term in τ of the matrices C_k we have to study the asymptotics of the integral

$$\mathcal{I}(z,\tau) = \int_{\frac{1}{2}}^{z}\mu(\eta,\tau)d\eta$$

as $\tau \to \infty$. This asymptotic must be uniform in z for large $|z|$.

Fix arbitrary ε, $0 < \varepsilon < \frac{1}{6}$, and split the integral $\mathcal{I}$ into two parts, keeping $|z| > C$, $\arg z = \frac{\pi}{3}j_k$,

$$\mathcal{I}(z,\tau) = \mathcal{I}_1(z^*,\tau) + \mathcal{I}_2(z^*,z,\tau), \quad z^* = \frac{1}{2} + \tau^{-1/2+\varepsilon},$$

$$\mathcal{I}_1(z,\tau) = \int_{\frac{1}{2}}^{z^*}\mu(\eta,\tau)\,d\eta, \qquad \mathcal{I}_2(z,\tau) = \int_{z^*}^{z}\mu(\eta,\tau)\,d\eta.$$

Denoting

$$a^2 = -\frac{w^2}{4x^2} + \frac{u^2}{4x}$$

we find

$$a^2 = O(\tau^{-1}), \quad \frac{u^4}{x^2} = O(\tau^{-2}). \tag{9.2.21}$$

Then, expanding $\mu(\eta,\tau)$ to leading order, we obtain the following asymptotics of $\mathcal{I}_1$:

$$\mathcal{I}_1 = 4i\int_{\frac{1}{2}}^{z^*}\sqrt{\Big(\eta^2-\frac{1}{4}\Big)^2+a^2+O(\tau^{-2})}\,d\eta$$

$$= 4i\int_0^{z^*-\frac{1}{2}}\sqrt{\xi^2(1+\xi)^2+a^2+O(\tau^{-2})}\,d\xi$$

$$= 4i\int_0^{z^*-\frac{1}{2}}\sqrt{\xi^2+a^2+O(\tau^{-3/2+3\varepsilon})}\,d\xi$$

$$= 4i\int_0^{z^*-\frac{1}{2}}\sqrt{\xi^2+a^2}\,d\xi+O(\tau^{-3/2+3\varepsilon})$$

$$= \Big(2iy\sqrt{y^2+a^2}+2ia^2\log\big(y+\sqrt{y^2+a^2}\big)\Big)\Big|_{y=0}^{y=z^*-\frac{1}{2}}+O\big(\tau^{-3/2+3\varepsilon}\big).$$

This yields

$$\mathcal{I}_1 = 2i\big(z^*-\frac{1}{2}\big)^2+ia^2-ia^2\log a^2+2ia^2\log\big(z^*-\frac{1}{2}\big)+2ia^2\log 2$$
$$+O\big(\tau^{-1-2\varepsilon}\big)+O\big(\tau^{-3/2+3\varepsilon}\big). \tag{9.2.22}$$

The second integral $\mathcal{I}_2$ can be estimated in a similar way:

$$\mathcal{I}_2 = 4i\int_{z^*}^{z}\Big(\eta^2-\frac{1}{4}\Big)\sqrt{1+\frac{a^2+O(\tau^{-2})}{(\eta^2-\frac{1}{4})^2}}\,d\eta$$

$$= 4i\int_{z^*}^{z}\Big(\eta^2-\frac{1}{4}+\frac{a^2}{\eta^2-\frac{1}{4}}+O\big(\tau^{-2}(\eta^2-\frac{1}{4})^{-3}\big)\Big)\,d\eta$$

$$= \Big[i\frac{4}{3}\eta^3-i\eta+2ia^2\log\frac{\eta-\frac{1}{2}}{\eta+\frac{1}{2}}\Big]_{\eta=z^*}^{\eta=z}$$
$$+O\big(\tau^{-1-2\varepsilon}\big)+O\big(\tau^{-2}z^{-1}\big). \tag{9.2.23}$$

Hence,

$$\mathcal{I}_2 = i\tfrac{4}{3}z^3-iz+2ia^2\log\tfrac{z-\frac{1}{2}}{z+\frac{1}{2}}+\tfrac{i}{3}-2i\big(z^*-\tfrac{1}{2}\big)^2-2ia^2\log\big(z^*-\tfrac{1}{2}\big)$$
$$+O\big(\tau^{-\frac{3}{2}+3\varepsilon}\big)+O\big(\tau^{-1-2\varepsilon}\big)+O\big(\tau^{-2}z^{-1}\big). \tag{9.2.24}$$

The sum of the two integrals (9.2.22) and (9.2.24) yields the asymptotics

$$\mathcal{I}(z,\tau) = \frac{4}{3}iz^3-iz+\frac{i}{3}+b+2ia^2\log\frac{z-\frac{1}{2}}{z+\frac{1}{2}}+o(\tau^{-1}),$$
$$b = ia^2-ia^2\log a^2+2ia^2\log 2,\quad \tau\to\infty. \tag{9.2.25}$$

The asymptotics (9.2.25) is uniform in z as $|z| > C$, $\arg z = \frac{\pi}{3}(k-1)$. Substituting this asymptotics into equation (9.2.20), we find the following leading terms of the matrices C_k:

$$\mathrm{C}_k = \exp\left\{\frac{i\tau}{3} + \tau b\right\}\left(1 + o(1)\right), \qquad \tau \to \infty, \quad k = 1, 2. \tag{9.2.26}$$

We note that C_k appear to be independent of k. This is due to the way the branch of $\mu(\eta)$ was fixed (see the end of Subsection 1).

IV. Calculation of the matrices N_k

The estimate (9.2.9) implies that for the solutions Ψ^k_{WKB} in the matching domain Ω^ε_+, $\left(z - \frac{1}{2} \simeq \tau^{-1/2+\varepsilon}\right)$, the following asymptotics is valid:

$$\Psi^k_{WKB}(z) = \left(I + o(1)\right)\exp\left\{-\tau\sigma_3\int_{\frac{1}{2}}^{z}\mu(\eta,\tau)\,d\eta\right\},$$

$$\tau \to \infty, \quad z \in \Omega^\varepsilon_+, \quad \arg\left(z - \frac{1}{2}\right) = \frac{\pi}{2}(k-1), \quad k = 1, 2.$$

Thus, we need to compute the asymptotic expansion of integral $\mathcal{I}(z,\tau)$ as $\tau \to \infty$, with z belonging to Ω^ε_+. However, this computation was already performed in (9.2.22). Assuming $0 < \varepsilon < \frac{1}{6}$, we find

$$\int_{\frac{1}{2}}^{z}\mu(\eta,\tau)\,d\eta = 2i\left(z - \frac{1}{2}\right)^2 + 2ia^2\log\left(z - \frac{1}{2}\right) + b + o(\tau^{-1}),$$

$$\tau \to \infty, \quad z \in \Omega^\varepsilon_+, \quad \arg\left(z - \frac{1}{2}\right) = \frac{\pi}{2}(k-1), \quad k = 1, 2.$$

Introducing in the right-hand side the variables (9.2.11), we find the following expression for Ψ^k_{WKB},

$$\Psi^k_{WKB}(z) = \left(I + o(1)\right)\exp\left\{-\sigma_3\left(\frac{\zeta^2}{4} - (\nu+1)\log\zeta\right)\right\}\mathrm{N}_0,$$

$$\tau \to \infty, \quad \zeta \in \Omega^\varepsilon_+, \quad \arg\left(z - \frac{1}{2}\right) = \frac{\pi}{2}(k-1), \quad k = 1, 2, \tag{9.2.27}$$

where

$$\mathrm{N}_0 = \exp\left\{\sigma_3\left(\frac{1}{2}(\nu+1)\log 8\tau - i(\nu+1)\frac{\pi}{4} - \tau b\right)\right\}.$$

We now consider the turning point solution $\breve{\Psi}_+(z)$. Taking the same z as in (9.2.27) we can derive the asymptotics of $\breve{\Psi}_+(z)$ using the well-known asymptotic expansions for the parabolic cylinder functions (see [**BE**]; see also Chapter 2). The final result is given below:

$$\breve{\Psi}_+(z) = \big(I + o(1)\big)\exp\Big\{-\sigma_3\Big(\frac{\zeta^2}{4} - (\nu+1)\log\zeta\Big)\Big\}$$

$$\times\begin{pmatrix} e^{-(1+\nu)i\pi/2} & 0 \\ 0 & -\dfrac{1}{v}\end{pmatrix}, \quad \tau\to\infty, \tag{9.2.28}$$

where $\zeta\in\Omega^{\varepsilon}_{+}$, $\arg\Big(z-\frac{1}{2}\Big)=0$.

$$\breve{\Psi}_+(z) = \big(I + o(1)\big)\exp\Big\{\sigma_3\big(\frac{\zeta^2}{4} - 9\nu+1)\log\zeta\big)\Big\} \tag{9.2.29}$$

$$\times\begin{pmatrix} e^{-i\pi(\nu+1)/2} & 0 \\ -\dfrac{i\sqrt{2\pi}}{v\Gamma(\nu+1)}e^{-i\pi(\nu+1)/2} & -\dfrac{1}{v}\end{pmatrix}, \quad \tau\to\infty,$$

where $\zeta\in\Omega^{\varepsilon}_{+}$, $\arg\Big(z-\frac{1}{2}\Big)=\frac{\pi}{2}$.

The leading order term of the matrices

$$\mathsf{N}_k = \big(\breve{\Psi}_+\big)^{-1}\Psi^k_{WKB}$$

can be derived by comparing equation (9.2.27) with (9.2.28)–(9.2.29); this yields

$$\mathsf{N}_1 = \begin{pmatrix} e^{-i\pi(\nu+1)/2} & 0 \\ 0 & -v\end{pmatrix}\big(I+o(1)\big)\mathsf{N}_0,$$

$$\mathsf{N}_2 = \begin{pmatrix} e^{-i\pi(\nu+1)/2} & 0 \\ -\dfrac{i\sqrt{2\pi}}{v\Gamma(\nu+1)} & -v\end{pmatrix}\big(I+o(1)\big)\mathsf{N}_0. \tag{9.2.30}$$

V. Calculation of the matrix S_1

Applying the asymptotics (9.2.30) and (9.2.26) to equation (9.2.4), which defines matrix S_1, we obtain

$$\mathsf{S}_1 = \begin{pmatrix} 1 & 0 \\ \dfrac{i\sqrt{2\pi}}{v\Gamma(\nu+1)}\exp\Big\{\dfrac{2i}{3}\tau + (\nu+1)\log 8\tau - \dfrac{\pi i}{2}(\nu+1)\Big\} + o(1) & 1\end{pmatrix},$$

$\tau\to\infty$.

The quantity s enters the nondiagonal element of the matrix S_6; recalling that $\tau=(-x)^{3/2}$ and that $1+\nu=i|v|^2$ we obtain equation (9.1.20), which concludes the proof of Theorem 9.3.

3. The direct monodromy problem approach. The proof of Theorem 9.4

Making a rescaling similar to the one used in the previous section

$$z = x^{-1/2}\lambda,$$

the system on Ψ-equation takes the form

$$\frac{d\Psi}{dz} = \tau \mathrm{A}_0(z,\tau), \tag{9.3.1}$$

where

$$\begin{aligned}
&\tau = x^{3/2},\\
&\mathrm{A}_0 = -(4iz^2 - c)\sigma_3 - z(12\sqrt{2} - 4ir)\sigma_2 - \gamma\sigma_1,\\
&c = 2\sqrt{\frac{2}{x}} - \frac{2}{x}iu_0^2 = O(\tau^{-1/2}),\\
&r = iu_0x^{-1/2} = O(\tau^{-1/2}),\\
&\gamma = 2w_0x^{-1} + \frac{i}{\sqrt{2}}x^{-3/2} = O(\tau^{-1/2}).
\end{aligned} \tag{9.3.2}$$

We will now apply to equation (9.3.1) the WKB-technique following the steps identical to those used in Section 2, thus some details are omitted.

The WKB ansatz has again the structure (9.2.2) (note that $\tau = x^{3/2}$), where the matrices T and Λ are defined as follows:

$$\begin{aligned}
&\Lambda = -\mu\sigma_3,\\
&\mu = 4i\sqrt{z^4 + \frac{1}{2}z^2 - \frac{c^2+\gamma^2}{16}},\\
&T = \frac{1}{8i\sqrt{z^4+z^2/2}}\Big[(4iz^2 - c + \mu)\sigma_0 + (4zr - 2\sqrt{2}z)\sigma_1 - i\gamma\sigma_2\Big].
\end{aligned} \tag{9.3.3}$$

We fix the branches of the functions $\sqrt{z^4+z^2/2}$ and $\mu(z)$ by the conditions

$$\sqrt{z^4+z^2/2} \geq 0, \quad z \in \mathbb{R},$$

$$\mu(z) = 4i\sqrt{z^4+z^2/2} + o(1), \quad \tau \to \infty.$$

The following statements follow directly from (9.3.3) and the estimates (9.1.21):

1) There exist two simple turning points

$$z_\pm = \pm\frac{i}{\sqrt{2}},$$

and one double turning point

$$z_0 = 0.$$

2) In the domain where

$$\sqrt{\tau}\left|z\sqrt{z^2+\frac{1}{2}}\right| \to \infty,$$

the following estimate holds for the matrix T,

$$T = \Big(\frac{1}{2}+\frac{1}{2}z^2\big(z^4+\frac{z^2}{2}\big)^{-1/2}\Big)\sigma_0 - \frac{\sqrt{2}}{4i}z\big(z^4+\frac{z^2}{2}\big)^{-1/2}\sigma_1 + O\Big(\frac{z+1}{\sqrt{\tau}\sqrt{z^4+z^2/2}}\Big).$$

In particular, it yields the asymptotics

(9.3.4) $$T \to I, \quad z \to \infty,$$

and

$$T = \left[\begin{pmatrix} 1 & 1 \\ i & -i \end{pmatrix} + o(1)\right] P_k, \quad z = O\Big(\tau^{-1/2+\varepsilon}\Big),$$

$$\arg z = \gamma_k, \quad k = 1, 4, \quad \varepsilon > 0,$$

where

(9.3.5) $$P_1 = \begin{pmatrix} \frac{1}{2} & 0 \\ 0 & \frac{i}{2} \end{pmatrix}, \quad P_4 = \begin{pmatrix} 0 & -\frac{i}{2} \\ \frac{i}{2} & 0 \end{pmatrix},$$

$$\gamma_k = \begin{cases} 0, & k = 1, \\ \pi, & k = 4. \end{cases}$$

3) The Stokes lines tend asymptotically to the real axis, to the interval $\left[-\frac{i}{\sqrt{2}}, \frac{i}{\sqrt{2}}\right]$ and to the branches of the hyperbola

$$(\mathrm{Re}\ z)^2 - \frac{1}{3}(\mathrm{Im}\ z)^2 + \frac{1}{4} = 0;$$

see Figure 9.2.

The statements 1)–3) imply the proper construction of two WKB-solutions, Ψ^1_{WKB} and Ψ^2_{WKB} with the following asymptotic behavior in both τ and z:

(9.3.6) $$\Psi^k_{WKB}(z) = \big[T(z) + o(1)\big]\exp\Big\{-\tau\sigma_3\int_0^z \mu(\eta)d\eta\Big\}$$

$$\sqrt{\tau}\Big|z\sqrt{z^2+\tfrac{1}{2}}\Big| \to \infty, \quad \arg z = \gamma_k = \begin{cases} 0, & k = 1, \\ \pi, & k = 4. \end{cases}$$

It is clear from (9.3.4) and (9.3.6) that the solutions Ψ^1_{WKB} and Ψ^4_{WKB} are related by a right matrix multiplier with the canonical solutions Ψ_1 and Ψ_4. Moreover, the

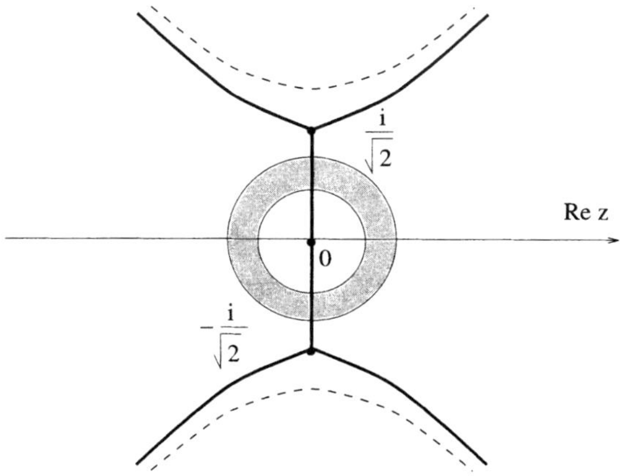

FIGURE 9.2. The Stokes lines are drawn in solid; broken lines denote the branches of the hyperbola. The shaded domain indicates the matching area of Ω_0^ε

relevant matrices can be calculated in closed form,

$$\text{(9.3.7)} \qquad \mathrm{C}_k = \left[\Psi_{WKB}^k\right]^{-1}\Psi_k$$

$$= \lim_{\substack{cz\to\infty \\ \arg z=\gamma_k}} \left\{\tau\sigma_3\int_0^z \mu(\eta)d\eta - \frac{4i}{3}\tau z^3\sigma_3 - i\tau z\sigma_3\right\}.$$

At the neighborhood of the double turning point $z_0 = 0$, equation (9.3.1) simplifies to the system

$$\text{(9.3.8)} \qquad \frac{d\Psi_0}{dz} = \tau\begin{pmatrix} c & -2\sqrt{2}z-\gamma \\ 2\sqrt{2}z-\gamma & -c \end{pmatrix}\Psi_0.$$

Letting

$$\Psi_0 = \begin{pmatrix} 1 & 1 \\ i & -i \end{pmatrix}\Phi,$$

the system (9.3.8) takes the standard form of (9.2.11), where

$$v = \mathrm{e}^{i\pi/4}\Big(u_0(2x)^{1/4} + iw_0(2x)^{-1/4}\Big) = O(1),$$

$$v^* = \mathrm{e}^{i\pi/4}\Big(u_0(2x)^{1/4} - iw_0(2x)^{-1/4}\Big) \equiv -i\overline{v} = O(1),$$

$$\text{(9.3.9)} \qquad \zeta = 2^{5/4}x^{3/4}z\mathrm{e}^{-i\pi/4}.$$

Therefore, under the usual assumption that $|v| \geq c > 0$, in the domain

$$\mathcal{D}_0^\varepsilon = \left\{z\,:\,|z| \leq \tau^{-1/2+\varepsilon},\quad 0<\varepsilon<\frac{1}{6}\right\},$$

it follows that there exists a solution $\Psi_0(z)$ of equation (9.3.1) with the following asymptotic behavior:

$$\Psi_0(z) = \begin{pmatrix} 1 & 1 \\ i & -i \end{pmatrix} \begin{pmatrix} D_{-1-\nu}(i\zeta) & D_\nu(\zeta) \\ \dot{D}_{-1-\nu}(i\zeta) & \dot{D}_\nu(\zeta) \end{pmatrix} (I + o(1)),$$

$$\tau \to \infty, \quad z \in \mathcal{D}_0^\varepsilon, \quad \arg z = 0, \pi, \tag{9.3.10}$$

where

$$\nu = -vv^* - 1 = -i\Big(u_0(2x)^{1/4} - iw_0(2x)^{-1/4}\Big) - 1.$$

The asymptotic expansion of the integral

$$\int_0^z \mu(\eta)d\eta,$$

both at infinity and in the circular domain

$$\Omega_0^\varepsilon = \Big\{z : |z| \simeq \tau^{-1/2+\varepsilon}, \quad 0 < \varepsilon < \frac{1}{6}\Big\},$$

can be obtained as in Section 2. As a result, we find

$$\int_0^z \mu(\eta)d\eta = \frac{4i}{3}\Big(z^2 + \frac{1}{2}\Big)^{3/2} - i\sqrt{2}a^2 \log \frac{\sqrt{z^2 + \frac{1}{2}} - \frac{1}{\sqrt{2}}}{\sqrt{z^2 + \frac{1}{2}} + \frac{1}{\sqrt{2}}}$$

$$-i\frac{\sqrt{2}}{3} + b + o(\tau^{-1}), \tag{9.3.11}$$

$$\tau \to \infty, \quad |z| > C, \quad \arg z = 0,$$

and

$$\int_0^z \mu(\eta)d\eta = -\frac{1}{\tau}\Big(\frac{\zeta^2}{4} - (\nu+1)\log\zeta + \frac{\nu+1}{2}\log 8\sqrt{2}x^{3/2}$$

$$-\frac{\pi i}{4}(\nu+1) + b + o(\tau^{-1})\Big), \tag{9.3.12}$$

$$\tau \to \infty, \quad z \in \Omega_0^\varepsilon, \quad \arg z = 0;$$

in the above expressions

$$a^2 = \frac{u_0^2}{2x} + \frac{w_0^2}{4x^2},$$

$$b = 4i\sqrt{2}\Big[\frac{a^2}{2}\log(-a^2)^{1/2} - \frac{a^2}{4} - \frac{a^2}{2}\log 2\Big]. \tag{9.3.13}$$

Estimates similar to (9.3.11) and (9.3.12) in the case that $\arg z = \pi$, can be derived using the symmetry

$$\mu(-z) = \mu(z), \quad z \in \mathbb{R}.$$

Hence, letting $\arg z = \pi$ we find that

$$\int_0^z \mu(\eta)d\eta = -\int_0^z \mu(\eta)d\eta, \quad \arg(-z) = 0,$$

which yields the asymptotic formulas

$$\begin{aligned}\int_0^z \mu(\eta)d\eta &= -\frac{4i}{3}\Big((-z)^2 + \frac{1}{2}\Big)^{3/2} + i\sqrt{2}a^2 \log\frac{\sqrt{(-z)^2+\frac{1}{2}} - \frac{1}{\sqrt{2}}}{\sqrt{(-z)^2+\frac{1}{2}} + \frac{1}{\sqrt{2}}} \\ &\quad + i\frac{\sqrt{2}}{3} - b + o(\tau^{-1}), \\ &\quad \tau \to \infty, \quad |z| > C, \quad \arg z = \pi,\end{aligned} \tag{9.3.14}$$

and

$$\begin{aligned}\int_0^z \mu(\eta)d\eta &= -\frac{1}{\tau}\Big(\frac{\zeta^2}{4} - (\nu+1)\log\zeta + \frac{\nu+1}{2}\log 8\sqrt{2}x^{3/2} \\ &\quad + \frac{3\pi i}{4}(\nu+1) - b + o(\tau^{-1})\Big), \\ &\quad \tau \to \infty, \quad z \in \Omega_0^{\varepsilon}, \quad \arg z = \pi.\end{aligned} \tag{9.3.15}$$

Finally the asymptotics (9.3.11) and (9.3.14) yield the following expressions for C_1 and C_4,

$$C_1 = C_4^{-1} = \exp\Big\{\Big(-ix^{3/2}\frac{\sqrt{2}}{3} + x^{3/2}b\Big)\sigma - 3\Big\} + o(1). \tag{9.3.16}$$

On the other hand, the asymptotic expansions (9.3.12) and (9.3.15) together with (9.3.5), (9.3.6) and (9.3.10), yield asymptotic representations for the matrices $N_k = \Psi_0^{-1}\Psi_{WKB}^k$:

$$\begin{aligned}N_1 &= \begin{pmatrix} \frac{1}{2}e^{\pi i(\nu+1)/2} & 0 \\ 0 & -\frac{i}{2}v \end{pmatrix} \exp\Big\{\sigma_3\Big(\frac{\nu+1}{2}\log 8\sqrt{2}x^{3/2} \\ &\quad -\frac{\pi i}{4}(\nu+1)\Big) - \tau b\sigma_3\Big\}(I + o(1)), \\ N_4^{-1} &= \exp\Big\{\sigma_3\Big(\frac{\nu+1}{2}\log 8\sqrt{2}x^{3/2} + \frac{3\pi i}{4}(\nu+1)\Big) - \tau b\sigma_3\Big\} \\ &\quad \times \begin{pmatrix} -\frac{2i\sqrt{2\pi}}{v\Gamma(1+\nu)}e^{-\pi i(\nu+1)/2} & -\frac{2}{v} \\ 2ie^{3\pi i(\nu+1)/2} & \frac{2i\sqrt{2\pi}}{\Gamma(-\nu)}e^{-\pi i(\nu+1)} \end{pmatrix}(I + o(1)).\end{aligned} \tag{9.3.17}$$

The monodromy data for the system (9.1.1), can be reconstructed through the matrix

$$S^{-1} = \frac{1}{1+|s|^2}\begin{pmatrix} 1+\overline{s}^2 & \overline{s}-s \\ \overline{s}-s & 1+s^2 \end{pmatrix} \equiv S_3^{-1}S_2^{-1}S_1^{-1}$$

$$= \Psi_4^{-1}\Psi_1 = C_4^{-1}N_4^{-1}N_1C_1. \tag{9.3.18}$$

On the other hand, the asymptotic expression of S^{-1} is given by formulas (9.3.16)–(9.3.18):

$$S^{-1} =$$

$$= \begin{pmatrix} -\frac{i\sqrt{2\pi}}{v\Gamma(1+\nu)}e^{-\frac{2i\sqrt{2}}{3}x^{3/2}+(\nu+1)\log 8\sqrt{2}x^{3/2}+\frac{\pi i}{2}(\nu+1)} & ie^{\pi i(\nu+1)} \\ ie^{\pi i(\nu+1)} & \frac{i\sqrt{2\pi}}{v\Gamma(-\nu)}e^{-\frac{2i\sqrt{2}}{3}x^{3/2}-(\nu+1)\log 8\sqrt{2}x^{3/2}+\frac{\pi i}{2}(\nu+1)} \end{pmatrix}$$

$$+o(1).$$

This concludes the proof of Theorem 9.4.

4. The inverse monodromy problem approach. The proof of Theorem 9.1.

In this and in the following section we will present an alternative way of deriving and proving the asymptotic formulae, for the purely imaginary solutions of the second Painlevé equation (9.0.1). This approach is based on the direct analysis of the relevant Riemann-Hilbert via the Deift-Zhou nonlinear steepest descent method, which was already used in Chapter 8.

We will start with formulating, once again, the Riemann-Hilbert problem associated with equation (9.0.1). As in Chapter 8 we will present its "Ψ-form ", which consists of finding the piecewise holomorphic 2×2 matrix-valued function $\Psi(\lambda)$ of the complex variable λ such that:

1) $\Psi(\lambda)$ is holomorphic for $\lambda \in \mathbb{C}\setminus\cup\{\gamma_k\}$, where γ_k are the rays $\gamma_k = \{\lambda \in \mathbb{C}\colon\ \arg\lambda = \frac{\pi}{6}+\frac{\pi}{3}(k-1)\}$, $k=1,\dots,6$, oriented from zero to infinity.
2) On the rays γ_k the following jump conditions are valid:

$$\Psi_+(\lambda) = \Psi_-(\lambda)S_k, \quad \lambda\in\gamma_k, \tag{9.4.1}$$

where

$$S_{2k-1} = \begin{pmatrix} 1 & 0 \\ s_{2k-1} & 1 \end{pmatrix}, \quad S_{2k} = \begin{pmatrix} 1 & s_{2k} \\ 0 & 1 \end{pmatrix}; \tag{9.4.2}$$

the parameters s_k, which do not depend either on x or on λ, satisfy the constraints

$$s_{k+3} = -s_k, \quad s_1 - s_2 + s_3 + s_1s_2s_3 = 0. \tag{9.4.3}$$

Here, $\Psi_+(\lambda)$ and $\Psi_-(\lambda)$ are the respective limits of $\Psi(\lambda)$ as λ approaches γ from the left and from the right.

3) The behavior of $\Psi(\lambda)$ at $\lambda = \infty$ is described by the asymptotic condition

$$\Psi(\lambda)e^{\theta(\lambda,x)\sigma_3} \to I, \quad \lambda \to \infty, \quad \theta(\lambda, x) = i(\frac{4}{3}\lambda^3 + x\lambda), \tag{9.4.4}$$

$$\sigma_3 = \begin{pmatrix} 1 & 0 \\ 0 & -1 \end{pmatrix}.$$

The above RH problem is depicted in Figure 8.2 (see Chapter 8). We recall that for any given set of the parameters $s \equiv (s_1, s_2, s_3)$ satisfying (9.4.3), the above Riemann-Hilbert problem is meromorphically (with respect to x) solvable and it determines the solution $u(x)$ of the Painlevé equation via the relation,

$$u(x) = 2 \lim_{\lambda \to \infty} \big(\lambda(\Psi(\lambda)e^{\theta(\lambda)})_{12}\big). \tag{9.4.5}$$

We also recall that every solution of (9.4.3) admits the Riemann-Hilbert representation (9.4.5), and that the selection of the purely imaginary solutions $u(x)$, x real, corresponds to the addition reduction,

$$s_1 = -\overline{s}_3.$$

In our analysis in this section, we will in fact assume that the monodromy parameters s_k satisfy a somewhat weaker conditions than the one above. We will assume that

$$\arg\big(1 - s_1 s_3\big) \in (-\pi, \pi), \quad s_1 s_3 \neq 0, \tag{9.4.6}$$

which includes the purely imaginary reduction, $s_1 = -\overline{s}_3$, as well as the real reduction, $s_1 = \overline{s}_3$ under the additional restriction $|s_1| < 1$.

In analogy with the classical steepest descent method we perform the natural change of variable,

$$\lambda = (-x)^{1/2}z, \quad x \to t = (-x)^{3/2}, \quad \Psi(\lambda(z)) \equiv \Psi(z).$$

Then, the characteristic exponent, $\exp(\theta(\lambda))$, is replaced by the exponent

$$e^{t\theta(z)}, \qquad \theta(z) = \frac{4i}{3}z^3 - iz.$$

Thus the stationary phase points are $z_\pm = \pm 1/2$. The anti-Stokes lines are depicted in Figure 9.3.

Our aim is to deform the original RH problem to a RH problem formulated on these anti-Stokes lines. We first note that the original RH problem can be immediately rewritten as the problem on the solid lines of Figure 9.4.

Since the jump matrix across the segment $[z_-, z_+]$ is nonsingular, it admits a unique Gauss LU-decomposition,

$$\big(S_3 S_4 S_5\big)^{-1} = \begin{pmatrix} 1 - s_1 s_3 & s_1 \\ s_1 & 1 + s_1 s_2 \end{pmatrix} = S_L S_D S_U \tag{9.4.7}$$

$$= \begin{pmatrix} 1 & 0 \\ s_1(1 - s_1 s_3)^{-1} & 1 \end{pmatrix} \begin{pmatrix} 1 - s_1 s_3 & 0 \\ 0 & (1 - s_1 s_3)^{-1} \end{pmatrix} \begin{pmatrix} 1 & s_1(1 - s_1 s_3)^{-1} \\ 0 & 1 \end{pmatrix},$$

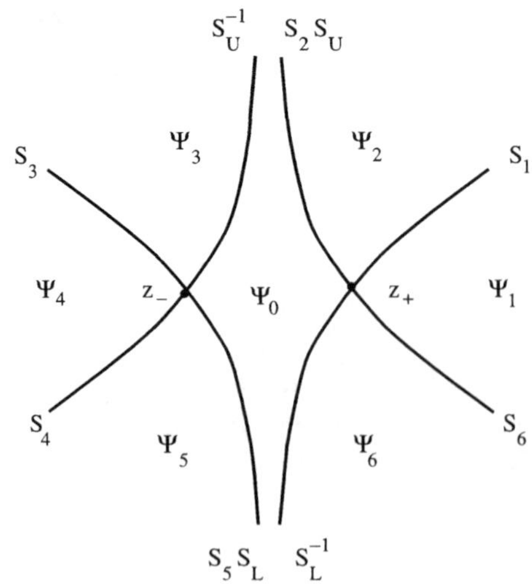

FIGURE 9.3. Anti-Stokes lines of the canonical exponent $\theta(z) = i\frac{4}{3}z^3 - iz$

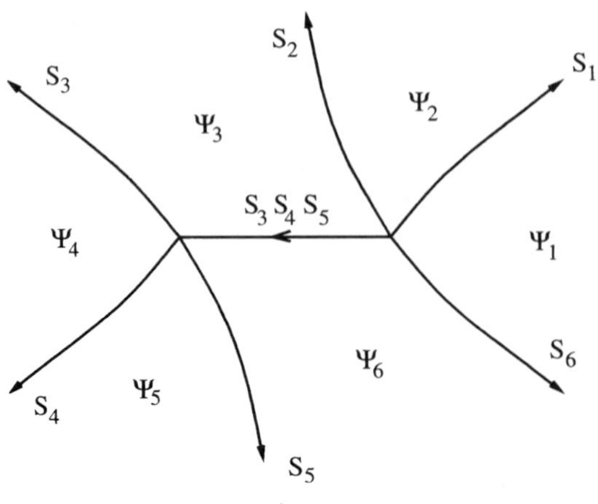

FIGURE 9.4. Deformation of the jump contour of the original RH problem

where we have taken into account the monodromy surface equation, $s_1 - s_2 + s_3 + s_1s_2s_3 = 0$. Therefore, it is possible to split the jump segment $[z_+, z_-]$ into the three contours shown in Figure 9.5.

Inflating the pockets shown in Figure 9.5, we arrive at the jump contour depicted in Figure 9.6. One can easily verify the triviality of the local monodromies around the crossing points $z = \pm 1/2$ which implies that the RH problem is smooth at these points (see [**BDT**], [**Zh**]; see also Section 1 of Chapter 3).

Taking into account the triangular structure of the jump matrices across the infinite branches of the RH jump contour in Figure 9.6, it follows that all jump matrices for the function $Y = \Psi e^{t\theta\sigma_3}$, approach the identity exponentially fast (away from the turning points $\pm 1/2$) as $t \to \infty$. Thus, in analogy with the classical steepest descent method (see also Chapter 8), one expects that the main contribution to

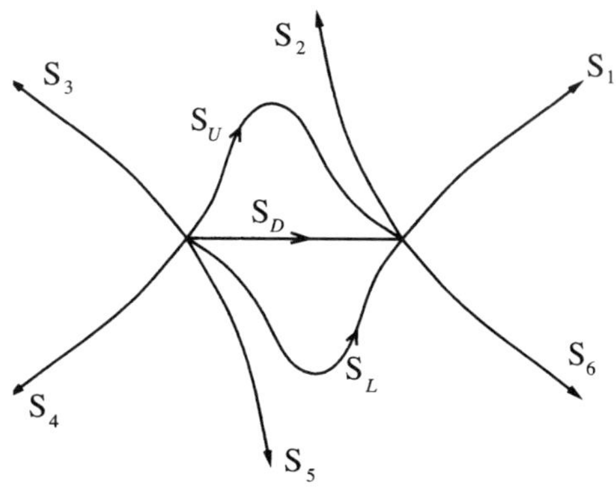

FIGURE 9.5. Splitting of the jump segment $[z_-, z_+]$

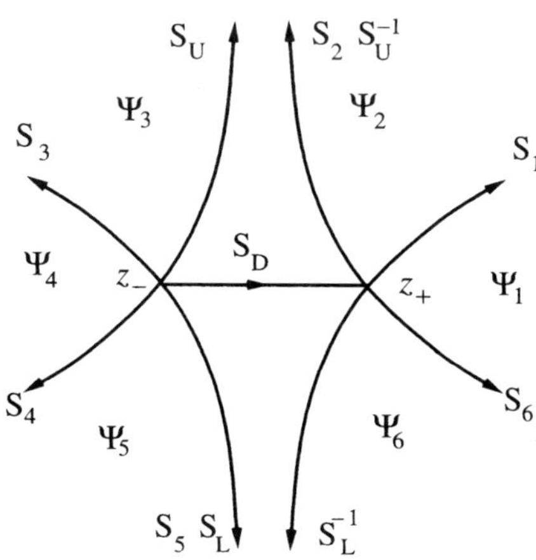

FIGURE 9.6. The final graph of the RH problem

the asymptotics of $Y(z)$ as $t \to \infty$, comes from the segment $[z_-, z_+]$ and from the neighborhood of the points $z_\pm$.

THE MODEL RH PROBLEM ON THE SEGMENT $[z_-, z_+]$

We first consider the RH problem on $[z_-, z_+]$, $z_\pm = \pm 1/2$. The problem consists of finding a piecewise holomorphic function $\hat{\Psi}^D(z)$ with the following properties:

1) $\hat{\Psi}^D(z)e^{t\theta(z)\sigma_3} \to I, \quad z \to \infty, \quad \theta(z) = i\frac{4}{3}z^3 - iz.$

2) On the segment $[z_-, z_+]$ oriented from the left to the right, the following jump condition is valid:

$$\hat{\Psi}^D_+(z) = \hat{\Psi}^D_-(z)S_D, \quad S_D = \left(1 - s_1 s_3\right)^{\sigma_3}.$$

3) At the end points $z = z_\pm$ of the segment, $\hat{\Psi}^D(z)$ does not have a singularity which is stronger than a square integrable singularity.

Since the RH problem for $\hat{\Psi}^D(z)$ is diagonal, it can be solved in a closed form, namely,

$$\hat{\Psi}^D(z) = \left(\frac{z-z_-}{z-z_+}\right)^{\nu\sigma_3} e^{-t\theta(z)\sigma_3}, \tag{9.4.8}$$
$$\nu = -\frac{1}{2\pi i}\ln(1-s_1s_3), \quad \arg\bigl(1-s_1s_3\bigr) \in (-\pi,\pi),$$

where the function $\left(\frac{z-z_-}{z-z_+}\right)^{\nu}$ is defined on the complex z-plane cut along the segment $[z_-, z_+]$, with its branch fixed by the asymptotic condition

$$\left(\frac{z-z_-}{z-z_+}\right)^{\nu} \to 1, \qquad z \to \infty.$$

The model RH problem near $z_\pm$

The exponent $\theta(z)$ in the neighborhood of the node point z_+ is quadratic,

$$\theta(z) = \theta(z_+) + 2i(z-z_+)^2 + \dots.$$

The quadratic local form of the exponent $\theta(z)$ suggests the *parabolic cylinder RH problem* presented in detail in Section 2.5 of Chapter 2 as the candidate for the model RH problem in the neighborhood of the point z_+. This intuitive idea can be implemented as follows.

Introduce (cf. Section 2.5 of Chapter 2) the Wronskian matrix of the parabolic cylinder functions $D_\nu(\zeta)$ and $D_{-\nu-1}(i\zeta)$,

$$Z_0(\zeta) = 2^{-\sigma_3/2}\begin{pmatrix} D_{-\nu-1}(i\zeta) & D_\nu(\zeta) \\ \frac{d}{d\zeta}D_{-\nu-1}(i\zeta) & \frac{d}{d\zeta}D_\nu(\zeta)\end{pmatrix}\begin{pmatrix} e^{i\frac{\pi}{2}(\nu+1)} & 0 \\ 0 & 1\end{pmatrix}. \tag{9.4.9}$$

Furthermore, introduce the auxiliary matrix functions

$$\begin{aligned} Z_1(\zeta) &= Z_0(\zeta)H_0, \quad Z_2(\zeta) = Z_1(\zeta)H_1,\\ Z_3(\zeta) &= Z_2(\zeta)H_2, \quad Z_4(\zeta) = Z_3(\zeta)H_3, \end{aligned} \tag{9.4.10}$$

where the triangular matrices H_k satisfy the equations

$$H_{k+2} = e^{i\pi(\nu+\frac{1}{2})\sigma_3}H_k e^{-i\pi(\nu+\frac{1}{2})\sigma_3}, \quad H_0 = \begin{pmatrix} 1 & 0 \\ h_0 & 1\end{pmatrix}, \quad H_1 = \begin{pmatrix} 1 & h_1 \\ 0 & 1\end{pmatrix}, \tag{9.4.11}$$
$$h_0 = -i\frac{\sqrt{2\pi}}{\Gamma(\nu+1)}, \quad h_1 = \frac{\sqrt{2\pi}}{\Gamma(-\nu)}e^{i\pi\nu}, \quad 1+h_0h_1 = e^{2\pi i\nu}.$$

All the functions $Z_k(\zeta)$ are solutions of a matrix equation with polynomial coefficients,

$$\frac{dZ}{d\zeta} = \begin{pmatrix} 0 & 1/2 \\ \frac{\zeta^2}{2} - 2\nu - 1 & 0\end{pmatrix} Z, \tag{9.4.12}$$

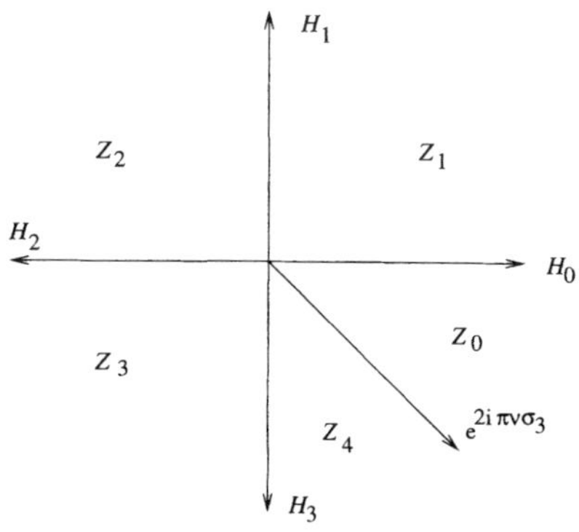

FIGURE 9.7. The model RH problem which can be solved in terms of the parabolic cylinder functions

thus they are entire functions of ζ. Note, that the symmetry of (9.4.12) with respect to the transformation $\zeta \to -\zeta$ allows one to derive the following identity:

$$\sigma_3 Z_{k+2}(e^{i\pi}\zeta)\sigma_3 = Z_k(\zeta)e^{-i\pi(\nu+1)\sigma_3}. \tag{9.4.13}$$

Let us now assemble the following piecewise holomorphic function $Z^{RH}(\zeta)$; see Figure 9.7:

$$Z^{RH}(\zeta) = \begin{cases} Z_0(\zeta), & \arg\zeta \in \left(-\frac{\pi}{4}, 0\right), \\ Z_1(\zeta), & \arg\zeta \in \left(0, \frac{\pi}{2}\right), \\ Z_2(\zeta), & \arg\zeta \in \left(\frac{\pi}{2}, \pi\right), \\ Z_3(\zeta), & \arg\zeta \in \left(\pi, \frac{3\pi}{2}\right), \\ Z_4(\zeta), & \arg\zeta \in \left(\frac{3\pi}{2}, \frac{7\pi}{4}\right). \end{cases} \tag{9.4.14}$$

Using the asymptotic properties of the parabolic cylinder functions [**BE**], we find that $Z^{RH}(\zeta)$ solves the following RH problem (cf. again Section 2.5 of Chapter 2):

1) as $\zeta \to \infty$,

$$Z^{RH}(\zeta) = \zeta^{-\sigma_3/2}\frac{1}{\sqrt{2}}\begin{pmatrix} 1 & 1 \\ 1 & -1 \end{pmatrix}(I + O(\zeta^{-2}))e^{(\frac{\zeta^2}{4} - (\nu+\frac{1}{2})\ln\zeta)\sigma_3}. \tag{9.4.15}$$

2) The following jump conditions are valid:

$$Z_+^{RH}(\zeta) = Z_-^{RH}(\zeta)H_k, \quad \arg\zeta = \frac{\pi}{2}k, \quad k = 0, 1, 2, 3, \tag{9.4.16}$$

and

$$Z_+^{RH}(\zeta) = Z_-^{RH}(\zeta)e^{2\pi i\nu\sigma_3}, \quad \arg\zeta = -\frac{\pi}{4}. \tag{9.4.17}$$

The cyclic identity $e^{2\pi i\nu\sigma_3}H_0H_1H_2H_3 = I$ implies the triviality of the monodromy at the origin, which in turn ensures the continuity of the model RH problem.

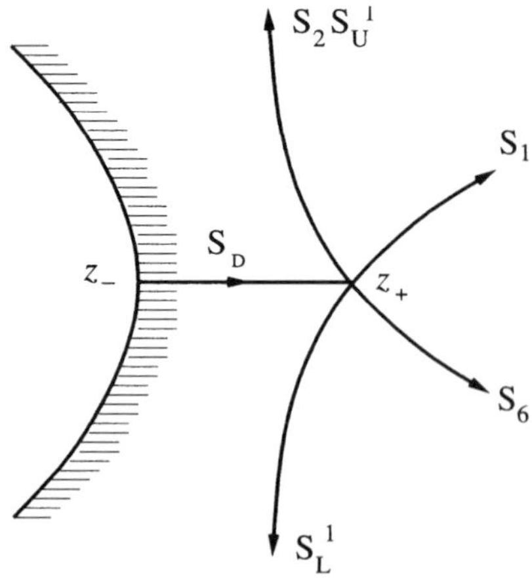

FIGURE 9.8. Transformation of the jump graph of the model RH problem and the jump properties of $\hat{\Psi}^r(z)$

The jump graph shown on Figure 9.7, if rotated by $-3\pi/4$, approximate the jump graph near a node point $z = z_+ \equiv 1/2$ on Figure 9.6. Comparison of the quadratic terms in the exponentials in (9.4.8) and (9.4.15) suggests the relation $\zeta \sim e^{3\pi i/4}2\sqrt{2t}\,(z - \frac{1}{2})$. In fact, we define the local change of the independent variable, $\zeta = \zeta(z)$, by the equation,

$$\zeta(z) = 2\sqrt{-\frac{4it}{3}z^3 + itz - \frac{it}{3}} \equiv 2\sqrt{-t\theta(z) + t\theta(z_+)}. \tag{9.4.18}$$

Since z_+ is a stationary point, i.e., $\theta'(z_+) = 0$, equation (9.4.18) defines a holomorphic change of variable in some neighborhood,

$$|z - z_+| < r < 1, \tag{9.4.19}$$

of the point $z = z_+$. The branch of the square root in (9.4.18) is defined by the condition mentioned above,

$$\zeta \sim e^{3\pi i/4}2\sqrt{2t}\,\left(z - \frac{1}{2}\right).$$

The parametrix $\hat{\Psi}^r(z)$ near $z = z_+ \equiv \frac{1}{2}$, can be defined as follows:

$$\hat{\Psi}^r(z) = \mathsf{B}(z)\big(-h_1/s_3\big)^{-\sigma_3/2}e^{\frac{it}{3}\sigma_3}2^{-\sigma_3/2} \times \begin{pmatrix} \zeta(z) & 1 \\ 1 & 0 \end{pmatrix} Z^{RH}\big(\zeta(z)\big)\big(-h_1/s_3\big)^{\sigma_3/2}, \tag{9.4.20}$$

where $\zeta = \zeta(z)$ is the holomorphic change of the independent variable introduced in (9.4.18), h_1 is the Stokes multiplier for the parabolic cylinder functions defined in (9.4.11), and the diagonal matrix $\mathsf{B}(z)$ is given by the equations,

$$\mathsf{B}(z) = \left(\zeta(z)\frac{z - z_-}{z - z_+}\right)^{\nu\sigma_3}, \quad \mathsf{B}(z_+) = (8t)^{\frac{\nu}{2}\sigma_3}e^{\frac{3\pi i}{4}\nu\sigma_3}. \tag{9.4.21}$$

It is important to note that the matrix function $\mathsf{B}(z)$ is holomorphic in neighborhood (9.4.19) of $z = z_+$.

By construction, the function $\hat{\Psi}^r(z)$ has jumps across the curves emanating from $z = z_+$ and depicted in Figure 9.8. These curves can be always be made to coincide in the disk (9.4.19) (for r sufficiently small, e.g., less than 1/2) with the corresponding jump curves of the final RH graph (see Figure 9.6); indeed, one can take the latter to be the pre-images of the former. Moreover, the definition of the function $\hat{\Psi}^r(z)$ ensures that its jumps are described by *exactly the same* Stokes matrices as the ones describing the solution $\Psi(z)$ of the original RH problem (see Figure 9.8 and Figure 9.6). Therefore, in a finite neighborhood of the node point $z = z_+$, $\Psi(\lambda)$ and $\hat{\Psi}^r(z)$ can be related by a left holomorphic multiplier $\mathsf{N}_r(z)$,

$$\Psi(\lambda) = \mathsf{N}_r(z)\hat{\Psi}^r(z). \tag{9.4.22}$$

Finally, we observe that

$$\mathsf{B}(z)\mathrm{e}^{(\frac{\zeta^2}{4}-\nu\ln\zeta)\sigma_3} = \mathrm{e}^{-\frac{it}{3}\sigma_3}\left(\frac{z-z_-}{z-z_+}\right)^{\nu\sigma_3}\mathrm{e}^{-t\theta(z)\sigma_3} \equiv \hat{\Psi}^D(z).$$

This relation, together with the use of appropriate asymptotic formulae of the parabolic cylinder functions (see e.g. [**BE**]), implies that,

$$\begin{aligned}
\hat{\Psi}^r(z) &= \mathsf{B}(z)\big(-h_1/s_3\big)^{-\sigma_3/2}\mathrm{e}^{\frac{it}{3}\sigma_3}\begin{pmatrix}1 & \frac{\nu}{\zeta}\\ \frac{1}{\zeta} & 1\end{pmatrix}\\
&\quad\times\big(I+O(\zeta^{-2})\big)\mathrm{e}^{(\frac{\zeta^2}{4}-\nu\ln\zeta)\sigma_3}\big(-h_1/s_3\big)^{\sigma_3/2}\\
&= \big(-h_1/s_3\big)^{-\sigma_3/2}\mathrm{e}^{\frac{it}{3}\sigma_3}\mathsf{B}(z)\begin{pmatrix}1 & \frac{\nu}{\zeta}\\ \frac{1}{\zeta} & 1\end{pmatrix}\mathsf{B}^{-1}(z)\\
&\quad\times\mathsf{B}(z)\big(I+O(\zeta^{-2})\big)\mathsf{B}^{-1}(z)\left(\mathsf{B}(z)\mathrm{e}^{(\frac{\zeta^2}{4}-\nu\ln\zeta)\sigma_3}\right)\big(-h_1/s_3\big)^{\sigma_3/2}\\
&= \begin{pmatrix}1 & -\frac{\nu s_3}{h_1}e^{\frac{2it}{3}}\beta(z)\frac{1}{\zeta}\\ -\frac{h_1}{s_3}e^{-\frac{2it}{3}}\beta^{-1}(z)\frac{1}{\zeta} & 1\end{pmatrix}\\
&\quad\times\left(I+O(\zeta^{-2+2|\operatorname{Re}\nu|})\right)\hat{\Psi}^D(z),
\end{aligned}\tag{9.4.23}$$

as $t\to\infty$ and $0 < r_1 \le |z-z_+| \le r_2 < 1$ (and hence $|\zeta|\to\infty$!). Here we use the notation

$$\beta(z) = \left(\zeta(z)\frac{z-z_-}{z-z_+}\right)^{2\nu}.$$

Near the other node point $z = z_- = -z_+ = -1/2$, the parametrix $\hat{\Psi}^l(z)$ can be introduced in a similar way. Alternatively, we can use the symmetry of the RH problem in Figure 9.6, and define

$$\hat{\Psi}^l(z) = \sigma_2\hat{\Psi}^r(-z)\sigma_2. \tag{9.4.24}$$

In analogy to (9.4.22), the function $\hat{\Psi}^l(z)$ differs from $\Psi(\lambda)$ by a left holomorphic multiplier $\mathrm{N}_l(z)$,

$$\Psi(\lambda) = \mathrm{N}_l(z)\hat{\Psi}^l(z), \quad \mathrm{N}_l(z) = \sigma_2 \mathrm{N}_r(-z)\sigma_2. \tag{9.4.25}$$

Also, $\hat{\Psi}^l(z)$ matches the "external" function $\hat{\Psi}^D(z)$ on the boundary of the disc $|z - z_-| < r < 1$ via the relation similar to (9.4.23).

The approximate solution of the original RH problem as $x \to -\infty$

Let us assemble the piecewise holomorphic matrix function $\hat{\Psi}(z)$ as follows:

$$\hat{\Psi}(z) = \begin{cases} \hat{\Psi}^D(z), & |z - z_\pm| > R, \\ \hat{\Psi}^r(z), & |z - z_+| < R, \\ \hat{\Psi}^l(z), & |z - z_-| < R, \end{cases} \tag{9.4.26}$$

where $0 < R < \frac{1}{2}$ is constant. The function $\hat{\Psi}(z)$ is discontinous across the graph shown in Figure 9.9 where C_r and C_l denote the clockwise oriented circles of radius R and centered at z_+ and z_-, respectively.

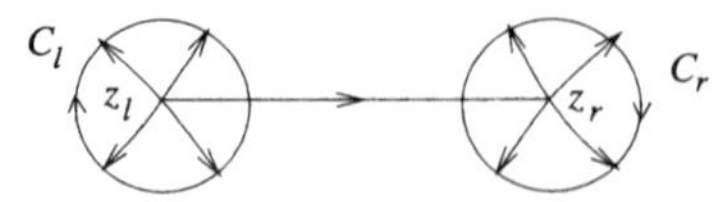

FIGURE 9.9. The RH problem for the combined function $\hat{\Psi}(z)$

We look for the exact solution $\Psi(z)$ of the RH problem of Figure 9.6 in the form

$$\Psi(\lambda) = \chi(z)\hat{\Psi}(z). \tag{9.4.27}$$

The error function $\chi(z) = \Psi(z)\hat{\Psi}^{-1}(z)$ solves the following RH problem.

1) $\chi(z) \to I$ as $z \to \infty$.

2) On the graph depicted in Figure 9.10, the function $\chi(z)$ has jumps described by the following equations:

On the infinite parts of the anti-Stokes lines γ_k,

$$\chi_+(z) = \chi_-(z)G(z), \quad G_k(z) = \hat{\Psi}^D(z)\mathrm{S}_k\hat{\Psi}^{D-1}(z), \quad z \in \gamma_k, \quad k = 1, \ldots, 6, \tag{9.4.28}$$

On the clockwise oriented circles,

$$\chi_+(z) = \chi_-(z)G_m(z), \quad G_m(z) = \hat{\Psi}^m(z)\hat{\Psi}^{D-1}(z), \quad z \in C_m, \quad m \in \{r, l\}. \tag{9.4.29}$$

We emphasize one of the key properties of our construction: by virtue of equations (9.4.22) and (9.4.25), the function $\chi(z)$ *does not* have any jumps inside the circles C_r and C_l.

Due to their triangular structure, the jump matrices $G_k(z)$ for the infinite parts of the anti-Stokes lines approach the unit matrix exponentially fast. Indeed, these matrices satisfy

$$\|G_k(z) - I\| \leq ce^{-4t|z-\frac{1}{2}|^2}, \quad k = 1, \ldots, 6, \tag{9.4.30}$$

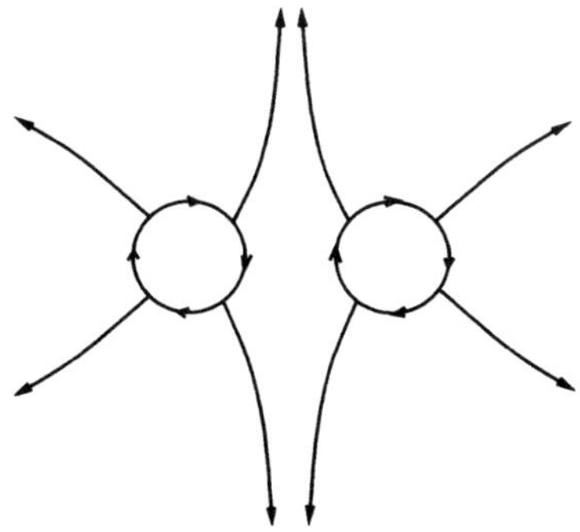

FIGURE 9.10. The RH problem for the error function $\chi(z) = \Psi(z)\hat{\Psi}^{-1}(z)$

for some positive constant c whose value is not important here. At the same time, the relation (9.4.23) leads to the following asymptotic representations for the jump matrices $G_r(z)$ and $G_l(z) = \sigma_2 G_r(-z)\sigma_2$:

$$G_r(z) = \begin{pmatrix} 1 & -\frac{\nu s_3}{h_1} e^{\frac{2it}{3}} \beta(z) \frac{1}{\zeta} \\ -\frac{h_1}{s_3} e^{-\frac{2it}{3}} \beta^{-1}(z) \frac{1}{\zeta} & 1 \end{pmatrix}$$

$$\times \left(I + O(\zeta^{-2+2|\operatorname{Re}\nu|})\right), \tag{9.4.31}$$

and

$$G_l(z) = \begin{pmatrix} 1 & \frac{h_1}{s_3} e^{-\frac{2it}{3}} \beta^{-1}(-z) \frac{1}{\zeta(-1)} \\ \frac{\nu s_3}{h_1} e^{\frac{2it}{3}} \beta(-z) \frac{1}{\zeta(-z)} & 1 \end{pmatrix}$$

$$\times \left(I + O(\zeta^{-2+2|\operatorname{Re}\nu|})\right). \tag{9.4.32}$$

Because $\zeta \geq c|t|^{1/2}$ on $C_r \cup C_l$ for some positive constant c, the following estimate for the jump matrices is valid:

$$\left\|G_{r,l}(z) - I\right\| \leq c_1 t^{-\frac{1}{2}+|\operatorname{Re}\nu|} \tag{9.4.33}$$

with some positive constant whose value is not important here. Thus, the jump matrices $G_r(z)$ and $G_l(z)$ approach the unit matrix uniformly as $t \to \infty$, provided that the following inequality is valid

$$\left|\operatorname{Re}\nu\right| < \frac{1}{2} - \delta, \quad \text{or, equivalently,} \quad \left|\arg(1 - s_1 s_3)\right| < \pi - 2\pi\delta, \tag{9.4.34}$$

where $0 < \delta < 1/2$. In the imaginary reduction, $s_1 = -\bar{s}_3$, we have $1 - s_1 s_3 = 1 + |s_1|^2 > 0$ and hence the condition (9.4.34) is satisfied.

Condition (9.4.34) implies the unique solvability of the RH problem (9.4.28)–(9.4.29) for the function $\chi(z)$. Indeed, the RH problem is equivalent (cf. Chapters 3 and 8) to the singular integral equation,

$$\chi_-(z) = I + \frac{1}{2\pi i}\int_\gamma \chi_-(z')(G(z') - I)\frac{dz'}{z' - z_-}, \tag{9.4.35}$$

where the contour γ is shown in Figure 9.10, and the jump matrix $G(z)$ is described by equations (9.4.28) and (9.4.29). In virtue of the inequalities (9.4.30) and (9.4.33) we have

$$\|G - I\|_{L^2(\gamma)\cap L^\infty(\gamma)} \le ct^{-\frac{1}{2}+|\mathrm{Re}\,\nu|}. \tag{9.4.36}$$

Therefore, by applying Theorem 8.1 of Chapter 8 we conclude that for sufficiently large t the relevant integral operator is contracting, and equation (9.4.35) can be solved in $L^2(\gamma)$ by iterations. Moreover, its unique solution satisfies

$$\|\chi_- - I\|_{L^2(\gamma)} \le ct^{-\frac{1}{2}+|\mathrm{Re}\,\nu|}. \tag{9.4.37}$$

Estimate (9.4.37) allows us to extract the asymptotics of the solution of PII, which is related to the solution $\chi(z)$ of the RH problem (9.4.28)–(9.4.29), via the equation (cf. (9.4.5)),

$$u(x) = 2\sqrt{-x}\lim_{z\to\infty}\big(z(\Psi(z)e^{t\theta(z)})_{12}\big).$$

Taking into account (9.4.27), (9.4.26), and (9.4.8)), we can rewrite this representation in the form,

$$u(x) = 2\sqrt{-x}\lim_{z\to\infty}\big(z\chi_{12}(z)\big). \tag{9.4.38}$$

In order to determine the entry $\chi_{12}(z)$, we use the integral representation for $\chi(z)$,

$$\chi(z) = I + \frac{1}{2\pi i}\int_\gamma \chi_-(z')(G(z') - I)\frac{dz'}{z' - z}, \quad z \notin \gamma, \tag{9.4.39}$$

which implies the relation

$$\lim_{z\to\infty}\big(z\chi_{12}(z)\big) = -\frac{1}{2\pi i}\int_\gamma \big(\chi_-(z)(G(z) - I)\big)_{12}\,dz. \tag{9.4.40}$$

Note that the estimates (9.4.36)and (9.4.37) allow as to take the limit inside the integral in (9.4.39). Actually, using the same estimates, and neglecting the exponentially small contribution of the infinite parts of γ, we can transform (9.4.40) into the equation,

$$\lim_{z\to\infty}\big(z\chi_{12}(z)\big) = -\frac{1}{2\pi i}\int_{C_r\cup C_l} G_{12}(z)\,dz + O\big(t^{-1+2|\mathrm{Re}\,\nu|}\big). \tag{9.4.41}$$

On the circles C_r and C_l the jump matrix $G(z)$ coincide with the matrices $G_r(z)$ and $G_l(z)$, respectively. Therefore, by using the estimates (9.4.31) and (9.4.32), we can transform (9.4.41) into the asymptotic formula,

$$\lim_{z\to\infty}\big(z\chi_{12}(z)\big) = \frac{1}{2\pi i}\frac{\nu s_3}{h_1}e^{\frac{2it}{3}}\int_{C_r}\frac{\beta(z)}{\zeta(z)}\,dz$$

$$+\frac{1}{2\pi i}\frac{h_1}{s_3}e^{-\frac{2it}{3}}\int_{C_r}\frac{\beta^{-1}(z)}{\zeta(z)}\,dz + O\big(t^{-1+2|\mathrm{Re}\,\nu|}\big). \tag{9.4.42}$$

The functions $\zeta(z)$, $\beta(z)$, and $\beta^{-1}(z)$ are holomorphic inside the circle C_r. The first terms of the relevant Taylor series about the center point $z = z_+ = 1/2$ are:

$$\zeta(z) = \mathrm{e}^{3i\pi/4}(8t)^{1/2}(z - \frac{1}{2}) + \dots,$$

$$\beta(z) = \mathrm{e}^{i\frac{3\pi\nu}{2}}(8t)^{\nu} + \dots, \quad \beta^{-1}(z) = \mathrm{e}^{-i\frac{3\pi\nu}{2}}(8t)^{-\nu} + \dots$$

Moreover, the point $z = 1/2$ is the only zero of $\zeta(z)$ inside of C_r. Therefore, the contour integrals in (9.4.42) can be easily evaluated:

$$\frac{1}{2\pi i}\oint_{C_r}\frac{\beta^{\pm 1}(z)}{\zeta(z)}\,dz = -\mathrm{Res}_{z=1/2}\left(\frac{\beta^{\pm 1}(z)}{\zeta(z)}\right)$$
$$= \mathrm{e}^{i\pi/4}\mathrm{e}^{\pm i\frac{3\pi\nu}{2}}(8t)^{\pm\nu - 1/2},$$

where we have taken into account the clockwise orientation of C_r. Combining this with equations (9.4.42) and (9.4.38), we obtain the following asymptotic representation for the Painlevé function $u(x)$,

$$u(x) = 2\mathrm{e}^{i\pi/4}\sqrt{-x}\Big(\frac{\nu s_3}{h_1}\mathrm{e}^{\frac{2it}{3}+i\frac{3\pi\nu}{2}}(8t)^{\nu-1/2} + \frac{h_1}{s_3}\mathrm{e}^{-\frac{2it}{3}-i\frac{3\pi\nu}{2}}(8t)^{-\nu-1/2}\Big)$$
$$+O\big(|x|^{-1+3|\mathrm{Re}\,\nu|}\big), \tag{9.4.43}$$

as $x \to -\infty$ and $t = (-x)^{3/2}$. Recalling the definition (9.4.11) of the quantity h_1, and using the classical identity,

$$\Gamma(-\nu)\Gamma(\nu) = -\frac{\pi}{\nu\sin\pi\nu}, \tag{9.4.44}$$

we can rewrite (9.4.43) in the final symmetric form,

$$u(x) = (-x)^{-1/4}\Big\{\frac{\sqrt{\pi}\,\mathrm{e}^{-i\frac{\pi\nu}{2}}}{s_1\Gamma(\nu)}\mathrm{e}^{i\frac{2}{3}(-x)^{3/2}+\nu\ln 8(-x)^{3/2}-i\frac{\pi}{4}} \tag{9.4.45}$$
$$+\frac{\sqrt{\pi}\,\mathrm{e}^{-i\frac{\pi\nu}{2}}}{s_3\Gamma(-\nu)}\mathrm{e}^{-i\frac{2}{3}(-x)^{3/2}-\nu\ln 8(-x)^{3/2}+i\frac{\pi}{4}}\Big\} + O\big(|x|^{-1+3|\mathrm{Re}\,\nu|}\big),$$

where

$$\nu = -\frac{1}{2\pi i}\ln\big(1 - s_1s_3\big), \quad \arg\big(1 - s_1s_3\big) \in \big(-\pi + 2\pi\delta, \pi - 2\pi\delta\big), \quad \delta > 0.$$

Under the purely imaginary reduction,

$$s_1 = -\overline{s}_3 \equiv s,$$

formula (9.4.45) can be simplified further. Indeed, in this case

$$u(x) = (-x)^{-1/4}2i\,\mathrm{Im}\left\{\frac{\sqrt{\pi}\,\mathrm{e}^{\frac{\pi d^2}{4}}}{s\Gamma\left(i\frac{d^2}{2}\right)}\mathrm{e}^{i\frac{2}{3}(-x)^{3/2}+\frac{i}{2}d^2\ln 8(-x)^{3/2}-i\frac{\pi}{4}}\right\} \tag{9.4.46}$$
$$+O\big(|x|^{-1}\big), \quad x \to -\infty,$$

where we have introduced the notation (cf. (9.1.7)),

$$d^2 = -2i\nu \equiv \frac{1}{\pi}\log(1 + |s|^2).$$

Using again the gamma-function identity (9.4.44) we find

$$\left|\frac{1}{\Gamma\left(i\frac{d^2}{2}\right)}\right|^2 = -\frac{i}{2}d^2\frac{\sin\frac{\pi i}{2}d^2}{\pi} = \frac{d^2}{4\pi}\left(\mathrm{e}^{\frac{\pi}{2}d^2} - \mathrm{e}^{-\frac{\pi}{2}d^2}\right) = \frac{d^2}{4\pi}\mathrm{e}^{-\frac{\pi}{2}d^2}|s|^2.$$

Therefore, equation (9.4.46) can be transformed into the formula,

$$
\begin{aligned}
u(x) &= i(-x)^{-1/4} d \sin\left\{\frac{2}{3}(-x)^{3/2} + \frac{3}{4} d^2 \log(-x) + \varphi\right\} \\
&\quad + O(|x|^{-1}), \quad x \to -\infty,
\end{aligned} \tag{9.4.47}
$$

where

$$
d^2 = \frac{1}{\pi}\log(1+|s|^2), \quad d > 0, \tag{9.4.48}
$$

$$
\varphi = \frac{3}{2} d^2 \log 2 - \frac{\pi}{4} - \arg\Gamma\left(i\frac{d^2}{2}\right) - \arg s, \tag{9.4.49}
$$

which is exactly equation (9.1.6). We emphasized that, since the parameter s can be any complex number, we have proved that *any* purely imaginary solution of the second Painlevé equation (9.0.1), has the asymptotics described by (9.4.47).

Condition (9.4.34) is also satisfied for the real solutions, i.e., when,

$$
s_1 = \overline{s}_3 \equiv s,
$$

provided that the following additional constraint is valid:

$$
|s_1| \equiv |s| < 1.
$$

In this case, following the same steps used to transform (9.4.43) into (9.4.47), we arrive at the following asymptotic formula for the *subclass* of the real solutions of (9.0.1) satisfying $|s| < 1$,

$$
\begin{aligned}
u(x) &= (-x)^{-1/4} d \cos\left\{\frac{2}{3}(-x)^{3/2} - \frac{3}{4} d^2 \log(-x) + \varphi\right\} \\
&\quad + O(|x|^{-1}), \quad x \to -\infty,
\end{aligned} \tag{9.4.50}
$$

where

$$
d^2 = -\frac{1}{\pi}\log(1-|s|^2), \quad d > 0, \tag{9.4.51}
$$

$$
\varphi = -\frac{3}{2} d^2 \log 2 - \frac{\pi}{4} + \arg\Gamma\left(i\frac{d^2}{2}\right) - \arg s. \tag{9.4.52}
$$

We summarize the results of this section in the following theorem.

THEOREM 9.5. *Let the parameters s_1 and s_3 satisfy the inequality,*

$$
\arg(1 - s_1 s_3) \in (-\pi + 2\pi\delta, \pi - 2\pi\delta), \quad s_1 s_3 \neq 0, \tag{9.4.53}
$$

for an arbitrary $0 < \delta < 1/2$. Then there exists a constant $x_0 < 0$, such that the RH problem (9.4.1)–(9.4.4) shown in Figure 8.2 is uniquely solvable for all $x < x_0$. Its solution is described by the formulae (9.4.27), (9.4.26), (9.4.20), (9.4.8). The asymptotics of the relevant solution of the second Painlevé equation PII($\alpha = 0$) as $x \to -\infty$ is given by (9.4.45). In the imaginary reduction, $s_1 = -\overline{s}_3 \equiv s$, we obtain (9.1.6)–(9.1.7).

REMARK 9.2. The special case when $s_1 s_3 = 0$ will be analyzed in Chapter 11.

REMARK 9.3. It turns out that the leading order asymptotics of the Painlevé function (9.1.6)–(9.1.7) (and (9.4.45) in the more general case) remains valid even if $\alpha \neq 0$ in PII. In fact, the leading order contribution of the parameter α to the asymptotics (9.4.45) is the additive term α/x.

REMARK 9.4. The error bound in (9.4.45) can be improved: The leading order terms presented in Theorem 9.1, together with the term α/x provide the asymptotic behavior of the Painlevé function, with an error bound equal to $O\big(x^{-\frac{7}{4}+\frac{9}{2}|\mathrm{Re}\,\nu|}\big)$.

5. The inverse monodromy problem approach. The proof of Theorem 9.2

We begin with the usual scalings

$$\lambda = x^{1/2}z, \quad x \mapsto t = x^{3/2}, \quad \Psi(\lambda(z)) \equiv \Psi(z),$$

which transform the canonical exponent to the expression,

$$i\Big(\frac{4}{3}\lambda^3 + x\lambda\Big) = it\Big(\frac{4}{3}z^3 + z\Big) \equiv t\theta(z).$$

The saddle points now are $z = \pm i/2$, and we have that

$$\mathrm{e}^{t\theta(\pm i/2)} = \mathrm{e}^{\mp\frac{t}{3}}.$$

Therefore, similarly with the situation discussed in the beginning of Section 3 of Chapter 8, half of the jump matrices for the function $Y(z) \equiv \Psi(z)\mathrm{e}^{t\theta(z,t)\sigma_3}$, blow up exploding in the neighborhoods of the saddle points as $t \to \infty$ (assuming the generic condition, $s_2 \neq 0$). To overcome this problem we introduce the "g"-function by the equations,

$$g(z) = i\frac{4}{3}\left(z^2 + \frac{1}{2}\right)^{3/2}, \quad g(z) \sim i\frac{4}{3}z^3, \quad z \to \infty, \tag{9.5.1}$$

which define $g(z)$ as an analytic function outside the vertical segment $[-i/\sqrt{2}, i/\sqrt{2}]$. We note that

$$g(z) = \theta(z) + O(z^{-2}), \quad z \to \infty, \tag{9.5.2}$$

and

$$g\left(\pm\frac{i}{\sqrt{2}}\right) = 0, \quad g(0) = \pm i\frac{t\sqrt{2}}{3}, \tag{9.5.3}$$

where $z = 0, \pm\frac{i}{\sqrt{2}}$, are the saddle points of the exponential function,

$$\mathrm{e}^{tg(z)} \equiv \mathrm{e}^{i\frac{4}{3}\left(z^2+\frac{1}{2}\right)^{3/2}}.$$

By virtue of (9.5.3), the functions $\mathrm{e}^{\pm tg(z)}$ are bounded in the neighborhoods of the relevant stationary phase points as $t \to \infty$ (cf. Chapter 8). We note that the g-function (9.5.1) is the limiting case, as $\arg x \to 0$, of the elliptic g-function of Chapter 8 (see (8.3.4)). More precisely,

$$\frac{4}{3}\left(z^2 + \frac{1}{2}\right)^{3/2} = 4\int_{z_3}^{z} w|_{\arg x \to 0}(z')dz', \tag{9.5.4}$$

where $w(z) = \sqrt{(z^2 - z_1^2)(z^2 - z_1^2)}$ is the elliptic curve whose branch points $\pm z_1$, $\pm z_3$ are defined, as functions of $\varphi \equiv \arg x$, by equations (8.3.1) and (8.3.6). The limit $\varphi \to 0$ implies that $z_1 \to 0$ and $z_3 \to -i/\sqrt{2}$, and then (9.5.4) follows.

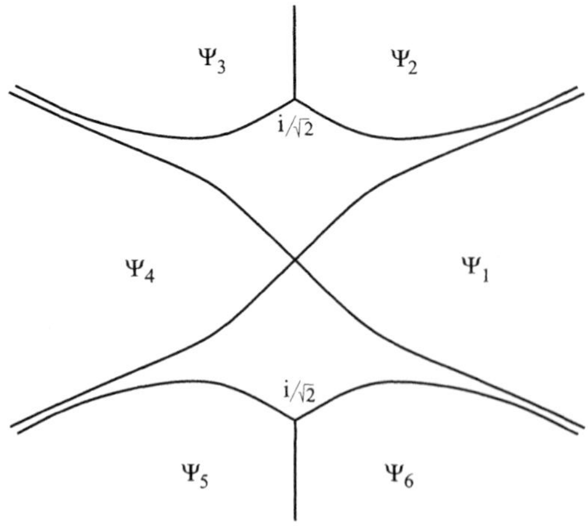

FIGURE 9.11. The anti-Stokes curves for $\exp\left\{it\frac{4}{3}(z^2+\frac{1}{2})^{3/2}\right\}$

Following the same procedure as in Chapter 8, we will deform the original Riemann-Hilbert problem (9.4.1)–(9.4.4) to one posed on the steepest-descent (or anti-Stokes) curves of the exponents $e^{\pm tg(z)}$ depicted in Figure 9.11. For this purpose we factorize the jump matrices S_2 and S_5 on the segments $(0, i/\sqrt{2})$ and $(0, -i/\sqrt{2})$, respectively. The standard LDU decomposition (with a diagonal matrix D, a lower L and an upper U triangular matrices with the unit diagonal entries) is not suitable, we use the modified forms LPL and UPU, where P is a quasi-permutation matrix (cf. Chapter 8). Namely,

$$S_2=\begin{pmatrix}1 & s_2\\ 0 & 1\end{pmatrix}=\begin{pmatrix}1 & 0\\ 1/s_2 & 1\end{pmatrix}\begin{pmatrix}0 & s_2\\ -1/s_2 & 0\end{pmatrix}\begin{pmatrix}1 & 0\\ 1/s_2 & 1\end{pmatrix}=S_LS_{P_1}S_L, \tag{9.5.5}$$

and

$$S_5=\begin{pmatrix}1 & 0\\ -s_2 & 1\end{pmatrix}=\begin{pmatrix}1 & -1/s_2\\ 0 & 1\end{pmatrix}\begin{pmatrix}0 & 1/s_2\\ -s_2 & 0\end{pmatrix}\begin{pmatrix}1 & -1/s_2\\ 0 & 1\end{pmatrix}=S_US_{P_2}^{-1}S_U. \tag{9.5.6}$$

Using this factorization, the segments $(0, i/\sqrt{2})$ and $(0, -i/\sqrt{2})$ can be split as shown in Figure 9.12. Inflating the pockets shown in Figure 9.12, we obtain the final jump graph shown in Figure 9.13.

Since the jump matrices for $\Psi\exp\{it\frac{4}{3}(z^2+\frac{1}{2})^{3/2}\sigma_3\}$ approach the unit matrix exponentially fast along the infinite parts of the anti-Stokes curves, we expect that the main contribution to the Ψ-function comes from the union of the segments $(-i/\sqrt{2}, 0)\cup(0, i/\sqrt{2})$, as well as from the neighborhoods of the turning points $z=0, \pm i/\sqrt{2}$.

REMARK 9.5. Throughout this chapter we assume the generic condition

$$s_2\neq 0, \tag{9.5.7}$$

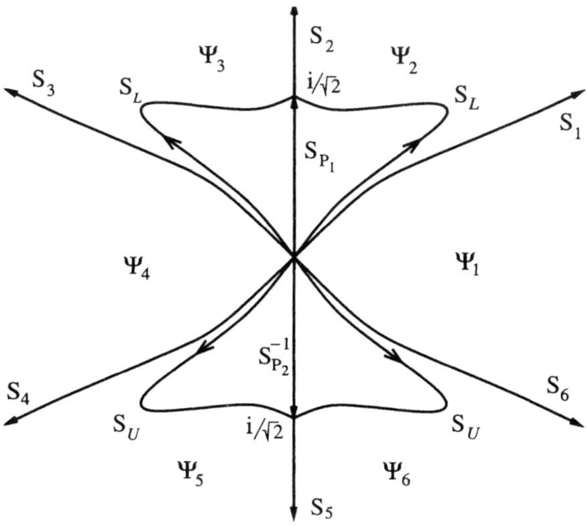

FIGURE 9.12. Deformation of the curves of the basic RH problem as a result of the factorization of S_2 and S_5

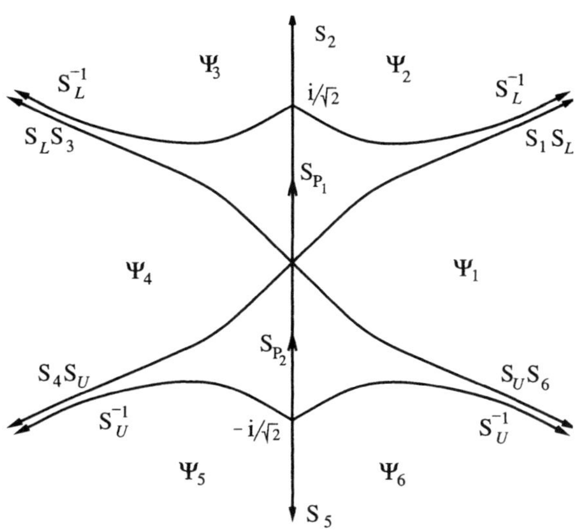

FIGURE 9.13. The graph of the final RH problem for the case $x \to +\infty$

which implies the nontriviality of the matrices S_2 and S_5. This in turn implies that the jump matrix $G(z) = e^{-t\theta(z)}S_k e^{t\theta(z)}$ blows up in the neighborhoods of the saddle points $\pm i/2$ of the function $\theta(z)$, while the modified jump matrices, i.e., $e^{-tg(z)}S_k e^{tg(z)}$, remain bounded in the neighborhoods of the saddle points of the exponent $g(z)$. In the case $s_2 = 0$, the transition to $g(z)$ is not needed and the analysis is simplified dramatically. We will discuss this case in detail in Chapter 11.

THE MODEL RH PROBLEM ON THE SEGMENT $[-i/\sqrt{2}, i/\sqrt{2}]$

In addition to the generic condition, $s_2 \neq 0$, we will assume from now on that

$$-\frac{\pi}{2} < \arg(i\sigma s_2) < \frac{\pi}{2}, \tag{9.5.8}$$

where the parameter σ equals either 1 or -1. Note, that this restriction is consistent with the purely imaginary reduction, $\overline{s}_2 = -s_2$, but *excludes* the real case, i.e., $\overline{s}_2 = s_2$. Actually, in the purely imaginary case we have

$$\sigma = \operatorname{sign} is_2. \tag{9.5.9}$$

The model RH problem on the segment $[-i/\sqrt{2}, i/\sqrt{2}]$ consists of finding a piecewise holomorphic function $\breve{\Psi}(z)$ satisfying the following conditions:

1) $\breve{\Psi}(z)e^{t\theta(z)} \to I \quad \text{as} \quad z \to \infty$.

2) $\breve{\Psi}(z)$ is discontinues across the segments $(-i/\sqrt{2}, 0)$ and $(0, i/\sqrt{2})$ oriented as is in Figure 9.13, where it satisfies the jump conditions

$$\begin{aligned} &\breve{\Psi}_+(z) = \breve{\Psi}_-(z)S_{P_1}, \quad z \in (0, i/\sqrt{2}), \\ &\breve{\Psi}_+(z) = \breve{\Psi}_-(z)S_{P_2}, \quad z \in (-i/\sqrt{2}, 0), \\ &S_{P_1} = \begin{pmatrix} 0 & s_2 \\ -1/s_2 & 0 \end{pmatrix}, \quad S_{P_2} = \begin{pmatrix} 0 & -1/s_2 \\ s_2 & 0 \end{pmatrix}. \end{aligned} \tag{9.5.10}$$

3) At the endpoints $0, \pm i/\sqrt{2}$, the function $\breve{\Psi}(z)$ has no singularities which are stronger than square integrable singularities.

The above quasi-permutation RH problem can be solved by elementary means. This can be achieved as follows (cf. [**FoIK4**]):

Consider first the auxiliary scalar RH problem consisting of finding the scalar function $\delta(z)$ analytic outside the vertical interval $[-i/\sqrt{2}, i/\sqrt{2}]$ and such that:

- $\delta(z) \to 1 \quad \text{as} \quad z \to \infty$;
- $\delta_+(z)\delta_-(z) = \frac{1}{i\sigma s_2}, \quad z \in (0, i/\sqrt{2})$;
- $\delta_+(z)\delta_-(z) = i\sigma s_2. \quad z \in (-i/\sqrt{2}, 0)$.

The solution of this scalar problem is

$$\delta(z) = \Delta^{-\frac{\nu}{2}}(z), \tag{9.5.11}$$

where

$$\nu := \frac{1}{i\pi}\ln(i\sigma s_2), \quad -\pi < \arg(i\sigma s_2) < \pi, \tag{9.5.12}$$

and

$$\Delta(z) := \frac{\left(z^2 + \frac{1}{2}\right)^{1/2} - \frac{1}{\sqrt{2}}}{\left(z^2 + \frac{1}{2}\right)^{1/2} + \frac{1}{\sqrt{2}}}. \tag{9.5.13}$$

The branches of the functions $\left(z^2 + \frac{1}{2}\right)^{1/2}$ and $\Delta^{-\frac{\nu}{2}}$ are fixed by the conditions $\left(z^2 + \frac{1}{2}\right)^{1/2} > 0$ as $z > 0$ and $-\pi < \arg\Delta < \pi$, respectively.

We introduce now the function $\breve{Y}$,

$$\breve{Y}(z) = \breve{\Psi}(z)e^{tg(z)\sigma_3}\delta^{-\sigma_3}(z). \tag{9.5.14}$$

Due to (9.5.2), this function is normalized at infinity,

$$\breve{Y}(z) \to I, \quad z \to \infty, \tag{9.5.15}$$

and has the following jumps across $(-i/\sqrt{2}, 0)$, $(0, i/\sqrt{2})$:

- $z \in (-i/\sqrt{2}, 0)$:

$$\breve{Y}_+(z) = \breve{Y}_-(z)\breve{S}_{P_2}, \tag{9.5.16}$$

$$\breve{S}_{P_2} = e^{-tg_-(z)}\delta_-^{\sigma_3}(z)S_{P_2}\delta_+^{-\sigma_3}(z)e^{tg_+(z)}$$

$$= \begin{pmatrix} 0 & -s_2^{-1}\delta_+\delta_- \\ s_2\delta_+^{-1}\delta_-^{-1} & 0 \end{pmatrix};$$

- $z \in (0, i/\sqrt{2})$:

$$\breve{Y}_+(z) = \breve{Y}_-(z)\breve{S}_{P_1}, \tag{9.5.17}$$

$$\breve{S}_{P_1} = e^{-tg_-(z)}\delta_-^{\sigma_3}(z)S_{P_1}\delta_+^{-\sigma_3}(z)e^{tg_+(z)}$$

$$= \begin{pmatrix} 0 & s_2\delta_+\delta_- \\ -s_2^{-1}\delta_+^{-1}\delta_-^{-1} & 0 \end{pmatrix},$$

where we have taken into account the relation $g_+(z) + g_-(z) = 0$. Using the auxiliary scalar RH problem for $\delta(z)$, we see that the jump matrices $\breve{S}_{P_{1,2}}$ become *equal*. Indeed, we have

$$\breve{S}_{P_1} = \breve{S}_{P_2} = -i\sigma\sigma_1 = \begin{pmatrix} 0 & -i\sigma \\ -i\sigma & 0 \end{pmatrix}. \tag{9.5.18}$$

Thus the permutation RH problem for $\breve{Y}(z)$, reduces to a permutation RH problem defined on a *single* interval, which can be diagonalized and hence can be solved in terms of elementary functions (cf. [**DeIZ**] and [**DeIKZ**]). Namely, let

$$\beta(z) = \left(\frac{z + i/\sqrt{2}}{z - i/\sqrt{2}}\right)^{1/4}, \tag{9.5.19}$$

which is an analytic function defined on the complex z-plane cut along the segment $[-i/\sqrt{2}, i/\sqrt{2}]$. Its branch is fixed by the asymptotic condition

$$\beta(z) \to 1, \quad z \to \infty. \tag{9.5.20}$$

This function satisfies the jump condition

$$\beta_+ = -i\beta_- \tag{9.5.21}$$

across the interval $[-i/\sqrt{2}, i/\sqrt{2}]$. Therefore, the matrix function $\breve{Y}(z)$, defined by

$$\breve{Y}(z) = \frac{1}{2}\begin{pmatrix} \beta + \beta^{-1} & \sigma(\beta - \beta^{-1}) \\ \sigma(\beta - \beta^{-1}) & \beta + \beta^{-1} \end{pmatrix}$$

$$= \frac{1}{2}\sigma^{-\sigma_3/2}\begin{pmatrix} 1 & 1 \\ 1 & -1 \end{pmatrix}\beta^{\sigma_3}\begin{pmatrix} 1 & 1 \\ 1 & -1 \end{pmatrix}\sigma^{\sigma_3/2}, \tag{9.5.22}$$

is normalized at infinity,

$$\breve{Y}(z) \to I, \quad z \to \infty,$$

and satisfies the desired jump condition

$$\breve{Y}_+(z) = \breve{Y}_-(z)(-i\sigma\sigma_1), \quad z \in (-i/\sqrt{2}, i/\sqrt{2}). \tag{9.5.23}$$

In summary, the solution of the model RH problem on the segment $[-i/\sqrt{2}, i/\sqrt{2}]$, is given by the formula

$$\breve{\Psi}(z) = \breve{Y}(z)\mathrm{e}^{-tg(z)\sigma_3}\delta^{\sigma_3}(z), \tag{9.5.24}$$

where the functions $\breve{Y}(z)$, $g(z)$, $\delta(z)$, and the parameter ν are defined by equations (9.5.22), (9.5.1), (9.5.11), and (9.5.12), respectively.

The model RH problems near $z = \pm i/\sqrt{2}$

Introduce the matrix function $A_0(\zeta)$,

$$A_0(\zeta) = \sqrt{2\pi}\,\mathrm{e}^{-i\pi/4}\begin{pmatrix} v_2(\zeta) & v_1(\zeta) \\ \frac{d}{d\zeta}v_2(\zeta) & \frac{d}{d\zeta}v_1(\zeta) \end{pmatrix}\mathrm{e}^{-i\frac{\pi}{4}\sigma_3}, \tag{9.5.25}$$

where

$$v_1(\zeta) = \mathrm{Ai}\,(\zeta), \quad v_2(\zeta) = \mathrm{e}^{i2\pi/3}\,\mathrm{Ai}\,(\mathrm{e}^{i2\pi/3}\zeta). \tag{9.5.26}$$

Using the well-known asymptotic properties of the Airy functions [**BE**], we find that this function has the following asymptotics as $|\zeta| \to \infty$, $\arg\zeta \in \left(-\pi, \frac{\pi}{3}\right)$,

$$A_0(\zeta) = \zeta^{-\sigma_3/4}\frac{1}{\sqrt{2}}\begin{pmatrix} 1 & 1 \\ 1 & -1 \end{pmatrix}(I + O(\zeta^{-3/2}))\mathrm{e}^{\frac{2}{3}\zeta^{3/2}\sigma_3}. \tag{9.5.27}$$

Also, introducing the matrix functions

$$A_1(\zeta) := A_0(\zeta)G_0, \quad A_2(\zeta) := A_1(\zeta)G_1, \quad A_3(\zeta) := A_2(\zeta)G_2,$$

$$G_0 = G_2 = \begin{pmatrix} 1 & 0 \\ -i & 1 \end{pmatrix}, \quad G_1 = \begin{pmatrix} 1 & -i \\ 0 & 1 \end{pmatrix}, \tag{9.5.28}$$

we find that the functions $A_j(\zeta)$ have the asymptotics (9.5.27) as $|\zeta| \to \infty$, $\arg\zeta \in \left(-\pi + \frac{2\pi}{3}j, \frac{\pi}{3} + \frac{2\pi}{3}j\right)$, $j = 0, 1, 2, 3$.

Define a piecewise holomorphic function $A^{RH}(\zeta)$,

$$A^{RH}(\zeta) = \begin{cases} A_0(\zeta)(is_2)^{-\sigma_3/2}, & \arg z \in \left(-\frac{\pi}{3}, 0\right), \\ A_j(\zeta)(is_2)^{-\sigma_3/2}, & \arg\zeta \in \left(-\frac{2\pi}{3} + \frac{2\pi}{3}j, \frac{2\pi}{3}j\right), \quad j = 1, 2, \\ A_3(\zeta)(is_2)^{-\sigma_3/2}, & \arg\zeta \in \left(\frac{4\pi}{3}, \frac{5\pi}{3}\right). \end{cases} \tag{9.5.29}$$

By construction, $A^{RH}(\zeta)$ has the uniform asymptotics
(9.5.30)

$$A^{RH}(\zeta) = \zeta^{-\sigma_3/4}\frac{1}{\sqrt{2}}\begin{pmatrix} 1 & 1 \\ 1 & -1 \end{pmatrix}\left(I + O(\zeta^{-3/2})\right)e^{\frac{2}{3}\zeta^{3/2}\sigma_3}(is_2)^{-\sigma_3/2}, \quad \zeta \to \infty,$$

as well as the jump properties indicated in Figure 9.14.

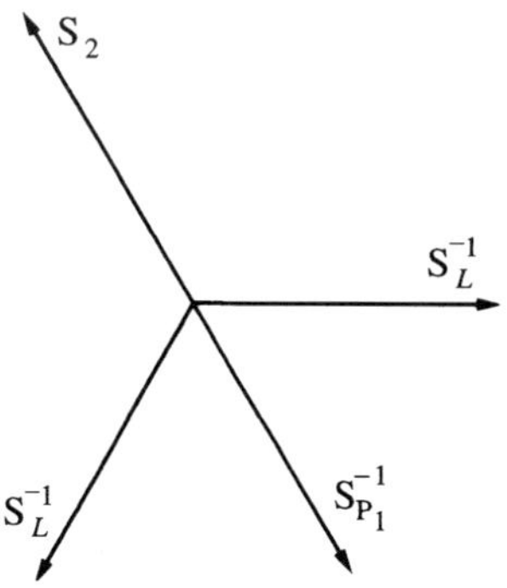

FIGURE 9.14. Jump graph for $A^{RH}(\zeta)$

Let us now introduce in the neighborhood of the saddle point $i/\sqrt{2}$, the change of the independent variable $\zeta = \zeta(z)$ by the equation

$$\zeta(z) = (-2it)^{2/3}\left(z^2 + \frac{1}{2}\right) \equiv 2^{2/3}t^{2/3}e^{-\pi/3}\left(z^2 + \frac{1}{2}\right). \tag{9.5.31}$$

Note, that

$$\zeta(z) = 2^{7/6}t^{2/3}e^{i\pi/6}(z - i/\sqrt{2})\left(1 + O(z - i/\sqrt{2})\right), \quad z \to i/\sqrt{2}, \tag{9.5.32}$$

and

$$tg(z) + \tfrac{2}{3}\zeta^{3/2} = 0. \tag{9.5.33}$$

The pre-image of the jump graph of Figure 9.14, of the model RH problem, under the map $\zeta = \zeta(z)$ is shown in Figure 9.15; we will assume that it coincides with the jump graph of Figure 9.13 in the neighborhood of the point $i/\sqrt{2}$.

The parametrix $\Psi^u(z)$ near $z = \frac{i}{\sqrt{2}}$ is defined by the equation,

$$\Psi^u(z) = \mathsf{B}_u(z)A^{RH}(\zeta(z)), \tag{9.5.34}$$

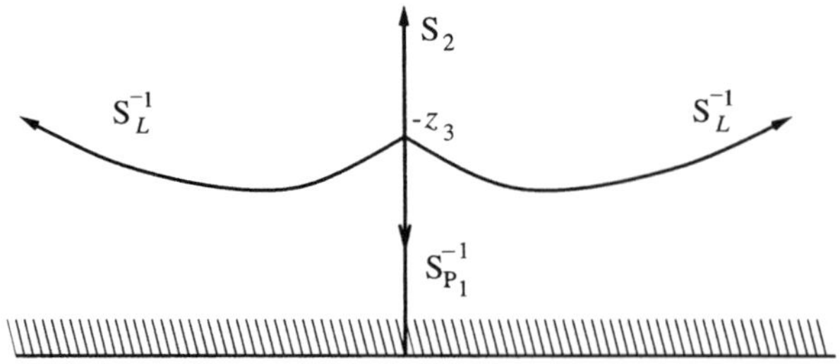

FIGURE 9.15. Transformation of the jump graph of the model RH problem and the jump properties of $\Psi^u(z)$

where $\zeta(z)$ is the holomorphic change of variables (9.5.31), and

$$\mathrm{B}_u(z) := \breve{Y}(z)\delta^{\sigma_3}(z)\big(i\sigma s_2\big)^{\sigma_3/2} \tag{9.5.35}$$

$$\times\sigma^{-\frac{\sigma_3}{2}}\frac{1}{\sqrt{2}}\begin{pmatrix}1 & 1\\ 1 & -1\end{pmatrix}\zeta^{\sigma_3/4}(z), \quad -\frac{\pi}{3} < \arg\zeta < \frac{5\pi}{3}.$$

Observe that on the interval $(0, i/\sqrt{2})$, we have

$$\zeta_+^{\sigma_3/4}(z) = \zeta_-^{\sigma_3/4}(z)\mathrm{e}^{\frac{i\pi}{2}\sigma_3},$$

where the interval is oriented toward its upper end. Recalling the jump properties of the functions $\breve{Y}(z)$ and $\delta(z)$ across this interval, one can easily verify that the matrix function $\mathrm{B}_u(z)$ is holomorphic in the deleted neighborhood of the point $z = i/\sqrt{2}$. A direct calculation shows that the function $\mathrm{B}_u(z)$ is bounded near the point $z = i/\sqrt{2}$. Indeed, we have

$$\mathrm{B}_u\Big(\frac{i}{\sqrt{2}}\Big) = \sigma^{-\frac{\sigma_3}{2}}\frac{1}{\sqrt{2}}\begin{pmatrix}1 & 1\\ 1 & -1\end{pmatrix}$$

$$\times\begin{pmatrix}\mathrm{e}^{i\pi/6}t^{1/6}2^{5/12} & \nu\mathrm{e}^{i\pi/3}t^{-1/6}2^{19/12}\\ 0 & \mathrm{e}^{-i\pi/6}t^{-1/6}2^{-5/12}\end{pmatrix}.$$

Hence, we conclude that the matrix-valued function $\mathrm{B}_u(z)$ is a holomorphic function which does not vanish in a neighborhood of $z = \frac{i}{\sqrt{2}}$ (compare with the function $\mathrm{B}_u(z)$ in (9.4.20)–(9.4.21)). Therefore, locally the function $\Psi^u(z)$ is related with the exact solution $\Psi(z)$ of the original RH problem by a left holomorphic matrix multiplier $\mathrm{N}_u(z)$ (cf. (9.4.22)),

$$\Psi(z) = \mathrm{N}_u(z)\Psi^u(z). \tag{9.5.36}$$

At the same time, in analogy with the analysis of the previous section, the introduction of the matrix $\mathrm{B}_u(z)$ allows us to match the parametrix solution $\Psi^u(z)$ with the solution $\breve{\Psi}(z)$ of the model RH problem posed on the interval $[-i/\sqrt{2}, i/\sqrt{2}]$.

Indeed, using the asymptotic formulae (9.5.30), we can derive the relation,

$$\Psi^u(z) = \breve{Y}(z)\delta^{\sigma_3}(z)\big(is_2\big)^{\sigma_3/2}\big(I + O(\zeta^{-3/2})\big)e^{\frac{2}{3}\zeta^{3/2}\sigma_3}\big(is_2\big)^{-\sigma_3/2}$$

$$= \big(I + O(\zeta^{-3/2})\big)\breve{Y}(z)\delta^{\sigma_3}(z)e^{\frac{2}{3}\zeta^{3/2}\sigma_3}$$

$$= \big(I + O(\zeta^{-3/2})\big)\breve{\Psi}(z), \tag{9.5.37}$$

$$t \to \infty, \quad 0 < r_1 \le |z - i/\sqrt{2}| \le r_2 < 1/\sqrt{2},$$

where we have also used the identity (9.5.33) and the definition (9.5.24) of the function $\breve{\Psi}(z)$. We also note that the function $\zeta(z)$ is of order $t^{2/3}$, uniformly in the annular domain $0 < r_1 \le |z - i/\sqrt{2}| \le r_2 < 1/\sqrt{2}$.

The parametrix solution $\Psi^d(z)$ near the point $z = -\frac{i}{\sqrt{2}}$, can be constructed using the symmetry relation $z \mapsto -z$,

$$\Psi^d(z) = \sigma_2\Psi^u(-z)\sigma_2. \tag{9.5.38}$$

The model RH problem near $z = 0$

Consider again the set of Wronskian matrices $Z_k(\zeta)$ of the parabolic cylinder functions described by (9.4.9)–(9.4.11), but with the parameter ν defined by equation (9.5.12), i.e.,

$$\nu = \frac{1}{i\pi}\ln(i\sigma s_2). \tag{9.5.39}$$

Introduce the auxiliary function $Z_r^{RH}(\zeta)$ defined in a complex half-plane $\arg\zeta \in \big(-\frac{3\pi}{4}, \frac{\pi}{4}\big)$,

$$Z_r^{RH}(\zeta) := Z_k(\zeta)Q^{\sigma_3/2}, \tag{9.5.40}$$

where

$$k = \begin{cases} -1, & \arg\zeta \in \big(-\frac{3\pi}{4}, -\frac{\pi}{2}\big), \\ 0, & \arg\zeta \in \big(-\frac{\pi}{2}, 0\big), \\ 1, & \arg\zeta \in \big(0, \frac{\pi}{4}\big), \end{cases} \qquad Q = \frac{1 + s_1 s_2}{s_2 h_0}.$$

Define a piecewise holomorphic function $Z^{RH}(\zeta)$ on the whole complex ζ-plane by the relations,

$$Z^{RH}(\zeta) = \begin{cases} Z_r^{RH}(\zeta), & \arg\zeta \in \big(-\frac{3\pi}{4}, \frac{\pi}{4}\big), \\ -\sigma\sigma_3 Z_r^{RH}(e^{-i\pi}\zeta)\sigma_2, & \arg\zeta \in \big(\frac{\pi}{4}, \frac{5\pi}{4}\big), \end{cases} \tag{9.5.41}$$

where $\sigma = \pm 1$ is the same as in (9.5.22) and (9.5.18). Taking into account (9.5.39), we find that the function $Z^{RH}(\zeta)$ has the following jumps across the rays oriented

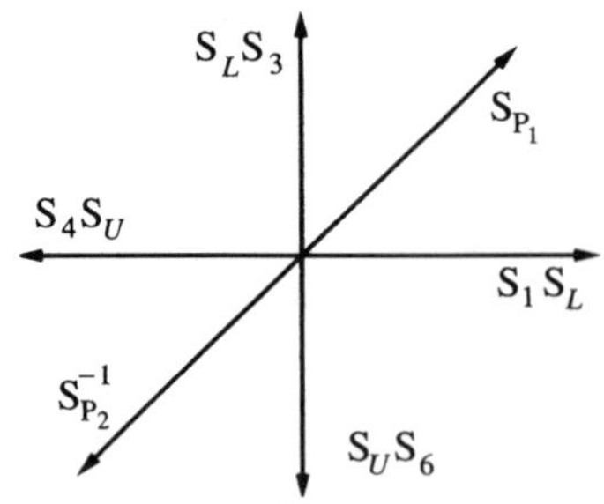

FIGURE 9.16. The model RH problem which can be solved in terms of the parabolic cylinder functions

toward infinity,

$$Z_+^{RH}(\zeta) = Z_-^{RH}(\zeta)\mathrm{S},$$

$$\arg\zeta = -\frac{\pi}{2}: \qquad \mathrm{S} = \mathrm{S}_U\mathrm{S}_6 = \begin{pmatrix} 1 & -(1+s_2s_3)s_2^{-1} \\ 0 & 1 \end{pmatrix},$$

$$\arg\zeta = 0: \qquad \mathrm{S} = \mathrm{S}_1\mathrm{S}_L = \begin{pmatrix} 1 & 0 \\ (1+s_1s_2)s_2^{-1} & 1 \end{pmatrix},$$

$$\arg\zeta = \frac{\pi}{2}: \qquad \mathrm{S} = \mathrm{S}_L\mathrm{S}_3 = \begin{pmatrix} 1 & 0 \\ (1+s_2s_3)s_2^{-1} & 1 \end{pmatrix},$$

$$\arg\zeta = \pi: \qquad \mathrm{S} = \mathrm{S}_4\mathrm{S}_U = \begin{pmatrix} 1 & -(1+s_1s_2)s_2^{-1} \\ 0 & 1 \end{pmatrix},$$

$$\arg\zeta = \frac{\pi}{4}: \qquad \mathrm{S} = \mathrm{S}_{P_1} = \begin{pmatrix} 0 & s_2 \\ -s_2^{-1} & 0 \end{pmatrix},$$

$$\arg\zeta = -\frac{3\pi}{4}: \qquad \mathrm{S} = \mathrm{S}_{P_2}^{-1} = \begin{pmatrix} 0 & s_2^{-1} \\ -s_2 & 0 \end{pmatrix}; \tag{9.5.42}$$

see Figure 9.16.

Indeed, the jump matrix across the ray $\arg\zeta = 0$ in (9.5.42) is simply the conjugation of the Stokes matrix H_0 in (9.4.11) for the parabolic cylinder functions. The jump matrix across the ray $\arg\zeta = -\pi/2$ is the conjugation of the matrix H_{-1} (see (9.4.11)), where we use the constraint on the Stokes multipliers (9.4.3), i.e., $s_1 - s_2 + s_3 + s_1s_2s_3 = 0$, and formula (9.5.39). The jump matrices across $\arg\zeta = \pi/2, \pi$ can be obtained from the jump matrices for $\arg\zeta = -\pi/2, 0$ with the help of the symmetry $S \mapsto \sigma_2 S \sigma_2$. The jump matrices across $\arg\zeta = \pi/4, -3\pi/4$, can be computed using the symmetry (9.4.13).

From the asymptotics of the parabolic cylinder functions (see [**BE**], see also Section 2.5 of Chapter 2), we derive for $Z^{RH}(\zeta)$ the following asymptotic relations as $\zeta \to \infty$,

$$Z^{RH}(\zeta) = \frac{\zeta^{-\sigma_3/2}}{\sqrt{2}}$$

$$\times \begin{pmatrix} 1 + \frac{(\nu+1)(\nu+2)}{2\zeta^2} + O(\zeta^{-4}) & 1 - \frac{\nu(\nu-1)}{2\zeta^2} + O(\zeta^{-4}) \\ 1 + \frac{(\nu-2)(\nu+1)}{2\zeta^2} + O(\zeta^{-4}) & -1 + \frac{\nu(\nu+3)}{2\zeta^2} + O(\zeta^{-4}) \end{pmatrix}$$

$$\times e^{(\frac{\zeta^2}{4} - (\nu+\frac{1}{2})\ln\zeta)\sigma_3} Q^{\sigma_3/2}$$

$$= \begin{pmatrix} 0 & 1 \\ 1 & -\zeta \end{pmatrix} 2^{\sigma_3/2} Q^{\sigma_3/2} \begin{pmatrix} 1 & \frac{\nu}{Q\zeta} \\ \frac{Q}{\zeta} & 1 \end{pmatrix}$$

$$\times \left(I + O(\zeta^{-2})\right) e^{(\frac{\zeta^2}{4} - \nu\ln\zeta)\sigma_3}, \qquad \arg\zeta \in \left(-\tfrac{3\pi}{4}, \tfrac{\pi}{4}\right), \tag{9.5.43}$$

$$Z^{RH}(\zeta) = \begin{pmatrix} 0 & 1 \\ 1 & -\zeta \end{pmatrix} 2^{\sigma_3/2} Q^{\sigma_3/2} \begin{pmatrix} 1 & \frac{\nu}{Q\zeta} \\ \frac{Q}{\zeta} & 1 \end{pmatrix} \times$$

$$\times \left(I + O(\zeta^{-2})\right) e^{(\frac{\zeta^2}{4} - \nu\ln\zeta)\sigma_3} S_{P_1}, \qquad \arg\zeta \in \left(\tfrac{\pi}{4}, \tfrac{5\pi}{4}\right). \tag{9.5.44}$$

In order to construct the parametrix near $z = 0$, we also need a suitable local change of the independent variable, $\zeta = \zeta(z)$. For the latter we use the formula,

$$\zeta(z) = 2\sqrt{it\frac{\sqrt{2}}{3} - itg_0(z)}\,, \tag{9.5.45}$$

where the function $g_0(z)$ is defined by the equations,

$$g_0(z) = \text{sign}\,(\text{Re}\ z)\, g(z) \equiv \text{sign}\,(\text{Re}\ z)\, \frac{4}{3}\left(z^2 + \frac{1}{2}\right)^{3/2}, \qquad z \neq 0, \tag{9.5.46}$$

$$g_0(0) = 0.$$

Observe that, since the function $g(z)$ is defined on the z-plane cut along the vertical interval $[-i/\sqrt{2}, i/\sqrt{2}]$, the function $g_0(z)$ is holomorphic in a neighborhood of $z = 0$. Moreover, in this neighborhood we have

$$it\frac{\sqrt{2}}{3} - itg_0(z) = -it\sqrt{2}z^2 + \dots,$$

and hence equation (9.5.45) defines $\zeta(z)$ as a genuine local holomorphic change of independent variables. In fact, $\zeta(z)$ admits a biholomorphic continuation into the domain shown in Figure 9.17. The branch of the square root in (9.5.45) is fixed by the condition,

$$\zeta(z) = e^{-i\pi/4}\sqrt{t}2^{5/4} z(1 + 0(z)), \quad z \sim 0. \tag{9.5.47}$$

We assume that the near-zero-part of the graph shown in Figure 9.17, is the pre-image of the near-zero-part of the graph of Figure 9.16 under the map $\zeta = \zeta(z)$. We also assume (as we have done in the case of the stationary points $\pm i/\sqrt{2}$) that

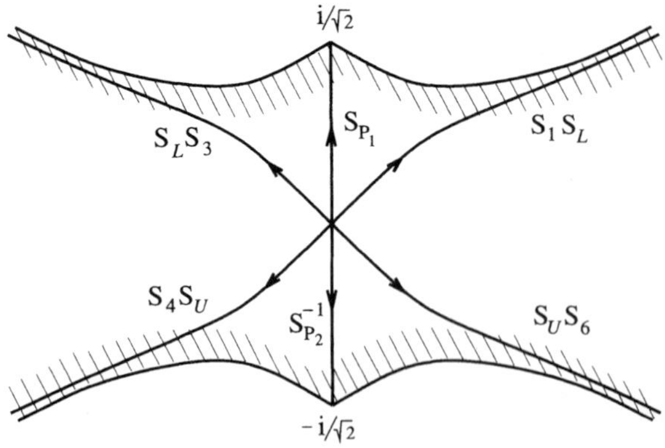

FIGURE 9.17. Transformation of the jump graph of the model RH problem and the jump properties of $\Psi^0(z)$

the graph of Figure 9.17 coincides locally with the original jump graph depicted in Figure 9.13 in the neighborhood of the point $z = 0$. Finally, we note that, by definition, the change of variable function $\zeta(z)$ satisfies the relations,

$$\frac{\zeta^2}{4} - it\frac{\sqrt{2}}{3} = -itg(z), \quad \text{Re } z > 0, \tag{9.5.48}$$

$$\frac{\zeta^2}{4} - it\frac{\sqrt{2}}{3} = itg(z). \quad \text{Re } z < 0, \tag{9.5.49}$$

The parametrix $\Psi^0(z)$ at $z = 0$ is defined by (compare with the construction (9.4.20)–(9.4.21) of the relevant parametrix in Section 4)

$$\Psi^0(z) = \mathrm{B}_0(z)\mathrm{e}^{-it\frac{\sqrt{2}}{3}\sigma_3}(2Q)^{-\frac{\sigma_3}{2}}$$

$$\times \begin{pmatrix} \zeta(z) & 1 \\ 1 & 0 \end{pmatrix} Z^{RH}(\zeta(z)), \tag{9.5.50}$$

where the matrix $\mathrm{B}_0(z)$ is defined by the relations,

$$\mathrm{B}_0(z) = \check{Y}(z)\Big(\zeta^{\nu}(z)\delta(z)\Big)^{\sigma_3}, \tag{9.5.51}$$

$$\text{Re } z > 0, \quad -\frac{3\pi}{4} - \epsilon < \arg\zeta < \frac{\pi}{4} + \epsilon,$$

$$\mathrm{B}_0(z) = \check{Y}(z)i\sigma\sigma_1(i\sigma s_2)^{-\sigma_3}\Big(\zeta^{\nu}(z)\delta^{-1}(z)\Big)^{\sigma_3}, \tag{9.5.52}$$

$$\text{Re } z < 0, \quad \frac{\pi}{4} - \epsilon < \arg\zeta < \frac{5\pi}{4} + \epsilon.$$

The matrix functions $\check{Y}(z)$ and $\delta(z)$ are given by formulae (9.5.22) and (9.5.11), respectively. The above choice of branches of the function $\zeta^{\nu}(z)$ implies that this function satisfies the following jump conditions across the interval $[-i/\sqrt{2}, i/\sqrt{2}]$:

- $\zeta^{\nu}_+(z) = \zeta^{\nu}_-(z), \quad z \in (0, i/\sqrt{2})$;
- $\zeta^{\nu}_+(z) = \zeta^{\nu}_-(z)\mathrm{e}^{2\pi i\nu} \equiv \zeta^{\nu}_-(z)(i\sigma s_2)^2, \quad z \in (-i/\sqrt{2}, 0)$.

Recalling the jump properties of the function $\delta(z)$ across the same interval we conclude that

$$\Big(\zeta^{\nu}(z)\delta^{-1}(z)\Big)_{+} = \Big(\zeta^{\nu}(z)\delta(z)\Big)_{-} i\sigma s_2, \quad z \in (-i/\sqrt{2}, i/\sqrt{2}).$$

This, together with equation (9.5.23), imply that the matrix function $\mathrm{B}_0(z)$ is holomorphic in the deleted neighborhood of the point $z=0$. It is also holomorphic at the point $z=0$, since $\mathrm{B}_0(z)$ is bounded near $z=0$. Indeed,

$$\lim_{z\to 0,\ \mathrm{Re}\ z>0} \zeta^2(z)\frac{1}{\Delta(z)} = \lim_{z\to 0,\ \mathrm{Re}\ z<0} \zeta^2(z)\Delta(z) = -it2^{7/2},$$

and taking into account the choices made for the branches of the multivalued functions involved, we arrive at the equations

$$\begin{aligned} \mathrm{B}_0(0) &= \check{Y}_{+}(0)\mathrm{e}^{-i\frac{\pi\nu}{4}\sigma_3} t^{\frac{\nu}{2}\sigma_3} 2^{\frac{7\nu}{4}\sigma_3} \\ &= \frac{1}{\sqrt{2}}\begin{pmatrix} 1 & i\sigma \\ i\sigma & 1 \end{pmatrix} \mathrm{e}^{-i\frac{\pi\nu}{4}\sigma_3} t^{\frac{\nu}{2}\sigma_3} 2^{\frac{7\nu}{4}\sigma_3}. \end{aligned} \tag{9.5.53}$$

As before, the analyticity of the matrix $\mathrm{B}_0(z)$ yields the "parametrix" equation,

$$\Psi(z) = \mathrm{N}_0(z)\Psi^0(z), \tag{9.5.54}$$

where $\Psi(z)$ is the solution of the original RH problem and $\mathrm{N}_0(z)$ is holomorphic in the neighborhood of $z=0$.

Equations (9.5.48) and (9.5.49), together with the definition (9.5.24) of the solution $\breve{\Psi}(z)$ of the model RH problem on the segment $[-i/\sqrt{2}, i/\sqrt{2}]$, imply

$$\check{Y}(z)\Big(\zeta^{\nu}(z)\delta(z)\Big)^{\sigma_3} \mathrm{e}^{-it\frac{\sqrt{2}}{3}\sigma_3} \mathrm{e}^{(\frac{\zeta^2(z)}{4} - \nu\ln\zeta(z))\sigma_3} = \breve{\Psi}(z), \tag{9.5.55}$$

$$\mathrm{Re}\ z > 0, \quad \arg\zeta \in \left(-\tfrac{3\pi}{4}, \tfrac{\pi}{4}\right)$$

and

$$\begin{aligned} &\check{Y}(z) i\sigma\sigma_1 (i\sigma s_2)^{-\sigma_3}\Big(\zeta^{\nu}(z)\delta^{-1}(z)\Big)^{\sigma_3} \mathrm{e}^{-it\frac{\sqrt{2}}{3}\sigma_3} \mathrm{e}^{(\frac{\zeta^2(z)}{4} - \nu\ln\zeta(z))\sigma_3} S_{P_1} \\ &\equiv \check{Y}(z) S_{P_1}^{-1}\Big(\zeta^{\nu}(z)\delta^{-1}(z)\Big)^{\sigma_3} \mathrm{e}^{-it\frac{\sqrt{2}}{3}\sigma_3} \mathrm{e}^{(\frac{\zeta^2(z)}{4} - \nu\ln\zeta(z))\sigma_3} S_{P_1} = \breve{\Psi}(z), \end{aligned} \tag{9.5.56}$$

$$\mathrm{Re}\ z < 0, \quad \arg\zeta \in \left(\tfrac{\pi}{4}, \tfrac{5\pi}{4}\right).$$

These relations, together with the asymptotic formulae (9.5.43) and (9.5.44), ensure the basic asymptotic property of the parametrix $\Psi^0(z)$, namely the matching relation,

$$\Psi^0(z) = \check{G}_0(z)\big(I + O(\zeta^{-2+2|\mathrm{Re}\,\nu|})\big)\breve{\Psi}(z), \tag{9.5.57}$$

$$\begin{aligned} \check{G}_0(z) &= \mathrm{B}_0(z)\begin{pmatrix} 1 & \frac{\nu}{Q\zeta}\mathrm{e}^{-i\frac{2\sqrt{2}}{3}t} \\ \frac{Q}{\zeta}\mathrm{e}^{i\frac{2\sqrt{2}}{3}t} & 1 \end{pmatrix}\mathrm{B}_0^{-1}(z) \\ &= I + O(\zeta^{-1+2|\mathrm{Re}\,\nu|}), \end{aligned} \tag{9.5.58}$$

as $t\to\infty$ and $0 < r_1 \le |z| \le r_2 < 1/\sqrt{2}$ (and hence $\zeta\to\infty$!).

The approximate solution of the original RH problem as $x \to +\infty$

Let us introduce the following piecewise holomorphic function $\hat{\Psi}(z)$,

$$\hat{\Psi}(z) = \begin{cases} \breve{\Psi}(z), & |z \pm z_3| > R_3 \quad \text{and} \quad |z| > R_0, \\ \Psi^u(z), & |z + z_3| < R_3, \\ \Psi^d(z), & |z - z_3| < R_3, \\ \Psi^0(z), & |z| < R_0, \end{cases} \tag{9.5.59}$$

where R_3 and R_0 are some positive constants. The jump graph of this function is depicted in Figure 9.18.

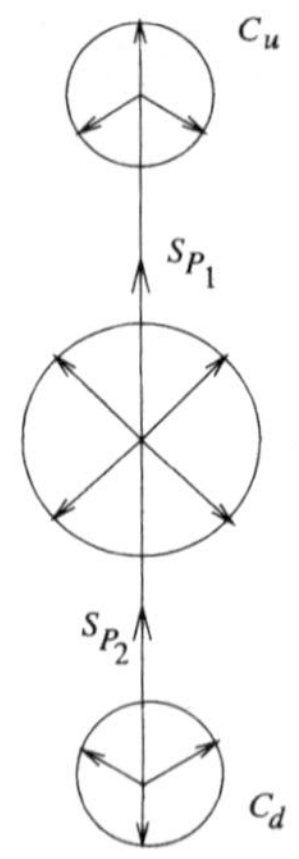

FIGURE 9.18. The RH problem for the combined function $\hat{\Psi}(z)$

We look for the exact solution $\Psi(z)$ of the original RH problem of Figure 9.13 in the form of

$$\Psi(z) = \chi(z)\hat{\Psi}(z). \tag{9.5.60}$$

The error function $\chi(z) = \Psi(z)\hat{\Psi}^{-1}(z)$ solves the following RH problem.

1) $\chi(z) \to I$ as $z \to \infty$.

2) On the graph γ depicted in Figure 9.19, the correction function $\chi(z)$ has jumps described by the following equations:

On the infinite parts of the anti-Stokes lines γ_k,

$$\begin{gathered} \chi_+(z) = \chi_-(z)G(z), \quad G(z) = G_k(z), \quad z \in \gamma_k, \quad k = 1, \dots, 6, \\ G_k(z) = \breve{Y}(z)e^{-tg(z)\sigma_3}\delta^{\sigma_3}(z)S_k\delta^{-\sigma_3}(z)e^{tg(z)\sigma_3}\breve{Y}^{-1}(z), \end{gathered} \tag{9.5.61}$$

On the clockwise oriented circles,

$$\begin{gathered} \chi_+(z) = \chi_-(z)G(z), \quad G(z) = G_m(z), \quad z \in C_m, \quad m \in \{0, u, d\}, \\ G_m(z) = \Psi^m(z)\breve{\Psi}^{-1}(z). \end{gathered} \tag{9.5.62}$$

We note that due to (9.5.36) and (9.5.54), there exist *no* jumps inside the circles C_m and across the vertical segments between the circles.

Due to their triangular structure, the jump matrices $G_k(z)$ approach the unit matrix exponentially fast on the infinite tails of the anti-Stokes lines. Indeed,

$$\|G_k(z) - I\| \le ce^{-2^{15/4}3^{-1}|t||z \pm i/\sqrt{2}|^{3/2}}, \tag{9.5.63}$$

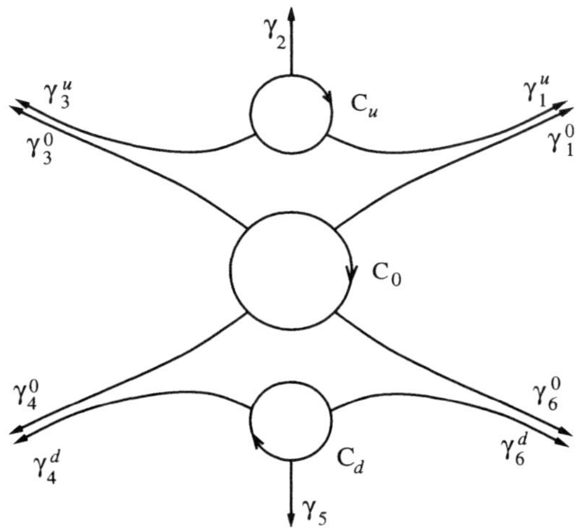

FIGURE 9.19. The RH problem for the correction function $\chi(z) = \Psi(z)\hat{\Psi}^{-1}(z)$

for the anti-Stokes lines $\gamma_k^{u,d}$ emanating from the circles $C_{u,d}$, and

$$\big\|G_k(z) - I\big\| \le c e^{-2^{3/2}\,|t|\,|z|^2}, \tag{9.5.64}$$

for the anti-Stokes lines γ_k^0 emanating from the circle C_0; see Figure 9.19. Here, c is a positive constant whose precise value is not important here.

The jump matrices $G_{u,d}(z)$ are given by equations (9.5.62). Using the estimate (9.5.37) and the fact that $\zeta(z) \ge ct^{2/3}$ as $z \in C_{u,d}$, we conclude that,

$$\big\|G_{u,d}(z) - I\big\| \le c|t|^{-1}, \tag{9.5.65}$$

for some positive constant c.

The estimate for the jump matrix $G_0(z)$ across the circle C_0 around the origin, follows from equations (9.5.57) and (9.5.58). Indeed, taking into account that $\zeta(z) \ge c\sqrt{t}$ as $z \in C_0$, we have

$$\begin{aligned} G_0(z) &= \breve{G}_0(z)\big(I + O(t^{-1+|\mathrm{Re}\,\nu|})\big) \\ &= \big(I + O(t^{-1/2+|\mathrm{Re}\,\nu|})\big). \end{aligned} \tag{9.5.66}$$

Hence, the jump matrix $G_0(z)$ satisfies the uniform estimate,

$$\big\|G_0(z) - I\big\| \le c|t|^{-\frac{1}{2}+|\mathrm{Re}\,\nu|} \tag{9.5.67}$$

for a positive constant c whose value is not important here.

Condition (9.5.8) implies

$$|\,\mathrm{Re}\ \nu| < \frac{1}{2} - \delta, \quad 0 < \delta < \frac{1}{2}. \tag{9.5.68}$$

Hence, by virtue of the estimates (9.5.63)–(9.5.65) and (9.5.67), we find the key inequality (cf. (9.4.36)),

$$\|G - I\|_{L^2(\gamma)\cap L^\infty(\gamma)} \le c\,t^{-\frac{1}{2}+|\mathrm{Re}\,\nu|} \tag{9.5.69}$$

for the jump matrix $G(z)$ of the RH problem (9.5.61)–(9.5.62). Similarly with the previous sections, this inequality (using again Theorem 8.1 of Chapter 8) implies the existence of a unique solution $\chi(z)$ of the RH problem (9.5.61)–(9.5.62); Figure 9.19. Moreover, the function $\chi(z)$ admits the Cauchy integral representation,

$$\chi(z) = I + \frac{1}{2\pi i}\int_\gamma \chi_-(\xi)\big(G(\xi) - I\big)\frac{d\xi}{\xi - z}, \tag{9.5.70}$$

where the contour γ is shown in Figure 9.19 and the boundary value $\chi_-(z)$ satisfies the L^2-estimate,

$$\big\|\chi_- - I\big\|_{L^2(\gamma)} < c|t|^{-\frac{1}{2}+|\mathrm{Re}\,\nu|}, \tag{9.5.71}$$

where c is a positive constant.

In order to find the asymptotics of the corresponding Painlevé function we note that for large z,

$$\Psi(z) = \chi(z)\hat{\Psi}(z) = \chi(z)\breve{\Psi}(z) = \chi(z)\breve{Y}(z)e^{-tg(z)\sigma_3}\delta^{\sigma_3}(z).$$

Therefore, the basic representation (9.4.5) takes the following form:

$$\begin{aligned} u(x) &= 2\sqrt{x}\lim_{z\to\infty}\big(z(\Psi(z)e^{t\theta(z)})_{12}\big) \\ &= i\sigma\sqrt{\frac{x}{2}} + 2\sqrt{x}\lim_{z\to\infty}\big(z\chi_{12}(z)\big). \end{aligned} \tag{9.5.72}$$

The first term in (9.5.72) comes from the Laurent series expansion of $\breve{Y}(z)$ at $z = \infty$ (see (9.5.22) and (9.5.19)),

$$\breve{Y}(z) = I + \frac{i\sigma}{2\sqrt{2}}\frac{1}{z}\sigma_1 + O(z^{-2})$$

(the diagonal factors $e^{-tg(z)\sigma_3}$ and $\delta^{\sigma_3}(z)$ do not contribute to the 12 entry of the z^{-1} coefficient of $\Psi(z)e^{t\theta(z)}$). The second term in (9.5.72) can be evaluated using the Cauchy representation (9.5.70). Indeed, using (9.5.70) we find (cf. (9.4.40))

$$\lim_{z\to\infty}\big(z\chi_{12}(z)\big) = -\frac{1}{2\pi i}\int_\gamma \big(\chi_-(z)(G(z) - I)\big)_{12}\,dz. \tag{9.5.73}$$

Taking into account the estimates derived earlier, we can transform (9.5.73), as follows:

$$\begin{aligned} &\lim_{z\to\infty}\big(z\chi_{12}(z)\big) \\ &= -\frac{1}{2\pi i}\int_\gamma (G(z) - I)_{12}\,dz - \frac{1}{2\pi i}\int_\gamma \big((\chi_-(z) - I)(G(z) - I)\big)_{12}\,dz \\ &= -\frac{1}{2\pi i}\int_\gamma (G(z) - I)_{12}\,dz + O(t^{-1+2|\mathrm{Re}\,\nu|}) \\ &= -\frac{1}{2\pi i}\oint_{C_0} \big(G_0(z)\big)_{12}\,dz + O(t^{-1+2|\,\mathrm{Re}\,\nu|}), \end{aligned} \tag{9.5.74}$$

where in the last equation we have replaced the contour γ by just the circle C_0 which yields an error of order $O(t^{-1})$ (see estimates (9.5.63)–(9.5.65)) and hence can be included into the general error term, $O(t^{-1+2|\mathrm{Re}\,\nu|})$. Within the same error, we can replace the matrix $G_0(z)$ by the matrix $\breve{G}_0(z)$ according to (9.5.66), thus we can rewrite (9.5.74) in the form,

$$\lim_{z\to\infty}\big(z\chi_{12}(z)\big) = -\frac{1}{2\pi i}\oint_{C_0} \big(\breve{G}_0(z)\big)_{12}\,dz + O(t^{-1+2|\,\mathrm{Re}\,\nu|}). \tag{9.5.75}$$

The matrix function $\breve{G}_0(z)$ given by the explicit formula (9.5.58), is analytic inside the circle C_0, except for its center, $z=0$, where $\breve{G}_0(z)$ has a simple pole. Therefore, the integral in (9.5.75) can be evaluated explicitly. Indeed, taking into account relations (9.5.47) and (9.5.53), as well as the clockwise orientation of the circle C_0, we find

$$-\frac{1}{2\pi i}\oint_{C_0}\breve{G}_0(z)\,dz = \mathrm{Res}_{z=0}\breve{G}_0(z)$$

$$= \mathrm{B}_0(0)\begin{pmatrix} 1 & \frac{\nu}{Q\zeta'(0)}\mathrm{e}^{-i\frac{2\sqrt{2}}{3}t} \\ \frac{Q}{\zeta'(0)}\mathrm{e}^{i\frac{2\sqrt{2}}{3}t} & 1 \end{pmatrix}\mathrm{B}_0^{-1}(0)$$

(9.5.76)
$$= \frac{1}{2}\begin{pmatrix} 1 & i\sigma \\ i\sigma & 1 \end{pmatrix}\begin{pmatrix} 1 & \frac{p}{t^{1/2-\nu}}\mathrm{e}^{-i\frac{2\sqrt{2}}{3}t} \\ \frac{q}{t^{1/2+\nu}}\mathrm{e}^{i\frac{2\sqrt{2}}{3}t} & 1 \end{pmatrix} \times\begin{pmatrix} 1 & -i\sigma \\ -i\sigma & 1 \end{pmatrix},$$

where

(9.5.77)
$$p = \frac{\nu}{Q}\mathrm{e}^{i\frac{\pi}{4}-i\frac{\pi\nu}{2}}2^{-\frac{5}{4}+\frac{7\nu}{2}} = i\sigma\frac{2^{-3/4}\sqrt{\pi}\,\mathrm{e}^{\frac{i\pi\nu}{2}}}{(1+s_1s_2)\Gamma(\nu)}\mathrm{e}^{i\frac{3\pi}{4}}2^{\frac{7\nu}{2}},$$

and

(9.5.78)
$$q = Q\mathrm{e}^{i\frac{\pi}{4}+i\frac{\pi\nu}{2}}2^{-\frac{5}{4}-\frac{7\nu}{2}} = i\sigma\frac{2^{-3/4}\sqrt{\pi}\,\mathrm{e}^{\frac{i\pi\nu}{2}}}{(1+s_3s_2)\Gamma(\nu)}\mathrm{e}^{-i\frac{3\pi}{4}}2^{-\frac{7\nu}{2}}.$$

The 12 entry of (9.5.76) is given by the expression

$$\frac{1}{2\sqrt{t}}\big(p\mathrm{e}^{-i\frac{2\sqrt{2}}{3}t+\nu\ln t} + q\mathrm{e}^{i\frac{2\sqrt{2}}{3}t-\nu\ln t}\big),$$

which together with (9.5.72) and (9.5.75) yield the following leading terms for the asymptotics of the Painlevé function $u(x)$, as $x\to+\infty$:

(9.5.79)
$$\begin{aligned}
&u(x) = i\sigma\sqrt{\frac{x}{2}} + i\sigma x^{-1/4}\Big\{c\mathrm{e}^{ih} + d\mathrm{e}^{-ih}\Big\} \\
&+O(x^{-1+3|\,\mathrm{Re}\,\nu|}), \\
&cd = -\frac{i\nu}{4\sqrt{2}},\quad h = \frac{2\sqrt{2}}{3}x^{3/2} + i\nu\ln 8\sqrt{2}x^{3/2} - \frac{3\pi}{4}, \\
&\mathrm{Re}\,\nu\in\big(-\tfrac{1}{2}+\delta,\tfrac{1}{2}-\delta\big),\quad \delta>0.
\end{aligned}$$

In the above expressions the asymptotic parameters ν, c, d are related to the Stokes multipliers by the following equations:

$$\nu = \frac{1}{i\pi}\ln(i\sigma s_2),\quad \arg(i\sigma s_2)\in\big(-\tfrac{\pi}{2}+\pi\delta,\tfrac{\pi}{2}-\pi\delta\big),\quad \delta>0,$$

(9.5.80)
$$c = \frac{2^{-3/4}\sqrt{\pi}\,\mathrm{e}^{\frac{i\pi\nu}{2}}}{(1+s_2s_3)\Gamma(-\nu)},\quad d = \frac{2^{-3/4}\sqrt{\pi}\,\mathrm{e}^{\frac{i\pi\nu}{2}}}{(1+s_1s_2)\Gamma(\nu)}.$$

Under the purely imaginary reduction,

$$s_1 = -\overline{s}_3 \equiv s$$

we find

$$s_2 = \frac{s-\overline{s}}{1+|s|^2} = \frac{2i\,\mathrm{Im}\ s}{1+|s|^2}, \qquad \sigma = -\mathrm{sign}\ \mathrm{Im}\ s,$$

$$\nu = \frac{i}{\pi}\ln\frac{1+|s|^2}{2|\,\mathrm{Im}\ s|} \equiv i\rho^2, \quad \rho > 0,$$

$$d = \overline{c}, \quad |c| = \frac{\rho}{2}2^{-1/4}, \quad \arg c = \arg\Gamma(i\rho^2) + \arg(1+s^2).$$

Therefore, equation (9.5.79) can be reduced to the formula,

$$u(x) = i\sigma\sqrt{\frac{x}{2}} + i\sigma(2x)^{-1/4}\rho\cos\left\{\frac{2\sqrt{2}}{3}x^{3/2} - \frac{3}{2}\rho^2\log x + \theta\right\}$$

$$+O(x^{-1}), \quad x \to +\infty, \tag{9.5.81}$$

where

$$\rho^2 = \frac{1}{\pi}\log\frac{1+|s|^2}{2|\,\mathrm{Im}\ s|}, \quad \rho > 0, \sigma = -\mathrm{sign}\ \mathrm{Im}\ s, \tag{9.5.82}$$

$$\theta = -\frac{3\pi}{4} - \frac{7}{2}\rho^2\log 2 + \arg\Gamma(i\rho^2) + \arg(1+s^2), \tag{9.5.83}$$

which is exactly equation (9.1.9).

We summarize the results of this section in the following theorem.

THEOREM 9.6. *Let the parameters $s_2 \in \mathbb{C}\setminus\{0\}$ and $\sigma \in \{+1,-1\}$ satisfy the conditions*

$$\arg(i\sigma s_2) \in \left(-\tfrac{\pi}{2}+\delta, \tfrac{\pi}{2}-\delta\right), \quad (1+s_1s_2)(1+s_2s_3) \neq 0, \tag{9.5.84}$$

where δ is an arbitrary small positive constant. Then, there exists a constant $x_0 > 0$ such that the RH problem (9.4.1)–(9.4.4) shown in Figure 8.2 is uniquely solvable for all $x > x_0$. Its solution is described by the formulae (9.5.60), (9.5.59), (9.5.50), (9.5.34), (9.5.14) and (9.5.22). The asymptotics of the relevant solution of the second Painlevé equation PII($\alpha = 0$), as $x \to +\infty$, is given by (9.5.79), (9.5.80). In the imaginary reduction, $s_1 = -\overline{s}_3$, $s_2 = -\overline{s}_2$, these expressions turn into (9.1.9), (9.1.10) with $\rho^2 = -i\nu$.

REMARK 9.6. In the imaginary reduction, the inequality $|s_2| \leq 1$ is valid. This is consistent with the inequality $-i\nu \geq 0$.

REMARK 9.7. The special cases when $s_2 = 0$ and $(1+s_1s_2)(1+s_2s_3) = 0$, will be analyzed in Chapter 11.

REMARK 9.8. The error bounds in the above asymptotics can be improved, however, we will not discuss this standard issue here.

REMARK 9.9. The asymptotics of the Painlevé function as $x \to +\infty$ (and hence as $x \to e^{\pm i\frac{2\pi}{3}}\infty$) depends on the value of the parameter α in the Painlevé equation $u_{xx} = 2u^3 + xu - \alpha$. Indeed, generically, the local Ψ-function in a neighborhood of the origin cannot be expressed in terms of the parabolic cylinder functions, but in term of the Whittaker functions; see [**K8**]. This is in contrast to the asymptotics in all other directions.

CHAPTER 10

PII Asymptotics on the Canonical Six-rays. Real-valued Case

0. Introduction

Here we study the asymptotics on the six rays, $\arg x = \frac{\pi}{3}k$, $k = 0, 1, 2, \ldots, 5$ in the case that the solution u of PII,

$$u_{xx} = xu + 2u^3, \tag{10.0.1}$$

is real-valued for real x. This implies the "real-valued reduction" (4.2.47)

$$s_1 = \bar{s}_3, \quad s_2 = \bar{s}_2$$

of the space of monodromy data.

Due to the symmetry relations (7.2.10) and (7.2.11), it is enough to investigate the asymptotic behaviour of $u(x)$ along just two rays, $\arg x = 0$ and $\arg x = \pi$, i.e., along the real line,

$$u_{xx} = xu + 2u^3, \quad x \in \mathbb{R}. \tag{10.0.2}$$

Similar to the case of the pure-imaginary solutions, studied earlier in Chapter 9, it is impossible to get asymptotics on the real line, as well as on any other of the six-rays, by taking the limit $\arg x \to \pi k/3$ in the asymptotics (7.2.7). Also, the relevant asymptotics of $u(x)$ is much more sophisticated than that of Chapter 9. The function $u(x)$ behaves like a damped harmonic oscillation as $x \to -\infty$ if $|s_1| < 1$, and like an exponentially decreasing Airy function as $x \to +\infty$ if $s_2 = 0$; furthermore, there exist the following additional cases:

- $|s_2| \neq 0$: $\quad u(x) \sim \pm\sqrt{x}\tan\left(\frac{\sqrt{2}}{3}x^{3/2} + O(\ln x)\right), \quad x \to +\infty.$
- $|s_1| = 1$: $\quad u(x) \sim u_{\pm}(x) + O\left(x^{-1/4}e^{-\frac{2\sqrt{2}}{3}(-x)^{3/2}}\right),$
  $u_{\pm}(x) = \pm\sqrt{-\frac{x}{2}} + O(x^{-1}), \quad x \to -\infty.$
- $|s_1| > 1$: $\quad u(x) \sim \frac{\sqrt{-x}}{\sin\left(\frac{2}{3}(-x)^{3/2} + O(\ln(-x))\right)}, \quad x \to -\infty.$

The more complicated structure of the asymptotics is due to the more complicated structure of the monodromy data manifold. Unlike the pure-imaginary case, this manifold cannot be parametrized by a single complex parameter. The interplay of the two monodromy parameters s_1, s_2, constrained by the equation $s_1 + \bar{s}_1 - s_2(1 - |s_1|^2) = 0$, gives rise to the several cases listed above. This shows the rich structure of the asymptotics of the Painlevé transcendents, even for the simplest case of PII equation.

The main tool used in this chapter is the direct monodromy problem approach, used in Chapter 7. However, here a more advanced version of the WKB method is used. Indeed, if one uses the approach of Chapter 7, one must invent various gauge transformations in order to construct a local Ψ-function in the neighborhood of a turning point for all the cases listed above; cf. [**K1**]. The second disadvantage of the approach used in Chapter 7 is the difficulty for finding error bounds for the approximate Ψ-functions and hence for the asymptotics of the Painlevé function. Below, by using an advanced version of the WKB method developed in [**K7**], we overcome both of these difficulties[1].

1. Formulation of results. Main theorems and discussion

Recall the λ-equation corresponding to the second Painlevé equation (see Chapter 4)

$$\frac{d\Psi}{d\lambda} = \left[-(4i\lambda^2 + ix + 2iu^2)\sigma_3 - 4u\lambda\sigma_2 - 2w\sigma_1\right]\Psi, \tag{10.1.1}$$

where x is real and u, w are complex parameters. The set of the monodromy data of this equation consists of the six Stokes matrices,

$$S_{2k} = \begin{pmatrix} 1 & s_{2k} \\ 0 & 1 \end{pmatrix}, \quad S_{2k+1} = \begin{pmatrix} 1 & 0 \\ s_{2k+1} & 1 \end{pmatrix}, \quad k = 1, 2, 3,$$

defined in the equations

$$S_k = \Psi_k^{-1}\Psi_{k+1}, \quad k = 1, \ldots, 6,$$

where $\Psi_1(\lambda), \ldots, \Psi_7(\lambda)$ are the canonical solutions of the equation (10.1.1). These solutions are fixed by the asymptotic condition (Chapter 2, (2.1.88))

$$\Psi_k(\lambda) = (I + O(\lambda^{-1}))\exp\left\{-\frac{4}{3}i\lambda^3\sigma_3 - ix\lambda\sigma_3\right\}, \tag{10.1.2}$$

$$\lambda \to \infty, \quad \frac{\pi}{3}(k-2) < \arg\lambda < \frac{\pi}{3}k, \quad k = 1, \ldots, 6.$$

The Stokes matrices satisfy the constraints

$$S_{k+3} = \sigma_2 S_k \sigma_2, \quad k = 1, 2, 3, \tag{10.1.3}$$

$$S_1 S_2 \ldots S_6 = I.$$

Assuming $u, w \in \mathbb{R}$, we find (see (4.2.47))

$$s_1 = \overline{s}_3, \quad s_2 = \overline{s}_2. \tag{10.1.4}$$

Therefore, all six Stokes matrices can be parametrized by the complex s_1 and the real s_2 parameters coupled by the constraint equation

$$s_1 + \overline{s}_1 - s_2(1 - |s_1|^2) = 0.$$

For $|s_1| \neq 1$ this leads to the one-parameter representation

[1]Somewhat similar to this version, although still different in several important points, is the *method of uniform asymptotics* suggested in [**BCLM**].

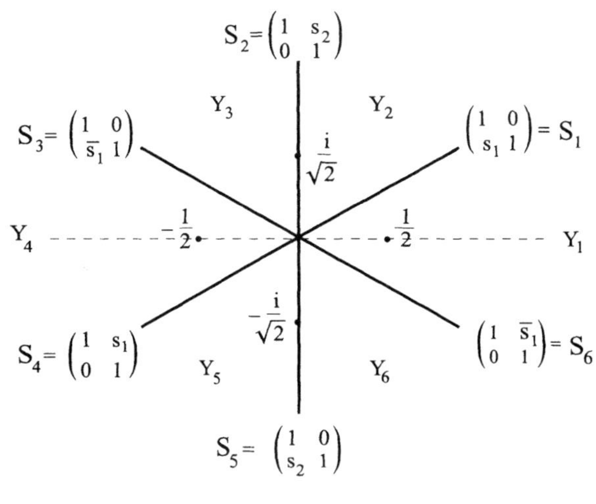

FIGURE 10.1. The basic RH problem

$$S_1 = \begin{pmatrix} 1 & 0 \\ s_1 & 1 \end{pmatrix}, \quad S_2 = \begin{pmatrix} 1 & \frac{s_1+\bar{s}_1}{1-|s_1|^2} \\ 0 & 1 \end{pmatrix}, \quad S_3 = \begin{pmatrix} 1 & 0 \\ \bar{s}_1 & 1 \end{pmatrix},$$

$$S_4 = \begin{pmatrix} 1 & -s_1 \\ 0 & 1 \end{pmatrix}, \quad S_5 = \begin{pmatrix} 1 & 0 \\ \frac{s_1+\bar{s}_1}{1-|s_1|^2} & 1 \end{pmatrix}, \quad S_6 = \begin{pmatrix} 1 & -\bar{s}_1 \\ 0 & 1 \end{pmatrix}.$$

For $|s_1| = 1$, which implies Re $s_1 = 0$ and therefore $s_1 = i\epsilon$, $\epsilon = \operatorname{sign}(\operatorname{Im} s_1)$, the relevant representation becomes

$$S_1 = \begin{pmatrix} 1 & 0 \\ i\epsilon & 1 \end{pmatrix}, \quad S_2 = \begin{pmatrix} 1 & s_2 \\ 0 & 1 \end{pmatrix}, \quad S_3 = \begin{pmatrix} 1 & 0 \\ -i\epsilon & 1 \end{pmatrix},$$

$$S_4 = \begin{pmatrix} 1 & -i\epsilon \\ 0 & 1 \end{pmatrix}, \quad S_5 = \begin{pmatrix} 1 & 0 \\ s_2 & 1 \end{pmatrix}, \quad S_6 = \begin{pmatrix} 1 & i\epsilon \\ 0 & 1 \end{pmatrix}.$$

We take the one-dimensional complex manifold

$$s = \left\{(s_1, s_2)\middle| s_1 + \bar{s}_1 - s_2(1 - |s_1|^2) = 0\right\} \tag{10.1.5}$$

as the parametrization of the real-valued solutions of PII:

$$u = u(x; s).$$

The main results for the asymptotics of the real-valued solutions of the second Painlevé transcendent are stated below in Theorems 10.1 and 10.2.

THEOREM 10.1. *The real-valued solutions of PII equation (10.0.2) have the following asymptotics as $x \to +\infty$:*

1) *If $s_2 \neq 0$, then*

$$u(x;s) = \sigma\sqrt{\frac{x}{2}}\cot\Big(\frac{\sqrt{2}}{3}x^{3/2} + \frac{\gamma}{2}\ln\big(8\sqrt{2}\,x^{3/2}\big) + \chi\Big) + O(x^{-1}), \tag{10.1.6}$$

where

$$\begin{aligned} &\sigma = \operatorname{sgn}(s_2) = \pm 1,\\ &\gamma = \frac{1}{\pi}\ln(\sigma s_2),\\ &\chi = -\frac{1}{2}\arg\Gamma\big(\frac{1}{2} + i\gamma\big) - \frac{1}{2}\arg(1 + s_2 s_3) + \frac{\pi}{2} + O(\tau^{-1/2}), \end{aligned} \tag{10.1.7}$$

and $\Gamma(z)$ is the Euler gamma-function. The asymptotics is uniform outside some neighborhoods of the singularities of the trigonometric functions appearing in the right-hand side of (10.1.6).

2) *If $s_2 = 0$, then*

$$u(x;s) = a x^{-1/4} e^{-\frac{2}{3}x^{3/2}}\left(1 + O\left(x^{-3/2}\right)\right), \tag{10.1.8}$$

where

$$a = -\frac{\operatorname{Im} s_1}{2\sqrt{\pi}}. \tag{10.1.9}$$

THEOREM 10.2. *The real-valued solutions of PII equation (10.0.2) have the following asymptotics as $x \to -\infty$:*

1) *If $|s_1| < 1$, then*

$$\begin{aligned} u(x;s) &= (-x)^{-1/4}\sqrt{-2\beta}\cos\Big(\frac{2}{3}(-x)^{3/2} + \beta\ln\big(8(-x)^{3/2}\big) + \phi\Big)\\ &+ O\big((-x)^{-7/10}\big), \end{aligned} \tag{10.1.10}$$

where

$$\begin{aligned} &\beta = \frac{1}{2\pi}\ln\big(1 - |s_1|^2\big) < 0,\\ &\phi = -\frac{\pi}{4} - \arg\Gamma(i\beta) - \arg s_1. \end{aligned} \tag{10.1.11}$$

2) *If $|s_1| > 1$, then*

$$u(x;s) = \frac{\sqrt{-x}}{\sin\Big(\frac{2}{3}(-x)^{3/2} + \beta\ln\big(8(-x)^{3/2}\big) + \phi\Big) + O\big((-x)^{-1/5}\big)}, \tag{10.1.12}$$

where

$$\begin{aligned} &\beta = \frac{1}{2\pi}\ln\left(|s_1|^2 - 1\right),\\ &\phi = -\arg\Gamma\big(\frac{1}{2} + i\beta\big) - \arg s_1. \end{aligned} \tag{10.1.13}$$

The asymptotics is uniform outside some neighborhoods of the singularities of the trigonometric functions appearing in the right-hand side of (10.1.12).

3) *If* $|s_1| = 1$, *i.e.*, $s_1 = i\epsilon$, *then*

$$u(x;s) = u_-(x) + b(-x)^{-1/4}e^{-\frac{2\sqrt{2}}{3}(-x)^{3/2}}\left(1 + O\left(x^{-3/2}\right)\right), \tag{10.1.14}$$

where

$$u_-(x) = -\epsilon\sqrt{-\frac{x}{2}} + O(x^{-5/2}), \quad b = -\frac{s_2}{2^{7/4}\sqrt{\pi}}. \tag{10.1.15}$$

The second statement of Theorem 10.1 and the third statement of Theorem 10.2 describe the so-called separatrix. This degenerated asymptotics can be derived heuristically from the asymptotics presented in the first assertions of the above theorems. Thus, we will concentrate on the asymptotics of the Painlevé transcendents in the generic case. The separatrix cases will be rigorously investigated in the next chapter.

In what follows, we will actually derive the relevant asymptotics under the more general assumptions than the real-valued constraint (10.1.4). The proof of Theorems 10.1 and 10.2 will then be obtained by imposing the real-valued constraint on the monodromy data.

2. Proof of Theorem 10.1: Asymptotics of the Painlevé functions as $x \to +\infty$

Preliminary assumptions and transformations

Here we prove the first statement of the theorem. The second statement will be proven in the part of Chapter 11 devoted to the quasi-linear Stokes phenomenon.

Using in (10.1.1) the change of the independent variable

$$\lambda = x^{1/2}z, \quad u = \sqrt{x}\,v, \quad w = xy, \tag{10.2.1}$$

we obtain the linear equation $\frac{d\Psi}{d\lambda} = \tau A_0(z,\tau)$, where $\tau = x^{3/2}$ and

$$A_0 = -i\big(4z^2 + 1 + 2v^2\big)\sigma_3 - 4vz\sigma_2 - 2y\sigma_1. \tag{10.2.2}$$

The eigenvalues $\mu_{1,2} = \pm\mu$ of the matrix $A_0(z,\tau)$ satisfy the equation

$$\mu^2 = -16z^2\big(z^2 + \frac{1}{2}\big) + 4D, \tag{10.2.3}$$
$$D = y^2 - \big(v^2 + \frac{1}{2}\big)^2.$$

Our principal assumption on the coefficients of $A_0(z,\tau)$ is $D = O(\tau^{-1})$, or, more precisely, that there exist some positive constants c_1 and c_2, such that

$$c_1\tau^{-1} \le |D| \le c_2\tau^{-1}. \tag{10.2.4}$$

As we will see, almost all Painlevé transcendents satisfy (10.2.4) as $x \to +\infty$. The assumption (10.2.4) means that the asymptotic location of the turning points and the Stokes graph structure coincide with those described in Chapter 9; see Figure 9.2. There exist two simple turning points $z_{1,2} = \pm i/\sqrt{2} + O(\tau^{-1})$ and two turning points approaching the origin, $z_{3,4} = O(\tau^{-1/2})$. As in Chapter 9, we will compute the asymptotics of the product

$$S_1S_2S_3 = \begin{pmatrix} 1+s_2s_3 & s_2 \\ s_1+s_3+s_1s_2s_3 & 1+s_1s_2 \end{pmatrix} = \Psi_1^{-1}(\lambda)\Psi_4(\lambda). \tag{10.2.5}$$

The behavior of the coefficients v and y is now different from the case of Chapter 9. The assumption (10.2.4) is consistent with the following asymptotics of the coefficients v, y:

$$y = O(\tau^{-\delta}), \quad v - \frac{i\varepsilon_1}{\sqrt{2}} = O(\tau^{-\delta}), \tag{10.2.6}$$

$$\begin{aligned} &y - \varepsilon_2\big(v^2 + \frac{1}{2}\big) = O(\tau^{-\delta}), \\ &y + \varepsilon_2\big(v^2 + \frac{1}{2}\big) = O(\tau^{-1+\delta}), \\ &\delta \in [0, \frac{1}{2}], \quad \varepsilon_{1,2} \in \{+,-\}. \end{aligned} \tag{10.2.7}$$

Observe that two-sided estimate (10.2.4) for D, implies that the estimates (10.2.7) are also two-sided.

The parameter δ introduced in (10.2.6), (10.2.7), stratifies the Painlevé transcendents according to their decrement velocity [**K8**]. Pure imaginary solutions discussed in Chapter 9 correspond to $\delta = 1/2$. As we will see, generic real solutions of PII correspond to the value $\delta = 0$, thus in what follows, we concentrate on this case.

Making use of a gauge transformation similar to (9.3.4),

$$\Psi(z) = P\Phi(z), \quad P = \frac{1}{\sqrt{2}}\begin{pmatrix} 1 & 1 \\ i\varepsilon_2 & -i\varepsilon_2 \end{pmatrix}, \tag{10.2.8}$$

we obtain a new matrix equation whose coefficients coincide with the quantities presented in (10.2.7):

$$\frac{d\Phi}{dz} = \tau B\Phi, \quad B = \begin{pmatrix} b_3 & b_+ \\ b_- & -b_3 \end{pmatrix}, \tag{10.2.9}$$

where

$$\begin{aligned} b_3 &= -\varepsilon_2 4zv, \\ b_+ &= -4iz^2 + 2i\varepsilon_2\Big(y - \varepsilon_2\big(v^2 + \frac{1}{2}\big)\Big), \\ b_- &= -4iz^2 - 2i\varepsilon_2\Big(y + \varepsilon_2\big(v^2 + \frac{1}{2}\big)\Big). \end{aligned} \tag{10.2.10}$$

In the subsequent calculation, we assume that the function $\mu(z)$ is defined on a Riemann surface glued using two copies of the complex z-plane both cut along the segments $[z_1, z_3]$ and $[z_2, z_4]$. On the upper sheet of the Riemann surface, the branch of $\mu(z)$ is fixed by the asymptotics at infinity,

$$\mu(z) = i(4z^2 + 1) + O(z^{-2}), \quad z \to \infty. \tag{10.2.11}$$

Now, we are ready to implement a scheme, similar to the one used in Sections 2 and 3 of Chapter 9, in order for compute the Stokes multipliers. This involves using several approximate solutions of (10.2.9), (10.2.10), which are valid in different overlapping asymptotic domains of the z-complex plane.

WKB APPROXIMATION

Following [**W1**], introduce the WKB (or, alternatively, the Liouville-Green) expansion to (10.2.9). Let T and Λ_0 be a matrix of eigenvectors and the diagonal matrix of eigenvalues to $\mathrm{B}(z)$, respectively,

$$T^{-1}\mathrm{B}T = \Lambda_0 = \operatorname{diag}(\mu, -\mu). \tag{10.2.12}$$

Let $\{\Lambda_n\}_{n=1}^{\infty}$ and $\{T_n\}_{n=1}^{\infty}$ be the sequences of diagonal and off-diagonal matrices, respectively, determined by the recursion

$$\begin{aligned} &[\Lambda_0, T_1] - \Lambda_1 = T^{-1}\frac{dT}{dz}, \\ &[\Lambda_0, T_n] - \Lambda_n = \sum_{m=1}^{n-1} T_{n-m}\Lambda_m + T^{-1}\frac{d(TT_{n-1})}{dz}, \quad n \geq 2. \end{aligned} \tag{10.2.13}$$

REMARK 10.1. From (10.2.13), it follows that $\Lambda_1 = -\operatorname{diag}\big(T^{-1}T_z\big)$. Since the substitution $T(z) \mapsto T(z)D(z)$, $D(z)$ a diagonal matrix, turns the product $T(z)\exp\Big\{\int_{z_0}^{z}\Lambda_1\,dz\Big\}$ into the same expression but multiplied from the right with the *constant* matrix $D(z_0)$, it is convenient to absorb the exponential factor $\exp\Big\{\int_{z_0}^{z}\Lambda_1\,dz\Big\}$ into the eigenvector matrix $T(z)$. We call the eigenvector matrix $T(z)$ defined in such a way, a *canonical* eigenvector matrix. Obviously, for the canonical matrix $T(z)$, the diagonal matrix $\Lambda_1(z)$ vanishes. In what follows, we assume that the eigenvector matrix is canonical.

Using the above observation, we specify the eigenvector matrix $T(z)$ in (10.2.12) as follows:

$$T(z) = \frac{1}{\sqrt{2b_+\mu}}\begin{pmatrix} b_+ & b_+ \\ \mu - b_3 & -\mu - b_3 \end{pmatrix} \exp\Big\{\frac{1}{2}\int_{z_0}^{z}\frac{1}{\mu}\big(b_3' - b_3\frac{b_+'}{b_+}\big)\,dz\,\sigma_3\Big\}, \tag{10.2.14}$$

where $(\ )' := \frac{d}{dz}$ and z_0 is some initial point. The expression (10.2.14) implies that the matrices T, T^{-1}, and therefore T_n and Λ_n are holomorphic and bounded outside some finite neighborhoods of zeros of $\mu^2(z)$ and $b_+(z)$.

We look for $\Phi(z)$ in the following form:

$$\begin{aligned} &\Phi = T^{(m)}(z)\chi^{(m)}(z)\exp\Big\{\tau\int_{z_0}^{z}\Lambda^{(m)}(z)\,dz\Big\}, \\ &T^{(m)}(z) = T\Big(I + \sum_{n=1}^{m}\tau^{-n}T_n\Big), \quad \Lambda^{(m)}(z) = \sum_{n=0}^{m}\tau^{-n}\Lambda_n(z). \end{aligned} \tag{10.2.15}$$

Equation (10.2.15), with the correction function $\chi^{(m)}(z)$ replaced by the unit matrix, defines the so-called WKB-approximation.

In order to describe the domain of validity of the WKB-approximation, we consider the Volterra integral equation for the correction function $\chi^{(m)}(z)$,

$$\chi^{(m)}(z) = I + \int_{\gamma(z)} e^{\tau\int_\zeta^z \Lambda^{(m)}(\xi)\,d\xi} R^{(m)}(\zeta)\chi^{(m)}(\zeta)e^{-\tau\int_\zeta^z \Lambda^{(m)}(\xi)\,d\xi}\,d\zeta, \tag{10.2.16}$$

where

$$\begin{aligned} R^{(m)}(z) &= \tau\Big(\big(T^{(m)}(z)\big)^{-1}\mathrm{B}(z)T^{(m)}(z) - \Lambda^{(m)}(z)\Big) - (T^{(m)}(z))^{-1}\frac{d}{dz}T^{(m)}(z)\\ &= \Big(I + \sum_{n=1}^{m}\tau^{-n}T_n\Big)^{-1}\Big\{-\sum_{n=1}^{m}\sum_{k=m-n+1}^{m}\tau^{1-n-k}T_n\Lambda_k - \tau^{-m}T^{-1}\frac{d}{dz}(TT_m)\Big\}, \end{aligned} \tag{10.2.17}$$

and $\gamma(z) = \big(\gamma_1(z), \gamma_2(z)\big)$ is a row-vector of contours with the common endpoint at z.

Outside some bounded neighborhoods of the turning points z_k,

$$\begin{aligned} R^{(m)} &= O(\tau^{-m}), \quad |z - z_k| > c_1, \quad |z| < c_2,\\ R^{(m)} &= O(\tau^{-m}z^{-2-3m}), \quad z \to \infty. \end{aligned} \tag{10.2.18}$$

A contour $\gamma_j(a,b)$ connecting the points a and b is called *j-canonical*, if

$$\operatorname{Re}\int_z^b \mu_j(\xi)\,d\xi \geq 0 \quad \forall z \in \gamma(a,b). \tag{10.2.19}$$

For instance, for $b > C > 0$ where C is a constant, the point a can be placed at infinity, the contour $\gamma_1(\infty, b)$ can be chosen so that it approaches the ray $\arg z = -\frac{\pi}{6}$, and $\gamma_2(\infty, b)$ can be chosen so that it approaches the ray $\arg z = \frac{\pi}{6}$. Similarly, for $b < C < 0$, the infinite contour $\gamma_1(\infty, b)$ approaches the ray $\arg z = \frac{7\pi}{6}$, while $\gamma_2(\infty, b)$ approaches the ray $\arg z = \frac{5\pi}{6}$.

A simply connected domain $\mathcal{D}$ is called *j-canonical* if there exists a point $\zeta_j \in \mathcal{D}$, such that the contour $\gamma(\zeta_j, z) \subset \mathcal{D}$ is homotopy equivalent to a j-canonical path for $\forall z \in \mathcal{D}$. The domain $\mathcal{D}$ is *canonical* if it is j-canonical for both $j = 1, 2$.

In order to construct a canonical domain containing a given point z_0, we first cut out a simply connected domain outside some neighborhoods of all the turning points, $\mathcal{L} \subset \mathbb{C}\backslash\{z_1, z_2, z_3, z_4\}$; see [**K9**]. Then we introduce a segment $\ell \subset \mathcal{L}$ of the *anti-Stokes line* passing trough z_0,

$$\ell = \Big\{z \in \mathcal{L}\colon \operatorname{Im}\int_{z_0}^z \mu(\xi)\,d\xi = 0\Big\}. \tag{10.2.20}$$

By definition, it is always possible to number the (finite or infinite) endpoints ζ_1 and ζ_2 of ℓ in such a way that for $\forall z \in \ell$ the part $\gamma(\zeta_j, z) \subset \ell$, $j = 1, 2$, is a j-canonical path.

Finally, introduce a segment ℓ^* of a *Stokes line* passing through a point $z^* \in \ell$,

$$\ell^* = \Big\{z \in \mathcal{L}\colon \operatorname{Re}\int_{z^*}^z \mu(\xi)\,d\xi = 0\Big\}. \tag{10.2.21}$$

The union of all ℓ^* is a canonical domain containing the point z_0. The construction of a canonical domain containing the point $b > 0$ is illustrated in Figure 10.2. Near infinity, this domain coincides with the canonical sector Ω_1 (see Figure 2.5), thus

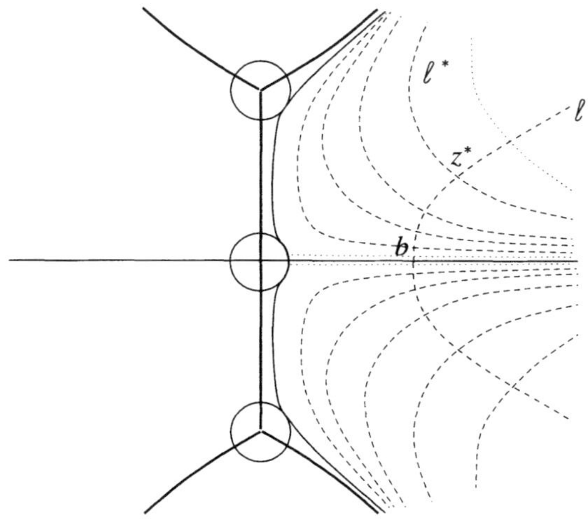

FIGURE 10.2. The canonical domain $\mathcal{D}_1$ containing the point $b > 0$, the anti-Stokes line ℓ passing through $z = b$, and the Stokes lines ℓ^* passing through the points $z^* \in \ell$

we denote it by $\mathcal{D}_1$. The canonical domain $\mathcal{D}_4$ containing the point $b < 0$ and corresponding to Ω_4, is the image of $\mathcal{D}_1$ under reflection with respect to the origin.

LEMMA 10.1. *Let $\mathcal{D}$ be a canonical domain. Let the contours $\gamma_j(z)$, $j = 1, 2$, be j-canonical. Let $z \in \mathcal{D}$ be located outside the finite neighborhoods of the zeros of μ and of b_+, and let contours $\gamma_j(z)$ avoid the finite neighborhoods of these zeros. Then, there exists a positive constant C such that*

$$\begin{aligned} &\|\chi^{(m)}(z) - I\| \le C\tau^{-m}, \quad z \in \mathcal{D}, \quad c_1 < |z| < c_2, \\ &\|\chi^{(m)}(z) - I\| \le C\tau^{-m}|z|^{-1-3m}, \quad z \in \mathcal{D}, \quad z \to \infty. \end{aligned} \tag{10.2.22}$$

PROOF. Using the definition of the canonical domain as well as the canonical paths with the estimates (10.2.18) for the matrix $R^{(m)}(z)$, it follows that the kernel of the Volterra integral equation (10.2.16) is bounded and integrable at infinity. Thus this equation is solvable by iterations, and the assertion is proven. □

We note that, if $m = 1$, then

$$\|\chi^{(1)} - I\| \le c_1\tau^{-1}, \quad |z| < const, \qquad \|\chi^{(1)} - I\| \le c_1\tau^{-1}|z|^{-4}, \quad z \to \infty.$$

For the considered canonical domain,

$$\begin{aligned} &T^{(1)}(z) = T^{(0)}(z)\big(I + O(\tau^{-1})\big), \quad |z| < const, \\ &T^{(1)}(z) = T^{(0)}(z)\big(I + O(\tau^{-1}z^{-4})\big) \quad z \to \infty. \end{aligned}$$

Since $\Lambda_1(z) \equiv 0$, this implies an improved estimate for the correction function $\chi^{(0)}$ for the zero-order WKB approximation,

$$\begin{aligned} &\|\chi^{(0)} - I\| \le c_1\tau^{-1}, \quad |z| < \text{const}, \\ &\|\chi^{(0)} - I\| \le c_1\tau^{-1}|z|^{-4}, \quad z \to \infty. \end{aligned} \tag{10.2.23}$$

Thus, we arrive at the following statement:

LEMMA 10.2. *Let $\mathcal{D}$ be a canonical domain. Let the contours $\gamma_j(z)$, $j = 1, 2$, be j-canonical. Let $z \in \mathcal{D}$ be located outside the finite neighborhoods of the zeros of μ and of b_+, and let the contours $\gamma_j(z)$ avoid the finite neighborhoods of these zeros. Then*

$$\Psi(z) = \Psi^{\mathcal{D}}_{WKB}(z) = P\Phi^{\mathcal{D}}_{WKB}(z) := PT(z)\chi_0(z)\exp\Big\{\tau \int_{z_0}^{z} \mu(\xi)\, d\xi\, \sigma_3\Big\}, \tag{10.2.24}$$

where P is defined in (10.2.8), $T(z)$ is defined in (10.2.14) and $\chi_0(z)$ satisfies (10.2.23).

Near infinity, the canonical domains $\mathcal{D}_j$, $j = 1, 4$, coincide with the sectors Ω_j, $j = 1, 4$, thus we can match the canonical solutions $\Psi_j(\lambda)$ with the solutions (10.2.24). By a straightforward computation,

$$\mathrm{C}_1^{-1} := \Psi_1^{-1}(\lambda)\Psi^{\mathcal{D}_1}_{WKB}(z) \tag{10.2.25}$$

$$= \begin{pmatrix} 0 & -i \\ \varepsilon_2 & 0 \end{pmatrix} \lim_{z\to+\infty} \exp\Big\{\tau\Big[\int_{z_0}^{z}\Big(\mu + \tfrac{1}{2\tau\mu}\big(b_3' - b_3\tfrac{b_+'}{b_+}\big)\Big)dz - i\big(\tfrac{4}{3}z^3 + z\big)\Big]\sigma_3\Big\}.$$

The matrix C_4^{-1} is given by the same expression with the limit $z \to +\infty$ replaced by $z \to -\infty$.

LOCAL SOLUTION IN A NEIGHBORHOOD OF THE COUPLING TURNING POINTS

In a neighborhood of two coupling turning points $z_{3,4} = O(\tau^{-1/2})$, the function $\Phi(z)$ can be expressed in terms of the parabolic cylinder functions. Indeed, in the entry b_+ defined in (10.2.10), the term $-4iz^2$ can be neglected in the asymptotic region $z = O(\tau^{-\frac{1}{2}+\epsilon})$, $0 < \epsilon < 1/2$, since the difference $y - \epsilon_2(v^2 + 1/2)$ is of order $O(1)$ (using (10.2.7) with $\delta = 0$). The model equation with the entry $b_+(z)$ replaced by its constant term is solvable exactly via the parabolic cylinder functions.

In order to justify rigorously the approximation in terms of the parabolic cylinder functions, let us introduce the parabolic cylinder equation for the entries of the first row of the Φ-function. In more in detail, following [**K7, K8**], we introduce the changes of the variables,

$$\xi = \varphi(z), \quad \Phi = VY, \tag{10.2.26}$$

where

$$V = \begin{pmatrix} \sqrt{\frac{b_+}{\varphi'}} & 0 \\ -\frac{1}{\sqrt{\varphi' b_+}}\Big(b_3 - \frac{1}{2\tau}\big(\frac{b_+'}{b_+} - \frac{\varphi''}{\varphi'}\big)\Big) & \sqrt{\frac{\varphi'}{b_+}} \end{pmatrix}, \quad (\,)' := \frac{d}{dz}. \tag{10.2.27}$$

The new function $Y(\xi)$ satisfies the equation

$$\frac{dY}{d\xi} = \mathcal{J}Y, \quad \mathcal{J} = \frac{1}{\varphi'}\big(\mathrm{V}^{-1}\mathrm{B}\mathrm{V} - \frac{1}{\tau}\mathrm{V}^{-1}\frac{d\mathrm{V}}{dz}\big) = \begin{pmatrix} 0 & 1 \\ \frac{\nu^2}{(\varphi')^2} & 0 \end{pmatrix}, \tag{10.2.28}$$

where

$$\nu^2 = \nu_0^2 + \frac{1}{2\tau^2}\{\varphi, z\}, \quad \{\varphi, z\} = \frac{\varphi'''}{\varphi'} - \frac{3}{2}\Big(\frac{\varphi''}{\varphi'}\Big)^2, \tag{10.2.29}$$

$$\nu_0^2 = \mu^2 + \frac{1}{\tau}\big(b_3' - b_3\frac{b_+'}{b_+}\big) - \frac{1}{2\tau^2}\Big(\frac{b_+''}{b_+} - \frac{3}{2}\big(\frac{b_+'}{b_+}\big)^2\Big). \tag{10.2.30}$$

Using (10.2.3), equation (10.2.30) yields

$$\nu_0^2 = \hat{w}_0^2(z)Q_0(z), \quad \hat{w}_0^2(z) = (z - \hat{z}_3)(z - \hat{z}_4), \tag{10.2.31}$$

with the function $Q_0(z)$, is holomorphic and does not vanish in some neighborhood of the *secondary* turning points $\hat{z}_j = O(\tau^{-1/2})$, $j = 3, 4$. Since we are aiming at the parabolic cylinder equation, we impose the following differential condition which determines the desired change of the independent variable,

$$\frac{\nu^2}{(\varphi')^2} = w^2(\xi) := \xi^2 + 2\rho, \tag{10.2.32}$$

or, equivalently,

$$\Big(\frac{d\xi}{dz}\Big)^2 w^2(\xi) = \nu_0^2 + \frac{1}{2\tau^2}\{\xi, z\}. \tag{10.2.33}$$

Neglecting the second term in the right-hand side of (10.2.33), we find the zero-order approximation of the change of the independent variable,

$$\xi' w_0(\xi) = \nu_0(z), \quad w_0^2(\xi) = \xi^2 + 2\rho_0, \quad \xi = \varphi_0(z), \quad \xi' := \frac{d\xi}{dz}. \tag{10.2.34}$$

We are looking for a solution $\xi = \varphi(z)$, which is holomorphic in the neighborhood of both secondary turning points $\hat{z}_3, \hat{z}_4 = O(\tau^{-1/2})$, thus we arrive at the integral representation

$$\int_{-i\sqrt{2\rho_0}}^{\xi} w_0(\xi)\, d\xi = \int_{\hat{z}_3}^{z} \nu_0(z)\, dz, \tag{10.2.35}$$

where $w_0(\xi)$ is determined by the equations

$$w_0^2(\xi) = \xi^2 + 2\rho_0, \quad \int_{-i\sqrt{2\rho_0}}^{i\sqrt{2\rho_0}} w_0(\xi)\, d\xi = i\pi\rho_0 = \int_{\hat{z}_3}^{\hat{z}_4} \nu_0(z)\, dz. \tag{10.2.36}$$

Here we assume that $\operatorname{Re}\sqrt{2\rho_0} \geq 0$, $w_0(\xi)$ is defined on the plane cut along two infinite rays $(-i\infty, -i\sqrt{2\rho_0}]$ and $[i\sqrt{2\rho_0}, +i\infty)$, and the branch of $w_0(\xi)$ is defined by the asymptotics $w_0(\xi) \simeq \xi$ as $\xi \to +\infty$.

Indeed, if the two coupling secondary turning points, i.e., the zeros $\hat{z}_j$ of $\nu_0(z)$, $\hat{z}_j = O(\tau^{-1/2})$, $j = 3, 4$, are simple, then the parameter $\rho_0 \neq 0$, and the two zeros of $w_0^2(\xi)$ are simple as well. Then, in some neighborhoods of $\hat{z}_3$ and ξ_1, equation (10.2.35) takes the form $(\xi - \xi_1)^{3/2}L(\xi) = (z - \hat{z}_3)^{3/2}R(z)$, for some holomorphic functions $L(\xi)$ and $R(z)$, which do not vanish at $\xi = \xi_1$ and $z = \hat{z}_3$, respectively. Then taking any branch of degree $2/3$ of both sides, and using the inverse function theorem, the above equation determines the holomorphic functions $\xi(z)$ and $z(\xi)$.

Using the condition (10.2.36) and applying the same arguments, we obtain an extension of the holomorphic functions $\xi(z)$ and $z(\xi)$ to a neighborhood of the points $\hat{z}_4$ and ξ_2, respectively. We note that the condition (10.2.36) is not satisfied for every choice of the branch of the cubic root. The appropriate branch is homotopy equivalent to the identity function $z(\xi) = \xi$ under a deformation with respect to a parameter $\epsilon \in [0,1]$ between a monic square polynomial and the function $\nu_0(z)$ defined by

$$\nu_0^2(z,\epsilon) = \hat{w}_0^2(z)\Big(1+\epsilon\big(Q_0(z)-1\big)\Big). \tag{10.2.37}$$

Note, the inverse function theorem prevents extending $\xi(z)$ to zeros and singularities of $\nu_0(z)$ except for $\hat{z}_3$ and $\hat{z}_4$. In particular, $\xi(z)$ cannot be continued at the zeros $z_\pm = O(\tau^{-\delta/2}) = O(1)$ of the coefficient $b_+(z)$.

Now, we look for a change of the independent variable $\varphi(z)$ in the form of the composition,

$$\varphi(z) = f_1\big((\varphi_0(z)\big), \tag{10.2.38}$$

which, using the Cayley identity, turns equation (10.2.33) into the equation

$$\Big(\frac{df_1}{dz}\Big)^2 w^2(f_1) = \nu_0^2 + \frac{1}{2\tau^2}\{f_1, z\} = \nu_0^2 + \frac{1}{2\tau^2}\Big(\{\varphi_0, z\} + \{f_1, \varphi_0\}(\varphi_0')^2\Big). \tag{10.2.39}$$

This equation is equivalent to the equation

$$(f_1')^2 w^2(f_1) = \nu_1^2(\varphi_0) + \frac{1}{2\tau^2}\{f_1, \varphi_0\}, \quad f_1' := \frac{df_1}{d\varphi_0}, \tag{10.2.40}$$

with

$$\begin{aligned} \nu_1^2(\varphi_0) &= \frac{1}{(\varphi_0')^2}\Big(\nu_0^2 + \frac{1}{2\tau^2}\{\varphi_0, z\}\Big) \\ &= w_0^2(\varphi_0) + \frac{\{\varphi_0, z\}}{2\tau^2(\varphi_0')^2} = \hat{w}_1^2(\varphi_0)Q_1(\varphi_0), \\ &\hat{w}^2(\zeta) = (\zeta-\zeta_1)(\zeta-\zeta_2), \quad \zeta_j = O(\tau^{-1/2}). \end{aligned} \tag{10.2.41}$$

Expanding the last two equalities of (10.2.41) into a Taylor series, we find, that $Q_1(\xi)$ satisfies uniformly the estimate,

$$Q_1(\xi) = 1 + O(\tau^{-2}). \tag{10.2.42}$$

An approximation $\varphi_1(\varphi_0)$ to the function $f_1(\varphi_0)$ can be found using the reduced equation,

$$\varphi_1' w_1(\varphi_1) = \nu_1(\varphi_0), \quad w_1(\xi) = \xi^2 + 2\rho_1, \quad \varphi_1' := \frac{d\varphi_1}{d\varphi_0}. \tag{10.2.43}$$

Integrating (10.2.43) and using the above arguments, we find the holomorphic change of the independent variable $\varphi_1(\varphi_0)$, which is valid in a finite neighborhood of the origin,

$$\int_{-i\sqrt{2\rho_1}}^{\xi} w_1(\xi)\,d\xi = \int_{\zeta_1}^{z} \nu_1(\zeta)\,d\zeta, \tag{10.2.44}$$

where $w_1(\zeta)$ is determined by the equations

$$w_1^2(\xi) = \xi^2 + 2\rho_1, \quad \int_{-i\sqrt{2\rho_1}}^{i\sqrt{2\rho_1}} w_1(\xi)\,d\xi = i\pi\rho_1 = \int_{\zeta_1}^{\zeta_2} \nu_1(\zeta)\,d\zeta. \tag{10.2.45}$$

Here we assume that Re $\sqrt{2\rho_1} \geq 0$, $w_1(\xi)$ is defined on the plane cut along the two infinite rays $(-i\infty, -i\sqrt{2\rho_1}]$ and $[i\sqrt{2\rho_1}, +i\infty)$, and the branch of $w_1(\xi)$ is defined by the asymptotics $w_1(\xi) \simeq \xi$ as $\xi \to +\infty$.

Elementary but rather long computations show that the estimate (10.2.42) implies that $\varphi_1(\varphi_0)$ satisfies uniformly the estimates [K7, K8],

$$\varphi_1 = \varphi_0 + O(\tau^{-2}), \quad \varphi_1' = 1 + O(\tau^{-2}), \quad \{\varphi_1, \varphi_0\} = O(\tau^{-2}). \tag{10.2.46}$$

The procedure described above can be continued giving rise to the iterative sequence of change of variables [Bl, K7, K8],

$$\xi = \varphi_n\big(\varphi_{n-1}\big(\dots \varphi_1\big(\varphi_0(z)\big)\dots\big)\big). \tag{10.2.47}$$

However, for our purposes, it is sufficient to consider $n = 1$ in (10.2.47).

Using in (10.2.26) the transformation (10.2.47), we obtain equation (10.2.28) with the matrix $\mathcal{J}$,

$$\mathcal{J} = \begin{pmatrix} 0 & 1 \\ w_n^2(\xi) + \frac{1}{2\tau^2}\frac{\{\varphi_n, \varphi_{n-1}\}}{(\varphi_n')^2} & 0 \end{pmatrix}. \tag{10.2.48}$$

The model equation

$$\frac{dZ^{(n)}}{d\xi} = \tau \begin{pmatrix} 0 & 1 \\ w_n^2(\xi) & 0 \end{pmatrix} Z^{(n)}, \quad w_n^2(\xi) = \xi^2 + 2\rho_n, \tag{10.2.49}$$

is exactly solvable in terms of parabolic cylinder functions,

$$Z^{(n)}(\xi) = \tau^{\sigma_3/4} \begin{pmatrix} D_{-\tau\rho_n - \frac{1}{2}}(\sqrt{2\tau}\,\xi) & D_{\tau\rho_n - \frac{1}{2}}(i\sqrt{2\tau}\,\xi) \\ D'_{-\tau\rho_n - \frac{1}{2}}(\sqrt{2\tau}\,\xi) & D'_{\tau\rho_n - \frac{1}{2}}(i\sqrt{2\tau}\,\xi) \end{pmatrix}, \tag{10.2.50}$$

where the prime means differentiation with respect to its argument. Since we need to analyze the associated Stokes phenomenon, we fix the canonical solutions of (10.2.49) by their asymptotics at infinity,

$$Z_k^{(n)}(\xi) = \xi^{-\frac{1}{2}\sigma_3} \frac{1}{\sqrt{2}} \begin{pmatrix} 1 & 1 \\ 1 & -1 \end{pmatrix} \Big(I + O(\tau^{-1}\xi^{-2})\Big) e^{\tau\theta_n\sigma_3}, \tag{10.2.51}$$

$$\begin{aligned} &\theta_n = \frac{\xi^2}{2} + \rho_n \ln\big(\tau^{1/2}\xi\big), \\ &\tau \to \infty, \quad |\xi| \geq c\tau^{-\frac{1}{2}+\varepsilon}, \quad c, \varepsilon > 0, \\ &\xi \in \omega_k = \Big\{\xi \in \mathbb{C}\colon \quad \arg\xi \in \big(\frac{\pi}{2}(k - \frac{3}{2}), \frac{\pi}{2}(k + \frac{1}{2})\big)\Big\}. \end{aligned}$$

The canonical solutions $Z_k^{(n)}(\xi)$ are related by the Stokes matrices G_k (cf. Section 2.5 of Chapter 2),

$$Z_{k+1}^{(n)}(\xi) = Z_k^{(n)}(\xi)G_k^{(n)}, \quad G_{k+2}^{(n)} = e^{-i\pi\rho\sigma_3}G_k^{(n)}e^{i\pi\rho\sigma_3}, \tag{10.2.52}$$

$$G_0^{(n)} = \begin{pmatrix} 1 & 0 \\ h_0^{(n)} & 1 \end{pmatrix}, \quad G_1^{(n)} = \begin{pmatrix} 1 & h_1^{(n)} \\ 0 & 1 \end{pmatrix},$$

$$h_0^{(n)} = -i\frac{\sqrt{2\pi}}{\Gamma(\frac{1}{2}-\rho)}2^{-\rho}, \quad h_1^{(n)} = -i\frac{\sqrt{2\pi}}{\Gamma(\frac{1}{2}+\rho)}2^{\rho}e^{-i\pi\rho}, \quad \rho := \tau\rho_n.$$

All solutions of (10.2.49) are oscillating along the rays $\arg\xi = \frac{\pi}{4} + \frac{\pi}{2}m$, $m \in \mathbb{Z}$.

We look for solutions of (10.2.28), (10.2.48) in the form of the product

$$Y(\xi) = \chi^{(n)}(\xi)Z^{(n)}(\xi). \tag{10.2.53}$$

The function $\chi^{(n)}(\xi)$ satisfies the integral Volterra equation

$$\chi^{(n)}(\xi) = I + \tau\int_{\xi_0}^{\xi} Z^{(n)}(\xi)Z^{(n)}(\zeta)^{-1}R^{(n)}(\zeta)\chi^{(n)}(\zeta)Z^{(n)}(\zeta)Z^{(n)}(\xi)^{-1}\,d\zeta,$$

$$R^{(n)}(\xi) = \begin{pmatrix} 0 & 0 \\ \frac{1}{2\tau^2}\frac{\{\varphi_n,\varphi_{n-1}\}}{(\varphi_n')^2} & 0 \end{pmatrix},$$

where ξ_0 is a normalization point, e.g., $\xi_0 = 0$. In a finite domain satisfying the condition

$$\big|\mathrm{Re}\,(\tau\theta)\big| < const, \tag{10.2.54}$$

it is easy to see, using (10.2.50), that

$$\big\|\chi^{(n)}(\xi) - I\big\| \le c\Big|\frac{\{\varphi_n,\varphi_{n-1}\}}{(\varphi_n')^2}\Big|, \quad c = const. \tag{10.2.55}$$

For $n = 1$, using (10.2.46), we find

$$\chi^{(1)}(\xi) = I + O(\tau^{-2}). \tag{10.2.56}$$

Furthermore, observing that

$$Z_k^{(1)}(\varphi_1(\varphi_0(z)))\big(Z_k^{(0)}(\varphi_0(z))\big)^{-1} = I + O(\tau^{-1}),$$

we find the estimate

$$\chi^{(0)}(\xi) = I + O(\tau^{-1}). \tag{10.2.57}$$

Thus we obtain the following statement:

LEMMA 10.3. *Let D_0 be a disk centered at the origin and such that the turning points $z_{1,2} = \pm i/\sqrt{2} + O(\tau^{-1})$, as well as the zeros of the coefficient $b_+(z)$ lie outside this disk. Let $z \in \mathrm{D}_0$, and let $\xi(z)$ be determined by equations (10.2.35) and (10.2.36). Suppose that z satisfies the condition $|\,\mathrm{Re}\,(\tau\xi^2(z)+2\tau\rho_0\ln(\tau^{1/2}\xi(z)))| < const$. Then*

$$\Phi(z) = \Phi_{loc}(z) := V(z)\chi^{(0)}(\xi(z))Z^{(0)}(\xi(z)), \tag{10.2.58}$$

where $V(z)$ is defined in (10.2.27), $Z^{(0)}(\xi)$ is defined in (10.2.50), and the function $\chi^{(0)}$ satisfies in D_0 uniformly the estimate (10.2.57).

The canonical domains $\mathcal{D}_j$, $j = 1, 4$, and the disk D_0 are allowed to intersect, thus it is possible to match the approximations valid in these domains. For this purpose, we find the ratio of the solutions $\Phi(z)$ defined in (10.2.24) and (10.2.58). A straightforward computation yields

$$\begin{aligned} \mathrm{N}_{\mathcal{D}_1} &:= \Phi_{loc}^{-1}(z)\Phi_{WKB}^{\mathcal{D}_1}(z) \\ &= \Big(I + O(\tau^{-1})\Big)\exp\Big\{\tau\Big[\int_{z_0}^{z}\Big(\mu + \frac{1}{2\tau\mu}(b_3' - b_3\frac{b_+'}{b_+})\Big)dz \\ &\qquad -\frac{\xi^2(z)}{2} - \rho_0\ln\big(\tau^{1/2}\xi(z)\big)\Big]\sigma_3\Big\}. \end{aligned} \tag{10.2.59}$$

Here we take into account that, for $z \in \mathrm{D}_0 \cap \mathcal{D}_1$, up to terms $O(\tau^{-1})$, $\mu(z)$ and ξ satisfy $\mu(z) \simeq e^{i\pi/2}2^{3/2}z$ and $\xi = \varphi_0(z) \simeq e^{i\pi/4}2^{3/4}z$. Furthermore, for $z \in \mathrm{D}_0 \cap \mathcal{D}_4$, $\mu(z)$ and ξ satisfy $\mu(z) \simeq e^{-i\pi/2}2^{3/2}z$ and $\xi = \varphi_0(z) \simeq e^{i\pi/4}2^{3/4}z$. This difference yields

$$\begin{aligned} \mathrm{N}_{\mathcal{D}_4} &:= \Phi_{loc}^{-1}(z)\Phi_{WKB}^{\mathcal{D}_4}(z) \\ &= i\sigma_1\Big(I + O(\tau^{-1})\Big)\exp\Big\{\tau\Big[\int_{z_0}^{z}\Big(\mu + \frac{1}{2\tau\mu}(b_3' - b_3\frac{b_+'}{b_+})\Big)dz \\ &\qquad +\frac{\xi^2(z)}{2} + \rho_0\ln\big(\tau^{1/2}\xi(z)\big)\Big]\sigma_3\Big\}. \end{aligned} \tag{10.2.60}$$

Thus, the difference of the canonical solutions $\Psi_j(\lambda)$ from the local canonical solutions is determined by the matrices $\mathrm{N}_{\mathcal{D}}$ multiplied from the right with the matrices C_j defined in (10.2.25),

$$\mathrm{N}_{\mathcal{D}_1}\mathrm{C}_1 = \Big(I + O(\tau^{-1})\Big)e^{\tau F_1\sigma_3}\begin{pmatrix} 0 & \varepsilon_2 \\ i & 0 \end{pmatrix}, \tag{10.2.61}$$

$$\begin{aligned} F_1 = \lim_{\substack{z\to+\infty \\ \hat{z}\in \mathrm{D}_0\cap\mathcal{D}_1}} \Big\{&\int_{z}^{\hat{z}}\Big(\mu + \frac{1}{2\tau\mu}(b_3' - b_3\frac{b_+'}{b_+})\Big)dz \\ &-\frac{\xi^2(\hat{z})}{2} - \rho_0\ln\big(\tau^{1/2}\xi(\hat{z})\big) + i\big(\frac{4}{3}z^3 + z\big)\Big\}, \end{aligned}$$

$$\mathrm{N}_{\mathcal{D}_4}\mathrm{C}_4 = i\sigma_1\Big(I + O(\tau^{-1})\Big)e^{\tau F_4\sigma_3}\begin{pmatrix} 0 & \varepsilon_2 \\ i & 0 \end{pmatrix}, \tag{10.2.62}$$

$$\begin{aligned} F_4 = \lim_{\substack{z\to-\infty \\ \hat{z}\in \mathrm{D}_0\cap\mathcal{D}_4}} \Big\{&\int_{z}^{\hat{z}}\Big(\mu + \frac{1}{2\tau\mu}(b_3' - b_3\frac{b_+'}{b_+})\Big)dz \\ &+\frac{\xi^2(\hat{z})}{2} + \rho_0\ln\big(\tau^{1/2}\xi(\hat{z})\big) + i\big(\frac{4}{3}z^3 + z\big)\Big\}. \end{aligned}$$

Computation of the Stokes matrices

Since τ is large and z is finite, the new local variable $\xi(z)$ defined by (10.2.34) is approximated by $\xi \simeq 2^{3/4}e^{i\pi/4}z$. Thus, the model function $Z_1^{(0)}$ for $\arg z = 0$, as well as the function for $\arg z = \pi$, have the canonical asymptotics (10.2.51). Thus (10.2.5) takes the following form,

$$\begin{aligned}
(10.2.63)\quad \mathrm{S}_1\mathrm{S}_2\mathrm{S}_3 &= \begin{pmatrix} 1+s_2s_3 & s_2 \\ s_1+s_3+s_1s_2s_3 & 1+s_1s_2 \end{pmatrix} \\
&= \left(\mathrm{N}_{\mathcal{D}_1}\mathrm{C}_1\right)^{-1} G_1^{(0)} G_2^{(0)} \mathrm{N}_{\mathcal{D}_4}\mathrm{C}_4 \\
&= \begin{pmatrix} 0 & -i \\ \varepsilon_2 & 0 \end{pmatrix} e^{-\tau F_1\sigma_3} G_1^{(0)} G_2^{(0)} i\sigma_1 e^{\tau F_4\sigma_3} \begin{pmatrix} 0 & \varepsilon_2 \\ i & 0 \end{pmatrix} (I+O(\tau^{-1})) \\
&= \begin{pmatrix} ih_2^{(0)} e^{\tau(F_1-F_4)} & \varepsilon_2 e^{\tau(F_1+F_4)} \\ -\varepsilon_2\left(1+h_1^{(0)}h_2^{(0)}\right)e^{-\tau(F_1+F_4)} & ih_1^{(0)} e^{-\tau(F_1-F_4)} \end{pmatrix} (I+O(\tau^{-1})),
\end{aligned}$$

where

$$h_1^{(0)} = -i\frac{\sqrt{2\pi}}{\Gamma(\frac{1}{2}+\rho)} 2^{\rho} e^{-i\pi\rho}, \quad h_2^{(0)} = -i\frac{\sqrt{2\pi}}{\Gamma(\frac{1}{2}-\rho)} 2^{-\rho} e^{2\pi i\rho},$$

$$(10.2.64)\qquad 1+h_1^{(0)}h_2^{(0)} = -e^{2\pi i\rho}, \quad \rho = \tau\rho_0.$$

REMARK 10.2. It is shown in [**K8**], that in the case of general $\delta \in \left[0, \frac{1}{2}\right]$, the same expression (10.2.63) is valid for the product of the Stokes matrices, but the error estimate must be replaced by $O(\tau^{-1+\delta})$.

COMPUTATION OF THE PHASE INTEGRALS $F_{1,4}$

The formula (10.2.63) yields the Stokes multipliers with an accuracy of $O(\tau^{-1})$. Therefore, the asymptotics of the phase integrals $F_{1,4}$ is not needed beyond the terms of order $O(\tau^{-2})$. In fact, in order to find the leading order asymptotics of the Stokes multipliers as $x \to +\infty$, it is sufficient to find the phase integrals with an accuracy of $O(\tau^{-1-\epsilon})$, $\epsilon > 0$.

It is not surprising that the computation of the phase integrals F_1, F_4 defined in (10.2.61), (10.2.62), depends on the value of $\delta \in \left[0, \frac{1}{2}\right]$ introduced in (10.2.6), (10.2.7). However, it is perhaps surprising that the form of the asymptotic ansatz of the Painlevé transcendent depends crucially on δ. In particular, as we will see, the asymptotics of the Painlevé function for $\delta = 0$, is quite different from that studied in Chapter 9 which corresponds to $\delta = 1/2$.

Using (10.2.36), we find

$$(10.2.65)\qquad \rho_0 = -\frac{i}{\sqrt{2}}\left(D - \varepsilon_2\frac{v}{\tau}\right) + O(\tau^{-2}) = O(\tau^{-1}).$$

The terms in F_1 (10.2.61) which involve $\xi(\hat{z})$ can be transformed as follows:

$$\begin{aligned}
\frac{\xi^2}{2} + \rho_0 \ln(\tau^{1/2}\xi) &= \int_0^{\xi} w_0(\xi)\,d\xi + \frac{\rho_0}{2}\ln\frac{\tau\rho_0}{2e} + O(\rho_0^2) \\
(10.2.66)\qquad &= \int_0^{\hat{z}} \nu_0(z)\,dz + \frac{\rho_0}{2}\ln\frac{\tau\rho_0}{2e} + O(\tau^{-2}),
\end{aligned}$$

where we have used (10.2.35) and have taken into account that $w_0(\xi)$ and $\nu_0(z)$ are even functions. The analogous terms in F_4 (10.2.62) are transformed as follows:

$$-\frac{\xi^2}{2} - \rho_0 \ln(\tau^{1/2}\xi) = \int_0^{\hat{z}} \nu_0(z)\,dz - \frac{\rho_0}{2}\ln\frac{\tau\rho_0}{2e} - i\pi\rho + O(\tau^{-2}). \tag{10.2.67}$$

Observing that

$$\mu + \frac{1}{2\tau\mu}\big(b_3' - b_3\frac{b_+'}{b_+}\big) = \nu_0 + O(\tau^{-2})$$

as $|z| \geq const$, we can rewrite the phase integral F_1 and F_4 in (10.2.61), (10.2.62) as follows:

$$F_1 = \lim_{z\to+\infty}\Big\{-\int_0^z \nu_0(z)\,dz + i\big(\frac{4}{3}z^3 + z\big) - c_1 + O(\tau^{-2})\Big\}, \tag{10.2.68}$$

$$F_4 = \lim_{z\to-\infty}\Big\{-\int_0^z \nu_0\,dz + i\big(\frac{4}{3}z^3 + z\big) + c_1 + i\pi\rho + O(\tau^{-2})\Big\}, \tag{10.2.69}$$

$$c_1 = \frac{\rho_0}{2}\ln\frac{\tau\rho_0}{2e}.$$

Because $\nu_0(z)$ is an even function of z, F_4 and F_1 are related by the equation

$$F_4 = -F_1 + i\pi\rho_0 + O(\tau^{-2}). \tag{10.2.70}$$

Using this equation in the nondiagonal entries of (10.2.63), we find the Stokes multiplier s_2,

$$s_2 = \varepsilon_2 e^{i\pi\tau\rho_0}\big(1 + O(\tau^{-1})\big). \tag{10.2.71}$$

To find the asymptotics as $\tau\to\infty$ of the phase integral F_1 with the accuracy $O(\tau^{-1-\epsilon})$, $\epsilon > 0$, we divide the path of integration into two segments: $(0,+\infty) = (0,\tilde{z}] \cup (\tilde{z},+\infty)$, where $\tilde{z} = c'\tau^{-\frac{1}{2}+\varepsilon}$ for some constants $c' > 0$ and $\varepsilon \in (0,1/2)$.

In the first segment, $z \in (0,\tilde{z})$, the integrand can be represented as the product (10.2.31). Indeed, introducing the notation $\pm z_+$ for the zeros of the coefficient b_+ in (10.2.10), we find

$$b_+(z) = -4i(z^2 - z_+^2), \quad z_+^2 = \frac{\varepsilon_2}{2}\Big(y - \varepsilon_2\big(v^2 + \frac{1}{2}\big)\Big), \tag{10.2.72}$$
$$\frac{b_+'}{b_+} = \frac{2z}{z^2 - z_+^2} = \frac{1}{z - z_+} + \frac{1}{z + z_+}.$$

Therefore, using (10.2.30) it follows that

$$\nu_0^2 = -8(z^2 - \hat{z}_3^2)\big(1 + O(z^2) + O(\tau^{-1})\big), \tag{10.2.73}$$

where

$$\hat{z}_3^2 = \frac{1}{2}\big(D - \varepsilon_2\frac{v}{\tau}\big) + O(\tau^{-2}) = i\frac{\rho_0}{\sqrt{2}} + O(\tau^{-2}). \tag{10.2.74}$$

Integrating estimate (10.2.73) from 0 to $\tilde{z} = c'\tau^{-1/2+\epsilon}$ yields

$$\int_0^{\tilde{z}} \nu_0(z)\,dz = i\sqrt{2}\,\tilde{z}^2 + \rho_0\ln\tilde{z} - \frac{\rho_0}{2}\ln\big(-i\frac{\rho_0}{e4\sqrt{2}}\big) + O(\tilde{z}^4) + O(\tau^{-2}\tilde{z}^{-2}). \tag{10.2.75}$$

In the region, $|z| \geq c'\tau^{-\frac{1}{2}+\epsilon}$, the function $\nu_0(z)$ (10.2.30) can be expanded into a series in degrees of $\rho_0 = O(\tau^{-1})$. Thus, integrating $\nu_0(z)$ along $(\tilde{z}, +\infty)$, we find

$$\int_{\tilde{z}}^{z} \nu_0(z)\,dz = i\Big(\frac{4}{3}z^3 + z + O(z^{-1})\Big) - i\frac{\sqrt{2}}{3} - i\sqrt{2}\tilde{z}^2 - \rho_0 \ln \tilde{z}$$

$$(10.2.76) \qquad +\frac{\rho_0}{2}\ln 2 + \frac{b_3(z_+)}{2\tau\mu(z_+)} \ln \frac{\frac{1}{\sqrt{2}} - \sqrt{z_+^2 + \frac{1}{2}}}{\frac{1}{\sqrt{2}} + \sqrt{z_+^2 + \frac{1}{2}}} + O(\tilde{z}^4) + O(\tau^{-1}\tilde{z}^2).$$

Combining (10.2.75) and (10.2.76), and using (10.2.68), we find

$$F_1 = i\frac{\sqrt{2}}{3} - \frac{\rho_0}{2}\ln 4\sqrt{2}\,\tau - i\frac{\pi\rho_0}{4} - \frac{b_3(z_+)}{2\tau\mu(z_+)} \ln \frac{\frac{1}{\sqrt{2}} - \sqrt{z_+^2 + \frac{1}{2}}}{\frac{1}{\sqrt{2}} + \sqrt{z_+^2 + \frac{1}{2}}}$$

$$(10.2.77) \qquad +O(\tau^{-1}\tilde{z}^2) + O(\tau^{-1}\tilde{z}^{-2}).$$

Choosing $\tilde{z} = c'\tau^{-1/4}$, it follows that the error in (10.2.77) does not exceed $O(\tau^{-3/2})$.

Asymptotics of the Stokes multipliers and of the Painlevé function as $x \to +\infty$

Since $\mu^2 = b_3^2 + b_+ b_-$, we have that at any zero of b_+, $\mu^2(z_+) = b_3^2(z_+)$. Let us put

$$(10.2.78) \qquad \hat{\sigma} = \frac{b_3(z_+)}{\mu(z_+)} \in \{-1, +1\}, \quad \arg z_+ \in \Big(-\frac{\pi}{2}, \frac{\pi}{2}\Big].$$

Thus, using (10.2.63), (10.2.64), (10.2.70) and (10.2.77), we find the following asymptotics for the Stokes multipliers,

$$1 + s_2 s_3 = \left(\frac{\frac{1}{\sqrt{2}} - \sqrt{z_+^2 + \frac{1}{2}}}{\frac{1}{\sqrt{2}} + \sqrt{z_+^2 + \frac{1}{2}}}\right)^{-\hat{\sigma}} \frac{\sqrt{2\pi}\, e^{i\frac{\pi\rho}{2}}}{\Gamma(\frac{1}{2} - \rho)} e^{i\frac{2\sqrt{2}}{3}\tau - \rho\ln 8\sqrt{2}\,\tau}\big(1 + O(\tau^{-1/2})\big),$$

(10.2.79)

$$1 + s_1 s_2 = \left(\frac{\frac{1}{\sqrt{2}} - \sqrt{z_+^2 + \frac{1}{2}}}{\frac{1}{\sqrt{2}} + \sqrt{z_+^2 + \frac{1}{2}}}\right)^{\hat{\sigma}} \frac{\sqrt{2\pi}\, e^{i\frac{\pi\rho}{2}}}{\Gamma(\frac{1}{2} + \rho)} e^{-i\frac{2\sqrt{2}}{3}\tau + \rho\ln 8\sqrt{2}\,\tau}\big(1 + O(\tau^{-1/2})\big),$$

$$s_2 = \varepsilon_2 e^{i\pi\rho}\big(1 + O(\tau^{-1})\big), \quad \rho = \tau\rho_0.$$

The fact that s_2 does not depend on τ implies the asymptotics $\rho = const + O(\tau^{-1})$ which, by virtue of equation (10.2.65), is consistent with the estimate (10.2.7). Also, by virtue of equation (10.2.72), the independence of $1 + s_2 s_3$ and $1 + s_1 s_2$ on τ is consistent with the estimate (10.2.7) with $\delta = 0$, provided that $\operatorname{Re}\rho = O(\tau^{-1})$. In turn, this estimate implies $\arg(\varepsilon_2 s_2) = O(\tau^{-1})$ and, since s_2 does not depend on τ, it follows that

$$(10.2.80) \qquad \arg(\varepsilon_2 s_2) = 0.$$

In fact, other values of $\arg(\varepsilon_2 s_2)$ correspond to other values of the parameter δ; see [**K5, K8**].

To find the asymptotics of the Painlevé function, we solve (10.2.79) for v, y assuming that the Stokes multipliers are the *constants* and that $s_2(1+s_1s_2)(1+s_2s_3) \neq 0$. First, we rewrite (10.2.79) in the form,

$$\left(\frac{\frac{1}{\sqrt{2}} - \sqrt{z_+^2 + \frac{1}{2}}}{\frac{1}{\sqrt{2}} + \sqrt{z_+^2 + \frac{1}{2}}}\right)^{\hat{\sigma}} = ce^{ih}(1 + O(\tau^{-1/2})),$$

$$\left(\frac{\frac{1}{\sqrt{2}} + \sqrt{z_+^2 + \frac{1}{2}}}{\frac{1}{\sqrt{2}} - \sqrt{z_+^2 + \frac{1}{2}}}\right)^{\hat{\sigma}} = de^{-ih}(1 + O(\tau^{-1/2})),$$

$$c = \frac{\sqrt{2\pi}\, e^{\frac{\pi\gamma}{2}}}{(1+s_2s_3)\Gamma\left(\frac{1}{2}+i\gamma\right)}, \quad d = \frac{\sqrt{2\pi}\, e^{\frac{\pi\gamma}{2}}}{(1+s_1s_2)\Gamma\left(\frac{1}{2}-i\gamma\right)}, \quad cd = 1,$$

$$h = \frac{2\sqrt{2}}{3}\tau + \gamma \ln 8\sqrt{2}\,\tau,$$

$$\gamma = i\rho = \frac{1}{\pi}\ln(\varepsilon_2 s_2) + O(\tau^{-1}), \quad \varepsilon_2 s_2 > 0. \tag{10.2.81}$$

Next we compute the sum $ce^{ih} + de^{-ih}$,

$$\frac{\frac{1}{\sqrt{2}} - \sqrt{z_+^2 + \frac{1}{2}}}{\frac{1}{\sqrt{2}} + \sqrt{z_+^2 + \frac{1}{2}}} + \frac{\frac{1}{\sqrt{2}} + \sqrt{z_+^2 + \frac{1}{2}}}{\frac{1}{\sqrt{2}} - \sqrt{z_+^2 + \frac{1}{2}}} = -2 - \frac{2}{z_+^2}.$$

Taking into account (10.2.72) and the second of equations (10.2.7) with $\delta = 0$, we find

$$z_+^2 = \varepsilon_2 y + O(\tau^{-1}) = -v^2 - \frac{1}{2} + O(\tau^{-1}) - \frac{1}{1 + \frac{c}{2}e^{ih} + \frac{d}{2}e^{-ih} + O(\tau^{-1/2})}. \tag{10.2.82}$$

The last equation in (10.2.82) together with the equation $cd = 1$, yield

$$v^2 = -\frac{1}{2}\left(\frac{ce^{ih} - 1 + O(\tau^{-1/2})}{ce^{ih} + 1 + O(\tau^{-1/2})}\right)^2 + O(\tau^{-1}),$$

which determines v up to a sign. Recalling that $w = du/dx$, the definition of y in (10.2.1) implies $y = \frac{3}{2}\frac{dv}{d\tau} + \frac{v}{2\tau}$, which allows us to fix the sign of v,

$$v = \varepsilon_2 i \frac{ce^{ih} - 1 + O(\tau^{-1/2})}{\sqrt{2}(ce^{ih} + 1) + O(\tau^{-1/2})} + O(\tau^{-1}), \tag{10.2.83}$$

where

$$h = \frac{2\sqrt{2}}{3}\tau + \gamma \ln 8\sqrt{2}\,\tau,$$

and the parameters $\varepsilon_2 = \pm 1$, $c \in \mathbb{C}$ and $\gamma \in \mathbb{R}$ are expressed in terms of the Stokes multipliers $s_2 \in \mathbb{R}$ and $s_3 \in \mathbb{C}$ of the associated linear system by the equations

$$c = \frac{\sqrt{2\pi}\, e^{\frac{\pi\gamma}{2}}}{(1+s_2s_3)\Gamma\left(\frac{1}{2}+i\gamma\right)}, \quad \gamma = \frac{1}{\pi}\ln(\varepsilon_2 s_2), \quad \varepsilon_2 = \operatorname{sgn} s_2.$$

Repeating the Brouwer Fixed Point Theorem arguments used in Section 6.1 of Chapter 7, we can prove that equation (10.2.83) indeed provides the rigorous asymptotics of the relevant Painlevé function.

The real reduction

The asymptotics as $x \to +\infty$ of the case that u is real can be extracted from the asymptotics (10.2.83) by applying the reductions $s_2 = \overline{s}_2$ which is already implied by (10.2.80), as well as $s_1 = \overline{s}_3$. The latter condition gives rise to the condition $d = \overline{c}$ in (10.2.81), and therefore to the equation

$$|c| = 1. \tag{10.2.84}$$

Thus (10.2.83) takes the form

$$v = \frac{\varepsilon_2}{\sqrt{2}} \cot\Big(\frac{\sqrt{2}}{3}\tau + \frac{\gamma}{2}\ln 8\sqrt{2}\,\tau + \chi\Big) + O(\tau^{-1}), \tag{10.2.85}$$

where

$$\chi = -\frac{1}{2}\arg\Gamma\big(\frac{1}{2} + i\gamma\big) - \frac{1}{2}\arg(1 + s_2 s_3) + \frac{\pi}{2} + O(\tau^{-1/2}),$$
$$\gamma = \frac{1}{\pi}\ln(\varepsilon_2 s_2), \quad \varepsilon_2 = \operatorname{sgn} s_2.$$

Using (10.2.1) in the above expressions completes the proof of the first assertion of Theorem 10.1.

Remark 10.3. The error estimate of the phase of the cotangent can be improved, however, we will not discuss this issue here.

Remark 10.4. Comparing the expression v of (10.2.83) with $\varepsilon_2 y = \frac{3}{2}v' + O(\tau^{-1})$ of (10.2.82), we see that the leading asymptotic term for $v(\tau)$ is differentiable and that the error estimate remains unchanged.

3. Proof of Theorem 10.2: Asymptotics of the Painlevé functions as $x \to -\infty$

Preliminary assumptions and transformations

Here we prove the first two statements of the theorem. The third assertion will be proved in Chapter 11 where the quasi-linear Stokes phenomenon will be discussed.

Using in (10.1.1) the change of the variables

$$\lambda = (-x)^{1/2} z, \quad u = \sqrt{-x}\, v, \quad w = -xy, \tag{10.3.1}$$

we obtain the linear equation $\frac{d\Psi}{d\lambda} = \tau \mathrm{A}_0(z,\tau)$, where $\tau = (-x)^{3/2}$ and

$$\mathrm{A}_0 = -i(4z^2 - 1 + 2v^2)\sigma_3 - 4vz\sigma_2 - 2y\sigma_1. \tag{10.3.2}$$

The eigenvalues $\mu_{1,2} = \pm\mu$ of the matrix $\mathrm{A}_0(z,\tau)$ satisfy the equation

$$\mu^2 = -16\big(z^2 - \frac{1}{4}\big)^2 + 4D, \tag{10.3.3}$$
$$D = y^2 + v^2(1 - v^2).$$

Our main assumption for the coefficients of $\mathrm{A}_0(z,\tau)$ is $D = O(\tau^{-1})$ similar to (10.2.4), or more precisely,

$$c_1\tau^{-1} \le |D| \le c_2\tau^{-1}. \tag{10.3.4}$$

We observe that (10.3.4) is consistent with the following asymptotics of the coefficients v, y:

$$
\begin{aligned}
&(10.3.5) \qquad && y = O(\tau^{-\delta}), \quad v = O(\tau^{-\delta}),\\
&(10.3.6) \qquad && y - i\sigma v\sqrt{1-v^2} = O(\tau^{-\delta}),\\
& && y + i\sigma v\sqrt{1-v^2} = O(\tau^{-1+\delta}),\\
& && \delta \in [0, \frac{1}{2}], \quad \sigma \in \{+,-\}.
\end{aligned}
$$

The parameter δ controls the solutions of the Painlevé equation as $x \to -\infty$. The real regular and the purely imaginary cases correspond to $\delta = 1/2$. Real singular solutions correspond to $\delta = 0$. Below, we concentrate on both cases $\delta = 1/2$ and $\delta = 0$ (for general δ; see [**K5, K8**]).

We introduce the notations

$$
\begin{aligned}
& a_3 = -i\big(4z^2 - 1 + 2v^2\big),\\
& a_+ = i4vz - 2y, \quad a_- = -i4vz - 2y,\\
(10.3.7) \qquad & z_+ = \frac{y}{2iv}.
\end{aligned}
$$

The assumption (10.3.4) means that the asymptotic location of the turning points, as well as the Stokes graph structure coincide with those described in Chapter 9 (see Figure 9.1): four turning points form two pairs approaching $\pm 1/2$, i.e., $z_{1,2} = \frac{1}{2} + O(\tau^{-1/2})$, $z_{3,4} = -\frac{1}{2} + O(\tau^{-1/2})$. The function $\mu(z)$ is defined on the complex z-plane cut along the two small segments $[z_1, z_2]$ and $[z_3, z_4]$, and the branch is fixed by the asymptotics $\mu(z) \to i(4z^2 - 1)$ as $z \to \infty$.

Asymptotics of the Stokes matrices

Using the approach described in detail in the previous section, we match the WKB approximations with the local approximate Ψ-functions near the coupling turning points $z_{1,2} = \frac{1}{2} + O(\tau^{-1/2})$. In this way we find the following asymptotic expression for the product of the Stokes matrices,

$$
(10.3.8) \qquad \mathrm{S}_0\mathrm{S}_1 = \begin{pmatrix} 1 - s_1 s_3 & -s_3 \\ s_1 & 1 \end{pmatrix} = \Psi_0^{-1}(\lambda)\Psi_2(\lambda)
$$

$$
= \begin{pmatrix} -e^{-2\pi i\rho} & -i\sqrt{2\pi}\frac{2^{-\rho}}{\Gamma(\frac{1}{2}-\rho)}e^{-2\tau F_1} \\ -i\sqrt{2\pi}\frac{2^{\rho}e^{-i\pi\rho}}{\Gamma(\frac{1}{2}+\rho)}e^{2\tau F_1} & 1 \end{pmatrix}\big(I + O(\tau^{-1})\big),
$$

where

$$
\begin{aligned}
& F_1 = \lim_{z \to +\infty}\Big\{\int_{\hat{z}_1}^{z} \nu_0\, dz + c_1 - i\big(\frac{4}{3}z^3 - z\big) + \frac{1}{2\tau}\ln\frac{2\mu}{a_+}\Big\},\\
& \nu_0^2 = \mu^2 + \frac{1}{\tau}\big(a_3' - a_3\frac{a_+'}{a_+}\big) - \frac{1}{2\tau^2}\Big(\frac{a_+''}{a_+} - \frac{3}{2}\big(\frac{a_+'}{a_+}\big)^2\Big),\\
(10.3.9) \qquad & c_1 = \frac{\rho_0}{2}\ln\big(e^{-i\pi}\frac{\tau\rho_0}{2e}\big),
\end{aligned}
$$

and $\hat{z}_1 = \frac{1}{2} + O(\tau^{-1/2})$ is a secondary turning point, i.e., a zero of $\nu_0(z)$, $\rho = \tau\rho_0$ and ρ_0 is defined by equation (10.2.36) with $\hat{z}_3$ and $\hat{z}_4$ replaced by $\hat{z}_1$ and $\hat{z}_2$, respectively.

ASYMPTOTICS OF THE PHASE INTEGRAL F_1

To compute the asymptotics as $\tau \to \infty$ of the phase integral F_1, we divide the integration path in two segments, $(\hat{z}_1, \tilde{z}] \cup (\tilde{z}, +\infty)$, where $\tilde{z} = \frac{1}{2} + c'\tau^{-\frac{1}{2}+\epsilon}$ for some $\epsilon \in \left(0, \frac{1}{2}\right)$. For ν_0 on the first segment of the integration path, we use the representation (10.2.31),

$$\nu_0^2(z) = -16(z - \hat{z}_1)(z - \hat{z}_2)\big(1 + O(z - \hat{z}_1) + O(\tau^{-1})\big). \tag{10.3.10}$$

Assuming $z_+ \neq 1/2$ for $\delta = 1/2$ (if $\delta = 0$ the assumption $z_+ = \frac{1}{2}$ contradicts the estimates (10.3.4), (10.3.5), (10.3.6)) we find

$$\hat{z}_{1,2} = \frac{1}{2} \pm \frac{1}{2}\sqrt{D + \frac{i}{\tau}(z_+ - \frac{1}{2}) + \frac{a_3(z_+)}{4\tau(z_+ - \frac{1}{2})}} + O(\tau^{-1}). \tag{10.3.11}$$

Using (10.3.11) in (10.2.36), we find

$$\rho_0 = -\frac{i}{2}(\hat{z}_2 - \hat{z}_1)^2 + O(\tau^{-3/2}) = -\frac{i}{2}\Big(D + \frac{i}{\tau}(z_+ - \frac{1}{2}) + \frac{a_3(z_+)}{4\tau(z_+ - \frac{1}{2})}\Big) + O(\tau^{-3/2}), \tag{10.3.12}$$

and

$$\int_{\hat{z}_1}^{\tilde{z}} \nu_0\,dz = 2i\zeta^2 + \rho_0 \ln(\tau^{1/2}\zeta) - \frac{\rho_0}{2}\ln(e^{-i3\pi/2}\frac{\tau\rho_0}{8e}) + O(\tau^{-1}\zeta) + O(\tau^{-2}\zeta^{-2}), \tag{10.3.13}$$

where

$$\zeta = \tilde{z} - \frac{1}{2}.$$

Expanding with respect to $\rho_0 = O(\tau^{-1})$ the integral of ν_0 along $(\tilde{z}, z)$ as $z \to \infty$, we find

$$\begin{aligned}\textstyle\int_{\tilde{z}}^{z} \nu_0\,dz &= i\big(\tfrac{4}{3}z^3 - z\big) - \tfrac{1}{2\tau}\ln z - 2i\zeta^2 - \rho_0 \ln\zeta \\ &+ \tfrac{i}{3} + \tfrac{a_3(z_+)}{2\tau\mu(z_+)}\ln\big(\tfrac{1}{2} - z_+\big) + O(\tau^{-1}z^{-1}) + O(\tau^{-1}\zeta) + O(\zeta^3) + O(\tau^{-2}\zeta^{-2}).\end{aligned} \tag{10.3.14}$$

Up to this point, the case $\delta = 0$ does not differ from the case $\delta = 1/2$. The difference emerges from the different choice of the sign in the equation $\mu(z_+) = \pm a_3(z_+)$. Namely,

$$\frac{a_3(z_+)}{\mu(z_+)} = \begin{cases} 1, & \delta = 0, \\ -1, & \delta = 1/2. \end{cases} \tag{10.3.15}$$

Using (10.3.13) and (10.3.14) in (10.3.9), and taking into account (10.3.15), we find

$$F_1 = \frac{i}{3} + \frac{\rho_0}{2}\ln(4\tau) + i\frac{\pi\rho_0}{4} + H + O(\tau^{-1}\zeta) + O(\zeta^3) + O(\tau^{-2}\zeta^{-2}),$$

$$H = \begin{cases} \frac{1}{2\tau}\ln\big(\frac{1}{v} + \frac{iy}{v^2}\big), & \delta = 0, \\ -\frac{1}{2\tau}\ln\frac{1}{4}(v + iy), & \delta = 1/2. \end{cases} \tag{10.3.16}$$

The least error estimate $O(\tau^{-6/5})$ in the above expression correspond to the choice $\epsilon = 1/5$, i.e., $\zeta = c'\tau^{-3/10}$.

ASYMPTOTICS OF THE PAINLEVÉ FUNCTION AS $x \to -\infty$

First, consider the case of $\delta = 1/2$. In this case, the parameter ρ_0 (10.3.12) can be expressed as

$$\rho_0 = -\frac{iD}{2} - \frac{1}{2\tau} + O(\tau^{-2}).$$

Hence

$$(v+iy)(v-iy) = D + O(\tau^{-2}) = \frac{2i}{\tau}\big(\tau\rho_0 + \frac{1}{2}\big). \tag{10.3.17}$$

Using elementary transformations and taking into account (10.3.17), equation (10.3.8) implies that

$$s_1 = \frac{2\sqrt{\pi}\, e^{-i\frac{\pi}{2}(\rho+\frac{1}{2})}}{\sqrt{\tau}(v+iy)\Gamma(\frac{1}{2}+\rho)} e^{i\frac{2}{3}\tau+(\rho+\frac{1}{2})\ln 8\tau - i\frac{\pi}{4}}\big(1+O(\tau^{-1/5})\big),$$

$$s_3 = \frac{2\sqrt{\pi}\, e^{-i\frac{\pi}{2}(\rho+\frac{1}{2})}}{\sqrt{\tau}(v-iy)\Gamma(-\frac{1}{2}-\rho)} e^{-i\frac{2}{3}\tau-(\rho+\frac{1}{2})\ln 8\tau + i\frac{\pi}{4}}\big(1+O(\tau^{-1/5})\big),$$

$$1 - s_1 s_3 = e^{-2\pi i(\rho+\frac{1}{2})}\big(1+O(\tau^{-1})\big), \quad \rho = \tau\rho_0. \tag{10.3.18}$$

Since the Stokes multipliers do not depend on τ, we obtain from the last of equations (10.3.18), $\rho = const + O(\tau^{-1})$. Then, the first two equations of the same system yield $\mathrm{Re}\,(\rho+\frac{1}{2}) = O(\tau^{-1})$. Therefore, the third equation together with the isomonodromy condition imply

$$\arg(1 - s_1 s_3) = 0. \tag{10.3.19}$$

Solving the system (10.3.18) for v we find,

$$v = \tau^{-1/2}\big(ae^{ig} + be^{-ig} + O(\tau^{-1/5})\big), \tag{10.3.20}$$

$$a = \frac{\sqrt{\pi}\, e^{\frac{\pi\beta}{2}}}{s_1\Gamma(i\beta)}, \qquad b = \frac{\sqrt{\pi}\, e^{\frac{\pi\beta}{2}}}{s_3\Gamma(-i\beta)}, \qquad ab = -\beta/2,$$

$$g = \frac{2\tau}{3} + \beta\ln 8\tau - \frac{\pi}{4},$$

$$\beta = -i\big(\rho + \frac{1}{2}\big) = \frac{1}{2\pi}\ln(1 - s_1 s_3) \in \mathbb{R}. \tag{10.3.21}$$

Regarding the above asymptotics we note the following: First, by solving the same system for y, we that the asymptotics (10.3.20) is differentiable with the error estimate remains unchanged. Second, the error estimate in the above asymptotics can be significantly improved, using standard tools such as nonlinear integral equations.

Let us now consider the case $\delta = 0$. Equation (10.3.8) implies

$$s_1 = \frac{\sqrt{2\pi}\, e^{-i\frac{\pi\rho}{2}}}{(\frac{1}{v} - \frac{iy}{v^2})\Gamma(\frac{1}{2}+\rho)} e^{i\frac{2}{3}\tau+\rho\ln 8\tau - i\frac{\pi}{2}}\big(1+O(\tau^{-1/5})\big),$$

$$s_3 = \frac{\sqrt{2\pi}\, e^{-i\frac{\pi\rho}{2}}}{(\frac{1}{v} + \frac{iy}{v^2})\Gamma(\frac{1}{2}-\rho)} e^{-i\frac{2}{3}\tau-\rho\ln 8\tau + i\frac{\pi}{2}}\big(1+O(\tau^{-1/5})\big),$$

$$s_1 s_3 - 1 = e^{-2\pi i\rho}\big(1+O(\tau^{-1})\big), \quad \rho = \tau\rho_0, \tag{10.3.22}$$

where we have used the fact that if $\delta = 0$, then

$$\big(\frac{1}{v} + \frac{iy}{v^2}\big)\big(\frac{1}{v} - \frac{iy}{v^2}\big) = 1 + O(\tau^{-1}).$$

The last equation of (10.3.22) together with the fact that $1-s_1s_3$ is independent of τ, imply that $\rho = const + O(\tau^{-1})$. Since s_1 and s_3 are independent of τ it follows that $\operatorname{Re}\rho = O(\tau^{-1})$. Hence $\arg(s_1s_3-1) = O(\tau^{-1})$, and since Stokes multipliers are independent of τ, we find

$$\arg(s_1s_3-1) = 0. \tag{10.3.23}$$

In order to find the asymptotics of the Painlevé function, all that remains is to solve (10.3.22) for v:

$$\frac{1}{v}-\frac{iy}{v^2} = ae^{ig}+O(\tau^{-1/5}), \quad \frac{1}{v}+\frac{iy}{v^2} = be^{-ig}+O(\tau^{-1/5}), \tag{10.3.24}$$

where

$$a = \frac{\sqrt{2\pi}\,e^{\frac{\pi\beta}{2}}}{s_1\Gamma(\frac{1}{2}+i\beta)}, \quad b = \frac{\sqrt{2\pi}\,e^{\frac{\pi\beta}{2}}}{s_3\Gamma(\frac{1}{2}-i\beta)}, \quad ab=1,$$

$$g = \frac{2}{3}\tau+\beta\ln 8\tau-\frac{\pi}{2},$$

$$\beta = -i\rho = \frac{1}{2\pi}\ln(s_1s_3-1)\in\mathbb{R}. \tag{10.3.25}$$

Adding the two equations (10.3.24), we obtain

$$v = \frac{2}{ae^{ig}+be^{-ig}+O(\tau^{-1/5})}. \tag{10.3.26}$$

By subtracting equations (10.3.24), it is easy to establish the differentiability of the leading asymptotic term (with the error estimate remains unchanged).

Repeating the Brouwer Fixed Point Theorem arguments used in Section 6.1 of Chapter 7, we can prove that equation (10.3.26) indeed provides the rigorous asymptotics of the relevant Painlevé function.

The real reduction

The real solutions are characterized by the condition $s_3=\overline{s}_1$ (if $|s_1|\neq 1$, then the condition $s_2=\overline{s}_2$ follows from the equation $s_1-s_2+s_3+s_1s_2s_3=0$). In terms of the parameters a,b defined in (10.3.21) and (10.3.25), the reality condition takes the form $b=\overline{a}$. Conditions (10.3.19) and (10.3.23) turn into

$$\begin{cases} |s_1|<1, & \delta = 1/2, \\ |s_1|>1, & \delta = 0, \end{cases} \tag{10.3.27}$$

respectively. The asymptotics of the real Painlevé function is given by the following equations:

$$|s_1|<1: \qquad v = \tau^{-1/2}\sqrt{-2\beta}\cos\left(\frac{2\tau}{3}+\beta\ln 8\tau+\phi\right)+O(\tau^{-\frac{7}{10}}),$$

$$\beta = \frac{1}{2\pi}\ln(1-|s_1|^2)<0, \quad \phi = -\frac{\pi}{4}-\arg\Gamma(i\beta)-\arg s_1. \tag{10.3.28}$$

$$|s_1|>1: \qquad v = \frac{1}{\sin\left(\frac{2}{3}\tau+\beta\ln 8\tau+\phi\right)+O(\tau^{-1/5})},$$

$$\beta = \frac{1}{2\pi}\ln(|s_1|^2-1), \quad \phi = -\arg\Gamma\left(\frac{1}{2}+i\beta\right)-\arg s_1. \tag{10.3.29}$$

Using the transformations (10.3.1), we complete the proof of the first and second assertions of Theorem 10.2.

CHAPTER 11

PII Quasi-linear Stokes Phenomenon

0. Introduction

In this chapter we will study those particular asymptotic properties of the second Painlevé transcendent which can be viewed as a generalization to the nonlinear case of the classical Stokes phenomenon. This means that we will analyze certain classes of solutions of the second Painlevé equation, which share the same asymptotic behavior in specific sectors of the complex x-plane. Our goal will be to describe the jump that occurs on the Stokes lines of the asymptotic form of the solution. The main result of this chapter is the following: The parametrization of the solutions of the Painlevé equation via the relevant Riemann-Hilbert monodromy data, allow us to control the exponentially vanishing terms of the asymptotics and even to compute the relevant coefficients of these terms explicitly.

Another purpose of this chapter is to study the general case of PII equation, namely

$$u_{xx} = xu + 2u^3 - \alpha, \qquad \alpha = \text{const}.$$

The associated RH problem is more complicated (cf. Chapter 5); it includes the complex number α as an additional parameter to the set of Stokes multipliers, $s_1, \ldots, s_6$, which parametrize the solution $u(x)$. This leads to a number of new technical issues which will be discussed in this chapter.

The outline of the chapter is: We will first recall the classical Stokes phenomenon using the Airy equation as an illustrative example. Then we will explain how this phenomenon manifests itself into a *quasi-linear* Stokes phenomenon for a family of classical solutions of equation PII. Finally, we will describe rigorously the quasi-linear Stokes phenomenon for a special type of degenerate but nonclassical solutions of equation PII. Our main tool will be the nonlinear steepest descent method. We will show that for the solutions analyzed, the associated Riemann-Hilbert jump graph can be decomposed into a union of two or more branches; one of them generates the dominant asymptotic contribution while the others generate exponentially small perturbations. In this way the recessive term can be effectively distinguished from the dominant background. Once again, we want to emphasize the methodological similarity of the Riemann-Hilbert approach to its linear counterpart — the method of contour integrals; indeed, it is the decomposition of the contours of integrations which form the basis of the study of the Stokes phenomenon in the classical special functions.

As a by-product of the analysis of this chapter we will also obtain the proof of the statement 2 of Theorem 10.1 and the proof of the statement 3 of Theorem 10.2 of the preceding chapter and hence complete the proofs of these theorems.

1. Linear Stokes phenomenon revisited

We recall that the classical Airy function $\mathrm{Ai}\,(x)$ is defined as the unique solution of the Airy equation,

$$u_{xx} = xu, \tag{11.1.1}$$

which satisfies the asymptotic condition,

$$\mathrm{Ai}\,(x) \sim \tfrac{1}{2\sqrt{\pi}} x^{-1/4} e^{-\frac{2}{3}x^{3/2}}, \quad x \to \infty, \quad \arg x = 0. \tag{11.1.2}$$

The Airy function admits the integral representation,

$$\mathrm{Ai}\,(x) = \frac{1}{2\pi i} \int_{L_1} \exp\left(\frac{1}{3}\lambda^3 - x\lambda\right) d\lambda, \tag{11.1.3}$$

where L_1 is any contour which begins at infinity in the sector $-\pi/2 < \arg\lambda < -\pi/6$ and ends at infinity in the sector $\pi/6 < \arg\lambda < \pi/2$. Using this representation, one can derive the following (differentiable in x) asymptotics of the function $\mathrm{Ai}\,(x)$ at the neighborhood of $x = \infty$,

$$\begin{aligned} \mathrm{Ai}\,(x) = \tfrac{1}{2\sqrt{\pi}} x^{-1/4} e^{-\frac{2}{3}x^{3/2}} \Big[&\textstyle\sum_{n=0}^{N} (-1)^n 3^{-2n} \frac{\Gamma(3n+\frac{1}{2})}{\Gamma(\frac{1}{2})(2n)!} x^{-\frac{3n}{2}} \\ &+ O\big(x^{-\frac{3}{2}(N+1)}\big)\Big], \\ & x \to \infty, \quad -\pi < \arg x < \pi; \end{aligned} \tag{11.1.4}$$

$$\begin{aligned} \mathrm{Ai}\,(x) = \tfrac{1}{2\sqrt{\pi}} x^{-1/4} e^{-\frac{2}{3}x^{3/2}} \Big[&\textstyle\sum_{n=0}^{N} (-1)^n 3^{-2n} \frac{\Gamma(3n+\frac{1}{2})}{\Gamma(\frac{1}{2})(2n)!} x^{-\frac{3n}{2}} \\ &+ O\big(x^{-\frac{3}{2}(N+1)}\big)\Big] \\ - \tfrac{i}{2\sqrt{\pi}} x^{-1/4} e^{\frac{2}{3}x^{3/2}} \Big[&\textstyle\sum_{n=0}^{N} 3^{-2n} \frac{\Gamma(3n+\frac{1}{2})}{\Gamma(\frac{1}{2})(2n)!} x^{-\frac{3n}{2}} \\ &+ O\big(x^{-\frac{3}{2}(N+1)}\big)\Big] \\ & x \to \infty, \quad -\frac{5\pi}{3} < \arg x < -\frac{\pi}{3}. \end{aligned} \tag{11.1.5}$$

Equation (11.1.4) implies, in particular, that

$$\mathrm{Ai}\,(x) \sim \tfrac{1}{2\sqrt{\pi}} x^{-1/4} e^{-\frac{2}{3}x^{3/2}}, \tag{11.1.6}$$

as

$$x \to \infty, \quad -\pi < \arg x \le -\frac{\pi}{3}.$$

At the same time, equation (11.1.5) yields

$$\mathrm{Ai}\,(x) \sim \tfrac{1}{2\sqrt{\pi}} x^{-1/4} e^{-\frac{2}{3}x^{3/2}} - \tfrac{i}{2\sqrt{\pi}} x^{-1/4} e^{\frac{2}{3}x^{3/2}}, \tag{11.1.7}$$

as

$$x \to \infty, \quad -\pi \le \arg x < -\frac{\pi}{3}.$$

These formulae imply that the coefficient of the recessive in the sector $(-\pi, -\pi/3)$ exponential term $x^{-1/4}e^{\frac{2}{3}x^{3/2}}$, jumps from $-\frac{i}{2\sqrt{\pi}}$ to 0 as the argument of x changes continuously from $-\pi$ to $-\pi/3$. Note that along the lines $\arg x = -\pi$ and $\arg x = -\pi/3$ the asymptotics is oscillatory, and the coefficients of both the exponential terms of the asymptotics are uniquely and unambiguously determined. These lines are called *Stokes' lines*, and the jump of the coefficient of the recessive term is referred to as the *(linear) Stokes phenomenon.* An equivalent way to describe the Stokes' phenomenon is as follows.

Denote

$$\mathrm{Ai}\,(x) \equiv u_1(x),$$

and let us introduce two more solutions of the Airy equation (11.1.1), which we will denote as $u_2(x)$ and $u_3(x)$, and which are uniquely defined by the asymptotic conditions,

$$u_2(x) \sim \tfrac{1}{2\sqrt{\pi}} x^{-1/4} e^{\frac{2}{3}x^{3/2}}, \quad x \to \infty, \quad \arg x = -\tfrac{2\pi}{3}, \tag{11.1.8}$$

and

$$u_3(x) \sim \tfrac{1}{2\sqrt{\pi}} x^{-1/4} e^{-\frac{2}{3}x^{3/2}}, \quad x \to \infty, \quad \arg x = -\tfrac{4\pi}{3}. \tag{11.1.9}$$

The solutions $u_{2,3}(x)$ admit the following integral representations,

$$u_2(x) = -\frac{1}{2\pi}\int_{L_2} \exp\left(\frac{1}{3}\lambda^3 - x\lambda\right) d\lambda, \tag{11.1.10}$$

where L_2 is any contour which begins at infinity in the sector $\pi/6 < \arg\lambda < \pi/2$ and ends at infinity in the sector $5\pi/6 < \arg\lambda < 7\pi/6$, and

$$u_3(x) = -\frac{1}{2\pi i}\int_{L_3} \exp\left(\frac{1}{3}\lambda^3 - x\lambda\right) d\lambda, \tag{11.1.11}$$

where L_3 is any contour which begins at infinity in the sector $5\pi/6 < \arg\lambda < 7\pi/6$ and ends at infinity in the sector $3\pi/2 < \arg\lambda < 11\pi/6$. Using these integral representations, one can derive for the solutions $u_{2,3}(x)$ the following full asymptotic series similar to the series (11.1.4):

$$\begin{aligned} u_2(x) = \tfrac{1}{2\sqrt{\pi}} x^{-1/4} e^{\frac{2}{3}x^{3/2}} \Big[&\textstyle\sum_{n=0}^{N} 3^{-2n} \frac{\Gamma(3n+\frac{1}{2})}{\Gamma(\frac{1}{2})(2n)!} x^{-\frac{3n}{2}} \\ &+ O\big(x^{-\frac{3}{2}(N+1)}\big)\Big], \\ x \to \infty, \quad -\frac{5\pi}{3} &< \arg x < \frac{\pi}{3}, \end{aligned} \tag{11.1.12}$$

and

$$\begin{aligned} u_3(x) = \tfrac{1}{2\sqrt{\pi}} x^{-1/4} e^{-\frac{2}{3}x^{3/2}} \Big[&\textstyle\sum_{n=0}^{N} (-1)^n 3^{-2n} \frac{\Gamma(3n+\frac{1}{2})}{\Gamma(\frac{1}{2})(2n)!} x^{-\frac{3n}{2}} \\ &+ O\big(x^{-\frac{3}{2}(N+1)}\big)\Big], \\ x \to \infty, \quad -\frac{7\pi}{3} &< \arg x < -\frac{\pi}{3}. \end{aligned} \tag{11.1.13}$$

Note that the asymptotic series for the solution $u_3(x)$ is *exactly* the same as the series (11.1.4) for the solution $u_1(x)$. Since

$$\int_{L_1+L_2+L_3} \exp\left(\frac{1}{3}\lambda^3 - x\lambda\right) d\lambda = 0,$$

we have the following relation between the solutions $u_1(x)$, $u_2(x)$, and $u_3(x)$,

$$u_1 - u_3 + iu_2 = 0. \tag{11.1.14}$$

The Stokes' phenomenon can be now identified with the following relation,

(11.1.15)
$$u_1(x) - u_3(x) = -\frac{i}{2\sqrt{\pi}}x^{-1/4}e^{\frac{2}{3}x^{3/2}}\Big(1 + O(x^{-\frac{3}{2}})\Big),$$
$$x \to \infty, \quad -\pi < \arg x < -\frac{\pi}{3}.$$

This equation gives an explicit form of the exponentially small jump occurring between the solutions $u_1(x)$ and $u_3(x)$, which have *exactly the same* exponentially increasing asymptotics in the sector $(-\pi, -\pi/3)$,

$$u_{1,3}(x) \sim u_f(x), \quad x \to \infty, \quad -\pi < \arg x < -\frac{\pi}{3}.$$

Here, $u_f(x)$ is the formal series,

(11.1.16)
$$u_f(x) = \frac{1}{2\sqrt{\pi}}x^{-1/4}e^{-\frac{2}{3}x^{3/2}}\sum_{n=0}^{\infty}(-1)^n 3^{-2n}\frac{\Gamma(3n+\frac{1}{2})}{\Gamma(\frac{1}{2})(2n)!}x^{-\frac{3n}{2}}.$$

We also note that the equation,

(11.1.17)
$$u(x;s) = u_1(x) + su_2(x),$$

defines a one-parameter family of the solutions of (11.1.1), all of which have the same asymptotic expansion in the sector $(-\pi, -\pi/3)$:

(11.1.18)
$$u(x;s) \sim u_f(x),$$
$$x \to \infty, \quad -\pi < \arg x < -\frac{\pi}{3}, \quad \forall s.$$

The particular values of the parameter s, 0 and i, correspond to the unique solutions, i.e., $u_1(x)$ and $u_3(x)$, which have the asymptotic representation (11.1.18) in the large sectors, i.e., in $-\pi < \arg x < \pi$ and $-7\pi/3 < \arg x < -\pi/3$, respectively.

REMARK 11.1. We have based our exposition of the Stokes' phenomenon on the integral representations for Airy functions. At the same time, in order to obtain rigorously all the asymptotic results described above, we need only the existence of the solutions $u_1(x)$, $u_2(x)$, and $u_3(x)$ (which have the asymptotic expansions (11.1.4), (11.1.12), and (11.1.13), respectively), as well as the explicit relation (11.1.14) between them. The existence result follows from the general analysis of the integral Volterra equations equivalent to (11.1.1) (cf. the discussion of the canonical solutions of the generic linear system of ODEs in Section 1.6 of Chapter 1). The connection formulae (11.1.14), which we have derived using the integral representations (11.1.3), (11.1.10) and (11.1.11), can be alternatively derived as follows. Using the invariance of equation (11.1.1) with respect to the transformation $x \to e^{i\frac{2\pi}{3}}x$, as well as uniqueness, it follows that

(11.1.19)
$$u_2(x) = e^{\frac{i\pi}{6}}u_1(e^{i\frac{2\pi}{3}}x) \quad \text{and} \quad u_3(x) = e^{\frac{i\pi}{3}}u_1(e^{i\frac{4\pi}{3}}x).$$
$$(u_1(z) = \operatorname{Ai}(x))$$

The solutions $u_2(x)$ and $u_3(x)$ are linearly independent (in the common sectors one grows exponentially while the another decay exponentially), hence we have

(11.1.20)
$$u_1(x) + c_2u_2(x) + c_3u_3(x) \equiv 0,$$

for some constants c_2 and c_3. The key fact now is that the symmetry relations (11.1.19) allow us to transform the identity (11.1.20) into the algebraic system,

(11.1.21)
$$1 + c_2e^{i\frac{\pi}{6}} + c_3e^{i\frac{\pi}{3}} = 0,$$

(11.1.22)
$$1 + c_2e^{i\frac{5\pi}{6}} + c_3e^{i\frac{5\pi}{3}} = 0,$$

and hence (11.1.14) follows. We note that a similar invariance is valid for any equation of the form $u_{xx} = x^n u$, $n \in \mathbb{N}$, but not for more general special functions of the confluent hypergeometric type. For the latter functions it seems that the use of the relevant integral representations is unavoidable (if one wants to evaluate the Stokes's multipliers explicitly). For instance, in the case of the parabolic cylinder functions,

$$u_{xx} = \left(\frac{x^2}{4} - \nu - \frac{1}{2}\right) u,$$

the canonical at infinity solutions (which are the analogs of the solutions $u_1(x)$, $u_2(x)$, and $u_3(x)$) are the functions $D_\nu(x)$, $D_{-\nu-1}(ix)$, and $D_\nu(-x)$. Hence the system (11.1.21)–(11.1.22) is replaced by the system,

$$D_\nu(0) + c_2 D_{-\nu-1}(0) + c_3 D_\nu(0) = 0, \tag{11.1.23}$$

$$D'_\nu(0) + ic_2 D'_{-\nu-1}(0) - c_3 D'_\nu(0) = 0. \tag{11.1.24}$$

In order to determine from this system the connection constants c_2 and c_3 we need $D_\nu(0)$ (it does not cancel out anymore), which can be obtained using the relevant contour integral representation[1].

Stokes phenomenon and the coefficient asymptotics

The Stokes phenomenon is intimately related to the fact that the series in the expansion (11.1.4) is divergent. Indeed, the asymptotics of the coefficients of this series and the Stokes phenomenon are very closely connected. In what follows we will explain this connection by evaluating the asymptotic behavior of the coefficients of the series (11.1.4) without using their explicit expressions, but by *using only the Stokes' phenomenon.* Later we will see that the same technique also works for the nonlinear case of the Painlevé equations, where there does not exist an explicit expression for the coefficients of the relevant asymptotic series.

Let us introduce a piecewise holomorphic function $\hat{u}(t)$,

$$\hat{u}(t) = 2\sqrt{\pi}\,(-1)^n u_{1+2n}(t^4) e^{\frac{2}{3}t^6}, \tag{11.1.25}$$

$$\arg t \in \left(-\tfrac{\pi}{6} - \tfrac{\pi}{3}n, \tfrac{\pi}{6} - \tfrac{\pi}{3}n\right), \quad n = 0, \pm 1, \pm 2, -3,$$

where the sequence of the solutions $u_n(x)$ of the Airy equation is defined by the relation (cf (11.1.19)),

$$u_{n+1}(x) = e^{i\frac{2\pi n}{3}} u_1\left(e^{i\frac{2\pi n}{3}} x\right).$$

This function has the uniform asymptotics expansion at infinity,

$$\hat{u}(t) \sim t^{-1} \sum_{n=0}^{\infty} a_n t^{-6n},$$

[1] We refer the reader to Section 2.4 of Chapter 2 for more details on the Stokes phenomenon for the parabolic cylinder functions and their integral representation. In particular, for the constants c_2 and c_3 we have the expressions,

$$c_2 = -\frac{\sqrt{2\pi}}{\Gamma(-\nu)} e^{-\frac{1}{2}(\nu+1)\pi i}, \quad c_3 = -e^{-\nu\pi i}.$$

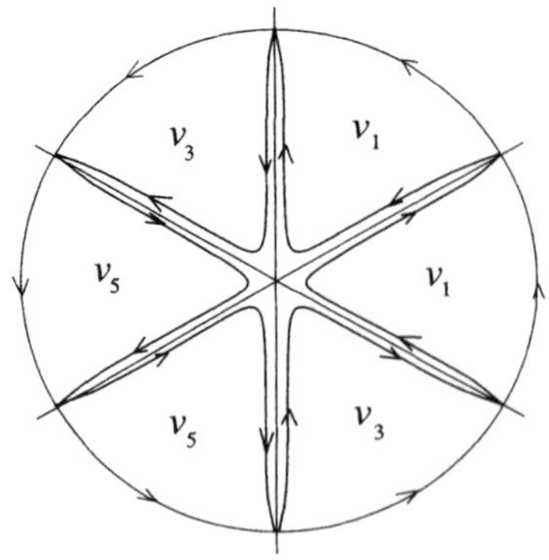

FIGURE 11.1. Contours of integration for $w_N(t)$ and for the Airy functions used in the construction of $\hat{u}(t)$

thus the coefficients a_n are the moments of this function. To compute their asymptotics we introduce the reduced sum, $u^{(N)}(t)$,

$$u^{(N)}(t) = t^{-1} \sum_{n=0}^{N-1} a_n t^{-6n},$$

and the function $w_N(t)$,

$$w_N(t) := \frac{t^{6N}}{2\pi i} \big(\hat{u}(t) - u^{(N)}(t) \big) = \frac{a_N}{2\pi i t} + O(t^{-7}).$$

Integrating this function over a circle around infinity counterclockwise, yields the coefficient a_N. On the other hand, the integral over the circle of large radius can be replaced by the sum of the integrals along the rays $\arg z = \frac{\pi}{6} + \frac{\pi}{3}k$ (see Figure 11.1):

$$\begin{aligned}
a_N &= \frac{1}{i\sqrt{\pi}} \sum_{k=0}^{5} (-1)^{k+1} \int\limits_{e^{-i\frac{5\pi}{6}+i\frac{\pi}{3}k}(0,\infty)} e^{\frac{2}{3}t^6} t^{6N} \big(u_{7-2k}(t^4) + u_{5-2k}(t^4) \big)\, dt \\
&= \frac{1}{i\sqrt{\pi}} \sum_{k=0}^{5} (-1)^{k} \int\limits_{e^{-i\frac{5\pi}{6}+i\frac{\pi}{3}k}(0,\infty)} e^{\frac{2}{3}t^6} t^{6N} u_{6-2k}(t^4)\, dt \\
&= 6\frac{(-1)^N}{\sqrt{\pi}} \int\limits_{(0,\infty)} e^{-\frac{2}{3}t^6} t^{6N} u_1(t^4)\, dt \\
&= \frac{1}{2\pi} (-1)^N (3/4)^N \Gamma(N) \big(1 + O(N^{-1}) \big).
\end{aligned} \tag{11.1.26}$$

In the latter equation, we have only used the leading term of the asymptotics of the Airy function (11.1.4).

REMARK 11.2. The asymptotics (11.1.26) is consistent with the exact value in (11.1.4),

$$a_n = (-1)^n 3^{-2n} \frac{\Gamma(3n+\frac{1}{2})}{\Gamma(\frac{1}{2})(2n)!}. \tag{11.1.27}$$

In fact, evaluating (11.1.26), we find the following integral identity

$$\int_0^\infty e^{-\frac{2}{3}z^{3/2}} z^{\frac{3}{2}(n-\frac{1}{2})} \operatorname{Ai}(z)\, dz = 3^{-2n-1} 2 \frac{\Gamma(3n+\frac{1}{2})}{(2n)!}. \tag{11.1.28}$$

2. General remarks on the Stokes phenomenon for PII

As it was shown in Chapters 7 and 8, the asymptotic behavior as $|x| \to \infty$ of generic solution of PII is given by the Jacobi sine, with its modulus depending on $\arg x$. Along the rays $\arg x = \frac{\pi}{3}k$, $k \in \mathbb{Z}$, the elliptic functions degenerate and the asymptotics are given in terms of trigonometric functions. It was found in [**JoKr1, N1**] that the phase shifts in the arguments of the elliptic functions representing the asymptotics of PII in the sectors between the above rays, are linear combinations of the elliptic periods. The connection formulae relating the coefficients of these linear combinations corresponding to different sectors were obtained in Theorem 7.1 (see also [**K3, N1, KK2, Ki7**]). Similar results for Painlevé IV and Painlevé III equations were published in [**K6, N4**], and for the case of Painlevé III equation, they are reproduced in Chapter 16.

In the seminal paper of Boutroux [**Bou**] (which was mentioned in Chapters 7 and 8), it was found that the Painlevé equations admit a one-parameter family of solutions represented by the sum

$$u = (\text{power series}) + (\text{exponential terms}). \tag{11.2.1}$$

The leading term $u_0(x)$ of the power series satisfies a certain algebraic equation,[2] while the exponential terms decrease (they form so-called transseries). Such solutions can be obtained using a standard perturbation analysis, and are referred to as truncated, or separatrix, or degenerate, or quasi-stationary, or instanton solutions.

In our study of the global asymptotics of the Painlevé functions we will accept the point of view that any discontinuity in the asymptotic form of an analytic function will be considered as a *Stokes phenomenon.* Following this convention, the nontrivial jump across the rays $\arg x = \frac{\pi}{3}n$, $n \in \mathbb{Z}$, in the phase shift of the elliptic asymptotic ansatz for PII, will be called a *nonlinear Stokes phenomenon.* Accordingly, an exponentially small jump in the asymptotics of the degenerate Painlevé function (which will be analyzed below), will be identified as a *quasi-linear Stokes phenomenon.*

The first, heuristic, description of a quasi-linear Stokes phenomenon was obtained in [**K2, K5**]. Important progress in the rigorous analysis of quasi-linear Stokes phenomena in context of Borel summation and resurgent analysis can be found in [**C1**]. These techniques are effective for the analysis of certain nonlinear asymptotic problems. For instance, in [**C2**], O. Costin applied the resurgent approach to find the asymptotic location of the first real pole of the degenerate solution of PI when the amplitude of its exponentially decreasing perturbation becomes large.

[2] In the case of PII, this equation is $2u_0^3 + xu_0 = \alpha$.

Y. Takei in [**Ta1**] proposed two approaches for using the Borel summation for the description of the quasi-linear Stokes phenomenon in PI. The first approach is similar to one of [**C1**] and is based on the study of the Borel transform of the leading power series. Assuming that the Borel series has certain analytic structure on the boundary of its convergence, Takei obtains a simple formula which relates the jump of the recessive term in (11.2.1) with the asymptotics of the Taylor coefficients of the Borel transform. However, the assumption on the analytic structure of the Borel transform remains unjustified. In his second approach, Takei studies the linear differential equation associated to PI in the framework of the isomonodromy method [**JMU**], using the so-called "exact" WKB analysis. This approach is based on the assumption of the Borel summability of the formal "instanton" series for the Painlevé functions which, for many cases, is still an open problem; see [**Ta2**].

The Borel summability of a formal series depends on the asymptotics of its coefficients. The asymptotics of the coefficients of the power series in (11.2.1) for the fifth Painlevé equation, were first found by Basor and Tracy in [**BasT**] up to a common constant factor. The approximate value for the latter was conjectured in [**BasT**] using numerical computations. The exact value of the Basor-Tracy constant, as well as the proof of their asymptotic formula were obtained by Andreev and Kitaev in [**AnK1**], via a combination of the isomonodromy method and of the Borel summation technique.

We emphasize the importance of the papers [**Ta1, AnK1**] as the first exact description of the relation between the asymptotics of the coefficients in the Painlevé transcendent expansions and the quasi-linear Stokes phenomenon.

In [**JoK**], Joshi and Kitaev studied the asymptotics of the leading power series coefficients for the degenerate solution of PI and obtained results about the asymptotics of the transseries coefficients. Their analysis is based on the use of recursion relations and leaves a common factor for asymptotic coefficients undetermined. The authors, however, have proposed the value of this factor indicating that it could be found using the technique presented in [**AnK1**].

A common tool of the approaches mentioned above is the use of the Borel transform. However, a disadvantage of this approach is that one must make certain assumptions about the analytic properties of the Borel transform of the solution.

In this chapter, we will study the quasi-linear Stokes phenomenon for PII,

$$u_{xx} = xu + 2u^3 - \alpha, \quad \alpha = \text{const}, \tag{11.2.2}$$

using the isomonodromy deformation method [**JMU, FN1, IN1**]. Our goal is twofold: (a) to prove rigorously the formal equation (11.2.1) (providing it, simultaneously, with an exact mathematical meaning), and (b) to evaluate the asymptotics of the coefficients of the leading power series in (11.2.1).

We emphasized that we *will not* use the Borel summation at any stage of our investigation. Instead, similarly to [**AnK1**], we fix the solutions of (11.2.2) with the power series asymptotic behavior, by choosing appropriately the monodromy data of the associated linear equation. However, this is the only similarity with the method of [**AnK1**]: The approach of [**AnK1**] is based on the asymptotic analysis of the direct monodromy problem, while our approach is based on the asymptotic analysis of the inverse monodromy problem, i.e., on the relevant Riemann-Hilbert problem (RH) and the nonlinear steepest descent method [**DeZ1**]. In this approach, we associate with the solution (11.2.1) a disjoint RH graph. This allows us to separate effectively the exponentially small terms from the power-like background.

In turn, the explicit evaluation of the exponentially small terms will allow us to calculate the asymptotics of the coefficients of the power-like dominant asymptotic background. In other words, we will follow a nonlinear analogue of the approach explained earlier for the case of Airy function.

3. The RH problem for PII

Here we modify slightly the RH formulation given in Chapter 5.

Consider on the complex plane λ the contour $\gamma = C \cup \rho_+ \cup \rho_- \cup_{k=1}^{6} \gamma_k$ which is the union of the rays $\gamma_k = \{\lambda \in \mathbb{C}\colon\ |\lambda| > r,\ \arg\lambda = \frac{\pi}{6} + \frac{\pi}{3}(k-1)\}$, $k = 1, \ldots, 6$, oriented to infinity, of the clockwise oriented circle $C = \{\lambda \in \mathbb{C}\colon\ |\lambda| = r\}$, and of the two radii $\rho_+ = \{\lambda \in \mathbb{C}\colon\ |\lambda| < r,\ \arg\lambda = \frac{\pi}{6}\}$ and $\rho_- = \{\lambda \in \mathbb{C}\colon\ |\lambda| < r,\ \arg\lambda = -\frac{5\pi}{6}\}$ oriented to the origin. The contour γ divides the complex λ-plane into 8 regions Ω_k, $k \in \{\text{left}, \text{right}, 1, \ldots, 6\}$ which are depicted, together with the contour γ itself, in Figure 11.2.

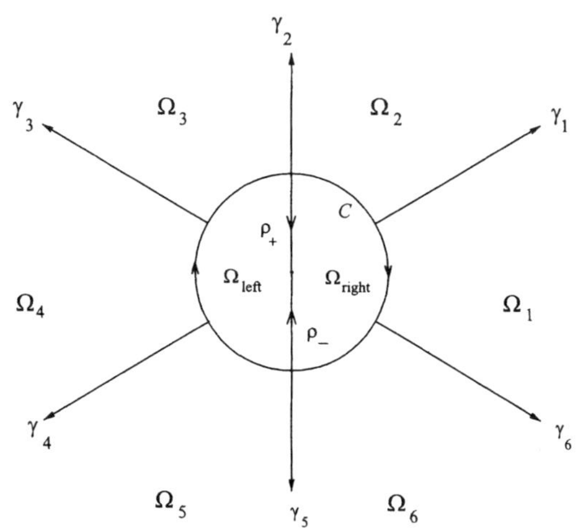

FIGURE 11.2. The Riemann-Hilbert problem graph

Let α be the constant appearing in (11.2.2), and let $s = \{S_1, S_2, S_3, S_4, S_5, S_6, E\}$ be the collection of seven 2×2 complex matrices satisfying the following conditions:

(1) All these matrices are unimodular.

(2) The matrices S_{2k-1} are low triangular while the matrices S_{2k} are upper triangular, i.e.,

$$S_{2k-1} = I + s_{2k-1}\sigma_-, \qquad S_{2k} = I + s_{2k}\sigma_+,$$

where we denote $\sigma_+ = \begin{pmatrix} 0 & 1 \\ 0 & 0 \end{pmatrix}$, and $\sigma_- = \begin{pmatrix} 0 & 0 \\ 1 & 0 \end{pmatrix}$. Moreover, the complex constants s_k are assumed to satisfy the constraints

$$s_{k+3} = -s_k, \qquad s_1 - s_2 + s_3 + s_1 s_2 s_3 = -2\sin\pi\alpha. \tag{11.3.1}$$

(3) Let $\sigma_3 = \begin{pmatrix} 1 & 0 \\ 0 & -1 \end{pmatrix}$, $\sigma_2 = \begin{pmatrix} 0 & -i \\ i & 0 \end{pmatrix}$, $\sigma_1 = \begin{pmatrix} 0 & 1 \\ 1 & 0 \end{pmatrix}$ denote the usual Pauli matrices. Then the following equation takes place,

$$E S_1 S_2 S_3 = \sigma_2 M^{-1} E \sigma_2, \tag{11.3.2}$$

where

$$M = \left(e^{i\pi(\alpha-\frac{1}{2})\sigma_3} + J_\pm\sigma_\pm\right)(-i\sigma_2), \tag{11.3.3}$$

$$J_+ = 0 \quad \text{if} \quad \tfrac{1}{2}+\alpha \notin \mathbb{N}, \quad \text{i.e.} \quad \alpha \notin \left\{\tfrac{1}{2}, \tfrac{3}{2}, \dots\right\},$$

$$J_- = 0 \quad \text{if} \quad \tfrac{1}{2}-\alpha \notin \mathbb{N}, \quad \text{i.e.} \quad \alpha \notin \left\{-\tfrac{1}{2}, -\tfrac{3}{2}, \dots\right\}.$$

We note that the matrix form of the first relation in (11.3.1) is the equation

$$\mathsf{S}_{k+3} = \sigma_2 \mathsf{S}_k \sigma_2, \tag{11.3.4}$$

which, together with (11.3.2), implies the *cyclic* equation

$$\mathsf{S}_1\mathsf{S}_2 \dots \mathsf{S}_6 = \mathsf{E}^{-1}e^{-2\pi i\alpha\sigma_3}\mathsf{E}. \tag{11.3.5}$$

We also note that the second of the relations in (11.3.1) is in fact a consequence of matrix equation (11.3.2).

The RH problem we are concerned with consists of finding a 2×2 matrix-valued function $\Psi(\lambda)$, which is piecewise analytic on the λ-plane, such that:

i) For every $k = 1, 2, \dots, 6$, the restriction, $\Psi_k(\lambda)$, of the function $\Psi(\lambda)$ on the set Ω_k is holomorphic and is continues up to the boundaries of Ω_k, i.e.,

$$\Psi_k(\lambda) := \Psi(\lambda)|_{\Omega_k} \in H(\Omega_k) \cap C(\overline{\Omega}_k), \quad k = 1, 2, \dots, 6. \tag{11.3.6}$$

Also, for any $\epsilon > 0$,

$$\Psi_{\text{left, right}} := \Psi(\lambda)|_{\Omega_{\text{left, right}}} \in H(\Omega_k) \cap C(\overline{\Omega}^{\epsilon}_{\text{left, right}}), \tag{11.3.7}$$

where

$$\Omega^{\epsilon}_{\text{left, right}} := \Omega_{\text{left, right}} \setminus \{\lambda : |\lambda| \le \epsilon\}.$$

ii) The limit $\Psi_-(\lambda)$ of $\Psi(\lambda)$ from the right side of γ and the limit $\Psi_+(\lambda)$ from the left side of γ are related by the equations,

$$\Psi_+(\lambda) = \Psi_-(\lambda)\mathsf{S}(\lambda), \quad \lambda \in \gamma \setminus \{0\}, \tag{11.3.8}$$

where the jump matrix $\mathsf{S}(\lambda)$ is constructed from the matrices of the collection s. Namely, on the rays γ_k the matrix $\mathsf{S}(\lambda)$ is given by the equation

$$\mathsf{S}(\lambda) = \mathsf{S}_k, \quad \lambda \in \gamma_k, \quad k = 1, 2, \dots, 6, \tag{11.3.9}$$

while on the circle C and on the radii $\rho_\pm$, the matrix $\mathsf{S}(\lambda)$ can be obtained from the following jump relations,

$$\begin{aligned} &\Psi_1(\lambda) = \Psi_{\text{right}}(\lambda)\mathsf{E}, \quad \Psi_2(\lambda) = \Psi_{\text{left}}(\lambda)\sigma_2\mathsf{E}\sigma_2\mathsf{S}_3^{-1}\mathsf{S}_2^{-1}, \\ &\Psi_3(\lambda) = \Psi_{\text{left}}(\lambda)\sigma_2\mathsf{E}\sigma_2\mathsf{S}_3^{-1}, \quad \Psi_4(\lambda) = \Psi_{\text{left}}(\lambda)\sigma_2\mathsf{E}\sigma_2, \\ &\Psi_5(\lambda) = \Psi_{\text{right}}(\lambda)\mathsf{E}\mathsf{S}_6^{-1}\mathsf{S}_5^{-1}, \quad \Psi_6(\lambda) = \Psi_{\text{right}}(\lambda)\mathsf{E}\mathsf{S}_6^{-1}, \end{aligned} \tag{11.3.10}$$

and

$$\begin{aligned} &\lambda \in \rho_- \setminus \{0\}: \ \Psi_{\text{left}}(\lambda) = \Psi_{\text{right}}(\lambda)M, \\ &\lambda \in \rho_+ \setminus \{0\}: \ \Psi_{\text{right}}(\lambda) = \Psi_{\text{left}}(\lambda)\sigma_2 M\sigma_2, \end{aligned} \tag{11.3.11}$$

respectively, where the matrix M is defined in equation (11.3.3) above.

iii) At $\lambda = 0$ the function $\Psi(\lambda)$ has a singularity which is specified by the inequality

$$\|\Psi_{\text{right}}(\lambda)\lambda^{-\alpha\sigma_3}\| \le \text{const}, \qquad \lambda \to 0, \tag{11.3.12}$$

where the branch of λ^α is chosen arbitrarily.

iv) At $\lambda = \infty$, the function $\Psi(\lambda)$ has an exponential behavior which is normalized by the asymptotic condition,

$$\Psi(\lambda)e^{\theta\sigma_3} \to I, \qquad \lambda \to \infty, \qquad \theta = i\left(\tfrac{4}{3}\lambda^3 + x\lambda\right). \tag{11.3.13}$$

Let

$$Y(\lambda) := \Psi(\lambda)e^{\theta\sigma_3}. \tag{11.3.14}$$

Following the techniques developed in Chapters 3, 4, and 5 and in Appendix A, one can show that the Riemann-Hilbert problem i)–iv) is uniquely and meromorphically (with respect to x) solvable for any choice of the data s. Furthermore, the solution of the second Painlevé equation (11.2.2) is defined by the relation ([**FN1**]),

$$u(x) = 2\lim_{\lambda\to\infty} \lambda Y_{12}(\lambda). \tag{11.3.15}$$

Moreover, the map

$$\{(s_1, s_2, \dots, s_6) \in \mathbb{C}^3 : \quad s_{k+3} = -s_k, \quad s_1 - s_2 + s_3 + s_1 s_2 s_3 = -2\sin\pi\alpha\}$$

$$\mapsto \{\text{solutions of P2}\} \tag{11.3.16}$$

is a bijection. Therefore, we will use the notation,

$$u(x) \equiv u(x; s|\alpha),$$

to indicate the second Painlevé transcendent parametrized by the RH monodromy data $s \equiv (s_1, s_2, \dots, s_6)$.

Before we proceed further we make the following remarks.

- For the bijectivity of the map (11.3.16) we need to assume that a point of the monodromy surface $s_1 - s_2 + s_3 + s_1 s_2 s_3 = -2\sin\pi\alpha$ is not special, i.e., $(s_1, s_2, s_3, \alpha) \neq \left((-1)^{n+1}, (-1)^n, (-1)^{n+1}, \tfrac{1}{2} + n\right)$, $n \in \mathbb{Z}$. In particular, the point is not special if

$$\alpha - \frac{1}{2} \notin \mathbb{Z}. \tag{11.3.17}$$

  At the end of this section we will outline the extension of the algebraic manifold (11.3.1) needed in order to incorporate its special point.
- Under the condition (11.3.17),

$$M = ie^{i\pi\alpha\sigma_3}\sigma_1, \tag{11.3.18}$$

  and equation (11.3.2) defines the matrix E via the parameters s_k, modulo a left diagonal multiplier. The transformation $\mathsf{E} \mapsto \Lambda\mathsf{E}$ induces the following transformations of the functions $\Psi_{\text{left}}(\lambda)$ and $\Psi_{\text{right}}(\lambda)$: $\Psi_{\text{right}}(\lambda) \mapsto \Psi_{\text{right}}(\lambda)\Lambda^{-1}$ and $\Psi_{\text{left}}(\lambda) \mapsto \Psi_{\text{left}}(\lambda)\sigma_2\Lambda^{-1}\sigma_2$, and this does not affect the Painelvé function $u(x)$.
- The conditions imposed on the data of the RH problem i)–iv), imply that the matrix $\Psi(\lambda)$ is unimodular for all λ, i.e.,

$$\det\Psi(\lambda) = 1, \quad \forall\lambda,$$

  and it satisfies the symmetry relation,

$$\Psi(\lambda) = \sigma_2\Psi(-\lambda)\sigma_2. \tag{11.3.19}$$

- It follows from equations (11.3.7), (11.3.11), and (11.3.12) that the functions $\Psi_{\text{right}}(\lambda)$ and $\Psi_{\text{left}}(\lambda)$ admit the following representations in the neighborhood of the origin,

(11.3.20) $$\Psi_{\text{right}}(\lambda) = \hat{\Psi}_0(\lambda)\lambda^{\alpha\sigma_3}P(\lambda), \quad -\frac{5\pi}{6} \le \arg\lambda \le \frac{\pi}{6},$$
$$\Psi_{\text{left}}(\lambda) = \hat{\Psi}_0(\lambda)\lambda^{\alpha\sigma_3}P(\lambda)\sigma_2 M^{-1}\sigma_2, \quad \frac{\pi}{6} \le \arg\lambda \le \frac{7\pi}{6},$$

where

(11.3.21) $$\hat{\Psi}_0(\lambda) \in H(\Omega_0) \cap C(\overline{\Omega}_0), \quad \Omega_0 = \{\lambda \in \mathbb{C} : \ |\lambda| < r\},$$
$$P(\lambda) = \bigl(I + j_\pm\sigma_\pm \ln\lambda\bigr)e^{\int^x u(x)\,dx\,\sigma_3}, \quad j_\pm = \pm\frac{1}{\pi}e^{\mp i\pi\alpha}e^{\pm 2\int^x u(x)\,dx}J_\pm.$$

In addition, the symmetry (11.3.19) yields the relation,

(11.3.22) $$\sigma_2\hat{\Psi}_0(-\lambda)\sigma_2 = \hat{\Psi}_0(\lambda)i\sigma_1.$$

- It can be easily checked that the jumps (11.3.10) on the circle C and on the radii $\rho_\pm$, are defined in such a way that each of the functions $\Psi_k(\lambda)$ has a holomorphic continuation in the neighborhoods of the relevant node points, i.e., the monodromies at each of the node points on the circle C are all trivial. The point $\lambda = 0$ is the only point where the RH problem is discontinous, and as the result at this point the function $\Psi(\lambda)$ has a power-like singularity described by the equations (11.3.20). This singularity can be actually replaced by an extra jump, so that the resulting RH graph has trivial monodromy at its *all* node points. Indeed, let us replace the parts of the radii $\rho_\pm$ in the neighborhood of the origin by the circle $C_{1/2} = \{\lambda \in \mathbb{C} : |\lambda| = r/2\}$, with the relevant jump matrices given by the equations,

(11.3.23) $$\mathsf{S}(\lambda) = \mathsf{E}_0\lambda^{\alpha\sigma_3}P(\lambda)\sigma_2 M^{-1}\sigma_2,$$

for λ belonging to the left half of $C_{1/2}$, and

(11.3.24) $$\mathsf{S}(\lambda) = \mathsf{E}_0\lambda^{\alpha\sigma_3}P(\lambda),$$

for λ belonging to the right half of $C_{1/2}$. Here,

$$\mathsf{E}_0 = \begin{pmatrix} 1 & 1 \\ -i & i \end{pmatrix},$$

and the branch of the function $\lambda^{\alpha\sigma_3}$ is defined by the condition,

$$-\frac{5\pi}{6} < \arg\lambda < \frac{7\pi}{6}.$$

This modified RH problem is depicted in Figure 11.3. The restriction of the function $\Psi(\lambda)$ on the new domain of its analyticity, i.e., on the interior ω_0 of the circle $C_{1/2}$, is given by the equation,

(11.3.25) $$\Psi(\lambda)|_{\omega_0} = \hat{\Psi}_0(\lambda)\mathsf{E}_0{}^{-1},$$

where $\hat{\Psi}_0(\lambda)$ is the holomorphic factor of the representation (11.3.20). We note that the introduction of the auxiliary matrix multiple E_0 in the equations (11.3.23)–(11.3.25), guarantees the symmetry relation (11.3.19) for

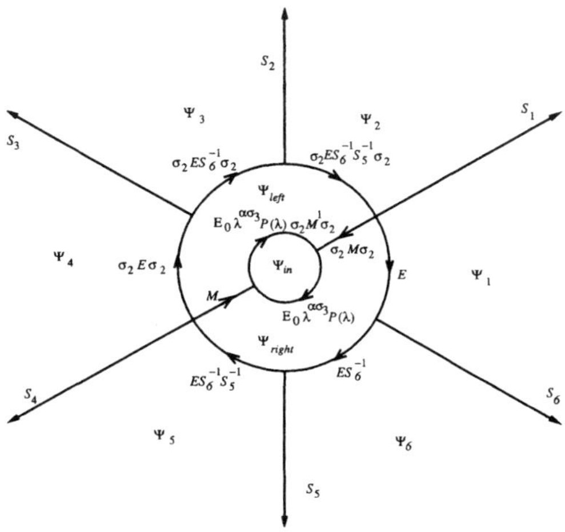

FIGURE 11.3. An equivalent RH setting

the modified Ψ-function. The reformulation of the original RH problem, as a RH problem where all node points have trivial monodromy, implies its meromorphic (with respect to the extra parameter x) solvability (see [**Zh**]; see also Chapters 3 and 5).

- The second Painlevé transcendent $u(x; s|\alpha)$ satisfies the following symmetry relations:

$$u(x;s|\alpha) = -u(x;s^-|-\alpha,), \quad u(x;s|\alpha) = \overline{u}(\overline{x};s^*|\overline{\alpha}),$$

$$u(x;s|\alpha) = e^{i\frac{2\pi}{3}n}u(e^{i\frac{2\pi}{3}n}x;s^{2n}|\alpha), \tag{11.3.26}$$

where bar means the complex conjugation. The induced automorphisms, $s \mapsto s^-$, $s \mapsto s^*$, and $s \mapsto s^m$ of the set of monodromy data, can be easily calculated by lifting the maps $(u,x,\alpha) \to (-u,x,-\alpha)$, $(u,x,\alpha) \to (\overline{u},\overline{x},\overline{\alpha})$, and $(u,x,\alpha) \to (e^{i\frac{2\pi}{3}n}u, e^{i\frac{2\pi}{3}n}x,\alpha)$ to the Ψ-function. Indeed, one finds that (see [**K3**] and Proposition 7.1 of Chapter 7),

$$(s^-)_k = -s_k,$$
$$(s^*)_k = \overline{s}_{4-k},$$
$$(s^m)_k = s_{k+m}, \quad m \in \mathbb{Z}.$$

- The solution $\Psi(\lambda, x)$ of the RH problem i)–iv) satisfies the system of linear differential equations (cf. Chapters 4 and 5),

$$\Psi_\lambda = \mathsf{A}(\lambda)\Psi, \quad \Psi_x = \mathsf{U}(\lambda)\Psi, \tag{11.3.27}$$

where

$$\mathsf{A}(\lambda) = -i(4\lambda^2 + x + 2u^2)\sigma_3 - \left(4u\lambda + \frac{\alpha}{\lambda}\right)\sigma_2 - 2u_x\sigma_1,$$

and

$$\mathsf{U}(\lambda) = -i\lambda\sigma_3 - u\sigma_2.$$

The Painlevé equation (11.2.2) is equivalent to the compatibility condition of the system (11.3.27),

$$\mathsf{A}_x(\lambda) - \mathsf{U}_\lambda(\lambda) + [\mathsf{A}(\lambda), \mathsf{U}(\lambda)] = 0.$$

This implies (compare again Chapters 4 and 5) that the set s above, is the set of monodromy data (S_k are the Stokes' matrices) of the first equation in (11.3.27), and that the Painlevé equation (11.2.2) describes isomonodromy deformations of this equation. It is this setting that allows one to prove that the inverse monodromy map (11.3.16) is indeed a surjection.

- The monodromy parametrization of PII used in this chapter, differs from the one of chapter 5 by the gauge transformation

$$\Psi(\lambda) \to e^{-\frac{i\pi}{4}\sigma_3}\Psi(\lambda).$$

More precisely, let $\Psi_k^{(\infty)}(\lambda)$ and $\Psi^{(0)}(\lambda)$ be the canonical solutions introduced in Chapter 5. Then we have the relations

$$\Psi_k^{(\infty)}(\lambda) = e^{-\frac{i\pi}{4}\sigma_3}\Psi_k(\lambda)e^{\frac{i\pi}{4}\sigma_3}, \quad \Psi^{(0)}(\lambda) = e^{-\frac{i\pi}{4}\sigma_3}\Psi_{\text{right}}(\lambda)\sigma_1, \tag{11.3.28}$$

which, in turn, generate the following relations between the respective monodromy data,

$$\mathsf{S}_k{}^{(\mathrm{V})} = e^{-\frac{i\pi}{4}\sigma_3}\mathsf{S}_k e^{\frac{i\pi}{4}\sigma_3}, \quad \mathsf{E}^{(\mathrm{V})} = \sigma_1\mathsf{E}e^{\frac{i\pi}{4}\sigma_3}, \tag{11.3.29}$$

where the data of Chapter 5 are marked by the superscript (v).

- Since the jump matrix $\mathsf{S}(\lambda)$ is a piecewise constant, each of the functions $\Psi_k(\lambda)$, $k \in \{\text{left}, \text{right}, 1, \ldots, 6\}$ admits an analytic continuation in the neighborhood of the respective domain Ω_k. Therefore, the RH problem above can be transformed to an equivalent RH problem with a rather arbitrary choice of the jump contour. The only condition which must be preserved is that the infinite branches of the new contour must be asymptotic to the rays γ_k. Once again, we emphasize that this rich flexibility in the choice of the jump graph, will play a crucial role in the forthcoming asymptotic analysis.

We conclude this section by discussing in more detail the connection matrix E for a special case of the monodromy data, which will be of a particular interest in this chapter; and which is characterized by the equation,

$$1 - s_3 s_1 = 0. \tag{11.3.30}$$

The constraints (11.3.1) imply $s_1 + s_3 = -2\sin\pi\alpha$, so that

$$s_1 = e^{-i\pi\sigma(\alpha+\frac{1}{2})}, \quad s_3 = 1/s_1 = e^{i\pi\sigma(\alpha+\frac{1}{2})}, \quad \sigma = \pm 1, \tag{11.3.31}$$

while the parameter s_2 remains arbitrary. Accordingly, for the connection matrix E we have

$$\alpha - \tfrac{1}{2} \notin \mathbb{Z}, \quad J_+ = J_- = 0, \quad M = e^{i\pi\alpha\sigma_3}i\sigma_1: \tag{11.3.32}$$

$$\begin{cases} \sigma = -1, \quad s_1 = -e^{i\pi(\alpha-\frac{1}{2})}: \quad \mathsf{E}\mathsf{S}_6^{-1}\mathsf{S}_1^{-1} = p^{\sigma_3}\begin{pmatrix} 1 & 0 \\ \frac{s_1^2(s_1+s_2)}{1-s_1^2} & 1 \end{pmatrix}, \\ \sigma = +1, \quad s_1 = -e^{-i\pi(\alpha-\frac{1}{2})}: \quad \mathsf{E}\mathsf{S}_6^{-1}\mathsf{S}_1^{-1} = p^{\sigma_3}i\sigma_2\begin{pmatrix} 1 & 0 \\ \frac{s_1^2(s_1+s_2)}{1-s_1^2} & 1 \end{pmatrix}, \end{cases}$$

where p is a nonzero constant;

$$\alpha - \tfrac{1}{2} = n \in \mathbb{Z}, \quad s_1 = s_3 = (-1)^{n+1}, \quad M\, i\sigma_2 = (-1)^n I + J_\pm \sigma_\pm :$$

$$(11.3.33)\quad \begin{cases} n \in \mathbb{Z}_{<0}, \quad J_+ = 0, \quad J_- \neq 0: \quad \mathsf{E}\mathsf{S}_6^{-1}\mathsf{S}_1^{-1} = p^{\sigma_3}\begin{pmatrix} 1 & 0 \\ q & 1 \end{pmatrix}, \\ n \in \mathbb{Z}_{\geq 0}, \quad J_+ \neq 0, \quad J_- = 0: \quad \mathsf{E}\mathsf{S}_6^{-1}\mathsf{S}_1^{-1} = p^{\sigma_3} i\sigma_2 \begin{pmatrix} 1 & 0 \\ q & 1 \end{pmatrix}, \end{cases}$$

where p and q are constants, and $p \neq 0$.

If $\alpha - \frac{1}{2} = n \in \mathbb{Z}$ and $s_1 = s_3 = (-1)^{n+1}$, then the condition $J_+ = J_- = 0$ is equivalent to the equation $s_2 = (-1)^n$. These points, which we will call *special points* of the monodromy surface (11.3.1), have important geometrical meaning. Indeed, they are the singular points of the algebraic manifold (11.3.1), i.e., the points where the complex gradient of the l.h.s. of equation (11.3.1) is zero. At the special points $\alpha = \frac{1}{2} + n$, $n \in \mathbb{Z}$, $s_3 = s_1 = -s_2 = (-1)^{n+1}$, equation (11.3.2) does not provide any restriction to the connection matrix E, hence E becomes arbitrary. Taking into account that, for $\alpha - \frac{1}{2} = n \in \mathbb{Z}$, the matrix E is determined up to an upper or lower (depending on the sign of n) triangular left multiplier, we conclude that the solutions of the Painlevé equation (11.2.2) corresponding to a special point of the monodromy surface form a one-parameter family, parametrized by the ratio of the column entries of the matrix E (cf. [**FN1**]). Geometrically, this means that in order to represent adequately the manifold of the RH monodromy data of PII in the case $\alpha - \frac{1}{2} = n \in \mathbb{Z}$, one has to attach a copy of the projective line $\mathbb{CP}^1$ to the special point of the surface (11.3.1). Note also, that the surface (11.3.1) has singular points only for half-integer α's .

In summary, if a point of the monodromy surface (11.3.1) is not special, the connection matrix can be eliminated from the RH problem data, while, at the special points, it must be retained. In fact, it then becomes a parameter of the Painlevé function.

4. Special points of the monodromy surface

Flaschka and Newell in [**FN1**] constructed the Ψ-function for the case of a half-integer α and $J_\pm = 0$ in terms of the Airy functions. There analysis was based on the direct monodromy problem for the λ-equation of the Lax pair (11.3.27). There result indicates that at the special points of the monodromy surface $\alpha - \frac{1}{2} = n \in \mathbb{Z}$, $s_3 = s_1 = -s_2 = (-1)^{n+1}$, the associated RH problem, although still non-Abelian, nevertheless admits an explicit solution. In this section, we will establish this remarkable fact using the direct analysis of the RH problem. As an immediate consequence, we will construct a one-parameter family of classical solutions of PII expressible in terms of the Airy functions. Historically, this family was first discovered by Gambier [**Gam**]. Later in the section we will analyze the quasi-linear Stokes phenomenon associated with this family and will calculate the relevant asymptotics of the coefficient of the power-like dominant asymptotic background.

Let us begin with the case $\alpha = \frac{1}{2}$, $s_1 = s_3 = -1$, $s_2 = 1$. In this case, $\mathsf{S}_1 = \mathsf{S}_5$, the monodromy matrix M equals $-i\sigma_2$, and a simple direct calculation shows that

$$\mathsf{S}_4^{-1}\mathsf{S}_3^{-1}\mathsf{S}_2^{-1} = \mathsf{S}_1^{-1}\mathsf{S}_6^{-1}\mathsf{S}_5^{-1} = -i\sigma_2.$$

Therefore, the general RH problem depicted in Figure 11.2 reduces to the one

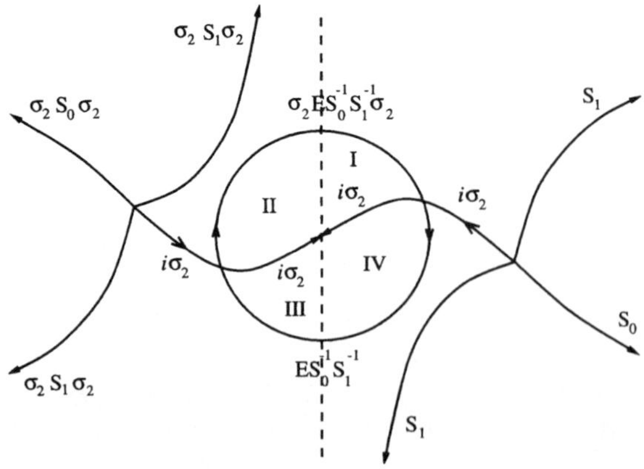

FIGURE 11.4. The RH problem graph for $\alpha = \frac{1}{2}$ and $s_1 = s_3 = -s_2 = -1$

shown in bold lines in Figure 11.4. We also assume that $x \neq 0$, $\arg x \in \left(\frac{2\pi}{3}, \pi\right)$, $r^2 < \frac{|x|}{2}$, and that those node points of the jump graph, which do not belong to the circle C, are located at $\lambda = \pm\sqrt{-\frac{x}{2}}$.

Introduce the matrix function $Z_0(z)$,

$$Z_0(z) = \sqrt{2\pi}\, e^{-i\pi/4} \begin{pmatrix} v_2(z) & v_1(z) \\ \frac{d}{dz} v_2(z) & \frac{d}{dz} v_1(z) \end{pmatrix} e^{-i\frac{\pi}{4}\sigma_3}, \tag{11.4.1}$$

where

$$v_1(z) = \operatorname{Ai}(z), \quad v_2(z) = e^{i2\pi/3} \operatorname{Ai}(e^{i2\pi/3} z). \tag{11.4.2}$$

Using (11.1.4), we see that this function has the following asymptotics as $|z| \to \infty$, $\arg z \in \left(-\pi, \frac{\pi}{3}\right)$,

$$Z_0(z) = z^{-\sigma_3/4} \frac{1}{\sqrt{2}} \begin{pmatrix} 1 & 1 \\ 1 & -1 \end{pmatrix} (I + O(z^{-3/2})) e^{\frac{2}{3} z^{3/2} \sigma_3}. \tag{11.4.3}$$

Also, introduce the matrix functions

$$Z_1(z) := Z_0(z) G_0, \quad Z_2(z) := Z_1(z) G_1, \quad Z_3(z) := Z_2(z) G_2,$$

$$G_0 = G_2 = \begin{pmatrix} 1 & 0 \\ -i & 1 \end{pmatrix}, \quad G_1 = \begin{pmatrix} 1 & -i \\ 0 & 1 \end{pmatrix}. \tag{11.4.4}$$

The matrix functions $Z_j(z)$ have the asymptotics (11.4.3) as $|z| \to \infty$, $\arg z \in \left(-\pi + \frac{2\pi}{3} j, \frac{\pi}{3} + \frac{2\pi}{3} j\right)$, $j = 0, 1, 2, 3$.

Define a piecewise holomorphic function $Z^{RH}(z)$ by

(11.4.5) $$Z^{RH}(z) = \begin{cases} Z_0(z)e^{-i\frac{\pi}{4}\sigma_3}, & \arg z \in (-\frac{\pi}{3}, 0), \\ Z_j(z)e^{-i\frac{\pi}{4}\sigma_3}, & \arg z \in (-\frac{2\pi}{3} + \frac{2\pi}{3}j, \frac{2\pi}{3}j), \quad j = 1, 2, \\ Z_3(z)e^{-i\frac{\pi}{4}\sigma_3}, & \arg z \in (\frac{4\pi}{3}, \frac{5\pi}{3}). \end{cases}$$

By construction, $Z^{RH}(z)$ has the uniform asymptotics

(11.4.6) $$Z^{RH}(z) = z^{-\sigma_3/4}\frac{1}{\sqrt{2}}\begin{pmatrix} 1 & 1 \\ 1 & -1 \end{pmatrix}(I + O(z^{-3/2}))e^{\frac{2}{3}z^{3/2}\sigma_3}e^{-i\frac{\pi}{4}\sigma_3}, \quad z \to \infty,$$

as well as the jump properties indicated in Figure 11.5.

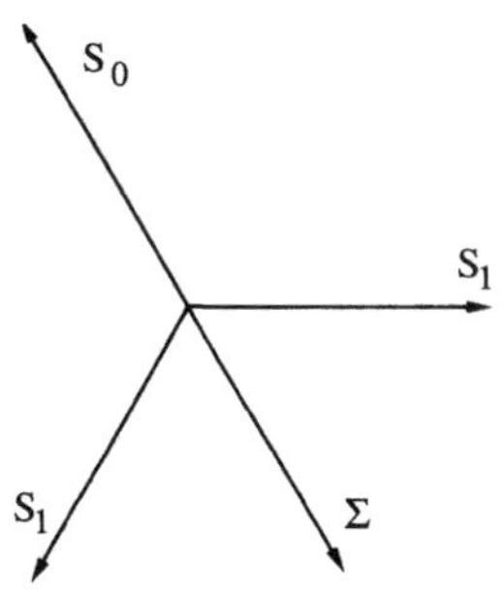

FIGURE 11.5. The jump graph for $Z^{RH}(z)$

Using the change of the independent variable

(11.4.7) $$z = -2^{2/3}(\lambda^2 + \frac{x}{2}),$$

introduce the piecewise holomorphic function $\Psi_r(\lambda)$ as follows:

(11.4.8) $$\Psi_r(\lambda) = \tfrac{1}{\sqrt{2}}(I - i\sigma_1)\lambda^{\sigma_3/2}(I + u\sigma_-)2^{\sigma_3/6}iZ^{RH}(z(\lambda)),$$
$$\operatorname{Re}\lambda > 0,$$

where the square root of λ is defined by the condition,

$$-\frac{\pi}{2} < \arg \lambda < \frac{\pi}{2},$$

and the parameter u will be determined later. We redefine slightly the function $Z^{RH}(z)$ using the domains indicated in Figure 11.6. This figure is the image of the right half of the Figure 11.4 under the map (11.4.7) (in the figure, for simplicity, we do not indicate the right diagonal factors $e^{-i\frac{\pi}{4}\sigma_3}$). The redefined function $Z^{RH}(z)$ still enjoys the canonical asymptotic behavior (11.4.6).

Using (11.4.6), the asymptotics of the function $\Psi_r(\lambda)$ at infinity have the canonical form

(11.4.9) $$\Psi_r(\lambda)e^{\theta\sigma_3} = I + \lambda^{-1}\Big[-i(\frac{x^2}{8} + \frac{u}{2})\sigma_3 + \frac{u}{2}\sigma_1\Big] + O(\lambda^{-2}).$$

At the origin, this function has an algebraic singularity,

(11.4.10) $$\Psi_r(\lambda + 0)) = \tfrac{1}{\sqrt{2}}(I - i\sigma_1)\lambda^{\sigma_3/2}C(x)\big(I + O(\lambda^2)\big),$$
$$C(x) = \big(I + u\sigma_-\big)2^{\sigma_3/6}iZ^{RH}(z(0)), \quad z(0) = e^{-i\pi}2^{-1/3}x.$$

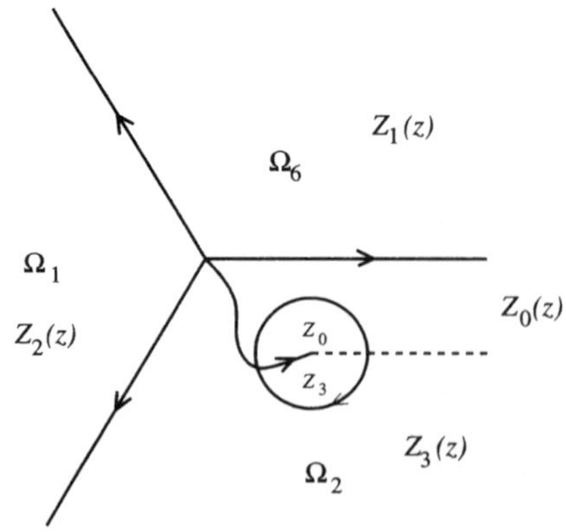

FIGURE 11.6. The definition of $Z^{RH}(z)$.

As a candidate for the solution of the RH problem we propose the piecewise holomorphic function $\Psi^{RH}(\lambda)$ defined by the relations,

$$\Psi^{RH}(\lambda) = \begin{cases} \Psi_r(\lambda), & \arg\lambda \in (-\frac{\pi}{2}, \frac{\pi}{2}), \quad |\lambda| > r, \\ \sigma_2\Psi_r(-\lambda)\sigma_2, & \arg\lambda \in (\frac{\pi}{2}, \frac{3\pi}{2}), \quad |\lambda| > r, \\ \Psi_r(\lambda)\sigma_2 \mathsf{S}_1\mathsf{S}_6\mathsf{E}^{-1}\sigma_2, & \lambda \text{ in domain I}, \\ \sigma_2\Psi_r(-\lambda)\mathsf{S}_1\mathsf{S}_6\mathsf{E}^{-1}\sigma_2, & \lambda \text{ in domain II}, \\ \sigma_2\Psi_r(-\lambda)\sigma_2\mathsf{S}_1\mathsf{S}_6\mathsf{E}^{-1}, & \lambda \text{ in domain III}, \\ \Psi_r(\lambda)\mathsf{S}_1\mathsf{S}_6\mathsf{E}^{-1}, & \lambda \text{ in domain IV}; \end{cases} \tag{11.4.11}$$

see Figure 11.4.

The crucial observation is that $\Psi^{RH}(\lambda)$ is continues across the imaginary axis. For instance, if $\arg\lambda = \frac{\pi}{2}$, $|\lambda| > r$, then the value of $\Psi^{RH}(\lambda)$ to the left of the imaginary axis is

$$\Psi^{RH}(\lambda - 0) = \sigma_2\Psi_r(-\lambda)\sigma_2$$
$$= \sigma_2\frac{1}{\sqrt{2}}(I - i\sigma_1)(e^{-i\pi}\lambda)^{\sigma_3/2}(I + u\sigma_-)2^{\sigma_3/6}iZ_0(z(\lambda))e^{-i\frac{\pi}{4}\sigma_3}\sigma_2$$
$$= \frac{1}{\sqrt{2}}(I - i\sigma_1)\lambda^{\sigma_3/2}(I + u\sigma_-)2^{\sigma_3/6}iZ_3(z(\lambda))e^{-i\frac{\pi}{4}\sigma_3} = \Psi^{RH}(\lambda + 0),$$

where we have used the identities $\sigma_2(I - i\sigma_1)\sigma_3 = -(I - i\sigma_1)$, $-\sigma_1 e^{-\frac{i\pi}{4}\sigma_3}\sigma_2 = e^{-\frac{i\pi}{4}\sigma_3}$ and $Z_0(z) = Z_3(z)i\sigma_1$. Continuity of $\Psi^{RH}(\lambda)$ across other segments of the imaginary axis can be checked similarly.

In order to establish that the function $\Psi^{RH}(\lambda)$ is the solution of the RH problem (11.3.13)–(11.3.2) for $\alpha = \frac{1}{2}$, $s_1 = s_3 = -s_2 = -1$, we must show that the parameter u in (11.4.8) can be chosen so that the condition (11.3.12) is satisfied. This condition is equivalent to the boundedness of $\Psi^{RH}(\lambda)\lambda^{-\sigma_3/2}$ as $\lambda \to 0$ in the domain III. Using (11.4.10), it follows that the boundedness of the product

$$\Psi_r(\lambda)\mathsf{S}_1\mathsf{S}_6\mathsf{E}^{-1}\lambda^{-\sigma_3/2}$$

as $\lambda \to 0$, is equivalent to the boundedness of

$$\lambda^{\sigma_3/2} C(x) \mathsf{S}_1 \mathsf{S}_6 \mathsf{E}^{-1} \lambda^{-\sigma_3/2},$$

or equivalently, to the upper triangularity of $C(x)\mathsf{S}_1\mathsf{S}_6\mathsf{E}^{-1}$. This triangularity condition uniquely determines the parameter u as the off-diagonal entry in the lower triangular matrix in the conventional Gauss LU decomposition,

$$2^{\sigma_3/6} i Z_0(z(0)) e^{-i\frac{\pi}{4}\sigma_3} \mathsf{S}_1 \mathsf{S}_6 \mathsf{E}^{-1} = (I - u\sigma_-)U, \quad (U)_{21} = 0. \tag{11.4.12}$$

In view of (11.4.9) and (11.3.15), the function $u \equiv u(x)$ becomes the solution of the Painlevé equation (11.2.2).

Classical solutions of PII for $\alpha = \frac{1}{2}$

Using (11.4.1) in (11.4.12), and introducing the notation,

$$\mathsf{S}_1 \mathsf{S}_6 \mathsf{E}^{-1} = \begin{pmatrix} a & b \\ c & d \end{pmatrix}, \tag{11.4.13}$$

we find after some elementary computations, that u satisfies

$$u = -2^{-1/3} \frac{v'(z_0)}{v(z_0)}, \quad v(z_0) = a v_2(z_0) - c v_1(z_0), \quad z_0 = -2^{-1/3} x, \tag{11.4.14}$$

where $(\ )' := \frac{d}{dz_0}$.

For this derivation we had assumed that x is restricted in a certain sector, but the formula (11.4.14) admits a meromorphic continuation (defined by itself) to the entire x-plane and thus defines a one-parameter family of explicit solutions of PII[3]. This reproduces the classical result of Gambier [**Gam**]. At the same time, we conclude that the analysis of the Stokes phenomenon for the Painlevé transcendents at a special point of the monodromy surface, can be obtained by using only the classical tools.

Let us choose in (11.4.13) $a = 0$. This choice of the connection matrix yields a particular Painlevé function, which we will denote $u_1^c(x|\frac{1}{2})$,

$$u_1^c(x|\tfrac{1}{2}) = -2^{-1/3} \tfrac{v_1'(z_0)}{v_1(z_0)}, \quad z_0 = -2^{-1/3} x. \tag{11.4.15}$$

Using (11.1.4), we derive the asymptotic expansion

$$u_1^c(x|\tfrac{1}{2}) \sim \sqrt{-\tfrac{x}{2}} \textstyle\sum_{n=0}^{\infty} c_n (-x)^{-\frac{3n}{2}}, \quad c_0 = 1, \quad \arg(-x) \in (-\pi, \pi). \tag{11.4.16}$$

The coefficient c_n, $n \geq 1$, in the expansion (11.4.16), can be found recursively.

Since $\mathrm{Ai}\,(z)$ is subordinate for $\arg z \in \left(-\frac{\pi}{3}, \frac{\pi}{3}\right)$, all other classical solutions (11.4.14) corresponding to the generic choice of the connection matrix E, i.e., to $a \neq 0$, are asymptotic to $u \sim -\sqrt{-\frac{x}{2}}$. More precisely, they have the asymptotic

[3]The fact that formula (11.4.14) defines a solution of the second Painlevé equation (11.2.2) with $\alpha = 1/2$, can be also verified by a direct substitution.

form given by (11.2.1),

$$\begin{aligned} u &= -2^{-1/3}\frac{v_2'}{v_2}\Big(1 - \tfrac{cv_1'}{av_2'} + \textstyle\sum_{n=1}^{\infty}\left(\frac{c}{a}\right)^n\left(\frac{v_1}{v_2}\right)^n - \sum_{n=1}^{\infty}\left(\frac{c}{a}\right)^{n+1}\frac{v_1'}{v_2'}\left(\frac{v_1}{v_2}\right)^n\Big) \\ &= -2^{-1/3}\frac{v_2'}{v_2} - i2^{-4/3}\frac{c/a}{\pi v_2^2} + O\big(x^{1/2}v_1^2v_2^{-2})\big) \\ &= -\sqrt{-\frac{x}{2}}\sum_{n=0}^{\infty}(-1)^n c_n(-x)^{-\frac{3n}{2}} \qquad (11.4.17) \\ &\quad + i\sqrt{2}\,\frac{c}{a}(-x)^{1/2}e^{-\frac{2\sqrt{2}}{3}(-x)^{3/2}}\big(1 + O(x^{-3/2})\big), \\ |x| &\to \infty, \quad \arg(-x) \in \left(-\tfrac{\pi}{3}, \tfrac{\pi}{3}\right). \end{aligned}$$

The *formal* expression (11.4.17) describes the 1-parametric family of the classical solutions of PII with $\alpha = \frac{1}{2}$, asymptotic to $-\sqrt{-\frac{x}{2}}$, $\arg x \in (-\frac{\pi}{3}, \frac{\pi}{3})$. The rigorous meaning of (11.4.17) is established by the asymptotic relation,

$$u(x) = u_2^c(x|\tfrac{1}{2}) + i\sqrt{2}\,\tfrac{c}{a}(-x)^{1/2}e^{-\frac{2\sqrt{2}}{3}(-x)^{3/2}}\big(1 + O(x^{-3/2})\big), \tag{11.4.18}$$

$$|x| \to \infty, \quad \arg(-x) \in \left(-\tfrac{\pi}{3}, \tfrac{\pi}{3}\right),$$

where $u_2^c(x|\frac{1}{2})$ is a particular solution from this family, which is defined by the equation

$$u_2^c(x|\tfrac{1}{2}) = -2^{-1/3}\tfrac{v_2'(z_0)}{v_2(z_0)}, \quad z_0 = -2^{-1/3}x.$$

It is important to stress that the classical solution of PII with $\alpha = \frac{1}{2}$, asymptotic to $+\sqrt{-\frac{x}{2}}$ in the same sector, i.e., the solution $u_1^c(x|\frac{1}{2})$, is unique. We will discuss further this interesting fact in more detail in the end of Section 5 of this chapter.

QUASI-LINEAR STOKES PHENOMENON AND THE ASYMPTOTICS OF THE COEFFICIENTS FOR THE CLASSICAL SOLUTIONS OF PII WITH $\alpha = 1/2$

Using the set of the Airy functions,

$$v_{n+1}(x) = e^{i\frac{2\pi}{3}n}v_1\left(e^{i\frac{2\pi}{3}n}x\right),$$

we define the corresponding classical solutions of PII,

$$u_{n+1}^c(x|\tfrac{1}{2}) = e^{i\frac{2\pi}{3}n}u_1^c(e^{i\frac{2\pi}{3}n}x|\tfrac{1}{2}) = -2^{-1/3}\tfrac{v_{n+1}'(z_0)}{v_{n+1}(z_0)}, \quad z_0 = -2^{-1/3}x, \tag{11.4.19}$$

where the prime means differentiation with respect to z_0. Similarly to u_1^c, the functions u_3^c and u_{-1}^c are also asymptotic to $\sqrt{-\frac{x}{2}}$, but in the following sectors,

$$\begin{aligned} u_3^c(x|\tfrac{1}{2}) &= e^{i\frac{4\pi}{3}}u_1^c(e^{i\frac{4\pi}{3}}x|\tfrac{1}{2}) \sim \sqrt{-\tfrac{x}{2}}, \quad \arg(-x) \in \left(-\tfrac{7\pi}{3}, -\tfrac{\pi}{3}\right), \\ u_{-1}^c(x|\tfrac{1}{2}) &= e^{-i\frac{4\pi}{3}}u_1(e^{-i\frac{4\pi}{3}}x|\tfrac{1}{2}) \simeq \sqrt{-\tfrac{x}{2}}, \quad \arg x \in \left(\tfrac{\pi}{3}, \tfrac{7\pi}{3}\right). \end{aligned} \tag{11.4.20}$$

Since the above different sectors overlap, the classical solutions of PII exhibit the quasi-linear Stokes phenomenon. The description of this phenomenon is based on the identity (cf.(11.1.14)), $v_n + v_{n+1} + v_{n+2} \equiv 0$. Indeed, consider the relations

$$u_3^c(x|\tfrac{1}{2}) - u_1^c(x|\tfrac{1}{2}) = -2^{-1/3}\Big(\frac{v_3'}{v_3} - \frac{v_1'}{v_1}\Big) = \frac{i2^{-4/3}}{\pi v_3 v_1}, \tag{11.4.21}$$

$$u_1^c(x|\tfrac{1}{2}) - u_{-1}^c(x|\tfrac{1}{2}) = -2^{-1/3}\Big(\frac{v_1'}{v_1} - \frac{v_{-1}'}{v_{-1}}\Big) = \frac{i2^{-4/3}}{\pi v_1 v_{-1}}, \tag{11.4.22}$$

$$u_{-1}^c(x|\tfrac{1}{2}) - u_3^c(x|\tfrac{1}{2}) = -2^{-1/3}\Big(\frac{v_{-1}'}{v_{-1}} - \frac{v_3'}{v_3}\Big) = \frac{i2^{-4/3}}{\pi v_{-1} v_3}. \tag{11.4.23}$$

In the sectors where the denominators are exponentially large, the differences in the left-hand side are exponentially small; this provides the description of the Stokes phenomenon in a form similar to the Airy relation (11.1.15),

$$\begin{aligned} &u_1^c(x|\tfrac{1}{2}) - u_3^c(x|\tfrac{1}{2}) = i\sqrt{2}(-x)^{1/2}e^{\frac{2\sqrt{2}}{3}(-x)^{3/2}}\big(1 + O(x^{-3/2})\big),\\ &|x| \to \infty, \quad \arg(-x) \in \big(-\pi, -\tfrac{\pi}{3}\big), \end{aligned} \tag{11.4.24}$$

$$\begin{aligned} &u_{-1}^c(x|\tfrac{1}{2}) - u_1^c(x|\tfrac{1}{2}) = i\sqrt{2}(-x)^{1/2}e^{\frac{2\sqrt{2}}{3}(-x)^{3/2}}\big(1 + O(x^{-3/2})\big),\\ &|x| \to \infty, \quad \arg(-x) \in \big(\tfrac{\pi}{3}, \pi\big), \end{aligned} \tag{11.4.25}$$

$$\begin{aligned} &u_3^c(x|\tfrac{1}{2}) - u_{-1}^c(x|\tfrac{1}{2}) = i\sqrt{2}(-x)^{1/2}e^{\frac{2\sqrt{2}}{3}(-x)^{3/2}}\big(1 + O(x^{-3/2})\big),\\ &|x| \to \infty, \quad \arg(-x) \in \big(-\tfrac{7\pi}{3}, -\tfrac{5\pi}{3}\big). \end{aligned} \tag{11.4.26}$$

Introduce a piecewise holomorphic function $\hat{u}(t|\frac{1}{2})$ as follows (cf. (11.1.25)),

$$\hat{u}\big(t|\tfrac{1}{2}\big) = \begin{cases} u_3^c\big(-t^2|\tfrac{1}{2}\big), & \arg t \in \big(-\pi, -\tfrac{\pi}{3}\big),\\ u_1^c\big(-t^2|\tfrac{1}{2}\big), & \arg t \in \big(-\tfrac{\pi}{3}, \tfrac{\pi}{3}\big),\\ u_{-1}^c\big(-t^2|\tfrac{1}{2}\big), & \arg t \in \big(\tfrac{\pi}{3}, \pi\big). \end{cases} \tag{11.4.27}$$

This function has the uniform asymptotic expansion,

$$\hat{u}(t|\tfrac{1}{2}) \sim \tfrac{t}{\sqrt{2}} \textstyle\sum_{n=0}^{\infty} c_n t^{-3n}, \quad t \to \infty. \tag{11.4.28}$$

Let $u^{(N)}(x)$ be the partial sum

$$u^{(N)}(t) = \frac{t}{\sqrt{2}} \sum_{n=0}^{N-1} c_n t^{-3n}, \tag{11.4.29}$$

and $w^{(N)}(t)$ be the product

$$w^{(N)}(t) = t^{3N-2}\big(\hat{u}(t) - u^{(N)}(t)\big) \sim \frac{1}{t\sqrt{2}} \sum_{n=0}^{\infty} c_{n+N} t^{-3n}. \tag{11.4.30}$$

Integrating (11.4.30) over a circle of large radius yields (cf. derivation of (11.1.26))

$$\begin{aligned} 0 &= \oint_{|t|=R} w^{(N)}(t)\,dt + \sum_{n=0}^{2} \int_{e^{-\pi + i\frac{2\pi}{3}n}(0,\infty)} \big(w_+^{(N)}(t) - w_-^{(N)}(t)\big)\,dt\\ &= \pi i\sqrt{2}\, c_N + (-1)^N i\sqrt{2}\big(\tfrac{3}{2\sqrt{2}}\big)^N \Gamma(N)\big(1 + O(N^{-1})\big). \end{aligned} \tag{11.4.31}$$

Thus,

$$c_n = (-1)^{n-1}\pi^{-1}\big(\tfrac{3}{2\sqrt{2}}\big)^n \Gamma(n)\big(1 + O(n^{-1})\big), \quad n \to \infty. \tag{11.4.32}$$

CLASSICAL SOLUTIONS FOR $\alpha = \frac{1}{2} + n$, $n \in \mathbb{Z}$

In order to describe the Painlevé functions for other values of the parameter α one can use the Bäcklund-Schlesinger transformations discussed in Chapter 6. We recall that the Schlesinger transformation of the Ψ-function,

(11.4.33) $\tilde{\Psi}(\lambda) = R(\lambda)\Psi(\lambda),$

$$R(\lambda) = I - \frac{q_\epsilon}{2i\lambda}(\sigma_3 - i\epsilon\sigma_1), \quad q_\epsilon = \frac{\alpha + \frac{\epsilon}{2}}{u_x - \epsilon(u^2 + \frac{x}{2})}, \quad \epsilon \in \{+1, -1\},$$

yields the Ψ-function of a new Painlevé transcendent, defined as follows,

$$\tilde{u} = u + \epsilon q_\epsilon, \quad \tilde{u}_x = u_x + \epsilon(\tilde{u}^2 - u^2), \quad \tilde{\alpha} = -\alpha - \epsilon; \tag{11.4.34}$$

but this new Ψ-function has the *same* Stokes matrices.

The classical solutions (11.4.14) of PII with $\alpha = \frac{1}{2}$, satisfy the Riccati equation,

$$u_x + u^2 + \frac{x}{2} = 0, \tag{11.4.35}$$

and therefore the above Bäcklund transformation to (11.4.14), as well as the associated Schlesinger transformation of the Ψ-function do not exist for $\epsilon = -1$. The transformation for $\epsilon = +1$ yields classical solutions of PII with $\alpha = -\frac{3}{2}$. Then, a sequence of iterated Bäcklund transformations with altered values of $\epsilon = \pm 1$, yields a sequence of classical solutions of PII corresponding to the parameter $\alpha = \alpha_n = (-1)^n\big(\frac{1}{2} + n\big)$, $n \in \mathbb{Z}_{\geq 0}$, and to the Stokes multipliers $s_1 = s_3 = -s_2 = -1$.

Applying the first of the symmetries (11.3.26) $u \mapsto -u$, $\alpha \mapsto -\alpha$, $s_k \mapsto -s_k$, to the above sequence of the classical solutions, we find the sequence corresponding to $\alpha = -\alpha_n = (-1)^{n+1}\big(\frac{1}{2} + n\big)$ and $s_1 = s_3 = -s_2 = 1$.

An elementary introduction to the analytic properties of this family of classical solutions of PII can be found in [**GroLS**]. Determinant formulas for any solution of this family are presented in [**Ai**]; see also [**FN1**].

5. Nonspecial points of the monodromy surface

CASE $\alpha - \frac{1}{2} \notin \mathbb{Z}$ AND $1 - s_1 s_3 = 0$

In the case of nonspecial points of the monodromy surface, it is convenient to deform the jump graph of the RH problem,to the one depicted in Figure 5, where we have introduced the notation

$$\Sigma := \mathrm{S}_1^{-1}\mathrm{S}_6^{-1}\mathrm{S}_1^{-1}. \tag{11.5.1}$$

The constraint $1 - s_1 s_3 = 0$, yields the equations

$$\Sigma = \begin{pmatrix} 0 & 1/s_1 \\ -s_1 & 0 \end{pmatrix}, \quad \text{and} \quad \mathrm{S}_2\mathrm{S}_4^{-1} = \begin{pmatrix} 1 & s_1 + s_2 \\ 0 & 1 \end{pmatrix}. \tag{11.5.2}$$

We also recall that the constraint $1 - s_1 s_3 = 0$, implies that (see (11.3.31))

$$s_1 = e^{-i\pi\sigma(\alpha + \frac{1}{2})}, \quad s_3 = 1/s_1 = e^{i\pi\sigma(\alpha+\frac{1}{2})}, \quad \sigma = \pm 1, \tag{11.5.3}$$

while the parameter s_2 remains arbitrary.

Using the flexibility in choosing the branches of the jump graph, we put the node points of the jump graph at the points $\lambda_0 = 0$ and $\lambda_{1,2} = \pm\sqrt{-\frac{x}{2}}$, and we assume that $\arg x \in \big[\frac{2\pi}{3}, \pi\big]$. Also, we choose the following curves: the curved segment

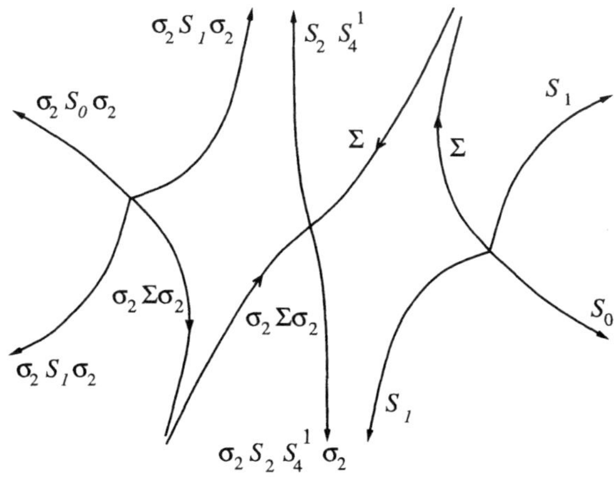

FIGURE 11.7. RH problem graph for nonspecial points of the monodromy surface

$[-\sqrt{-\frac{x}{2}}, \sqrt{-\frac{x}{2}}]$, indicated Figure 11.4, and the level curves Im $g(\lambda) = \text{const}$ of the function

$$g(\lambda) = i\tfrac{4}{3}\left(\lambda^2 + \tfrac{x}{2}\right)^{3/2}, \tag{11.5.4}$$

emanating from its critical points $\lambda = \pm\sqrt{-\frac{x}{2}}$ and $\lambda = 0$.

Regarding the choice of the jump graph we note the following:

Denote by γ_+ and γ_- the level curves emanating from $-\sqrt{-\frac{x}{2}}$ and $\sqrt{-\frac{x}{2}}$, respectively; these curves are asymptotic to the upper half of the imaginary axis and to the lower half of the imaginary axis, respectively. Strictly speaking, they are the level curves only if $\arg x \in \left[\frac{2\pi}{3}, \pi\right)$. However, we will use them in the graph, even for the case $\arg x = \pi$, noting that even in this case we have the following important property: Re $g(\lambda) < 0$ and Re $g(\lambda) > 0$ along the whole curves γ_- and γ_+, respectively, and $|\operatorname{Re} g(\lambda)| \to \infty$ as $|\lambda| \to \infty$, $\lambda \in \gamma_\pm$.

The function $\Psi(\lambda)$ is normalized at infinity by the asymptotic condition (11.3.13). Since we have eliminated from the jump graph the circle around the origin, the asymptotics for $\Psi(\lambda)$ at the origin (11.3.12) should be replaced by the condition,

$$\left\|\Psi(\lambda)\mathrm{S}_1\mathrm{S}_6\mathrm{E}^{-1}\lambda^{-\alpha\sigma_3}\right\| \le \text{const}, \quad \lambda \to 0, \quad \lambda \in \omega_-, \tag{11.5.5}$$

where ω_- is the domain bounded by the lower half of the level curve emanating from $\lambda = 0$, by the curved segment $[0, \sqrt{-\frac{x}{2}}]$, and by the curve γ_-.

Recalling formulae (11.3.32) and (11.3.33), we see that in the case of nonspecial points of the monodromy surface and even for a half-integer α, the parameters p and q, which are the only free parameters in the connection matrix E, do not affect the solution of Painlevé equation. Since the multipliers s_1, s_3 are fixed by (11.5.3) up to a choice of $\sigma = \pm 1$, the only essential free parameter of the RH problem is the Stokes parameter s_2, which determines the jump (namely the matrices $\mathrm{S}_2\mathrm{S}_4^{-1}$ and $\sigma_2\mathrm{S}_2\mathrm{S}_4^{-1}\sigma_2$) across the line Im $g(\lambda) = \text{const}$, emanating from the origin.

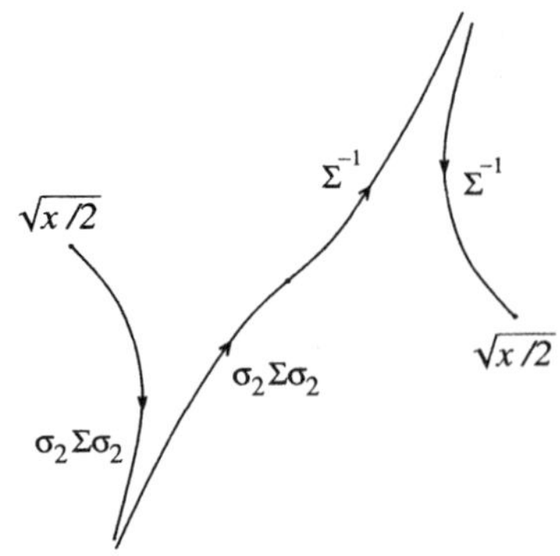

FIGURE 11.8. Graph of the RH model problem 1

THE REDUCED RH PROBLEM IN THE CASE OF $1 - s_1 s_3 = s_1 + s_2 = 0$

Consider the particular case of PII corresponding to $1-s_1s_3 = s_1+s_2 = 0$. The associated RH problem jump graph coincides with the one depicted in Figure 5, except for the curve lines emanating from the origin and approaching the vertical direction. These curves can be deleted from the graph since now the relevant jumps are the identity matrix,

$$\mathrm{S}_2\mathrm{S}_4^{-1} = \sigma_2\mathrm{S}_2\mathrm{S}_4^{-1}\sigma_2 = I.$$

The reduced RH problem is formulated as follows: find a piecewise holomorphic function $\hat{\Psi}(\lambda)$ such that

(1) $\hat{\Psi}(\lambda)e^{\theta\sigma_3} \to I, \quad \lambda \to \infty;$

(2) $\|\hat{\Psi}_-(\lambda)\mathrm{S}_1\mathrm{S}_6\mathrm{E}^{-1}\lambda^{-\alpha\sigma_3}\| \le \text{const}, \quad \lambda \to 0,$ where E is the connection matrix defined in (11.3.32) corresponding to $s_1 + s_2 = 0$;

(3) $\hat{\Psi}(\lambda)$ is discontinous across the contour shown in Figure 5 where the vertical curves emanating from the origin have been deleted.

THEOREM 11.1. *Let* $\arg x \in [\frac{2\pi}{3}, \pi]$, $\alpha - \frac{1}{2} \notin \mathbb{Z}$ *and* $1 - s_1s_3 = s_1 + s_2 = 0$. *Then, for sufficiently large* $|x|$, *there exists a unique solution of the reduced RH problem. If, additionally, the parameter* $\sigma \in \{+1, -1\}$ *is defined in such a way that* $s_1 = e^{-i\pi\sigma(\alpha+\frac{1}{2})}$, *then the corresponding Painlevé function has the asymptotics*

(11.5.6)
$$u(x) \equiv u_1(x|\alpha, \sigma) = \sigma\sqrt{-\tfrac{x}{2}} + O(x^{-2/5}), \quad |x| \to \infty, \quad \arg(-x) \in \left[-\tfrac{\pi}{3}, 0\right].$$

PROOF. It is sufficient to proof existence. Consider a RH problem on the segment $[-\sqrt{-\frac{x}{2}}, \sqrt{-\frac{x}{2}}]$, where $\arg\sqrt{-\frac{x}{2}} \in \left[-\frac{\pi}{6}, 0\right]$ (see Figure 11.8), defined as follows.

THE MODEL RH PROBLEM 1.

(1) $\Phi^0(\lambda)e^{\theta\sigma_3} \to I, \quad \lambda \to \infty;$

(2) $\|\Phi^0_-(\lambda)\mathrm{S}_1\mathrm{S}_6\mathrm{E}^{-1}\lambda^{-\alpha\sigma_3}\| \le \text{const}, \quad \lambda \to 0;$

(3) The matrix function Φ satisfies the following jump conditions:

$$\Phi^0_+(\lambda) = \Phi^0_-(\lambda)\sigma_2\Sigma\sigma_2, \quad \lambda \in \left(-\sqrt{-\frac{x}{2}}, 0\right),$$

$$\Phi^0_+(\lambda) = \Phi^0_-(\lambda)\Sigma^{-1}, \quad \lambda \in \left(0, \sqrt{-\frac{x}{2}}\right),$$

$$\sigma_2\Sigma\sigma_2 = \begin{pmatrix} 0 & s_1 \\ -1/s_1 & 0 \end{pmatrix}, \quad \Sigma^{-1} = \begin{pmatrix} 0 & -1/s_1 \\ s_1 & 0 \end{pmatrix}.$$

The solution to this RH problem is found in [**IFoK**] and [**DeZ2**] (see also Chapter 8) but in slightly different notations. On the complex λ-plane cut along $[-\sqrt{-\frac{x}{2}}, \sqrt{-\frac{x}{2}}]$, define the following scalar and matrix functions β, Y, δ:

$$\beta(\lambda) = \left(\frac{\lambda - \sqrt{-\frac{x}{2}}}{\lambda + \sqrt{-\frac{x}{2}}}\right)^{1/4}, \tag{11.5.7}$$

$$Y(\lambda) = \frac{1}{2}\begin{pmatrix} \beta + \beta^{-1} & -\sigma(\beta - \beta^{-1}) \\ -\sigma(\beta - \beta^{-1}) & \beta + \beta^{-1} \end{pmatrix}$$

$$= \frac{1}{\sqrt{2}}\begin{pmatrix} 1 & 1 \\ 1 & -1 \end{pmatrix}\beta^{-\sigma\sigma_3}\frac{1}{\sqrt{2}}\begin{pmatrix} 1 & 1 \\ 1 & -1 \end{pmatrix}, \tag{11.5.8}$$

$$\delta(\lambda) = \left(\frac{\sqrt{\lambda^2 + \frac{x}{2}} - i\sqrt{-\frac{x}{2}}}{\sqrt{\lambda^2 + \frac{x}{2}} + i\sqrt{-\frac{x}{2}}}\right)^{\nu}, \quad \nu = -\frac{1}{2\pi i}\ln(i\sigma s_1) = \frac{\sigma\alpha}{2}, \tag{11.5.9}$$

where the branch of the function $\beta(\lambda)$ is fixed by the condition

$$\beta \to 1 \quad \text{as} \quad \lambda \to \infty;$$

the function $\sqrt{\lambda^2 + \frac{x}{2}}$ is fixed by the asymptotics $\sqrt{\lambda^2 + \frac{x}{2}} = \lambda + O(\lambda^{-1})$ as $\lambda \to +\infty$, and the function ξ^ν is defined on the ξ-plane cut along the negative part of the real axis and the principal branch is choosen.

It is easy to check that the above definitions of the branches imply the relation,

$$\delta(-\lambda)\delta(\lambda) = 1. \tag{11.5.10}$$

Furthermore, the above functions satisfy the conditions:

$$Y_+(\lambda) = Y_-(\lambda)(-i\sigma)\sigma_1, \quad \lambda \in \left(-\sqrt{-\frac{x}{2}}, \sqrt{-\frac{x}{2}}\right), \tag{11.5.11}$$

$$\delta_+(\lambda)\delta_-(\lambda) = e^{2\pi i\nu}, \quad \lambda \in \left(-\sqrt{-\frac{x}{2}}, 0\right), \tag{11.5.12}$$

$$\delta_+(\lambda)\delta_-(\lambda) = e^{-2\pi i\nu}, \quad \lambda \in \left(0, \sqrt{-\frac{x}{2}}\right),$$

$$g_+(\lambda) = -g_-(\lambda), \quad \lambda \in \left(-\sqrt{-\frac{x}{2}}, \sqrt{-\frac{x}{2}}\right). \tag{11.5.13}$$

Therefore the function

$$\Phi^0 = Y\delta^{\sigma_3}e^{-g\sigma_3}, \tag{11.5.14}$$

satisfies the jump conditions of the model RH problem 1.

The asymptotics as $\lambda \to \infty$,

$$\begin{aligned}
\beta(\lambda) &= 1 - \frac{1}{2\lambda}\sqrt{-\frac{x}{2}} + O(\lambda^{-2}),\\
Y(\lambda) &= I + \frac{1}{2\lambda}\sigma\sqrt{-\tfrac{x}{2}}\,\sigma_1 + O(\lambda^{-2}),\\
\delta(\lambda) &= 1 + O(\lambda^{-2}),\\
g(\lambda) &= i\left(\tfrac{4}{3}\lambda^3 + x\lambda\right) + i\tfrac{x^2}{8\lambda} + O(\lambda^{-3}),
\end{aligned} \tag{11.5.15}$$

imply

$$\Phi^0(\lambda) = \Big(I + \tfrac{1}{2\lambda}\big(-\tfrac{ix^2}{4}\sigma_3 + \sigma\sqrt{-\tfrac{x}{2}}\,\sigma_1\big) + O(\lambda^{-2})\Big)e^{-\theta\sigma_3}. \tag{11.5.16}$$

The asymptotics as $\lambda \to 0$,

$$\begin{aligned}
\beta_-(\lambda) &= e^{-i\pi/4} + O(\lambda),\\
Y_-(\lambda) &= \frac{1}{\sqrt{2}}\Big(I + i\sigma\sigma_1 + O(\lambda)\Big),\\
\delta_-(\lambda) &= e^{-i\pi\nu}(-2x)^{\nu}\lambda^{-2\nu}\big(1 + O(\lambda^2)\big),\\
g_-(\lambda) &= -\tfrac{\sqrt{2}}{3}(-x)^{3/2} + \sqrt{2}(-x)^{1/2}\lambda^2 + O(\lambda^4 x^{-1/2}),
\end{aligned} \tag{11.5.17}$$

imply

$$\Phi^0_-(\lambda) = C\big(I + O(\lambda)\big)\lambda^{-\sigma\alpha\sigma_3} \tag{11.5.18}$$

for some constant matrix C. Therefore, using (11.3.32), we verify the condition $\left\|\Phi^0_-(\lambda)\mathsf{S}_1\mathsf{S}_6\mathsf{E}^{-1}\lambda^{-\alpha\sigma_3}\right\| \le \text{const}$, at the origin. Thus the function $\Phi^0(\lambda)$ (11.5.14) solves the model RH problem 1.

REMARK 11.3. The RH problem solved above is, perhaps, the simplest one in the class of the so-called *quasi-permutation RH problems.* A quasi-permutation RH problem for the elliptic case was solved in (8.4.17)–(8.4.18). More general quasi-permutation RH problems are solved in [**Ko2**].

We note that

$$\sigma_2\Phi^0(-\lambda)\sigma_2 = \Phi^0(\lambda). \tag{11.5.19}$$

The function $\Phi^0(\lambda)$ is singular at $\lambda = \pm\sqrt{-\frac{x}{2}}$. In order to eliminate this singularity we have to introduce a suitable parametrix in the neighborhoods of the points $\lambda = \pm\sqrt{-\frac{x}{2}}$. For this purpose we consider the following model RH problem.

THE MODEL RH PROBLEM 2.

Find a piecewise holomorphic function $\Phi_r(\lambda)$ with the jumps indicated in Figure 11.9.

Since we have not normalized $\Phi_r(\lambda)$, this function is determined up to a left multiplication by an entire matrix function. If we use this solution in a domain different than the entire complex plane, then $\Phi_r(\lambda)$ is determined up to a left multiplication by a matrix function holomorphic in this domain.

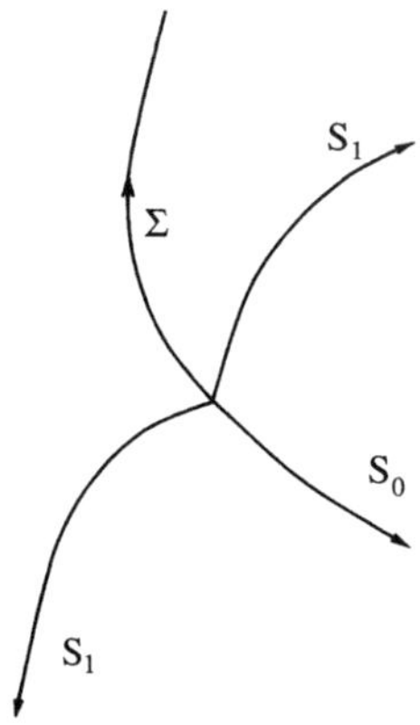

FIGURE 11.9. The model RH problem near $\lambda = \sqrt{-\frac{x}{2}}$

In analogy with the procedure of Chapter 8, $\Phi_r(\lambda)$ can be constructed in terms of the classical Airy functions. Indeed, the function $\Phi_r(\lambda)$ is almost identical to the function $\Psi_r(\lambda)$ introduced in the previous section in connection with the special points of the monodromy surface. More precisely, we take exactly the same matrix functions $Z_j(\lambda)$, $j = 0,1,2,3$ as in (11.4.1) and (11.4.4) but we slightly modify the definition (11.4.5) of the piecewise holomorphic function $Z^{RH}(z)$, in order to account for the fact that now $s_1 \neq -1$:

$$Z^{RH}(z) = \begin{cases} Z_0(z)(is_1)^{\sigma_3/2}, & \arg z \in \left(-\frac{\pi}{3}, 0\right), \\ Z_j(z)(is_1)^{\sigma_3/2}, & \arg z \in \left(-\frac{2\pi}{3} + \frac{2\pi}{3}j, \frac{2\pi}{3}j\right), \quad j = 1,2, \\ Z_3(z)(is_1)^{\sigma_3/2}, & \arg z \in \left(\frac{4\pi}{3}, \frac{5\pi}{3}\right). \end{cases} \tag{11.5.20}$$

This definition includes (11.4.5) as a special case (corresponding $s_1 = -1$). By construction, $Z^{RH}(z)$ has the uniform asymptotics
(11.5.21)

$$Z^{RH}(z) = z^{-\sigma_3/4}\frac{1}{\sqrt{2}}\begin{pmatrix} 1 & 1 \\ 1 & -1 \end{pmatrix}\Big(I + O(z^{-3/2})\Big)e^{\frac{2}{3}z^{3/2}\sigma_3}(is_1)^{\sigma_3/2}, \quad z \to \infty,$$

as well as the jump relations indicated in Figure 11.5 with $-i\sigma_2$ replaced by Σ.

Using the change of the independent variable (11.4.7),

$$z = -2^{2/3}\left(\lambda^2 + \frac{x}{2}\right), \tag{11.5.22}$$

and the notation

$$\zeta := \lambda - \sqrt{-\frac{x}{2}}, \tag{11.5.23}$$

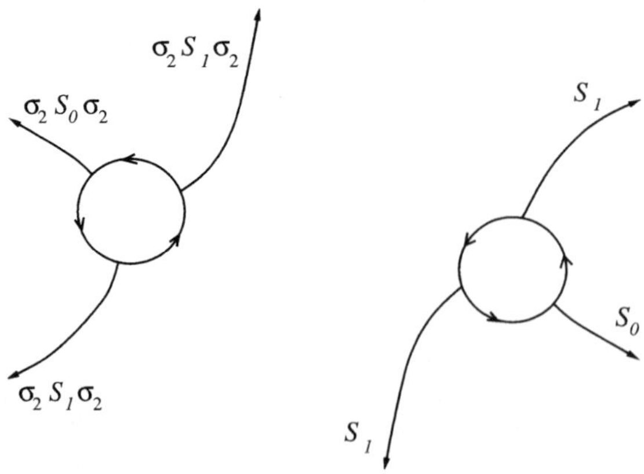

FIGURE 11.10. The RH problem graph for the correction function $\chi(\lambda)$

we compute the "fraction" of Φ^0 and $Z^{RH}(z(\lambda))$ in the annular domain $c_1|x|^{-\frac{1}{2}+\epsilon} \leq |\zeta| \leq c_2|x|^{-\frac{1}{2}+\epsilon}$, where $0 < c_1 < c_2$ and $\frac{1}{4} < \epsilon < \frac{1}{2}$ are some constants:

$$\Phi^0(\lambda)\big(Z^{RH}(z(\lambda))\big)^{-1} = \Big(I + O(x^{-1/4}\zeta^{1/2}) + O(x^{-1/2}\zeta^{-2})\Big)$$

$$\times\tfrac{1}{\sqrt{2}}\begin{pmatrix} 1 & 1 \\ 1 & -1 \end{pmatrix}\Big(\tfrac{1-\sigma}{2} i\sigma_1 + \tfrac{1+\sigma}{2} I\Big) e^{i\frac{\pi}{4}\sigma_3} 2^{\frac{2}{3}\sigma_3} (-x/2)^{\sigma_3/4}. \tag{11.5.24}$$

This relation allows us to introduce the local parametrix $\Phi_r(\lambda)$ near the endpoint $\lambda = \sqrt{-\frac{x}{2}}$ by the equation (cf. (11.4.8)),

$$\Phi_r(\lambda) = \tfrac{1}{\sqrt{2}}\begin{pmatrix} 1 & 1 \\ 1 & -1 \end{pmatrix}\Big(\tfrac{1-\sigma}{2} i\sigma_1 + \tfrac{1+\sigma}{2} I\Big) e^{i\frac{\pi}{4}\sigma_3} 2^{\frac{2}{3}\sigma_3} (-x/2)^{\sigma_3/4} Z^{RH}\big(z(\lambda)\big). \tag{11.5.25}$$

As a candidate for the global approximation of the solution of the reduced RH problem, we propose the piecewise holomorphic function $\Phi(\lambda)$,

$$\Phi(\lambda) = \begin{cases} \Phi_r(\lambda), & |\lambda - \sqrt{-\frac{x}{2}}| < R, \\ \sigma_2\Phi_r(e^{-i\pi}\lambda)\sigma_2, & |\lambda + \sqrt{-\frac{x}{2}}| < R, \\ \Phi^0(\lambda), & |\lambda \pm \sqrt{-\frac{x}{2}}| > R, \end{cases} \tag{11.5.26}$$

where $R = c|x|^{-\frac{1}{2}+\epsilon}$ for a constant $c > 0$ and $\frac{1}{4} < \epsilon < \frac{1}{2}$. We seek for the exact solution of the reduced RH problem in the form of the product

$$\hat{\Psi}(\lambda) = \chi(\lambda)\Phi(\lambda). \tag{11.5.27}$$

The correction function $\chi(\lambda)$ solves the RH problem with the jump graph shown in Figure 11.10. The jump matrix across the circle centered at $\lambda = \sqrt{-\frac{x}{2}}$ is given by the ratio $\Phi^0(\lambda)\Phi_r^{-1}(\lambda)$ which, due to (11.5.24), (11.5.25) and with the choice of $\epsilon = 2/5$, satisfies the estimate

$$\big\|\Phi^0(\lambda)\Phi_r^{-1}(\lambda) - I\big\| \leq c|x|^{-3/10}. \tag{11.5.28}$$

The jump matrices across the exterior parts of infinite contours approach the unit matrices exponentially fast,

$$\|\Phi^0(\lambda)S_j(\Phi^0(\lambda))^{-1} - I\| \le C|x|^{3/20}e^{-\frac{2^{15/4}}{3}|x|^{3/4}|\zeta|^{3/2}}. \tag{11.5.29}$$

Now, the solvability of the reduced RH problem is straightforward. Indeed, we consider the system of singular integral equations $\chi_- = I + K\chi_-$ equivalent to the RH problem for $\chi(\lambda)$,

$$\chi_-(\lambda) = I + \frac{1}{2\pi i}\int_\gamma \chi_-(\xi)\big(G(\xi) - I\big)\frac{d\xi}{\xi - \lambda_-},$$

where γ is the contour shown in Figure 11.10 and $G(\lambda) = \Phi_-(\lambda)S_j\Phi_+^{-1}(\lambda)$ is the relevant jump matrix. Using the estimates (11.5.28), (11.5.29), as well as the boundedness of the Cauchy operator in $L^2(\gamma)$, the singular integral operator K, which is a superposition of the operator of the right multiplication in $G(\lambda) - I$ and of the Cauchy operator C_-, satisfies the estimate, $\|K\|_{L^2(\gamma)} \le c|x|^{-3/10}$ (the precise value of the positive constant is not important at this point). Thus K is a contracting operator in $L^2(\gamma)$ for large enough $|x|$. Since KI is a square integrable function on γ, and since $\rho = \chi_- - I$ satisfies the singular integral equation $\rho = KI + K\rho$, it follows that $\rho \in L^2(\gamma)$ and therefore $\chi_- = I + \rho$ by iterations.

In addition, the solution found by iterations implies the following estimate of $\chi(\lambda)$ as $\lambda, |x| \to \infty$ (cf. Theorem 8.1 of Chapter 8),

$$\|\chi(\lambda) - I\| \le \frac{c}{1 + |\lambda|}|x|^{-2/5}. \tag{11.5.30}$$

Using (11.5.30) and (11.5.16) in (11.5.27), and taking into consideration the formula (11.3.15), we find that the Painlevé function corresponding to the above reduced RH problem has the following asymptotics as $|x| \to \infty$, $\arg(-x) \in \left[-\frac{\pi}{3}, 0\right]$,

$$u(x) = \sigma\sqrt{-\frac{x}{2}} + O(x^{-2/5}).$$

This completes the proof. □

Using the iterative series $\chi_- = \sum_{n=0}^{\infty} K^n I$, we can also establish the existence of the asymptotic series over the negative powers of $x^{1/2}$ representing the function $u_1(x|\alpha, \sigma)$ in the whole sector $\arg x \in [\frac{2\pi}{3}, \pi]$ (cf. [**DeZ2**]). The meromorphicity of $u_1(x|\alpha, \sigma)$ (the Painlevé property) allows us to differentiate this series and hence to use the differential equation (11.2.2) to determine its coefficients. An elementary analysis along these arguments leads to the following asymptotic formula for the solution $u_1(x|\alpha, \sigma)$,

$$u_1(x|\alpha, \sigma) \sim u_f(x|\alpha, \sigma), \quad x \to \infty, \quad \arg(-x) \in \left[-\frac{\pi}{3}, 0\right], \tag{11.5.31}$$

where $u_f(x|\alpha, \sigma)$ is the formal series,

$$u_f(x|\alpha, \sigma) = \sigma\sqrt{-\frac{x}{2}}\sum_{n=0}^{\infty} b_n(-x)^{-3n/2}, \tag{11.5.32}$$

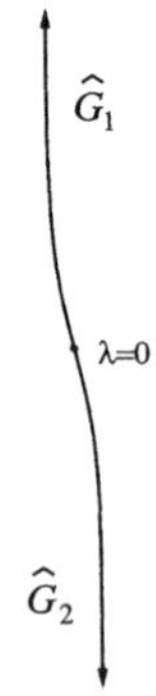

FIGURE 11.11. The RH problem graph for the correction function $\hat{\chi}(\lambda)$

whose coefficients b_n are determined uniquely by the recurrence relation,

$$b_0 = 1, \quad b_1 = \frac{\sigma\alpha}{\sqrt{2}}, \tag{11.5.33}$$

$$b_{n+2} = \frac{9n^2-1}{8} b_n - \sum_{m=1}^{n+1} b_m b_{n+2-m} - \frac{1}{2}\sum_{l=1}^{n+1}\sum_{m=1}^{n+2-l} b_l b_m b_{n+2-l-m}.$$

THE RH PROBLEM WITH $1 - s_1 s_3 = 0$ AND $s_1 + s_2 \neq 0$.

We now consider the case of $1 - s_1 s_3 = 0$, and arbitrary s_2, so that $s_1 + s_2 \neq 0$; see Figure 5. We look for the solution $\Psi(\lambda)$ in the form of the product

$$\Psi(\lambda) = \hat{\chi}(\lambda)\hat{\Psi}(\lambda), \tag{11.5.34}$$

where $\hat{\Psi}(\lambda)$ is the solution of the reduced RH problem, i.e., of the RH problem associated with $s_1 + s_2 = 0$. The correction function $\hat{\chi}(\lambda)$ satisfies the RH problem with jumps on the two curves $\operatorname{Im} g(\lambda) = \text{const}$, emanating from the origin and approaching the vertical direction (see Figure 11.11),

(1) $\hat{\chi}(\lambda) \to I, \quad \lambda \to \infty;$

(2) $\|\hat{\chi}(\lambda)\| \leq \text{const}, \quad \lambda \to +0;$

(3) $\hat{\chi}_+(\lambda) = \hat{\chi}_-(\lambda)\hat{G}(\lambda),$
$\hat{G}(\lambda) := \hat{G}_1(\lambda) = \hat{\Psi}_-(\lambda) S_2 S_4^{-1} \hat{\Psi}_-^{-1}(\lambda), \quad \lambda \in (0, +i\infty),$
$\hat{G}(\lambda) := \hat{G}_2(\lambda) = \hat{\Psi}_-(\lambda)\sigma_2 S_2 S_4^{-1}\sigma_2 \hat{\Psi}_-^{-1}(\lambda), \quad \lambda \in (0, -i\infty),$
$\sigma_2 \hat{G}_1(-\lambda)\sigma_2 = \hat{G}_2(\lambda).$

THEOREM 11.2. *If* $\alpha - \frac{1}{2} \notin \mathbb{Z}$, $\operatorname{Re}(\sigma\alpha) > -\frac{1}{2}$, $1 - s_1 s_3 = 0$, $\arg x \in [\frac{2\pi}{3}, \pi]$ *and* $|x|$ *is sufficiently large, then the RH problem* (11.3.6)*–*(11.3.13) *is uniquely solvable. The asymptotics of the corresponding Painlevé function as* $x \to \infty$ *is given by the equation*

$$\begin{aligned} u &= u_1(x|\alpha, \sigma) \\ &- \frac{s_1 + s_2}{\pi} 2^{-\frac{5}{2}\sigma\alpha - \frac{7}{4}} \Gamma(\tfrac{1}{2} + \sigma\alpha)(-x)^{-\frac{3}{2}\sigma\alpha - \frac{1}{4}} e^{-\frac{2\sqrt{2}}{3}(-x)^{3/2}} \\ &\times \big(1 + O(x^{-1/4})\big), \quad x \to \infty, \quad \arg(-x) \in \Big[-\frac{\pi}{3}, 0\Big], \end{aligned} \tag{11.5.35}$$

where $u_1(x|\alpha,\sigma) \sim \sigma\sqrt{-\frac{x}{2}}$ *is the solution of the Painlevé equation introduced by Theorem* 11.1 *(this solution corresponds to* $1 - s_1 s_3 = s_1 + s_2 = 0$*). The parameter* $\sigma \in \{+1, -1\}$ *is defined by the equation* $s_1 = e^{-i\pi\sigma(\alpha+\frac{1}{2})}$.

PROOF. It is enough to prove the solvability of the RH problem defining $\hat{\chi}(\lambda)$. In the neighborhood of the imaginary axis, $\hat{\Psi}(\lambda) = \chi(\lambda)\Phi^0(\lambda)$; see (11.5.27) and (11.5.26). Hence, using (11.5.14), we find the following expression for the jump matrix across $(0, +i\infty)$,

$$\hat{G}_1(\lambda) = I + (s_1 + s_2)\delta^2 e^{-2g}\chi Y\sigma_+ Y^{-1}\chi^{-1}, \tag{11.5.36}$$

$$\hat{G}_2(\lambda) = I - (s_1 + s_2)\delta^{-2} e^{2g}\chi Y\sigma_- Y^{-1}\chi^{-1}. \tag{11.5.37}$$

The jump contour coincides with the level line $\operatorname{Im} g(\lambda) = \text{const}$, thus the jump matrix $\hat{G}_1(\lambda)$ approaches the unit matrix exponentially, as $\lambda \to +i\infty$. Similarly, $\hat{G}_2(\lambda) - I$, decreases exponentially as $\lambda \to -i\infty$. For the exponential function $e^{g(\lambda)}$, the origin is the saddle point (see (11.5.17)).

The jump matrices have an algebraic singularity at $\lambda = 0$,

$$\begin{aligned}\hat{G}_1(\lambda) = I &+ (s_1 + s_2)e^{-2\pi i\nu}(-2x)^{-2\nu}\lambda^{4\nu}e^{-\frac{2\sqrt{2}}{3}(-x)^{3/2}+2\sqrt{2}(-x)^{1/2}\lambda^2} \\ &\times \tfrac{1}{2}(i\sigma\sigma_3 + \sigma_1)\big(1 + O(\lambda) + O(x^{-9/10})\big), \\ \hat{G}_2(\lambda) = I &- (s_1 + s_2)e^{2i\pi\nu}(-2x)^{-2\nu}\lambda^{4\nu}e^{-\frac{2\sqrt{2}}{3}(-x)^{3/2} + 2\sqrt{2}(-x)^{1/2}\lambda^2)} \\ &\times \tfrac{1}{2}(i\sigma\sigma_3 + \sigma_1)\big(1 + O(\lambda) + O(x^{-9/10})\big),\end{aligned}$$

where $\nu = \frac{\sigma\alpha}{2}$ (cf. (11.5.9)). Consider the model RH problem with the jump matrices $\tilde{G}_j(\lambda)$, $j = 1, 2$, of the form (11.5.36), (11.5.37), but with the matrix $\chi(\lambda)Y(\lambda)$ replaced by its asymptotics at $\lambda = 0$, i.e.,

$$\begin{aligned}\tilde{G}_1(\lambda) &= I + (s_1 + s_2)\delta^2 e^{-2g}\tfrac{1}{2}(i\sigma\sigma_3 + \sigma_1), \\ \tilde{G}_2(\lambda) &= I - (s_1 + s_2)\delta^{-2} e^{2g}\tfrac{1}{2}(i\sigma\sigma_3 + \sigma_1).\end{aligned}$$

Assume for the moment that

$$\operatorname{Re}(\sigma\alpha) \geq 0, \quad \text{i.e.,} \quad \operatorname{Re}\nu \geq 0. \tag{11.5.38}$$

Since the matrices $\tilde{G}_j(\lambda) - I$ are constant nilpotent matrices multiplied by scalar functions, the solution for $\tilde{\chi}$ is given by the Cauchy integral,

$$\begin{aligned}\tilde{\chi}(\lambda) &= I + \tfrac{s_1+s_2}{2\pi i}\Big\{\int_0^{+i\infty}\delta^2 e^{-2g}\tfrac{d\zeta}{\zeta-\lambda} + \int_{-i\infty}^0 \delta^{-2}e^{2g}\tfrac{d\zeta}{\zeta-\lambda}\Big\}\tfrac{1}{2}(i\sigma\sigma_3 + \sigma_1) \\ &= I + \tfrac{s_1+s_2}{2\pi i}\int_0^{+i\infty}\delta^2 e^{-2g}\tfrac{\lambda\, d\zeta}{\zeta^2-\lambda^2}(i\sigma\sigma_3 + \sigma_1).\end{aligned} \tag{11.5.39}$$

The condition (11.5.38) ensures the convergence of the above integrals. Obviously, $\tilde{\chi}(\lambda) \to I$ as $\lambda \to \infty$. Regarding the behavior at the origin, the same condition (11.5.38) ensures that $\|\tilde{\chi}(\lambda)\| \leq \text{const}$.

We will look for $\hat{\chi}(\lambda)$ in the form of the product

$$\hat{\chi}(\lambda) = X(\lambda)\tilde{\chi}(\lambda). \tag{11.5.40}$$

For the correction function $X(\lambda)$, we have the RH problem with the jump contour shown in Figure 11.11, but with different jump matrices:

(1) $X(\lambda) \to I, \quad \lambda \to \infty$;
(2) $\|X(\lambda)\| \leq \text{const}, \quad \lambda \to +0$;
(3) $X_+(\lambda) = X_-(\lambda)H(\lambda), \quad H(\lambda) = \tilde{\chi}_-(\lambda)\hat{G}(\lambda)\tilde{G}^{-1}(\lambda)\tilde{\chi}_-^{-1}(\lambda)$.

Consider the equivalent singular integral equation,

$$X_-(\lambda) = I + \frac{1}{2\pi i}\int_\gamma X_-(\zeta)\big(H(\zeta)-I\big)\frac{d\zeta}{\zeta-\lambda_-}, \tag{11.5.41}$$

where γ is the jump graph in Figure 11.11 and λ_- is the right limit of λ on the jump contour. This equation is of the form, $X_- = I + KX_-$, where the operator K is the superposition of the right multiplication by $H-I$ and of the Cauchy operator C_-. Using the boundedness of C_- in $L^2(\gamma)$, we find
(11.5.42)

$$\|K\|_{L^2(\gamma)} \le c\|H-I\|_{L^\infty(\gamma)} \le c'|x|^{-\frac12-5\nu}e^{-\frac{2\sqrt2}{3}|x|^{3/2}\cos\frac32(\arg x-\pi)} \le c''|x|^{-1/2},$$

where c, c' and c'' are positive constants. Taking into account the assumption $\mathrm{Re}\,\nu \ge 0$ (11.5.38), the singular integral operator K, for large enough $|x|$, $\arg x \in \left[\frac{2\pi}{3},\pi\right]$, defines a construction and therefore equation (11.5.41) is solvable by iterations.

To find the asymptotics of the relevant Painlevé function, it is enough to expand the function $\tilde\chi(\lambda)$ (11.5.39) at the vicinity of $\lambda=\infty$,

$$\begin{aligned}\tilde\chi(\lambda) &= I - \frac{s_1+s_2}{2\pi i\lambda}\int_0^{+i\infty}\delta^2 e^{-2g}\,d\zeta\big(1+O(\lambda^{-2})\big)(i\sigma\sigma_3+\sigma_1)\\ &= I + \lambda^{-1}h(i\sigma\sigma_3+\sigma_1) + O(\lambda^{-3}), \end{aligned}\tag{11.5.43}$$
$$h = -\frac{s_1+s_2}{2\pi}2^{-5\nu-\frac74}\Gamma(2\nu+\tfrac12)(-x)^{-3\nu-\frac14}e^{-\frac{2\sqrt2}{3}(-x)^{3/2}}\big(1+O(x^{-1/2})\big).$$

Thus, for $u(x)$ we have,

$$\begin{aligned}u(x) &= 2\lim_{\lambda\to\infty}\lambda\big(\Psi(\lambda)e^{\theta\sigma_3}\big)_{12} = 2\lim_{\lambda\to\infty}\lambda\big(X\tilde\chi\hat\Psi e^{\theta\sigma_3}\big)_{12}\\ &= u_1 + 2\hat h_1,\end{aligned}\tag{11.5.44}$$

where $\hat h_1 = h\big(1+O(x^{-2\nu-1/4})\big)$, involves the contribution of the correction function $X(\lambda)$. Recalling that $\nu=\sigma\alpha/2$, we conclude that the relation (11.5.44) yields the asymptotics (11.5.35) for $\mathrm{Re}\,(\sigma\alpha)\ge 0$, $\alpha-\frac12\notin\mathbb{Z}$.

The asymptotic formula (11.5.35) can be extended to the parameters α satisfying the condition $\mathrm{Re}\,(\sigma\alpha) > -\frac12$, with the help of the Bäcklund transformations. Indeed, the Schlesinger transformation of the Ψ-function (cf. (11.4.33)),

$$\tilde\Psi(\lambda) = R(\lambda)\Psi(\lambda),$$
$$R(\lambda) = I - \frac{q_\epsilon}{2i\lambda}(\sigma_3 - i\epsilon\sigma_1), \quad q_\epsilon = \frac{\alpha+\frac{\epsilon}{2}}{u_x - \epsilon(u^2+\frac{x}{2})}, \quad \epsilon\in\{+1,-1\},$$

yields the Ψ-function associated with the Painlevé transcendent (cf. (11.4.34))

$$\tilde u = u + \epsilon q_\epsilon, \quad \tilde u_x = u_x + \epsilon(\tilde u^2 - u^2), \quad \tilde\alpha = -\alpha-\epsilon. \tag{11.5.45}$$

The function $\tilde\Psi(\lambda)$ has the same Stokes multipliers, thus the definition of the parameter σ via $s_1 = e^{-i\pi\sigma(\alpha+\frac12)} = \tilde s_1 = e^{-i\pi\tilde\sigma(\tilde\alpha+\frac12)}$, implies the transformation, $\tilde\sigma = -\sigma$, for either value of $\epsilon=\pm1$, and hence the relation,

$$\tilde\sigma\tilde\alpha = \sigma\alpha + \sigma\epsilon. \tag{11.5.46}$$

Assume that $\sigma\epsilon = -1$. By a straightforward calculation, one checks that the asymptotic formula (11.5.35) is invariant with respect to the Bäcklund transformation (11.5.45), provided that $\mathrm{Re}\,(\sigma\alpha) > \frac12$. By virtue of (11.5.46), this observation establishes the validity of (11.5.35) for $\mathrm{Re}\,(\sigma\alpha) > -\frac12$. □

REMARK 11.4. The use of the Bäcklund transformation (11.5.45) allows one to obtain the asymptotics of the relevant Painlevé functions without any restrictions on the real part of the product $\sigma\alpha$. In fact, one can see that the asymptotic formula (11.5.35) is still valid inside the sector $[\frac{2\pi}{3}, \pi]$ for arbitrary values of $\operatorname{Re}(\sigma\alpha)$, while on the boundary, $\arg x = \frac{2\pi}{3}$ the behavior of the function $u(x)$ changes quite significantly if $\operatorname{Re}(\sigma\alpha) \le -\frac{1}{2}$.

THE INCREASING DEGENERATE (TRONQUÉE) PAINLEVÉ FUNCTIONS

Applying the second of the symmetries (11.3.26) to equation (11.5.35), we obtain the following result.

THEOREM 11.3. *If* $\alpha - \frac{1}{2} \notin \mathbb{Z}$, $\operatorname{Re}(\sigma\alpha) > -\frac{1}{2}$, $1 - s_3 s_1 = 0$, *and* $\arg x \in [\pi, \frac{4\pi}{3}]$, *then the asymptotics of the corresponding Painlevé function as* $x \to \infty$ *is given by the equation*

$$(11.5.47)\qquad u = u_0(x|\alpha,\sigma) \\ - \frac{s_3+s_2}{\pi} 2^{-\frac{5}{2}\sigma\alpha - \frac{7}{4}} \Gamma(\tfrac{1}{2}+\sigma\alpha)(-x)^{-\frac{3}{2}\sigma\alpha - \frac{1}{4}} e^{-\frac{2\sqrt{2}}{3}(-x)^{3/2}} \big(1 + O(x^{-1/4})\big), \\ x \to \infty, \quad \arg(-x) \in \left[0, \frac{\pi}{3}\right],$$

where $u_0(x|\alpha,\sigma) = \overline{u_1(\overline{x}|\overline{\alpha},\sigma)} \sim \sigma\sqrt{-\frac{x}{2}}$ *is the solution of the Painlevé equation corresponding to* $1 - s_3 s_1 = s_3 + s_2 = 0$. *The parameter* $\sigma \in \{+1,-1\}$ *is defined via the equation* $s_1 = e^{-i\pi\sigma(\alpha+\frac{1}{2})}$.

In analogy with the solution $u_1(x|\alpha,\sigma)$, the solution $u_0(x|\alpha,\sigma)$ admits the same asymptotic expansion (11.5.31), but in a different sector, namely

$$(11.5.48)\qquad u_0(x|\alpha,\sigma) \sim u_f(x|\alpha,\sigma), \quad x \to \infty, \quad x \in \left[\pi, \frac{4\pi}{3}\right], \\ \arg(-x) \in \left[0, \frac{\pi}{3}\right],$$

where $u_f(x|\alpha,\sigma)$ is the formal series defined in (11.5.32).

The solutions $u_1(x|\alpha,\sigma)$ and $u_0(x|\alpha,\sigma) = \overline{u_1(\overline{x}|\overline{\alpha},\sigma)}$ are meromorphic functions of $x \in \mathbb{C}$, and thus they can be continued beyond the sectors indicated in Theorems 11.2 and 11.3. To find the asymptotics of the solution $u_0(x|\alpha,\sigma)$ in the interior of the sector $\arg x \in [\frac{2\pi}{3}, \pi]$, we apply (11.5.35) with $s_2 = -s_3$. Similarly, using (11.5.47) with $s_2 = -s_1$, we find the asymptotics of the solution $u_1(x|\alpha,\sigma)$ in the interior of the sector $\arg x \in [\pi, \frac{4\pi}{3}]$. Either expression implies that, if $|x| \to \infty$, $\arg x \in [\frac{2\pi}{3}, \frac{4\pi}{3}]$, $\arg(-x) \in [-\frac{\pi}{3}, \frac{\pi}{3}]$, then

$$(11.5.49)\qquad u_0(x,\alpha,\sigma) - u_1(x,\alpha,\sigma) \\ = i\sigma \frac{2^{-\frac{5}{2}\sigma\alpha - \frac{3}{4}}}{\Gamma(\frac{1}{2}-\sigma\alpha)} (-x)^{-\frac{3}{2}\sigma\alpha - \frac{1}{4}} e^{-\frac{2\sqrt{2}}{3}(-x)^{3/2}} \big(1 + O(x^{-1/4})\big),$$

where we have used the relation

$$(s_1 - s_3)\Gamma(\tfrac{1}{2}+\sigma\alpha) = -\frac{2i\pi\sigma}{\Gamma(\frac{1}{2}-\sigma\alpha)}.$$

Due to the exponential decay of the difference (11.5.49) in the interior of the sector $\arg x \in \left(\frac{2\pi}{3}, \frac{4\pi}{3}\right)$, we have from (11.5.49) that in fact the solutions $u_1(x|\alpha,\sigma)$ and

$u_0(x|\alpha,\sigma)$ have the asymptotics given by the formal series $u_f(x|\alpha,\sigma)$ in the semi-closed sectors, which are bigger than the original sectors. Indeed, we have that

(11.5.50) $$u_1(x|\alpha,\sigma) \sim u_f(x|\alpha,\sigma), \quad |x| \to \infty, \quad \arg x \in \left[\tfrac{2\pi}{3}, \tfrac{4\pi}{3}\right), \quad \arg(-x) \in \left[-\tfrac{\pi}{3}, \tfrac{\pi}{3}\right),$$

and

(11.5.51) $$u_0(x|\alpha,\sigma) \sim u_f(x|\alpha,\sigma), \quad |x| \to \infty, \quad \arg x \in \left(\tfrac{2\pi}{3}, \tfrac{4\pi}{3}\right], \quad \arg(-x) \in \left(-\tfrac{\pi}{3}, \tfrac{\pi}{3}\right].$$

In view of these asymptotic relations we conclude that formula (11.5.49) constitutes the quasi-linear Stokes phenomenon for the increasing degenerate solution of PII (cf. (11.1.15).

Theorems 11.2 and 11.3 imply that the condition

(11.5.52) $$1 - s_1 s_3 = 0 \quad \Leftrightarrow \quad s_1 = e^{-i\pi\sigma(\alpha+\frac{1}{2})}, \quad s_3 = e^{i\pi\sigma(\alpha+\frac{1}{2})},$$

singles out a one-parameter family of solutions of PII, given by $u(x) \equiv u(x; s_2|\alpha,\sigma)$ and characterized by the asymptotic condition,

(11.5.53) $$u(x; s_2|\alpha,\sigma) \sim u_f(x|\alpha,\sigma), \quad |x| \to \infty, \quad \arg x \in \left(\tfrac{2\pi}{3}, \tfrac{4\pi}{3}\right), \quad \arg(-x) \in \left(-\tfrac{\pi}{3}, \tfrac{\pi}{3}\right).$$

The *unique* member of this family, which has the asymptotics (11.5.53) up to the upper boundary of the sector (i.e., $\arg x = \frac{4\pi}{3}$), is exactly the function $u_0(x|\alpha,\sigma)$. This solution corresponds to the choice of the free parameter

$$s_2 = -s_3 = -e^{i\pi\sigma(\alpha+\frac{1}{2})}.$$

In other words,

(11.5.54) $$u_0(x|\alpha,\sigma) = u\left(x; -e^{i\pi\sigma(\alpha+\frac{1}{2})}|\alpha,\sigma\right).$$

Similarly, the unique member of the family which has the asymptotics (11.5.53) up to the lower boundary of the sector (i.e., $\arg x = \frac{2\pi}{3}$), is exactly the function $u_1(x|\alpha,\sigma)$. This solution corresponds to the choice of the free parameter

$$s_2 = -s_1 = -e^{-i\pi\sigma(\alpha+\frac{1}{2})}.$$

In other words,

(11.5.55) $$u_1(x|\alpha,\sigma) = u\left(x; -e^{-i\pi\sigma(\alpha+\frac{1}{2})}|\alpha,\sigma\right).$$

REMARK 11.5. The parameter s_2 of the family (11.5.53), can be uniquely determine without any reference to the associated monodromy problem. Indeed, Theorems 11.2 and 11.3 imply that for any member of family (11.5.53), the following asymptotic formulae are valid:

(11.5.56) $$u = \sigma\sqrt{-\frac{x}{2}}\sum_{n=0}^{N} b_n(-x)^{-3n/2} + O\big(x^{-3N/2-1}\big) + c_+(-x)^{-\frac{3}{2}\sigma\alpha-\frac{1}{4}}e^{-\frac{2\sqrt{2}}{3}(-x)^{3/2}}\big(1+O(x^{-1/4})\big),$$
$$x \to \infty, \quad \arg x \in \left(\frac{2\pi}{3}, \frac{4\pi}{3}\right] \quad \arg(-x) \in \left(-\frac{\pi}{3}, \frac{\pi}{3}\right],$$

and

$$u = \sigma\sqrt{-\frac{x}{2}}\sum_{n=0}^{N} b_n(-x)^{-3n/2} + O\big(x^{-3N/2}\big) \tag{11.5.57}$$

$$+c_-(-x)^{-\frac{3}{2}\sigma\alpha-\frac{1}{4}}e^{-\frac{2\sqrt{2}}{3}(-x)^{3/2}}\big(1+O(x^{-1/4})\big),$$

$$x\to\infty,\quad \arg x\in\Big[\frac{2\pi}{3},\frac{4\pi}{3}\Big)\quad \arg(-x)\in\Big[-\frac{\pi}{3},\frac{\pi}{3}\Big),$$

On the boundaries of the sector the exponential term oscillates, thus the coefficients c_+ and c_- are determined uniquely by the above asymptotics, and each of them can be taken as an "internal" parameter of the family (11.5.53). The "external" parameter s_2 can be recovered from either of the following two relations,

$$c_+ = -\frac{e^{i\pi\sigma(\alpha+\frac{1}{2})}+s_2}{\pi}2^{-\frac{5}{2}\sigma\alpha-\frac{7}{4}}\Gamma(\tfrac{1}{2}+\sigma\alpha), \tag{11.5.58}$$

or

$$c_- = -\frac{e^{-i\sigma(\alpha+\frac{1}{2})}+s_2}{\pi}2^{-\frac{5}{2}\sigma\alpha-\frac{7}{4}}\Gamma(\tfrac{1}{2}+\sigma\alpha). \tag{11.5.59}$$

Equations (11.5.58) and (11.5.59) imply the nontrivial relation,

$$c_+ - c_- = -\frac{2\pi i\sigma}{\Gamma(\frac{1}{2}-\sigma\alpha)}2^{-\frac{5}{2}\sigma\alpha-\frac{7}{4}}. \tag{11.5.60}$$

This relation constitutes an explicit description of the quasi-linear Stokes phenomenon for the family (11.5.53).

The existence of the one-parameter family of Painlevé functions characterized by the asymptotic condition (11.5.53) was due to Boutroux. He called such solutions *tronquée*. However, to our knowledge, the exact Stokes relation (11.5.60) is absent in the classical literature. We believe that for the derivation of (11.5.60), as well as for the derivation of the general connection formulae for Painlevé equation, the use of the Riemann-Hilbert technique is essential.

REMARK 11.6. The third assertion of Theorem 10.2 is now straightforward. Indeed, the Painlevé function $u(x)$, which is real for x real, corresponds to the choice $s_1 = \overline{s}_3$, $s_2 = \overline{s}_2$, $\alpha = \overline{\alpha}$. Under these conditions, for $\alpha - \frac{1}{2} \notin \mathbb{Z}$, neither $u_1(x|\alpha,\sigma)$ nor $u_0(x|\alpha,\sigma)$ in Theorems 11.2 and 11.3 appear to be real. However, since the Painlevé function u is real, the imaginary parts of both $u_1(x|\alpha,\sigma)$ and $u_0(x|\alpha,\sigma)$ cancel with the imaginary part of the exponentially small correction that comes from the explicit expressions of the parameters s_1 and s_3 (e.g. $s_1 = \cos\pi(\alpha+\frac{1}{2}) - i\sigma\sin\pi(\alpha+\frac{1}{2})$).

Thus we find that, for x on the real line,

$$\begin{aligned}\mathrm{Im}\ u_0(x|\alpha,\sigma) &= -\,\mathrm{Im}\ u_1(x|\alpha,\sigma)\\ &= \frac{1}{2}i\sigma\frac{2^{-\frac{5}{2}\sigma\alpha-\frac{3}{4}}}{\Gamma(\frac{1}{2}-\sigma\alpha)}(-x)^{-\frac{3}{2}\sigma\alpha-\frac{1}{4}}e^{-\frac{2\sqrt{2}}{3}(-x)^{3/2}}\big(1+O(x^{-1/4})\big),\end{aligned}$$

which is just half of the difference (11.5.49), and

$$\begin{aligned}u &= \mathrm{Re}\ u_1(x|\alpha,\sigma)\\ &\quad -\frac{s_2+\cos\pi(\alpha+\frac{1}{2})}{\pi}2^{-\frac{5}{2}\sigma\alpha-\frac{7}{4}}\Gamma(\tfrac{1}{2}+\sigma\alpha)(-x)^{-\frac{3}{2}\sigma\alpha-\frac{1}{4}}e^{-\frac{2\sqrt{2}}{3}(-x)^{3/2}}\\ &\quad \times\big(1+O(x^{-1/4})\big).\end{aligned} \tag{11.5.61}$$

This expression can be interpreted as the half-sum of the representations (11.5.35) and (11.5.47). Putting $\alpha = 0$ in (11.5.61), we arrive at the third assertion of Theorem 10.2.

Applying the rotational symmetry of (11.3.26) to u_1 and u_0, we find the solutions

$$(11.5.62)\qquad u = u_{2n+1}(x|\alpha,\sigma) := e^{i\frac{2\pi}{3}n} u_1(e^{i\frac{2\pi}{3}n}x|\alpha,\sigma) \sim u_f(x|\alpha,\sigma)$$

$$|x| \to \infty, \quad \arg x \equiv \pi + \arg(-x) \in \left[\tfrac{2\pi}{3} - \tfrac{2\pi}{3}n, \tfrac{4\pi}{3} - \tfrac{2\pi}{3}n\right),$$

corresponding to

$$s_{2n} = -e^{i\pi\sigma(\alpha+\frac{1}{2})}, \quad s_{2n+1} = e^{-i\pi\sigma(\alpha+\frac{1}{2})}, \quad s_{2n+2} = -e^{-i\pi\sigma(\alpha+\frac{1}{2})};$$

and

$$(11.5.63)\qquad u = u_{2n}(x|\alpha,\sigma) := e^{i\frac{2\pi}{3}n} u_0(e^{i\frac{2\pi}{3}n}x|\alpha,\sigma) \sim u_f(x|\alpha,\sigma),$$

$$|x| \to \infty, \quad \arg x \in \left(\tfrac{2\pi}{3} - \tfrac{2\pi}{3}n, \tfrac{4\pi}{3} - \tfrac{2\pi}{3}n\right],$$

corresponding to

$$s_{2n} = -e^{i\pi\sigma(\alpha+\frac{1}{2})}, \quad s_{2n+1} = e^{-i\pi\sigma(\alpha+\frac{1}{2})}, \quad s_{2n+2} = -e^{i\pi\sigma(\alpha+\frac{1}{2})}.$$

Here we allow n to be any integer number and the Stokes multipliers s_n are defined modulo the relations

$$s_{n+3} = -s_n.$$

Comparing the Stokes multipliers defined in (11.5.62) and (11.5.63), we observe the following symmetries,

$$(11.5.64)\qquad u_n(x|\alpha,\sigma) = u_{n+6}(x|\alpha,\sigma), \quad u_{2n-1}(x|\alpha,\sigma) = u_{2n}(x|\alpha,-\sigma).$$

In other words, among the above solutions there exist only three different functions:

$$u_0(x|\alpha,\sigma), \quad u_1(x|\alpha,\sigma). \quad \text{and} \quad u_3(x|\alpha,\sigma).$$

Equation (11.5.49) and the definitions (11.5.62) and (11.5.63) imply that

$$(11.5.65)\qquad u_{2n}(x|\alpha,\sigma) - u_{2n+1}(x|\alpha,\sigma)$$
$$= i\sigma\frac{2^{-\frac{5}{2}\sigma\alpha-\frac{3}{4}}}{\Gamma(\frac{1}{2}-\sigma\alpha)} e^{i\frac{2\pi}{3}n}(e^{-i\pi}e^{i\frac{2\pi}{3}n}x)^{-\frac{3}{2}\sigma\alpha-\frac{1}{4}} e^{-\frac{2\sqrt{2}}{3}(e^{-i\pi}e^{i\frac{2\pi}{3}n}x)^{3/2}}\left(1 + O(x^{-1/4})\right),$$
$$|x| \to \infty, \quad \arg x \in \left[\tfrac{2\pi}{3} - \tfrac{2\pi}{3}n, \tfrac{4\pi}{3} - \tfrac{2\pi}{3}n\right].$$

These differences, which are exponentially small in the interior of the relevant sectors, represent the quasi-linear Stokes relations for the corresponding pair of the solutions.

Using the second of the equations (11.5.64), we can establish the following extension of the asymptotics of the solutions $u_n(x)$ as $|x| \to \infty$:

$$(11.5.66)\qquad u_{2n-1}(x|\alpha,(-1)^{n-1}\sigma) = u_{2n}(x|\alpha,(-1)^n\sigma) \sim u_f(x|\alpha,\sigma),$$
$$|x| \to \infty, \quad \arg x \equiv \pi + \arg(-x) \in \left(\tfrac{2\pi}{3}(1-n), \tfrac{2\pi}{3}(3-n)\right).$$

In particular, this means that

$$
\begin{aligned}
& u_0(x|\alpha,\sigma) \sim u_f(x|\alpha,\sigma), \quad |x| \to \infty, \\
& \qquad \arg x \equiv \pi + \arg(-x) \in \big(\frac{2\pi}{3}, 2\pi\big), \\
& u_1(x|\alpha,\sigma) \sim u_f(x|\alpha,\sigma), \quad |x| \to \infty, \\
& \qquad \arg x \equiv \pi + \arg(-x) \in \big(0, \frac{4\pi}{3}\big), \\
& u_3(x|\alpha,-\sigma) \sim u_f(x|\alpha,\sigma), \quad |x| \to \infty, \\
& \qquad \arg x \equiv \pi + \arg(-x) \in \big(-\frac{2\pi}{3}, \frac{2\pi}{3}\big).
\end{aligned}
\tag{11.5.67}
$$

The solutions $u_0(x|\alpha,\sigma)$, $u_1(x|\alpha,\sigma)$, $u_3(x|\alpha,\sigma)$, are uniquely defined by the asymptotics (11.5.67). In Boutroux's terminology, these are the *tri-tronquée* solutions of PII. The contribution to the classical theory of these solutions, is the exact evaluation of the associated quasi-linear Stokes phenomenon (see (11.5.65)).

By using the same rotational symmetry, it is possible to obtain two more families of the increasing tronquée solutions from the original family (11.5.53).

ASYMPTOTICS OF THE COEFFICIENTS FOR $\alpha - \frac{1}{2} \notin \mathbb{Z}$

Using the recurrence relations (11.5.33), we can find the first few terms of the formal series $u_f(x|\alpha,\sigma)$:

$$
\begin{aligned}
u_f(x|\alpha,\sigma) = \sigma\sqrt{-\frac{x}{2}}\Big\{1 &+ \frac{\sigma\alpha}{\sqrt{2}(-x)^{3/2}} - \frac{1+6\alpha^2}{8(-x)^3} + \frac{\sigma\alpha(11+16\alpha^2)}{8\sqrt{2}(-x)^{9/2}} \\
&- \frac{73+708\alpha^2+420\alpha^4}{128(-x)^6} + \frac{\sigma\alpha(1021+2504\alpha^2+768\alpha^4)}{64\sqrt{2}(-x)^{15/2}} \\
&- \frac{10657+129918\alpha^2+132060\alpha^4+24024\alpha^6}{1024(-x)^9} \\
&+ \frac{\sigma\alpha(248831+786304\alpha^2+416400\alpha^4+49152\alpha^6)}{512\sqrt{2}(-x)^{21/2}} + O\big(x^{-12}\big)\Big\}.
\end{aligned}
\tag{11.5.68}
$$

To find the asymptotics of the coefficients b_n in (11.5.32) as $n \to \infty$, let us construct a piecewise holomorphic function $\hat{u}(t)$,

$$
\begin{aligned}
& \hat{u}(t) = u_{2n-1}(e^{i\pi}t^2|\alpha,(-1)^{n-1}\sigma) = u_{2n}(e^{i\pi}t^2|\alpha,(-1)^n\sigma), \\
& \arg t \in \big(-\tfrac{\pi}{3}n, \tfrac{\pi}{3} - \tfrac{\pi}{3}n\big).
\end{aligned}
\tag{11.5.69}
$$

The function $\hat{u}(t)$ has a finite number of simple poles and, by construction, enjoys the uniform asymptotics near infinity, $\hat{u}(t) \simeq \sigma t/\sqrt{2}$ as $|t| \to \infty$. Using (11.5.32), it follows that $\hat{u}(t)$ has the following asymptotic series expansion,

$$
\hat{u}(t) \sim \sigma\frac{t}{\sqrt{2}}\sum_{n=0}^{\infty} b_n t^{-3n}. \tag{11.5.70}
$$

Let $u^{(N)}(x)$ be a partial sum

$$
u^{(N)}(t) = \sigma\frac{t}{\sqrt{2}}\sum_{n=0}^{N-1} b_n t^{-3n},
$$

and let $w^{(N)}(t)$ be defined by

$$w^{(N)}(t) = t^{3N-2}(\hat{u}(t) - u^{(N)}(t)) = \frac{\sigma}{t\sqrt{2}} \sum_{n=0}^{\infty} b_{n+N} t^{-3n}.$$

The function $t^{3N-2}u^{(N)}(t)$ is polynomial and the function $\hat{u}(t)/t$ is bounded as $|t| \geq \rho$, thus the integral of $v^{(N)}(t)$ along the circle of radius $|t| = \rho$ containing all the pole singularities of $\hat{u}(t)$, satisfies the estimate

$$\left| \oint_{|t|=\rho} w^{(N)}(t)\, dt \right| \leq \rho^{3N-2} \oint_{|t|=\rho} |\hat{u}(t)|\, dl \leq 2\pi\rho^{3N} \max_{|t|=\rho} |\hat{u}(t)/t| = C\rho^{3N}. \tag{11.5.71}$$

On the other hand,

$$\oint_{|t|=\rho} w^{(N)}(t)\, dx = \oint_{|t|=R} w^{(N)}(t)\, dt + \sum_{n=-3}^{2} \int_{e^{i\frac{\pi}{3}n}(\rho,R)} \left(w_+^{(N)}(t) - w_-^{(N)}(t)\right) dt. \tag{11.5.72}$$

Since $w^{(N)}(t) = \frac{\sigma}{t\sqrt{2}} b_N + O(t^{-3})$, the first of the integrals in the r.h.s. of (11.5.72) is

$$\oint_{|x|=R} w^{(N)}(t)\, dt = \pi i \sigma \sqrt{2}\, b_N + O(R^{-2}). \tag{11.5.73}$$

The last six integrals in (11.5.72) can be computed using (11.5.69), (11.5.62), (11.5.63), (11.5.64) and (11.5.49):

$$\begin{aligned}
&\sum_{n=-3}^{2} \int_{e^{i\frac{\pi}{3}n}(\rho,R)} \left(w_+^{(N)}(t) - w_-^{(N)}(t)\right) dt \\
&= 3\int_\rho^R t^{3N-2}\left(u_0(e^{i\pi}t^2, \alpha, \sigma) - u_1(e^{i\pi}t^2, \alpha, \sigma)\right) dt \\
&+3(-1)^{N-1} \int_\rho^R t^{3N-2}\left(u_0(e^{i\pi}t^2, \alpha, -\sigma) - u_1(e^{i\pi}t^2, \alpha, -\sigma)\right) dt \\
&= 3i\sigma \frac{2^{-\frac{5}{2}\sigma\alpha - \frac{3}{4}}}{\Gamma(\frac{1}{2} - \sigma\alpha)} \int_\rho^R t^{3(N-\sigma\alpha)-\frac{5}{2}} e^{-\frac{2\sqrt{2}}{3}t^3} \left(1 + O(t^{-1/2})\right) dt \\
&+3(-1)^N i\sigma \frac{2^{\frac{5}{2}\sigma\alpha - \frac{3}{4}}}{\Gamma(\frac{1}{2} + \sigma\alpha)} \int_\rho^R t^{3(N+\sigma\alpha)-\frac{5}{2}} e^{-\frac{2\sqrt{2}}{3}t^3} \left(1 + O(t^{-1/2})\right) dt \\
&= i\sigma 2^{-\frac{3}{2}N} 3^{N-\frac{1}{2}} \Big\{ 6^{-\sigma\alpha} \frac{\Gamma(N - \sigma\alpha - \frac{1}{2})}{\Gamma(\frac{1}{2} - \sigma\alpha)} \left(1 + O(N^{-1/6})\right) \\
&+(-1)^N 6^{\sigma\alpha} \frac{\Gamma(N + \sigma\alpha - \frac{1}{2})}{\Gamma(\frac{1}{2} + \sigma\alpha)} \left(1 + O(N^{-1/6})\right) \Big\} \\
&+O(\rho^{3(N+|\mathrm{Re}\alpha|)-\frac{3}{2}}) + O\big(e^{-\frac{2\sqrt{2}}{3}R^3} R^{3(N+|\mathrm{Re}\alpha|)-\frac{9}{2}}\big).
\end{aligned} \tag{11.5.74}$$

Thus, letting $R = \infty$, we find the asymptotics as $N \to \infty$, of the coefficient b_N in (11.5.32),

$$b_N = -\frac{1}{\pi\sqrt{6}}\Big(\frac{3}{2\sqrt{2}}\Big)^N \Big\{6^{-\sigma\alpha}\frac{\Gamma(N-\sigma\alpha-\frac{1}{2})}{\Gamma(\frac{1}{2}-\sigma\alpha)}(1+O(N^{-1/6}))$$

$$+(-1)^N 6^{\sigma\alpha}\frac{\Gamma(N+\sigma\alpha-\frac{1}{2})}{\Gamma(\frac{1}{2}+\sigma\alpha)}(1+O(N^{-1/6}))\Big\} + O(\rho^{3N}). \tag{11.5.75}$$

In particular, for $\alpha = \frac{1}{2}$, the two-term asymptotics (11.5.75) reduces into one-term asymptotics for the classical solutions (11.4.32) found earlier. Furthermore, it is possible to prove the validity of the same asymptotic formula (11.5.75), for any half-integer α.

In the case $\alpha = 0$, the recurrence (11.5.33) implies that all the odd coefficients vanish, $b_{2n-1} = 0$. The asymptotics (11.5.75) is consistent with this fact. At the same time, for the even coefficients the asymptotics (11.5.75) reduces to the one-term expression. In other words, if $\alpha = 0$, then

$$b_{2N-1} = 0,$$

$$b_{2N} = -\frac{\sqrt{2}}{\pi^{3/2}\sqrt{3}}\Big(\frac{3}{2\sqrt{2}}\Big)^{2N}\Gamma(2N-\tfrac{1}{2})(1+O(N^{-1/6})) + O(\rho^{3N}).$$

REMARK 11.7. The error in (11.5.35), (11.5.47) and (11.5.49) can be improved to $O(x^{-3/2})$. This implies the error $O(N^{-1})$ for (11.5.75).

THE CASE $\alpha - \frac{1}{2} = n \in \mathbb{Z}$ AND $1 - s_1 s_3 = 0$ REVISITED

If $\alpha - \frac{1}{2} = n \in \mathbb{Z}$, then the assumption (11.3.30) yields equations (11.3.31), which become

$$s_1 = s_3 = (-1)^{n+1}. \tag{11.5.76}$$

Thus the values of the Stokes multipliers cannot be used to determine the value of the parameter $\sigma = \pm 1$.

At the nonspecial points of the monodromy surface (11.3.1), and for α equal to half-integer, the connection matrix E satisfies the condition (11.3.33). Then, the inequality $s_1 + s_2 \neq 0$ is equivalent to the choice of the nonspecial point of the monodromy surface.

Choose the parameter $\sigma = \pm 1$ according to the inequality (11.5.38), i.e.,

$$\sigma\alpha > 0.$$

Then the construction above provides us with the solution of the RH problem (11.3.6)–(11.3.13) for $\alpha - \frac{1}{2} = n \in \mathbb{Z}$, $1 - s_1 s_3 = 0$ and arbitrary s_2. This, in turn, provides us with the tri-tronquée Painlevé function $u_1(x|\alpha,\sigma)$ (11.5.6) for α equal to half-integer,

$$u_1(x|\alpha,\sigma) \sim u_f(x|\alpha,\sigma), \quad \sigma\alpha > 0, \quad s_1 + s_2 = 0, \quad |x| \to \infty,$$
$$\arg x = \pi + \arg(-x) \in \big[\tfrac{2\pi}{3}, \pi\big],$$

and with the corresponding one-parameter family of tronquée solutions (11.5.35). Since the second of the symmetries (11.3.26) does not affect the Stokes multipliers, $s_1 = s_3 = (-1)^{n+1}$ and $s_2 = (-1)^n$, the solution $u_1(x|\alpha,\sigma)$ defined in (11.5.6) coincides with the solution $u_0(x|\alpha,\sigma) = \overline{u_1(\overline{x}|\overline{\alpha},\sigma)}$. This observation is consistent with the absence, for $\alpha - \frac{1}{2} = n \in \mathbb{Z}$ and $\sigma\alpha > 0$, of the quasi-linear Stokes phenomenon described by (11.5.49).

Furthermore, if $\alpha = \frac{1}{2}$, the uniqueness of the RH problem solution implies that $u_1(x|\frac{1}{2},1)$ must coincide with the classical solution $u_1^c(x|\frac{1}{2})$ (11.4.15),

$$u_1\left(x|\frac{1}{2},1\right) \equiv u_1^c\left(x|\frac{1}{2}\right).$$

This is nothing but the tri-tronquée background of the 1-parameter family of (non-classical!) tronquée solutions $u(x|\frac{1}{2},1,s_2)$ of PII with $\alpha = \frac{1}{2}$.

Similar analysis can be performed for the other members of the family of the classical solutions.

For the other choice of the sign of the product $\sigma\alpha$, i.e., for

$$\sigma\alpha < 0,$$

the unique solvability of the RH problem implies that all Painlevé functions coincide with the above classical solutions corresponding to the special points of the monodromy surface. Thus, for half-integer α, the sign of the parameter $\sigma\alpha$ distinguishes classical from nonclassical tronquée solutions of PII.

6. Asymptotic solution for $s_2 = 0$

Throughout this section, we assume that $\alpha - \frac{1}{2} \notin \mathbb{Z}$.

Let us put $s_2 = 0$ in the original RH problem i)–iv) of Section 3, and let us assume that $|x| \to \infty$ in the sector $\arg x \in [0, \frac{\pi}{3}]$. Equations (11.3.1) imply that $s_5 = 0$ and hence the function $\Psi(\lambda)$ has no jumps across two rays $\arg\lambda = \frac{\pi}{2}, \frac{3\pi}{2}$, $|\lambda| > r$; see Figure 11.2. The remaining parameters s_1 and s_3 satisfy the constraint

$$s_1 + s_3 = -2\sin\pi\alpha. \tag{11.6.1}$$

A direct calculation using (11.3.2) shows that the matrix E has the form

$$\mathsf{E} = \begin{pmatrix} p & \\ & q \end{pmatrix}\begin{pmatrix} 1 & ie^{-i\pi\alpha} \\ 1 & -ie^{i\pi\alpha} \end{pmatrix}, \qquad pq = -\frac{1}{2i\cos\pi\alpha}. \tag{11.6.2}$$

Recall the piecewise holomorphic function $Y(\lambda)$ introduced in (11.3.14), i.e.,

$$Y(\lambda) = \Psi(\lambda)e^{\theta(\lambda)\sigma_3}, \quad \theta(\lambda) = i\big(\frac{4}{3}\lambda^3 + x\lambda\big).$$

This function satisfies the following RH problem (cf. (11.3.20), (11.3.13), and (11.3.8)):

i) $\quad Y(\lambda) \to I$ as $\lambda \to \infty$, $\quad Y_{\text{right}}(\lambda)\lambda^{-\alpha\sigma_3} \in H(\Omega_0) \cap C(\overline{\Omega}_0)$;

ii) $\quad Y_+(\lambda) = Y_-(\lambda)G(\lambda)$, $\quad G(\lambda) = e^{-\theta\sigma_3}\mathsf{S}(\lambda)e^{\theta\sigma_3}$, $\quad \lambda \in \gamma \setminus \{0\}$.

If $\mathsf{S}(\lambda) = I + s\sigma_\pm$, then $G(\lambda) = I + se^{\mp 2\theta}\sigma_\pm$, and, according to the nonlinear steepest descend methodology, our first task is to transform the jump contour γ to the contour of the steepest descent for the matrix $G(\lambda) - I$.

Let us denote by γ_+ the level curve $\operatorname{Im}\theta(\lambda) = \text{const}$, passing through the stationary phase point $\lambda_+ = \frac{i}{2}x^{1/2}$ and asymptotic to the rays $\arg\lambda = \frac{\pi}{6}, \frac{5\pi}{6}$. This is the steepest descent curve for $e^{2\theta}$. Similarly, let us denote by γ_-, the level curve passing through the stationary phase point $\lambda_- = -\frac{i}{2}x^{1/2}$ and asymptotic to the

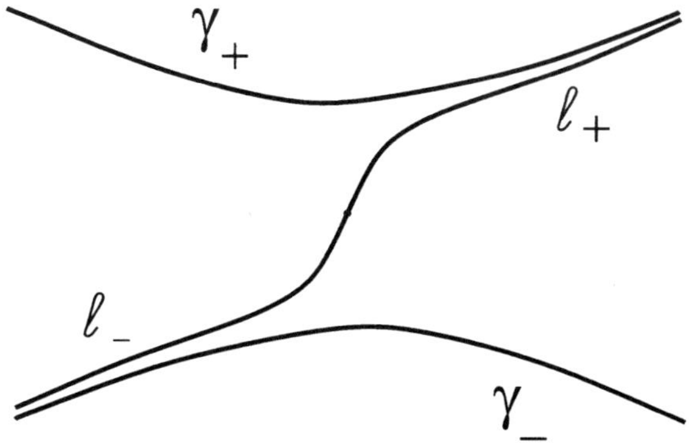

FIGURE 11.12. The level curves Im $\theta(\lambda) = \text{const}$, for $\arg x \in (0, \frac{\pi}{3}]$

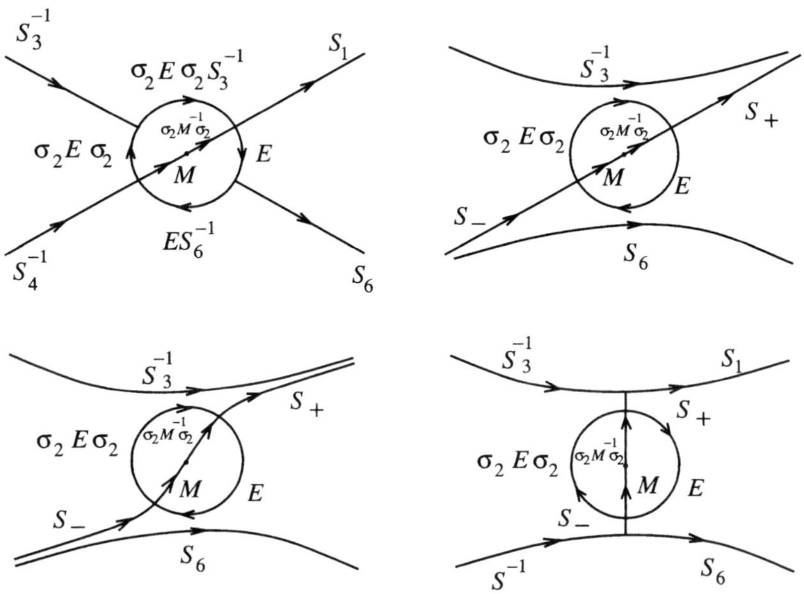

FIGURE 11.13. The graphs for the equivalent RH problems ($s_2 = 0$) with the associated jump matrices.

rays $\arg \lambda = -\frac{\pi}{6}, -\frac{5\pi}{6}$. This is the steepest descent curve for $e^{-2\theta}$. The curves γ_+ and γ_- are depicted in Figure 11.12.

Let us also denote by ℓ_+ and ℓ_- the level curves emanating from the origin and asymptotic to the respective rays $\arg \lambda = \frac{\pi}{6}$ and $\arg \lambda = -\frac{5\pi}{6}$. The curves ℓ_+ and ℓ_- are the steepest descent curves for $e^{2\theta}$ and $e^{-2\theta}$, respectively. For $\arg x \in (0, \frac{\pi}{3}]$, the curves ℓ_+ and ℓ_- are depicted in Figure 11.12

If $\arg x = 0$, the curves ℓ_+ and ℓ_- are the vertical segments connecting the origin $\lambda = 0$ with the respective stationary phase points λ_+ and λ_-.

We will also assume that

$$|x| > 16r^2, \tag{11.6.3}$$

where we recall that r is the radius of the circle C which forms the bounded part of the jump contour γ, depicted in Figure 11.2.

The original RH problem can be replaced by a sequence of equivalent RH problems, indicated in Figure 11.13, where the left and the right graphs in the bottom row correspond to the case $\arg x \in (0, \frac{\pi}{3}]$ and $\arg x = 0$ respectively (these RH problems are given in the Ψ-representation).

In the case when $\arg x \in (0, \frac{\pi}{3}]$, the jump contour of the new RH problem is the union of the curves γ_+, γ_-, ℓ_+, ℓ_- oriented from the left to the right, and the circle $C = C_{\text{right}} \cup C_{\text{left}}$ divided by $\ell_\pm$ into the right C_{right} and left C_{left} arcs, both with clockwise orientation. The jump matrices are as follows (cf. Figure 11.13):

$$\begin{aligned}
&\lambda \in \gamma_+: && \mathsf{S}(\lambda) = \mathsf{S}_3^{-1} = \begin{pmatrix} 1 & 0 \\ -s_3 & 1 \end{pmatrix}, \\
&\lambda \in \gamma_-: && \mathsf{S}(\lambda) = \mathsf{S}_6 = \begin{pmatrix} 1 & -s_3 \\ 0 & 1 \end{pmatrix}, \\
&\lambda \in \ell_+, \quad |\lambda| > r: && \mathsf{S}(\lambda) = \mathsf{S}_1\mathsf{S}_3 = \begin{pmatrix} 1 & 0 \\ -2\sin\pi\alpha & 1 \end{pmatrix} = \mathsf{S}_+, \\
&\lambda \in \ell_-, \quad |\lambda| > r: && \mathsf{S}(\lambda) = \mathsf{S}_6^{-1}S_4^{-1} = \begin{pmatrix} 1 & -2\sin\pi\alpha \\ 0 & 1 \end{pmatrix} = \mathsf{S}_-, \\
&\lambda \in \ell_+, \quad |\lambda| < r: && \mathsf{S}(\lambda) = \sigma_2 M^{-1}\sigma_2, \\
&\lambda \in \ell_-, \quad |\lambda| < r: && \mathsf{S}(\lambda) = M, \\
&\lambda \in C_{\text{right}}: \quad \mathsf{S}(\lambda) = \mathsf{E}, && \lambda \in C_{\text{left}}: \quad \mathsf{S}(\lambda) = \sigma_2\mathsf{E}\sigma_2.
\end{aligned} \tag{11.6.4}$$

Observe that the jump contour of this RH problem is decomposed into the disjoint sum of the lines γ_+, γ_- and $\gamma_0 = \ell_+ \cup \ell_- \cup C$.

In the case that $\arg x = 0$, the jump contour of the new RH problem is the union of the curves γ_+, γ_- oriented from the left to the right, the curves ℓ_+ and ℓ_- oriented from λ_- to λ_+, and the clockwise oriented circle $C = C_{\text{right}} \cup C_{\text{left}}$. The jump matrices are as follows (cf. Figure 11.13):

$$\lambda \in \gamma_+: \ \mathsf{S}(\lambda) = \begin{cases} \mathsf{S}_3^{-1}, \ \text{Re}\,\lambda < 0, \\ \mathsf{S}_1, \ \text{Re}\,\lambda > 0, \end{cases} \qquad \lambda \in \gamma_-: \ \mathsf{S}(\lambda) = \begin{cases} \mathsf{S}_4^{-1}, \ \text{Re}\,\lambda < 0, \\ \mathsf{S}_6, \ \text{Re}\,\lambda > 0, \end{cases}$$

$$\lambda \in \ell_+, \ |\lambda| > r: \ \mathsf{S}(\lambda) = \mathsf{S}_1\mathsf{S}_3 = \begin{pmatrix} 1 & 0 \\ -2\sin\pi\alpha & 1 \end{pmatrix} = \mathsf{S}_+,$$

$$\lambda \in \ell_-, \ |\lambda| > r: \ \mathsf{S}(\lambda) = \mathsf{S}_6^{-1}\mathsf{S}_4^{-1} = \begin{pmatrix} 1 & -2\sin\pi\alpha \\ 0 & 1 \end{pmatrix} = \mathsf{S}_-,$$

(11.6.5)

$$\lambda \in \ell_+, \ |\lambda| < r: \ \mathsf{S}(\lambda) = \sigma_2 M^{-1}\sigma_2, \quad \lambda \in \ell_-, \ |\lambda| < r: \ \mathsf{S}(\lambda) = M,$$
$$\lambda \in C_r: \ \mathsf{S}(\lambda) = \mathsf{E}, \quad \lambda \in C_l: \ \mathsf{S}(\lambda) = \sigma_2 \mathsf{E} \sigma_2.$$

We note that the problem (11.6.5) is a limiting case, $\arg x \to 0$, of (11.6.4).

We assume now that, in addition to the vanishing of s_2, the parameter s_3 is also zero, i.e., we let

(11.6.6) $$s_2 = s_3 = 0.$$

This leads to a *reduced* RH problem which is the same for all $\arg x \in [0, \frac{\pi}{3}]$. We will use the notation $\Phi(\lambda)$ for the solution of this RH problem. This problem, which is posed on the contour $\gamma_0 = \ell_+ \cup \ell_- \cup C$ oriented as above, is formulated as follows:

(11.6.7)

i) $\Phi(\lambda)e^{\theta\sigma_3} \to I, \quad \lambda \to \infty, \qquad \Phi_{\text{right}}(\lambda)\lambda^{-\alpha\sigma_3} \in H(\Omega_0) \cap C(\overline{\Omega}_0),$

ii) $\Phi_+(\lambda) = \Phi_-(\lambda)\mathsf{S}(\lambda), \quad \lambda \in \gamma_0 \setminus \{0\},$

where the jump matrix $\mathsf{S}(\lambda)$ is described in (11.6.4); see also Figure 11.14. The assumption (11.6.6) implies $s_1 = -2\sin\pi\alpha$ and hence $\mathsf{S}_1 = \mathsf{S}_+$ and $\mathsf{S}_4^{-1} = \mathsf{S}_-$.

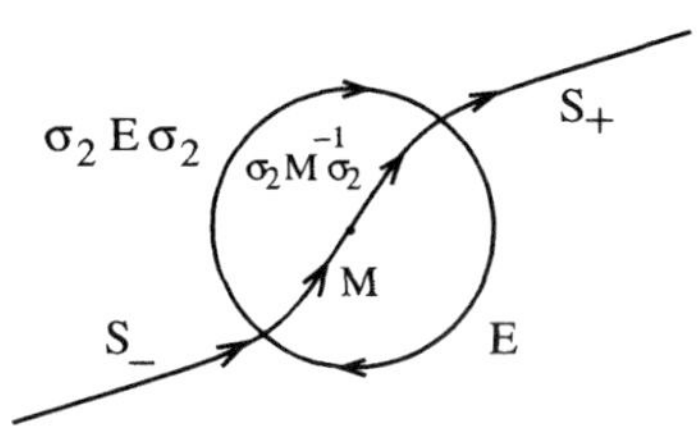

FIGURE 11.14. The graph for the reduced RH problem ($s_2 = s_3 = 0$) with the respective jump matrices

THEOREM 11.4. *Suppose that $\alpha - \frac{1}{2} \notin \mathbb{Z}$. Then, for all $\arg x \in [0, \frac{\pi}{3}]$ and $|x|$ sufficiently large, there exists a unique solution of the RH problem (11.6.7).*

PROOF. Since $\det \mathsf{S}(\lambda) \equiv 1$, we have $\det \Phi_+ = \det \Phi_-$, and hence $\det \Phi(\lambda)$ is an entire function. Furthermore, because of the normalization of $\Phi(\lambda)$ at infinity, $\det \Phi(\lambda) \equiv 1$. Let $\tilde{\Phi}$ and Φ be two solutions of (11.6.7). Then $\chi(\lambda) = \tilde{\Phi}(\lambda)\Phi^{-1}(\lambda)$ is a rational function of λ with the only possible pole at $\lambda = 0$. However, due to the boundedness of $\Phi_{\text{right}}(\lambda)\lambda^{-\alpha\sigma_3}$ and $\tilde{\Phi}_{\text{right}}(\lambda)\lambda^{-\alpha\sigma_3}$ as $\lambda \to 0$, $\chi(\lambda)$ is bounded at $\lambda = 0$ and hence, using the Liouville theorem, $\chi(\lambda) \equiv \text{const}$. Thus, because of the normalization of Φ and $\tilde{\Phi}$ at infinity, $\Phi(\lambda) \equiv \tilde{\Phi}(\lambda)$, and the uniqueness follows.

To prove the existence of the solution $\Phi(\lambda)$, we introduce the auxiliary function $\hat{\Phi}^0(z)$,

$$\hat{\Phi}^0(z) = B(z)\begin{pmatrix} v_1(z) & v_2(z) \\ v_1'(z) & v_2'(z) \end{pmatrix}, \tag{11.6.8}$$

where

$$B(z) = \tfrac{1}{2}e^{-i\frac{\pi}{4}\sigma_3}\begin{pmatrix} 1 & 1 \\ -1 & 1 \end{pmatrix}\begin{pmatrix} 1 & 0 \\ -\alpha/z & 1 \end{pmatrix},$$

the prime means differentiation with respect to z, and the functions $v_1(z)$ and $v_2(z)$ are defined as follows:

$$v_1(z) = \sum_{k=0}^{\infty}\frac{\Gamma(\alpha+\frac{1}{2})z^{\alpha+2k}}{4^k k!\Gamma(\alpha+\frac{1}{2}+k)} = 2^{\alpha-\frac{1}{2}}\Gamma(\alpha+\frac{1}{2})e^{i\frac{\pi}{2}(\alpha-\frac{1}{2})}z^{\frac{1}{2}}J_{\alpha-\frac{1}{2}}(e^{-\pi i/2}z),$$

$$v_2(z) = \sum_{k=0}^{\infty}\frac{\Gamma(\frac{3}{2}-\alpha)z^{1-\alpha+2k}}{4^k k!\Gamma(\frac{3}{2}-\alpha+k)} = 2^{\frac{1}{2}-\alpha}\Gamma(\frac{3}{2}-\alpha)e^{i\frac{\pi}{2}(\frac{1}{2}-\alpha)}z^{\frac{1}{2}}J_{\frac{1}{2}-\alpha}(e^{-\pi i/2}z).$$

Here $J_\nu(z)$ denotes the classical Bessel function [**BE**]. The function $\hat{\Phi}^0(z)$ satisfies the linear differential equation

$$\hat{\Phi}^0_z = \big(\sigma_3 - \frac{\alpha}{z}\sigma_2\big)\hat{\Phi}^0.$$

Also,

$$\hat{\Phi}^0(z)z^{-\alpha\sigma_3} \in H(\mathbb{C}).$$

Using the standard asymptotic properties of the Bessel functions, we find that the products

$$\hat{\Phi}_1(z) := \hat{\Phi}^0(z)\hat{\mathsf{E}}, \qquad \hat{\Phi}_2(z) := \hat{\Phi}_1(z)\hat{\mathsf{S}}_1, \qquad \hat{\Phi}_3(z) := \hat{\Phi}_2(z)\hat{\mathsf{S}}_2,$$

where

$$\hat{\mathsf{E}} = \frac{\sqrt{\pi}}{2\cos\pi\alpha}\begin{pmatrix} \frac{2^{1-\alpha}}{\Gamma(\frac{1}{2}+\alpha)} & \\ & \frac{2^{\alpha}}{\Gamma(\frac{3}{2}-\alpha)} \end{pmatrix} e^{i\frac{\pi}{4}\sigma_3}\begin{pmatrix} e^{-i\pi\alpha} & i \\ ie^{i\pi\alpha} & 1 \end{pmatrix},$$

$$\hat{\mathsf{S}}_1 = \begin{pmatrix} 1 & 2\sin\pi\alpha \\ 0 & 1 \end{pmatrix}, \qquad \hat{\mathsf{S}}_2 = \begin{pmatrix} 1 & 0 \\ -2\sin\pi\alpha & 1 \end{pmatrix},$$

satisfy the asymptotic relation

$$\hat{\Phi}_k(z) = (I - \frac{i\alpha}{2z}\sigma_1 + O(\frac{1}{z^2}))e^{z\sigma_3}, \tag{11.6.9}$$

as $z \to \infty$, $\arg z \in \big(\pi(k-\frac{3}{2}), \pi(k+\frac{1}{2})\big)$.

The following symmetry relations are valid,

$$\sigma_2\hat{\Phi}^0(e^{i\pi}z)\sigma_2 = \hat{\Phi}^0(z)M, \quad M = ie^{i\pi\alpha\sigma_3}\sigma_1, \quad \sigma_2\hat{\Phi}_{k+1}(e^{i\pi}z)\sigma_2 = \hat{\Phi}_k(z). \tag{11.6.10}$$

Furthermore,

$$\hat{\mathsf{E}}\hat{\mathsf{S}}_1 = D\mathsf{E},$$

where

$$D = \frac{\sqrt{\pi}\, e^{i\frac{\pi}{4}}}{2\cos\pi\alpha}\begin{pmatrix} \frac{2^{1-\alpha}e^{-i\pi\alpha}}{p\Gamma(\frac{1}{2}+\alpha)} & \\ & \frac{2^{\alpha}e^{i\pi\alpha}}{q\Gamma(\frac{3}{2}-\alpha)} \end{pmatrix}, \quad \mathsf{E} = \begin{pmatrix} p & \\ & q \end{pmatrix}\begin{pmatrix} 1 & ie^{-i\pi\alpha} \\ 1 & -ie^{i\pi\alpha} \end{pmatrix}.$$

Note that we have choosen p and q exactly in the same as in (11.6.2).

Strictly speaking, the functions $\hat{\Phi}^0(z)$ and $\hat{\Phi}_k(z)$ are defined on the universal covering of the complex z-plane punctured at $z = 0$. In particular, in virtue of (11.6.10) we have

$$\hat{\Phi}^0(e^{2\pi i}z) = \hat{\Phi}^0(z)e^{2\pi i\alpha\sigma_3}, \quad \hat{\Phi}_3(e^{2\pi i}z) = \hat{\Phi}_1(z). \tag{11.6.11}$$

However, we are going to use these functions, to define a piecewise holomorphic function $\hat{\Phi}(z)$ back on the z-plane. The function $\hat{\Phi}(z)$, in turn, will be used to construct a parametrix for the solution of the RH problem (11.6.7) in the neighborhood of $\lambda = 0$.

Let us introduce in the complex z-plane the contour $\hat{\gamma}_0 = \mathbb{R} \cup \hat{C}$, consisting of the real axis oriented from left to right, and the clockwise oriented circle $\hat{C} = \{z \in \mathbb{C}\colon\ |z| = \hat{r}\}$, divided by the real axis in the lower $\hat{C}_{\text{down}}$ and upper $\hat{C}_{\text{up}}$ arcs. This contour, which is depicted in Figure 11.15, divides the complex z-plane into four regions: $\hat{\Omega}_2$, which is the exterior of the circle in the lower half of the complex z-plane, $\hat{\Omega}_3$, which is the exterior of the circle in the upper half of the complex z-plane, $\hat{\Omega}_d$, which is the lower half of the interior of the circle and $\hat{\Omega}_u$, which is the upper half of the interior of the circle. We are now ready to define the function $\hat{\Phi}(z)$, mentioned above:

$$\hat{\Phi}(z) = \begin{cases} \hat{\Phi}_2(z), & \pi \le \arg z \le 2\pi, \quad \text{if} \quad z \in \hat{\Omega}_2, \\ \hat{\Phi}_3(z), & 2\pi \le \arg z \le 3\pi, \quad \text{if} \quad z \in \hat{\Omega}_3, \\ \hat{\Phi}^0(z)D, & \pi \le \arg z \le 2\pi, \quad \text{if} \quad z \in \hat{\Omega}_d, \\ \sigma_2\hat{\Phi}^0(e^{-i\pi}z)D\sigma_2, & 2\pi \le \arg z \le 3\pi, \quad \text{if} \quad z \in \hat{\Omega}_u. \end{cases} \tag{11.6.12}$$

We emphasize, once again, that the function $\hat{\Phi}(z)$ is defined as a piecewise analytic function on $\mathbb{C} \setminus \{0\}$ (and in particular, $\hat{\Phi}(e^{2\pi i}z) = \hat{\Phi}(z)$).

By construction, the function $\hat{\Phi}(z)$ solves the following RH problem (see Figure 11.15):

i) $\quad \hat{\Phi}(z)e^{-z\sigma_3} \to I, \quad z \to \infty, \qquad \hat{\Phi}(z)|_{\hat{\Omega}_d} z^{-\alpha\sigma_3} \in H(\{z : |z| \le \hat{r}\});$

ii) $\quad \hat{\Phi}_+(z) = \hat{\Phi}_-(z)\hat{\mathsf{S}}(z), \quad \lambda \in \hat{\gamma}_0 \setminus \{0\},$

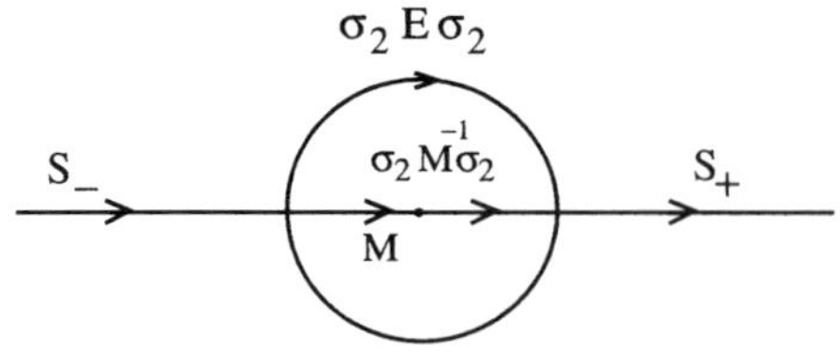

FIGURE 11.15. The graph for the model RH problem with the associated jump matrices

where the jump matrix $\hat{\mathrm{S}}(z)$ is given by the equations

$$\begin{aligned} z > \hat{r}:&\quad \hat{\mathrm{S}}(z) = \hat{\mathrm{S}}_2 = \begin{pmatrix} 1 & 0 \\ -2\sin\pi\alpha & 1 \end{pmatrix} = \mathrm{S}_+, \\ z < -\hat{r}:&\quad \hat{\mathrm{S}}(z) = \hat{\mathrm{S}}_1^{-1} = \begin{pmatrix} 1 & -2\sin\pi\alpha \\ 0 & 1 \end{pmatrix} = \mathrm{S}_-, \\ -\hat{r} < z < 0:&\quad \hat{\mathrm{S}}(z) = M, \qquad 0 < z < \hat{r}: \quad \hat{\mathrm{S}}(z) = \sigma_2 M^{-1}\sigma_2, \\ z \in \hat{C}_{\mathrm{down}}:&\quad \hat{\mathrm{S}}(z) = \mathrm{E}, \qquad z \in \hat{C}_{\mathrm{up}}: \quad \hat{\mathrm{S}}(z) = \sigma_2 \mathrm{E} \sigma_2. \end{aligned} \tag{11.6.13}$$

In other words, the function $\hat{\Phi}(z)$ has precisely the jump properties of the function $\Phi(\lambda)$, and also has the same behavior at $z = 0$. Hence, in order to construct a parametrix for the solution $\Phi(\lambda)$ of the RH problem (11.6.7), we only need to transfer the function $\hat{\Phi}(z)$ from the z-plane to the λ-plane in such way that the large z behavior of $\hat{\Phi}(z)$ is consistent with the large λ behavior specified in the problem (11.6.7). For this purpose, we introduce the conformal mapping

$$z(\lambda) = -\theta(\lambda) = -ix\lambda - i\frac{4}{3}\lambda^3. \tag{11.6.14}$$

For $|\lambda| \le R = \frac{1}{4}|x|^{1/2}$, the mapping (11.6.14) defines a holomorphic change of the independent variable.

Set

$$\tilde{\Phi}(\lambda) = \begin{cases} \hat{\Phi}(z(\lambda)), & |\lambda| < R, \\ e^{-\theta(\lambda)\sigma_3}, & |\lambda| > R, \end{cases} \tag{11.6.15}$$

and assume that in the definition of the function $\hat{\Phi}(z)$ the circle $\hat{C}$ is replaced by the image of the circle C under the mapping (11.6.14). Then, the part of the contour γ_0, which is inside the circle $\mathcal{L}$ of radius R, maps onto the part of the contour $\hat{\gamma}_0$, which is inside of the analytic curve $\hat{\mathcal{L}} \equiv z\{\mathcal{L}\}$. Therefore, on the parts of γ_0 which are inside the circle $\mathcal{L}$, the functions $\tilde{\Phi}(\lambda)$ and $\Phi(\lambda)$ have exactly the same jumps, and hence the former provides a parametrix for the latter inside the circle $\mathcal{L}$. (We note that due to condition (11.6.3) the circle C lies inside of the circle $\mathcal{L}$.)

We now look for the solution of the RH problem (11.6.7) in the form of the product

$$\Phi(\lambda) = \chi(\lambda)\tilde{\Phi}(\lambda).$$

The function $\chi(\lambda)$ is holomorphic inside the circle $\mathcal{L}$, it approaches the identity matrix as $\lambda \to \infty$, and it has jumps on the contour ℓ which is the union of the (clockwise oriented) circle $\mathcal{L}$ and the outer parts of the curves ℓ_+, and ℓ_-. In fact, the function $\chi(\lambda)$ solves the following RH problem:

$$\text{(11.6.16)} \qquad \begin{aligned} &i) && \chi(\lambda) \to I, \quad \lambda \to \infty; \\ &ii) && \chi_+(\lambda) = \chi_-(\lambda) H(\lambda), \quad \lambda \in \ell, \\ & && \lambda \in \ell_+,\ |\lambda| > R: \quad H(\lambda) = e^{-\theta\sigma_3} S_+ e^{\theta\sigma_3}, \\ & && \lambda \in \ell_-,\ |\lambda| > R: \quad H(\lambda) = e^{-\theta\sigma_3} S_- e^{\theta\sigma_3}, \\ & && |\lambda| = R,\ \lambda \in \mathcal{L}_r: \quad H(\lambda) = \hat{\Phi}_2(z(\lambda)) e^{\theta\sigma_3}, \end{aligned}$$

$$\text{(11.6.17)} \qquad |\lambda| = R,\ \lambda \in \mathcal{L}_l: \quad H(\lambda) = \hat{\Phi}_3(z(\lambda)) e^{\theta\sigma_3}.$$

We denote by $\mathcal{L}_l$ ($\mathcal{L}_r$) the arc of the circle $\mathcal{L}$ which is to the left (right) of the curve $\ell_+ \cup \ell_-$.

Taking into account the equations $S_+ = \hat{S}_2$, $S_- = \hat{S}_1^{-1}$ (see (11.6.13)), it is straightforward to establish the smoothness of the RH problem at the node points $\ell_+ \cap \mathcal{L}$ and $\ell_- \cap \mathcal{L}$. Furthermore, the jump matrix $H(\lambda)$ satisfies the estimates

$$\text{(11.6.18)} \qquad \|H(\lambda) - I\| \leq \begin{cases} c_1 e^{-c_2|x|\,|\lambda|} & \lambda \in \ell_\pm, \quad |\lambda| \geq R, \\ c_1 R^{-3} & |\lambda| = R, \end{cases}$$

(the precise values of the positive constants c_j, $j = 1, 2$, are not important for our considerations). The estimates (11.6.18) follow directly from (11.6.9). The solvability of the RH problem (11.6.16), and therefore of (11.6.7), follows from the estimates (11.6.18) and from the fact that $R = \frac{1}{4}|x|^{1/2}$. Indeed, consider the system of singular integral equations for the limiting value $\chi_-(\lambda)$,

$$\text{(11.6.19)} \qquad \chi_-(\lambda) = I + \frac{1}{2\pi i} \int_\ell \frac{\chi_-(\zeta)(H(\zeta) - I)}{\zeta - \lambda_-}\, d\zeta,$$

which is equivalent to the RH problem (11.6.16). In (11.6.19), λ_- means, as before, the limit of the integral,

$$\int_\ell \frac{\chi_-(\zeta)(H(\zeta) - I)}{\zeta - \lambda'}\, d\zeta,$$

as λ' approaches $\lambda \in \ell$ from the right side of ℓ. Equation (11.6.19) can also be written as the following singular integral equation for the function $\psi_- := \chi_- - I$,

$$\text{(11.6.20)} \qquad \psi_- = KI + K\psi_-.$$

Here, the operator K is the composition of the operator of the right multiplication by $H - I$ and of the Cauchy operator C_-. The Cauchy operator C_- is bounded in $L^2(\ell)$ (see e.g. [**LiS**]; see also [**BDT**], [**Zh**], and Chapter 3). Moreover, although the contour ℓ depends on x, the bound for $\|C_-\|$ can be choosen independently of x. This fact together with the L^∞-estimates (11.6.18) imply $\|K\|_{L_2(\ell)} \leq c|x|^{-3/2}$, and hence $I - K$ is invertible in $L^2(\ell)$. Since, in addition, $KI \in L^2(\ell)$, equation (11.6.20) for ψ_- is solvable in $L^2(\ell)$, and the solution $\chi(\lambda)$ of the RH problem (11.6.16) is determined via the integral representation[4],

$$\text{(11.6.21)} \qquad \chi(\lambda) = I + \frac{1}{2\pi i} \int_\ell \frac{(I + \psi_-(\zeta))(H(\zeta) - I)}{\zeta - \lambda}\, d\zeta, \quad \lambda \notin \ell.$$

[4]For more detail on the general L^2-theory of the RH problems see [**DeZ4, DeZ5**] and Chapter 3); see also similar constructions in the Chapters 8, 9 and in the preceding section of this chapter.

The function ψ_- is given by the convergent iterative series, $\psi_- = \sum_{n=1}^{\infty} K^n I$. This, and the L^∞-estimates (11.6.18), lead to the following asymptotic formula for the function $\chi(\lambda)$, valid for all $\arg x \in [0, \frac{\pi}{3}]$:

$$\|I - \chi^-\|_{L_2(\ell)} = \|\psi^-\|_{L_2(\ell)} \leq c\|KI\|_{L_2(\ell)} \leq c|x|^{-5/4}. \tag{11.6.22}$$

We also notice that

$$\|I - \chi(\lambda)\| \leq \frac{c}{|x|(1+|\lambda|)}, \quad \forall\lambda. \tag{11.6.23}$$

Indeed, if λ lies outside of a sectorial neighborhood of ℓ, then the estimate (11.6.23) is a direct consequence of integral representation (11.6.21). The domain of validity of (11.6.23) can be extended to the whole λ-plane, because we can always slightly deform contour ℓ and slightly rotate its infinite parts.

It follows from (11.6.15), that

$$Y(\lambda) \equiv \Phi(\lambda)e^{\theta\sigma_3} = \chi(\lambda),$$

for all $|\lambda| > R$. Therefore, the Painlevé function which corresponds to the choice of the monodromy data, $s_2 = s_3 = 0$, $s_1 = -2\sin\pi\alpha$, and which we will denote by $u_1(x) \equiv u_1(x|\alpha)$, is determined by the equation (cf. (11.3.15)),

$$u_1(x) = 2\lim_{\lambda\to\infty} \lambda\chi_{12}(\lambda) = -\frac{1}{\pi i}\int_\ell \big(\chi_-(\zeta)(H(\zeta) - I)\big)_{12}\,d\zeta.$$

This equation, together with (11.6.22), leads to the following asymptotic representation for the function $u_1(x)$,

$$u_1(x) = \frac{i}{\pi}\int_\ell \big((H(\zeta) - I)\big)_{12}\,d\zeta + O\big(x^{-5/2}\big), \tag{11.6.24}$$

as

$$x \to \infty, \quad \arg x \in \left[0, \frac{\pi}{3}\right].$$

Using the expressions (11.6.16) for the jump matrix $H(\lambda)$, the asymptotics (11.6.9), and the definition (11.6.14) of the change of variable function $z(\lambda)$, we obtain from (11.6.24) the leading asymptotic term of $u_1(x)$,

$$u_1(x) = \alpha/x + O(x^{-5/2}), \qquad \arg x \in \left[0, \frac{\pi}{3}\right].$$

Moreover, using the iterative series $\chi_- = \sum_{n=0}^{\infty} K^n I$, we can establish the existence of the asymptotic series over negative powers of $x^{1/2}$ representing the function $u_1(x)$ in the whole sector $\arg x \in [0, \frac{\pi}{3}]$ (cf. [**DeZ2**]). The meromorphicity of $u_1(x)$ (the Painlevé property) allows us to differentiate this series and hence to use the differential equation (11.2.2) to determine its coefficients. An elementary analysis using these arguments leads to the following asymptotic formula for the solution $u_1(x)$,

$$u_1(x|\alpha) \sim u_f(x|\alpha), \quad x \to \infty, \quad \arg x \in \left[0, \frac{\pi}{3}\right], \tag{11.6.25}$$

where $u_f(x|\alpha)$ is the formal series,

$$u_f(x|\alpha) := \frac{\alpha}{x}\sum_{n=0}^{\infty} a_n x^{-3n}, \tag{11.6.26}$$

whose coefficients a_n are determined uniquely by the recurrence relation

$$a_0 = 1, \qquad a_{n+1} = (3n+1)(3n+2)a_n - 2\alpha^2 \sum_{\substack{k,l,m=0\\ k+l+m=n}}^{n} a_k a_l a_m. \tag{11.6.27}$$

REMARK 11.8. In all the above considerations we have implicitly assumed that $\alpha \neq 0$. If $\alpha = 0$, then the reduced Riemann-Hilbert problem becomes elementary. In this case, condition (11.6.6) implies that the Stokes parameter s_1 is also zero, and there is no singularity at $\lambda = 0$. For the jump matrices of the problem (11.6.7) we have the relations

$$\mathsf{S}_+ = \mathsf{S}_- = I,$$
$$M = i\sigma_1 = \sigma_2 M^{-1}\sigma_2, \quad \text{and} \quad \mathsf{E}^{-1}M = \sigma_2 \mathsf{E}^{-1}\sigma_2.$$

These equations yield the following explicit solution of the problem (11.6.7)

$$\Phi(z) = \begin{cases} e^{-\theta(\lambda)\sigma_3}, & \text{if } \ |z| > r, \\ e^{-\theta(\lambda)\sigma_3}\mathsf{E}^{-1}, & \text{if } \ z \in \Omega_{\text{right}}, \\ e^{-\theta(\lambda)\sigma_3}\mathsf{E}^{-1}M, & \text{if } \ z \in \Omega_{\text{left}}. \end{cases} \tag{11.6.28}$$

In fact, in the case $\alpha = 0$ the function $\tilde{\Phi}(\lambda)$ defined in (11.6.15) reduces to the above expression. From (11.6.28) we obtain that $Y(\lambda) \equiv I$ for all $|\lambda| > r$ and hence

$$u_1(x|0) \equiv 0.$$

Consider now the generic case of the RH problem (11.6.4)–(11.6.5), i.e., we now assume that $s_3 \neq 0$. We look for the solution $\Psi(\lambda)$ in the form of the product

$$\Psi(\lambda) = X(\lambda)\Phi(\lambda), \tag{11.6.29}$$

where $\Phi(\lambda)$ is the solution of the reduced RH problem (11.6.7). Observe that the function $X(\lambda)$ has jumps only across the curve $\gamma_+ \cup \gamma_-$. In fact, in terms of $X(\lambda)$, the RH problem (11.6.4)–(11.6.5), for all $x \in [0, \frac{\pi}{3}]$, takes the following form:

$$\begin{aligned} &i) \quad X(\lambda) \to I, \quad \lambda \to \infty, \\ &ii) \quad X_+(\lambda) = X_-(\lambda)\mathcal{H}(\lambda), \quad \lambda \in \gamma_+ \cup \gamma_-, \\ &\qquad \lambda \in \gamma_+: \quad \mathcal{H}(\lambda) = \Phi_-(\lambda)\mathsf{S}_3^{-1}\Phi_-^{-1}(\lambda), \\ &\qquad \lambda \in \gamma_-: \quad \mathcal{H}(\lambda) = \Phi_-(\lambda)\mathsf{S}_6\Phi_-^{-1}(\lambda), \end{aligned} \tag{11.6.30}$$

where $\Phi_-(\lambda)$ denotes the right limit of $\Phi(\lambda)$ on $\gamma_+ \cup \gamma_-$. We note that, in virtue of (11.6.23), for all $|\lambda| > R$ we have that $\|\Phi(\lambda)e^{\theta\sigma_3}\| = \|\chi(\lambda)\| <$ const. Then, the following estimate for the jump matrix $\mathcal{H}(\lambda)$ on $\gamma_\pm$ is valid,

$$\|\mathcal{H}(\lambda) - I\| \le c_1 e^{-\frac{2}{3}|x|^{3/2}\cos(\frac{3}{2}\arg x)} e^{-c_2|x|^{1/2}|\lambda-\lambda_0|^2},$$

where $c_{1,2}$ are positive constants and $\lambda_0 = \pm\frac{i}{2}x^{1/2}$ is the stationary phase point. This estimate in turn yields the estimate,

$$\|\mathcal{K}\|_{L^2(\gamma_+\cup\gamma_-)} \le c e^{-\frac{2}{3}|x|^{3/2}\cos(\frac{3}{2}\arg x)}, \tag{11.6.31}$$

for the norm of the singular integral operator $\mathcal{K}$ in the integral equations $X_- = I + \mathcal{K}X_-$. It follows then, that if $|x|$ is sufficiently large and $\arg x \in [0, \frac{\pi}{3} - \epsilon]$, $\epsilon > 0$, the operator $\mathcal{K}$ is contracting and the system $X_- = I + \mathcal{K}X_-$ is solvable by iterations in $L^2(\gamma_+ \cup \gamma_-)$, i.e., $X_- = \sum_{n=0}^{\infty} \mathcal{K}^n I = I + \mathcal{K}I + O(e^{-\frac{4}{3}x^{3/2}})$. In fact, in what follows we will use a more refined analysis which will allow us to include the oscillating direction, $\arg x = \frac{\pi}{3}$, in our considerations.

THEOREM 11.5. *If $s_2 = 0$ and $|x| \to \infty$, $\arg x \in [0, \frac{\pi}{3}]$, then the asymptotics of the second Painlevé transcendent is given by*

$$u = u_1(x|\alpha) - \frac{is_3}{2\sqrt{\pi}} x^{-1/4} e^{-\frac{2}{3}x^{3/2}} (1 + O(x^{-3/4})) + O(s_3^2 x^{-7/4} e^{-\frac{4}{3}x^{3/2}}), \tag{11.6.32}$$

where $u_1(x|\alpha) \sim \alpha/x$ is the solution of the Painlevé equation for $s_2 = s_3 = 0$, $s_1 = -2\sin\pi\alpha$. If $\alpha = 0$, then the function $u_1(x|\alpha)$ is identically zero.

PROOF. Since $\lambda \in \gamma_+ \cup \gamma_-$ implies $|\lambda| > R$, we can use for $\Phi_-(\lambda)$ the representation,

$$Y_-(\lambda) = \chi_-(\lambda) e^{-\theta(\lambda)\sigma_3} = (I + \psi_-(\lambda)) e^{-\theta(\lambda)\sigma_3}.$$

From the estimate (11.6.23), it follows that

$$\|\psi_-(\lambda)\| \le \frac{c}{|x|(1 + |\lambda|)} < \frac{c}{|x|}, \qquad \lambda \in \gamma_\pm, \tag{11.6.33}$$

which in turn implies the following asymptotics of the jump matrix $\mathcal{H}(\lambda)$, as $x \to \infty$ and $\arg x \in [0, \frac{\pi}{3}]$,

$$\mathcal{H}(\lambda) = \begin{cases} I - s_3 e^{2\theta(\lambda)} \sigma_- + O\left(s_3 \frac{e^{2\theta(\lambda)}}{|x|}\right), & \lambda \in \gamma_+, \\ I - s_3 e^{-2\theta(\lambda)} \sigma_+ + O\left(s_3 \frac{e^{-2\theta(\lambda)}}{|x|}\right), & \lambda \in \gamma_-. \end{cases} \tag{11.6.34}$$

We also note that, due to the flexibility in choosing contours for the RH problem (11.6.16), the function $\psi_-(\lambda)$, and hence the jump matrix $\mathcal{H}(\lambda)$, can be analytically continued in a sectorial neighborhood of the contour $\gamma_+ \cup \gamma_-$, where they also satisfy the same estimates (11.6.33) and (11.6.34), respectively.

Let us introduce the triangular matrices $P(\lambda)$, $Q(\lambda)$,

$$\begin{aligned} P(\lambda) &= I - \frac{s_3}{2\pi i} \int_{\gamma_+} e^{2\theta(\zeta)} \frac{d\zeta}{\zeta - \lambda} \sigma_-, \\ Q(\lambda) &= I - \frac{s_3}{2\pi i} \int_{\gamma_-} e^{-2\theta(\zeta)} \frac{d\zeta}{\zeta - \lambda} \sigma_+, \end{aligned} \tag{11.6.35}$$

which solve the following triangular RH problems:

i) $P(\lambda) \to I$ as $\lambda \to \infty$, and $P_+(\lambda) = P_-(\lambda)(I - s_3 e^{2\theta}\sigma_-)$ as $\lambda \in \gamma_+$,

ii) $Q(\lambda) \to I$ as $\lambda \to \infty$, and $Q_+(\lambda) = Q_-(\lambda)(I - s_3 e^{-2\theta}\sigma_+)$ as $\lambda \in \gamma_-$.

The asymptotics (11.6.34) suggests to look for the solution $X(\lambda)$ of the RH problem (11.6.30) in the form of the product

$$X(\lambda) = Z(\lambda) Q(\lambda) P(\lambda). \tag{11.6.36}$$

The function $Z(\lambda)$ satisfies the RH problem

$$\begin{aligned} &i) \quad Z(\lambda) \to I, \quad \lambda \to \infty, \\ &ii) \quad Z_+(\lambda) = Z_-(\lambda) V(\lambda), \quad \lambda \in \gamma_+ \cup \gamma_-, \end{aligned} \tag{11.6.37}$$

where

$$\begin{aligned} V(\lambda) &= Q(\lambda) P_-(\lambda) \mathcal{H}(\lambda) (I + s_3 e^{2\theta}\sigma_-) P_-^{-1}(\lambda) Q^{-1}(\lambda), \quad \lambda \in \gamma_+, \\ V(\lambda) &= Q_-(\lambda) P(\lambda) \mathcal{H}(\lambda) P^{-1}(\lambda) (I + s_3 e^{-2\theta}\sigma_+) Q_-^{-1}(\lambda), \quad \lambda \in \gamma_-. \end{aligned}$$

Using the first of the relations in (11.6.34), we find the estimate for the jump matrix $V(\lambda)$ on γ_+,

$$V(\lambda) = I + O\left(\frac{s_3 e^{2\theta(\lambda)}}{|x|}\right), \qquad \lambda \in \gamma_+. \tag{11.6.38}$$

The product $P\mathcal{H}P^{-1}$ on γ_- admits the representation,

$$P(\lambda)\mathcal{H}(\lambda)P^{-1}(\lambda) = I - s_3 e^{-2\theta}\sigma_- - \frac{s_3^2}{2\pi i}\varphi e^{-2\theta}\sigma_3 + O\left(\frac{s_3 e^{-2\theta}}{|x|}\right), \qquad \lambda \in \gamma_-,$$

where the function $\varphi(\lambda) = \int_{\gamma_+} e^{2\theta(\zeta)}(\zeta - \lambda)^{-1}\,d\zeta$ is square integrable and holomorphic in a neighborhood of γ_-. It satisfies the estimate

$$|\varphi(\lambda)| \le c|x|^{-3/4} e^{-\frac{2}{3}|x|^{3/2}\cos(\frac{3}{2}\arg x)}, \qquad \lambda \in \gamma_-,$$

for some constant c. Thus,

$$V(\lambda) = I + O\left(\frac{s_3 e^{-2\theta(\lambda)}}{|x|^{3/4}}\right), \qquad \lambda \in \gamma_-. \tag{11.6.39}$$

We keep the parameter s_3 in the error terms to indicate explicitly that these terms disappear if $s_3 = 0$. We can also combine the estimates (11.6.38) and (11.6.39) into the single inequality,

$$\|V(\lambda) - I\| \le c\left|s_3 \frac{e^{\pm 2\theta(\lambda)}}{x^{3/4}}\right|, \qquad \lambda \in \gamma_\pm. \tag{11.6.40}$$

Observe that the inequality (11.6.40) implies the following L_∞- and L_2-estimates for $V(\lambda)$:

$$\|V - I\|_{L_\infty(\gamma_+ \cup \gamma_-)} \le c|s_3| \frac{e^{-\frac{2}{3}|x|^{3/2}\cos(\frac{3}{2}\arg x)}}{|x|^{3/4}}, \tag{11.6.41}$$

$$\|V - I\|_{L_2(\gamma_+ \cup \gamma_-)} \le c|s_3| \frac{e^{-\frac{2}{3}|x|^{3/2}\cos(\frac{3}{2}\arg x)}}{|x|^{7/8}}. \tag{11.6.42}$$

Consider now the system of singular integral equations for $Z_-(\lambda)$, $Z_- = I + \mathsf{K}Z_-$, which is equivalent to the RH problem (11.6.37). Here the singular integral operator K is the superposition of the multiplication by $V - I$ and of the Cauchy operator C_-. From (11.6.41) and from the boundedness of the Cauchy operator in $L^2(\gamma_+ \cup \gamma_-)$, we have that

$$\|\mathsf{K}\|_{L^2(\gamma_+ \cup \gamma_-)} \le c|s_3| \frac{e^{-\frac{2}{3}|x|^{3/2}\cos(\frac{3}{2}\arg x)}}{|x|^{3/4}},$$

for all sufficiently large $|x|$ and $\arg x \in [0, \frac{\pi}{3}]$. Thus, the equation $\zeta_- = \mathsf{K}I + \mathsf{K}\zeta_-$ for the difference $\zeta_- := Z_- - I$, is solvable by iterations in the space $L^2(\gamma_+ \cup \gamma_-)$ and its solution, in view of (11.6.42), satisfies the estimate,

$$\|\zeta_-\|_{L_2(\gamma_+ \cup \gamma_-)} \le c|s_3| \frac{e^{-\frac{2}{3}|x|^{3/2}\cos(\frac{3}{2}\arg x)}}{|x|^{7/8}}. \tag{11.6.43}$$

The solution of the RH problem (11.6.37) is given by the integral formula

$$Z(\lambda) = I + \frac{1}{2\pi i}\int_{\gamma_+ \cup \gamma_-} \frac{V(\xi) - I}{\xi - \lambda}\,d\xi$$

$$+\frac{1}{2\pi i}\int_{\gamma_+ \cup \gamma_-} \frac{\zeta_-(\xi)(V(\xi) - I)}{\xi - \lambda}\,d\xi, \quad \lambda \notin \gamma_+ \cup \gamma_-.$$

This formula, together with (11.6.36), (11.6.35), (11.6.29), and (11.3.15), yields the representation

$$u(x) = u_1(x) + \frac{s_3}{\pi i}\int_{\gamma_-} e^{-2\theta(\lambda)}\,d\lambda$$

$$-\frac{1}{2\pi i}\int_{\gamma_+\cup\gamma_-}(V(\lambda)-I)\,d\lambda - \frac{1}{2\pi i}\int_{\gamma_+\cup\gamma_-}\zeta_-(\lambda)(V(\lambda)-I)\,d\lambda,$$

for the Painlevé function $u(x) \equiv u(x; s_2 = 0|\alpha)$. This representation, taking into account the estimates (11.6.40), (11.6.42), and (11.6.43), implies that

$$u(x) = u_1(x) + \frac{s_3}{\pi i}\int_{\gamma_-} e^{-2\theta(\lambda)}\,d\lambda$$

$$+O(s_3x^{-1}e^{-\frac{2}{3}x^{3/2}}) + O(s_3^2x^{-7/4}e^{-\frac{4}{3}x^{3/2}}). \tag{11.6.44}$$

Evaluating the integral in the r.h.s. of (11.6.44) by Laplace's method, we arrive at (11.6.32).

THE DECREASING DEGENERATE (TRONQUÉE) PAINLEVÉ FUNCTIONS

Applying the second of the symmetries (11.3.26) to (11.6.32), we obtain the following result.

THEOREM 11.6. *If $s_2 = 0$ and $|x| \to \infty$, $\arg x \in [-\frac{\pi}{3}, 0]$, then the asymptotics of PII is given by*

$$u = u_3(x|\alpha) + \frac{is_1}{2\sqrt{\pi}}x^{-1/4}e^{-\frac{2}{3}x^{3/2}}(1 + O(x^{-3/4})) + O(s_1^2x^{-7/4}e^{-\frac{4}{3}x^{3/2}}), \tag{11.6.45}$$

where $u_3(x|\alpha) = \overline{u_1(\overline{x}|\overline{\alpha})} \sim \alpha/x$ is the solution of the Painlevé equation for $s_2 = s_1 = 0$, $s_3 = -2\sin\pi\alpha$.

The solutions $u_1(x|\alpha)$ and $u_3(x|\alpha) = \overline{u_1(\overline{x}|\overline{\alpha})}$ are meromorphic functions of $x \in \mathbb{C}$ and thus they can be continued beyond the sectors indicated in Theorems 11.5 and 11.6. Using (11.6.32), we can find the asymptotics of the solution $u_3(x|\alpha)$ in the interior of the sector $\arg x \in [0, \frac{\pi}{3}]$. Similarly, using (11.6.45), we can find the asymptotics of the solution $u_1(x|\alpha)$ in the interior of the sector $\arg x \in [-\frac{\pi}{3}, 0]$. These expressions imply that, if $|x| \to \infty$ and $\arg x \in [-\frac{\pi}{3}, \frac{\pi}{3}]$, then

$$u_3(x|\alpha) - u_1(x|\alpha) = i\frac{\sin\pi\alpha}{\sqrt{\pi}}x^{-1/4}e^{-\frac{2}{3}x^{3/2}}(1 + O(x^{-3/4})) + O(x^{-7/4}e^{-\frac{4}{3}x^{3/2}}). \tag{11.6.46}$$

REMARK 11.9. The same asymptotic relation can be obtained using the analysis of the RH problem satisfied by the product $\hat{\Psi}(\lambda)\Psi^{-1}(\lambda)$, where $\hat{\Psi}(\lambda)$ is the solution of the RH problem for $s_2 = s_1 = 0$, and $\Psi(\lambda)$ is the solution of the RH problem for $s_2 = s_3 = 0$.

REMARK 11.10. The proof of the second assertion of Theorem 10.1 is now straightforward. Indeed, for the Painlevé function u(x), which is real for real x, we have $s_1 = \overline{s}_3$, $s_2 = \overline{s}_2$, $\alpha = \overline{\alpha}$. Under these conditions, neither $u_1(x|\alpha)$ nor $u_3(x|\alpha)$ in Theorems 11.5 and 11.6 appear real. However, since the Painlevé function u is real, the imaginary parts of both $u_1(x|\alpha)$ and $u_3(x|\alpha)$ cancel with the imaginary part of the exponentially small correction which comes from the explicit expressions of the parameters s_1 and $s_3 = \overline{s}_1$, $s_1 = -\sin \pi\alpha + i \operatorname{Im} s_1$. Thus,

$$\operatorname{Im} u_3(x|\alpha) = -\operatorname{Im} u_1(x|\alpha) = i\frac{\sin \pi\alpha}{2\sqrt{\pi}} x^{-1/4} e^{-\frac{2}{3}x^{3/2}} (1 + O(x^{-3/4})),$$

which is just half of the right-hand side of (11.6.46), and

$$u = \operatorname{Re} u_1(x|\alpha) - \frac{\operatorname{Im} s_1}{2\sqrt{\pi}} x^{-1/4} e^{-\frac{2}{3}x^{3/2}} (1 + O(x^{-3/4})).$$

This expression can be interpreted as the half-sum of the representations (11.6.32) and (11.6.45). Putting $\alpha = 0$ in (11.6.47) yields $u_1(x|0) \equiv 0$, and hence we arrive at the second assertion of Theorem 10.1.

In addition to $u_1(x|\alpha)$ and $u_3(x|\alpha)$, let us introduce the solution $u_2(x|\alpha)$ corresponding to the Stokes multipliers $s_1 = s_3 = 0$, $s_2 = 2\sin \pi\alpha$. Due to the second of the symmetries (11.3.26), we find $u_2(x|\alpha) = \overline{u_2(\overline{x}|\overline{\alpha})}$.

The last of the symmetries (11.3.26) yields the following relations:

$$\begin{aligned} u_1(x|\alpha) &= e^{i\frac{2\pi}{3}} u_2\left(e^{i\frac{2\pi}{3}} x|\alpha\right) = e^{-i\frac{2\pi}{3}} u_3\left(e^{-i\frac{2\pi}{3}} x|\alpha\right), \\ u_2(x|\alpha) &= e^{i\frac{2\pi}{3}} u_3\left(e^{i\frac{2\pi}{3}} x|\alpha\right) = e^{-i\frac{2\pi}{3}} u_1\left(e^{-i\frac{2\pi}{3}} x|\alpha\right), \\ u_3(x|\alpha) &= e^{i\frac{2\pi}{3}} u_1\left(e^{i\frac{2\pi}{3}} x|\alpha\right) = e^{-i\frac{2\pi}{3}} u_2\left(e^{-i\frac{2\pi}{3}} x|\alpha\right). \end{aligned} \tag{11.6.47}$$

These equations imply the decreasing asymptotics of the functions $u_k(x|\alpha)$,

$$\begin{aligned} u_1(x|\alpha) &\sim \alpha/x, \qquad |x| \to \infty, \qquad \arg x \in \left(-\frac{\pi}{3}, \pi\right), \\ u_2(x|\alpha) &\sim \alpha/x, \qquad |x| \to \infty, \qquad \arg x \in \left(\frac{\pi}{3}, \frac{5\pi}{3}\right), \\ u_3(x|\alpha) &\sim \alpha/x, \qquad |x| \to \infty, \qquad \arg x \in \left(-\pi, \frac{\pi}{3}\right). \end{aligned} \tag{11.6.48}$$

The same symmetry (11.3.26) applied to the equation (11.6.46), yields the following relations:

$$\begin{aligned} &u_2(x|\alpha) - u_3(x|\alpha) = -\frac{\sin \pi\alpha}{\sqrt{\pi}} x^{-1/4} e^{\frac{2}{3}x^{3/2}} \left(1 + O(x^{-3/4})\right) + O\left(x^{-7/4} e^{\frac{4}{3}x^{3/2}}\right), \\ &\arg x \in \left[-\pi, -\frac{\pi}{3}\right]; \\ &u_1(x|\alpha) - u_2(x|\alpha) = \frac{\sin \pi\alpha}{\sqrt{\pi}} x^{-1/4} e^{\frac{2}{3}x^{3/2}} \left(1 + O(x^{-3/4})\right) + O\left(x^{-7/4} e^{\frac{4}{3}x^{3/2}}\right), \\ &\arg x \in \left[\frac{\pi}{3}, \pi\right]. \end{aligned} \tag{11.6.49}$$

REMARK 11.11. It follows from the known general asymptotics of the second Painlevé transcendent in the complex domain (see e.g. [**K5**], [**Ki7**] and Chapters 7 and 8), that if the condition $s_2 = s_3 = 0$ is violated, then the Painlevé function has an elliptic asymptotic behavior in a subsector of the sector $\arg x \in [0, \frac{2\pi}{3}]$, unless $1 + s_2 s_3 = 0$. In the latter case, as we have seen in the previous section, the solution behaves as $\sqrt{-\frac{x}{2}}$. Therefore, the solution u_1 is uniquely characterized by the asymptotic behavior indicated in (11.6.48). Similar results are valid for the solutions u_2 and u_3.

The solutions $u_1(x|\alpha)$, $u_2(x|\alpha)$, and $u_3(x|\alpha)$ are the *decreasing tri-tronquée* solutions of equation (11.2.2). Equations (11.6.46) and (11.6.49) express the manifestation of the quasi-linear Stokes phenomenon.

The solutions $u_1(x|\alpha)$, $u_2(x|\alpha)$, and $u_3(x|\alpha)$, together with the increasing solutions $u_0(x|\alpha,\sigma)$, $u_1(x|\alpha,\sigma)$, and $u_3(x|\alpha,\sigma)$, which were studied in the previous section, form the complete set of tri-tronquée solutions of PII. Their defining property is that they have a power-like asymptotics in the maximal sectors discussed above.

Theorems 11.5 and 11.6 imply, that the condition

$$s_2 = 0 \tag{11.6.50}$$

singles out a one-parameter family of the *decreasing tronquée* solutions having the asymptotic series expansion (11.6.25) in the open sector $\left(-\frac{\pi}{3}, \frac{\pi}{3}\right)$. At the same time, these theorems provide an explicit description of the corresponding quasi-linear Stokes' phenomenon. The latter is expressed by the exact equation

$$s_1 + s_3 = -2\sin\pi\alpha,$$

which connects the coefficients of the oscillatory terms appear on the boundary rays $\arg x = \pm\frac{\pi}{3}$ (cf. Remark 11.5)

Applying the last of the symmetries (11.3.26), to equations (11.6.32), (11.6.45), and using (11.6.47), we can construct two additional families of decreasing tronquée solutions. The complete description of the quasi-linear Stokes' phenomena for each of the three families of decreasing tronquée solutions of PII can then be represented as follows:

i) If $s_1 = 0$, then $s_3 - s_2 = -2\sin\pi\alpha$, and

$$u = \begin{cases} u_2(x|\alpha) - \frac{s_3}{2\sqrt{\pi}}\, x^{-1/4} e^{\frac{2}{3}x^{3/2}}(1 + O(x^{-3/4})), & \arg x \in [-\pi, -\frac{2\pi}{3}], \\ u_3(x|\alpha) - \frac{s_2}{2\sqrt{\pi}}\, x^{-1/4} e^{\frac{2}{3}x^{3/2}}(1 + O(x^{-3/4})), & \arg x \in [-\frac{2\pi}{3}, -\frac{\pi}{3}]. \end{cases}$$

ii) If $s_2 = 0$, then $s_1 + s_3 = -2\sin\pi\alpha$ and

$$u = \begin{cases} u_3(x|\alpha) + \frac{is_1}{2\sqrt{\pi}}\, x^{-1/4} e^{-\frac{2}{3}x^{3/2}}(1 + O(x^{-3/4})), & \arg x \in [-\frac{\pi}{3}, 0], \\ u_1(x|\alpha) - \frac{is_3}{2\sqrt{\pi}}\, x^{-1/4} e^{-\frac{2}{3}x^{3/2}}(1 + O(x^{-3/4})), & \arg x \in [0, \frac{\pi}{3}]. \end{cases}$$

iii) If $s_3 = 0$, then $s_1 - s_2 = -2\sin\pi\alpha$ and

$$u = \begin{cases} u_1(x|\alpha) - \frac{s_2}{2\sqrt{\pi}}\, x^{-1/4} e^{\frac{2}{3}x^{3/2}}(1 + O(x^{-3/4})), & \arg x \in [\frac{\pi}{3}, \frac{2\pi}{3}], \\ u_2(x|\alpha) - \frac{s_1}{2\sqrt{\pi}}\, x^{-1/4} e^{\frac{2}{3}x^{3/2}}(1 + O(x^{-3/4})), & \arg x \in [\frac{2\pi}{3}, \pi]. \end{cases}$$

The asymptotics of the coefficients

It follows from (11.6.25), with the help of the relevant symmetries, that the asymptotic relations (11.6.48) can in fact be extended to the full asymptotic power series, i.e.

$$
\begin{aligned}
&u_1(x|\alpha) \sim u_f(x|\alpha), && |x| \to \infty, && \arg x \in \left(-\frac{\pi}{3}, \pi\right),\\
(11.6.51)\quad &u_2(x|\alpha) \sim u_f(x|\alpha), && |x| \to \infty, && \arg x \in \left(\frac{\pi}{3}, \frac{5\pi}{3}\right),\\
&u_3(x|\alpha) \sim u_f(x|\alpha), && |x| \to \infty, && \arg x \in \left(-\pi, \frac{\pi}{3}\right),
\end{aligned}
$$

where $u_f(x|\alpha)$ is the formal power series defined by equations (11.6.26) and (11.6.27). The first few terms of this expansion are given by

$$
\begin{aligned}
u_f(x|\alpha) &= \frac{\alpha}{x}\Big\{1 + \frac{2(1-\alpha^2)}{x^3} + \frac{4(10 - 13\alpha^2 + 3\alpha^4)}{x^6}\\
&+ \frac{8}{x^9}(280 - 397\alpha^2 + 129\alpha^4 - 12\alpha^6)\\
&+ \frac{16}{x^{12}}(15400 - 22736\alpha^2 + 8427\alpha^4 - 1146\alpha^6 + 55\alpha^8) + O(x^{-15})\Big\}.
\end{aligned}
$$

Note that for $\alpha \to 0$ (linear limit), the recurrence (11.6.27) is exactly solvable,

$$
a_n = \frac{\Gamma(3n)}{3^{n-1}\Gamma(n)}, \qquad \alpha \to 0. \tag{11.6.52}
$$

The goal of this section is to determine the asymptotics of the coefficients a_n in (11.6.26) as $n \to \infty$, for arbitrary α. For this purpose, following the scheme developed earlier, we define the following sectorial analytic function $\hat{u}(x)$:

$$
\hat{u}(x) = \begin{cases} u_3(x|\alpha), & \arg x \in (-\frac{2\pi}{3}, 0),\\ u_1(x|\alpha), & \arg x \in (0, \frac{2\pi}{3}),\\ u_2(x|\alpha), & \arg x \in (\frac{2\pi}{3}, \frac{4\pi}{3}). \end{cases} \tag{11.6.53}
$$

The function $\hat{u}(x)$ has the uniform asymptotic expansion (11.6.26) near infinity. In particular, this means that $\hat{u}(x)$ is bounded for $|x| \geq \rho$. Let $u^{(N)}(x)$ be the partial sum

$$
u^{(N)}(x, \alpha) = \frac{\alpha}{x}\sum_{n=0}^{N-1} a_n x^{-3n},
$$

and let $w^{(N)}(x)$ be defined by

$$
w^{(N)}(x) = x^{3N}(\hat{u}(x) - u^{(N)}(x, \alpha)) \sim \frac{\alpha}{x}\sum_{n=0}^{\infty} a_{n+N} x^{-3n}. \tag{11.6.54}
$$

The function $x^{3N}u^{(N)}(x, \alpha)$ is a polynomial, thus the integral of $w^{(N)}(x)$ along the circle of the radius $|x| = \rho$, satisfies the estimate

(11.6.55)

$$
\left|\oint_{|x|=\rho} w^{(N)}(x)\,dx\right| \leq \rho^{3N}\oint_{|x|=\rho} |\hat{u}(x)|\,dl \leq 2\pi\rho^{3N+1}\max_{|x|=\rho}|\hat{u}(x)| = C\rho^{3N+1}.
$$

On the other hand,

$$\oint_{|x|=\rho} w^{(N)}(x)\,dx = \oint_{|x|=R} w^{(N)}(x)\,dx$$

(11.6.56)

$$+\int_{\rho}^{R} x^{3N}(u_1-u_3)\,dx + \int_{e^{i\frac{2\pi}{3}}\rho}^{e^{i\frac{2\pi}{3}}R} x^{3N}(u_2-u_1)\,dx + \int_{e^{-i\frac{2\pi}{3}}\rho}^{e^{-i\frac{2\pi}{3}}R} x^{3N}(u_3-u_2)\,dx.$$

Noting that $w^{(N)}(x) = \frac{\alpha}{x}a_N + O(x^{-4})$, it follows that the first of the integrals in the r.h.s. of (11.6.56) can be estimated by

$$\oint_{|x|=R} w^{(N)}(x)\,dx = 2\pi i\alpha a_N + O(R^{-3}).$$

The last three integrals in (11.6.56) can be computed using (11.6.46) and (11.6.49):

$$\int_{\rho}^{R} x^{3N}(u_1-u_3)\,dx + \int_{e^{i\frac{2\pi}{3}}\rho}^{e^{i\frac{2\pi}{3}}R} x^{3N}(u_2-u_1)\,dx + \int_{e^{-i\frac{2\pi}{3}}\rho}^{e^{-i\frac{2\pi}{3}}R} x^{3N}(u_3-u_2)\,dx$$

$$= 3\int_{\rho}^{R} x^{3N}(u_1-u_3)\,dx = -3i\frac{\sin\pi\alpha}{\sqrt{\pi}}\int_{\rho}^{R} x^{3N-\frac{1}{4}}e^{-\frac{2}{3}x^{3/2}}(1+O(x^{-1/2}))\,dx$$

$$= -3i\frac{\sin\pi\alpha}{\sqrt{\pi}}(\frac{3}{2})^{2N-\frac{1}{2}}\Big[\Gamma(2N+\frac{1}{2})(1+O(N^{-1/3})) + O(\rho^{3N}) + O(R^{3N}e^{-\frac{2}{3}R^{3/2}})\Big].$$

Thus, letting $R=\infty$, we find the asymptotics of the coefficient a_N in (11.5.31),

(11.6.57)
$$a_N = \frac{\sin\pi\alpha}{\alpha\pi^{3/2}}(3/2)^{2N+\frac{1}{2}}\Gamma(2N+\frac{1}{2})(1+O(N^{-1/3})) + O(\rho^{3N}), \qquad N\to\infty.$$

In the linear limit, $\alpha\to 0$, this equation is consistent with the equation (11.6.52). For $\alpha\in\mathbb{Z}\backslash\{0\}$, the asymptotics (11.6.57) is reduced to $a_N = O(\rho^{3N})$, which is consistent with the expansion of the rational solution of PII determined by the conditions $s_1=s_2=s_3=0$, and $\alpha\in\mathbb{Z}$.

7. The Hastings-McLeod solution

Assume that

$$\alpha-\frac{1}{2}\notin\mathbb{Z},\quad \operatorname{Re}(\sigma\alpha) > -\frac{1}{2},$$

and consider the family (11.5.53) of increasing tronquée solutions. The corresponding monodromy parameters s_1 and s_3 are given by the equations (see (11.5.52)),

(11.7.1)
$$s_1 = e^{-i\pi\sigma(\alpha+\frac{1}{2})}\quad\text{and}\quad s_3 = e^{i\pi\sigma(\alpha+\frac{1}{2})}.$$

Therefore, there exists only one solution in the family (11.5.53), which belongs to the decreasing tronquée solutions. This solution corresponds to the choice $s_2=0$ and has a canonical power-like asymptotics in the sector $\left(-\frac{\pi}{3},\frac{\pi}{3}\right)$. In more details, collecting the relevant results of the previous sections, we arrive at the following theorem.

THEOREM 11.7. *Let a point of the monodromy data manifold (11.3.1) be specified by the equations,*

$$s_1 = e^{-i\pi\sigma(\alpha+\frac{1}{2})}, \quad s_3 = e^{i\pi\sigma(\alpha+\frac{1}{2})} \quad \text{and} \quad s_2 = 0, \tag{11.7.2}$$

where $\sigma \in \{1, -1\}$. Then the Riemann-Hilbert problem (11.3.6)-(11.3.13) has a meromorphic with respect to x, unique solution. The corresponding solution of the second Painlevé equation (11.2.2), which we will denote $u_{HM}(x)$, has the following asymptotic behavior:

$$u_{HM}(x) \sim \sigma\sqrt{-\frac{x}{2}}\sum_{n=0}^{\infty} b_n(-x)^{-3n/2}, \quad b_0 = 1 \tag{11.7.3}$$

as

$$x \to \infty, \quad \arg(-x) \in \left(-\frac{\pi}{3}, \frac{\pi}{3}\right),$$

and

$$u_{HM}(x) \sim \frac{\alpha}{x}\sum_{n=0}^{\infty} a_n x^{-3n}, \quad a_0 = 1, \tag{11.7.4}$$

as

$$x \to \infty, \quad \arg x \in \left(-\frac{\pi}{3}, \frac{\pi}{3}\right),$$

where the coefficients b_n and a_n are defined by the recurrence relations (11.5.33) and (11.6.27). If $\alpha = 0$, then the asymptotic series (11.7.4) must be replaced by the asymptotic relation,

$$u_{HM}(x) = \sigma\frac{1}{2\sqrt{\pi}}x^{-1/4}e^{-\frac{2}{3}x^{3/2}}\left(1 + O(x^{-3/4})\right), \tag{11.7.5}$$

$$x \to +\infty, \quad \arg x = 0.$$

Appropriately modified, the asymptotics (11.7.3) and (11.7.4) can be continued up to the boundaries of the relevant sectors. Indeed,

$$\begin{aligned} u_{HM}(x) &= \sigma\sqrt{-\frac{x}{2}}\sum_{n=0}^{N} b_n(-x)^{-3n/2} + O\big(x^{-3N/2-1}\big) \\ &+c_+(-x)^{-\frac{3}{2}\sigma\alpha-\frac{1}{4}}e^{-\frac{2\sqrt{2}}{3}(-x)^{3/2}}\big(1 + O(x^{-1/4})\big), \\ & x \to \infty, \quad \arg x \in \left(\frac{2\pi}{3}, \frac{4\pi}{3}\right], \quad \arg(-x) \in \left(-\frac{\pi}{3}, \frac{\pi}{3}\right]; \end{aligned} \tag{11.7.6}$$

$$\begin{aligned} u_{HM}(x) &= \sigma\sqrt{-\frac{x}{2}}\sum_{n=0}^{N} b_n(-x)^{-3n/2} + O\big(x^{-3N/2-1}\big) \\ &+c_-(-x)^{-\frac{3}{2}\sigma\alpha-\frac{1}{4}}e^{-\frac{2\sqrt{2}}{3}(-x)^{3/2}}\big(1 + O(x^{-1/4})\big), \\ & x \to \infty, \quad \arg x \in \left[\frac{2\pi}{3}, \frac{4\pi}{3}\right), \quad \arg(-x) \in \left[-\frac{\pi}{3}, \frac{\pi}{3}\right); \end{aligned} \tag{11.7.7}$$

$$\begin{aligned} u_{HM}(x) &= \frac{\alpha}{x}\Big(\sum_{n=0}^{N} a_n x^{-3n} + O\big(x^{-3N-4}\big)\Big) \\ &+d_+x^{-\frac{1}{4}}e^{-\frac{2}{3}x^{3/2}}\big(1 + O(x^{-3/4})\big), \\ & x \to \infty, \quad \arg x \in \left(-\frac{\pi}{3}, \frac{\pi}{3}\right], \end{aligned} \tag{11.7.8}$$

and

$$(11.7.9)\qquad \begin{aligned} u_{HM}(x) = \frac{\alpha}{x}\Big(\sum_{n=0}^{N} a_n x^{-3n} + O\big(x^{-3N-4}\big)\Big) \\ +d_- x^{-\frac{1}{4}} e^{-\frac{2}{3}x^{3/2}}\big(1 + O(x^{-3/4})\big), \\ x \to \infty, \quad \arg x \in \big[-\frac{\pi}{3}, \frac{\pi}{3}\big). \end{aligned}$$

In the above expressions, N is an arbitrary positive integer number, and the quasi-linear Stokes' multipliers, i.e., the constants c_+, c_-, d_+, and d_-, are given by the following equations:

$$(11.7.10)\qquad c_+ = -\frac{e^{i\pi\sigma(\alpha+\frac{1}{2})}}{\pi} 2^{-\frac{5}{2}\sigma\alpha-\frac{7}{4}}\Gamma(\tfrac{1}{2}+\sigma\alpha),$$

$$(11.7.11)\qquad c_- = -\frac{e^{-i\pi\sigma(\alpha+\frac{1}{2})}}{\pi} 2^{-\frac{5}{2}\sigma\alpha-\frac{7}{4}}\Gamma(\tfrac{1}{2}+\sigma\alpha),$$

$$(11.7.12)\qquad d_+ = -i\frac{e^{i\pi\sigma(\alpha+\frac{1}{2})}}{2\sqrt{\pi}},$$

$$(11.7.13)\qquad d_- = i\frac{e^{-i\pi\sigma(\alpha+\frac{1}{2})}}{2\sqrt{\pi}}.$$

The subscript "HM" in the notation $u_{HM}(x)$ are abbreviations for *Hastings-McLeod*, since S.P. Hastings and J.B. McLeod were the first to discover this solution in the case $\alpha = 0$ [**HasM**]. Furthermore, using an earlier version of the isomonodromy approach, they were able to prove existence and uniqueness and also prove that the solution has no singularities for real x. The behavior for complex x was not discussed in [**HasM**]. The case of nonzero α, presented in this chapter, was analyzed in [**IK5**] and [**K10**].

In addition to the results collected in Theorem 11.7, we should mention that the function $u_{HM}(x)$ has no real poles even in the case of nonzero α. This fact has been proven by T. Claeys, A. Kuijlaars, and M. Vanlessen in [**CK**] with the help of a Riemann-Hilbert technique, similar to the one developed in [**FoZ**] and presented in Chapter 5 of this book (see the Section *Global solutions of the Painlevé equations* in that chapter).

REMARK 11.12. The solution $u_{HM}(x)$ is uniquely characterized by the asymptotic conditions (11.7.3) and (11.7.4). Indeed, $u_{HM}(x)$ is the unique solution of the second Painlevé equation (11.2.2), such that

$$(11.7.14)\qquad u_{HM}(x) \sim \sigma\sqrt{-\frac{x}{2}}, \quad x \to -\infty, \quad \arg(-x) = 0,$$

and, simultaneously,

$$(11.7.15)\qquad u_{HM}(x) \sim \frac{\alpha}{x}, \quad x \to +\infty, \qquad \arg x = 0.$$

If $\alpha = 0$, then the asymptotic condition (11.7.15) must be replaced [**HasM**] by the asymptotic relation,

$$(11.7.16)\qquad u_{HM}(x) \sim \sigma\frac{1}{2\sqrt{\pi}} x^{-1/4} e^{-\frac{2}{3}x^{3/2}},$$

$$x \to +\infty, \quad \arg x = 0,$$

which by itself is sufficient, in this case, for the characterization of the solution $u_{HM}(x)$.

REMARK 11.13. Each of the four asymptotic relations (11.7.6)–(11.7.9) can be taken as a characterization of the solution $u_{HM}(x)$; in fact, it is enough to specify the asymptotics just along any of the boundary rays where the solution is oscillatory.

REMARK 11.14. Each of the following four asymptotic conditions provides yet another, alternative, characterization of the solution $u_{HM}(x)$:

$$u_{HM}(x) = u_0(x|\alpha,\sigma) + c_+(-x)^{-\frac{3}{2}\sigma\alpha-\frac{1}{4}}e^{-\frac{2\sqrt{2}}{3}(-x)^{3/2}}\big(1+O(x^{-1/4})\big),$$
$$x \to \infty, \quad \arg(-x) = 0; \tag{11.7.17}$$

$$u_{HM}(x) = u_1(x|\alpha,\sigma) + c_-(-x)^{-\frac{3}{2}\sigma\alpha-\frac{1}{4}}e^{-\frac{2\sqrt{2}}{3}(-x)^{3/2}}\big(1+O(x^{-1/4})\big),$$
$$x \to \infty, \quad \arg(-x) = 0; \tag{11.7.18}$$

$$u_{HM}(x) = u_1(x|\alpha) + d_+x^{-\frac{1}{4}}e^{-\frac{2}{3}x^{3/2}}\big(1+O(x^{-3/4})\big),$$
$$x \to \infty, \quad \arg x = 0, \tag{11.7.19}$$

and

$$u_{HM}(x) = u_3(x|\alpha) + d_-x^{-\frac{1}{4}}e^{-\frac{2}{3}x^{3/2}}\big(1+O(x^{-3/4})\big),$$
$$x \to \infty, \quad \arg x = 0, \tag{11.7.20}$$

where $u_{0,1}(x|\alpha,\sigma)$ and $u_{1,3}(x|\alpha)$ are tri-tronquée solutions uniquely determined by the conditions (11.5.67) and (11.6.51); in the case $\alpha = 0$, the decreasing solutions $u_{1,3}(x|\alpha)$ are identically zero. The constants $c_\pm$ and $d_\pm$ are defined by (11.7.10)–(11.7.13).

The statements of the above remarks follow by combining the results obtained in this chapter with the asymptotic results obtained in [**K5**] and [**K8**].

We conclude this chapter with three more remarks.

REMARK 11.15. The asymptotic relations (11.7.6)–(11.7.9) describe the quasi-linear Stokes phenomenon for the Hastings-McLeod solution $u_{HM}(x)$. The formulae (11.7.10)–(11.7.13) supplement this description, by providing the exact evaluation of the relevant Stokes' multipliers.

REMARK 11.16. The asymptotic behavior of the function $u_{HM}(x)$ in the complementary sectors, i.e., in $\left(\frac{\pi}{3},\frac{2\pi}{3}\right)$ and $\left(-\frac{2\pi}{3},-\frac{\pi}{3}\right)$, is given in terms of elliptic functions. The exact form of the relevant asymptotics can be extracted from the results of Chapters 7 and 8. Indeed, applying the symmetry (7.2.10) with $l = 2$ to the asymptotics (7.2.7), we find the following formulae:

$$u(x;s) = -\frac{i\varkappa}{\sqrt{1+\varkappa^2}}\,x^{1/2}\,\mathrm{sn}\left(\frac{2i}{3}\,\frac{x^{3/2}}{\sqrt{1+\varkappa^2}} - 2\sigma K\big(\alpha+\tfrac{1}{2}\big) + O\big(x^{-3/2}\big)\big|\varkappa\right),$$
$$x \to \infty, \quad \arg x = \varphi + O\big(x^{-3/2}\big), \quad \varphi \in \big(-\tfrac{2\pi}{3},-\tfrac{\pi}{3}\big), \tag{11.7.21}$$

and

$$u(x;s) = -\frac{i\overline{\varkappa}}{\sqrt{1+\overline{\varkappa}^2}}\, x^{1/2}\, \mathrm{sn}\left(\frac{2i}{3}\, \frac{x^{3/2}}{\sqrt{1+\overline{\varkappa}^2}} + 2\sigma\overline{K}\big(\alpha+\tfrac{1}{2}\big) + O\big(x^{-3/2}\big)\big|\overline{\varkappa}\right),$$

$$x \to \infty, \quad \arg x = \varphi + O\big(x^{-3/2}\big), \quad \varphi \in \big(\tfrac{\pi}{3}, \tfrac{2\pi}{3}\big). \tag{11.7.22}$$

In these expressions, $\varkappa = \varkappa(\varphi)$ satisfies Boutroux equations (7.1.7) with the parameter $\phi = \varphi + \frac{4\pi}{3} \in \big(\frac{2\pi}{3}, \pi\big)$ for (11.7.21), and $\phi = -\varphi + \frac{4\pi}{3} \in \big(\frac{2\pi}{3}, \pi\big)$ for (11.7.22), and $K = K(\varkappa)$ is the complete elliptic integral.

We note that these are sectors containing the poles of $u_{HM}(x)$. The distribution of these poles near the point $x = \infty$, can be extracted from the asymptotic formulae (11.7.21) and (11.7.22).

REMARK 11.17. Applying the symmetries (11.3.26) to the solution $u_{HM}(x)$, we can obtain two additional Hastings-McLeod type solutions, namely the ones associated with the pairs of the sectors $\big(\frac{\pi}{3}, \pi\big) \cup \big(\frac{4\pi}{3}, 2\pi\big)$ and $\big(\pi, \frac{5\pi}{3}\big) \cup \big(0, \frac{2\pi}{3}\big)$.

Part 3

Asymptotics of the Third Painlevé Transcendent

The third Painlevé equation (PIII)

$$u'' = \frac{(u')^2}{u} - \frac{u'}{x} + \frac{1}{x}(\alpha u^2 + \beta) + \gamma u^3 + \frac{\delta}{u} \tag{PIII}$$

represents in full the features of integrable nonlinear ODE. In contrast to PII, it contains two essential singularities at $x = 0$ and $x = \infty$, as well as four arbitrary coefficients $\alpha, \beta, \gamma, \delta$. This adds new phenomena, which are typical of the higher Painlevé equations:

(a) Complicated manifold of generic solutions with different asymptotics at singular points.
(b) Rich group of Bäcklund transformations.
(c) Several classes of rational and algebraic solutions

Following the scheme of Part 2 we will concentrate on (a) above; some results on (c) will be presented in Chapter 12. Because of complexity of generic solutions of PIII, we will not present an exhaustive treatment of all cases, but we will concentrate on the most important solutions. Those solutions are expressed by simple RH problems and their asymptotics are useful in applications. They include the self-similar reductions of the sine-Gordon equation

$$w_{xx} + \frac{1}{x} w_x + \sin w = 0, \tag{SG-PIII}$$

which is PIII with $\alpha = \beta = 0$, $\gamma = -\delta = 1$ and $u(x) = \exp(iw(x))$.

In Chapters 13–15 we will describe the RH approach to this particular PIII case. The isomonodromic technique is slightly different than that of Chapter 5, and the proper adaptation is discussed in Chapter 13. The asymptotics at $x = 0$ is studied in Chapter 14, where the necessary parametrization of the monodromy data is derived. The connection formulas are discussed in Chapter 15 (asymptotics at the boundaries of characteristic sectors) and Chapter 16 (asymptotics in the complex plane).

In our treatment of equation (SG-PIII) we concentrate on the *construction* of connection formulae rather than on the complete proofs. The existence of nonsingular real-valued solutions of equation (SG-PIII) can be obtained with the use of apriori estimates (see Theorem 14.1). Thus for almost all cases considered we can use the direct monodromy problem, assuming a certain type of asymptotic behavior of the solution $w(x)$. The monodromy data calculated in this way is explicitly invertible in the same way as for PII. For some degenerate cases, such as exponentially vanishing solutions, we have used the inverse monodromy problem (see Chapter 14) in a way similar to the Deift-Zhou approach.

CHAPTER 12

PIII Equation, an Overview

0. Introduction. Generic transforms and elementary solutions

The third Painlevé equation

$$u'' = \frac{(u')^2}{u} - \frac{u'}{x} + \frac{1}{x}(\alpha u^2 + \beta) + \gamma u^3 + \frac{\delta}{u} \tag{12.0.1}$$

admits a large group of transformations acting on its solutions $u\,(x|\alpha, \beta, \gamma, \delta)$. We will first describe some obvious continuous transformations, including scalings, which will enables us to reduce the number of parameters.

A generic acting on solutions of PIII has the form

$$\hat{u}\left(x|\hat{\alpha}, \hat{\beta}, \hat{\gamma}, \hat{\delta}\right) = \mathcal{F}\left(u\left(\phi\left(x\right)|\alpha, \beta, \gamma, \delta\right), x\right), \tag{12.0.2}$$

where $\hat{u}\,(x)$ satisfies (12.0.1) with parameters $\hat{\alpha}, \hat{\beta}, \hat{\gamma}, \hat{\delta}$, while $u\,(\phi\,(x))$ satisfies the same equation it with parameters $\alpha, \beta, \gamma, \delta$.

Since u'' and the first two terms in the right-hand side of equation (12.0.1) must remain invariant under the transformation (12.0.2), we get immediately the generic form of the function $\mathcal{F}\,(u, x)$,

$$\begin{gathered} \mathcal{F}\,(u, x) = C_2\,(x)\, u^{C_1(x)}, \quad u = u\left(\phi\left(x\right)|\alpha, \beta, \gamma, \delta\right), \\ \phi\,(x) = \exp\left(C_3 \int \tfrac{dx}{xC_1(x)}\right), \end{gathered} \tag{12.0.3}$$

where $C_1\,(x)$, $C_2\,(x)$ are arbitrary functions and C_3 is an arbitrary constant. Substituting equations (12.0.3) into (12.0.1) one can obtain four generators of the transformation group (12.0.2):

$$Q: \quad u\,(x|\alpha, \beta, \gamma, \delta) \mapsto \hat{u}\left(x|\hat{\alpha}, \hat{\beta}, \hat{\gamma}, \hat{\delta}\right) = qu\left(x|\alpha q^{-1}, \beta q, \gamma q^{-2}, \delta q^2\right), \tag{12.0.4}$$

$$P: \quad u\,(x|\alpha, \beta, \gamma, \delta) \mapsto \hat{u}\left(x|\hat{\alpha}, \hat{\beta}, \hat{\gamma}, \hat{\delta}\right) = u\left(px|\alpha p, \beta p, \gamma p^2, \delta p^2\right), \tag{12.0.5}$$

$$J: \quad u\,(x|\alpha, \beta, \gamma, \delta) \mapsto \hat{u}\left(x|\hat{\alpha}, \hat{\beta}, \hat{\gamma}, \hat{\delta}\right) = \frac{1}{u\,(x|-\alpha, -\beta, -\gamma, -\delta)}, \tag{12.0.6}$$

$$S: \quad u\,(x|0, 0, \gamma, \delta) \mapsto \hat{u}\left(x|\hat{\alpha}, \hat{\beta}, \hat{\gamma}, \hat{\delta}\right) = u^2\left(\sqrt{x}\,\Big|\,\frac{\alpha}{2}, \frac{\beta}{2}, 0, 0\right). \tag{12.0.7}$$

The first three substitutions are applicable to arbitrary solutions, while the fourth, S, acts on solutions with $\alpha = \beta = 0$ and the inverse transformation S^{-1} acts on solutions with $\gamma = \delta = 0$. In fact, S gives rise to two different transformations,

$$\mathrm{S}_{\pm}^{-1}: \quad u\,(x|\alpha, \beta, 0, 0) \mapsto \hat{u}\left(x|\hat{\alpha}, \hat{\beta}, \hat{\gamma}, \hat{\delta}\right) = \pm\sqrt{u\left(x^2\,|\,0, 0, 2\gamma, 2\delta\right)}, \tag{12.0.8}$$

since either choice of the branch leaves the parameters in equation (12.0.1) invariant.

The commutation relations of the transformations (12.0.3)–(12.0.8) are given by

$$
\begin{aligned}
&PQ = QP, \qquad JP = PJ, \qquad S_{\mp}^{-1}J = JS_{\pm}^{-1},\\
&JQ = Q^{-1}J, \quad SP = P^2S, \quad PS_{\pm}^{-1} = S_{\pm}^{-1}P^2,\\
&SQ = Q^2S, \qquad SJ = JS, \qquad PS_{\mp}^{-1} = S_{\mp}^{-1}P^2.
\end{aligned}
\tag{12.0.9}
$$

For a concrete choice of the parameters $\alpha, \beta, \gamma, \delta$ equation (12.0.1) can be integrated through elementary or classic transcendent functions:

I. If $\alpha = \beta = \gamma = \delta = 0$, then $u = C_1 x^{C_2}$.

II. If $\beta = \delta = 0$, then after $x = \exp z$ one has

$$\left(\frac{u'}{u}\right)^2 + 2\left(\frac{u'}{u}\right) - 2\alpha u \exp z - \gamma u^2 \exp 2z = C_1, \quad u' = \frac{du}{dz}.$$

III. If $\alpha = \gamma = 0$, then

$$\left(\frac{u'}{u}\right)^2 + 2\left(\frac{u'}{u}\right) + \frac{2\beta}{u}\exp z + \frac{\delta}{u^2}\exp 2z = C_2.$$

IV. For $\gamma = a^2, \delta = -c^2, a\beta + c(\alpha - 2a) = 0, a \neq 0$, equation (12.0.1) reduces to the Riccati equation

$$u' = au^2 + \frac{\alpha - a}{ax}u + c. \tag{12.0.10}$$

The first order equations in II–III above, can be integrated in terms of elementary functions. For example, deploying

$$u = \frac{1}{v\exp(-z)},$$

we find

$$(v')^2 = 2\alpha v + \gamma + (1 + C_1)v^2. \tag{12.0.11}$$

If $1 + C_1 = 0,\ \beta = \delta = 0$, the general solution of (12.0.1) has the form

$$u = \frac{2\alpha}{x(\alpha^2 \log^2 x + 2\alpha C \log x + C^2 - \gamma)}.$$

For the degenerate case of $\alpha = 0,\ \ 1 + C_1 = 0$, the solution reduces to

$$u = \frac{1}{x(\sqrt{\gamma}\log x + C)}.$$

If $1 + C_1 = 0$, the solution becomes

$$u = \frac{x^{k-1}}{Cx^{2k} + K_1 x^k + K_2},$$

where

$$C \neq 0, \quad K_1 = -\frac{\alpha}{1 + C_1}, \quad K_2 = \frac{\alpha^2 - \gamma(1 + C_1)}{4C(1 + C_1)^2}, \quad k^2 = 1 + C_1.$$

The Riccati equation (12.0.10) can be transformed into the linear equation

$$w'' + \frac{a-\alpha}{a\tau} w' + w = 0, \tag{12.0.12}$$

with $x = \lambda\tau$, $\lambda^2 = (ac)^{-1}$, $u = -w'/aw$, and its general solution is expressed through the Bessel functions.

1. Rational solutions of PIII

We will apply the Bäcklund transformations, derived in Chapter 6, to describe a set of rational solutions of PIII equation (12.0.1). These solutions were first found in [**GroL1, GroL2**] and [**FoA2**] later in more compact form in [**Bor2**]. The results of Chapter 6 for PIII can be summarized in the following two theorems.

THEOREM 12.1. *Let $u(x|\alpha,\beta,\gamma,\delta)$ be a solution of equation (12.0.1) for $\gamma\delta \neq 0$. Then*

$$\begin{aligned} \hat{u}\left(x|\hat{\alpha},\hat{\beta},\hat{\gamma},\hat{\delta}\right) &= u(x|\alpha,\beta,\gamma,\delta) \\ &\times \sqrt{\frac{\gamma}{\hat{\gamma}}}\left(1 + \frac{2 + \frac{\beta}{\sqrt{-\delta}} + \frac{\alpha}{\sqrt{\gamma}}}{x\left(\frac{u'}{u} + \sqrt{\gamma}u + \frac{\sqrt{-\delta}}{u}\right) - 1 - \frac{\beta}{\sqrt{-\delta}}}\right) \end{aligned} \tag{12.1.1}$$

is a solution of equation (12.0.1) with parameters

$$\begin{aligned} \hat{\alpha} &= -\left(2 + \tfrac{\beta}{\sqrt{-\delta}}\right)\sqrt{\hat{\gamma}}, \qquad \hat{\gamma} \neq 0, \\ \hat{\beta} &= -\left(2 + \tfrac{\alpha}{\sqrt{\gamma}}\right)\tfrac{\sqrt{-\gamma\delta}}{\sqrt{\hat{\gamma}}}, \quad \hat{\delta} = \tfrac{\gamma\delta}{\hat{\gamma}}. \end{aligned} \tag{12.1.2}$$

THEOREM 12.2. *If $\gamma = 0$, then the transformed solution is given by*

$$\hat{u}\left(x|\hat{\alpha},\hat{\beta},\hat{\gamma},\hat{\delta}\right) = \frac{xu' - \left(1 + \frac{\beta}{\sqrt{-\delta}}\right)u + x\sqrt{-\delta}}{u^2}, \tag{12.1.3}$$

with parameters

$$\begin{aligned} \hat{\alpha} &= \sqrt{-\delta}, \qquad \hat{\beta} = \alpha\left(2 + \frac{\beta}{\sqrt{-\delta}}\right), \\ \hat{\gamma} &= 0, \qquad \hat{\delta} = -\alpha^2. \end{aligned} \tag{12.1.4}$$

When the numerator in (12.1.1) vanishes

$$2 + \frac{\beta}{\sqrt{-\delta}} + \frac{\alpha}{\sqrt{\gamma}} = 0,$$

there exists a one-parameter family of solutions characterized by the requirement that the denominator in (12.1.1) also vanishes:

$$x\left(\frac{u'}{u} + \sqrt{\gamma}u + \frac{\sqrt{-\delta}}{u}\right) - 1 - \frac{\beta}{\sqrt{-\delta}} = 0.$$

The transformation (12.1.1) is still valid for this case but with the altered signs of the roots

$$\sqrt{\gamma} \mapsto -\sqrt{\gamma}, \qquad \sqrt{-\delta} \mapsto -\sqrt{-\delta}.$$

All possible Bäcklund transformations can be derived from the composition of (12.1.1) and (12.1.3) with elementary transformations P, Q, J, S, defined by equations (12.0.4)–(12.0.9).

We are now ready to describe all rational solutions of PIII. These solutions can be parametrized by two integers which specify the coefficients α, β, γ, δ.

THEOREM 12.3 ([**Bor2**]). *The PIII equation (*12.0.1*) posesses rational solutions iff* $\gamma\delta \neq 0$ *and*

$$\frac{\alpha}{\sqrt{\gamma}} + \frac{\beta}{\sqrt{-\delta}} = 4n, \tag{12.1.5}$$
$$\frac{\alpha}{\sqrt{\gamma}} - \frac{\beta}{\sqrt{-\delta}} = 4m,$$

*All rational solutions can be obtained from the Bäcklund transformations (*12.1.1*), (*12.1.3*) and (*12.0.4*)-(*12.0.9*) applied to the constants*

$$u_0(x) = \left(-\frac{\delta}{\gamma}\right)^{1/4} \exp\left(\frac{i\pi k}{2}\right), \quad k = 0, 1, 2, 3, \tag{12.1.6}$$

which are solutions of PIII for

$$\alpha\sqrt{-\delta} + \beta\sqrt{\gamma} = 0.$$

For a fixed set of n, m *and* k*, PIII has four rational solutions, if both conditions (*12.1.6*) are satisfied and it has two rational solutions, if only one of (*12.1.6*) is valid.*

PROOF. We first prove the necessity of condition (12.1.5) for the existence of rational solutions of PIII equation. Let $u(x|\alpha, \beta, \gamma, \delta)$ be a rational solution, then at the neighborhood of any of its poles $x_0 \neq 0$, it has the power series expansion

$$u(x|\alpha, \beta, \gamma, \delta) = \frac{\epsilon}{\sqrt{\gamma}(x - x_0)} - \frac{\alpha + \epsilon\sqrt{\gamma}}{3x_0} + a_1(x - x_0) + \dots, \quad x \to x_0,$$

where $\epsilon^2 = 1$ and a_1 is arbitrary. At infinity, the analogous expansion is

$$u(x|\alpha, \beta, \gamma, \delta) = \epsilon\left(-\frac{\delta}{\gamma}\right)^{1/4} + \frac{1}{4\sqrt{\gamma}}\left(-\frac{\alpha}{\sqrt{\gamma}} + \frac{\beta}{\sqrt{-\delta}}\right)x^{-1} + \dots, \quad x \to \infty. \tag{12.1.7}$$

The residue theorem applied to $u(x|\alpha, \beta, \gamma, \delta)$, immediately implies the condition (12.1.5). Let $x = 0$ be a simple pole; applying the transformation J (12.0.6) to $u(x|\alpha, \beta, \gamma, \delta)$, implies that $u^{-1}(x| - \beta, -\alpha, -\gamma, -\delta)$ is also a solution of (12.0.1). Since $x = 0$ is a simple zero, this also leads to condition (12.1.5).

The sufficiency of (12.1.5) can be verified in a constructive way. For $n = 0$, the constants (12.1.6) are solutions of PIII. Applying the Bäcklund transformations (12.1.1) with alternating signs of radicals, we obtain rational solution with arbitrary integers n and m satisfying equations (12.1.5).

On the other hand, for any rational solution $u(x|\alpha, \beta, \gamma, \delta)$, the constraints (12.1.5) are satisfied for some integers n and m. Applying to $u(x|\alpha, \beta, \gamma, \delta)$ the Bäcklund transformations (12.1.1) with the proper signs of radicals, we find the rational solution $\hat{u}(x|\hat{\alpha}, \hat{\beta}, \hat{\gamma}, \hat{\delta})$ with

$$\hat{\alpha}\sqrt{-\hat{\delta}} + \hat{\beta}\sqrt{\hat{\gamma}} = 0.$$

This solution, due to (12.1.7), has the asymptotics at infinity

$$\hat{u}(x|\hat{\alpha},\hat{\beta},\hat{\gamma},\hat{\delta}) \to \epsilon\left(-\frac{\hat{\delta}}{\hat{\gamma}}\right)^{1/4}, \quad x\to\infty. \tag{12.1.8}$$

But the constant $\epsilon\left(-\frac{\hat{\delta}}{\hat{\gamma}}\right)^{1/4}$ is a solution of the same equation,

$$\hat{u}(x|\hat{\alpha},\hat{\beta},\hat{\gamma},\hat{\delta}) = \epsilon\left(-\frac{\hat{\delta}}{\hat{\gamma}}\right)^{1/4},$$

which proves that $u(x|\alpha,\beta,\gamma,\delta)$ is a composition of Bäcklund transformations applied to the constant.

The number of rational solutions can be estimated as follows: consider rational solutions of PIII for large x. A simple analysis implies that if the first equation (12.1.5) is valid, then those solutions possess the asymptotics (12.1.8). For the second equation (12.1.5), the relevant asymptotics is

$$u_{3,4}(x) \to \pm i\left(-\frac{\delta}{\gamma}\right)^{1/4}, \quad x\to\infty.$$

This leads to 2 distinct solutions when only one constraint (12.1.5) holds, and to 4 solutions when both constrains (12.1.5) are valid. □

As an illustration of the above theorems we construct simple rational solutions starting from some elementary solutions. Take the coefficients

$$\hat{\alpha}=4, \quad \hat{\beta}=0, \quad \hat{\gamma}=1, \quad \hat{\delta}=-1,$$

then equation (12.0.1) has 4 rational solutions. Solving equations (12.1.2), we find

$$\alpha=-2, \quad \beta=2, \quad \gamma=1, \quad \hat{\delta}=-1.$$

Starting with $u(x|\alpha,\beta,\gamma,\delta)=\pm 1$ and applying the Bäcklund transformation (12.1.1), find the first pair of solutions

$$u(x|\hat{\alpha},\hat{\beta},\hat{\gamma},\hat{\delta}) = \frac{2x\pm 1}{\pm 2x+3}.$$

The second pair follows from the choice of

$$\alpha=-2, \quad \beta=-2, \quad \gamma=1, \quad \hat{\delta}=-1,$$

and of the constant solutions $u(x|\alpha,\beta,\gamma,\delta)=\pm i$:

$$u(x|\hat{\alpha},\hat{\beta},\hat{\gamma},\hat{\delta}) = i\frac{2ix\pm 1}{\pm 2ix+3}.$$

2. Algebraic solutions of PIII

Algebraic solutions can be found in a similar way as the rational solutions. They have the form

$$u(x) = R(\sqrt[3]{x}), \quad R - \text{rational function},$$

and can be constructed from the elementary solution

$$u(x) = \sqrt[3]{-\frac{\delta x}{\alpha}}, \quad \alpha\delta\neq 0. \tag{12.2.1}$$

Now, the coefficients of equation (12.0.1) satisfy one of the following constraints,

$$\gamma = 0, \quad \alpha\delta \neq 0, \quad \frac{\beta}{\sqrt{-\delta}} = 2n, \tag{12.2.2}$$

$$\delta = 0, \quad \beta\gamma \neq 0, \quad \frac{\alpha}{\sqrt{\gamma}} = 2m, \tag{12.2.3}$$

for some integers n and m. We will present a concise proof of this result, following [**GroL2**] and [**Bor2**].

THEOREM 12.4. *All algebraic solutions of PIII equation (12.0.1) can be derived using the Bäcklund transformations (12.1.2), (12.1.3) and the transformations J and Q starting from the elementary solution (12.2.1). These solutions can be obtained assuming that either (12.2.2) or (12.2.3) are valid.*

For each set of coefficients satisfying (12.2.2) or (12.2.3), there exist three algebraic solutions of the form

$$u(x) = \tau y R(\tau y^2), \quad y = \sqrt[3]{x},$$

$$\tau = \exp\left(\tfrac{2\pi i k}{3}\right), \quad k = 0, 1, 2,$$

for some rational function R.

PROOF. By constructing power series expansions at $x = 0$, $x = \text{const}$, and $x = \infty$, we can derive the necessary condition $\gamma\delta = 0$ for the existence of algebraic (non-rational) solutions.

If $\gamma = 0$, let $u(x) = yw(y)$, $x = y^3$. Then PIII takes the form

$$\left(\frac{yw'(y)}{w(y)}\right)' \frac{1}{9y^2} = y\left(\alpha w(y) + \frac{\delta}{w^2(y)}\right) + \frac{\beta}{yw(y)}.$$

A rational solution $w(y)$ can be constructed by analyzing Laurent series at $y = 0$, $y = \text{const}$ and $y = \infty$. This construction shows that all zeros of w are simple and all poles are of the second order. Furthermore, the asymptotics at infinity have the form

$$\begin{gathered} w(y) = w_0 + \frac{w_2}{y^2} + \frac{w_4}{y^4} + \dots, \quad y \to \infty, \\ w_0^3 = -\frac{\delta}{\alpha}, \quad w_2 = -\frac{\beta}{3\alpha w_0}, \quad \dots, \end{gathered} \tag{12.2.4}$$

where all w_{2k} are uniquely found from w_0. This leads to the representation

$$w(y) = w_0 \frac{y^p \prod_{i=1}^{2n-p/2} (y^2 + a_i)}{\prod_{i=1}^{n} (y^2 + b_i)^2}, \quad p = 0, -2.$$

We now consider a contour in the complex plane y enclosing all the zeroes of $w(y)$ and compute the integral $\oint yw^{-1}(y)dy$. The asymptotics (12.2.4) leads to (12.2.2). The assumption $\delta = 0$ yields (12.2.3).

Thus, algebraic and nonrational solutions of (12.0.1) exist only under conditions (12.2.2) or (12.2.3), and have the form $yR(y^2)$. The equation $w_0^3 = -\frac{\delta}{\alpha}$ has three roots for $\alpha\delta \neq 0$, thus for a fixed set of $\alpha, \beta, \gamma, \delta$, there exist three distinct solutions $w(x)$.

Finally, we must prove that the Bäcklund transformation (12.1.2) does not change the form $u(x) = yR(y^2)$, $y = \sqrt[3]{x}$ of the solution:

$$\hat{u}(x|\hat{\alpha}, \hat{\beta}, \hat{\gamma}, \hat{\delta}) = 2y\frac{y^2 R'(y^2) - -(1+3n)R(y^2) + \frac{3}{2}\sqrt{-\delta}y^2}{3y^2 R^2(y^2)} = y\hat{R}(y^2).$$

Since the trivial algebraic solution has the above form, with $R(y^2)$ = const, a composition of Bäcklund transformations starting with this solution, will lead to an algebraic solution $u(x)$. For a given set of $\alpha, \beta, \gamma, \delta$, three of these solutions can be constructed from $u(x0$ by the transformations P and Q with parameters $p = q = \exp(i\pi k)$, $k = 0, 1$. This yields the solutions $\tau^2 R(\tau y^2)$ with $\tau = \exp\left(\frac{2\pi i k}{3}\right)$, $k = 0, 1, 2$. □

CHAPTER 13

Sine-Gordon Reduction of PIII

0. Introduction

In Chapters 13–16 we will study a special case of PIII, namely the equation

$$u_{xx} + \frac{1}{x}u_x + \sin u = 0. \tag{13.0.1}$$

This equation can be obtained from (12.0.1), the general case of

$$w'' = \frac{(w')^2}{w} - \frac{w'}{x} + \frac{1}{x}(\alpha w^2 + \beta) + \gamma w^3 + \frac{\delta}{w} \tag{13.0.2}$$

with the substitutions

$$w(x) = \exp\left(\frac{iu(x)}{2}\right), \quad \alpha = \beta = 0, \quad \gamma = -\delta = -\frac{1}{4}. \tag{13.0.3}$$

In applications, equation (13.0.1) is the most common case of PIII. Its appearances include self-similar reduction of the sine-Gordon equation, as well as the mathematical formalization of the pendulum with a friction.

In connection with the, so-called sine-Gordon or Liouville type equations, we note the following. Let $U(z, \overline{z})$ satisfy the above equations, namely

$$U_{z\overline{z}} = a\mathrm{e}^{\varepsilon U} + b\mathrm{e}^{-\varepsilon U}, \quad z \in \mathbb{C}. \tag{13.0.4}$$

where a, b, ε are constants. If $ab \neq 0$ (sine-Gordon case), then the radial-symmetric self-similar solution of (13.0.4)

$$U = U(x), \quad x = \sqrt{z\overline{z}} = |z|,$$

leads to the ODE

$$U_{xx} + \frac{1}{x}U_x = a\mathrm{e}^{\varepsilon U} + b\mathrm{e}^{-\varepsilon U}. \tag{13.0.5}$$

This equation is a particular case of PIII,

$$w(x) = \exp\left(\frac{1}{2}\varepsilon U(x)\right), \quad \alpha = \beta = 0, \quad \gamma = 4a\varepsilon, \quad \delta = 4b\varepsilon.$$

The radial-symmetric substitution

$$U = U(x), \quad x = z\overline{z} = |z|^2,$$

yields the ODE

$$U_{xx} + \frac{1}{x}U_x = \frac{1}{x}\left(a\mathrm{e}^{\varepsilon U} + b\mathrm{e}^{-\varepsilon U}\right),$$

This ODE is also a particular case of PIII (12.0.1),

$$w(x) = \exp\left(\frac{1}{2}\varepsilon U(x)\right), \quad \alpha = a\varepsilon, \quad \beta = b\varepsilon, \quad \gamma = \delta = 0.$$

For the Liouville case $b = 0$, the associate PIII can be solved explicitly in quadra-

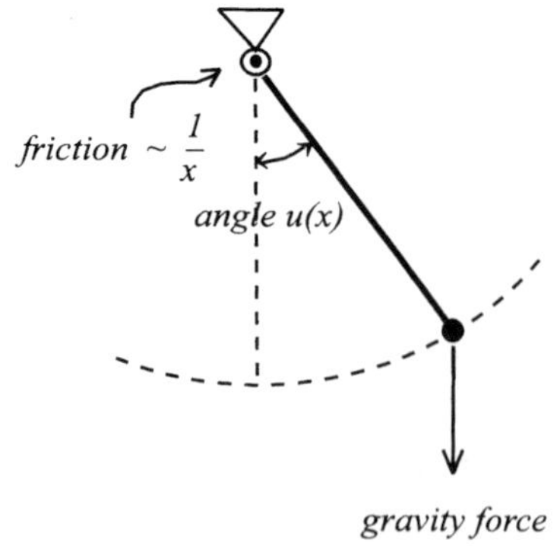

FIGURE 13.1. The mathematical formulation of the pendulum with friction, where x is time

tures (see Section 1 of Chapter 12), which is consistent with the fact that the Liouville equation itself can also be solved explicitly.

Among the many applications of the elliptic version of sine-Gordon equation (13.0.4), we note its appearance in the theory of surfaces of constant mean curvature (CMC). The associated Gauss equation describing these surfaces takes the invariant form (for details see, for example, [**BobEi1**])

$$v_{z\bar{z}} + 2\mathrm{e}^{v} - 2Q\overline{Q}\mathrm{e}^{-v} = 0.$$

Here z is a local complex variable on a Riemann surface, $\mathrm{e}^{v}dzd\bar{z}$ is the conformal metric of the surface in the Euclidean space, and $Q(z)dz^2$ is the holomorphic Hopf differential. For a simply connected domain the transformation

$$\frac{dw}{dz} = \sqrt{Q}, \quad u = v - \log|Q|,$$

reduces this equation to the elliptic sine-Gordon equation

$$u_{z\bar{z}} + 4\sinh u = 0.$$

The umbilic points of the above surface are the zeroes of Q. Obviously, if the metric $v(z,\bar{z})$ is smooth the solution $u(w,\overline{w})$ is singular at these points. A rotationally-symmetric CMC surface depends on only one variable $x = |w|$, thus its metric $u = u(x)$, satisfies PIII (see Chapter 14),

$$u_{xx} + \frac{1}{x}u_x + 4\sinh u = 0.$$

Concluding the discussion, we also note that the hyperbolic version of equation (13.0.4), namely the equation

$$U_{tt} - U_{yy} = a\mathrm{e}^{\varepsilon U} + b\mathrm{e}^{-\varepsilon U}, \tag{13.0.6}$$

also reduces to the PIII equation (13.0.5), under the reduction

$$U = U(x), \quad x = \sqrt{y^2 - t^2}.$$

The appearance of equation (13.0.1) in elementary mechanics can be used to explain certain basic properties of the solutions of equation (13.0.1). Recall that the equation

$$u_{xx} + \frac{g}{l}\sin u = 0, \tag{13.0.7}$$

describes the motion of the pendulum, where x is time, u is the angle to the vertical, and the ratio of the gravity constant g to the pendulum length l is normalized to 1. Adding to equation (13.0.7) a friction in the form of the term u_x with coefficient inversely proportional to time, we come to equation (13.0.1) (see Figure 13.1).

It is clear that due to the friction, periodic motions are no longer possible. Now, any solution goes to the stationary point $u = \pi n, \quad n \in \mathbb{Z}$, and the separatrix loops between $-\pi + 2\pi n$ and $\pi + 2\pi n$ break down. From the point of view of physics this is due to the infinite friction force at $x = 0$, one needs an infinite initial torque to make the pendulum pass from one stationary point to another. On the other hand, at a large time the fiction becomes negligible, and the motion is governed by the classical pendulum equation (13.0.7).

The above discussion suggests that on the phase plane the saddle at points $u = \pi + 2\pi n$ survive even after friction is switched on, but the centers around $u = 2\pi n$ become stable focuses. The separatrix loops now split into separatrices going to infinity and to the focuses.

A comparison of the phase portraits of the classical and of the damped pendulums is presented in Figure 13.2

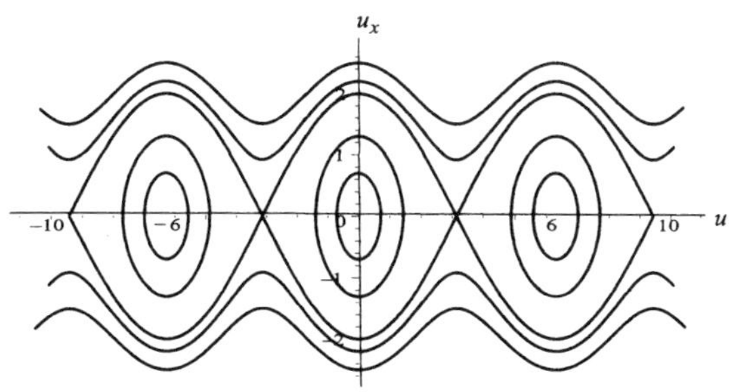

FIGURE 13.2. The phase plane of the classical pendulum

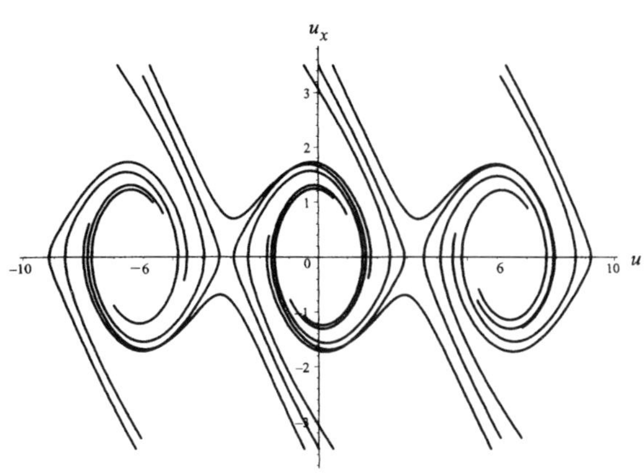

FIGURE 13.3. The phase plane of the SG-PIII equation (13.0.1)

The sine-Gordon reduction (13.0.1) of PIII cannot be solved in terms of quadratures, but instead it reveals features of the generic PIII. Note also that due to the transformation S (12.0.7), the SG-PIII equation includes the case of $\alpha\beta \neq 0$, $\gamma\delta = 0$.

1. Direct monodromy problem for the sine-Gordon PIII reduction

The Lax pair and the associated monodromy theory for the SG-PIII equation (13.0.1), is much simpler than the general case (cf. (5.3.5) and (5.3.6 in Chapter 5). Indeed, the Lax pair consists of the following linear equations,

$$\begin{aligned} &\frac{\partial\Psi}{\partial\lambda} = \mathsf{A}\Psi; \quad \mathsf{A} = A_0 + \frac{A_{-1}}{\lambda} + \frac{A_{-2}}{\lambda^2}, \\ &\frac{\partial\Psi}{\partial x} = \mathsf{U}\Psi; \quad \mathsf{U} = \lambda U_1 + U_0, \end{aligned} \tag{13.1.1}$$

where

$$A_0 = -\frac{ix}{16}\sigma_3, \quad A_{-1} = -\frac{ixu_x}{4}\sigma_1, \quad A_{-2} = i(\cos u\sigma_3 - \sin u\sigma_2),$$

and

$$U_0 = -\frac{iu_x}{2}\sigma_1, \qquad U_1 = -\frac{ix}{8}\sigma_3.$$

Hence the basic equation is

$$\frac{d\Psi}{d\lambda} = \left\{-i\frac{x^2}{16}\sigma_3 - \frac{1}{2\lambda}\frac{ixu_x}{2}\sigma_1 + \frac{i\cos u}{\lambda^2}\sigma_3 - \frac{i\sin u}{\lambda^2}\sigma_2\right\}\Psi, \tag{13.1.2}$$

where σ_k are the three Pauli matrices

$$\sigma_1 = \begin{pmatrix} 0 & 1 \\ 1 & 0 \end{pmatrix}, \quad \sigma_2 = \begin{pmatrix} 0 & -i \\ i & 0 \end{pmatrix}, \quad \sigma_3 = \begin{pmatrix} 1 & 0 \\ 0 & -1 \end{pmatrix},$$

and x, u and u_x are complex-valued parameters.

Equation (13.1.2) has two irregular singular points of the first rank, $\lambda = 0$ and $\lambda = \infty$. The structure of the essential singularities of the solution $\Psi(\lambda)$ at these points is fixed by the following asymptotics:

$$\Psi(\lambda) \quad \to \quad \exp\left(-\frac{ix^2\lambda}{16}\sigma_3\right), \quad \lambda \to \infty,$$

$$\Psi(\lambda) \quad \to \quad \exp\left(-\frac{i\sigma_3}{\lambda}\right), \quad \lambda \to 0.$$

The matrix $\mathsf{A}(\lambda)$ satisfies

$$\mathsf{A}(\lambda) = -\sigma_1\mathsf{A}(-\lambda)\sigma_1. \tag{13.1.3}$$

The coefficients of the formal series for Ψ,

$$\Psi_f^{(\infty)}(\lambda) = \left(I + \psi_1^{(\infty)}\lambda^{-1} + \ldots\right)\exp\left\{-\frac{ix^2}{16}\sigma_3\lambda\right\},$$

can be calculated in terms of the matrix elements of A,

$$A_0 = -i\frac{x^2}{16}\sigma_3, \tag{13.1.4}$$

$$A_{-1} = -\frac{ix^2}{16}\left[\psi_1^{(\infty)}, \sigma_3\right] = \begin{pmatrix} 0 & -\frac{ix}{4}w \\ -\frac{ix}{4}v & 0 \end{pmatrix},$$

where

$$w = -\frac{x}{2}\left(\psi_1^{(\infty)}\right)_{12},$$
$$v = \frac{x}{2}\left(\psi_1^{(\infty)}\right)_{21}.$$

The reduction (13.1.3) imposed on $\mathsf{A}(\lambda)$ is equivalent to $v = w = u_x$ and, consequently

$$A_{-1} = -\frac{ixu_x}{4}\sigma_1.$$

Similarly, the formal solution at $\lambda = 0$,

$$\Psi_f^{(0)}(\lambda) = P_0(I + \ldots)\exp\left\{-\frac{i\sigma_3}{\lambda}\right\},$$

yields

$$A_{-2} = iP_0\sigma_3P_0^{-1},$$

provided that

(13.1.5) $$\det A_{-2} = 1,$$
$$\operatorname{Tr} A_{-2} = 0.$$

Taking into account the reduction (13.1.3), there appears the third condition on A_{-2}, namely

(13.1.6) $$\sigma_1 A_{-2}\sigma_1 = -A_{-2}.$$

Equations (13.1.5) and (13.1.6) imply that A_{-2} is of the form

(13.1.7) $$A_{-2} = \begin{pmatrix} i\cos u & -\sin u \\ \sin u & -i\cos u \end{pmatrix}, \quad u \in \mathbb{C}.$$

The real-valued reduction of equation (13.1.2), i.e., relations $u = \overline{u}$, $w = \overline{w}$, imply the constraint equation

(13.1.8) $$\mathsf{A}(\lambda) = -\mathsf{A}^*\left(\overline{\lambda}\right).$$

It is possible to insert two arbitrary parameters c_1, c_2 in the matrix P_0,

$$P_0 = \begin{pmatrix} \cos\frac{u}{2} & -i\sin\frac{u}{2} \\ -i\sin\frac{u}{2} & \cos\frac{u}{2} \end{pmatrix}\begin{pmatrix} c_1 & 0 \\ 0 & c_2 \end{pmatrix}, \quad c_1, c_2 \in \mathbb{C}.$$

However, we will require that the matrix P_0 satisfies the identity

(13.1.9) $$\sigma_1 P_0 \sigma_1 = P_0,$$

which yields $c_1 = c_2 = 1$. If, in addition, u, w are real-valued, then P_0 satisfies[1] equation

(13.1.10) $$\left[P_0^*\right]^{-1} = P_0.$$

[1]The constraints (13.1.9), (13.1.10) fix P_0 with only one scalar multiplier of the form $e^{i\alpha}$, where α is an arbitrary constant.

The complete set of the monodromy data for equation (13.1.2) consists of four Stokes matrices $S_{1,2}^{(\infty)}$, $S_{1,2}^{(0)}$ and of one connection matrix $Q \equiv Q^{(0)}$. Assuming that $x \in \mathbb{R}$, the canonical sectors $\Omega_{1,2,3}^{(\infty,0)}$ in the neighborhoods of the origin and of infinity are defined as follows:

$$\Omega_1^{(\infty,0)} = \{\lambda : -\pi < \arg(\lambda) < \pi, \quad 0 \neq |\lambda| \lessgtr \rho > 0\},$$

$$\Omega_2^{(\infty,0)} = \{\lambda : 0 < \arg(\lambda) < 2\pi, \quad 0 \neq |\lambda| \lessgtr \rho > 0\},$$

$$\Omega_3^{(\infty,0)} = \{\lambda : \pi < \arg(\lambda) < 3\pi, \quad 0 \neq |\lambda| \lessgtr \rho > 0\}.$$

The sectors Ω_k are shown in Figure 13.4.

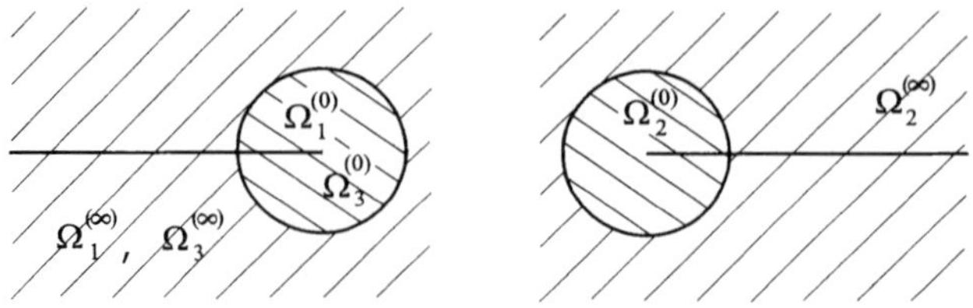

FIGURE 13.4. Sectors Ω_k for $\arg x = 0$

The canonical solutions $\Psi_k^{(\infty)}(\lambda)$ and $\Psi_k^{(0)}(\lambda)$ are fixed at the sectors $\Omega_k^{(\infty)}$ and $\Omega_k^{(0)}$ respectively by their asymptotics,

$$\Psi_k^{(\infty)} = \left(I + \psi_1^{(\infty)}\lambda^{-1} + \dots\right)\exp\left\{-\frac{ix^2}{16}\sigma_3\lambda\right\},$$

$$\lambda \in \Omega_k^{(\infty)}, \quad \lambda \to \infty,$$

(13.1.11) $$\Psi_k^{(0)} = P_0\left(I + \lambda\psi_1^{(\infty)} + \dots\right)\exp\left\{-\frac{i\sigma_3}{\lambda}\right\},$$

$$\lambda \in \Omega_k^{(0)}, \quad \lambda \to 0,$$

$$P_0 = \begin{pmatrix} \cos\frac{u}{2} & -i\sin\frac{u}{2} \\ -i\sin\frac{u}{2} & \cos\frac{u}{2} \end{pmatrix}, \quad k = 1, 2, 3.$$

The Stokes matrices are defined by

(13.1.12) $$S_k^{(\infty,0)} = \Psi_k^{-1}\Psi_{k+1}, \quad k = 1, 2,$$

while the connection matrix Q satisfies the equation

(13.1.13) $$\Psi_1^{(\infty)} = \Psi_1^{(0)}Q.$$

The function $\Psi_3^{(\infty,0)}(\lambda)$ is defined on the universal covering $\widetilde{\mathbb{C}P}^1$. This means

(13.1.14) $$\Psi_3^{(\infty,0)}(\lambda) = \Psi_1^{(\infty,0)}\left(\lambda e^{-2\pi i}\right),$$

where $\pi < \arg\lambda < 3\pi$.

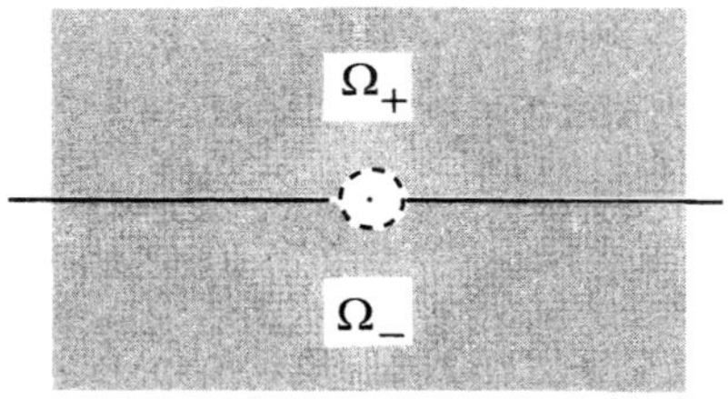

FIGURE 13.5. Domains for auxiliary Stokes matrices $S_\pm$

From this equation and the definition (13.1.12) of the Stokes matrices, it follows that

$$\Psi_1^{(\infty,0)}(\lambda)\, S_1^{(\infty,0)} S_2^{(\infty,0)} = \Psi_1^{(\infty,0)}\left(\lambda e^{-2\pi i}\right), \quad \lambda \in \Omega_3.$$

In other words, the monodromy group has the following two generators:

$$M^{(\infty)} = \left[S_2^{(\infty)}\right]^{-1}\left[S_1^{(\infty)}\right]^{-1},$$

$$M^{(0)} = Q^{-1}\cdot\left[S_1^{(0)} S_2^{(0)}\right]^{-1} Q.$$

In contrast to the general theory we have chosen here the inverse orientation by going around the point $\lambda = \infty$. Therefore, now the cyclic relation (2.0.31) takes the form

$$M^{(\infty)} = M^{(0)}. \tag{13.1.15}$$

It is convenient for future considerations to introduce, in addition to the solutions $\Psi_k^{(\infty,0)}$, the auxiliary pair $\Psi_\pm(\lambda)$ fixed by conditions

$$\Psi_\pm(\lambda) = \left(I + \psi_1^{\pm(\infty)}\lambda^{-1} + \dots\right)\exp\left\{-\frac{ix^2}{16}\sigma_3\lambda\right\},$$

$$\lambda \to \infty, \quad \lambda \in \Omega_\pm,$$

$$\Psi_\pm(\lambda) = P_\pm\left(I + \psi^{\pm(0)}\lambda + \dots\right)\exp\left\{-\frac{i\sigma_3}{\lambda}\right\}, \tag{13.1.16}$$

$$\lambda \to 0, \quad \lambda \in \Omega_\pm,$$

where (see Figure 13.4)

$$\Omega_+ = \{\lambda : 0 < \arg\lambda < \pi, \quad \lambda \neq 0\},$$

$$\Omega_- = \{\lambda : -\pi < \arg\lambda < 0, \quad \lambda \neq 0\}.$$

We note that the solutions $\Psi_\pm(\lambda)$ are defined uniquely by their asymptotics (13.1.16), in spite of absence of the rays, $\operatorname{Im}\lambda = 0$, in the domains $\Omega_\pm$. Indeed, if there exists another solution $\widetilde{\Psi}_+(\lambda)$ with the same asymptotic behaviour as $\Psi_+(\lambda)$, then the matrix

$$S = \widetilde{\Psi}_+^{-1}\Psi_+$$

has the following properties. According to the first equation (13.1.16) it is upper-triangular, whereas according to the second equation (13.1.16) it is lower-triangular, thus $S = I$.

We note that in general the matrices $P_\pm$ do not coincide with the matrix P_0,

$$P_0 = P_\pm\Lambda_\pm,$$

where $\Lambda_\pm = \operatorname{diag}\left\{\lambda_1^\pm, \lambda_2^\pm\right\}$, and

$$\det \Lambda_\pm = 1, \quad \lambda_1^\pm \lambda_2^\pm = 1. \tag{13.1.17}$$

Let us define the auxiliary Stokes matrices $S_\pm^{(\infty,0)}$, which connect the solutions $\Psi_\pm$ with that of $\Psi_1^{(\infty,0)}$:

$$\begin{aligned} \Psi_\pm(\lambda) &= \Psi_1^{(\infty)}(\lambda)\, S_\pm^{(\infty)}, \quad \lambda \in \Omega_\pm \cap \Omega_1^{(\infty)}, \\ \Psi_\pm(\lambda)\,\Lambda_\pm &= \Psi_1^{(0)}(\lambda)\, S_\pm^{(0)}, \quad \lambda \in \Omega_\pm \cap \Omega_1^{(0)}. \end{aligned} \tag{13.1.18}$$

The Stokes matrices $S_-^{(\infty)}$, $S_-^{(0)}$, together with $S_k^{(\infty,0)}$ and the connection matrix Q, can be expressed in terms of $S_+^{(\infty)}, \quad S_+^{(0)}$.

$$Q = \left[\Psi_1^{(0)}\right]^{-1} \Psi_1^{(\infty)} = \left[\Psi_+ \Lambda_+ S_+^{(0)^{-1}}\right]^{-1} \cdot \Psi_+(\lambda)\, S_+^{(\infty)^{-1}},$$

or

$$Q = S_+^{(0)} \Lambda_+^{-1} \left[S_+^{(\infty)}\right]^{-1}. \tag{13.1.19}$$

It was mentioned above that the solutions $\Psi_k^{(\infty,0)}(\lambda)$ and $\Psi_\pm(\lambda)$ are correctly defined only in the universal covering $\mathbb{C}\backslash\{0\}$. Taking into account the invariance of the essential singularity in the formal solutions $\Psi_f^{(\infty)}$, we obtain the relations

$$\Psi_-(\lambda) = \sigma_1 \Psi_+\left(e^{i\pi}\lambda\right)\sigma_1, \quad -\pi < \arg\lambda < 0, \tag{13.1.20}$$

$$\Psi_2^{(\infty)}(\lambda) = \sigma_1 \Psi_1^{(\infty)}\left(e^{-i\pi}\lambda\right)\sigma_1, \quad 0 < \arg\lambda < 2\pi. \tag{13.1.21}$$

These relations are consistent with the formal substitution

$$\Psi_-(\lambda) = \sigma_1 \Psi_+\left(e^{i\pi}\lambda\right)\sigma_1, \quad -\pi < \arg\lambda < 0, \tag{13.1.22}$$

which is a consequence of the real reduction.

Formulae (13.1.20) and (13.1.21) imply

$$\begin{aligned} \Psi_1^{(\infty)} S_+^{(\infty)} &= \Psi_+(\lambda) = \sigma_1 \Psi_-\left(e^{-i\pi}\lambda\right)\sigma_1 = \sigma_1 \Psi_1^{(\infty)}\left(e^{-i\pi}\lambda\right) S_-^{(\infty)} \sigma_1 \\ &= \Psi_2^{(\infty)}(\lambda)\,\sigma_1 S_-^{(\infty)} \sigma_1, \quad 0 < \arg\lambda < \pi. \end{aligned}$$

That yields immediately

$$S_1^{(\infty)} = S_+^{(\infty)} \sigma_1 \left[S_-^{(\infty)}\right]^{-1} \sigma_1. \tag{13.1.23}$$

Assuming then $\pi < \arg\lambda < 2\pi$ in (13.1.20) we find

$$\begin{aligned} \Psi_1^{(\infty)}\left(e^{-2\pi i}\lambda\right) S_-^{(\infty)} &= \Psi_-\left(e^{-2\pi i}\lambda\right) = \sigma_1 \Psi_+\left(\lambda e^{-\pi i}\right)\sigma_1 \\ &= \sigma_1 \Psi_1^{(\infty)}\left(\lambda e^{-\pi i}\right) S_+^{(\infty)} \sigma_1 = \Psi_2^{(\infty)}(\lambda)\,\sigma_1 S_+^{(\infty)} \sigma_1. \end{aligned} \tag{13.1.24}$$

This formula, together with equation (13.1.14) leads to the equation

$$\Psi_3^{(\infty)}(\lambda) = \Psi_2^{(\infty)}(\lambda)\,\sigma_1 S_+^{(\infty)} \sigma_1 \left[S_-^{(\infty)}\right]^{-1}, \tag{13.1.25}$$

where $\pi < \arg\lambda < 2\pi$.

Hence,

$$S_2^{(\infty)} = \sigma_1 S_+^{(\infty)} \sigma_1 \left[S_-^{(\infty)}\right]^{-1}. \tag{13.1.26}$$

Formulae (13.1.23) and (13.1.26) express the matrices $S_{1,2}^{(\infty)}$ in terms of the matrices $S_{\pm}^{(\infty)}$. The corresponding constraints between the Stokes matrices near the origin take the form

$$S_1^{(0)} = S_+^{(0)}\sigma_1 \left[S_-^{(0)}\right]^{-1}\sigma_1,$$

$$S_2^{(0)} = \sigma_1 S_+^{(0)}\sigma_1 \left[S_-^{(0)}\right]^{-1}.$$

The demonstration is similar to the case of infinity, with the additional use of the invariance of P_0 matrix (see (13.1.9)) under the automorphism and the following relation between Λ_+ and Λ_-,

$$\Lambda_- = \sigma_1\Lambda_+\sigma_1,$$

which follows from (13.1.20).

Formula (13.1.18) can be interpreted as providing the continuation of the functions $\Psi_\pm(\lambda)$ onto the half-line $\lambda > 0$. Thus, the following equations are valid,

$$\Psi_-(\lambda) = \Psi_+(\lambda)\,G_+^{(\infty)}, \quad G_+^{(\infty)} = \left[S_+^{(\infty)}\right]^{-1} S_-^{(\infty)}, \quad \lambda > 0, \tag{13.1.27}$$

$$\Psi_-(\lambda) = \Psi_+(\lambda)\,G_+^{(0)}, \quad G_+^{(0)} = \Lambda_+\left[S_+^{(0)}\right]^{-1} S_-^{(0)}\Lambda_-^{-1}, \quad \lambda > 0,$$

which lead to the equation

$$G_+^{(0)} = G_+^{(\infty)} = G_+. \tag{13.1.28}$$

The a priori triangular structure (see (5.5.15)) of $S_\pm^{(0,\infty)}$, allows us to present these matrices in the form

$$S_+^{(\infty)} = \begin{pmatrix} 1 & 0 \\ q & 1 \end{pmatrix}, \quad S_-^{(\infty)} = \begin{pmatrix} 1 & f \\ 0 & 1 \end{pmatrix}, \tag{13.1.29}$$

$$S_+^{(0)} = \begin{pmatrix} 1 & p \\ 0 & 1 \end{pmatrix}, \quad S_-^{(0)} = \begin{pmatrix} 1 & 0 \\ g & 1 \end{pmatrix}.$$

Substituting these equations into (13.1.28), and making use of (13.1.17), we find

$$f = -p, \quad g = -q, \quad pq \neq -1, \tag{13.1.30}$$

$$\lambda_1^+ = \frac{1}{\sqrt{1+pq}}, \quad \lambda_2^- = \sqrt{1+pq}.$$

Clearly these conditions guarantee that the cyclic relation (13.1.15) is satisfied identically for any p and q.

Summing up, the monodromy data can be expressed in the following way:

$$\mathrm{S}_1^{(\infty)} = \begin{pmatrix} 1 & 0 \\ p+q & 1 \end{pmatrix}, \quad \mathrm{S}_2^{(\infty)} = \begin{pmatrix} 1 & p+q \\ 0 & 1 \end{pmatrix},$$

$$\mathrm{S}_1^{(0)} = \begin{pmatrix} 1 & p+q \\ 0 & 1 \end{pmatrix}, \quad \mathrm{S}_2^{(0)} = \begin{pmatrix} 1 & 0 \\ p+q & 1 \end{pmatrix}, \tag{13.1.31}$$

$$\mathrm{Q} = (1+pq)^{-\frac{1}{2}} \begin{pmatrix} 1 & p \\ -q & 1 \end{pmatrix},$$

where p, $q \in \mathbb{C}$, $pq \neq -1$, are the nondiagonal parameters of the matrices $\mathrm{S}_+^{(\infty)}$ and $\mathrm{S}_+^{(0)}$,

$$\mathrm{S}_+^{(\infty)} = \begin{pmatrix} 1 & 0 \\ q & 1 \end{pmatrix}, \quad \mathrm{S}_+^{(0)} = \begin{pmatrix} 1 & p \\ 0 & 1 \end{pmatrix}.$$

The auxiliary Stokes matrices $\mathrm{S}_-^{(0,\infty)}$ have the form

$$\mathrm{S}_-^{(\infty)} = \begin{pmatrix} 1 & -p \\ 0 & 1 \end{pmatrix}, \quad \mathrm{S}_-^{(0)} = \begin{pmatrix} 1 & 0 \\ -q & 1 \end{pmatrix}.$$

Posing the real-valued reduction (13.1.8), imposes the additional constraints

$$\Psi_-(\lambda) = \left[\Psi_+^*\left(\overline{\lambda}\right)\right]^{-1}, \quad \Psi_1^{(\infty)} = \left[\Psi_1^*\left(\overline{\lambda}\right)\right]^{-1},$$
$$-\pi < \arg\lambda < 0, \quad \arg\overline{\lambda} = -\arg\lambda.$$

Finally, this yields

$$\mathrm{S}_-^{(\infty)} = \left[\mathrm{S}_+^{(\infty)*}\right]^{-1} \Longleftrightarrow p = \overline{q}. \tag{13.1.32}$$

REMARK 13.1. Pure imaginary solutions of equation (13.1.2) correspond to the monodromy data restriction

$$p = -\overline{q}.$$

2. Inverse monodromy problem

Let us now discuss the inverse monodromy problem for the system (13.1.2). For the general case of arbitrary finite monodromy data p and q we will formulate the RH problem on the real axis.

We define the pair of functions $\Phi_1(\lambda)$ and $\Phi_2(\lambda)$ in the open upper and lower half-planes respectively by the equations

$$\Phi_1(\lambda) = \Psi_+(\lambda), \quad \Phi_2(\lambda) = \Psi_-(\lambda), \tag{13.2.1}$$

where the functions $\Psi_+(\lambda)$ and $\Psi_-(\lambda)$ are defined in the domains $0 < \arg\lambda < \pi$ and $-\pi < \arg\lambda < 0$ respectively (see Section 1). The continuation of the functions

$\Phi_1(\lambda)$ and $\Phi_2(\lambda)$ onto the real axis can be achieved with the use of formulae (13.1.18), (13.1.24), as follows:

$$\lambda > 0: \quad \begin{aligned} \Phi_1(\lambda) &= \Psi_1^{(\infty)}(\lambda)\, S_+^{(\infty)}, \\ \Phi_2(\lambda) &= \Psi_1^{(\infty)}(\lambda)\, S_-^{(\infty)}, \end{aligned} \tag{13.2.2}$$

$$\lambda < 0: \quad \begin{aligned} \Phi_1(\lambda) &= \Psi_2^{(\infty)}(\lambda)\, \sigma_1 S_-^{(\infty)} \sigma_1, \\ \Phi_2(\lambda) &= \Psi_2^{(\infty)}(\lambda)\, \sigma_1 S_+^{(\infty)} \sigma_1. \end{aligned} \tag{13.2.3}$$

The functions $\Phi_{1,2}(\lambda)$ have an additional singularity at $\lambda = 0$. Nevertheless, the following equation is valid on the real axis

$$\Phi_2(\lambda) = \Phi_1(\lambda)\, G_0(\lambda), \quad \operatorname{Im} \lambda = 0,$$

where,

$$G_0(\lambda) = \begin{cases} \begin{pmatrix} 1 & -p \\ -q & 1+pq \end{pmatrix}, & \lambda > 0, \\ \begin{pmatrix} 1 & q \\ p & 1+pq \end{pmatrix}, & \lambda < 0. \end{cases} \tag{13.2.4}$$

Taking into account the asymptotic behaviour of $\Phi_{1,2}(\lambda)$ at the singular points $\lambda = 0$ and $\lambda = \infty$ (see formula (13.1.16)), we introduce the piecewise analytical function χ,

$$\chi(\lambda, x) = \begin{cases} \Phi_1(\lambda) \exp\left\{\frac{ix^2}{16}\lambda\sigma_3 + \frac{i}{\lambda}\sigma_3\right\}, & \operatorname{Im} \lambda \geq 0, \\ \Phi_2(\lambda) \exp\left\{\frac{ix^2}{16}\lambda\sigma_3 + \frac{i}{\lambda}\sigma_3\right\}, & \operatorname{Im} \lambda \leq 0. \end{cases}$$

This leads to the formulation of the inverse monodromy problem for the system (13.1.2) in terms of the following RH problem:

1°. $\chi(\lambda, x)$ is analytical function for $\operatorname{Im} \lambda > 0$ and $\operatorname{Im} \lambda < 0$.
2°. $\chi^-(\lambda, x) = \chi^+(\lambda, x)\, G(\lambda, x)$, $\operatorname{Im} \lambda = 0$, where

$$G(\lambda, x) = \exp\left\{-\frac{ix^2}{16}\lambda\sigma_3 - \frac{i}{\lambda}\sigma_3\right\} G_0(\lambda) \exp\left\{\frac{ix^2}{16}\lambda\sigma_3 + \frac{i}{\lambda}\sigma_3\right\},$$

and $G_0(\lambda)$ is defined by (13.2.4).
3°. $\chi(\lambda, x) \to I$ as $\lambda \to \infty$, $\operatorname{Im} \lambda \neq 0$.

This RH problem can be reformulated in terms of the matrix singular integral equation similar to equation (3.1.3) of Chapter 3. Due to the normalization 3°, the piecewise analytic function χ satisfies

$$\chi(\lambda,x)=I+\frac{1}{2\pi i}\int_{-\infty}^{\infty}\chi^+(\mu,x)\frac{I-G(\mu,x)}{\mu-\lambda}d\mu,$$

which yields the following integral equation

$$\chi^+(\lambda,x)=I+\frac{1}{2\pi i}\int_{-\infty}^{\infty}\chi^+(\mu,x)\frac{I-G(\mu,x)}{\mu-\lambda-i0}d\mu.$$

More explicitly, taking into account (13.2.4),

$$\begin{aligned}\chi^+(\lambda,x) &= I+\frac{1}{2\pi i}\left\{-\int_{-\infty}^{0}\frac{\chi^+(\mu,x)}{\mu-\lambda-i0}\begin{pmatrix}0 & q\,e^{-\frac{ix^2}{8}\mu-\frac{2i}{\mu}}\\ p\,e^{\frac{ix^2}{8}\mu+\frac{2i}{\mu}} & 1+pq\end{pmatrix}d\mu\right.\\ &\quad\left.+\int_{0}^{\infty}\frac{\chi^+(\mu,x)}{\mu-\lambda-i0}\begin{pmatrix}0 & p\,e^{-\frac{ix^2}{8}\mu-\frac{2i}{\mu}}\\ q\,e^{\frac{ix^2}{8}\mu+\frac{2i}{\mu}} & -1-pq\end{pmatrix}d\mu\right\}.\end{aligned}\tag{13.2.5}$$

In the general case of the RH problem 1°–3°, the conjugation matrix $G_0(\lambda)$, which is piecewise constant to the right and to the left of the origin, is discontinous. However, under the assumption $p=-q$, the RH problem 1°–3° becomes regular, and equation (13.2.5) simplifies to the equation

$$\chi^+(\lambda,x)=I+\frac{1}{2\pi i}\int_{-\infty}^{\infty}\frac{\chi^+(\mu,x)}{\mu-\lambda-i0}\begin{pmatrix}0 & p\,e^{-\frac{ix^2}{8}\mu-\frac{2i}{\mu}}\\ -p\,e^{\frac{ix^2}{8}\mu+\frac{2i}{\mu}} & p^2\end{pmatrix}d\mu.\tag{13.2.6}$$

The solution of the SG-PIII equation can be expressed in terms of the function χ^+ by the following equations:

$$\begin{aligned}u_x(x)&=-\frac{x}{2}\lim_{\lambda\to\infty}\lambda\chi^+_{12}(\lambda,x),\\ \sin\frac{u}{2}&=i\lim_{\lambda\to0}\chi^+_{12}(\lambda,x),\\ \cos\frac{u}{2}&=\lim_{\lambda\to0}\chi^+_{11}(\lambda,x),\end{aligned}\tag{13.2.7}$$

where χ^+_{12} and χ^+_{11} are the upper nondiagonal and the upper diagonal elements of the matrix $\chi^+(\lambda,x)$, respectively.

3. Exponential solutions of SG-PIII via the inverse monodromy problem

In order to obtain the asymptotics of $u(x)$ as $x\to\pm\infty$ we will analyze equation (13.2.6).

We first rescale the independent variables

$$z = \lambda x,$$
$$\xi = \mu x.$$

This implies the appearance of rapid oscillations in the kernel of equation (13.2.6), where the relevant phase is given by

$$x\left(\frac{\xi}{8} + \frac{2}{\xi}\right).$$

Hence the stationary-phase points are $\xi = \pm 4$; therefore, these points (independently of the sign of x) always lie on the contour of integration. In order to avoid singularities of $\chi^+(\lambda)$ near the stationary points, we need to implement "asymptotic undressing" of the RH problem 1°–3° (see the discussion at the beginning of Chapter 8).

However, for the case of purely imaginary parameter x, it is possible to avoid the above "undressing" and to obtain asymptotics in a more elementary manner. Letting

$$x \mapsto ix,$$

we find another form of SG-PIII equation, namely

$$u_{xx} + \frac{u_x}{x} - \sin u = 0, \quad x \in \mathbb{R}. \tag{13.3.1}$$

Let us replace

$$x^2 \mapsto -x^2$$

and put

$$q = 0 \tag{13.3.2}$$

in the RH[2] 1° - 3°. Then the problem can be reformulated as the RH problem well-posed on the imaginary axis. Indeed, reduction (13.3.2) yields

$$G_0(\lambda) = \begin{cases} \begin{pmatrix} 1 & -p \\ 0 & 1 \end{pmatrix}, & \lambda > 0, \\ \begin{pmatrix} 1 & 0 \\ p & 1 \end{pmatrix}, & \lambda < 0. \end{cases} \tag{13.3.3}$$

Following the Deift-Zhou method discussed in Chapter 8, we can reformulate the RH problem 1°–3°, by introducing a new piecewise analytic function $Y(\lambda, x)$ as shown in Figure 13.6. Then, the jump condition 2° for the matrix $\chi(\lambda, ix)$,

$$\chi(\lambda - i0, ix) = \chi(\lambda + i0, ix) e^{\frac{ix^2}{16}\lambda\sigma_3 - \frac{i}{\lambda}\sigma_3} G_0(\lambda) e^{-\frac{ix^2}{16}\lambda\sigma_3 + \frac{i}{\lambda}\sigma_3},$$

implies

$$Y(\lambda - i0, x) = Y(\lambda + i0, x),$$

[2]This reduction corresponds to the following special choice of the monodromy data:

$$S_1^{(\infty)} = S_1^{(0)} = \begin{pmatrix} 1 & p \\ 0 & 1 \end{pmatrix}, \quad S_2^{(\infty)} = S_2^{(0)} = \begin{pmatrix} 1 & 0 \\ p & 1 \end{pmatrix}, \quad Q = I.$$

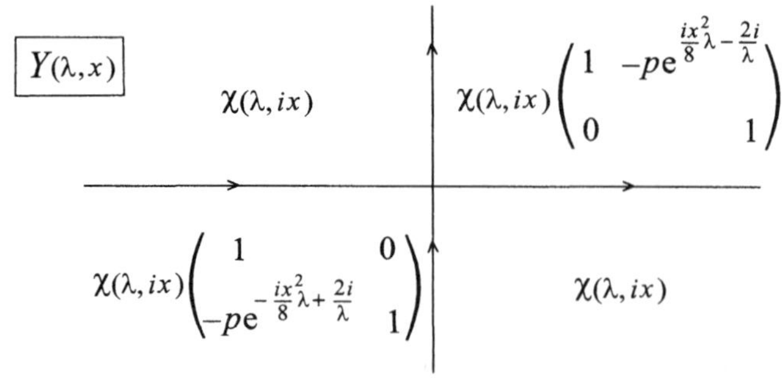

FIGURE 13.6. Definition of the $Y(\lambda, x)$ matrix

which provides the analyticity of Y both in the left and right half-planes. However, the jump condition for Y moves now to the imaginary axis. Note also that Y is bounded at $\lambda = 0$ and $\lambda \to \infty$, since χ is bounded and furthermore,

$$e^{\frac{ix^2}{8}\lambda - \frac{2i}{\lambda}} \to 0, \qquad \text{Im } \lambda > 0, \quad \lambda \to 0, \text{ or } \lambda \to \infty,$$

$$e^{-\frac{ix^2}{8}\lambda + \frac{2i}{\lambda}} \to 0, \qquad \text{Im } \lambda < 0, \quad \lambda \to 0, \text{ or } \lambda \to \infty.$$

Hence, the RH problem for the matrix-valued function Y becomes:

1°. $Y(\lambda, x)$ is an analytical function for $\text{Re } \lambda > 0$ and for $\text{Re } \lambda < 0$.
2°. $Y^-(\lambda, x) = Y^+(\lambda, x)\, G_Y(\lambda, x)\,, \quad \text{Re } \lambda = 0,$
where

$$G_Y(\lambda, x) = \begin{cases} \begin{pmatrix} 1 & -pe^{\frac{ix^2}{8}\lambda - \frac{2i}{\lambda}} \\ 0 & 1 \end{pmatrix}, & \text{Im } \lambda > 0, \\ \begin{pmatrix} 1 & 0 \\ pe^{-\frac{ix^2}{8}\lambda + \frac{2i}{\lambda}} & 1 \end{pmatrix}, & \text{Im } \lambda < 0. \end{cases}$$

3°. $Y(\lambda, x) \to I$ as $\lambda \to \infty$.

The solution of the SG-PIII equation (13.3.1) can be expressed in terms of the function Y by the following equations:

$$u_x(x) = \frac{x}{2} \lim_{\lambda \to \infty} \lambda Y_{12}(\lambda, x),$$

$$\sin \frac{u}{2} = i \lim_{\lambda \to 0} Y_{12}(\lambda, x), \tag{13.3.4}$$

$$\cos \frac{u}{2} = \lim_{\lambda \to 0} Y_{11}(\lambda, x).$$

The solution $Y(\lambda, x)$ of the above RH problem admits the following integral representation,

$$Y(\lambda, x) = I + \frac{1}{2\pi i} \int_{-i\infty}^{i\infty} \frac{\rho(\mu, x)}{\mu - \lambda} (I - G(\mu, x))\, d\mu, \quad \operatorname{Re} \lambda \neq 0, \tag{13.3.5}$$

where $\rho(\lambda, x) \equiv Y^{+}(\lambda, x)$ satisfies the integral equations

$$\rho(\lambda, x) = I + \frac{1}{2\pi i} \int_{-i\infty}^{i\infty} \frac{\rho(\mu, x)}{\mu - \lambda - i0} (I - G(\mu, x))\, d\mu \tag{13.3.6}$$

(cf. (3.1.2) and (3.1.3) of Chapter 3). Note that the matrix kernel in (13.3.6) is exponentially small on the imaginary axis

$$||I - G_Y(\lambda, x)||_{L^2 \cap L^\infty} < C e^{-|x|}. \tag{13.3.7}$$

Indeed, this follows from the definition of the matrix G, as well as from the application of the (classical) steepest descent method which yields the estimates

$$\int_0^{i\infty} e^{\frac{ix^2}{4}\lambda - \frac{4i}{\lambda}}\, d\lambda, \quad \int_0^{-i\infty} e^{-\frac{ix^2}{4}\lambda + \frac{4i}{\lambda}}\, d\lambda = O\left(x^{-3/2} e^{-2|x|}\right).$$

The conditions of Theorem 8.1 are now satisfied, and hence ρ exists (for sufficiently large x) and satisfies the estimate

$$||I - \rho(\lambda, x)||_{L^2} < C_1 e^{-|x|}. \tag{13.3.8}$$

Applying this to the right-hand side of (13.3.5), we find,

$$\begin{aligned} Y(\lambda, x) &= I + \frac{1}{2\pi i} \int_{-i\infty}^{i\infty} \frac{I - G_Y(\mu, x)}{\mu - \lambda} d\mu + \frac{1}{2\pi i} \int_{-i\infty}^{i\infty} \frac{(\rho(\mu, x) - I)(I - G_Y(\mu, x))}{\mu - \lambda} d\mu \\ &= I + \frac{1}{2\pi i} \int_{-i\infty}^{i\infty} \frac{I - G_Y(\mu, x)}{\mu - \lambda} d\eta + O\left(\frac{e^{-2|x|}}{1 + |\lambda|}\right), \quad x \to \pm\infty. \end{aligned}$$

Then, for the scalar components Y_{11} and Y_{12} we obtain

$$Y_{11}(\lambda, x) = 1 + O\left(\frac{e^{-2|x|}}{1 + |\lambda|}\right),$$

$$Y_{12}(\lambda, x) = -\frac{p}{2\pi i} \int_0^{i\infty} \frac{e^{\frac{ix^2}{8}\mu - \frac{2i}{\mu}}}{\lambda - \mu} d\mu + O\left(\frac{e^{-2|x|}}{1 + |\lambda|}\right), \quad x \to \pm\infty.$$

This asymptotics together with the formula (13.3.4) yield the following representation for the solution $u(x)$,

$$u_x(x) = \frac{ipx}{4\pi} \int_0^{i\infty} e^{\frac{ix^2}{8}\mu - \frac{2i}{\mu}}\, d\mu + O(e^{-2|x|}), \quad x \to \pm\infty. \tag{13.3.9}$$

Evaluating the integral in the right-hand side of this equation by the steepest descent method, we find

$$u_x(x) = -\mathrm{sign}(x)\, p\sqrt{\frac{2}{\pi|x|}}\, \mathrm{e}^{-|x|} + O(x^{-1}\,\mathrm{e}^{-|x|}), \quad x \to \pm\infty.$$

Integrating with respect to x, we obtain the following final expression for the leading term of the asymptotics of $u(x)$:

$$u(x) = 2\pi k + p\sqrt{\frac{2}{\pi|x|}}\, \mathrm{e}^{-|x|} + O(x^{-1}\,\mathrm{e}^{-|x|}), \quad x \to \pm\infty, \quad k \in \mathbb{Z}. \tag{13.3.10}$$

The integration constant $2\pi k$ comes from the second formula (13.3.4). The value of integer k will be discussed in Section 3.2 of Chapter 14.

Thus we have proved the following theorem.

THEOREM 13.1. *The set of the monodromy data*

$$\mathrm{S}_1^{(\infty)} = \mathrm{S}_1^{(0)} = \begin{pmatrix} 1 & p \\ 0 & 1 \end{pmatrix}, \quad \mathrm{S}_2^{(\infty)} = \mathrm{S}_2^{(0)} = \begin{pmatrix} 1 & 0 \\ p & 1 \end{pmatrix}, \quad \mathrm{Q} = I,$$

where p is arbitrary, parametrize the solution of SG-PIII equation (13.3.1) *which has the asymptotics* (13.3.10) *at infinity.*

CHAPTER 14

Canonical Four-rays. Real-valued Solutions of SG-PIII

0. Introduction

Here we describe the manifolds of the two-parameter solutions of the SG-PIII equation (13.0.1), in terms of their monodromy data p and q introduced in Chapter 13. The reductions $p = \overline{q}$ and $p = -\overline{q}$, specify the real-valued and pure imaginary solutions, respectively. In analogy with Part 2, we will calculate p and q explicitly in terms the asymptotics at $x = 0$ and at $x \to +\infty$. This will provide connection formulae for generic nonsingular solutions, describing the manifolds of solutions.

The four canonical rays

$$x = re^{\frac{i\pi k}{2}}, \quad r \in [0, +\infty), \quad k = 0, 1, 2, 3, \tag{14.0.1}$$

play for the SG-PIII the same role that the six canonical rays play for PII; see Chapters 5, 7–9. In some sense, these rays are dual to the Stokes lines in the λ-plane.

The simple automorphism (or Lie-point symmetry)

$$x \mapsto xe^{\frac{i\pi k}{2}}, \quad u \mapsto u \pm \pi, \quad k = 0, 1, 2, 3, \tag{14.0.2}$$

maps solutions of the SG-PIII equation

$$u_{xx} + \frac{u_x}{x} + \sin u = 0$$

along the four rays (14.0.1).

We now consider the action of the automorphism (14.0.2) on the monodromy data. The functions $\Psi_k^{(0)}$ and $\Psi_k^{(\infty)}$, satisfy the Lax pair equations (13.1.1), and normalize according to equations (13.1.11). The transformation (14.0.2), supplemented with the transformation

$$\lambda \mapsto -\lambda \tag{14.0.3}$$

maps the Lax pair onto itself. The normalization at infinity (13.1.11), is invariant under this transformation, hence $\Psi_k^{(\infty)}(\lambda, x) = \Psi_k^{(\infty)}(-\lambda, ix)$, and $S_k^{(\infty)}$ satisfy (13.1.31),

$$S_1^{(\infty)} = \begin{pmatrix} 1 & 0 \\ p+q & 1 \end{pmatrix}, \quad S_2^{(\infty)} = \begin{pmatrix} 1 & p+q \\ 0 & 1 \end{pmatrix},$$

The matrix P_0 transforms to $P_0 \mapsto -iP_0\sigma_1$, thus the commutation $\sigma_1 e^{\frac{i\sigma_3}{\lambda}} = e^{-\frac{i\sigma_3}{\lambda}}\sigma_1$ implies

$$\Psi_1^{(0)}(-\lambda, ix) = -i\Psi_1^{(0)}(\lambda, x)\sigma_1. \tag{14.0.4}$$

Therefore, the Stokes matrices $S_k^{(0)}$ become

$$S_1^{(0)} = \begin{pmatrix} 1 & 0 \\ p+q & 1 \end{pmatrix}, \quad S_2^{(0)} = \begin{pmatrix} 1 & p+q \\ 0 & 1 \end{pmatrix}.$$

and the connection matrix Q takes the form

$$Q = \left[\Psi_1^{(0)}(-\lambda, ix)\right]^{-1} \Psi_1^{(\infty)}(-\lambda, ix) = \frac{i}{\sqrt{1+pq}} \begin{pmatrix} -q & 1 \\ 1 & p \end{pmatrix}. \tag{14.0.5}$$

1. Monodromy data parametrization at $x = 0$.

The SG-PIII equation (13.0.1)

$$u_{xx} + \frac{u_x}{x} + \sin u = 0,$$

is distinguished among the general PIII equations (12.0.1) by the simple nonlinearity $F(u) = \sin u$. This makes it possible to prove existence and uniqueness of solutions near the origin $x = 0$, by traditional nonlinear ODEs methods without appeal to the isomonodromy method. In particular, the linearization as $u \to 0$ yields the Bessel equation

$$u_{xx} + \frac{u_x}{x} + u = 0$$

which possesses the known leading term asymptotics $u(x) \sim r \ln x + s$ as $x \to 0$.

Starting with this function as Cauchy data at the origin, we will establish the existence of solution of the SG-PIII equation. Then, from the knowledge of the data x, u, u_x for small x, we can solve the direct monodromy problem and connect them with the monodromy data p and q.

THEOREM 14.1. *For any given r, $s \in \mathbb{C}$ such that $|\text{Im}\, r| < 2$, there exists a solution of the SG-PIII equation (*13.0.1*) with the following asymptotic behaviour near the origin*

$$u\,(x) = r \ln x + s + O\left(|x|^{2-|Im\, r|}\right), \quad x \to 0. \tag{14.1.1}$$

PROOF. The formal series expansion

$$u(x) = r \ln x + s + \sum_{k=1}^{\infty} \sum_{|l|<k} \left(a_{kl} x^{2k+irl} + b_{kl} x^{2k-irl}\right), \quad x \to 0, \tag{14.1.2}$$

satisfies the SG-PIII equation (13.0.1), where all coefficients a_k and b_k can be recursively determined in terms of r and s. The series (14.1.2) is asymptotic for $|\,\text{Im}\; r| < 2$, and in addition it is convergent in the circle $|x| < \varepsilon$ for sufficiently small ε. This can be proven by estimating the error term $h_n(x)$ of the series (14.1.2),

$$u(x) = r \ln x + s + \sum_{k=1}^{n-1} \sum_{|l|<k} \left(a_{kl} x^{2k+irl} + b_{kl} x^{2k-irl}\right) + h_n(x). \tag{14.1.3}$$

Inserting this sum into the equation (13.0.1) and expanding e^{iu} and e^{-iu} into the Taylor series, we find the initial value problem for the error term h_n,

$$h_n'' + \frac{h_n'}{x} + F(h_n, x) = 0, \tag{14.1.4}$$

where

$$h_n(x) = O\left(|x|^{n(2-|\operatorname{Im} r|)}\right), \quad x \to 0, \tag{14.1.5}$$

and

$$F(h_n(x), x) = \tfrac{1}{2i}\left(x^{ir}e^{is} + x^{-ir}e^{-is} + O(x^{2-|\operatorname{Im} r|})\right)$$

$$\times\left(h_n(x) + O(h_n^2(x))\right) + O\left(|x|^{(n-1)(2-|\operatorname{Im} r|)}\right), \quad x \to 0.$$

The equation (14.1.4) with the initial condition (14.1.5) is equivalent to the integral equation

$$h_n(x) = \frac{1}{\sqrt{x}} \int_0^x \int_0^\xi \sqrt{\eta} F(h_n(\eta), \eta)\, d\eta\, d\xi. \tag{14.1.6}$$

The above integral operator here is a contraction on the class of functions h with asymptotics (14.1.5) on the small disk $|x| < \varepsilon$. This yields the existence and uniqueness of the error function $h_n(x)$. □

Now we are ready to calculate the monodromy data corresponding to the solution (14.1.1).

THEOREM 14.2. *The monodromy data p and q associated with the solution (14.1.1) can be expressed in the form*

$$p = \frac{Ae^{-\frac{\pi r}{4}} - Be^{\frac{\pi r}{4}}}{A+B}, \quad q = \frac{Be^{-\frac{\pi r}{4}} - Ae^{\frac{\pi r}{4}}}{A+B}, \tag{14.1.7}$$

where

$$A = 2^{\frac{3ir}{2}} e^{\frac{is}{2}} \Gamma^2\left(\frac{1}{2} + \frac{ir}{4}\right),$$

$$B = 2^{-\frac{3ir}{2}} e^{-\frac{is}{2}} \Gamma^2\left(\frac{1}{2} - \frac{ir}{4}\right).$$

PROOF. The monodromy data for the system (13.1.2) are reconstructed through the connection matrix (see Section 1),

$$\mathrm{Q} = \left[\Psi_1^{(0)}\right]^{-1} \Psi_1^{(\infty)}. \tag{14.1.8}$$

We recall that under the assumption $w = u_x$, where $u(x)$ is the solution of PIII equation (14.3.1), the matrix Q is independent of x. Therefore it is sufficient to compute the asymptotics of $\Psi_1^{(0)}(\lambda, x)$ and of $\Psi_1^{(\infty)}(\lambda, x)$ as $x \to 0$, assuming $w = u_x$ where the form of $u(x)$ in (13.1.2) is given by (14.1.1). We note that it is sufficient to know the asymptotics of the Ψ-function on the ray $\arg \lambda = 0$.

Let us begin with $\Psi_1^{(\infty)}$ and let us seek its asymptotics in the form

$$\Psi_1^{(\infty)}(\lambda, x) = \psi(\xi, xu_x)\left[1 + x^{2-|\operatorname{Im} r|}\Phi(\xi, x)\right], \tag{14.1.9}$$

where $\xi = \frac{x^2\lambda}{16}$ and the matrix ψ satisfies the "short-cut" system (13.1.2):

$$\frac{d\psi}{d\xi} = \left\{ -i\sigma_3 - \frac{ixu_x}{4\xi}\sigma_1 \right\}\psi. \tag{14.1.10}$$

The solution of equation (14.1.10) is normalized by the condition (13.1.16),

$$\psi \to \exp\{-i\sigma_3\xi\}, \quad \xi \to \infty, \quad \arg\xi = 0. \tag{14.1.11}$$

For the components of the matrix ψ we obtain the Bessel equations

$$\begin{aligned} \frac{d^2}{d\xi^2}v_1 + \left(1 + \frac{a_+}{\xi^2}\right)v_1 &= 0, \\ \frac{d^2}{d\xi^2}v_2 + \left(1 + \frac{a_-}{\xi^2}\right)v_2 &= 0, \end{aligned} \tag{14.1.12}$$

where

$$v_1 = \psi_{11} + \psi_{21}, \quad v_2 = \psi_{11} - \psi_{21}, \quad a_\pm = \frac{x^2u_x^2}{16} \pm i\frac{xu_x}{4}.$$

Thus, $v_{1,2} = \sqrt{\xi}Z_{\nu_{1,2}}(\xi)$, where $Z_{\nu_{1,2}}(\xi)$ are the Bessel functions with symbols

$$\nu_{1,2} = \frac{1}{2} \pm \frac{ixu_x}{4}.$$

Taking into consideration the normalization (14.1.11), as well as the known asymptotics of the Hankel function, we find

$$\psi(\xi) = \frac{1}{2}\sqrt{\frac{\pi}{2}}e^{-\frac{\pi x u_x}{8}}\sqrt{\xi} \tag{14.1.13}$$

$$\times \begin{pmatrix} -iH_\nu^{(2)}(\xi) + H_{\nu-1}^{(2)}(\xi), & iH_{-\nu+1}^{(1)}(\xi) - H_{-\nu}^{(1)}(\xi) \\ iH_\nu^{(2)}(\xi) + H_{\nu-1}^{(2)}(\xi), & iH_{-\nu+1}^{(1)}(\xi) + H_{-\nu}^{(1)}(\xi) \end{pmatrix},$$

where $\nu = \frac{1}{2} \pm \frac{ixu_x}{4}$.

Substituting (14.1.9) into equation (13.1.2), and scaling the terms of order $x^{2-|\operatorname{Im} r|}$ we find that the remainder term Φ in the asymptotics (14.1.9) satisfies

$$\Phi = O\left(\xi^{-1}\right), \quad \arg\xi = 0. \tag{14.1.14}$$

On the other hand, making use of equality $xu_x = r + O\left(x^{2-|\operatorname{Im} r|}\right)$, and using (14.1.13), we find

$$\Psi_1^{(\infty)} = \psi(\xi, r)\left(1 + O\left(x^{2-|\operatorname{Im} r|}\right)\right),$$

where $\psi(\xi, r)$ is defined by the formula (14.1.13) with xu_x replaced by r. Combining the latter estimate with that of (14.1.14), we obtain

$$\Psi_1^{(\infty)} = \psi(\xi, r)\{1 + O[x^{2-|\operatorname{Im} r|}(1 + \xi^{-1})]\}. \tag{14.1.15}$$

Let us pass now to the construction of the matrix $\Psi_1^{(0)}$. It is convenient to make the change of variables

$$\lambda = \zeta^{-1}$$

in equation (13.1.2), and to seek the solution in the form

$$\Psi_1^{(0)} = \varphi(\zeta, xu_x, u)\left(1 + x^2\varkappa(\zeta, x)\right). \tag{14.1.16}$$

Here the matrix ϕ satisfies the "short-cut" system (13.1.2), i.e.

$$\frac{d\varphi}{d\zeta} = \left\{-i\sigma_3 \cos u + i\sigma_2 \sin u + \frac{ixu_x}{4\zeta}\sigma_1\right\}\varphi, \tag{14.1.17}$$

normalized by the condition (13.1.16),

$$\varphi \to \left(\cos\frac{u}{2} - i\sigma_1 \sin\frac{u}{2}\right) \mathrm{e}^{-i\sigma_3\zeta}, \quad \zeta \to \infty, \quad \arg\zeta = 0. \tag{14.1.18}$$

It is easy to check in the same way as above, that equation (14.1.17) can be explicitly solved in terms the Bessel functions. Using the normalization (14.1.18), we find

$$\varphi(\zeta, xu_x, u) = \frac{1}{2}\sqrt{\frac{\pi}{2}}e^{-\frac{\pi x u_x}{8}}\zeta^{\frac{1}{2}} \tag{14.1.19}$$

$$\times \begin{pmatrix} -ie^{-\frac{iu}{2}}H_\nu^{(2)} + e^{\frac{iu}{2}}H_{\nu-1}^{(2)}, & -ie^{\frac{iu}{2}}H_{-\nu+1}^{(1)} + e^{-\frac{iu}{2}}H_{-\nu}^{(1)} \\ -ie^{-\frac{iu}{2}}H_\nu^{(2)} - e^{\frac{iu}{2}}H_{\nu-1}^{(2)}, & ie^{\frac{iu}{2}}H_{-\nu+1}^{(1)} + e^{-\frac{-iu}{2}}H_{-\nu}^{(1)} \end{pmatrix},$$

where $\nu = \frac{1}{2} + \frac{ixu_x}{4}$, $\quad H_\nu = H_\nu(\zeta)$.

Following the same reasoning as for the case of $\Psi_1^{(\infty)}$, we find the estimate

$$\Psi_1^{(0)} = \varphi(\zeta, x, r\ln x + s)\left\{1 + O\left(x^2\zeta^{-1} + x^{2-|\operatorname{Im} r|}\right)\right\}. \tag{14.1.20}$$

Expressing the estimates (14.1.15) and (14.1.20) in the variable λ, we find that in the domain

$$x^{-|\operatorname{Im} r|} \ll \lambda \ll x^{-2}, \tag{14.1.21}$$

the remainder terms in both estimates are small in x. Therefore, the matrix Q can be calculated as $x \to 0$, by the formula

$$\mathrm{Q} = \lim_{x\to 0} \varphi^{-1}\psi|_{\lambda = x^{-2+\varepsilon}}, \quad 0 < \varepsilon < 2 - |\operatorname{Im} r|. \tag{14.1.22}$$

At the point $\lambda = x^{-2+\varepsilon}$, both variables ξ and ζ tend to zero as $x \to 0$. Thus, we can use the well-known asymptotics of the Hankel functions near the origin to calculate Q through the formulae (14.1.22), (14.1.13), (14.1.19). In this way we find the following expression,

$$\mathrm{Q} = \frac{1}{2\pi}\begin{pmatrix} A + B & Ae^{-\frac{\pi r}{4}} - Be^{\frac{\pi r}{4}} \\ Ae^{\frac{\pi r}{4}} - Be^{-\frac{\pi r}{4}} & A + B \end{pmatrix}, \tag{14.1.23}$$

where

$$A = 2^{\frac{3ir}{2}}e^{\frac{is}{2}}\Gamma^2\left(\frac{1}{2} + \frac{ir}{4}\right),$$

$$B = 2^{-\frac{3ir}{2}}e^{-\frac{is}{2}}\Gamma^2\left(\frac{1}{2} - \frac{ir}{4}\right).$$

Finally, we recall the expression of Q in terms of the monodromy parameters p and q and this would conclude the proof of the theorem. □

2. Monodromy data parametrization at infinity

Here we prove for the system (13.1.2) the theorem analogous to Theorem 9.3.

THEOREM 14.3. *Let the parameters u and w of the system (13.1.2) depend on x in such a way that:*

a)

$$u = O\left(x^{-\frac{1}{2}}\right) + 2k\pi, \quad k \in \mathbb{Z},$$
$$w = O\left(x^{-\frac{1}{2}}\right), \quad x \to +\infty; \tag{14.2.1}$$

b) at least one of the values of v and v^*,

$$v = -\frac{\sqrt{x}}{4}\left(u_k + iw\right)e^{\frac{i\pi}{4}},$$

$$v^* = \frac{\sqrt{x}}{4}\left(u_k - iw\right)e^{\frac{i\pi}{4}}, \quad u_k = u - 2\pi k,$$

is different than zero as $x \to +\infty$;

c)

$$-1 \le \operatorname{Re} vv^* \le 0.$$

Then, the monodromy data p and q have the following asymptotic expansions:

$$p = -\frac{1}{v^*}\frac{\sqrt{2\pi}}{\Gamma(-i\delta)}\exp\left\{ix - i\delta\ln 4x + \delta\frac{\pi}{2}\right\} + o(1),$$

$$q = \frac{i}{v}\frac{\sqrt{2\pi}}{\Gamma(i\delta)}\exp\left\{-ix + i\delta\ln 4x + \delta\frac{\pi}{2}\right\} + o(1), \tag{14.2.2}$$

$$i\delta = -vv^*, \quad x \to +\infty.$$

PROOF. The proof follows closely the proof of the Theorem 9.3, thus we will omit most of the relevant calculations.

In order to apply the WKB method we transform the original system (13.1.2) by scaling the variable λ:

$$\frac{d\Psi}{dz} = xA_0(z,x)\psi,$$

$$z = \lambda x, \tag{14.2.3}$$

$$A_0 = \left(-\frac{i}{16} + \frac{i}{z^2}\cos u\right)\sigma_3 - \frac{iw}{4z}\sigma_1 - \frac{i}{z^2}\sin u\sigma_2.$$

The WKB-solution of equation (14.2.3) has the structure of equation (16.4.10) ($\tau \equiv x$!), with the matrices T and Λ, defined by the equations

$$\Lambda = -\mu\sigma_3, \quad \mu = i\sqrt{\left(\frac{1}{16} - \frac{1}{z^2}\right)^2 - \frac{w^2}{16z^2} + \frac{1-\cos u}{8z^2}},$$

(14.2.4)

$$T = \left[\left(\frac{i}{16} - \frac{i}{z^2}\cos u + \mu\right)\sigma_0 - \frac{1}{z^2}\sin u\sigma_1 + \frac{w}{4z}\sigma_2\right]\frac{8z^2}{i(z^2-16)}.$$

From equations (14.2.4) and from the estimates (14.2.1) it follows, that:

a) Equation (14.2.3) has four simple turning points, tending in pairs to the two double points:

$$z_{\pm} = \pm 4.$$

b) In the domain, where

$$\sqrt{x}\left|z - \frac{16}{z}\right| \to \infty,$$

the matrix T satisfies the estimate

$$T = \left[I - \frac{16}{z^2} P_0 \cos\frac{u}{2}\right] \frac{z^2}{z^2 - 16} + O\left(\frac{z}{\sqrt{x}\,(z^2 - 16)}\right).$$

This means, in particular, that

$$T \to I, \qquad z \to \infty, \qquad T \to P_0 \cos\tfrac{u}{2}, \qquad z \to 0. \tag{14.2.5}$$

c) The conjugate Stokes lines tend to either the real axis or to the circle $\xi^2 + \eta^2 = 16, \quad z = \xi + i\eta$ (see Figure 14.1).

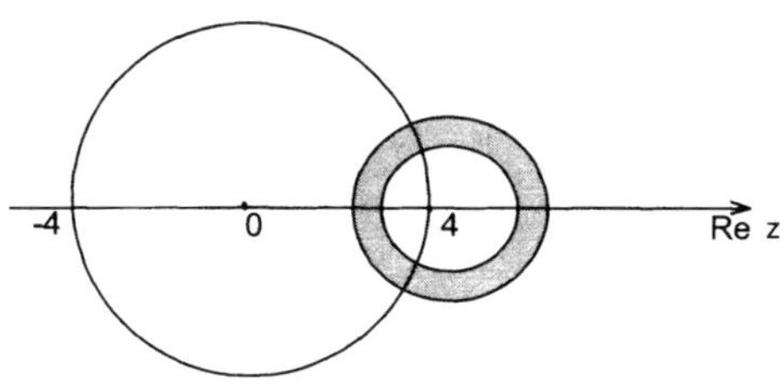

FIGURE 14.1. The shaded domain Ω_+^{ε} for the asymptotic matching

We denote the exterior of the circle $|z| = 4$ without the ray $(-\infty, -4)$ by Ω_1, and the open upper and lower half-planes by $\Omega_{\pm}$. Also, we assume that

$$-\pi < \arg z < \pi, \quad z \in \Omega_1 \cup \Omega_+ \cup \Omega_-. \tag{14.2.6}$$

The Propositions a)–c) allow us to define the WKB- solutions Ψ^1_{WKB} and $\Psi^{\pm}_{WKB}$ at the domains Ω_1, $\Omega_{\pm}$. These solutions have double asymptotics of the form

$$\Psi^1_{WKB} = \left[\left(I - \frac{16}{z^2} P_0 \cos\frac{u}{2}\right) \frac{z^2}{z^2 - 16} + O\left(\frac{z}{\sqrt{x}\,(z^2 - 16)}\right)\right] \times \exp\left\{-x\sigma_3 \int_4^z \mu(\eta, x)\, d\eta\right\}, \quad \sqrt{x}\,|z - 4| \to \infty, \tag{14.2.7}$$

$$\Psi^{\pm}_{WKB} = \left[\left(I - \frac{16}{z^2} P_0 \cos\frac{u}{2}\right) \frac{z^2}{z^2 - 16} + O\left(\frac{z}{\sqrt{x}\,(z^2 - 16)}\right)\right] \times \exp\left\{-x\sigma_3 \int_4^z \mu(\eta, x)\, d\eta\right\}, \quad \sqrt{x}\left|z - \frac{16}{z}\right| \to \infty. \tag{14.2.8}$$

The asymptotic expressions (14.2.5), (14.2.7), (14.2.8), show that the solutions Ψ^1_{WKB} and $\Psi^{\pm}_{WKB}$ are related by right diagonal matrices with the canonical solutions Ψ^{∞}_1 and $\Psi_{\pm}$, respectively. The diagonal matrices have the form

$$C_1 = \lim_{z\to\infty,\, \arg z=0} \exp\left\{x\sigma_3 \int_4^z \mu(\eta)\, d\eta - \frac{ixz}{16}\sigma_3\right\},$$

$$C_1 = \left[\Psi^1_{WKB}\right]^{-1} \Psi^{\infty}_1,$$

(14.2.9)
$$C_{\pm} = \lim_{z\to\infty,\, \arg z=\pm\gamma} \exp\left\{x\sigma_3 \int_4^z \mu(\eta)\, d\eta - \frac{ixz}{16}\sigma_3\right\},$$

$$C_{\pm} = \left[\Psi^{\pm}_{WKB}\right]^{-1} \Psi_{\pm},$$

$$\arg z = \pm\gamma, \quad 0<\gamma<\pi.$$

At the neighborhood of the turning point

$$D^{\varepsilon}_{+} = \left\{z : |z-4| < x^{-\frac{1}{2}+\varepsilon}, \quad 0<\varepsilon<\frac{1}{6}\right\},$$

the equation (14.2.3) reduces to the model equation

(14.2.10)
$$\frac{d\Psi_0}{d\zeta} = \begin{pmatrix} \zeta/2 & v \\ v^* & -\zeta/2 \end{pmatrix} \Psi_0$$

with the parameters

(14.2.11)
$$\begin{aligned} v &= -\frac{\sqrt{x}}{4}(u_k + iw)\, e^{\frac{i\pi}{4}}, \\ v^* &= \frac{\sqrt{x}}{4}(u_k - iw)\, e^{\frac{i\pi}{4}}, \end{aligned}$$

and the independent variable

(14.2.12)
$$\zeta = \frac{\sqrt{x}}{4}(z-4)\, e^{-\frac{i\pi}{4}}.$$

Thus, in the domain D^{ε}_{+} equation (14.2.3) has a solution Ψ_0 with the following asymptotic behaviour,

(14.2.13)
$$\Psi_0(z) = (I + o(1)) \begin{pmatrix} D_{-1-\nu}(i\zeta) & D_{\nu}(\zeta) \\ \dot{D}_{-1-\nu}(i\zeta) & \dot{D}_{\nu}(\zeta) \end{pmatrix},$$

where $\tau\to\infty$, $z\in D^{\varepsilon}_{+}$, $\arg(z-4) = 0,\ \pm\frac{\pi}{2}$,

(14.2.14)
$$\nu = -vv^* - 1 = i\frac{x}{16}\left(w^2 + u_k^2\right) - 1.$$

The investigation of the integral $\int_4^z \mu(\eta)\, d\eta$ at the neighborhood of infinity, and in the circular domain

$$\Omega^{\varepsilon}_{+} = \left\{z : |z-4| \approx x^{-\frac{1}{2}+\varepsilon}, \quad 0<\varepsilon<\frac{1}{6}\right\}$$

is similar to the investigation performed in Chapter 5. In this way, we obtain the following asymptotic formulae:

$$\int_4^z \mu(\eta)\, d\eta = \frac{i}{16}z + \frac{i}{z} - \frac{i}{2}$$

$$+b + ia^2 \ln\frac{z-4}{z+4} + o\left(\tau^{-1}\right), \tag{14.2.15}$$

$$x \to \infty, \quad |z| > C, \quad \arg z = 0,\ \pm\gamma,$$

and

$$\int_4^z \mu(\eta)\, d\eta = \frac{i}{64}(z-4)^2 + ia^2\ln(z-4)$$

$$+b - 3ia^2 \ln 2 + o\left(\tau^{-1}\right), \tag{14.2.16}$$

$$x \to \infty, \quad z \in \Omega_+^{\varepsilon}, \quad \arg(z-4) = 0,\ \pm\frac{\pi}{2},$$

where a and b are defined by

$$a^2 = \frac{w^2+u^2}{16}, \quad b = \frac{i}{2}a^2 - \frac{i}{2}a^2\ln 64a^2 + 4ia^2\ln 2.$$

The substitution of the asymptotics (14.2.15) into equation (14.2.9) implies that matrices C_1, $C_\pm$ are given by

$$C_1 = C_\pm = \exp\left\{\left(-\frac{xi}{2} + bx\right)\sigma_3\right\} + o(1). \tag{14.2.17}$$

On the other hand, the formulae (14.2.16) and (14.2.15) as well as the equations (14.2.7) and (14.2.8), can be used to obtain the leading term in x of the matrices $N_1 = \Psi_0^{-1}\Psi_{WKB}^1$ and $N_\pm = \Psi_0^{-1}\Psi_{WKB}^\pm$:

$$N_1 = \begin{pmatrix} e^{(1+\nu)\frac{i\pi}{2}} & 0 \\ 0 & -v \end{pmatrix}(I + o(1))\exp\left\{\left(\frac{\nu+1}{2}\ln 4x - \frac{\nu+1}{4}i\pi - bx\right)\sigma_3\right\},$$

$$N_+ = \begin{pmatrix} e^{(1+\nu)\frac{i\pi}{2}} & 0 \\ -\frac{i\sqrt{2\pi}}{\Gamma(1+\nu)} & -v \end{pmatrix}(I + o(1)) \tag{14.2.18}$$

$$\times \exp\left\{\left(\frac{\nu+1}{2}\ln 4x - \frac{\nu+1}{4}i\pi - bx\right)\sigma_3\right\},$$

$$N_- = \begin{pmatrix} e^{(1+\nu)\frac{i\pi}{2}} & \frac{v}{\Gamma(-\nu)}\sqrt{2\pi}e^{-\frac{i\pi}{2}(1+\nu)} \\ 0 & -v \end{pmatrix}(I + o(1))$$

$$\times \exp\left\{\left(\frac{\nu+1}{2}\ln 4x - i\frac{\nu+1}{2}\pi - bx\right)\sigma_3\right\}.$$

The monodromy data for the system (13.1.2) can be reconstructed through the associated Stokes matrices

$$S_+^{(\infty)} = \left[\Psi_1^\infty\right]^{-1}\Psi_+ = C_1^{-1}N_1^{-1}N_+C_+,$$

and

$$\mathrm{S}_-^{(\infty)} = [\Psi_1^\infty]^{-1}\,\Psi_- = C_1^{-1}N_1^{-1}N_-C_+.$$

Equations (14.2.17) and(14.2.18) provide the asymptotic expressions for the matrices $\mathrm{S}_\pm^{(\infty)}$:

$$\mathrm{S}_+^\infty = \begin{pmatrix} 1 & 0 \\ \dfrac{i}{v}\dfrac{\sqrt{2\pi}}{\Gamma(1+\nu)}\exp\left\{-ix+(1+\nu)\ln 4x - i\dfrac{\nu+1}{2}\pi\right\} & 1 \end{pmatrix} + o(1),$$

$$\mathrm{S}_-^\infty = \begin{pmatrix} 1 & \dfrac{v}{\Gamma(-\nu)}\sqrt{2\pi}\exp\left\{ix-(1+\nu)\ln 4x - i\dfrac{\nu+1}{2}\pi\right\} \\ 0 & 1 \end{pmatrix} + o(1),$$

$$x\to+\infty.$$

□

3. The asymptotics of nonsingular solutions

First we introduce the manifold $\mathbb{M}$ of nonsingular solutions of the SG-PIII equation

$$u_{xx}+\frac{1}{x}u_x+\sin u = 0,\quad x>0, \tag{14.3.1}$$

parametrized by the monodromy data (13.1.16),

$$\mathbb{M} = \left\{(p,q)\in\mathbb{C}^2:\ pq>-1,\right\}. \tag{14.3.2}$$

This manifold is naturally decomposed into four submanifolds,

$$\mathbb{M} = \mathbb{M}_r\cup\mathbb{M}_q\cup\mathbb{M}_p\cup\mathbb{M}_i, \tag{14.3.3}$$

where

$$\mathbb{M}_r = \left\{(p,q)\in\mathbb{C}^2:\quad \arg q = -\arg p,\quad p,q\neq 0\right\},$$
$$\mathbb{M}_i = \left\{(p,q)\in\mathbb{C}^2:\quad \arg q = -\arg p+\pi,\quad 0<|pq|<1\right\},$$
$$\mathbb{M}_p = \left\{(p,q)\in\mathbb{C}^2:\quad q=0\right\},\quad \mathbb{M}_q = \left\{(p,q)\in\mathbb{C}^2:\quad p=0\right\}.$$

THEOREM 14.4. *Let the monodromy data $(p,q)\in\mathbb{M}$, correspond to the decreasing as $x\to+\infty$ solution $u(x)$ of PIII equation*[1]*. Then, according to the decomposition (14.3.3), the leading term of the asymptotics of $u(x)$ is given by the following formulae:*

a)

$$(p,q)\in\mathbb{M}_r:$$

$$u(x) = 2\pi k + \frac{\alpha}{2}x^{-\frac{1}{2}}\Big[\left(\beta+\frac{1}{\beta}\right)\cos\left\{x-\frac{\alpha^2}{16}\ln x+\varphi\right\}$$

$$+i\left(\beta-\frac{1}{\beta}\right)\times\sin\left\{x-\frac{\alpha^2}{16}\ln x+\varphi\right\}\Big]+o(1),\quad x\to+\infty, \tag{14.3.4}$$

$$\alpha^2 = \frac{8}{\pi}\ln(1+|pq|),\quad \beta^2 = \left|\frac{q}{p}\right|,\quad \alpha,\beta>0,$$

[1]The solution $u(x)$ is defined $\bmod 2\pi$.

$$\varphi = -\frac{\alpha^2}{8}\ln 2 + \frac{3\pi}{4} - \arg\Gamma\left(-\frac{i\alpha^2}{16}\right) - \arg p.$$

b)

$$(p,q) \in \mathbb{M}_i :$$

$$u(x) = 2\pi k + \frac{\alpha}{2}x^{-\frac{1}{2}}\left[\left(\beta - \frac{1}{\beta}\right)\cos\left\{x + \frac{\alpha^2}{16}\ln x + \varphi\right\}\right.$$

(14.3.5) $$\left. + i\left(\beta + \frac{1}{\beta}\right) \times \sin\left\{x + \frac{\alpha^2}{16}\ln 2 + \varphi\right\}\right] + o(1), \quad x \to +\infty,$$

$$\alpha^2 = -\frac{8}{\pi}\ln(1 - |pq|), \quad \beta^2 = \left|\frac{q}{p}\right|, \quad \alpha, \beta > 0,$$

$$\varphi = \frac{\alpha^2}{8}\ln 2 + \frac{3\pi}{4} - \arg\Gamma\left(\frac{i\alpha^2}{16}\right) - \arg p.$$

c)

$$(p,q) \in \mathbb{M}_p :$$

(14.3.6) $$u(x) = 2\pi k + \alpha\sqrt{\frac{2}{\pi}}x^{-\frac{1}{2}}\exp\left\{-ix - \frac{i\pi}{4} + i\varphi\right\} + o(1),$$

$$x \to +\infty, \quad \alpha e^{i\varphi} = p.$$

d)

$$(p,q) \in \mathbb{M}_q :$$

(14.3.7) $$u(x) = 2\pi k + \alpha\sqrt{\frac{2}{\pi}}x^{-\frac{1}{2}}\exp\left\{ix + \frac{i\pi}{4} + i\varphi\right\} + o(1),$$

$$x \to +\infty, \quad \alpha e^{i\varphi} = q.$$

The asymptotics (14.3.4)*–*(14.3.7) *are differentiable in* x.

PROOF. The proof of the asymptotics (14.3.4)–(14.3.7) is similar to the proof of Theorem 7.1; this proof splits into three steps.

1. Define the approximate solution

$$u_{\text{asym}}(x; s) = 2\pi k + O(x^{-\frac{1}{2}}),$$

where $s = \{p, q\}$ is the monodromy data, and the right-hand side of $u_{\text{asym}}(x; s)$ is one of the formulae (14.3.4)–(14.3.7). Substitute this function into the system (13.1.2) and solve the resulting system for $|x| \to \infty$.

2. Since, by definition, the conditions a)–c) of Theorem 14.3 are true for u_{asym}, we can apply the WKB-analysis to calculate the monodromy data $s_{\text{asym}}(x; s)$. Theorem 14.3 guarantees that the approximate monodromy data satisfies the uniform estimate

$$|s - s_{\text{asym}}| < C|x|^{-\epsilon},$$

where C and $\epsilon > 0$ do not depend on x.

3. Allow s to depend on x and on the constant s_0 in such a way, that both

(14.3.8) $$s_{\text{asym}}(x, s(x, s_0)) = s_0$$

and hence

(14.3.9) $$s(x, s) \to s_0, \quad |x| \to \infty.$$

The existence of such a function, as well as the solvability of (14.3.8) follow from Theorem 14.3 and the Brouwer Fixed Point theorem (see [**Ki2**] and Chapter 4). Then, since its monodromy data s_0 is constant, $u_{\text{asym}}(x, s(x, s_0))$ is an exact solution of SG-PIII equation. Furthermore, equation (14.3.9) implies that this function has the desired asymptotics, and hence the estimates in (14.3.4)–(14.3.7) follow. □

REMARK 14.1. The system (13.1.2) is invariant under the transformation $u \mapsto u + 2k\pi$. Therefore the integer k in the formulae (14.3.4)–14.3.7) cannot be constructed through the monodromy data p and$q$. This is a consequence of the fact that the entire class of solutions $u(x) \bmod 2\pi$, have been associated to the same point $(p,q) \in \mathbb{M}$.

REMARK 14.2. The submanifold

$$\mathbb{M}_R = \{(p,q) \in \mathbb{M}_r : |p| = |q|\},$$

corresponds to the real-valued solutions of PIII equation. The asymptotics of those solutions has the form

$$\begin{aligned} u(x) &= 2\pi k + \alpha x^{-\frac{1}{2}} \cos\left\{x - \frac{\alpha^2}{16}\ln x + \varphi\right\} + o(1), \\ & \qquad x \to +\infty, \\ \alpha^2 &= \frac{8}{\pi}\ln\left(1 + |p^2|\right), \quad \alpha > 0, \\ \varphi &= -\frac{\alpha^2}{8}\ln 2 + \frac{3\pi}{4} - \arg\Gamma\left(-\frac{i\alpha^2}{16}\right) - \arg p. \end{aligned} \tag{14.3.10}$$

Similarly, the purely imaginary solutions, which decrease as $x \to +\infty$, are parametrized by the submanifold

$$\mathbb{M}_I = \{(p,q) \in \mathbb{M}_i : |p| = |q|\}.$$

The asymptotics of these solutions have the form

$$\begin{aligned} u(x) &= 2\pi k + \alpha x^{-\frac{1}{2}} i \sin\left\{x + \frac{\alpha^2}{16}\ln x + \varphi\right\} + o(1), \quad x \to +\infty, \\ \alpha^2 &= -\frac{8}{\pi}\ln\left(1 - |p^2|\right), \quad \alpha > 0, \\ \varphi &= \frac{\alpha^2}{8}\ln 2 + \frac{3\pi}{4} - \arg\Gamma\left(\frac{i\alpha^2}{16}\right) - \arg p. \end{aligned} \tag{14.3.11}$$

3.1. Connection formulae for nonsingular solutions. We will now construct explicit connection formulae for the solutions of PIII equation. There is an essential difference in the structure of these formulae in comparison to the corresponding formulae for PII. The solutions of PIII have two singular points, $x = 0$ and $x = \infty$), while the solutions of PII have only one, $x = \infty$. Therefore, the correct connection problem for PIII is posed through the Cauchy problem on the half-line $(0, \infty)$. More precisely, it is necessary to construct the asymptotic solutions as $x \to 0$ and as $x \to \infty$, and to connect the parameters of these asymptotics so that they represent the asymptotics of solution of the Cauchy problem.

The parameters r and s must satisfy certain conditions which will be derived below in order for the solution $u(x)$ to be in the manifold $\mathbb{M}$. Comparing the two formulae (14.1.23) and (13.1.31), which expresses the matrix Q through r, s and p, q respectively, yields the constraint[2]

$$1+pq=\frac{4\pi^2}{(A+B)^2}. \tag{14.3.12}$$

This means that the solution (14.1.1) lies in the manifold $\mathbb{M}$, if the parameters r and s are such that

$$\operatorname{Im}(A+B)=0. \tag{14.3.13}$$

The condition (14.3.13) is satisfied for all real-valued r and s (then $A=\overline{B}$), as well as for all purely imaginary r and s ($A, B\in\mathbb{R}$).

Thus under the constraint (14.3.13) we can describe the asymptotics of the solution $u(x)$ as $x\to+\infty$, with the initial condition (14.1.1). More precisely, the following theorem is valid:

THEOREM 14.5. *Suppose that the following statements are true:*

a) near the origin $u(x)$ satisfies the asymptotics (14.1.1) *where r and s satisfy the condition* (14.3.13)*;*

b) the solution $u(x)$ has no singularities for all $x>0$, and its asymptotics at infinity is described by one of the formulae (14.3.4)*–*(14.3.7) *with certain parameters α, β, φ.*

Then the following formulae hold, connecting the initial data r and s with the parameters of α, β, φ:

$$\alpha^2=\pm\frac{16}{\pi}\ln\frac{1}{2\pi}\left|2^{\frac{3ir}{2}}e^{\frac{is}{2}}\Gamma^2\left(\frac12+\frac{ir}{4}\right)+2^{-\frac{3ir}{2}}e^{-\frac{is}{2}}\Gamma^2\left(\frac12-\frac{ir}{4}\right)\right|, \tag{14.3.14}$$

$$\beta^2=\left|\frac{e^{\frac{\pi r}{4}}2^{\frac{3ir}{2}}e^{\frac{is}{2}}\Gamma^2\left(\frac12+\frac{ir}{4}\right)-e^{-\frac{\pi r}{4}}2^{-\frac{3ir}{2}}e^{-\frac{is}{2}}\Gamma^2\left(\frac12-\frac{ir}{4}\right)}{e^{\frac{is}{2}-\frac{\pi r}{4}}2^{\frac{3ir}{2}}\Gamma^2\left(\frac12+\frac{ir}{4}\right)-e^{-\frac{is}{2}+\frac{\pi r}{4}}2^{-\frac{3ir}{2}}\Gamma^2\left(\frac12-\frac{ir}{4}\right)}\right|,$$

$$\varphi=\pm\frac{\alpha^2}{8}\ln 2+\frac{3\pi}{4}-\arg\Gamma\left(\pm\frac{i\alpha^2}{16}\right)-\pi^{\frac{1-\varepsilon}{2}}$$

$$-\arg\left\{e^{\frac{is}{2}-\frac{\pi r}{4}}2^{\frac{3ir}{2}}\Gamma^2\left(\frac12+\frac{ir}{4}\right)-e^{-\frac{is}{2}+\frac{\pi r}{4}}2^{-\frac{3ir}{2}}\Gamma^2\left(\frac12-\frac{ir}{4}\right)\right\},$$

where the upper sign in "$\pm$" occurs in the asymptotics (14.3.4) *and the lower sign occurs in the asymptotics* (14.3.5)*. The value of ε is*

$$\varepsilon=\operatorname{sign}\left\{2^{\frac{3ir}{2}}e^{\frac{is}{2}}\Gamma^2\left(\frac12+\frac{ir}{4}\right)+2^{-\frac{3ir}{2}}e^{-\frac{is}{2}}\Gamma^2\left(\frac12-\frac{ir}{4}\right)\right\}.$$

We specify the "separatrix" case of the asymptotics (14.3.6), (14.3.7). In this case, the connection formulae for r, s and α, φ, become much simple. For example, the solution (14.1.1) belongs to the manifold $\mathbb{M}_p$ provided that

$$\exp(is)=2^{-3ir}\Gamma^2\left(\frac12-\frac{ir}{4}\right)\Gamma^{-2}\left(\frac12+\frac{ir}{4}\right)\exp\left(-\frac{\pi r}{2}\right). \tag{14.3.15}$$

Thus formulae (14.1.7) simplify to

$$p=2\sinh\frac{\pi r}{4},\quad q=0.$$

[2]Clearly, the constraint (14.3.12) may be obtained directly from the formulae (14.0.5).

This implies that the connection formulae for the solution (14.3.6) are given by

$$\alpha = 2\left|\sinh\frac{\pi r}{4}\right|, \quad \varphi = \arg\sinh\frac{\pi r}{4}, \tag{14.3.16}$$

where r and s are restricted by the constraints (14.3.13) and (14.3.15).

Similarly, for the solution (14.3.7) we find

$$\alpha = 2\left|\sinh\frac{\pi r}{4}\right|, \quad \varphi = -\pi + \arg\sinh\frac{\pi r}{4}, \tag{14.3.17}$$

where, in addition to (14.3.13), the following condition also holds:

$$\exp(is) = 2^{-3ir}\Gamma^2\left(\frac{1}{2} - \frac{ir}{4}\right)\Gamma^{-2}\left(\frac{1}{2} + \frac{ir}{4}\right)\exp\frac{\pi r}{2}. \tag{14.3.18}$$

In the particular case of $\operatorname{Re} r = 0$, the connection formulae (14.3.17) for the solution (14.3.7) were first obtained earlier in [**McTW**]. We note that for purely imaginary r, the conditions (14.3.18) and (14.3.13) become explicit expressions of s in terms of r:

$$\operatorname{Re} s = -\frac{i\pi r}{2} + 2\pi\ell, \quad \ell \in \mathbb{Z},$$

$$\operatorname{Im} s = 3ir - 2\ln\left|\frac{\Gamma\left(\frac{1}{2} - \frac{ir}{4}\right)}{\Gamma\left(\frac{1}{2} + \frac{ir}{4}\right)}\right|.$$

For real-valued solutions of SG-PIII it is possible to eliminate the assumption b) of Theorem 14.3. Indeed, the uniqueness of solution of the corresponding Riemann-Hilbert problem (which in turn is the trivial corollary of the Liouville theorem) implies uniqueness of the inverse problem. Therefore, if we have an independent proof of the global solvability of the Cauchy problem for SG-PIII equation, then we can eliminate assumption b). For example, such a proof exists for the case of real-valued solutions. Actually, any real-valued solution of SG-PIII equation is obviously nonsingular as $x > 0$. Furthermore, as mentioned above the condition (14.3.13) is satisfied identically for real-valued r and s. Also, the real-valued solutions with the asymptotics (14.3.10) exist for any α and φ. Thus, the connection formulae (14.3.4)–(14.3.7) can be inverted, to give expressions of p and q in terms of α and φ. Therefore, we can formulate a stronger version of Theorem 14.3.

THEOREM 14.6. *The real-valued solution of the PIII equation (*14.3.1*) with the asymptotics (*14.1.1*) at the origin, has the following asymptotics as $x \to +\infty$:*

$$u(x) = 2\pi k + \alpha x^{-\frac{1}{2}}\cos\left(x - \frac{\alpha^2}{16}\ln x + \varphi\right) + o(1), \tag{14.3.19}$$

where $\alpha > 0$,

$$\alpha^2 = -\frac{16}{\pi}\ln\frac{1}{\pi}\left|\operatorname{Re}\left\{2^{\frac{3ir}{2}}e^{\frac{is}{2}}\Gamma^2\left(\frac{1}{2} + \frac{ir}{4}\right)\right\}\right|,$$

$$\begin{aligned}\varphi = &-\frac{\alpha^2}{8}\ln 2 + \frac{3\pi}{4} - \arg\Gamma\left(-\frac{i\alpha^2}{16}\right)\\ &-\frac{\pi}{2}\left(1 - \mathit{sign}\left\{\operatorname{Re}\left[2^{\frac{3ir}{2}}e^{\frac{is}{2}}\Gamma^2\left(\frac{1}{2} + \frac{ir}{4}\right)\right]\right\}\right)\\ &-\arg\left\{e^{\frac{is}{2} - \frac{\pi r}{4}}2^{\frac{3ir}{2}}\Gamma^2\left(\frac{1}{2} + \frac{ir}{4}\right) - e^{-\frac{is}{2} + \frac{\pi r}{4}}2^{-\frac{3ir}{2}}\Gamma^2\left(\frac{1}{2} - \frac{ir}{4}\right)\right\}.\end{aligned} \tag{14.3.20}$$

All that remains is to calculate the integer k in the formulae (14.3.4)–(14.3.7):

THEOREM 14.7. *Under the conditions of Theorem* 14.2, *the integer* k *equals*

$$k = \left[\frac{1}{2\pi}\left\{\pi - s_1 + 3r_1 \ln 2 - 4 \arg \Gamma\left(\frac{1}{2} + \frac{ir}{4}\right)\right\}\right], \tag{14.3.21}$$

where $s = s_1 + is_r$, $r = r_1 + ir_2$ *and* $[A]$ *denotes the integer part of* A.

PROOF. Consider the phase space $(u,\ u_x)$ of the solutions of PIII equation. The points $u = 2\pi k,\ u_x = 0$, are stable focuses, whereas the points $u = \pi + 2\pi k,\ u_x = 0$, are saddles, which have separatrices which disjoint the focuses. The real-valued case is presented in Figure 14.2.

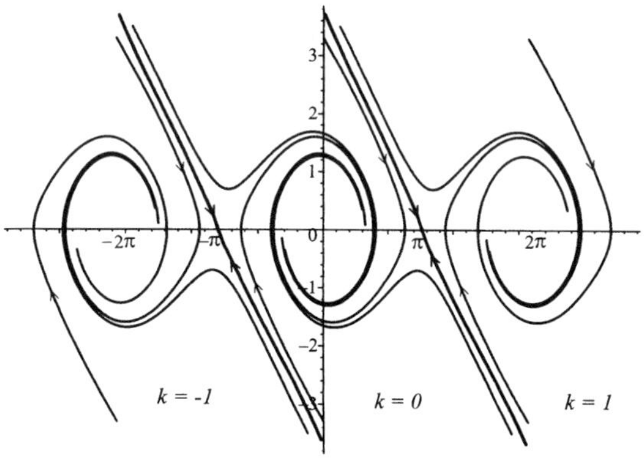

FIGURE 14.2. Indexing the separatrix domains with integers k

The asymptotic solutions (14.3.4)–(14.3.7) describe the behaviour of the phase curve in the neighborhood of the k-th focus, whereas the asymptotics defined in (14.1.1), i.e.,

$$\begin{aligned} u(x) &= r \ln x + s + O\left(x^{2-|\operatorname{Im} r|}\right), \\ u_x(x) &= \frac{r}{x} + O\left(x^{1-|\operatorname{Im} r|}\right), \quad x \to 0, \end{aligned}$$

characterizes the solution at infinity in the $(u,\ u_x)$ variables. The domains, divided by the separatrices, are indexed by k, because there is only one focus $u = 2\pi k$ in the k-th domain.

Let us derive equations for the separatrices in terms of the initial data r and s. The integral curves described by the asymptotics (14.3.4)–(14.3.7), which fill in the domain, have the property that curves passing closer to the boundary have the larger value of amplitude α in (14.3.4)–(14.3.7). Therefore, the boundaries, which by definition coincide with the separatrices, correspond to the case of $\alpha^2 = \infty$. Then, according to Theorem 14.2, we obtain from (14.3.14) (recalling that $|\operatorname{Im} r| < 2!$),

$$\alpha^2 = \infty \Leftrightarrow |A + B| = 0 \tag{14.3.22}$$

$$\Leftrightarrow 2^{3ir}\Gamma^2\left(\frac{1}{2} + \frac{ir}{4}\right)\Gamma^{-2}\left(\frac{1}{2} - \frac{ir}{4}\right) = -e^{-is}.$$

The real and the imaginary parts of this equation yield

$$\begin{aligned} s_2 &= -3r_2 \ln 2, \\ s_1 &= \pi - 2\pi k + 3r_1 \ln 2 - 4 \arg \Gamma\left(\frac{1}{2} + \frac{ir}{4}\right). \end{aligned} \tag{14.3.23}$$

The value of the integer k appearing in the second of equations (14.3.23) must be matched with the condition

$$k = 0 \Leftrightarrow r = s = 0; \tag{14.3.24}$$

this follows from the fact that equation (14.1.1) with $r = s = 0$ implies $u = 0$. Combining (14.3.23) and (14.3.24), we find (14.3.21). □

REMARK 14.3. It is clear from equation (14.3.22), that the real-valued dimension of the separatrix manifold is 2, whereas the dimension of the entire phase-space (u, u_x) is four. This apparent mismatch in dimensions is explained by the fact that the solutions $u(x)$ belong to the manifold $\mathbb{M}$ which is three-dimensional due to the condition (14.3.13). Thus, the codimension of separatrix on the manifold $\mathbb{M}$ is 1.

3.2. The structure of separatrix solution. Consider the asymptotic behavior of the separatrix solution defined by (14.3.22). In order to find the monodromy data parametrization of this solution, we first note that equation (14.3.22) implies that Q (14.1.23) is given by

$$\mathrm{Q} = \begin{pmatrix} 0 & i \\ i & 0 \end{pmatrix} \equiv i\sigma_1.$$

This is possible only if $p, q \to \infty$. Thus, we consider the limit case of the submanifold $\mathbb{M}$:

$$\mathbb{M}_\infty = \left\{(p, q) \in \mathbb{C}^2 : \quad q = -p + \Delta, \quad p, q \to \infty\right\},$$

where the only parameter, which enters the Stokes matrices (13.1.31) is Δ,

$$\mathrm{S}_1^{(\infty)} = \begin{pmatrix} 1 & 0 \\ \Delta & 1 \end{pmatrix}, \quad \mathrm{S}_2^{(\infty)} = \begin{pmatrix} 1 & \Delta \\ 0 & 1 \end{pmatrix},$$

$$\mathrm{S}_1^{(0)} = \begin{pmatrix} 1 & \Delta \\ 0 & 1 \end{pmatrix}, \quad \mathrm{S}_2^{(0)} = \begin{pmatrix} 1 & 0 \\ \Delta & 1 \end{pmatrix}, \tag{14.3.25}$$

It is clear that, since p and q are infinite, Theorems 14.4–14.6 are not valid for the separatrix solution. In order to get the asymptotics at infinity we use the RH problem of Section 3 of Chapter 13.

Introduce the piecewise analytical function Y,

$$Y(\lambda, x) = \begin{cases} \sigma_1 \Psi_1^{(\infty)}(\lambda) \sigma_1 \exp\left\{-\frac{ix^2}{16}\lambda\sigma_3 + \frac{i}{\lambda}\sigma_3\right\}, & \operatorname{Re}\lambda \geq 0, \\ \sigma_1 \Psi_2^{(\infty)}(\lambda) \sigma_1 \exp\left\{-\frac{ix^2}{16}\lambda\sigma_3 + \frac{i}{\lambda}\sigma_3\right\}, & \operatorname{Re}\lambda \leq 0. \end{cases}$$

Since $Q = i\sigma_1$, we have simultaneously that

$$Y(\lambda, x) = \begin{cases} i\sigma_1 \Psi_1^{(0)}(\lambda) \exp\left\{-\frac{ix^2}{16}\lambda\sigma_3 + \frac{i}{\lambda}\sigma_3\right\}, & \text{Re } \lambda \geq 0, \\ i\sigma_1 \Psi_2^{(0)}(\lambda) \exp\left\{-\frac{ix^2}{16}\lambda\sigma_3 + \frac{i}{\lambda}\sigma_3\right\}, & \text{Re } \lambda \leq 0. \end{cases}$$

It is easy to see that the function $Y(\lambda, x)$ satisfies the RH problem 1°–3° of Section 3 of Chapter 13, where p must now be replaced by Δ. Hence, in view of Theorem 13.1, we obtain at once the following result.

THEOREM 14.8. *The separatrix solution of the SG-PIII equation (13.0.1), is related to the manifold $\mathbb{M}_\infty$ and has the following asymptotics at infinity,*

$$u_{sep}(x) = 2\pi\left(k + \frac{1}{2}\right) + \Delta\sqrt{\frac{2}{\pi}}x^{-\frac{1}{2}}e^{-x}(1 + o(1)), \quad x \to +\infty, \tag{14.3.26}$$

where $\Delta = p + q$ is finite as $p, q \to \infty$.

Note that the appearence of the term $2\pi\left(k + \frac{1}{2}\right)$ instead of $2\pi k$ is due to the relation

$$Y(0, x) = i\sigma_1 P_0 = \begin{pmatrix} \sin\frac{u}{2} & i\cos\frac{u}{2} \\ i\cos\frac{u}{2} & \sin\frac{u}{2} \end{pmatrix}.$$

Now, the connection formula for the above separatrix solution is straightforward. According to Theorem 14.2

$$p + q = -2\sinh\frac{\pi r}{4},$$

and the initial conditions r and s connected by (14.3.15),

$$\exp(is) = -2^{-3ir}\Gamma^2\left(\frac{1}{2} - \frac{ir}{4}\right)\Gamma^{-2}\left(\frac{1}{2} + \frac{ir}{4}\right). \tag{14.3.27}$$

COROLLARY OF THEOREM 14.8. *The separatrix solution of the SG-PIII equation (13.0.1) is specified by the following asymptotics:*
(14.3.28)

$$u_{sep}(x) = \begin{cases} r\ln x + s + o(x), & x \to 0, \\ 2\pi\left(k + \frac{1}{2}\right) - 2\sinh\frac{\pi r}{4}\sqrt{\frac{2}{\pi}}x^{-\frac{1}{2}}e^{-x}(1 + o(1)), & x \to +\infty, \end{cases}$$

where constants r and s are connected by the constraint (14.3.27), and the integer k is defined by

$$k = \left[\frac{1}{2\pi}\left\{\pi - \text{Re } s + 3\,\text{Re } r\ln 2 - 4\arg\Gamma\left(\frac{1}{2} + \frac{ir}{4}\right)\right\}\right].$$

The fine tuning of the separatrix specified by the constraint (14.3.27), is illustrated by the numerical simulation shown in Figure 14.3. The three phase curves start from the asymptotics $u(x) = r\ln x + s$, $u_x(x) = r/x$ for small x (right lower lines), with $r = 4$. For this value of r, the value $s = 4.49774$ matches the separatrix constraint (14.3.27).

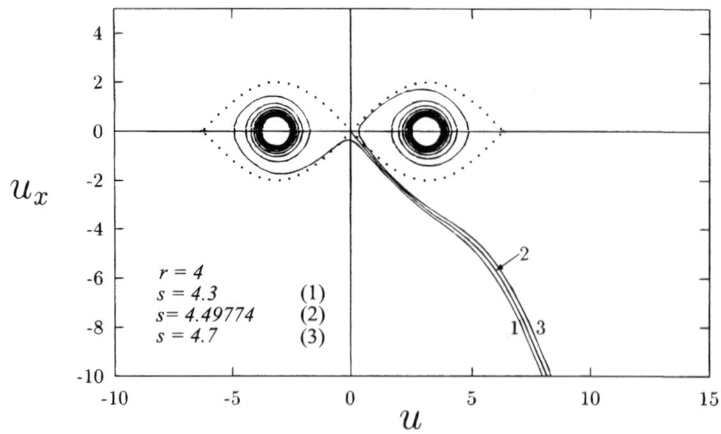

FIGURE 14.3. Separatrix (2) between two phase trajectories (1) and (3) going to neighboring focuses. Dashed lines are the separatrices of the classical pendulum

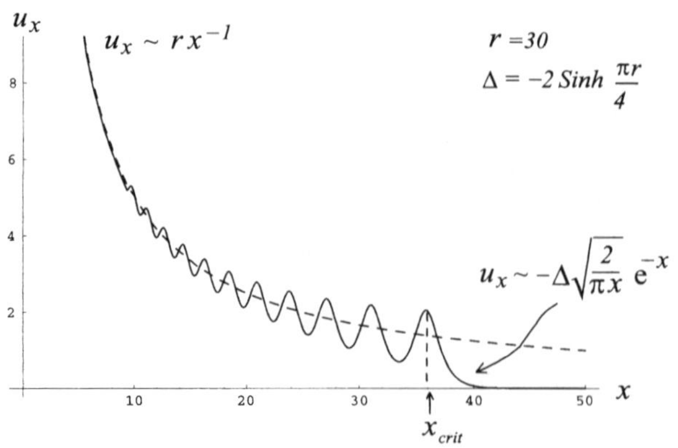

FIGURE 14.4. Intermediate "soliton train" asymptotics. Separatrix solution (solid line) vs. its asymptotics at the origin (dashed line)

The exponentially vanishing separatrix solution u_{sep} is rather complicated. Indeed, Besides logarithmic asymptotics at the origin and the exponential decay at infinity, it also possesses an intermediate asymptotic behavior, similar to a "soliton train" modes of the sine-Gordon equation. The intermediate oscillations are revealed for large values of r. They start near the initial asymptotic mode $u_x(x) = r/x$, and they end up at a certain value of x called x_{crit}, where the "front-end soliton" is situated (see Figure 14.4). For large values of r we see the clear-cut profile of the "soliton train" with a distinct front-end soliton. The existence of the above "wavefront boundary" at $x = x_{crit}$ plays an important role in applications (see Introduction and [**N6**]).

The value of x_{crit} can be estimated asymptotically for large r. Indeed, the right slope of the front-end soliton has the exponential decaying profile described by (14.3.28),

$$\frac{d}{dx}u_{sep}(x) = 2\sinh\frac{\pi r}{4}\sqrt{\frac{2}{\pi}}x^{-\frac{1}{2}}\mathrm{e}^{-x}\left(1+o(1)\right), \quad x \to +\infty.$$

Thus at the boundary of the front-end soliton, we have

$$2\sinh\frac{\pi r}{4}\sqrt{\frac{2}{\pi}}x_{crit}^{-\frac{1}{2}}\mathrm{e}^{-x_{crit}} = O(1), \quad r \to \infty.$$

This yields

$$x_{crit} = \frac{\pi}{4}r - \frac{1}{2}\ln r + O(1), \quad r \to \infty.$$

The existence of the "soliton train" mode is illustrated in Figure 14.5. The reason of its existence, from the point of view of physics, is the following. The larger the value of r the larger the number of revolutions that the pendulum must perform before it comes to an upper unstable position.

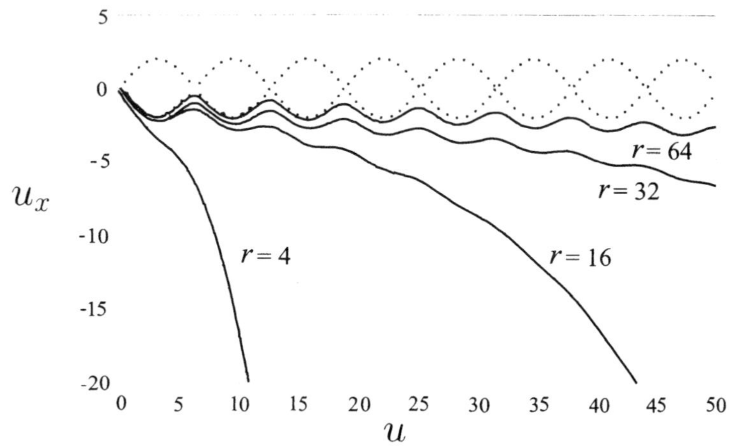

FIGURE 14.5. Four separatrix solutions (solid line) vs. the separatrix solutions of classical pendulum (dashed line)

REMARK 14.4. Separatrix solutions, vanishing exponentially at infinity, also exist for the sinh-Gordon version of SG-PIII equation, i.e., the equation

$$v_{xx} + \frac{1}{x}v_x - \sinh v = 0. \tag{14.3.29}$$

Indeed, the following one-parameter family of solutions was first found by B. M. McCoy, T. T. Wu and C. A. Tracy in [**McTW**],

$$v(x) = \begin{cases} r\ln x + s + o(x), & x \to 0, \\ -2\sin\frac{\pi r}{4}\sqrt{\frac{2}{\pi}}x^{-\frac{1}{2}}\mathrm{e}^{-x}\left(1+o(1)\right), & x \to +\infty, \end{cases} \tag{14.3.30}$$

under the condition

$$\exp(s) = 2^{-3r}\Gamma^2\left(\frac{1}{2}-\frac{r}{4}\right)\Gamma^{-2}\left(\frac{1}{2}+\frac{r}{4}\right).$$

This family can be obtained from (14.3.28) and (14.3.15) using the change of variables

$$v = iu, \quad r \mapsto ir, \quad s \mapsto is.$$

CHAPTER 15

Canonical Four-rays. Singular Solutions of the SG-PIII

0. Introduction

Consider another special type of the Painlevé III equation, namely the equation

$$(15.0.1)\qquad u_{xx} + \frac{1}{x}u_x = 4\sinh u, \quad x > 0.$$

This equation can be transformed to equation (13.0.1) by the simple change of variables $u \mapsto \pi + iu$. Equation (15.0.1) arises in applications of modern mathematical physics (see, for example, [**McTW, CeVa2, CeVa3, TW4, FLZZ**]) and geometry [**BobEi1, BobEi2**]. Some of these applications suggest the study, not only of smooth bounded solutions of equation (15.0.1), but also of the singularities.

We assume that the solution of equation (15.0.1), $u = u(x)$, is a function with the following asymptotics near the origin

$$(15.0.2)\qquad u(x) = r\ln x + s + O\left(x^{2-|r|}\right), \quad x \to 0, \quad |r| < 2.$$

It is shown in [**McTW**], that under a special choice of the initial data r and s, there exists a smooth solution of equation (15.0.1) with the asymptotics

$$(15.0.3)\qquad u(x) \sim \alpha x^{-\frac{1}{2}} e^{-2x}, \quad x \to \infty.$$

Exact connection formulae for the parameters r, s and α, are given in [**McTW**], (see also equations (15.0.12) and (15.0.11)). Taking the linear limit as $r, s \to 0$ in the solution (15.0.3), we find the Bessel function of imaginary argument $K_0(2x)$. The nonlinearity of equation (15.0.1) excludes the corresponding $I_0(2x)$ function, which has exponential growth at infinity. Indeed, it turns out that any solution, linearly independent to the solution satisfying (15.0.3) must have singularities at finite values of x. These singularities, correspond to the movable poles of the solution of PIII written in the canonical form

$$(15.0.4)\qquad w_{xx} = \frac{w_x^2}{w} - \frac{1}{x}w_x + w^3 - \frac{1}{w}.$$

This equation follows from equation (15.0.1) under the transformation

$$(15.0.5)\qquad u = 2\ln w.$$

The word "movable" used above means that the coordinates of the poles depend only of initial data.

The logarithmic transform (15.0.5), maps the poles of w into branching points of u. Therefore, near the singularity $x = a$, we seek u in the asymptotic form

$$u(x) = -2\ln(x-a) - \frac{x-a}{a} + \frac{b(x-a)^2}{a^2} + +O(x-a)^3, \quad x \to a, \tag{15.0.6}$$

where a and b are some fixed real parameters.

The main goal of this chapter is to find the connection formulas connecting the parameters a and b with the initial data r and s of equation (15.0.2).

Equation (15.0.4) can have a solution with an infinite number of poles. In analogy with the case of singular solutions of PII studied in Chapter 10, we will apply the Isomonodromic Deformation Method to obtain a unified description of such solutions. The monodromy data of the corresponding Ψ-function appear, as usual, the global invariants for the solution u. It was shown in Chapter 13, that if u satisfies the asymptotics (15.0.2) as $x \to 0$, then the system of Ψ-functions reduces to Bessel equations, and the related monodromy matrices can be calculated explicitly in terms of the initial data r and s. On the other hand, if we apply instead the asymptotics (15.0.6), it will turn out that the system for Ψ-functions as $x \to a$, will reduce to the Mathieu equation, and then the monodromy data can also be computed explicitly. The combination of the above two calculations gives the following exact functional equations for the singularity parameters a and b in equation (15.0.6):

$$\arg V(0,a,\nu) = -\frac{1}{2}\arg\frac{Ae^{\frac{i\pi r}{4}} + Be^{-\frac{i\pi r}{4}}}{A-B-1}, \tag{15.0.7}$$

$$\mu(a,\nu) = \frac{r}{4} - \frac{1}{2}, \tag{15.0.8}$$

where $\nu^2 = \frac{3}{4}b + \frac{3}{8}$,

$$A = 2^r e^{\frac{s}{2}}\Gamma^2\left(\frac{1}{2}+\frac{r}{4}\right), \quad B = 2^{-r}e^{-\frac{s}{2}}\Gamma^2\left(\frac{1}{2}-\frac{r}{4}\right). \tag{15.0.9}$$

In equation (15.0.7) the function V is the solution of modified Mathieu equation,

$$\frac{d^2V}{dz^2} - \left(\nu^2 - \frac{a^2}{2}\cosh 2z\right)V = 0,$$

with the asymptotics

$$V(z,a,\nu) \to e^{-\frac{z}{2}}\exp\left(\frac{ia}{2}e^z\right), \quad z \to +\infty.$$

The function μ in equation (15.0.8) is the Floquet exponent for the solution $V(iz,a,\nu)$ of the periodic Mathieu equation.

It seems that equations (15.0.7) and (15.0.8) have no exact solutions for almost any r and s. However, the asymptotic solutions as $a \to \infty$ can be calculated effectively. For example, the asymptotic distribution of poles at infinity are described by the following formula,

$$\begin{aligned} a_n &= \pi\left(n-\frac{1}{2}\right) - \frac{x}{2}\ln 8\pi n + \frac{1}{2}\arg\Gamma\left(\frac{1}{2}+ix\right) \\ &+\frac{1}{2}\arg\left(Ae^{\frac{i\pi r}{4}} + Be^{-\frac{i\pi r}{4}}\right)(A-B) + O\left(\frac{\ln n}{n}\right), \quad n \to \infty, \end{aligned} \tag{15.0.10}$$

where $x = \frac{1}{\pi}\ln\left|\frac{A-B}{2\pi}\right|$ and A, B are defined by equation (15.0.9).

It will be shown in Section 4 that there is no finite point $x \in \mathbb{R}$ where these poles accumulate. Combining this result with that of Chapter 13, it is possible to describe the asymptotic behavior of the general real-valued solution of equation (15.0.1). For almost all initial data r and s in equation (15.0.2) ($|r| < 2$), there is a meromorphic solution with infinitely many poles satisfying asymptotic distribution (15.0.10).

These poles vanish, and there exists a smooth for $x > 0$ one-parameter solution, having the asymptotics (15.0.3), where

$$\alpha = -\frac{2}{\sqrt{\pi}}\sin\frac{\pi r}{4}, \tag{15.0.11}$$

iff r and s satisfy the constraint

$$e^{\frac{s}{2}} = 2^{-r}\Gamma\left(\frac{1}{2}-\frac{r}{4}\right)\Gamma^{-1}\left(\frac{1}{2}+\frac{r}{4}\right), \tag{15.0.12}$$

(see [**McTW**] and also formula (14.3.16) in Chapter 8).

Finally we discuss briefly some additional results obtained in [**McTW**]. The regular solution (15.0.3) of PIII equation (15.0.1) was studied in [**McTW**] for all positive values of α. If $\alpha < \frac{2}{\sqrt{\pi}}$, then the solution can be continued smoothly up to the origin, having there the asymptotics (15.0.2), and the connection formula (15.0.11) holds. If $|\alpha| = \frac{2}{\sqrt{\pi}}$, then the solution remains smooth (see [**McTW**]), but near the origin the asymptotics is now given by

$$u(x) = \pm 2\ln\left[-x\left(\ln\frac{x}{2}+C\right)\right] + O\left(x^4\ln^2 x\right), \quad x \to 0,$$

where C is the Euler constant. This case corresponds to the limit case $r \to \pm 2$ in equation (15.0.2). Finally, if $\alpha > \frac{2}{\sqrt{\pi}}$, the solution becomes a meromorphic function with the poles a_n tending to zero ([**McTW**]),

$$a_n \sim 4\exp\left[-\frac{\pi n}{2\mu} - \frac{1}{\mu}\arg\Gamma(i\mu)\right], \quad n \to \infty,$$

where $\alpha = \frac{2}{\sqrt{\pi}}\cosh\pi\mu$.

We expect, that it should be possible to obtain the above results by the technique of the isomonodromic deformations described below.

1. Monodromy data and the Laurent series expansion

We recall here briefly the main facts of the isomonodromy deformation method, developed in Chapter 13 for PIII. Under the change of variables

$$u \mapsto \pi - iu, \quad x \mapsto \frac{x}{2},$$

equation (16.0.1) transforms into equation (15.0.1). The system (13.1.2) can be rewritten in the form

$$\frac{d\Psi}{d\lambda} = \left\{-\frac{ix^2}{4}\sigma_3 - \frac{xu_x}{4\lambda}\sigma_1 + \frac{1}{\lambda^2}\left(\sigma_2\sinh u - i\sigma_3\cosh u\right)\right\}\Psi, \tag{15.1.1}$$

where as usual $\Psi = \Psi(\lambda, x, u) - 2\times 2$ is a complex-valued matrix, σ_1, σ_2, σ_3 are the three Pauli matrices and x, $u = u(x)$ are real-valued parameters.

Near the irregular points $\lambda = 0$ and $\lambda = \infty$ we define two solutions Ψ and Φ of the system (15.1.1), specified by the following asymptotics:

$$\Psi \to \exp\left(-\frac{ix^2}{4}\sigma_3\lambda\right), \quad \lambda \to \infty, \tag{15.1.2}$$

$$\Phi \to i\left(\sinh\frac{u}{2} - \sigma_1\cosh\frac{u}{2}\right)\exp\left(-\frac{i\sigma_3}{\lambda}\right), \quad \lambda \to 0. \tag{15.1.3}$$

The matrix Q connects the solutions Ψ and Φ,

$$\mathrm{Q} = \Phi^{-1}\Psi, \tag{15.1.4}$$

while the Stokes matrices arise for each of the solutions Ψ and Φ as we circle around infinity and the origin respectively. Let $\Psi_1(\Psi_2)$ be the solution of equation (15.1.1) in the upper (lower) half-plane λ with asymptotics (15.1.2), while $\arg\lambda = 0$ ($\arg\lambda = \pi$). The pair of the solutions Φ_1, Φ_2 is defined in a similar way using the asymptotics (15.1.3). Then, (see Chapter 13)

$$\begin{aligned} \Psi_1(\lambda) &= \Psi_2(\lambda)\begin{pmatrix} 1 & 0 \\ \mathrm{T} & 1 \end{pmatrix}, \\ \Phi_1(\lambda) &= \Phi_2(\lambda)\begin{pmatrix} 1 & \mathrm{T} \\ 0 & 1 \end{pmatrix}, \end{aligned} \tag{15.1.5}$$

where T is a Stokes multiplier, which is obviously independent of λ. In our case it is also independent of x; the connection matrix Q is also independent of x.

THEOREM 15.1. *The function $u = u(x)$ satisfies the Painlevé III equation (15.0.1) if the connection matrix (15.1.4) Q, and the Stokes multiplier T, defined by equations (15.1.5), do not depend on x.*

For the proof see Chapter 5, also see [**FN1**] were this result was first established.

Knowing the monodromy data Q and T, we can solve the inverse problem, i.e., we can find the solution Ψ of equation (15.1.1), by solving a singular integral equation. Then, the derivative u_x of the Painlevé function can be obtained in terms of Ψ by the formula

$$u_x(x) = 2ix\lim_{\lambda\to\infty}\lambda\Psi_{12}(\lambda,x)\exp\left(-\frac{ix^2}{4}\lambda - \frac{i}{\lambda}\right), \tag{15.1.6}$$

which coincides with formula (13.2.7) in Chapter 5 (Ψ_{12} denotes, as usual, the "12" entry of matrix Ψ).

REMARK 15.1. The monodromy data Q and T represent the global invariants of the Painlevé function $u(x)$ even when x becomes complex-valued (see [**FN1**]). The function $w = \exp\frac{u}{2}$ must be meromorphic for $|x| > 0$, according to the Painlevé property of equation (15.0.4). Thus, by exponentiating the asymptotics (15.0.6), we obtain the first terms of the Laurent series at the pole $x = a$. The parameters a and b appearing in its coefficients, are uniquely determined by the data Q and T. This means that having Q and T, we can expand the solution u beyond the pole.

The obvious symmetries of equation (15.1.1) and the reality of $u(x)$ yield the constraints[1] on the complex-valued parameters T, Q_{ij}, $i,j=1,2$,

$$\mathsf{Q}_{11}=\mathsf{Q}_{22}=-\overline{\mathsf{Q}}_{11},\quad \mathsf{Q}_{12}=-\overline{\mathsf{Q}}_{21}, \tag{15.1.7}$$

$$\mathsf{T}=-\overline{\mathsf{T}},\quad \mathsf{Q}_{11}\mathsf{T}=\mathsf{Q}_{21}-\mathsf{Q}_{12}, \tag{15.1.8}$$

(compare with equations (13.1.31), (13.1.32) in Chapter 13). The constraints (15.1.7) and (15.1.8) imply that their exist only two arbitrary real parameters in Q and T. For example, we can choose $|\mathsf{Q}_{11}|$ and $\arg\mathsf{Q}_{12}$, or $|\mathsf{Q}_{11}|$ and T.

In fact we have calculated the matrix Q through the initial data (15.0.2), r and s of Chapter 13 (equation (14.1.23)). It is easy to prove that all calculations in Chapter 13 remain true, after the transform $u\mapsto\pi-iu$, $x\mapsto\frac{x}{2}$, where $|\mathrm{Re}\,r|<2$ in equation (15.0.2). Thus, we obtain the following the result for the connection matrix,

$$\mathsf{Q}=\frac{1}{2\pi i}\begin{pmatrix} A-B & Ae^{-\frac{i\pi r}{4}}+Be^{\frac{i\pi r}{4}} \\ Ae^{\frac{i\pi r}{4}}+Be^{-\frac{i\pi r}{4}} & A-B\end{pmatrix}, \tag{15.1.9}$$

where

$$A=2^{r}e^{\frac{s}{2}}\Gamma^{2}\left(\frac{1}{2}+\frac{r}{4}\right),\quad B=2^{-r}e^{-\frac{s}{2}}\Gamma^{2}\left(\frac{1}{2}-\frac{r}{4}\right).$$

For the solution $u(x)$ which is regular near the origin, ($u(0)=s$, $u_x(0)=0$, or $r=0$ in equation (15.0.2)) the matrix Q simplifies to

$$\mathsf{Q}=-i\begin{pmatrix}\sinh\frac{s}{2} & \cosh\frac{s}{2}\\ \cosh\frac{s}{2} & \sinh\frac{s}{2}\end{pmatrix}. \tag{15.1.10}$$

The Stokes multiplier T is determined from equations (15.1.8) and (15.1.9), and takes the form

$$\mathsf{T}=2i\sin\frac{\pi r}{4}. \tag{15.1.11}$$

Finally, let us write the integral equation, expressing the solution of equation (15.1.1). Denoting by Ψ_1 and Ψ_2 the columns of the matrix Ψ, we have

$$\Psi_1(\lambda,s)\,e^{\theta}=\begin{pmatrix}1\\0\end{pmatrix}+\frac{\mathsf{T}}{2\pi i}\int_{-\infty}^{0}\frac{\Psi_2(\xi,x)}{\lambda-\xi}e^{\theta}d\xi +\frac{\mathsf{Q}_{21}}{2\pi i\mathsf{Q}_{22}}\int_{-\infty}^{\infty}\frac{\Psi_2(\xi,x)}{\lambda-\xi}e^{\theta}d\xi, \tag{15.1.12}$$

where $\theta=\frac{ix^2}{4}\lambda+\frac{i}{\lambda}$, $\pi\leqslant\arg\lambda<2\pi$ (see (13.2.4)). Equation (15.1.12) can be obtained by the Cauchy formula to the function $\Psi_2 e^{\theta}$, which is analytic in the lower half-plane and by using equations (15.1.5), (15.1.7) and (15.1.8).

---

[1] In the notation of Chapter 13 we have that $\mathsf{T}=\overline{p}-p$, $\mathsf{Q}_{11}=1/\sqrt{1-|p|^2}$, $\mathsf{Q}_{12}=p/\sqrt{1-|p|^2}$. According to Remark 13.1 we are dealing with a submanifold of PIII solutions for which $p=-\overline{q}$, $|p|>1$.

2. Ψ-function asymptotics at a pole

We will study here the asymptotic expansions of the Ψ-function, the solution of equation (15.1.1), under the assumption that $u = u(x)$ has the following asymptotics,

$$u(x) = -2\ln\varepsilon - \frac{\varepsilon}{a} - \frac{b\varepsilon^2}{a^2} + O\left(\varepsilon^2\right), \tag{15.2.1}$$

where $\varepsilon = x - a$, $\varepsilon \to 0$.

We will first determine the leading term of the Ψ-function as $\varepsilon \to 0$. Let $(\psi_1, \psi_2)^T$ be the first column of the matrix Ψ. Introducing the new variables

$$w_1 = \psi_1 + \psi_2, \quad w_2 = \psi_2 - \psi_1,$$

then the system (15.1.1) takes the form

$$\begin{cases} \frac{dw_1}{d\lambda} = -\frac{xu_x}{4\lambda} w_1 + \left(\frac{ix^2}{4} + \frac{ie^{-u}}{\lambda^2}\right) w_2, \\ \frac{dw_2}{d\lambda} = \left(\frac{ix^2}{4} + \frac{ie^{u}}{\lambda^2}\right) w_1 + \frac{xu_x}{4\lambda} w_2. \end{cases} \tag{15.2.2}$$

Expanding the above coefficients in the power series of ε, we find

$$\frac{ix^2}{4} = \frac{ia^2}{4} + \frac{ia\varepsilon}{2} + \frac{i\varepsilon^2}{4},$$

$$\frac{xu_x}{4} = -\frac{a}{2\varepsilon} - \frac{3}{4} - \frac{\varepsilon(1+2b)}{4a} + O\left(e^2\right),$$

$$e^u = \frac{2}{\varepsilon^2} - \frac{1}{a\varepsilon} + \frac{1-2b}{2a^2} + O(\varepsilon),$$

$$e^{-u} = O\left(e^2\right).$$

Taking the derivative of the first equation (15.1.1) with respect to λ, and expressing w_2 through the second equation, we transform the system (15.1.1) into a second order scalar equation for w_1,

$$\frac{d^2w_1}{d\lambda^2} + \left(\frac{x^4}{16} + \frac{\frac{1}{4}-\nu^2}{\lambda^2} + \frac{1}{\lambda^4}\right) w_1 = O\left(\varepsilon\lambda^{-2}w_1 + \varepsilon^2\lambda^{-4}\frac{dw_1}{d\lambda}\right), \tag{15.2.3}$$

where $\nu^2 = \frac{3}{4}b + \frac{3}{8}$. The function w_2 satisfy the equation

$$w_2 = -\frac{4i}{a^2}\left[\frac{dw_1}{d\lambda} - \left(\frac{a}{2\varepsilon\lambda} + \frac{3}{4\lambda}\right) w_1\right] + O\left(\varepsilon\lambda^{-1}\right).$$

Define a pair of functions W and $\overline{W}$ as the basis of solutions for the equation

$$\frac{d^2W}{d\lambda^2} + \left(\frac{x^4}{16} + \frac{\frac{1}{4}-\nu^2}{\lambda^2} + \frac{1}{\lambda^4}\right) W = 0, \tag{15.2.4}$$

with the following asymptotics at infinity

$$W \to \exp\left(\frac{ix^2}{4}\lambda\right), \tag{15.2.5}$$

$$\overline{W} \to \exp\left(-\frac{ix^2}{4}\lambda\right), \quad \lambda \to \infty, \quad \arg\lambda = 0.$$

Then, for the components of the first column we find the asymptotic formulas

(15.2.6)
$$\begin{aligned}\psi_1 &= \frac{2i}{x^2}\frac{d\overline{W}}{d\lambda} + \left(\frac{3i}{2a^2\lambda} - \frac{i}{\varepsilon a\lambda} + \frac{1}{2}\right)\overline{W} + O\left(\varepsilon\lambda^{-1} + \varepsilon^2\lambda^{-3}\right),\\ \psi_2 &= \frac{2i}{x^2}\frac{d\overline{W}}{d\lambda} + \left(-\frac{3i}{2a^2\lambda} + \frac{i}{\varepsilon a\lambda} + \frac{1}{2}\right)\overline{W} + O\left(\varepsilon\lambda^{-1} + \varepsilon\lambda^{-3}\right).\end{aligned}$$

It is easy to prove that these components of the Ψ-function satisfy the asymptotics (15.1.2) as $\lambda \to \infty$. The second column of Ψ-function is constructed in a similar way. Finally, the leading term in ε for Ψ is obtained in the form

(15.2.7)
$$\Psi = \begin{pmatrix} \psi_1 & \overline{\psi}_2 \\ \psi_2 & \overline{\psi}_1 \end{pmatrix},$$

where ψ_1, ψ_2 are defined by equation (15.2.6).

Let us proceed now to the matrix Φ — the solution of equation (15.1.1) with asymptotics (15.1.3). We perform the same calculations, doing the transformation

$$\Phi(\lambda) = P\chi(\xi), \quad \xi = \frac{4}{x^2\lambda}, \quad P = i\left(\sinh\frac{u}{2} - \sigma_1\cosh\frac{u}{2}\right).$$

Then, for the matrix χ we find the system

$$\frac{d\chi}{d\xi} = \left\{-\frac{ix^2}{4}\sigma_3 + \frac{xu_x}{4\xi}\sigma_1 + \frac{1}{\xi^2}\left(\sigma_2\sinh u - i\sigma_3\cosh u\right)\right\}\chi,$$

which differs from equation (15.1.1) only by the sign in the second term of the right-hand side. Calculations similar to those performed above, yield the following asymptotic expression as $\varepsilon \to 0$,

(15.2.8)
$$\Phi = P\begin{pmatrix} \chi_1 & \overline{\chi}_2 \\ \chi_2 & \overline{\chi}_1 \end{pmatrix},$$

where

(15.2.9)
$$\begin{aligned}\chi_1 &= \frac{2i}{x^2}\frac{d\overline{W}}{d\xi} - \left(\frac{3i}{2a^2\xi} + \frac{i}{\varepsilon a\xi} - \frac{1}{2}\right)\overline{W} + O\left(\varepsilon\xi^{-1} + \varepsilon^2\xi^{-3}\right),\\ \chi_2 &= \frac{2i}{x^2}\frac{d\overline{W}}{d\xi} - \left(\frac{3i}{2a^2\xi} + \frac{i}{\varepsilon a\xi} + \frac{1}{2}\right)\overline{W} + O\left(\varepsilon\xi^{-1} + \varepsilon^2\xi^{-3}\right).\end{aligned}$$

Here $W = W(\xi)$ is the solution of equation (15.2.4) in the variable ξ, where $\xi = \frac{4}{x^2\lambda}$.

In what follows, we will establish the domains of applicability of the formal asymptotics (15.2.7) and (15.2.8).

THEOREM 15.2. *Let Ψ and Φ be the solutions of equation (*15.1.1*) with the asymptotics (*15.1.2*), (*15.1.3*). Let $u = u(x)$ satisfy the asymptotic condition (*15.2.1*). Then, Ψ and Φ have the asymptotics given by equations (*15.2.7*) and (*15.2.8*), where the error is of $O\left(\varepsilon\lambda^{-1} + \varepsilon^2\lambda^{-3}\right)$ for Ψ and of $O\left(\varepsilon\lambda + \varepsilon^2\lambda^3\right)$ for Φ.*

PROOF. Equation (15.2.3) is equivalent to the following integral equation,

$$\overline{w}_1 = \exp\left(-\frac{ix^2}{4}\lambda\right) + \frac{2}{x}\int_\lambda^\infty \sin\frac{x^2}{4}(\lambda-\zeta) \tag{15.2.10}$$

$$\times\left\{\left[p(\zeta)+\varepsilon\zeta^{-2}g(\zeta,\varepsilon)\right]\overline{w}_1(\zeta)+\varepsilon^2\zeta^{-4}f(\zeta,\varepsilon)\,w_{1_x}(\zeta)\right\}d\zeta,$$

where $p(\lambda) = \frac{\nu^2-\frac{1}{4}}{\lambda^2} - \frac{1}{\lambda^4}$, $|f|$, $|g| < \text{const}$.

Let $\overline{w}_1 = \overline{W} + \varepsilon M$, where W is the solution of equations (15.2.4), (15.2.5). Then $\overline{W}$ satisfies equation (15.2.10) when $\varepsilon = 0$, therefore for the remainder M satisfies the estimate

$$|M| < C\left(\varepsilon\lambda^{-1}+\varepsilon^2\lambda^{-3}\right).$$

Estimates for the components of Φ can be derived in a similar way. □

3. Mathieu equation

Consider the equation for the leading term of the Ψ-function

$$\frac{d^2W}{d\lambda^2} + \left(\frac{a^4}{16}+\frac{\frac{1}{4}-\nu^2}{\lambda^2}+\frac{1}{\lambda^4}\right)W = 0. \tag{15.3.1}$$

We will study the solutions of this equation in the sectors $\Omega'_j = \{\lambda : \pi j \leqslant \arg\lambda < \pi(j+1)\}$, $j = 0, 1$. We will note that the Stokes multiplier T_0, connecting its solutions in the sectors Ω'_j, and the matrix Q_0, connecting solutions around irregular points $\lambda = 0$ and $\lambda = \infty$, defined in the sectors Ω'_0 and Ω'_1 respectively, are actually the leading terms, as $\varepsilon \to 0$, of the corresponding parameters of the equation (15.1.1). Equation (15.3.1) is simply transformed into the Mathieu equation, therefore, the parameters T_0 and Q_0 will be expressed through the Mathieu functions.

It is convenient to perform the relevant calculation for the connection matrix Q_0 on the real axis $\arg\lambda = 0$.

The basis of solutions of equation (15.3.1), W and $\overline{W}$ is introduced by the asymptotic condition,

$$\begin{aligned} W &\to \exp\left(\frac{ia^2}{4}\lambda\right), \\ \overline{W} &\to \exp\left(-\frac{ia^2}{4}\lambda\right), \quad \lambda\to+\infty. \end{aligned} \tag{15.3.2}$$

Equation (15.3.1) is invariant under the transformation

$$W(\lambda)\to\xi^{-1}W(\xi), \quad \xi = \frac{4}{a^2\lambda}, \tag{15.3.3}$$

so that the pair of functions, $\xi^{-1}W(\xi)$ and $\xi^{-1}\overline{W}(\xi)$ also constitute a basis for equation (15.3.1). The old basis can be expressed in terms of the new basis by the equation

$$W(\lambda) = \frac{a}{2}\lambda\left\{p_0 W\left(\frac{4}{a^2\lambda}\right)+q_0\overline{W}\left(\frac{4}{a^2\lambda}\right)\right\}, \tag{15.3.4}$$

where p_0, $q_0 \in \mathbb{C}$.

The asymptotics (15.3.2), together with equation (15.3.4), imply the following representation of W near the origin,

$$W(\lambda)\to\frac{a}{2}\lambda\left(p_0 e^{\frac{i}{\lambda}}+q_0 e^{-\frac{i}{\lambda}}\right), \quad \lambda\to 0, \quad \arg\lambda = 0. \tag{15.3.5}$$

Let us define the matrix Q_0, by

$$Q_0 = \begin{pmatrix} -\overline{p}_0 & -q_0 \\ \overline{q}_0 & p_0 \end{pmatrix}, \tag{15.3.6}$$

then from equation (15.3.4) we obtain

$$\left(\lambda^{-1}\overline{W}(\lambda),\quad \lambda^{-1}W(\lambda)\right) = \left(-\overline{W}\left(\frac{4}{a^2\lambda}\right),\quad W\left(\frac{4}{a^2\lambda}\right)\right)\frac{a}{2}Q_0. \tag{15.3.7}$$

This means that the matrix (15.3.6) is the connection matrix for equation (15.3.1) (according to the definition given in Chapter 13, equation (15.1.4)). The following theorem establishes the properties of Q_0.

THEOREM 15.3. *The matrix* Q_0 *defined by equation (*15.3.6*), satisfies the following identities:*

1) $\det Q_0 = |q_0|^2 - |p_0|^2 = 1$,
2) $p_0 = -\overline{p}_0$,
3) $\arg W\left(\frac{2}{a}\right) = \frac{1}{2}\arg\frac{p_0 - 1}{q_0}$.

PROOF. The identity 1) follows directly from the Wronskian conservation theorem of equation (15.3.1).

Letting $\lambda = \frac{2}{a}$ in equation (15.3.4), implies $\overline{W}W^{-1} = (1-p_0)\,q_0^{-1}$, which yields 3). Taking the absolute value of both sides of the latter equation, we find $|1-p_0|^2 = |q_0|^2$. Solving this equation together with equation 1), the identity 2) follows, and the theorem is proved. □

Let us turn now to the calculation of the Stokes multiplier, which can be defined from the identity

$$\left(W\left(\lambda e^{-i\pi}\right), \overline{W}\left(\overline{\lambda}e^{i\pi}\right)\right) = \left(\overline{W}\left(\overline{\lambda}\right), W(\lambda)\right)\begin{pmatrix} 1 & 0 \\ -T_0 & 1 \end{pmatrix}, \quad \lambda \in \overline{\Omega}_0'. \tag{15.3.8}$$

It is convenient to introduce the new variables

$$\lambda = \frac{2}{a}e^{i\zeta}, \quad W(\lambda) = \sqrt{\lambda}\,v(\zeta). \tag{15.3.9}$$

The transformation (15.3.9) maps equation (15.3.1) into the Mathieu equation

$$\frac{d^2v}{d\zeta^2} + \left(\nu^2 - \frac{a^2}{2}\cos 2\zeta\right)v = 0. \tag{15.3.10}$$

Applying the Floquet theorem to equation (15.3.10), we can represent the function $W(\lambda)$ in the form

$$W(\lambda) = e^{\frac{ir}{2}}\left(e^{i\mu\zeta}y_1(\zeta) + e^{-i\mu\zeta}y_2(\zeta)\right), \tag{15.3.11}$$

where $y_{1,2}$ are π- periodic functions. From (15.3.11) we have that

$$W\left(\lambda e^{i\pi}\right) - W\left(\lambda e^{-i\pi}\right) = 2i\cos\mu\pi W(\lambda).$$

Comparing this equality with (15.3.8), we conclude that

$$T_0 = 2i\cos\pi\mu. \tag{15.3.12}$$

The last formula coincides with the result of [**McTW**], where the Stokes multipliers for the Mathieu equation were constructed.

We are now ready to compare the connection matrices (15.3.6) and (15.1.4), as well as the Stokes multipliers (15.3.8) and (15.1.5).

THEOREM 15.4. *Let* Q *and* T *be the connection matrix and the Stokes multiplier for equation* (15.1.1) *respectively, where* $u(x)$ *satisfies the asymptotics* (15.2.1). *Then, the following estimates are valid*

$$\mathsf{Q}-\mathsf{Q}_0=O(\varepsilon), \quad \mathsf{T}-\mathsf{T}_0=O(\varepsilon), \tag{15.3.13}$$

where Q_0, T_0 *are defined for the Mathieu equation* (15.3.1) *by equations* (15.3.6) *and* (15.3.12).

PROOF. According to the Theorem 15.2, there exists a circular domain $\varepsilon^{\frac{2}{3}+\delta}<|\lambda|<\varepsilon^{-\frac{2}{3}-\delta}$, $\delta>0$, where the solution Ψ is represented by equation (15.2.8). Then, equation $\Psi=\Phi\mathsf{Q}$ can be written in the form

$$\frac{i}{a\lambda}\begin{pmatrix} -\overline{W}(\lambda), & -W(\lambda) \\ \overline{W}(\lambda), & W(\lambda)\end{pmatrix}+O(\varepsilon) = \left\{-\frac{i}{2}\begin{pmatrix} -\overline{W}(\xi), & W(\xi) \\ \overline{W}(\xi), & -W(\xi)\end{pmatrix}+O(\varepsilon)\right\}\mathsf{Q}.$$

Expressing the left-hand side through the variable ξ by equation (15.3.4), and making use of equation (15.3.7), we find the estimate

$$\mathsf{Q}=\mathsf{Q}_0+O(\varepsilon),$$

which proves the first statement of the theorem.

In order to derive the second estimate (15.3.13), let us turn to the integral equation (15.2.10); the solution of this equation can be written in the form

$$\overline{w}_1(\lambda)=\left\{e^{-\frac{ia^2}{4}\lambda}+\theta_1(\lambda)\,\mathsf{T}(\varepsilon)\,e^{\frac{ia^2}{4}\lambda}\right\}(1+o(1)), \ |\lambda|\to\infty, \tag{15.3.14}$$

where

$$\theta_1(\lambda)=\begin{cases} 0, & 0\le\arg\lambda<\pi, \\ 1, & \pi\le\arg\lambda<2\pi. \end{cases}$$

Substituting the ansatz (15.3.14) into equation (15.2.10), we can compare terms multiplied by equal powers of λ and equal exponents. On the boundary of the sector Ω_0', the following asymptotics formula is valid:

$$\begin{aligned}&\int_\lambda^\infty\left(1-e^{\frac{ia^2}{2}(\lambda-\zeta)}\right)\zeta^{-k-1}d\zeta \\ &=\begin{cases} \frac{\lambda^{-k}}{k}(1+o(1)), & \arg\lambda\to 0, \\ \left[\frac{\lambda^{-k}}{k}+\frac{2\pi i}{k!}\left(\frac{a^2}{2i}\right)^k e^{\frac{ia^2}{2}\lambda}\right](1+o(1)), & \arg\lambda\to\pi. \end{cases}\end{aligned} \tag{15.3.15}$$

This can be derived by applying the integration by parts and deforming the contour of integration, so that the residue at the origin gives the nonvanishing term as $\arg\lambda \to \pi$, $|\lambda| \to \infty$.

The last two integrals in equation (15.2.10), using the asymptotics (15.3.15), are of order $O(\varepsilon)$, while $|\lambda| \to \infty$, $0 \leqslant \arg\lambda \leqslant \pi$. Thus, the Stokes multiplier $\mathsf{T}(\varepsilon)$ satisfies the equation

$$\mathsf{T}(\varepsilon) = \mathsf{T}_0 + O(\varepsilon),$$

where T_0 is the Stokes multiplier of equation (15.1.1) as $\varepsilon = 0$. However, the solution of equation (15.2.10) with $\varepsilon = 0$, coincides with $\overline{W}(\lambda)$, so T_0 is the Stokes multiplier for the Mathieu equation (15.3.1), and the theorem is proved. □

It is useful to transform equation (15.3.1) into the modified Mathieu equation. This can be achieved by a change of variables similar to that used for equation (15.3.9),

$$\lambda = \frac{2}{a}\mathrm{e}^{z}, \quad W(\lambda) = \sqrt{\lambda}V(z), \tag{15.3.16}$$

where $V = V(z, a, \nu)$, satisfies the modified Mathieu equation [**McTW**],

$$\frac{d^2V}{dz^2} - \left(\nu^2 - \frac{a^2}{2}\cosh 2z\right)V = 0. \tag{15.3.17}$$

The boundary conditions (15.3.2) and (15.3.5), under the transformation (15.3.16) are mapped into the following asymptotic conditions:

$$V(z) \to \mathrm{e}^{-\frac{z}{2}}\exp\left(\frac{ia}{\sqrt{2}}\mathrm{e}^{z}\right), \quad z \to +\infty, \tag{15.3.18}$$

$$V(z) \to \mathrm{e}^{\frac{z}{2}}p_0\exp\left(\frac{ia}{\sqrt{2}}\mathrm{e}^{-z}\right) + q_0\mathrm{e}^{\frac{z}{2}}\exp\left(-\frac{ia^2}{\sqrt{2}}\mathrm{e}^{-z}\right), \quad z \to -\infty. \tag{15.3.19}$$

The value $\lambda = \frac{2}{a}$ corresponds to $z = 0$, so that $\arg W\left(\frac{2}{a}\right) = \arg V(0, a, \nu)$.

THEOREM 15.5. *Let $V = V(z, a, \nu)$ be the solution of the modified Mathieu equation (*15.3.17*) with the asymptotics (*15.3.18*). Then the parameters a and ν satisfy the equations*

$$\arg V(o, a, \nu) = \frac{1}{2}\arg\frac{\mathsf{Q}_{11} - 1}{\mathsf{Q}_{12}}, \tag{15.3.20}$$

$$2i\cos\pi\mu = \mathsf{T}, \tag{15.3.21}$$

*where $\mu = \mu(a, \nu)$ is the Floquet exponent for the solution $V(iz, a, \nu)$ of equation (*15.3.10*), and $\mathsf{T}, \mathsf{Q} = (\mathsf{Q}_{ij})$, are respectively the Stokes multiplier and the connection matrix for equation (*15.1.1*).*

PROOF. The proof follows immediately from Theorem 15.4, equation (15.3.12), and the property 3) of the Theorem 15.3.

Finally, let us estimate the difference between the Painlevé function satisfying equations (15.0.1) and (15.0.2) and its asymptotics (15.0.6) as $\varepsilon = x - a \to 0$.

THEOREM 15.6. *Let $u = u(x)$ be the solution of equations (*15.0.1*) and (*15.0.2*) associated with the monodromy parameters T and Q (equations (*15.1.11*) and (*15.1.4*)). Then, there exist constants a and $b = \frac{4}{3}\nu^2 - \frac{1}{2}$, satisfying equations (*15.3.20*) and (*15.3.21*), such that $u(x)$ satisfies the asymptotics (*15.0.6*) holds.*

PROOF. Let us construct the solution of the Mathieu equation (15.3.17) with the boundary conditions (15.3.21) and (15.3.19), assuming $\mathrm{T} = \mathrm{T}_0$, $\mathrm{Q} = \mathrm{Q}_0$, where T_0 and Q_0 are the monodromy parameters defined in (15.3.6) and (15.3.12). The existence of such a solution V can be established by applying the Gelfand-Levitan-Marchenko equation to the original equation (15.3.1). Thus, together with V we have found the parameters a and ν satisfying the functional equations (15.3.20) and (15.3.21). As it was proved by Paul Painlevé in the classical work [**P**], there exists a solution $\hat{u}(x)$ of the PIII equation (15.2.1) with a and $b = \frac{4}{3}\nu^2 - \frac{1}{2}$. Substituting this solution into the system (15.1.1), and calculating the asymptotics of the Ψ-function as $\varepsilon \to 0$, we conclude, according to Theorem 15.4, that the corresponding monodromy data $\widehat{\mathrm{T}}$ and $\widehat{\mathrm{Q}}$ satisfy the estimates

$$\begin{aligned} \widehat{\mathrm{T}} - \mathrm{T} &= \widehat{\mathrm{T}}_0 - \mathrm{T}_0 = O(\varepsilon), \\ \widehat{\mathrm{Q}} - \mathrm{Q} &= \widehat{\mathrm{Q}}_0 - \mathrm{Q}_0 = O(\varepsilon). \end{aligned}$$

On the other hand, the isomonodromic condition implies that $\widehat{\mathrm{T}}$ and $\widehat{\mathrm{Q}}$ are independent of x, i.e., of $\varepsilon = x - a$. This yields immediately the equalities

$$\widehat{\mathrm{T}} = \mathrm{T}, \quad \widehat{\mathrm{Q}} = \mathrm{Q}.$$

Therefore, the newly constructed solution $\widehat{u}(x)$ has the prescribed asymptotics (15.2.1) with the parameters a and ν defined by equations (15.3.20) and (15.3.21). □

4. Asymptotic distribution of poles

The functional equations (15.3.20) and (15.3.21) satisfied by the parameters a and ν appear to be transcendent, i.e., apparently there is no explicit solution for any choice of T and Q. In fact, for any fixed Floquet exponent μ, there exist an equation on a and ν involving an infinite chain fraction (Hill's determinant) ([**BE**], p. 143). However, we can derive some qualitative properties from equations (15.3.20) and (15.3.21) concerning the distribution of poles, and we can calculate their asymptotics for large a. We will also prove that $x = \infty$ is the only accumulation point of the poles of PIII function with the initial condition (15.0.2).

Let us first calculate the asymptotic solution of the Mathieu equation (15.3.17) for large a. It is convenient to rewrite this equation in the form of a Sturm-Liouville equation,

$$\frac{d^2V}{dz^2} + a^2\left(\frac{1}{2}\cosh 2z - \frac{\nu^2}{a^2}\right)V = 0. \tag{15.4.1}$$

This equation, together with the boundary conditions (15.3.18) and(15.3.19) can be interpreted as a problem of one-dimensional scattering of short waves for the potential

$$U(z) = \frac{1}{2}\cosh 2z - \frac{\nu^2}{a^2}. \tag{15.4.2}$$

The asymptotic solution of this problem is well-known, and it is given by the classical WKB-method (see, for example, [**Fe2**]). The solution V can be represented in the form

$$V(z) \sim \begin{cases} U^{-\frac{1}{4}}(z)\exp\left\{ia\int_{z_0}^{z}\sqrt{U(\zeta)}d\zeta\right\} & z > z_0, \\ U^{-\frac{1}{4}}(z)\left[p_0\exp\left\{ia\int_{z}^{-z_0}\sqrt{U(\zeta)}d\zeta\right\}\right. & \\ \left.+q_0\exp\left\{-ia\int_{z}^{-z_0}\sqrt{U(\zeta)}d\zeta\right\}\right] & z < z_0. \end{cases} \tag{15.4.3}$$

The WKB-solutions (15.4.3) are valid everywhere outside the neighborhoods of turning point, which are the zeros of potential $U(z)$. Let us introduce the notation

$$\nu^2 = \frac{a^2}{2} + 2ax, \tag{15.4.4}$$

where the value x is assumed to be finite and independent of a. Then, the turning points denoted by $\pm z_0$ lie closely to each other:

$$z_0^2 = \frac{2x}{a} + O\left(a^{-3}\right), \quad a \to \infty.$$

The scattering data p_0 and q_0 in equation (15.4.3) are expressed explicitly through the values x an φ by the equations

$$p_0 = -ie^{\pi x}\left(1 + O\left(a^{-1}\right)\right), \quad \operatorname{Im} p_0 < 0, \tag{15.4.5}$$

$$q_0 = \frac{\sqrt{2\pi}\exp\left[\frac{\pi x}{2} + ix\left(\ln|x| - 1\right) - 2i\varphi\right]}{\Gamma\left(\frac{1}{2} + ix\right)}\left(1 + O\left(a^{-1}\right)\right), \tag{15.4.6}$$

where

$$\varphi = \frac{a}{2}\int_{z_0}^{\infty}\left\{\sqrt{2\cosh 2z - 1 - 4xa^{-1}} - e^{z}\right\}dz.$$

The formulae (15.4.5) and (15.4.6) remain true for any $x \in \mathbb{R}$. They can be obtained by matching the WKB-solutions (15.4.3) at the neighborhood of turning points $-z_0,\ z_0$, with a solution that can be expressed through the parabolic cylinder functions (see the Chapter 14).

The asymptotics of the integral φ as $a \to \infty$, can be calculated by reducing this integral to the elliptic integral of the second kind:

$$\varphi = -a + \frac{x}{2}\left[\ln\frac{2|x|}{a} - 1 - 4\ln 2\right] + O\left(a^{-1}\ln a\right).$$

Then, from equations (15.4.5) and (15.4.6) we find

$$x = \frac{1}{\pi}\ln|p_0|, \quad |p_0| \neq 0, \tag{15.4.7}$$

$$\arg q_0 = 2a - 2\pi n + x\ln a + 3x\ln 2 - \arg\Gamma\left(\frac{1}{2} + ix\right), \tag{15.4.8}$$

where n is an integer, $n \to \infty$.

From (15.4.8), we get the asymptotic formula for the distribution of poles,

$$a_n = \pi n - \frac{x}{2}\ln 8\pi n + \frac{1}{2}\arg\Gamma\left(\frac{1}{2} + ix\right) + \frac{1}{2}\arg q_0 + O\left(\frac{\ln n}{n}\right), \quad n \to \infty. \tag{15.4.9}$$

The asymptotics of ν^2 as $n \to \infty$ can be derived by substitution into (15.4.4) the right-hand side of equation (15.4.8).

Finally, we can express $p_0 = Q_{22}$ and $q_0 = -Q_{21}$ through the initial data r and s using to equation (15.1.9), and so the formula (15.0.10) is proved.

REMARK 15.2. There exists another approach to the asymptotics (15.4.9). Instead of equation (15.4.6), we can use the asymptotic expression for the phase-shift of the WKB-solution near the turning points $\pm z_0$. Then, the resulting formula (15.4.9) involves the value $\mathsf{T} = (q_0 + \overline{q}_0)\, p_0^{-1}$, instead of $\arg q_0$.

REMARK 15.3. The connection formula (15.4.5), fails when the imaginary part of p_0 becomes positive. However, this case can be reduced to the previous one by the simple transformation $u \mapsto -u$ which preserves equation (15.0.1). Clearly, from equations (15.1.9) and (15.3.6) we find the following the transformation for monodromy data,

$$p \mapsto -p,\ q \mapsto -q.$$

The formalism of the isomonodromic deformations method, applied to equation (15.0.1) with u replaced by $-u$, leads again to the Mathieu equation (15.4.1). The connection formulas (15.4.5) and (15.4.6) have now the proper signs, so the asymptotic (15.4.9) is valid. The points a_n in (15.4.9) are now the poles of the inverse function $w = \exp\left(\frac{u}{2}\right)$, which satisfies equation (15.0.4).

REMARK 15.4. For the limiting case[2] $p_0 \to 0$ equation (15.4.5) yields $x \to -\infty$, so the coordinates of poles a_n, with n fixed, tend to infinity. Equation $p_0 = 0$, is equivalent, according to (15.0.10), to the condition

$$e^{\frac{s}{2}} = 2^{-r}\Gamma\left(\frac{1}{2} - \frac{r}{4}\right)\Gamma^{-1}\left(\frac{1}{2} + \frac{r}{4}\right),$$

which shows that for $x > 0$, there exists a smooth solution $u = u(x)$ of equation (15.0.1) (see [**McTW**]). Thus, the limiting case $p_0 \to 0$, leads to the "deportation" of poles to infinity.

THEOREM 15.7. *The poles of Painlevé function* $w = \exp\frac{u}{2}$, *with the initial condition* (15.0.2), *have the asymptotic distribution* (15.4.9) *as* $x \to \infty$. *No other point* $x \in \mathbb{R}$ *can be the limit point of the set of poles of this function.*

PROOF. It is necessary to prove only the second statement of the theorem. Suppose that there exists a sequence of poles $a_n \to a^* < \infty$. Let us consider the spectrum points ν_n of the periodic Mathieu equation (15.3.9) at the fixed value of the Floquet exponent $\mu(\mathsf{T} = 2i\cos\pi\mu)$. The values $a = a_n$ are bounded, but the values of ν_n, satisfying the equation $\mu = \mu\,(a_n, \nu_n)$, go to infinity as $n \to \infty$ ([**BE**], p. 163). Then for the modified Mathieu equation (15.3.17),

$$\frac{d^2V}{dz} - \nu^2\left(\frac{a^2}{2\nu^2}\cosh 2z - 1\right)V = 0$$

we can write again the WKB-solutions (15.4.3), where a is replaced by ν and the potential satisfies $U(z) = \frac{a^2}{2\nu^2}\cosh 2z - 1$ as $\nu \to \infty$. It can be shown that the

[2] In terms of the monodromy data p, $q = -\overline{p}$, this corresponds to the case $|p| \to \infty$, $\operatorname{Im} p = 0$.

connection formulas (15.4.5) and (15.4.6) remain valid also in this case, where the turning points lie far apart,

$$z_0^2 \simeq \frac{2x}{a}, \quad x \simeq \nu^2,$$

according to (15.4.4).

Thus, equation (15.4.5) does not hold as $\nu_n \to \infty$ because the value $|p_0| = \exp \pi x$ remains fixed and finite. This contradiction proves the theorem. □

CHAPTER 16

Asymptotics in the Complex Plane of the SG-PIII Transcendent

0. Introduction

This chapter, which is similar to Chapter 7, presents the asymptotic study in the complex plane, of a particular case of PIII transcendent, namely of the SG-PIII (13.0.5). We start with the formalism of the Boutroux ansatz, and present an averaging procedure for determining the modulus and the phase shift of the elliptic function, appearing in the Boutroux ansatz. Next we use the direct monodromy problem to connect the phase shift with the monodromy data; this can be achieved by evaluating the zeroes of PIII and by comparing them with zeroes of the relevant elliptic function.

Consider the sine-Gordon reduction of the PIII equation (SG-PIII),

$$u_{xx} + \frac{1}{x}u_x + \sin u = 0, \quad x, u \in \mathbb{C}, \tag{16.0.1}$$

with the initial condition

$$u(x) = r\ln x + s + O\left(|x|^{2-|\operatorname{Im} r|}\right), \quad x \to 0, \quad |\operatorname{Im} r| < 2. \tag{16.0.2}$$

Let $u(x)$ be the solution of equation (16.0.1) satisfying (16.0.2). As it was shown in Chapter 14, this solution can be parametrized by the monodromy data p and q (see Theorem 14.2),

$$p = \frac{Ae^{-\frac{\pi r}{4}} - Be^{\frac{\pi r}{4}}}{A+B}, \quad q = \frac{Be^{-\frac{\pi r}{4}} - Ae^{\frac{\pi r}{4}}}{A+B}, \tag{16.0.3}$$

where

$$A = 2^{\frac{3ir}{2}} e^{\frac{is}{2}} \Gamma^2\left(\frac{1}{2} + \frac{ir}{4}\right),$$

$$B = 2^{-\frac{3ir}{2}} e^{-\frac{is}{2}} \Gamma^2\left(\frac{1}{2} - \frac{ir}{4}\right).$$

We will compute the asymptotics of $u(x)$ at infinity

$$|x| \to \infty, \quad 0 < \arg x < \frac{\pi}{2},$$

and will express this asymptotics in terms of the monodromy data (16.0.3). The asymptotic procedure, as well as the final answer here are similar to those used in Chapter 7 for PII. We assume that the monodromy data are nondegenerate, which is equivalent to the condition $pq + 1 \neq 0$ (see Chapter 14).

First let us discuss an elementary way for deriving the leading term of asymptotics directly from the SG-PIII equation (16.0.1) itself. For large $|x|$, the second term u_x/x appears to be small, so the leading term is described in terms of the pendulum equation

$$u_{xx} + \sin u = -\frac{u_x}{x}.$$

Fix an initial point x_0 and the Cauchy data

$$u_0 = u(x_0), \quad u_1 = u'(x_0).$$

Define the energy of the pendulum by

$$F(x) = \frac{u_x^2(x)}{2} - \cos u(x),$$

and construct the elliptic function $U(x, F(x_0))$ solving the Cauchy problem

$$\begin{aligned} &U_x^2 - 2\cos U = 2F(x_0), \\ &U(x_0) = u_0, \\ &U_x(x_0) = u_1. \end{aligned} \tag{16.0.4}$$

The solution U of (16.0.4) is the Jacobi sine function

$$\sin(\frac{U(t) - u_0}{2}) = \operatorname{sn}\left((x - x_0)\sqrt{2(F+1)}\middle|\sqrt{\frac{2}{F+1}}\right). \tag{16.0.5}$$

It is clear that

$$\Big|u(x) - U(t - t_0, F(x_0))\Big| = O(|x_0|^{-1})$$

for any x in a disc of finite-radius, which avoids the lattice of poles of U.

Introducing the standard uniformization,

$$\begin{aligned} &V = \cos U, \\ &U_x^2 = F + V, \\ &V_x^2 = 2(F+V)(1 - V^2), \end{aligned} \tag{16.0.6}$$

we choose a cycle $\mathcal{L}$ on a surface $W^2 = 2(F+V)(1-V^2)$, and we construct the sequence $\{x_n\}, \quad n = 0, 1, \dots$ by the recurrent formula

$$\omega_n = \oint_{\mathcal{L}} \frac{dV}{\sqrt{2(F+V)(1-V^2)}},$$

$$x_{n+1} = x_n + \omega_n,$$

$$F(x_{n+1}) = \left(\frac{V_x^2}{2(1-V^2)} - V\right)\Big|_{x=x_{n+1}}.$$

For a cycle $\mathcal{L}$ with nontrivial homology, this sequence, called by Boutroux a "line of periods", tends to the following finite limits:

$$\omega_n = \omega_\infty + O(x_n^{-1}),$$

$$F(x_n) = F_\infty + O(x_n^{-1}),$$

$$\omega_\infty = \oint_{\mathcal{L}} \frac{dV}{\sqrt{2(F_\infty+V)(1-V^2)}}, \tag{16.0.7}$$

where F_∞ satisfies the transcendental equation

$$\oint_{\mathcal{L}} \sqrt{\frac{F+V}{1-V^2}}\,dV = 0. \tag{16.0.8}$$

This result extends the estimate (16.0.5) to a large domain of order $O(|x|)$, $t \gg 1$. Now, the elliptic function (16.0.4) becomes "modulated" with respect to $\phi = \arg x$, which varies by $o(x_{n+1} - x_n)$ as $x_{n+1} - x_n = O(1)$:

$$V(x) = V\big(x - x_\infty(\phi)|F_\infty(\phi)\big). \tag{16.0.9}$$

Here for the modulus F_∞ of the elliptic function the equation (16.0.8) takes the form

$$\arg G_a(F) = \arg G_b(F) = -\phi - \frac{\pi}{2}, \tag{16.0.10}$$

where

$$\phi = \arg x,$$

$$G_a(F) \overset{def}{=} \oint_a \sqrt{\frac{F+V}{1-V^2}}\,dV, \quad G_b(F) \overset{def}{=} \oint_b \sqrt{\frac{F+V}{1-V^2}}\,dV, \tag{16.0.11}$$

and a, b are basic cycles of the elliptic integrals.

Equations (16.0.10) are called the *Boutroux equations.* They can be considered as the modulation equations for the modulus $\varkappa = \varkappa(\phi)$ of the Jacobi sine function (16.0.9),

$$\frac{dF}{d\phi} = -2i(F+1)\frac{E\left(\sqrt{\frac{2}{F+1}}\right)}{K\left(\sqrt{\frac{2}{F+1}}\right)}, \tag{16.0.12}$$

with the initial condition

$$F(0) = 1,$$

which comes from equations (16.0.10) as $\phi = 0$. Here $E(\varkappa)$ and $K(\varkappa)$ denote the standard complete elliptic integrals,

$$\begin{gathered} K = K(\varkappa) = \int_0^1 \frac{dz}{\sqrt{(1-z^2)(1-\varkappa^2 z^2)}}, \quad K' = K(\varkappa'), \\ E = E(\varkappa) = \int_0^1 \sqrt{\frac{1-\varkappa^2 z^2}{1-z^2}}\,dz, \quad E' = E(\varkappa'), \\ \varkappa = \sqrt{\frac{2}{F+1}}, \\ \varkappa' = \sqrt{1-\varkappa^2}, \quad 0 < \arg\varkappa' < \frac{\pi}{2}, \quad -\frac{\pi}{2} < \arg(1+\varkappa^2) < 0. \end{gathered} \tag{16.0.13}$$

One can show that in the first quadrant $0 < \phi < \frac{\pi}{2}$, the system (16.0.10) has a unique smooth solution. The real and the imaginary parts of the function $F = F(\phi)$ are shown in Figure 16.1.

The associated modulus of the elliptic function (16.0.5), is shown in Figure 16.2.

The above Boutroux ansatz will be proved below rigorously with the help of the Isomonodromic Deformation Method. Moreover, in addition to of equations (7.1.7) this will yield a representation for the phase shift (16.0.5).

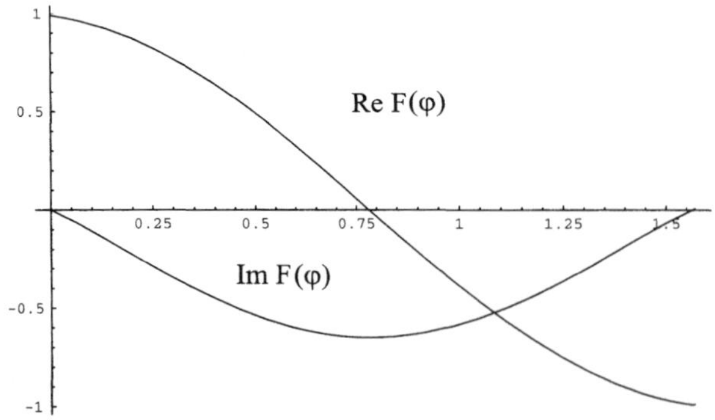

FIGURE 16.1. The energy function F in the first quadrant $0 < \arg x < \pi/2$

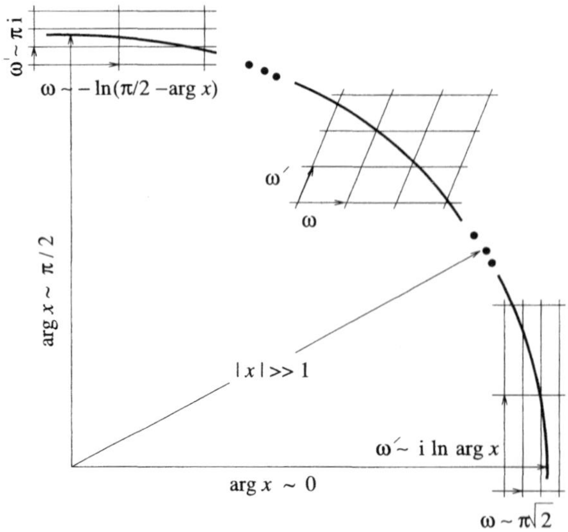

FIGURE 16.2. The deformation of the period lattice for elliptic function V in the first quadrant $0 < \arg x < \pi/2$

THEOREM 16.1. *Any solution $u(x|p,q)$ of the SG-PIII equation (16.0.1) parametrized by the monodromy data (16.0.3) with $p, q \neq 0$ and $pq \neq -1$ has the following asymptotics as $|x| \to \infty$, $0 < \arg x < \pi/2$:*

$$u(x) = 2 \arcsin \left\{ \operatorname{sn} \left(\frac{2(x - x_0(\phi))}{\varkappa(\phi)} \Big| \varkappa(\phi) \right) \right\} + o(1), \tag{16.0.14}$$

where the modulus $\varkappa = \varkappa(\phi)$ of the Jacobi sine satisfies the Whitham equation

$$\frac{d\varkappa}{d\phi} = i\varkappa \frac{E(\varkappa)}{K(\varkappa)}, \quad \varkappa(0) = 1. \tag{16.0.15}$$

Since

$$\varkappa^2 = \frac{2}{F+1},$$

*t he above representation is equivalent to the modulation equation (*16.0.10*) for the full energy $F = F(\phi)$ in (*16.0.4*).*

The phase shift x_0 is given by

$$x_0(\phi) = \frac{i\varkappa(\phi)}{2\pi}\left[K(\phi)\log(-p) - (K(\phi) - iK'(\phi))\log(-1-pq)\right]. \tag{16.0.16}$$

1. Direct monodromy problem

As it was shown in Chapter 13, the SG-PIII equation (16.0.1) is equivalent to

$$\mathsf{U}_\lambda - \mathsf{A}_x + [\mathsf{U}, \mathsf{A}] = 0, \tag{16.1.1}$$

where $\mathsf{U} = \mathsf{U}(u, u', x, \lambda) and \mathsf{A} = \mathsf{A}(u, u', x, \lambda)$ are the 2×2-matrices

$$\mathsf{A} = i\left(\frac{\cos u}{\lambda^2} - \frac{x^2}{16}\right)\sigma_3 - \frac{ixu'}{4\lambda}\sigma_1 + \frac{i\sin u}{\lambda^2}\sigma_2, \tag{16.1.2}$$

$$\mathsf{U} = -\frac{ix\lambda}{8}\sigma_3 - \frac{iu'}{2}\sigma_1, \tag{16.1.3}$$

and $\sigma_{1,2,3}$ are the three Pauli matrices

$$\sigma_1 = \begin{pmatrix} 0 & 1 \\ 1 & 0 \end{pmatrix}, \quad \sigma_2 = \begin{pmatrix} 0 & -i \\ i & 0 \end{pmatrix}, \quad \sigma_3 = \begin{pmatrix} 1 & 0 \\ 0 & -1 \end{pmatrix}.$$

The matrix equations (16.1.1) is the compatibility condition of the Lax pair

$$\begin{cases} \Psi_\lambda = \mathsf{A}\Psi, \\ \Psi_x = \mathsf{U}\Psi. \end{cases} \tag{16.1.4}$$

Given x, u, u_x, introduce the following two pairs of canonical solutions of the λ-equation (16.1.4), (similar to $\Psi_\pm$ in Chapter 13); see Figure 16.3:

$$\begin{aligned} \Psi_+^\infty &= [I + O(\lambda^{-1})]\exp\left(-\tfrac{ix^2}{16}\lambda\sigma_3\right), \quad \lambda \to \infty, \quad \lambda \in \Gamma_1 \cap \Omega_+, \\ \Psi_-^\infty &= [I + O(\lambda^{-1})]\exp\left(-\tfrac{ix^2}{16}\lambda\sigma_3\right), \quad \lambda \to \infty, \quad \lambda \in \Gamma_2 \cap \Omega_-, \end{aligned} \tag{16.1.5}$$

and

$$\begin{aligned} \Psi_+^0 &= P_0[I + O(\lambda)]\exp\left(-\tfrac{i}{\lambda}\sigma_3\right), \quad \lambda \to 0, \quad \operatorname{Im}\lambda > 0, \\ \Psi_-^0 &= P_0[I + O(\lambda)]\exp\left(-\tfrac{i}{\lambda}\sigma_3\right), \quad \lambda \to 0, \quad \operatorname{Im}\lambda < 0, \end{aligned} \tag{16.1.6}$$

where

$$P_0 = \begin{pmatrix} \cos\frac{u}{2} & i\sin\frac{u}{2} \\ i\sin\frac{u}{2} & \cos\frac{u}{2} \end{pmatrix}.$$

The monodromy data for the canonical Ψ-functions (16.1.5) and (16.1.6) are defined

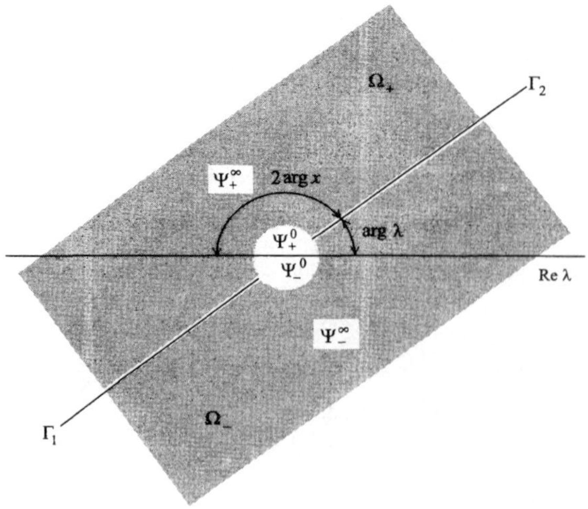

FIGURE 16.3. Domains of the canonical Ψ-functions

by (cf. definitions (13.1.16) in Chapter 13)

$$\Psi_-^\infty = \Psi_+^\infty S_j^\infty, \quad \lambda \in \Gamma_j, \quad j = 1, 2, \tag{16.1.7}$$

and

$$\Psi_\pm^0 = \Psi_\pm^\infty Q, \quad \lambda \in \Omega_\pm.$$

The matrices S_1, S_2 and Q are parametrized by the two scalar parameters p and q,

$$S_1 = \begin{pmatrix} 1 & q \\ p & 1+pq \end{pmatrix}, \quad S_2 = \begin{pmatrix} 1 & -p \\ -q & 1+pq \end{pmatrix}, \tag{16.1.8}$$

$$Q = \frac{1}{\sqrt{1+pq}} \begin{pmatrix} 1 & p \\ -q & 1 \end{pmatrix}.$$

Here p and q are the same as those in (16.0.3), which parametrize the initial data (16.0.2); this is a consequence of the fact that the same definitions of canonical solutions as in Chapter 14. For simplicity, we will restrict ourselves to the generic case of nondegenerate monodromy matrices, i.e., we will assume

$$1 + pq \neq 0.$$

2. WKB approximation of the Ψ-function

Now we will solve the direct monodromy problem (16.1.7), with the approximate "potential"

$$u = U, \quad u_x = U_x, \quad |x| \gg 1,$$

where

$$U_x = \sqrt{2\cos U + 2F}.$$

We introduce new variables, scaled with respect to $r = |x| \gg 1$,

$$X = x/r = \mathrm{e}^{i\phi}, \quad \phi = i \arg X, \quad \xi = r\lambda. \tag{16.2.1}$$

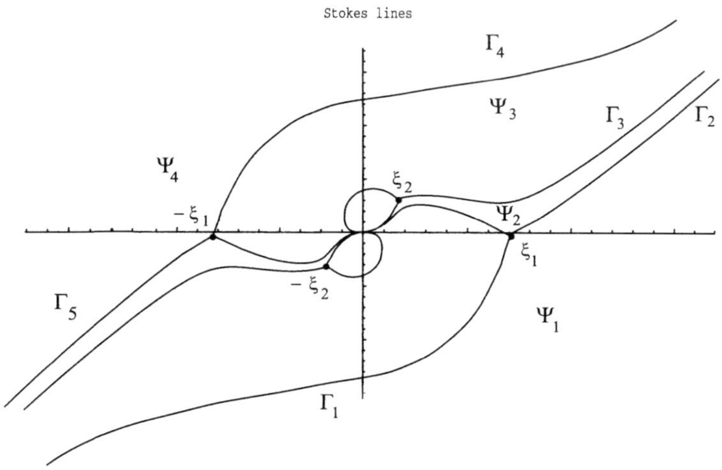

FIGURE 16.4. Stokes lines of equation (16.2.4)

In the new variables, equation (16.1.4) becomes

$$\Psi_\xi = \tilde{\mathsf{A}}\Psi, \tag{16.2.2}$$

$$\mathsf{A} = r\tilde{\mathsf{A}} = r\begin{pmatrix} -\dfrac{iX^2}{16} + \dfrac{i\cos U}{\xi^2} & -\dfrac{iXU_x}{4\xi} - \dfrac{i\sin U}{\xi^2} \\ -\dfrac{iXU_x}{4\xi} + \dfrac{i\sin U}{\xi^2} & \dfrac{iX^2}{16} - \dfrac{i\cos U}{\xi^2} \end{pmatrix}.$$

The large parameter r, allows us to construct the asymptotics of the Ψ-function explicitly by means of the WKB-approximation. For this purpose we calculate the determinant of $\tilde{\mathsf{A}}$,

$$\mu^2(\xi) = \det \mathsf{A} = \frac{X^4}{256} - \frac{X^2F}{8\xi^2} + \frac{1}{\xi^4}.$$

The eigenvalues $\pm i\mu(\xi)$ of $\tilde{\mathsf{A}}$, will form the phase integrals of the relevant WKB-solutions. Let $\pm\xi_1$ and $\pm\xi_2$ be the zeroes of $\mu(\xi)$,

$$\mu(\xi) = \frac{X^2}{16\xi^2}\sqrt{(\xi^2 - \xi_1^2)(\xi^2 - \xi_2^2)}, \tag{16.2.3}$$

$$\xi_1 = \frac{4i}{X}\sqrt{F + \sqrt{F^2 - 1}}, \quad \xi_2 = \frac{4i}{X}\sqrt{F - \sqrt{F^2 - 1}}.$$

The Stokes lines for the phase integrals $\exp\left\{\pm ir\int \mu(\zeta)d\zeta\right\}$ are defined as solutions of the equation

$$\mathrm{Re}\left[ir\int_{\xi_j}^{\xi} \mu(\zeta)d\zeta\right] = 0. \tag{16.2.4}$$

These lines form the Stokes graph shown on Figure 16.4. The asymptotic directions at infinity coincide with the straight line $\arg\lambda = \pi - 2\phi$ shown in Figure 16.3, thus the functions Ψ_4 and Ψ_1 represent Ψ_+^∞ and Ψ_-^∞ respectively. The two small domains in the center of Figure 16.4, accommodate the functions Ψ_+^0 and Ψ_-^0 from the small upper and lower half-circles in Figure 16.3.

The canonical solutions are normalized by

$$\Psi_j = [I + O(\xi^{-1})] \exp\left(-\frac{irX^2}{16}\xi\sigma_3\right), \quad \xi \to \infty, \quad \xi \in \Gamma_j, \quad j = 1, 4. \tag{16.2.5}$$

Since x and U are parameters in the λ-equation (16.2.2), we can specify these parameters in a suitable way, so that the monodromy data calculated via x and U, provide sufficient information to recover the elliptic function $U(x - x_0(\phi)|\varkappa(\phi))$. It appears that the simplest way is to concentrate on the zeroes of the elliptic function U. If the modulus $\varkappa$ is known, the zeroes of U uniquely determine the phase shift x_0. Indeed, choose x, U and U_x in the λ-equation (16.2.2) to be in the form

$$x = rX_{n,m}, \quad \cos U(x) = 1, \quad U_x = \sqrt{2\cos U + 2F} = \sqrt{2(F+1)}. \tag{16.2.6}$$

Then $\varkappa = \sqrt{\frac{2}{F+1}}$ and $X_{n,m}$ can be expressed in terms of the monodromy data p and q.

The matrix $\tilde{\mathrm{A}}$ in (16.2.2) now simplifies

$$\tilde{\mathrm{A}} = i\begin{pmatrix} -\dfrac{X^2}{16} + \dfrac{1}{\xi^2} & -\dfrac{X\sqrt{2(F+1)}}{4\xi} \\ -\dfrac{X\sqrt{2(F+1)}}{4\xi} & \dfrac{X^2}{16} - \dfrac{1}{\xi^2} \end{pmatrix}. \tag{16.2.7}$$

For the WKB approximation of equation (16.2.7) we diagonalize the matrix $\tilde{\mathrm{A}}$ by the left-hand multiplier $\Psi \mapsto T\Psi$, $\tilde{\mathrm{A}} \mapsto T^{-1}\tilde{\mathrm{A}}T$:

$$\Psi = \left[T + O(r^{-1}\mu^{-3/2})\right] \exp\left[-ir\sigma_3\int_{\xi_0}^{\xi} \mu(\zeta)d\zeta + i\int_{\xi_0}^{\xi} T^{-1}T_\zeta d\zeta\right] C, \tag{16.2.8}$$

where

$$T = \sqrt{\frac{d_1}{\mu}}\begin{pmatrix} 1 & \frac{d_1-\mu}{d_2} \\ \frac{\mu-d_1}{d_2} & 1 \end{pmatrix}, \quad d_1 = \frac{X^2}{16} - \xi^{-2}, \ d_2 = \frac{X\sqrt{2F+2}}{4\xi}, \tag{16.2.9}$$

and C is some, constant in ξ, matrix.

Equation (16.2.9) defines a Riemann surface $(\mu, \ \xi)$ with the canonical cycles defined as shown in Figure 16.5. The asymptotics of the phase integrals are given by the following lemma.

LEMMA 16.1. *The phase integrals in (16.2.8) satisfy the estimates*

$$\int_{\xi_2}^{\xi} \mu(\zeta)d\zeta = \frac{X^2}{16}\xi + \frac{1}{4}\oint_b \mu(\zeta)d\zeta + O(\xi^{-1}), \ \xi \in \Gamma_2, \ \xi \to \infty, \tag{16.2.10}$$

$$\int_{\pm\xi_1}^{\xi} \mu(\zeta)d\zeta = \frac{X^2}{16}\xi \mp \frac{1}{4}\oint_{2a-b} \mu(\zeta)d\zeta + O(\xi^{-1}), \ \xi \in \begin{cases}\Gamma_1 \\ \Gamma_4\end{cases}, \ \xi \to \infty. \tag{16.2.11}$$

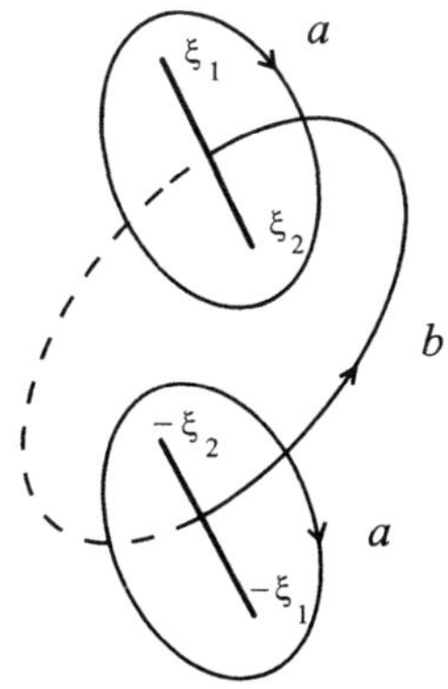

FIGURE 16.5. Canonical cycles on the Riemann surface (16.2.3)

PROOF. By the definition (16.2.3), $\mu(\xi) \to \frac{X^2}{16}$ at infinity. For large ξ, such that $\xi \in \Gamma_2$, (see Figure 16.4), the phase integral (16.2.10) is

$$\int_{\xi_2}^{\xi} \mu(\zeta)d\zeta = \frac{X^2}{16}(\xi - \xi_2) + \int_{\xi_2}^{\xi} \left(\mu(\zeta) - \frac{X^2}{16}\right) d\zeta$$

$$(16.2.12) \qquad = \frac{X^2}{16}(\xi - \xi_2) + \int_{\Gamma_2} \left(\mu(\zeta) - \frac{X^2}{16}\right) \zeta + O(\xi^{-1}).$$

The function $\mu(\xi)$ is even, thus the last integral can be written as

$$\int_{\Gamma_2} \left(\mu(\zeta) - \frac{X^2}{16}\right) d\zeta = \frac{1}{2} \int\limits_{\Gamma_2 \cup (-\Gamma_2)} \left(\mu(\zeta) - \frac{X^2}{16}\right) d\zeta = \frac{1}{4} \int\limits_{b} \mu(\zeta)d\zeta + \frac{1}{2} \int_{-\xi_2}^{\xi_2} \frac{X^2}{16} d\zeta.$$

The asymptotic formula (16.2.11) can be proved similarly. □

THEOREM 16.2. *The canonical solutions Ψ_j (16.2.5), have the following asymptotics as $r \to \infty$:*

(16.2.13)

$$\Psi_1(\xi) = T[1 + o(1)] \exp \left[-ir\sigma_3 \left(\int_{\xi_1}^{\xi} \mu(\zeta)d\zeta - \frac{1}{4} \oint_{2a-b} \mu(\zeta)d\zeta\right)\right], \quad \xi \in \Gamma_1;$$

(16.2.14)

$$\Psi_4(\xi) = T[1 + o(1)] \exp \left[-ir\sigma_3 \left(\int_{-\xi_1}^{\xi} \mu(\zeta)d\zeta + \frac{1}{4} \oint_{2a-b} \mu(\zeta)d\zeta\right)\right], \quad \xi \in \Gamma_4;$$

$$(16.2.15) \quad \Psi_2(\xi) = T[1 + o(1)] \exp \left[-ir\sigma_3 \left(\int_{\xi_2}^{\xi} \mu(\zeta)d\zeta - \frac{1}{4} \oint_{b} \mu(\zeta)d\zeta\right)\right], \quad \xi \in \Gamma_2.$$

PROOF. The proof follows immediately from the definition of the canonical solutions (16.2.5) and from the asymptotics of the phase integrals (16.2.10) and (16.2.11) stated in Lemma 16.1. □

We will apply the asymptotics (16.2.13)–(16.2.15) to the equation

$$\Psi_1 = \Psi_4 \begin{pmatrix} 1 & q \\ p & 1+pq \end{pmatrix}. \tag{16.2.16}$$

This equation follows from equation (16.1.7), since Ψ_-^∞ is now Ψ_1 and Ψ_+^∞ is Ψ_4. The conjugation equation (16.2.16) is valid in the domain where both Ψ-functions are defined. Thus we need to make a extension of Ψ_1 and Ψ_4 into a common domain within the lines Γ_2 and Γ_4. The best way to do this is analysis of WKB-solutions at the turning points ξ_j at the ends of the Stokes lines.

3. Asymptotics at the turning points

WKB-approximation (16.2.13)–(16.2.15) fails in a small neighborhood of the turning points ξ_j. In this neighborhood, we can use an Airy function approximation, since all the turning points are *simple* (we recall that we study nondegenerate case of the fourth-order polynomial $\mu^2(\xi)$). This approximation allows us to extend the Ψ-function to the neighboring $\frac{2\pi}{3}$-sector around simple turning point. In analogy with the procedure used in Chapter 7 this extension involves a right-hand multiplication by appropriate Stokes matrices (see [**W2**] for the generic theory).

LEMMA 16.2. *Any solution of the equation $\Psi_\xi = \tilde{A}\Psi$ where the matrix $\tilde{A}$ is defined in (16.2.7), has the following asymptotics at the turning points ξ_j, $j = 1, 2$:*

$$\Psi(\xi, r, X) = r^{1/6} \begin{pmatrix} \omega(\tau) & -i\overline{\omega}(\tau) \\ i\omega(\tau) & \overline{\omega}(\tau) \end{pmatrix} D + O\left[r^{-1/6}\tau^{3/4}\exp(\pm 4i\gamma_j\tau^{3/2})\right], \tag{16.3.1}$$

where

$$\xi \in \Gamma_j, \ \tau = r^{2/3}(\xi - \xi_j), \quad \gamma_j^2 = \frac{X^4(\xi_j^2 - \xi_{3-j}^2)}{4608\xi_j^3}, \quad j = 1, 2,$$

D is a constant in τ matrix, and the functions ω and $\overline{\omega}$ are linear independent solutions of the Airy equation

$$\omega_{\tau\tau} + 36\gamma^2\tau\omega = 0. \tag{16.3.2}$$

PROOF. It is easy to verify that near the point ξ_j the phase function μ has the asymptotics

$$\mu = 6\gamma_j r^{-1/3}\tau^{1/2} + O((\xi - \xi_j)^{3/2}). \tag{16.3.3}$$

Let ψ_1 and ψ_2 be the components of the first column of Ψ. The matrix equation (16.2.7) is equivalent to the system of the scalar equations

$$\begin{aligned} \chi_{1,\xi} &= -\tfrac{irX^2}{16\xi^2}(\xi - \xi_1)(\xi - \xi_2)\chi_2, \\ \chi_{2,\xi} &= -\tfrac{irX^2}{16\xi^2}(\xi + \xi_1)(\xi + \xi_2)\chi_1, \end{aligned} \tag{16.3.4}$$

where $\chi_{1,2} = \psi_1 \pm i\psi_2$. This system yields the second-order equation

$$\chi_{2,\xi\xi} = \frac{irX^2}{16\xi^2}\left(\xi_1 + \xi_2 + 2\frac{\xi_1\xi_2}{\xi}\right)\chi_1 - \frac{r^2X^4}{256\xi^4}(\xi^2 - \xi_1^2)(\xi^2 - \xi_2^2)\chi_2. \tag{16.3.5}$$

At the turning point ξ_j, we choose a solution of (16.3.5) with the leading term of asymptotics

$$\chi_{1,2} = \omega_{1,2} + O(r^{-1/3}).$$

This yields a model equation for the leading order,

$$\omega_{2,\tau\tau} + 36\gamma_j^2\tau\omega_2 = 0, \ \omega_1 = 0, \tag{16.3.6}$$

where $\tau = r^{2/3}(\xi - \xi_j)$. The remainder terms of the Taylor series imply the estimates

$$\psi_1 = \frac{1}{2}\omega_2 + O(r^{-1/3}), \ \psi_2 = \frac{i}{2}\omega_2 + O(r^{-1/3}),$$

which prove the Lemma. □

We will now use the well-known asymptotics of the Airy functions ([**BE**]). Let us normalize basic functions ω and $\overline{\omega}$ of the Airy equation (16.3.2), by their asymptotics $\arg\tau = -\frac{2}{3}\arg\gamma$:

$$\begin{aligned} \omega(\tau) &= \tau^{-1/4}\left[1 + O(\tau^{-3/2})\right]\exp\left(-4i\gamma\tau^{3/2} - \tfrac{\pi i}{4}\right), \\ \overline{\omega}(\tau) &= \tau^{-1/4}\left[1 + O(\tau^{-3/2})\right]\exp\left(4i\gamma\tau^{3/2} + \tfrac{\pi i}{4}\right). \end{aligned} \quad \tau \to \infty. \tag{16.3.7}$$

Then,

$$\begin{aligned} \overline{\omega}\left(\tau e^{-2\pi i/3}\right) &= e^{-\pi i/3}\left[\overline{\omega}(\tau) - \omega(\tau)\right], \quad \overline{\omega}\left(\tau e^{2\pi i/3}\right) = e^{\pi i/3}\omega(\tau), \\ \omega\left(\tau e^{2\pi i/3}\right) &= e^{\pi i/3}\left[\omega(\tau) - \overline{\omega}(\tau)\right], \quad \omega\left(\tau e^{-2\pi i/3}\right) = e^{-\pi i/3}\overline{\omega}(\tau). \end{aligned} \tag{16.3.8}$$

Now, we can extend the functions Ψ_1 and Ψ_4 to the common domain containing the points ξ_1, ξ_2 and $-\xi_1$.

THEOREM 16.3. *Let $\tilde{\Psi}_1$ and $\tilde{\Psi}_4$ denote the continuation of the canonical solutions Ψ_1 and Ψ_4 to the domain Ω_3 (see Figure 16.6). Then, these functions have the following asymptotics as $r \to \infty$:*

$$\begin{aligned} \tilde{\Psi}_1(\xi) \ &= \ T\left[1 + O(r^{-1}\mu^{-3/2})\right] \\ &\times \ \exp\left[-ir\sigma_3\int_{\xi_2}^{\xi}\mu(\zeta)d\zeta\right]\begin{pmatrix}1 & 1\\ 0 & 1\end{pmatrix} D_1 \begin{pmatrix}1 & 1\\ 0 & 1\end{pmatrix} D_2, \end{aligned} \tag{16.3.9}$$

$$\tilde{\Psi}_4(\xi) = T\left[1 + O(r^{-1}\mu^{-3/2})\right]\exp\left[-ir\sigma_3\int_{\xi_2}^{\xi}\mu(\zeta)d\zeta\right] D_3 \begin{pmatrix}1 & 0\\ 1 & 1\end{pmatrix} D_4, \tag{16.3.10}$$

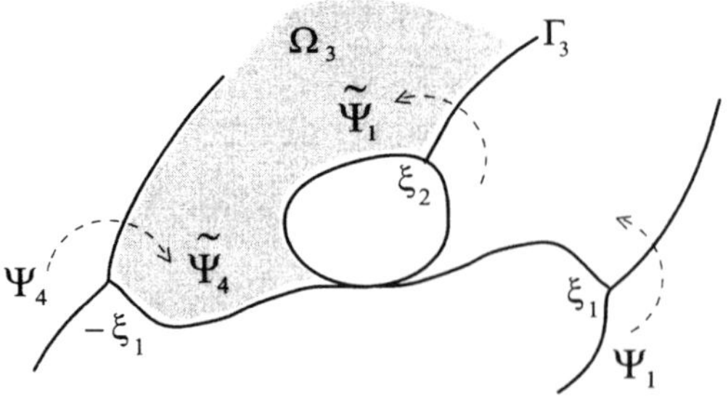

FIGURE 16.6. Extension of Ψ_1 and Ψ_4 into a common domain

(16.3.11)
$$D_1 = \exp\left[\frac{-ir\sigma_3}{4}\oint_{2a}\mu(\zeta)d\zeta\right], \quad D_2 = \exp\left[\frac{ir\sigma_3}{4}\oint_{2a-b}\mu(\zeta)d\zeta\right],$$
$$D_3 = \exp\left[\frac{ir\sigma_3}{4}\oint_{2a-2b}\mu(\zeta)d\zeta\right], \quad D_4 = \exp\left[-\frac{ir\sigma_3}{4}\oint_{2a-b}\mu(\zeta)d\zeta\right].$$

PROOF. First let us illustrate how to continue Ψ_1 around the point ξ_1 (see Figure 16.6). Consider the auxiliary WKB-solution

$$\hat{\Psi}_1 = T\,[1 + o(1))]\exp\left[-ir\sigma_3\int_{\xi_2}^{\xi}\mu(\zeta)d\zeta\right]C, \quad \xi \in \Gamma_3. \tag{16.3.12}$$

Since (16.2.13) and (16.3.12) are solutions of the same equation $\Psi_\xi = \tilde{A}\Psi$, in the common domain where $\xi - \xi_1$ is small and τ is large, these solutions coincide up to a certain constant multiplier. All we need now is to compare both Ψ_1 and $\hat{\Psi}_1$ with the Airy-type solution (16.3.1). First calculate the multiplier D in (16.3.1), using the asymptotics (16.2.13) of Ψ_1 as $\xi \to \xi_1$:

$$\begin{aligned}\Psi_1 &= \left[r^{1/6}\zeta^{-1/4}\sqrt{\frac{d_1}{6\gamma_1}}\begin{pmatrix}1 & -i\\ i & 1\end{pmatrix} + O(r^{-1/2}\zeta^{-3/4})\right]\\ &\times\ \exp\left\{-i\sigma_3\left[4\gamma_1\zeta^{3/2} + O(r^{-2/3}\zeta^{5/2})\right]\right\}D_2.\end{aligned} \tag{16.3.13}$$

Applying the asymptotics (16.3.7) of the Airy function to (16.3.1), and comparing the leading terms with (16.3.13), we find

$$D = \sqrt{\frac{d_1}{\gamma_1}}\exp\left(\frac{\pi i\sigma_3}{4}\right)D_2. \tag{16.3.14}$$

Next, we follow the same procedure for $\hat{\Psi}_1$ as $\xi \to \xi_1$. We first replace the lower limit of integration from ξ_2 to ξ_1:

$$\begin{aligned}\hat{\Psi}_1 &= T\left[1+o(1)\right]\exp\left[-ir\sigma_3\int_{\xi_1}^{\xi}\mu(\zeta)d\zeta\right]\exp\left[-ir\sigma_3\int_{\xi_2}^{\xi_1}\mu(\zeta)d\zeta\right]C\\ &= T\left[1+o(1)\right]\exp\left[-ir\sigma_3\int_{\xi_1}^{\xi}\mu(\zeta)d\zeta\right]D_1C,\end{aligned}$$

where we have used

$$\exp\left[-ir\sigma_3\int_{\xi_2}^{\xi_1}\mu(\zeta)d\zeta\right]=\exp\left[\frac{ir\sigma_3}{2}\int_a\mu(\zeta)d\zeta\right]=D_1.$$

Replace ζ by $\zeta e^{2\pi i/3}$ in the Airy-function representation (16.3.1), and using (16.3.8) we find

$$\begin{aligned}&\hat{\Psi}_1(\xi,r)|_{\xi\to\xi_1}=\hat{\Psi}_1(\xi_1+\zeta r^{-2/3}e^{-2\pi i/3},r)\\ &=r^{1/6}\begin{pmatrix}\omega\left(\zeta e^{-2\pi i/3}\right) & -i\overline{\omega}\left(\zeta e^{-2\pi i/3}\right)\\ i\omega\left(\zeta e^{-2\pi i/3}\right) & \overline{\omega}\left(\zeta e^{-2\pi i/3}\right)\end{pmatrix}D+r^{-1/6}R(\zeta,r)\\ &=r^{1/6}e^{-\pi i/3}\begin{pmatrix}\overline{\omega}(\zeta) & -i\left[\overline{\omega}(\zeta)-\omega(\zeta)\right]\\ i\omega(\zeta) & \overline{\omega}(\zeta)-\omega(\zeta)\end{pmatrix}D+r^{-1/6}R(\zeta,r)\\ &=r^{1/6}e^{-\pi i/3}\begin{pmatrix}\overline{\omega} & -i\omega\\ i\overline{\omega} & \omega\end{pmatrix}\begin{pmatrix}1 & i\\ 0 & -1\end{pmatrix}D+r^{-1/6}R(\zeta,r)\\ &=-ir^{1/6}\zeta^{-1/4}\begin{pmatrix}1 & -i\\ i & 1\end{pmatrix}\exp\left[i\sigma_3\left(4\gamma_1\zeta^{3/2}e^{\pi i}+\frac{\pi}{4}\right)\right]\begin{pmatrix}1 & -i\\ 0 & -1\end{pmatrix}D\end{aligned}$$

$$+r^{-1/6}R(\zeta,r), \tag{16.3.15}$$

where $R(\zeta,r)=O\left(\zeta^{3/4}\exp\left(\pm 4i\gamma_1\zeta^{3/2}\right)+r^{-1/3}\zeta^{5/4}\right)$.

Comparing the leading terms of (16.3.13) and (16.3.15), we find that the constant C is given by

$$C=D_1\begin{pmatrix}1 & 1\\ 0 & 1\end{pmatrix}D_2.$$

Finally, we need to pass from $\hat{\Psi}_1$ to $\tilde{\Psi}_1$, going around the point ξ_2. This can be achieved similarly with the case of ξ_1, but without changing the lower limit of integration. Thus, the connection here reads

$$\tilde{\Psi}_1 = \hat{\Psi}_1 \begin{pmatrix} 1 & 1 \\ 1 & 0 \end{pmatrix} = \Psi_1 \begin{pmatrix} 1 & 1 \\ 1 & 0 \end{pmatrix} D_1 \begin{pmatrix} 1 & 1 \\ 1 & 0 \end{pmatrix} D_2,$$

which proves formula (16.3.9).

The asymptotic formula (16.3.10) can be proved similarly, passing over $-\xi_1$ going in counterclockwise direction; this yields the matrix $\left(\begin{smallmatrix} 1 & 0 \\ 1 & 1 \end{smallmatrix}\right)$ multiplied on the left by D_3 (which comes from the change of the lower limit of integration in (16.2.14) from $-\xi_1$ to ξ_2). □

4. Zeros of the elliptic function ansatz

The Ψ-function representations (16.3.9) and (16.3.10) will allow us to solve the monodromy equation (16.2.16),

$$\tilde{\Psi}_1 = \tilde{\Psi}_4 S_1, \quad S_1 = \begin{pmatrix} 1 & q \\ p & 1+pq \end{pmatrix}, \tag{16.4.1}$$

in terms of the asymptotic parameters $x = re^{i\phi}$ and $F = F(\phi)$. This will prove rigorously the Boutroux equation (16.0.10) and will also determine the phase shift $x_0(\phi)$ of the Jacobi sine function $\operatorname{sn}(x - x_0(\phi)|\varkappa(\phi))$ (16.0.5). Actually, we need the asymptotic distribution of the zeroes $x_{m,n}$ of the Jacobi sine for $r \gg 1$, $0 < \phi < \pi/2$. The latter distribution will be found by inverting the of asymptotic formulae for p and q (recall that the monodromy parameters are constant in x, and they are fixed as the first integrals of SG-PIII equation (13.0.1).

Changing the variable of integration, $\xi = \frac{4}{X}\nu$, reduces the phase integrals to the form

(16.4.2)

$$ir\int \mu(\xi)d\xi = -\frac{ix_{m,n}}{4}\int \nu^{-2}\sqrt{(\nu^2-\nu_1^2)(\nu^2-\nu_2^2)}d\nu = -\frac{ix_{m,n}}{4}\int \tilde{\mu}(\xi)d\xi,$$

where

$$\nu_1^2 = -F + \sqrt{F^2-1}, \quad \nu_2^2 = -F - \sqrt{F^2-1}.$$

Calculating $\mathrm{S}_1 = \tilde{\Psi}_4^{-1}\tilde{\Psi}_1$ in the domain Ω_3, in terms of the WKB-approximation (16.3.9) and (16.3.10) we find the following:

THEOREM 16.4. *The Stokes matrix S_1 has the form*

$$\mathrm{S}_1 = \begin{pmatrix} 1 & \dfrac{1+\alpha^2}{\alpha^2\beta^2} \\ -\beta^2 & -\alpha^{-2} \end{pmatrix}, \tag{16.4.3}$$

where

$$\alpha = \exp\left[-\frac{ir}{2}\oint_a \mu(\xi)d\xi\right] = (D_1)_{11}, \quad \beta = \exp\left[-\frac{ir}{4}\oint_{2a-b} \mu(\xi)d\xi\right] = (D_4)_{11}.$$

PROOF. The proof follows by the direct multiplication of the matrices (16.3.9) and (16.3.10), and by the substitution of explicit expressions (16.3.11):

$$\Psi_4^{-1}\Psi_1 = D_4^{-1}\begin{pmatrix} 1 & 0 \\ -1 & 1 \end{pmatrix} D_3^{-1}\begin{pmatrix} 1 & 1 \\ 0 & 1 \end{pmatrix} D_1 \begin{pmatrix} 1 & 1 \\ 0 & 1 \end{pmatrix} D_2 = \begin{pmatrix} 1 & \dfrac{1+\alpha^2}{\alpha^2\beta^2} \\ -\beta^2 & -\alpha^{-2} \end{pmatrix}.$$

□

LEMMA 16.3. *The phase integrals (16.4.2) over the basic cycles, can be expressed as*

$$\oint_a \tilde{\mu}(\xi)d\xi = i\sqrt{2}G_b(F) = \frac{2}{\varkappa}\left[K'(\varkappa) - E'(\varkappa)\right], \tag{16.4.4}$$

$$\oint_b \tilde{\mu}(\xi)d\xi = \frac{i}{\sqrt{2}}G_a(F) = -\frac{4i}{\varkappa}E(\varkappa), \tag{16.4.5}$$

where $G_a(F)$ and $G_b(F)$ are the basic periods of the integral (16.0.11)

$$G_{a,b}(F) = \oint_{a,b}\sqrt{\frac{F+V}{1-V^2}}.$$

and $E(\varkappa)$, $K(\varkappa)$ denote the standard complete elliptic integrals (16.0.13):

$$K = K(\varkappa) = \int_0^1 \frac{dz}{\sqrt{(1-z^2)(1-\varkappa^2 z^2)}}, \qquad K' = K(\varkappa'),$$

$$E = E(\varkappa) = \int_0^1 \sqrt{\frac{1-\varkappa^2 z^2}{1-z^2}}\,dz, \quad E' = E(\varkappa'), \tag{16.4.6}$$

$$\varkappa = \sqrt{\frac{2}{F+1}}.$$

PROOF. The integrals (16.4.2) and $G(F)$ admit the standard uniformization

$$\int \mu(\zeta)d\zeta = \nu_2\int\sqrt{\frac{1-\kappa^2\nu^2}{1-\nu^2}}d\nu + \nu_1\int\sqrt{\frac{1-\kappa^{-2}\nu^2}{1-\nu^2}}d\nu,$$

$$\int\sqrt{\frac{F+V}{1-V^2}} = -\frac{2\sqrt{2}}{\varkappa}\int\sqrt{\frac{1-\varkappa^2\nu^2}{1-\nu^2}}d\nu,$$

$$\kappa = \frac{\nu_1}{\nu_2} = F - \sqrt{F^2-1}, \quad \varkappa = \frac{2\sqrt{\kappa}}{\kappa+1}.$$

The latter modular transformation, together with an integration over the basic cycles a and b, yield formulae (16.4.4) and (16.4.5). □

COROLLARY 16.1. *Suppose* $F = F_0 + r^{-1}F_1 + O(r^{-2})$, *then*

$$\oint_a \tilde{\mu}(\xi)d\xi = i\sqrt{2}G_b(F_0) - \frac{2\sqrt{2}i\varkappa K'(\varkappa)}{2r}F_1 + O(r^{-2}), \tag{16.4.7}$$

$$\oint_b \tilde{\mu}(\xi)d\xi = \frac{i}{\sqrt{2}}G_a(F_0) - \frac{\sqrt{2}\varkappa K(\varkappa)}{2r}F_1 + O(r^{-2}), \tag{16.4.8}$$

PROOF. This follows from the Taylor series expansion of $\sqrt{\frac{F+V}{1-V^2}}$, and from the identities

$$\oint_a \frac{dV}{\sqrt{(F+V)(1-V^2)}} = \sqrt{2}\varkappa K(\varkappa), \quad \oint_b \frac{dV}{\sqrt{(F+V)(1-V^2)}} = 2\sqrt{2}\varkappa K'(\varkappa).$$

PROOF OF THEOREM 16.1. Comparing the matrices (16.4.1) and (16.4.3) we find

$$\log(-p) = -\frac{ir}{2}\oint_{2a-b} \mu(\zeta)d\zeta + 2\pi iN = \frac{ix_{m,n}}{8}\oint_{2a-b} \tilde{\mu}(\zeta)d\zeta + 2\pi iN,$$

(16.4.9)

$$\log(-1-pq) = ir\oint_a \mu(\zeta)d\zeta + 2\pi iN = -\frac{ix_{m,n}}{4}\oint_a \tilde{\mu}(\zeta)d\zeta + 2\pi iN.$$

Since

$$|x_{m,n}| = r = O(N), \quad \arg x_{m,n} = \phi,$$

equations (16.4.9) yield the Whitham equations for the modulus $\varkappa = \varkappa(\phi)$

$$\oint_a \tilde{\mu}(\zeta)d\zeta = \oint_b \tilde{\mu}(\zeta)d\zeta = -\phi.$$

Due to Lemma 16.3, these equations are equivalent to the equations (16.0.10),

$$G_a(F) = G_b(F) = -\phi - \frac{\pi}{2}.$$

The zeroes $x_{m,n}$ can be found by their asymptotics in r^{-1} :

$$x_{m,n} = m\omega + n\omega' + x_0 + O(r^{-1}), \quad |m| + |n| = O(r), \tag{16.4.10}$$

where the periods of the Jacobi sine function (16.0.14) are given as

$$\omega = 2\varkappa K(\varkappa), \quad \omega' = i\varkappa K'(\varkappa).$$

Calculate now terms of order $O(1)$ in equations (16.4.9). These functions are formed by x_0 and $r^{-1}F_1$, which yield the following system:

$$\log(-p) = \frac{ix_0}{2\varkappa}(K' - E' + E) + 2\sqrt{2}\varkappa F_1(K + iK'),$$

$$\log(-1-pq) = \frac{ix_0}{2\varkappa}(K' - E') + 2\sqrt{2}F_1\varkappa(K).$$

The determinant of this system is $\pi/2$, using the Legendre equality $EK' + E'K - KK' = \pi/2$. Thus, the solution for x_0 is straightforward,

$$x_0(\phi) = \frac{i\varkappa(\phi)}{2\pi}\left[K(\phi)\log(-p) - (K(\phi) - iK'(\phi))\log(-1-pq)\right].$$

□

APPENDIX A

Proof of Theorem 3.4

This Appendix is based on the paper [**BolIK**].

Let us define the function $\Psi(\lambda, x)$ by the equations,

$$\Psi(\lambda, x)\big|_{\overline{\Omega_1^0}} = Y_1(\lambda, x)e^{-\theta(\lambda,x)\sigma_3}, \tag{A.1}$$

$$\Psi(\lambda, x)\big|_{\overline{\Omega_k^0}} = Y_k(\lambda, x)e^{-\theta(\lambda,x)\sigma_3}\mathrm{S}_{k-1}^{-1}\dots\mathrm{S}_1^{-1}, \quad k = 1, \dots, 6, \tag{A.2}$$

where the branch of the $\ln\lambda$ is fixed by the condition,

$$\left|\arg\lambda - \frac{k-1}{3}\pi\right| < \frac{\pi}{6} \quad \text{when} \quad \lambda \in \Omega_k^0.$$

Slightly abusing the terminology, we will call the function $\Psi(\lambda, x)$, along with the function $Y(\lambda, x)$, *the solution of the Riemann-Hilbert problem* (3.3.98)*–*(3.3.99). In terms of the function$\Psi(\lambda, x)$, Theorem 3.4 can be reformulated as follows.

THEOREM A.0 *Let $s \equiv (\mathrm{S}_1, \mathrm{S}_2, \dots, \mathrm{S}_6; \nu)$ be the given monodromy data, i.e., the complex number ν and the set of six 2×2 matrices satisfying equations* (3.3.117) *and* (3.3.118). *Then there exists the countable set $X_s = \{x_j \equiv x_j(s)\} \subset \mathbb{C}$, $x_j \to \infty$ and the unique matrix function $\Psi(\lambda, x)$ with the following properties.*

(a) *$\Psi(\lambda, x)$ is holomorphic in $\mathbb{C} \times (\mathbb{C} \setminus X_s)$ and meromorphic along $\mathbb{C} \times X_s$ (i.e., it has poles at $x_j \in X_s$ and the coefficients of the corresponding Laurent series are entire functions of λ).*

(b) *For every $x \in \mathbb{C} \setminus X_s$, the following asymptotic relation takes place,*

$$\Psi(\lambda, x)\mathrm{S}_1 \dots \mathrm{S}_{k-1} \sim \Big(I + \sum_{j=1}^{\infty} \frac{\Psi_j(x)}{\lambda^j}\Big)e^{-\theta(\lambda,x)\sigma_3}, \tag{A.3}$$

$$\lambda \to \infty, \quad \lambda \in \overline{\Omega_k^0}, \quad \left|\arg\lambda - \frac{k-1}{3}\pi\right| < \frac{\pi}{6},$$

where

$$\theta(\lambda, x) = \frac{4i}{3}\lambda^3 + ix\lambda + \nu\ln\lambda,$$

and the coefficient functions $\Psi_j(x)$ are meromorphic in x and have the set X_s as the set of their poles.

(c) *The asymptotics* (A.3) *is uniform, i.e., for any positive number N and for any compact subset $\mathbb{K} \in \mathbb{C} \setminus X_s$, there exists a constant $C > 0$ such that*

$$\Big\|\Psi(\lambda, x)\mathrm{S}_1 \dots \mathrm{S}_{k-1}e^{-\theta(\lambda,x)\sigma_3} - I - \sum_{j=1}^{N} \frac{\Psi_j(x)}{\lambda^j}\Big\| < \frac{C}{\lambda^{N+1}}, \tag{A.4}$$

for all $|\lambda| > 1$, $\lambda \in \overline{\Omega_k^0}$ and $x \in \mathbb{K}$.

(d) *The asymptotics* (A.3) *is differentiable with respect to* λ *and* x.
(e) $\det \Psi(\lambda, x) \equiv 1$.

REMARK A.1. Because of the "correct" triangular structure of the Stokes matrices S_k, the asymptotic condition (A.3) actually holds in the open sectors Ω_k defined in (2.1.90). Indeed, we have that for any positive number N, for any compact subset $\mathbb{K} \Subset \mathbb{C} \setminus X_s$, and for any closed subsector $\Omega_k' \subset \Omega_k$, there exists a constant $C > 0$, such that

$$\left\| \Psi(\lambda, x) S_1 \dots S_{k-1} e^{-\theta(\lambda,x)\sigma_3} - I - \sum_{j=1}^{N} \frac{\Psi_j^0}{\lambda^j} \right\| \le \frac{C}{\lambda^{N+1}},$$

$$\forall (\lambda, x) \in \Omega_k' \times \mathbb{K}, \quad |\lambda| > 1.$$

In what follows we will prove Theorem A.0. Theorem 3.4 will then follow by using equations (A.1) and (A.2) to define the function $Y(\lambda, x)$ via the already known function $\Psi(\lambda, x)$.

We split the proof into three steps. In the first step, we establish the existence of the local, in both λ and x, solution of the Riemann-Hilbert problem. This means we prove that, given a neighborhood Ω of $\lambda = \infty$ and a disk D on the x-plane, there exists a matrix function, which we denote by $M(\lambda, x)$, which is holomorphically invertible in $\Omega \times D$, and which satisfies the asymptotic condition (A.3) with all $\Psi_j(x)$ holomorphic in D (see Theorem A.1 below). We note, that for the standard case that the dependence of RH on a parameter is not addressed, this result follows from the Sibuya Theorem [**Si2**] concerning the local solvability of the general Riemann-Hilbert-Birkhoff problem for a linear system near its irregular singularity. Therefore, we call this step *the Sibuya Theorem with a parameter.* The principal construction used in this step, is the introduction of the explicit model functions $g_k(\lambda, x)$, such that each of them factorizes the corresponding Stokes matrix S_k. It is essentially this technique, together with the application of the Birkhoff-Grothendieck theorem (the second step), that allows us to avoid using the L^2-Fredholm theory of singular integral operators, which is the principal tool used in the works [**BS**] and [**FoZ**].

In the second step, we apply the classical Birkhoff-Grothendieck theorem, appropriately generalized to the case that the RH depends on a parameter, to the function $M(\lambda, x)$. This yields a global in λ, but still local in x, solution of the original RH problem (see Theorem A.3 below). This is the most important step, therefore for the convenience of the reader we will present a detailed proof of the Birkhoff-Grothendieck theorem with a parameter, following [**Bol4**], in Appendix B.

Finally, in the third step, we cover the x-plane by a countable number of small disks, and glue up the corresponding local solutions into a single, global in both λ and x, solution of the inverse monodromy problem, which completes the proof of Theorem A.0 and hence Theorem 3.4.

Before we proceed, we make one technical assumption concerning the formal monodromy exponent ν. We assume that

$$\nu \notin \mathbb{Z} \setminus \{0\}. \tag{A.5}$$

The following lemma shows that we can make this assumption, without loss of generality.

LEMMA A.1. *Let $\Psi(\lambda, x) \equiv \Psi_1(\lambda, x)$ be the solution of the RH problem with the monodromy data $s = (S_1, \ldots, S_6; \nu)$, and let $u(x)$ and $v(x)$ be the corresponding functional parameters of the manifold $\mathcal{A}$ of the singular data. Then, the functions*

$$\hat{\Psi}(\lambda, x) := \begin{pmatrix} \lambda + u_x/2iu & -u/2 \\ 2/u & 0 \end{pmatrix} \Psi(\lambda, x),$$

and

$$\tilde{\Psi}(\lambda, x) := \begin{pmatrix} 0 & 2/v \\ -v/2 & \lambda - v_x/2iv \end{pmatrix} \Psi(\lambda, x),$$

solve RH problems with the monodromy data

$$\hat{s} = (S_1, \ldots, S_6; \nu - 1),$$

and

$$\tilde{s} = (S_1, \ldots, S_6; \nu + 1),$$

respectively.

Note, that only the asymptotic property (3.3.125) needs to be proven. The proof can be obtained by a direct calculation using equations (3.3.132) and (3.3.133), relating the matrix coefficients Ψ_j of the asymptotic series for Ψ and the functions u, v, $w = u_x$, and $z = v_x$[1].

0.1. The Sibuya theorem with a parameter. Throughout this section we will use the following notations. For any set $\mathbb{M} \subset \mathbb{C}^N$, with nonempty interior, we denote $H(\mathbb{M})$ the Banach space of the 2×2 matrix functions holomorphic in the interior of $\mathbb{M}$ and continuous in $\overline{\mathbb{M}}$. We will denote by $H^0(\mathbb{M})$ the subset of $H(\mathbb{M})$ consisting of holomorphically invertible matrix functions. We shall call such functions holomorphic in $\mathbb{M}$ and holomorphically invertible in $\mathbb{M}$ respectively.

Let

$$\Omega_\rho = \{\lambda \in \mathbb{C}: \quad |\lambda| > \rho\}$$

be a neighborhood of $\lambda = \infty$. We define the following subsets of Ω_ρ:

$$\Gamma_k^\rho = \left\{\lambda \in \mathbb{C}: \quad |\lambda| \geq 2\rho, \quad \arg\lambda = \frac{2k-1}{6}\pi\right\}, \quad k = 1, \ldots, 6,$$

$$\Omega_k^\rho = \left\{\lambda \in \mathbb{C}: \quad |\lambda| > \rho, \quad \left|\arg\lambda - \frac{2k-1}{6}\pi\right| < \varepsilon\right\},$$

$$\mathcal{L}_k = \left\{\lambda \in \mathbb{C}: \quad |\lambda| \geq 3\rho, \quad \left|\arg\lambda - \frac{k-1}{3}\pi\right| \leq \frac{\pi}{6}\right\}, \quad k = 1, \ldots, 6.$$

---

[1]The transformations $\Psi \mapsto \hat{\Psi}$ and $\Psi \mapsto \tilde{\Psi}$ are examples (found by A. Kitaev) of the *Schlesinger transformations* (see e.g. **[JMU]**; see also Chapter 6). They induce the following *Bäcklund transformations*,

$$(u, v) \mapsto \left(-\frac{u}{4}(uv + x - u_x^2/u^2),\ \frac{4}{u}\right),$$

and

$$(u, v) \mapsto \left(\frac{4}{v},\ -\frac{v}{4}(uv + x - v_x^2/v^2)\right),$$

on the solution space of the system (2.1.86).

We assume that the contours Γ_k^ρ are oriented towards the ∞, and that ε is a sufficiently small positive number such that $\Omega_k^\rho \cap \Omega_{k+1}^\rho = \emptyset$. At the moment, we do not impose any restrictions on the positive number ρ. The contours Γ_k^ρ and the sets Ω_k^ρ are depicted in Figure A.1. The sectors $\mathcal{L}_k$ are shown in Figure A.2.

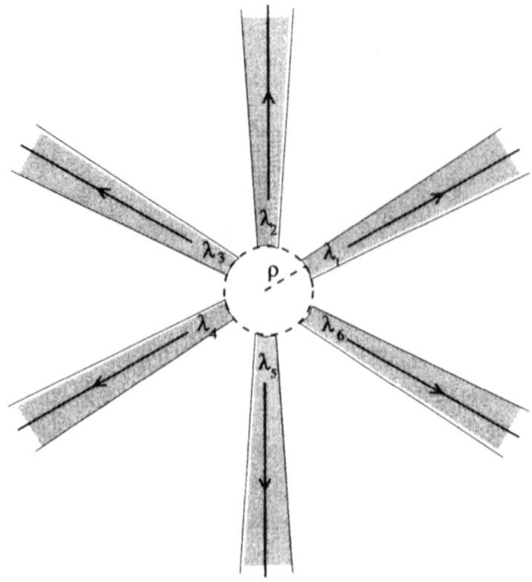

FIGURE A.1. The contours Γ_k^ρ and the sets Ω_k^ρ

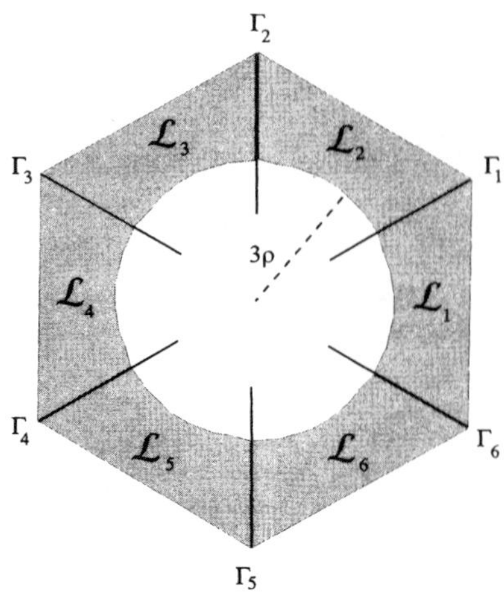

FIGURE A.2. The sectors $\mathcal{L}_k$

The goal of this section is to prove the following theorem.

THEOREM A.1 (The Sibuya theorem with a parameter). *Let* $s = (S_1, \ldots, S_6; \nu) \in \mathcal{M}$ *be the given monodromy data, i.e., the complex number* ν *and the set of six* 2×2 *matrices satisfying equations* (3.3.117) *and* (3.3.118). *Let* $\mathcal{K}$ *be a closed disk on the complex* x*-plane. Then, there exists a number* $\rho_0 \equiv \rho_0(\mathcal{K})$, *and a matrix function* $M(\lambda, x)$, *such that for all* $\rho \geq \rho_0$:

1. *$M(\lambda,x)\in H^0(\Omega_{3\rho}\times\mathcal{K})$.*
2. *For every $x\in\mathcal{K}$,*

$$M(\lambda,x)S_1\dots S_{k-1}\sim\Big(I+\sum_{j=1}^{\infty}\frac{\Psi_j^0(x)}{\lambda^j}\Big)e^{\{-\frac{4}{3}i\lambda^3\sigma_3-ix\lambda\sigma_3-\nu\ln\lambda\sigma_3\}},\tag{A.6}$$

$$\lambda\to\infty,\quad \lambda\in\mathcal{L}_k,\quad \Big|\arg\lambda-\frac{k-1}{3}\pi\Big|\le\frac{\pi}{6},$$

where the coefficients $\Psi_j^0(x)$ are holomorphic in $\mathcal{K}$.
3. *The asymptotics (A.6) is uniform, i.e., for any positive number l, there exists a constant $C>0$, such that*

$$\Big\|M(\lambda,x)S_1\dots S_{k-1}\exp\{\frac{4}{3}i\lambda^3\sigma_3+ix\lambda\sigma_3+\nu\ln\lambda\,\sigma_3\}-I-\sum_{j=1}^{l}\frac{\Psi_j^0}{\lambda^j}\Big\|\le\frac{C}{\lambda^{l+1}},$$

$$\forall(\lambda,x)\in\mathcal{L}_k\times\mathcal{K}.$$

4. *The asymptotics (A.6) is differentiable with respect to both x and λ.*

In an equivalent way, Theorem A.1 can be reformulated as follows.

THEOREM A.1′. *Let $s=(S_1,\dots,S_6;\nu)\in\mathcal{S}$ be the given monodromy data, i.e., the complex number ν and the set of six 2×2 matrices satisfying equations (3.3.117) and (3.3.118). Let $\mathcal{K}$ be a closed disk on the complex x-plane. Then, there exists a number $\rho_0\equiv\rho_0(\mathcal{K})$, and a collection of six matrix functions $Y_k(\lambda,x)$, $k=1,\dots,6$, such that for all $\rho\ge\rho_0$:*

1. *$Y_k(\lambda,x)\in H^0(\mathcal{L}_k\times\mathcal{K})$.*
2. *$Y_{k+1}(\lambda,x)=Y_k(\lambda,x)G_k(\lambda,x)$, $\quad\lambda\in\Gamma_k^\rho\cap\{|\lambda|\ge3\rho\}\quad k=1,\dots6$, where*

$$G_k(\lambda,x)=e^{-\theta(\lambda,x)\sigma_3}S_ke^{\theta(\lambda,x)\sigma_3},\quad \theta(\lambda,x)=i\frac{4}{3}\lambda^3+ix\lambda+\nu\ln\lambda,\tag{A.7}$$

$$\arg\lambda=\frac{2k-1}{6}\pi\quad\text{if}\quad\lambda\in\Gamma_k^\rho,$$

and $Y_7(\lambda,x)\equiv Y_1(\lambda,x)^2$.
3. *For every $x\in\mathcal{K}$, and $k=1,\dots,6$,*

$$Y_k(\lambda,x)\sim I+\sum_{j=1}^{\infty}\frac{\Psi_j^0(x)}{\lambda^j},\tag{A.8}$$

$$\lambda\to\infty,\quad\lambda\in\mathcal{L}_k,$$

where the coefficients $\Psi_j^0(x)$ are holomorphic in $\mathcal{K}$.
4. *The asymptotics (A.8) is uniform, i.e., for any positive number l, there exists a constant $C>0$, such that*

$$\Big\|Y_k(\lambda,x)-I-\sum_{j=1}^{l}\frac{\Psi_j^0}{\lambda^j}\Big\|\le\frac{C}{\lambda^{l+1}},$$

$$\forall(\lambda,x)\in\mathcal{L}_k\times\mathcal{K},\quad\text{and for all } k.$$

5. *The asymptotics (A.8) is differentiable with respect to both x and λ.*

---

[2]We emphasize that the sectors $\mathcal{L}_k$ belong to $\mathbb{C}$ but *not* to the universal covering of $\mathbb{C}\setminus\{0\}$.

To see that Theorem A.1′ implies Theorem A.1, let us define (cf. (A.1) and (A.2)) the function $M(\lambda, x)$ by the equations,

$$M(\lambda, x)\big|_{\mathcal{L}_1} = Y_1(\lambda, x)e^{-\theta(\lambda,x)\sigma_3},$$

$$M(\lambda, x)\big|_{\mathcal{L}_k} = Y_k(\lambda, x)e^{-\theta(\lambda,x)\sigma_3}\mathrm{S}_{k-1}^{-1}\cdots\mathrm{S}_1^{-1}, \quad k > 1,$$

where the branch of the $\ln\lambda$ is fixed by the condition,

$$\left|\arg\lambda - \frac{k-1}{3}\pi\right| \le \frac{\pi}{6} \quad \text{when} \quad \lambda \in \mathcal{L}_k.$$

The function $M(\lambda, x)$ defined above satisfies all the statements of the Theorem A.1. Indeed, because of the property 2 of $Y_k(\lambda, x)$, and of the cyclic constraint (3.3.118) on S_k, the function $M(\lambda, x)$ has no jumps across any of Γ_k^ρ and hence it is holomorphic in the whole $\Omega_{3\rho}$. The remaining of the properties stated in Theorem A.1 follows directly from the analogous properties of the functions $Y_k(\lambda, x)$.

In what follows, we prove Theorem A.4′. Our approach is based on replacing the jump conditions posed on the rays Γ_k^ρ, with some equivalent jump conditions posed on the horseshoe shape domains around the rays. To this end we introduce the following covering of the neighborhood Ω_ρ:

$$\Omega_\rho \subset \Omega_+ \cup \Omega_-,$$

where the sets Ω_+ and Ω_- are depicted in Figure A.3 and Figure A.4 respectively. These sets have the following properties:

- The set Ω_+ is obtained by deleting from the complex plane $\mathbb{C}$ a collection of thin neighborhoods, Ω_+^k, of the contours Γ_k^ρ. The closure of the neighborhood Ω_+^k is a subset of Ω_k^ρ. In other words,

$$\Omega_+ = \mathbb{C} \setminus \cup_{j=1}^6 \Omega_+^k, \quad \Gamma_k^\rho \subset \Omega_+^k, \quad \overline{\Omega_+^k} \subset \Omega_k^\rho.$$

- The set Ω_- is the union of the (slightly bigger) neighborhoods, Ω_-^k, of the contours Γ_k^ρ, whose closures are also the subsets of the respective Ω_k^ρ. In other words,

$$\Omega_- = \cup_{k=1}^6 \Omega_-^k, \quad \Gamma_k^\rho \subset \Omega_+^k, \quad \overline{\Omega_+^k} \subset \Omega_-^k, \quad \overline{\Omega_-^k} \subset \Omega_k^\rho.$$

- Let $\gamma^+ := \partial\Omega_+$ be the boundary of the set Ω_+. Then

$$\gamma^+ = \cup_k \gamma_k^+, \quad \gamma_k^+ = -\partial\Omega_+^k,$$

and we assume that each γ_k^+ is a simple analytic curve which encircles the ray Γ_k^ρ, and which is asymptotic to the rays $\Gamma_{(k;\ \pm\varepsilon/3)}$, where

$$\Gamma_{(k;\ \delta)} := \left\{\lambda \in \mathbb{C}: \quad \arg\lambda - \frac{2k-1}{6}\pi = \delta\right\}.$$

- Let $\gamma^- := \partial\Omega_-$ be the boundary of the set Ω_-. Then

$$\gamma^- = \cup_k \gamma_k^-, \quad \gamma_k^- = \partial\Omega_-^k,$$

and we assume that each γ_k^- is a simple analytic curve which encircles the ray Γ_k^ρ, and which is asymptotic to the rays $\Gamma_{(k;\ \pm\varepsilon/2)}$.

In addition to the sets Ω_+ and Ω_- we introduce their intersection,

$$\Omega_0 := \Omega_+ \cap \Omega_-.$$

The set Ω_0 is depicted in Figure A.5. Observe that the above listed properties of the sets Ω_+ and Ω_- imply the following properties of the set Ω_0.

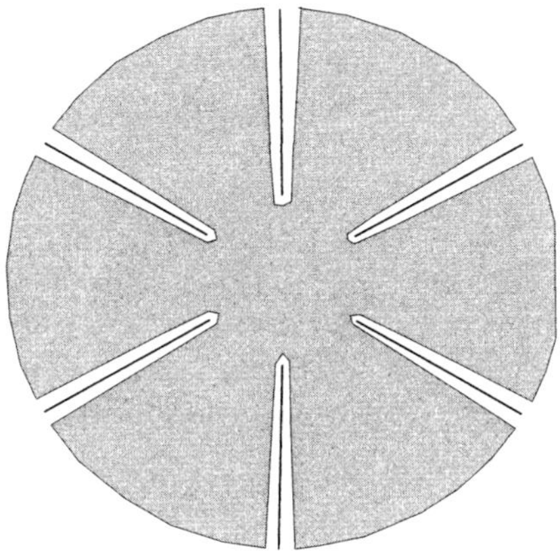

FIGURE A.3. The set $\Omega_+ = \mathbb{C} \setminus \cup_{j=1}^6 \Omega_+^k$

FIGURE A.4. the set $\Omega_- = \cup_{k=1}^6 \Omega_-^k$

- $\Omega_0 = \cup_{k=1}^6 \omega_k, \quad \overline{\omega_k} \subset \Omega_k^\rho \setminus \Gamma_k^\rho$, and each subset ω_k is an open and connected set.
- $\partial\omega_k = \gamma_k^+ + \gamma_k^-$.
- $\operatorname{dist}(\gamma_k^+, \gamma_k^-) \geq c_\varepsilon$.

We are now ready to introduce the main ingredient of our proof, namely the model functions $g_k(\lambda, x)$. Assuming that $\lambda \in \Omega_k^\rho \setminus \Gamma_k^\rho$, we define g_k by

$$g_k(\lambda, x) = \exp\Big\{-\frac{1}{2\pi i}\int_{\Gamma_k^\rho} \frac{\ln G_k(\mu, x)}{\mu - \lambda}\, d\mu\Big\}$$

$$(A.9) \qquad = \begin{cases} \begin{pmatrix} 1 & -\frac{s_k}{2\pi i}\int_{\Gamma_k^\rho} \frac{e^{-2\theta(\mu,x)}}{\mu-\lambda}\, d\mu \\ 0 & 1 \end{pmatrix}, & k = 2l, \\ \begin{pmatrix} 1 & 0 \\ -\frac{s_k}{2\pi i}\int_{\Gamma_k^\rho} \frac{e^{2\theta(\mu,x)}}{\mu-\lambda}\, d\mu & 1 \end{pmatrix}, & k = 2l-1, \end{cases}$$

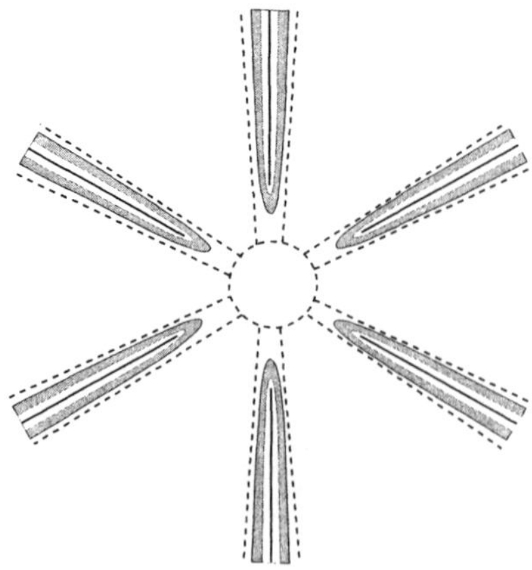

FIGURE A.5. The set $\Omega_0 = \Omega_- \cap \Omega_+ = \cup_{k=1}^{6} \omega_k$

where the jump matrices $G_k(\lambda, x)$ and the phase function $\theta(\lambda, x)$ are given in (A.7). The characteristic property of $g_k(\lambda, x)$ is the jump relation,

$$g_{k+}(\lambda, x) = g_{k-}(\lambda, x) G_k^{-1}(\lambda, x), \quad \lambda \in \Gamma_k^{\rho} \cap \{|\lambda| \geq 2\rho + \delta\}, \tag{A.10}$$

where $g_{k+}(\lambda, x)$ and $g_{k-}(\lambda, x)$ denote the boundary values of $g_k(\lambda, x)$ as λ approaches Γ_k^{ρ} from the left and from the right, respectively.

On the set $\Omega_0 \times \mathcal{K}$, the function $g(\lambda, x)$ can be defined by setting

$$g(\lambda, x)\big|_{\omega_k} = g_k(\lambda, x). \tag{A.11}$$

The following are the direct consequences of the *explicit* formulae (A.9) and (A.11), as well as of the geometric properties of the set Ω_0 indicated above.

(i) $g(\lambda, x)$ is holomorphic in both λ and x. Indeed,

$$g(\lambda, x) \in H^0(\Omega_0 \times \mathcal{K}).$$

(ii) The following uniform estimate is valid,

$$\sup_{\substack{\lambda \in \Omega_0 \\ x \in \mathcal{K}}} \|\lambda (g(\lambda, x) - I)\| \leq c_{\rho, \mathcal{K}}. \tag{A.12}$$

Moreover,

$$c_{\rho, \mathcal{K}} \to 0 \quad \text{as} \quad \rho \to \infty. \tag{A.13}$$

The next lemma, is one of the local versions of the Birkhoff-Grothendieck theorem with a parameter. This is the main technical step in our proof of the Sibuya Theorem A.1′. The basic idea is the uniform boundness of the relevant Cauchy operators (see Proposition A.1 below), which is the main reason for moving from the rays Γ_k^{ρ} to the horseshoe domains ω_k. This idea was first suggested in [**AnoBol**] (see also [**Bol4**] and Appendix B).

LEMMA A.2 (The basic factorization lemma). *For sufficiently large $\rho > 0$, there exist functions $\Psi_{\pm}(\lambda, x)$, such that:*

1. $\Psi_{\pm}(\lambda, x) \in H^0(\Omega_{\pm} \times \mathcal{K})$.
2. $\sup_{(\lambda,x)\in\Omega_{\pm}\times\mathcal{K}} \|\lambda(I - \Psi_{\pm}(\lambda, x))\| \le C.$
3. $\Psi_{+}(\lambda, x)g(\lambda, x) = \Psi_{-}(\lambda, x) \quad \forall\lambda \in \Omega_0, \quad \forall x \in \mathcal{K}.$

PROOF. Let us introduce three more Banach spaces.

$$H = \Big\{f \in H(\Omega_0 \times \mathcal{K}), \quad \sup_{(\lambda,x)\in\Omega_0\times\mathcal{K}} |\lambda f(\lambda, x)| < \infty\Big\},$$

$$H_{\pm} = \Big\{f \in H(\Omega_{\pm} \times \mathcal{K}), \quad \sup_{(\lambda,x)\in\Omega_{\pm}\times\mathcal{K}} |(1 + |\lambda|)f(\lambda, x)| < \infty\Big\} \subset H,$$

$$\|f\|_H \equiv \sup_{(\lambda,x)\in\Omega_0\times\mathcal{K}} |\lambda f(\lambda, x)|,$$

$$\|f\|_{H_{\pm}} \equiv \sup_{(\lambda,x)\in\Omega_{\pm}\times\mathcal{K}} |(1 + |\lambda|)f(\lambda, x)|.$$

Given

$$f \in H,$$

we can define the Cauchy operators:

$$(P_{\pm}f)(\lambda, x) = \frac{1}{2\pi i}\int_{\gamma^{\pm}} \frac{f(\mu, x)}{\mu - \lambda}\, d\mu, \quad \lambda \in \Omega_{\pm}.$$

We recall that

$$\gamma^{\pm} = \partial\Omega_{\pm},$$

and that

$$\partial\Omega_0 = \gamma^+ + \gamma^- = \sum_{k=1}^{6} \partial\omega_k = \sum_{k=1}^{6}(\gamma_k^+ + \gamma_k^-).$$

Note that the Cauchy formula,

$$f(\lambda, x) = (P_+f)(\lambda, x) + (P_-f)(\lambda, x), \quad \lambda \in \Omega_0, \tag{A.14}$$

is valied for the functions from H.

Put

$$g(\lambda, x) = I + \overset{o}{g}(\lambda, x), \quad \overset{o}{g}(\lambda, x) \in H,$$

and introduce the operators $P_{\pm}^g$ by the equation,

$$P_{\pm}^g(f) := P_{\pm}(f\, g^o).$$

PROPOSITION A.1. *$P_{\pm}^g$ are bounded operators from H to $H_{\pm}$.*

PROOF OF THE PROPOSITION (an adaptation of the technique of [**AnoBol**]). Each ω_k can be represented as the union of the domains ω_k^+ and ω_k^- according to the Figure 6. Assume that $\lambda \in \Omega_+ \setminus \cup_k \omega_k^+$ and hence

$$\frac{|\lambda|}{\operatorname{dist}\{\lambda, \gamma_+\}} \le C_\epsilon,$$

for some positive constant C_ϵ. Then

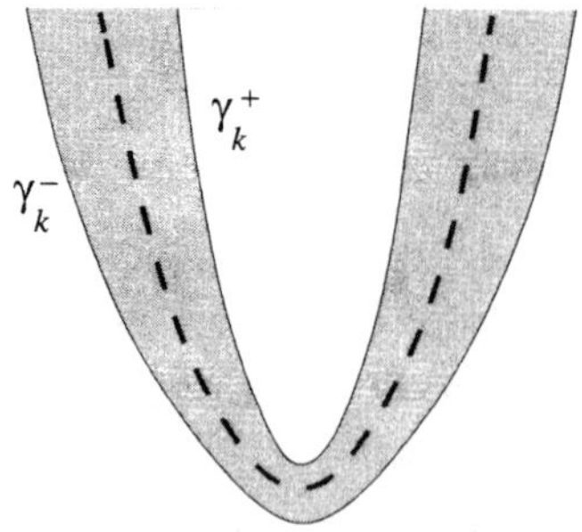

FIGURE A.6. $\omega_k = \omega_k^+ \cup \omega_k^-$

$$\lambda\big(P_+^g f\big)(\lambda,x) = \frac{\lambda}{2\pi i}\int_{\gamma^+}\frac{f(\mu,x)g^o(\mu,x)}{\mu-\lambda}\,d\mu,$$

$$\begin{aligned}\big|\lambda\big(P_+^g f\big)(\lambda,x)\big| &\le \frac{|\lambda|}{2\pi}\|f\|_H\|g^o\|_H\int_{\gamma^+}\frac{|d\mu|}{|\mu|^2|\mu-\lambda|}\\ &\le \frac{|\lambda|}{2\pi\,\mathrm{dist}\,\{\lambda,\gamma_+\}}\|f\|_H\|g^o\|_H\int_{\gamma^+}\frac{|d\mu|}{|\mu|^2}\\ &\le \frac{C_\epsilon}{2\pi}\|f\|_H\|g^o\|_H\int_{\gamma^+}\frac{|d\mu|}{|\mu|^2} \equiv C_+\cdot\|f\|_H\,.\end{aligned}$$

Similarly, if $\lambda \in \Omega_-\setminus \cup_k\,\omega_k^-$, then

$$\big|\lambda\big(P_-^g f\big)(\lambda,x)\big| \le C_-\cdot\|f\|_H\,. \tag{A.15}$$

Suppose now that $\lambda \in \omega_k^+$. Then, we can use identity (A.14) and rewrite $\lambda\big(P_+^g f\big)(\lambda,x)$ as

$$\lambda\big(P_+^g f\big)(\lambda,x) = \lambda f(\lambda,x)g^o(\lambda,x) - \lambda\big(P_-^g f\big)(\lambda,x). \tag{A.16}$$

On the other hand, $\lambda \in \omega_k^+ \Rightarrow \lambda\in\Omega_-\setminus\cup_k\,\omega_k^-$, and hence we can use (A.15) in the right-hand side of (A.16). Therefore (cf. [**AnoBol**] and the proof of of Lemma B.1 in Appendix B), we obtain,

$$\big|\lambda\big(P_+^g f\big)(\lambda,x)\big| \le C'_+\cdot\|f\|_H,$$

if $\lambda\in\omega_k^+$. In other words, we have the inequality

$$\big|\lambda\big(P_+^g f\big)(\lambda,x)\big| \le C''\cdot\|f\|_H,\quad C'' := \max\{C_+, C'_+\} \tag{A.17}$$

for all $(\lambda,x)\in\overline{\Omega_+}\times\mathcal{K}$.

Assuming that $|\lambda|\le\rho/2$ we obtain at once that

$$\begin{aligned}\big|\big(P_+^g f\big)(\lambda,x)\big| &\le \frac{1}{2\pi}\|f\|_H\|g^o\|_H\int_{\gamma^+}\frac{|d\mu|}{|\mu|^2|\mu-\lambda|}\\ &\le \frac{1}{2\pi\,\mathrm{dist}\,\{\lambda,\gamma_+\}}\|f\|_H\|g^o\|_H\int_{\gamma^+}\frac{|d\mu|}{|\mu|^2} \equiv \tilde{C}_+\cdot\|f\|_H\,.\end{aligned}$$

This estimate together with (A.17) imply

$$\big\|P_+^g f\big\|_{H_+} \le C\cdot\|f\|_H,\qquad C := \max\{C'',\tilde{C}_+\}$$

Similarly,

$$\left\| P_-^g f \right\|_{H_-} \leq C' \cdot \|f\|_H$$

and the proposition is proved. □

Two additional facts should be noted. First, all the C-constants above contain the norm of g^o, i.e., $\left\| g^o \right\|_H$, as a factor. Hence, in virtue of estimates (A.12) and (A.13), the norm of the operators $P_\pm^g$ can be made as small as we wish by choosing ρ large enough. Secondly, from the integral formula (A.9) we derive the existence of the asymptotic expansion of the function $g^0(\lambda, x)$ over λ^{-1} as $\lambda \to \infty$, $\lambda \in \Omega_0$. This, in particular, implies that there exist the matrix function $c(x)$ and the constant C_0 such that

$$g^o(\lambda, x) = \frac{c(x)}{\lambda} + \tilde{g}(\lambda, x), \quad \left|\tilde{g}(\lambda, x)\right| \leq \frac{C_0}{|\lambda|^2}, \tag{A.18}$$

for all $(\lambda, x) \in \overline{\Omega_0} \times \mathcal{K}$. This estimate allows us to claim that

$$P_+ g^o \in H_+. \tag{A.19}$$

To see (A.19), we note that

$$\left(P_+ g^o\right)(\lambda, x) = \frac{c(x)}{2\pi i} \int_{\gamma^+} \frac{d\mu}{\mu(\mu - \lambda)} + \frac{1}{2\pi i} \int_{\gamma^+} \frac{\tilde{g}(\mu, x)}{(\mu - \lambda)}\, d\mu. \tag{A.20}$$

Denote the second integral as $g_r(\lambda, x)$. In virture of the estimate (A.18) and repeating the same line of arguments as in the proof of Proposition A.1, we conclude that

$$g_r \in H_+, \quad \text{and moreover} \quad \left\| g_r \right\|_{H_+} \to 0, \quad \rho \to \infty.$$

Since the first integral in (A.20) is obviously zero for all $\lambda \in \Omega_+$, we arrive at (A.19). Moreover, we have that

$$\left\| P_+ g^0 \right\|_{H_+} \to 0, \quad \rho \to \infty. \tag{A.21}$$

We are ready now to complete the proof of Lemma A.2.

We are looking for

$$\Psi_\pm(\lambda, x) = I + \overset{o}{\Psi}_\pm(\lambda, x), \quad \overset{o}{\Psi}_\pm(\lambda, x) \in H_\pm,$$

such that

$$I + \overset{o}{\Psi}_- = \left(I + \overset{o}{\Psi}_+\right)\left(I + \overset{o}{g}\right),$$

or

$$\overset{o}{\Psi}_- = \overset{o}{\Psi}_+ + \overset{o}{\Psi}_+ \overset{o}{g} + \overset{o}{g}.$$

Applying P_+, we have

$$0 = \overset{o}{\Psi}_+ + P_+\left(\overset{o}{\Psi}_+ \overset{o}{g}\right) + P_+\left(\overset{o}{g}\right).$$

Hence, we need just to solve the following integral equation in the space H_+:

$$\overset{o}{\Psi}_+ + P_+^g\left(\overset{o}{\Psi}_+\right) = -P_+ \overset{o}{g}.$$

As we have already indicated, because of estimates (A.12) and (A.13), we conclude that the operator,

$$I + P_+^g$$

is invertible, for sufficiently large ρ, in the Banach space H_+, and, due to (A.21), the norm of $\Psi^o{}_+$ can be made small enough to ensure the invertability of matrix $\Psi_+(\lambda, x)$ for all $(\lambda, x) \in \Omega_+ \times \mathcal{K}$. This completes the proof of the Lemma. □

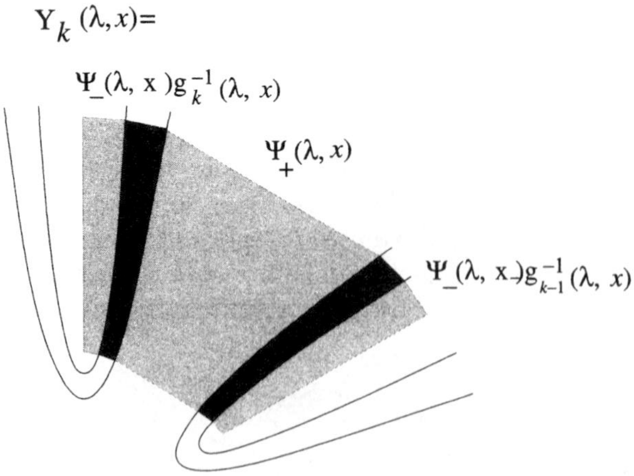

FIGURE A.7. Functions $Y_k(\lambda, x)$

PROOF OF THEOREM A.1$'$. Define the functions $Y_k(\lambda, x)$ as shown in Figure A.7. We claim that these functions satisfy all the properties stated in the theorem. The only properties which do not follow directly from the definition of the functions $Y_k(\lambda, x)$, and hence need a proof, are the asymptotic statements 3–5. The proof is based on the use of the appropriate integral representations for the functions $Y_k(\lambda, x)$; these representations can be obtained as follows (cf. [**Pa**]).

We first note that property 2 of the functions $\Psi_\pm(\lambda, x)$ implies the estimate

$$\sup_{(\lambda,x)\in\mathcal{L}_k\times\mathcal{K}} \|\lambda\big(I - Y_k(\lambda, x)\big)\| \le C, \tag{A.22}$$

which allows us to use the Cauchy formula in the sectors $\mathcal{L}_k$, and to obtain the equations,

$$Y_k(\lambda, x) = I + \int_{\partial\mathcal{L}_k} \frac{Y_k(\mu, x) - I}{\mu - \lambda}\frac{d\mu}{2\pi i}, \quad \lambda \in \overset{o}{\mathcal{L}}_k, \tag{A.23}$$

and

$$0 = \int_{\partial\mathcal{L}_k} \frac{Y_k(\mu, x) - I}{\mu - \lambda}\frac{d\mu}{2\pi i}, \quad \lambda \notin \mathcal{L}_k, \tag{A.24}$$

where $\mathcal{L}^o{}_k$ denotes the interior of $\mathcal{L}_k$. Denote by

$$\Gamma_k := \Gamma_k^\rho \cap \{|\lambda| \ge 3\rho\},$$

and let C_k be the arc of the circle $|\lambda| = 3\rho$, between the rays Γ_{k-1} and Γ_k (see Figure A.2). Assume that C_k is oriented counterclockwise. Then,

$$\partial\mathcal{L}_k = -\Gamma_k - C_k + \Gamma_{k-1},$$

and we can rewrite (A.23) as,

$$Y_k(\lambda, x) = I - \int_{C_k} \frac{Y_k(\mu, x) - I}{\mu - \lambda}\frac{d\mu}{2\pi i}$$

$$-\int_{\Gamma_k} \frac{Y_k(\mu, x) - I}{\mu - \lambda}\frac{d\mu}{2\pi i} + \int_{\Gamma_{k-1}} \frac{Y_k(\mu, x) - I}{\mu - \lambda}\frac{d\mu}{2\pi i}, \tag{A.25}$$

where $\lambda \in \mathcal{L}^o{}_k$. Observe that from property 2 of Y_k, we have that

$$Y_k(\lambda,x) - I = -Y_k(\lambda,x)\big(G_k(\lambda,x) - I\big) + Y_{k+1}(\lambda,x) - I, \quad \lambda \in \Gamma_k,$$

and

$$Y_k(\lambda,x) - I = Y_{k-1}(\lambda,x)\big(G_{k-1}(\lambda,x) - I\big) + Y_{k-1}(\lambda,x) - I, \quad \lambda \in \Gamma_{k-1}.$$

Therefore, (A.25) can be transformed into

$$Y_k(\lambda,x) = I - \int_{C_k} \frac{Y_k(\mu,x) - I}{\mu - \lambda}\frac{d\mu}{2\pi i}$$

$$+ \int_{\Gamma_k} \frac{Y_k(\mu,x)\big(G_k(\mu,x) - I\big)}{\mu - \lambda}\frac{d\mu}{2\pi i} + \int_{\Gamma_{k-1}} \frac{Y_{k-1}(\mu,x)\big(G_{k-1}(\mu,x) - I\big)}{\mu - \lambda}\frac{d\mu}{2\pi i}$$

$$(A.26) \qquad - \int_{\Gamma_k} \frac{Y_{k+1}(\mu,x) - I}{\mu - \lambda}\frac{d\mu}{2\pi i} + \int_{\Gamma_{k-1}} \frac{Y_{k-1}(\mu,x) - I}{\mu - \lambda}\frac{d\mu}{2\pi i}.$$

Using the second Cauchy relation (A.24), we can replace the last two integrals by

$$\int_{C_{k+1}} \frac{Y_{k+1}(\mu,x) - I}{\mu - \lambda}\frac{d\mu}{2\pi i} + \int_{\Gamma_{k+1}} \frac{Y_{k+1}(\mu,x) - I}{\mu - \lambda}\frac{d\mu}{2\pi i},$$

and

$$- \int_{C_{k-1}} \frac{Y_{k-1}(\mu,x) - I}{\mu - \lambda}\frac{d\mu}{2\pi i} + \int_{\Gamma_{k-2}} \frac{Y_{k-1}(\mu,x) - I}{\mu - \lambda}\frac{d\mu}{2\pi i},$$

respectively. Thus, equation (A.26) can be further transformed into the equation,

$$Y_k(\lambda,x) = I - \int_{C_{k+1}} \frac{Y_{k+1}(\mu,x) - I}{\mu - \lambda}\frac{d\mu}{2\pi i} - \int_{C_k} \frac{Y_k(\mu,x) - I}{\mu - \lambda}\frac{d\mu}{2\pi i}$$
$$- \int_{C_{k-1}} \frac{Y_{k-1}(\mu,x) - I}{\mu - \lambda}\frac{d\mu}{2\pi i}$$

$$+ \int_{\Gamma_k} \frac{Y_k(\mu,x)\big(G_k(\mu,x) - I\big)}{\mu - \lambda}\frac{d\mu}{2\pi i} + \int_{\Gamma_{k-1}} \frac{Y_{k-1}(\mu,x)\big(G_{k-1}(\mu,x) - I\big)}{\mu - \lambda}\frac{d\mu}{2\pi i}$$

$$(A.27) \qquad - \int_{\Gamma_{k+1}} \frac{Y_{k+1}(\mu,x) - I}{\mu - \lambda}\frac{d\mu}{2\pi i} + \int_{\Gamma_{k-2}} \frac{Y_{k-1}(\mu,x) - I}{\mu - \lambda}\frac{d\mu}{2\pi i}.$$

Comparing the two integrals in the last row of (A.27) with the analogous integrals of (A.25), we see that the effect of the manipulations that led us from (A.25) to (A.27) is, the increase of the index k in the first integral and its decrease in the second one. Therefore, repeating several times the above procedure, we arrive at the following final integral representation for $Y_k(\lambda,x)$,

$$Y_k(\lambda,x) = I - \sum_{l=1}^{6} \int_{C_l} \frac{Y_l(\mu,x) - I}{\mu - \lambda}\frac{d\mu}{2\pi i}$$

$$(A.28) \qquad + \sum_{j=1}^{6} \int_{\Gamma_l} \frac{Y_l(\mu,x)\big(G_l(\mu,x) - I\big)}{\mu - \lambda}\frac{d\mu}{2\pi i},$$

for all $\lambda \in \mathcal{L}^o{}_k$.

Each term of the first sum in the r.h.s. of (A.28) is holomorphic in the neighborhood of $\lambda = \infty$, while the integrand of each term in the second sum decays exponentially as $\lambda \to \infty$. In fact, there exist positive constants C and c such that

$$\|Y_l(\mu, x)\big(G_l(\mu, x) - I\big)\| < Ce^{-c|\lambda|^3}, \quad \mu \in \Gamma_l, \quad x \in \mathcal{K}.$$

This means that, we indeed have for $Y_k(\lambda, x)$, uniformly in $x \in \mathcal{K}$, the full asymptotic series (A.8) with coefficients given by the equation,

$$\Psi_j^0(x) = \sum_{l=1}^{6} \int_{C_l} \big(Y_l(\mu, x) - I\big)\mu^{j-1}\frac{d\mu}{2\pi i}$$

$$-\sum_{l=1}^{6} \int_{\Gamma_l} Y_l(\mu, x)\big(G_l(\mu, x) - I\big)\mu^{j-1}\frac{d\mu}{2\pi i}. \tag{A.29}$$

¿From this equation, in particular, it follows that all $\Psi_j^0(x)$ are holomorphic functions.

By standard arguments, the integral representation (A.28) implies the x and λ-differentiability of the asymptotics (A.8). In fact, the λ and x-differentiability of the asymptotics (A.8) is a direct consequence of the fact that it is uniformly valid (see, e.g. [**W1**]).

Finally, we note that the asymptotic expansion (A.8) is valid in the closed sector $\mathcal{L}_k$ (in fact, in some sector which includes $\mathcal{L}_k$). Indeed, because of property 2 each of the functions $Y_k(\lambda, x)$ can be analytically continued beyond the boundaries of $\mathcal{L}_k$, so that $Y_k(\lambda, x)$ becomes holomorphic and satisfies the estimate (A.22) in a sector which includes $\mathcal{L}_k$. Therefore, in the representation (A.28) we can slightly rotate the rays Γ_k and Γ_{k-1} so that we can make the representation, and hence the asymptotics (A.8) valid for all $\lambda \in \mathcal{L}_k$. □

0.2. The local solution of the RH problem. Denote by Ω_0, Ω_∞, and ω the following subsets of the Riemann sphere $\mathbb{C}P^1$:

$$\Omega_0 = \{\lambda \in \mathbb{C} : |\lambda| \le R\},$$
$$\Omega_\infty = \{\lambda \in \mathbb{C} \cup \infty : |\lambda| \ge 4\rho\},$$
$$\omega = \Omega_0 \cap \Omega_\infty,$$

where $R > 4\rho$. Also denote by $D_\delta(x_0)$ the closed disk in the x-plane of radius δ and centered at x_0, i.e.,

$$D_\delta(x_0) = \{x \in \mathbb{C} : |x - x_0| \le \delta\}.$$

Let $M(\lambda, x)$ be the Sibuya function of the previous section corresponding to the choice

$$\mathcal{K} = D_\delta(x_0).$$

Observe that

$$M(\lambda, x) \in H^0\Big(\big(\Omega_\infty \setminus \{\infty\}\big) \times D_\delta(x_0)\Big), \tag{A.30}$$

for all

$$\rho \ge \rho_0(D_\delta(x_0)) \equiv \rho_0(x_0, \delta).$$

In particular,

$$M(\lambda, x) \in H^0(\omega \times D_\delta(x_0)). \tag{A.31}$$

By a direct application to the function $M(\lambda, x)$ of the Birkhoff-Grothendieck theorem with a parameter, which is formulated and proven in Appendix B, we arrive at the following result.

THEOREM A.2. *Let $x_0 \in \mathbb{C}$. Then there exists a number $\delta_0 > 0$, the finite set $X^0 \equiv \{x_{0j}\} \subset D_{\delta_0}(x_0)$, and the matrix functions $T(\lambda, x)$ and $\Phi(\lambda, x)$ such that:*

- *$T(\lambda, x)$ and $T^{-1}(\lambda, x)$ are holomorphic in $\Omega_\infty \times \big(D_{\delta_0}(x_0) \setminus X^0 0\big)$ and meromorphic along $\Omega_\infty \times X^0$ (i.e., they have poles at $x_{0j} \in X^0$ and the coefficients of the corresponding Laurent series are holomorphic in D_∞).*
- *$\Phi(\lambda, x)$ and $\Phi^{-1}(\lambda, x)$ are holomorphic in $\Omega_0 \times \big(D_{\delta_0}(x_0) \setminus X^0\big)$ and meromorphic along $\Omega_0 \times X^0$.*
-

$$M(\lambda, x) = T^{-1}(\lambda, x)\lambda^{\begin{pmatrix} \kappa_1 & 0 \\ 0 & \kappa_2 \end{pmatrix}}\Phi(\lambda, x), \tag{A.32}$$

where $\kappa_1 \geq \kappa_2$ are some integer numbers, which are the same for all $x \in D_{\delta_0}(x_0) \setminus X^0$.

The specific features of our function $M(\lambda, x)$, yield the important additional properties of the objects participating in the factorization formula (A.32).

PROPOSITION A.2. *The function $\Phi(\lambda, x)$ is an entire function of λ; indeed,*

$$\Phi(\lambda, x) \in H^0\Big(\mathbb{C} \times \big(D_{\delta_0}(x_0) \setminus X^0\big)\Big),$$

and both $\Phi(\lambda, x)$ and $\Phi^{-1}(\lambda, x)$ are meromorphic along $D_0 \times X^0$. Moreover,

$$\Phi(\lambda, x)S_1 \ldots S_{k-1} \sim \lambda^{\begin{pmatrix} -\kappa_1 & 0 \\ 0 & -\kappa_2 \end{pmatrix}} B_0\Big(I + \sum_{j=1}^{\infty} \frac{\Psi_j(x)}{\lambda^j}\Big)e^{-\theta(\lambda, x)\sigma_3}, \tag{A.33}$$

$$\lambda \to \infty, \quad \lambda \in \mathcal{L}_k, \quad \det B_0 \neq 0,$$

where the coefficients $\Psi_j(x)$ are holomorphic in $D_{\delta_0}(x_0) \setminus X^0$ and meromorphic in $D_{\delta_0}(x_0)$. The asymptotics is uniform in $x \in D$, where D is an arbitrary compact subset of $D_{\delta_0}(x_0) \setminus X^0$, and it is differentiable with respect to both x and λ.

The proposition is an immediate consequence of equations (A.30) and (A.32), and of the asymptotic properties of $M(\lambda, x)$ (see (A.6)). We note also that the coefficients $\Psi_j(x)$ are finite combinations of the coefficients $\Psi_j^0(x)$ and the coefficients of the Taylor expansions of $T(\lambda, x)$ at $\lambda = \infty$.

PROPOSITION A.3. *The numbers κ_1 and κ_2 in (A.32) satisfy the equation,*

$$\kappa_1 + \kappa_2 = 0. \tag{A.34}$$

PROOF. The matrix equation (A.32) implies the following relation in the annular domain $\omega = \Omega_0 \cap \Omega_\infty$ for the corresponding determinants,

$$\det \Phi(\lambda, x) = \det T(\lambda, x) \det M(\lambda, x)\lambda^{-\kappa_1 - \kappa_2}. \tag{A.35}$$

At the same time, we observe that the asymptotic properties of the matrix function $M(\lambda, x)$ (see statement 3 of Theorem A.1) yield

$$\det M \to 1, \tag{A.36}$$

as $\lambda \to \infty$.

Assume that $\kappa_1 + \kappa_2 \neq 0$; let

$$\phi(\lambda, x) := \begin{cases} \det \Phi(\lambda, x), & \text{for} \quad \lambda \in \Omega_0, \\ \det T(\lambda, x) \det M(\lambda, x) \lambda^{-\kappa_1 - \kappa_2}, & \text{for} \quad \lambda \in \Omega_\infty \setminus \{\infty\} \end{cases}$$

($x \in D_{\delta_0}(x_0) \setminus X^0$). Equations (A.35) and (A.36) mean that, for every x, the (entire) function $\phi_x(\lambda) \equiv \phi(\lambda, x)$ is either identically zero ($\kappa_1 + \kappa_2 > 0$) or it is a polynomial of degree $-\kappa_1 - \kappa_2$ ($\kappa_1 + \kappa_2 < 0$). In both cases there are points on the finite complex plane λ where the matrix $\Phi(\lambda, x)$ is not invertible. This contradicts proposition 2 and hence the sum $\kappa_1 + \kappa_2$ must be zero. □

Equation (A.34) allows us to specify (A.32) as follows:

$$M(\lambda, x) = T^{-1}(\lambda, x)\lambda^{\kappa\sigma_3}\Phi(\lambda, x), \tag{A.37}$$

where $\kappa \geq 0$ is an integer number (the same for all $x \in D_{\delta_0}(x_0) \setminus X_s^0$), and σ_3 is defined by

$$\sigma_3 = \begin{pmatrix} 1 & 0 \\ 0 & -1 \end{pmatrix}.$$

PROPOSITION A.4. *The number κ in equation (A.37) is (identically) zero.*

PROOF. It is this statement that we use the assumption (A.5). Suppose that $\kappa > 0$. Then, the functions $T(\lambda, x)$ and $\Phi(\lambda, x)$ in (A.32) are defined up to the following transformations,

$$T(\lambda, x) \mapsto \tilde{T}(\lambda, x) = \begin{pmatrix} c_1 & 0 \\ \sum_{j=0}^{2\kappa} a_j \lambda^{j-2\kappa} & c_2 \end{pmatrix} T(\lambda, x),$$

$$\Phi(\lambda, x) \mapsto \tilde{\Phi}(\lambda, x) = \begin{pmatrix} c_1 & 0 \\ \sum_{j=0}^{2\kappa} a_j \lambda^{j} & c_2 \end{pmatrix} \Phi(\lambda, x),$$

where $c_1 c_2 \neq 0$ and $a_0, \dots, a_{2\kappa}$, are arbitrary. Using this freedom, we can adjust the matrix functions $T(\lambda, x)$ and $\Phi(\lambda, x)$ in such a way (perhaps adding a few more points to the singular set X^0 and slightly reducing the radius δ_0) so that asymptotic relation (A.33) can be transformed into the relation,

$$\Phi(\lambda, x) S_1 \dots S_{k-1} \sim \Big(I + \sum_{j=1}^{\infty} \frac{\Psi_j(x)}{\lambda^j}\Big) e^{-\left(\theta(\lambda, x) + \kappa \ln \lambda\right)\sigma_3}, \tag{A.38}$$

$$\lambda \to \infty, \quad \lambda \in \mathcal{L}_k,$$

with the meromorphic in $D_{\delta_0}(x_0)$ and holomorphic in $D_{\delta_0}(x_0) \setminus X^0$ coefficients $\Psi_k(x)$, such that

$$(\Psi_j)_{12} \equiv 0, \quad \forall j < 2\kappa. \tag{A.39}$$

The asymptotics (A.38) is still uniform and differentiable with respect to both x and λ.

Applying to the logarithmic derivatives $\Phi_\lambda \Phi^{-1}$ and $\Phi_x \Phi^{-1}$ the arguments based on the Liouville theorem, which we have already used in Chapters 3 and 4 when deriving the isomonodromy deformation equations (4.2.28), we conclude that the function $\Phi(\lambda, x)$ satisfies the isomonodromy Lax pair (4.2.24). The functional parameters $u(x)$, $v(x)$, $w(x)$, and $z(x)$ are meromorphic in $D_{\delta_0}(x_0)$ and holomorphic in $D_{\delta_0}(x_0) \setminus X^0$; furthermore, they satisfy the general relations,

$$w = u_x, \quad y = v_x, \tag{A.40}$$

$$\nu + \kappa = vw - uy. \tag{A.41}$$

Moreover,

$$u = 2(\Psi_1)_{12}, \quad \text{and} \quad v = 2(\Psi_1)_{21}.$$

¿From the first equation and (A.39) we conclude that

$$u(x) \equiv 0,$$

which in virtue of (A.40) means that

$$w(x) \equiv 0$$

as well, and hence the right-hand side of (A.41) is identically zero, but this cannot be, since ν is not an integer. This contradiction proves that κ must be zero[3]. □

Propositions A.2–A.4 yield the following corollary of Theorem A.2 which constitutes the *local* (with respect to x) solution of the original RH problem.

THEOREM A.3. *Let $x_0 \in \mathbb{C}$. Then, there exists a number $\delta_0 > 0$, the finite set $X_s^0 \equiv \{x_{0j}\} \subset D_{\delta_0}(x_0)$, and the unique matrix function $\Psi_0(\lambda, x)$, such that:*

- **(i)** *$\Psi_0(\lambda, x)$ and $\Psi_0^{-1}(\lambda, x)$ are holomorphic in $\mathbb{C} \times \left(D_{\delta_0}(x_0) \setminus X_s^0\right)$ and meromorphic along $\mathbb{C} \times X_s^0$.*
- **(ii)** $\det \Psi_0(\lambda, x) \equiv 1$.
- **(iii)** *For every $x \in D_{\delta_0}(x_0) \setminus X_s^0$,*

$$\Psi_0(\lambda, x)S_1 \dots S_{k-1} \sim \Big(I + \sum_{j=1}^{\infty} \frac{\Psi_j(x)}{\lambda^j}\Big)e^{-\theta(\lambda,x)\sigma_3},$$

$$\lambda \to \infty, \quad \lambda \in \mathcal{L}_k,$$

where the coefficients $\Psi_j(x)$ are holomorphic in $D_{\delta_0}(x_0) \setminus X_s^0$ and meromorphic in $D_{\delta_0}(x_0)$. These asymptotics are uniform in $x \in D_{\delta_0}(x_0) \setminus X_s^{0,\epsilon}$ and differentiable with respect to both x and λ. $X_s^{0,\epsilon}$ denotes an arbitrary ϵ-neighborhood of the set X_s^0.

PROOF. Let $\Phi(\lambda, x)$ and $T(\lambda, x)$ be the functions from Theorem A.2. Set

$$\Psi_0(\lambda, x) := T^{-1}(\infty, x)\Phi(\lambda, x).$$

We will argue that this function satisfies the conditions **(i)**–**(iii)** above. Indeed, condition **(i)** follows directly from Proposition A.2 and the analytic properties of $T(\lambda, x)$. By virtue of (A.37) and Proposition A.4, the following equation holds:

$$\Psi_0(\lambda, x) = T^{-1}(\infty, x)T(\lambda, x)M(\lambda, x).$$

Note that

$$T^{-1}(\infty, x)T(\lambda, x) = I + \sum_{j=1}^{\infty} T_j(x)\lambda^{-j},$$

---

[3]Strictly speaking, in the proof we have excluded the case where $T_{11}(\infty, x) \equiv 0$. If $T_{11}(\infty, x) \equiv 0$, then κ must be replaced by $-\kappa$ in (A.38), and "12" by "21" in (A.39). The rest of the proof will go along the same lines as before.

where the series converges in Ω_∞ uniformly for $x \in D$, $\overline{D} \subset D_{\delta_0}(x_0) \setminus \Theta_0$, and the coefficients $T_j(x)$ are holomorphic in $D_{\delta_0}(x_0) \setminus X_s^0$ and meromorphic in $D_{\delta_0}(x_0)$. Hence, taking into account the asymptotic properties of the Sibuya function $M(\lambda, x)$ (see statement 2 of Theorem A.4), we arrive to the asymptotic relation **(iii)**. To obtain **(ii)**, we note that for every fixed x,

$$\det \Psi_0(\lambda, x) \to 1, \quad \lambda \to \infty,$$

and hence **(ii)** follows via the Liouville theorem. Finally, to prove the uniqueness statement of the theorem, assume that $\tilde{\Psi}_0(\lambda, x)$ is another function satisfying **(i)**–**(iii)** and consider the ratio,

$$R(\lambda, x) := \tilde{\Psi}_0(\lambda, x)\Psi_0^{-1}(\lambda, x).$$

For every fixed x, we have that $R(\lambda, x)$ is an entire function of λ. Moreover, as $\lambda \to \infty$ and $\lambda \in \mathcal{L}_k$, we have from **(iii)**,

$$\begin{aligned} R(\lambda, x) &= \tilde{\Psi}_0(\lambda, x)\Psi_0^{-1}(\lambda, x) \\ &= \left[\tilde{\Psi}_0(\lambda, x)S_1 \dots S_{k-1}e^{\theta(\lambda,x)\sigma_3}\right]\left[\Psi_0(\lambda, x)S_1 \dots S_{k-1}e^{\theta(\lambda,x)\sigma_3}\right]^{-1} \\ &= I + O\left(\frac{1}{\lambda}\right). \end{aligned}$$

In other words, in the whole neighborhood of infinity we have

$$R(\lambda, x) \to I, \quad \lambda \to \infty,$$

and therefore

$$R(\lambda, x) \equiv I.$$

The uniqueness, and hence the theorem are proven.

0.3. The solution of the RH problem. By virtue of Theorem A.3, and the standard topological properties of the complex plane, there exists a countable covering,

$$\mathbb{C} = \bigcup_{n=1}^{\infty} D_{\delta_n}(x_n),$$

of the complex plane x by the disks $D_{\delta_n}(x_n)$, such that every disc $D_{\delta_n}(x_n)$ has nonempty intersection only with finitely many other discs from the collection, and such that for each $D_{\delta_n}(x_n)$ the statement of Theorem A.3 is valid. In other words, for every n there exists the finite set $X_s^n \equiv \{x_{nj}\} \subset D_{\delta_n}(x_n)$, and the unique matrix function $\Psi_n(\lambda, x)$ such that:

- **(i)** $\Psi_n(\lambda, x)$ and $\Psi_n^{-1}(\lambda, x)$ are holomorphic in $\mathbb{C} \times \left(D_{\delta_n}(x_n) \setminus X_s^n\right)$ and meromorphic along $\mathbb{C} \times X_s^n$.
- **(ii)** $\det \Psi_n(\lambda, x) \equiv 1$.
- **(iii)** For every $x \in D_{\delta_n}(x_n) \setminus X_s^n$,

$$\Psi_n(\lambda, x)S_1 \dots S_{k-1} \sim \left(I + \sum_{j=1}^{\infty} \frac{\Psi_j(x)}{\lambda^j}\right)e^{-\theta(\lambda,x)\sigma_3},$$

$$\lambda \to \infty, \quad \lambda \in \mathcal{L}_k,$$

where the coefficients $\Psi_j(x)$ are holomorphic in $D_{\delta_n}(x_n) \setminus X_s^n$ and meromorphic in $D_{\delta_n}(x_n)$. The asymptotics is uniform in $x \in D$, where D is an arbitrary compact subset of $D_{\delta_n}(x_n) \setminus X_s^n$, and it is differentiable with respect to both x and λ.

Our aim now is to glue this collection of the local (with respect to x) solutions of the original RH problem, into the global function $\Psi(\lambda, x)$ which would solve this problem for all complex $x \in \mathbb{C} \setminus X_s$, where

$$X_s := \bigcup_{n=1}^{\infty} X_s^n.$$

(Note that the set X_s is countable and has ∞ as its only limit point.) The following lemma plays a central role in this construction.

LEMMA A.3. *Suppose that*

$$\overset{o}{D}_{\delta_n}(x_n) \cap \overset{o}{D}_{\delta_m}(x_m) \neq \emptyset.$$

Then

$$\Psi_n(\lambda, x) = \Psi_m(\lambda, x), \quad \forall (\lambda, x) \in \mathbb{C} \times \Big(\big(D_{\delta_n}(x_n) \setminus X_s^n\big) \cap \big(D_{\delta_m}(x_m) \setminus X_s^m\big) \Big).$$

PROOF. The proof is identical with the proof used for the uniqueness part of Theorem A.3. The key point is that both functions has exactly the same asymptotic behaviour (with exactly the same Stokes matrices) as $\lambda \to \infty$.

Let us now set

$$\Psi(\lambda, x) := \Psi_n(\lambda, x), \quad (\lambda, x) \in \mathbb{C} \times \big(D_{\delta_n}(x_n) \setminus \Theta_n\big). \tag{A.42}$$

Because of Lemma A.3, this equation defines $\Psi(\lambda, x)$ as a holomorphically invertible function on the whole product $\mathbb{C} \times \big(\mathbb{C} \setminus X_s\big)$. Moreover, it inherits all the common analytic properties of the local functions $\Psi_n(\lambda, x)$, including uniqueness. In other words, the function $\Psi(\lambda, x)$ satisfies all the properties stated in Theorem A.0. □

REMARK A.2. The poles of the functions $u(x; s)$ and $v(x; s)$ are the values of the parameter x for which the generalized inverse monodromy problem with the given monodromy data,

$$s = \{S_1, \dots, S_6; \nu\},$$

is not solvable. At the same time, by using the Schlesinger transformations indicated in Lemma A.1, we see that if x is a pole of the functions $u(x; s)$ and $v(x; s)$, then the inverse monodromy problem is solvable for each of the following choices of the monodromy data

$$\hat{s} = \{S_1, \dots, S_6; \nu - 1\}$$

and

$$\tilde{s} = \{S_1, \dots, S_6; \nu + 1\}.$$

REMARK A.3. The Sibuya function $M(\lambda, x)$ defines a holomorphic vector bundle over $\mathbb{C}P^1$. The points x_j of the set X_s, i.e., the poles of the functions $u(x; s)$ and $v(x; s)$, are the points where this bundle is not trivial. In this case, the index κ in the representation (A.37) equals 1.

APPENDIX B

The Birkhoff-Grothendieck Theorem with a Parameter

The Appendix is based on the work of A. A. Bolibruch [**Bol4**].

Denote by Ω_0, Ω_∞, and ω the following subsets of the Riemann sphere $\mathbb{C}P^1$:

$$\Omega_0 = \{\lambda \in \mathbb{C} : |\lambda| \leq R\},$$
$$\Omega_\infty = \{\lambda \in \mathbb{C} \cup \infty : |\lambda| \geq r\},$$
$$\omega = \Omega_0 \cap \Omega_\infty,$$

where $R > r$. Also, as in Appendix A, we denote by $D_\delta(x_0)$ the closed disk in the x - plane of radius δ and centered at x_0, i.e.

$$D_\delta(x_0) = \{x \in \mathbb{C} : |x - x_0| \leq \delta\}.$$

Our goal is to present an elementary proof of the following (local) version of the classical Birkhoff-Grothendieck theorem with the parameter.

THEOREM B.1. *Let $M(\lambda, x)$ belongs to the set*[1] *$H^0(\omega \times D_\delta(x_0))$. Then there exists a number $0 < \delta_0 \leq \delta$, the finite set $X^0 \equiv \{x_{0j}\} \subset D_{\delta_0}(x_0)$, and the matrix functions $T(\lambda, x)$ and $\Phi(\lambda, x)$ such that*

- *$T(\lambda, x)$ and $T^{-1}(\lambda, x)$ are holomorphic in $\Omega_\infty \times \big(D_{\delta_0}(x_0) \setminus X^0\big)$ and meromorphic along $\Omega_\infty \times X^0$ (i.e., they have poles at $x_{0j} \in X^0$ and the coefficients of the corresponding Laurent series are holomorphic in D_∞).*
- *$\Phi(\lambda, x)$ and $\Phi^{-1}(\lambda, x)$ are holomorphic in $\Omega_0 \times \big(D_{\delta_0}(x_0) \setminus \Theta_0\big)$ and meromorphic along $\Omega_0 \times X^0$.*
-

$$M(\lambda, x) = T^{-1}(\lambda, x)\lambda^{\left(\begin{smallmatrix}\kappa_1 & 0\\ 0 & \kappa_2\end{smallmatrix}\right)}\Phi(\lambda, x), \tag{B.1}$$

  where $\kappa_1 \geq \kappa_2$ are some integer numbers, which are the same for all $x \in D_{\delta_0}(x_0) \setminus X^0$.

We shall prove the theorem in a series of lemmas, following closely the proof of [**Bol4**] where the case of $N \times N$ matrix and an arbitrary finite number of parameters is considered. As in [**Bol4**], the first two lemmas are just specification of the corresponding statements proven in [**AnoBol**] for the 2×2 case, as well as the modification of this result for the case that the matrix depends of a parameter x. Lemma 3 is a particular case of the Sauvage lemma (see [**H**]).

The Banach spaces $H(\omega \times D_\delta(x_0))$, $H(\Omega_\infty \times D_\delta(x_0))$, and $H(\Omega_0 \times D_\delta(x_0))$ are equipped by the standard norm $\|f\|_H = \max_{(\lambda,x)} \|f(\lambda, x)\|$. In addition to the subsets $H^0(\Omega_{\infty,0} \times D_\delta(x_0))$, we also introduce the subspace, $\hat{H}(\Omega_\infty \times D_\delta(x_0))$, of

[1] We recall that for any set $\mathbb{M} \subset \mathbb{C}^N$, with nonempty interior, we denote $H(\mathbb{M})$ the Banach space of the 2×2 matrix functions holomorphic in the interior of $\mathbb{M}$ and continuous in $\overline{\mathbb{M}}$. We also denote by $H^0(\mathbb{M})$ the subset of $H(\mathbb{M})$ consisting of holomorphically invertible matrix functions.

the space $H(\Omega_\infty \times D_\delta(x_0))$, consisting of all $f(\lambda, x) \in H(\Omega_\infty \times D_\delta(x_0))$ such that $f(\infty, x) \equiv 0$. We also note that the spaces $H(\Omega_{\infty,0} \times D_\delta(x_0))$ are the subspaces of $H(\omega \times D_\delta(x_0))$.

LEMMA B.1. *If under the conditions of the theorem, $\|M - I\|_H < \epsilon$ for sufficiently small $\epsilon > 0$, then there exists unique functions $T(\lambda, x) \in H^0(\Omega_\infty \times D_\delta(x_0))$ and $\Phi(\lambda, x) \in H^0(\Omega_0 \times D_\delta(x_0))$, such that $T(\infty, x) \equiv I$ and*

$$M(\lambda, x) = T^{-1}(\lambda, x)\Phi(\lambda, x). \tag{B.2}$$

PROOF. Similar to the proof of the basic factorization lemma in Appendix A (and, once again, following [**AnoBol**]), we introduce the Cauchy operators $P_\pm$: $H(\omega \times D_\delta(x_0)) \mapsto H(\Omega_{0,\infty} \times D_\delta(x_0))$, by the equations

$$(P_+f)(\lambda, x) = \frac{1}{2\pi i}\oint_{|\mu|=R} \frac{f(\mu, x)}{\mu - \lambda} d\mu, \quad (P_-f)(\lambda, x) = -\frac{1}{2\pi i}\oint_{|\mu|=r} \frac{f(\mu, x)}{\mu - \lambda} d\mu,$$

where the orientation, in both integrals, is counterclockwise. Let us now show that the operators $P_\pm$ are bounded. Once again, the crucial role is played by the Cauchy formula

$$f(\lambda, x) = (P_+f)(\lambda, x) + (P_-f)(\lambda, x), \tag{B.3}$$

which holds for any $f \in H(\omega \times D_\delta(x_0))$. Note also that

$$\operatorname{Im} P_- = \hat{H}(\Omega_\infty \times D_\delta(x_0)).$$

Denote by s the number $(R - r)/2$. If $0 \le |\lambda| < s + r$, then

$$\|(P_+f)(\lambda, x)\| \le \frac{1}{2\pi}\oint_{|\mu|=R} \left\|\frac{f(\mu, x)}{\mu - \lambda}\right\| |d\mu| \le ||f||_H \frac{1}{2\pi}\oint_{|\mu|=R} \frac{1}{s}|d\mu| \le \frac{R}{s}||f||_H.$$

If $s + r \le |\lambda| \le R$, using (B.3), we obtain

$$\|(P_+f)(\lambda, x)\| \le ||f||_H + \|(P_-f)(\lambda, x)\| \le$$

$$||f||_H + \frac{1}{2\pi}\oint_{|\mu|=r} \left\|\frac{f(\mu, x)}{\mu - \lambda}\right\| |d\mu| \le ||f||_H + ||f||_H \frac{1}{2\pi}\oint_{|\mu|=r} \frac{1}{s}|d\mu| \le \frac{s+r}{s}||f||_H.$$

Thus,

$$|(P_+f)(\lambda, x)| \le \max\left\{\frac{R}{s}, \frac{s+r}{s}\right\} ||f||_H,$$

for all $(\lambda, x) \in \Omega_0 \times D_\delta(x_0)$, and hence

$$||P_+||_H \le C := \max\left\{\frac{R}{s}, \frac{s+r}{s}\right\}.$$

Similar arguments show that the same estimate is also true for the norm of the operator P_-.

Let $N(\lambda, x) := M(\lambda, x) - I$, then $||N||_H < \epsilon$. To prove the existence statement of the lemma, it is sufficient to show the existence of a matrix function $X(\lambda, x) \in \hat{H}(\Omega_\infty \times D_\delta(x_0))$, such that $||X||_H < 1/2$ and

$$P_-\left((I + X)M\right) = 0. \tag{B.4}$$

Indeed, in this case we can put

$$T(\lambda, x) := I + X(\lambda, x) \in H^0(\Omega_\infty \times D_\delta(x_0))$$

and

$$\Phi(\lambda, x) := T(\lambda, x)M(\lambda, x) \in H^0(\Omega_0 \times D_\delta(x_0)).$$

Since (B.3), (B.4), it is clear that the function $\Phi(\lambda, x)$ is in $H(\Omega_0 \times D_\delta(x_0))$, and we only need to establish the holomorphic invertibility of $\Phi(\lambda, x)$. We have,

$$\Phi = I + X + (I + X)N,$$

or, using $\Phi = P_+(\Phi)$,

$$\Phi = I + P_+\big((I + X)N\big),$$

and the norm $||P_+\big((I + X)N\big)||_H < ||P_+||_H \frac{3\epsilon}{2}$, can be made smaller than 1/2 for sufficiently small ϵ.

Equation (B.4) can be rewritten as follows,

$$P_-\left((I + X)M\right) = P_-\left((I + X)(I + N)\right)$$
$$= P_-\left((I + X + XN + N)\right) = X + P_-(XN) + P_-(N) = 0,$$

where we have used the obvious facts that

$$P_-(X) = X, \quad \text{and} \quad P_-(I) = 0.$$

Therefore, the equation we need to solve, can be written as the integral equation,

$$X + P_-(XN) = -P_-(N),$$

in the space $\hat{H}(\Omega_\infty \times D_\delta(x_0))$. Since the operator P_- is bounded and $||N||_H < \epsilon$, for sufficiently small ϵ the operator $I + P_-[\cdot\, N]$ is invertible and the norm of the solution $X = \big(I + P_-[\cdot\, N]\big)^{-1}\big(-P_-(N)\big)$ can be made as small as needed. This proves the existence part of the lemma.

Suppose $\{\tilde{T}(\lambda, x), \tilde{\Phi}(\lambda, x)\}$ is another pair of functions with the same properties. Then, $M = T^{-1}\Phi = \tilde{T}^{-1}\tilde{\Phi}$, and the function R defined by

$$R(\lambda, x) := \begin{cases} \tilde{T}(\lambda, x)T^{-1}(\lambda, x), & \text{for} \quad \lambda \in \Omega_\infty, \\ \tilde{\Phi}(\lambda, x)\Phi^{-1}(\lambda, x), & \text{for} \quad \lambda \in \Omega_0, \end{cases}$$

is a holomorphically invertible matrix function in $\overline{\mathbb{C}} \times D_\delta(x_0)$. Thus, by Liouville theorem, $R(\lambda, x)$ does not depend on λ. On the other hand, from the above construction one has $R(\infty, x) \equiv I$, therefore $R(\lambda, x) \equiv I$ and hence, $\tilde{T}(\lambda, x) \equiv T(\lambda, x)$, $\tilde{\Phi}(\lambda, x) \equiv \Phi(\lambda, x)$.

The second step of the proof of the theorem is to replace the function $M(\lambda, x)$, by some matrix function rational in λ. □

LEMMA B.2. *For every matrix function $M(\lambda, x) \in H^0(\omega \times D_\delta(x_0))$, there exists a number $0 < \delta'' \leq \delta$, a matrix function $F(\lambda, x)$ rational in λ with holomorphic coefficients in x and matrix functions $T(\lambda, x)$, $\Phi(\lambda, x)$ such that:*

(1) *The only possible poles of $F(\lambda, x)$ are $\lambda = 0$ and $\lambda = \infty$. Moreover,*

$$\alpha_{0,\infty} := \sup_{x \in D_\delta(x_0)} \text{(order of the pole at } 0, \infty) < \infty.$$

(2) *$F(\lambda, x)$ is holomorphically invertible in $\omega \times D_{\delta''}(x_0)$, and, furthermore, the positions of the zeros of $\det F(\lambda, x)$ in $\big(\Omega_0 \setminus \omega\big) \times D_{\delta''}(x_0)$, do not depend on x.*

(3) *$T(\lambda, x)$ and $T^{-1}(\lambda, x)$ are holomorphic in $\Omega_\infty \times D_{\delta''}(x_0)$.*

(4) *$\Phi(\lambda, x)$ and $\Phi^{-1}(\lambda, x)$ are holomorphic in $\Omega_0 \times D_{\delta''}(x_0)$.*

(5) *$M(\lambda, x) = T^{-1}(\lambda, x)F(\lambda, x)\Phi(\lambda, x)$.*

PROOF. The function $M(\lambda, x_0)$ can be uniformly approximated in ω by a matrix function rational in λ with possible poles in 0 and ∞ only. For this one can just take the parts of the Laurent expansion of $M(\lambda, x_0)$ in the annular domain ω, cutting the infinite tails of the expansion in an appropriate way .

On the other hand, for every $\epsilon > 0$, one can choose a number $\delta' > 0$, such that $\|M^{-1}(\lambda, x_0)M(\lambda, x) - I\| < \epsilon$ in $\omega \times D_{\delta'}(x_0)$.

Thus, one can always find a number $0 < \delta' \le \delta$, and a holomorphically invertible in ω matrix function $B(\lambda)$ rational in λ with poles at 0 and ∞ only, such that $\|B^{-1}(\lambda)M(\lambda, x) - I\|$ is uniformly small in $\omega \times D_{\delta'}(x_0)$. Applying to the product $B^{-1}(\lambda)M(\lambda, x)$ lemma B.1, we obtain

$$B^{-1}M = T_1^{-1}\Phi_1, \quad M = BT_1^{-1}\Phi_1,$$

where $T_1(\lambda, x) \in H^0(\Omega_\infty \times D_{\delta'}(x_0))$ and $\Phi_1(\lambda, x) \in H^0(\Omega_0 \times D_{\delta'}(x_0))$. Consider the function $B(\lambda)T_1^{-1}(\lambda, x)$, and approximate it from the right in a similar way by a function $H(\lambda)$ with the same properties as B. Applying again lemma B.1, we find

$$BT_1^{-1}H^{-1} = T_2^{-1}\Phi_2,$$

where $T_2(\lambda, x)$ and $\Phi_2(\lambda, x)$ have the same properties as $T_1(\lambda, x)$ and $\Phi_1(\lambda, x)$ (for some $0 < \delta'' \le \delta'$). Consequently,

$$\Phi_2 = T_2BT_1^{-1}H^{-1}, \quad M = T_2^{-1}\Phi_2H\Phi_1 = T_2^{-1}F\Phi_1, \quad F := \Phi_2H.$$

From the first relation above it follows that for every $x \in D_{\delta''}(x_0)$, the matrix function $\Phi_2(\lambda, x)$ is meromorphic in Ω_∞ with poles independent of x, and with Laurent series coefficients holomorphic in x. On the other hand, Φ_2, by the construction, is holomorphically invertible in $\Omega_0 \times D_{\delta''}(x_0)$. Thus, Φ_2 is rational in λ with holomorphic coefficients in x. This implies that F possesses exactly the same properties. Moreover, since $\Phi_2(\lambda, x)$ and $H(\lambda)$ are holomorphically invertible in $\omega \times D_{\delta''}(x_0)$, $F(\lambda, x)$ has the same property.

Finally, we have

$$F = \begin{cases} T_2BT_1^{-1}, & \text{for } \lambda \in \Omega_\infty \\ \Phi_2H, & \text{for } \lambda \in \Omega_0, \end{cases}$$

and hence the only possible poles of F are 0 and ∞. Moreover, the corresponding maximal orders of the poles, α_0 and α_∞, do not exceed the orders of the corresponding poles of the (independent of x) matrix functions $H(\lambda)$ and $B(\lambda)$, respectively. Also, the position of zeros of the $\det F(\lambda, x)$ in $(\Omega_0 \setminus \omega) \times D_{\delta''}(x_0)$, do not depend on x, because $\det \Phi_2$ has no zeros in Ω_0 and H does not depend on x.

Now, denoting $T = T_2$ and $\Phi = \Phi_1$ we get the decomposition (5). □

The following lemma plays a crucial role in proving the theorem. In fact, it proves the theorem for a particular class of the rational matrices M without any small norm assumptions.

It is convenient to introduce one more notation: Let Ω be a subset of the extended complex λ plane, and let X be a finite subset of the interior of the disk $D_\delta(x_0)$ on the complex x plane. We will denote

$$H_m^0\Big(\Omega \times \big(D_\delta(x_0) \setminus X\big)\Big)$$

the set of matrix valued functions $f(\lambda, x)$ such that $f(\lambda, x)$ and $f^{-1}(\lambda, x)$ are holomorphic in $\Omega \times \big(D_\delta(x_0) \setminus X\big)$ and meromorphic along $\Omega \times X$.

LEMMA B.3. *Let* $M(\lambda, x) \in H_m^0\Big((\Omega_0 \setminus \{0\}) \times (D_\delta(x_0) \setminus X)\Big)$ *where* X *is a finite subset of the interior of* $D_\delta(x_0)$. *Moreover, we assume that* $M(\lambda, x)$ *is rational in* λ *with the coefficients meromorphic in* $D_\delta(x_0)$ *and holomorphic in* $D_\delta(x_0) \setminus X$. *Suppose in addition that the only possible poles of* M *as a function of* λ *are* $\lambda = 0$ *and* $\lambda = \infty$, *and that the corresponding maximal orders of these poles,* α_0 *and* α_∞, *are finite. Then, there exists a number* $\delta_0 \leq \delta$, *the finite set* $X^0 \subset D_{\delta_0}(x_0)$, $X \subset X^0$, *and the matrix functions* $T(\lambda, x)$ *and* $\Phi(\lambda, x)$, *such that:*

- $T(\lambda, x)$ *and* $T^{-1}(\lambda, x)$ *are holomorphic in* $(\overline{\mathbb{C}} \setminus \{0\}) \times (D_{\delta_0}(x_0) \setminus X^0)$ *and meromorphic along* $\overline{\mathbb{C}} \setminus \{0\} \times X^0$. *In other words,*
$$T(\lambda, x) \in H_m^0\Big((\overline{\mathbb{C}} \setminus \{0\}) \times (D_{\delta_0}(x_0) \setminus X^0)\Big).$$
- $\Phi(\lambda, x)$ *and* $\Phi^{-1}(\lambda, x)$ *are holomorphic in* $\Omega_0 \times (D_{\delta_0}(x_0) \setminus X^0)$ *and meromorphic along* $\Omega_0 \times X^0$. *In other words,*
$$\Phi(\lambda, x) \in H_m^0\Big(\Omega_0 \times (D_{\delta_0}(x_0) \setminus X^0)\Big).$$
Moreover, $\Phi(\lambda, x)$ *is a polynomial in* λ *with the coefficients meromorphic in* $D_{\delta_0}(x_0)$ *and holomorphic in* $D_{\delta_0}(x_0) \setminus X^0$, *and also* $\deg \Phi(\lambda, x) \leq \alpha_0 + \alpha_\infty \quad \forall x$.
-
$$M(\lambda, x) = T^{-1}(\lambda, x)\lambda^{\begin{pmatrix} \kappa_1 & 0 \\ 0 & \kappa_2 \end{pmatrix}}\Phi(\lambda, x), \tag{B.5}$$
where $\kappa_1 \geq \kappa_2$ *are some integer numbers, which are the same for all* $x \in D_{\delta_0}(x_0) \setminus X^0$.

PROOF. Since the only poles of M in the λ-plane are 0 and ∞, the matrix function M can be assumed to be of the form

$$M(\lambda, x) = \lambda^{\begin{pmatrix} \kappa_1 & 0 \\ 0 & \kappa_2 \end{pmatrix}} L(\lambda, x), \tag{B.6}$$

for some integers $\kappa_1 \geq \kappa_2$, where $L(\lambda, x)$ is a polynomial matrix,

$$L(\lambda, x) = L_0(x) + L_1(x)\lambda + \cdots + L_m(x)\lambda^m.$$

(One can just take $\kappa_1 = \kappa_2 = -\alpha_0$.) We note that

$$L(\lambda, x) \in H_m^0\Big((\Omega_0 \setminus \{0\}) \times (D_\delta(x_0) \setminus X)\Big). \tag{B.7}$$

The degree m of the polynomial $L(\lambda, x)$ can be assumed to be the same for all x (equals, e.g., $\alpha_0 + \alpha_\infty$), although for some x the coefficient for the highest degree term might be zero. We also note that all the coefficients $L_j(x)$ are meromorphic in $D_\delta(x_0)$ with possible poles at the points of the set X.

If for all $x \in D_\delta(x_0) \setminus X$ the inequality, $\det L(0, x) \neq 0$, holds, then one has nothing to prove: just put $\delta_0 \equiv \delta$, $X^0 \equiv X$, $T \equiv I$, and $\Phi \equiv L$.

Let $\det L(0, x_1) = 0$ for some $x_1 \in D_\delta(x_0) \setminus X$. We will then prove that

$$\det L(0, x) \equiv 0. \tag{B.8}$$

Indeed, we first observe that

$$\det L(\lambda, x) = l_0(x) + l_1(x)\lambda + \cdots + l_n(x)\lambda^n,$$

for some integer n (which can be choosen the same for all x). All the coefficients $l_k(x)$ are meromorphic functions of $x \in D_\delta(x_0)$ with possible poles belonging to the set X. According to our assumption, $l_0(x_1) = 0$. At the same time, for at least one

$0 < k \le n$ we must have that $l_k(x_1) \neq 0$, otherwise, $\det L(\lambda, x_1) \equiv \det L(0, x_1) = 0$ which contradicts (B.7). Denote by k_0 the maximal of such k's, and let

$$P_0(\lambda, x) := l_0(x) + l_1(x)\lambda + \cdots + l_{k_0}(x)\lambda^{k_0}.$$

Take $R_0 < R$ so that the circle $C_0 := \{\lambda : |\lambda| = R_0\}$ does not pass through the zeros of $P_0(\lambda, x_1)$. The choice of k_0 implies that $\det L(\lambda, x_1) = P_0(\lambda, x_1)$. Therefore,

$$\det L(\lambda, x) - P_0(\lambda, x) \to 0, \quad x \to x_1,$$

uniformly with respect to $\lambda \in C_0$. At the same time,

$$P_0(\lambda, x) \to P_0(\lambda, x_1), \quad x \to x_1,$$

uniformly with respect to $\lambda \in C_0$. The choice of R_0 implies that $|P_0(\lambda, x_1)| \ge c_0 > 0$, for all $\lambda \in C_0$. Hence, there exists $\delta' > 0$, such that

$$|\det L(\lambda, x) - P_0(\lambda, x)| < |P_0(\lambda, x)|, \tag{B.9}$$

for all $\lambda \in C_0$, and for all $x \in D_{\delta'}(x_1) \subset D_\delta(x_0) \setminus X$.

Suppose now that (B.8) is not true. Then we can find $0 < \delta'' \le \delta'$ such that

$$l_0(x) \neq 0, \quad \forall x \in D_{\delta''}(x_1) \setminus \{x_1\}. \tag{B.10}$$

Denote by $\lambda_j(x)$, $j = 1, \dots, k_0$, the zeros of the polynomial $P_x(\lambda) := P_0(\lambda, x)$. These zeros satisfy the relation

$$\prod_j \lambda_j(x) = \frac{l_0(x)}{l_{k_0}(x)}(-1)^{k_0}. \tag{B.11}$$

Note that

$$|l_{k_0}(x)| \ge c_0' > 0, \quad \forall x \in D_{\delta'''}(x_1),$$

for some $0 < \delta''' \le \delta''$. Taking into account this inequality, the fact that $l_0(x) \to 0$ as $x \to x_1$, and equation (B.11), we conclude that in any neighborhood of $\lambda = 0$ we can find at least one $\lambda_j(x)$ for some $x \in D_{\delta'''}(x_1) \setminus \{x_1\}$. In particular, we have that for some $y \in D_{\delta'''}(x_1) \setminus \{x_1\}$, at least one zero of $P_0(\lambda, y)$ is inside the circle C_0. In virtue of (B.9), using the Rouche theorem, we conclude that inside the circle C_0 the polynomial $Q_y(\lambda) := \det L(\lambda, y)$ has at least one zero. Denote this zero as μ. Since (B.10), $Q_y(0) \neq 0$ and therefore $\mu \neq 0$. Thus, there exist at least one pair (λ, x) in $(\Omega_0 \setminus \{0\}) \times D_\delta(x_0)$, namely (μ, y), for which $\det L(\lambda, x) = 0$. This contradicts (B.7), and hence proves (B.8).

Denote by $m_1(x)$ and $m_2(x)$ the rows of the matrix $L_0(x) \equiv L(0, x)$, and assume that

$$m_1(x) \neq 0 \quad \text{for some } x. \tag{B.12}$$

Then, the identity (B.8) implies that there exists a scalar function $s(x)$ rational in entries of $L_0(x)$ such that

$$m_2(x) = s(x)m_1(x) \tag{B.13}$$

(e.g., $s(x) = m_{21}(x)/m_{11}(x)$ if $m_{11}(x) \neq 0$ for some x). Define T_1 by

$$T_1(\lambda, x) := \begin{pmatrix} 1 & 0 \\ s(x)\lambda^{k_2 - k_1} & 1 \end{pmatrix}.$$

It is clear that $T_1(\lambda, x) \in H^0_m\Big((\overline{\mathbb{C}} \setminus \{0\}) \times (D_\delta(x_0) \setminus X_1)\Big)$, where X_1 is the set of poles of $s(x)$, i.e., the set X extended by the zeros of the relevant entry of $L_0(x)$). Restricting, if needed, the x-domain to $D_{\delta_1}(x_0)$ with $\delta_1 < \delta$, we can always make X_1 finite (and still containing X).

It follows immediately from (B.13) that

$$T_1(\lambda, x) M(\lambda, x) = \lambda^{\begin{pmatrix} \kappa_1 & 0 \\ 0 & \kappa_2+1 \end{pmatrix}} L_1(\lambda, x), \tag{B.14}$$

where

$$L_1(\lambda, x) = \begin{pmatrix} 1 & 0 \\ -s(x)\lambda^{-1} & \lambda^{-1} \end{pmatrix} L(\lambda, x).$$

Because of (B.13) the function $L_1(\lambda, x)$ is still a matrix polynomial in λ. Its coefficients are meromorphic in $D_{\delta_1}(x_0)$ and holomorphic in $D_{\delta_1}(x_0) \setminus X_1$, and furthermore, $\deg L_1 \leq \deg L \leq \alpha_0 + \alpha_\infty$. Also, equation (B.14) implies that L_1 inherits the basic analytic properties of M. Indeed,

$$L_1(\lambda, x) \in H^0_m\Big((\Omega_0 \setminus \{0\}) \times (D_{\delta_1}(x_0) \setminus X_1)\Big). \tag{B.15}$$

If (B.12) is not true, i.e., if

$$m_1(x) \equiv 0,$$

then we can directly increase the number κ_1 without the procedure presented above. In fact, in this case,

$$M(\lambda, x) = \lambda^{\begin{pmatrix} \kappa_1+1 & 0 \\ 0 & \kappa_2 \end{pmatrix}} L_1(\lambda, x), \tag{B.16}$$

where

$$L_1(\lambda, x) = \begin{pmatrix} \lambda^{-1} & 0 \\ 0 & 1 \end{pmatrix} L(\lambda, x)$$

is again a polynomial satisfying (B.15).

Multiplying the right hand side of (B.14) or of (B.16) by either the identity or by σ_1, we can rewrite these equations in the form

$$T'(\lambda, x) M(\lambda, x) = \lambda^{\begin{pmatrix} \kappa_1' & 0 \\ 0 & \kappa_2' \end{pmatrix}} L'(\lambda, x),$$

where $\kappa_1' \geq \kappa_2'$ and the matrices $T'(\lambda, x)$ and $L'(\lambda, x)$ have the same properties as the matrices $T_1(\lambda, x)$ and $L_1(\lambda, x)$, respectively. Since $\kappa_1' + \kappa_2' > \kappa_1 + \kappa_2$, iterating if necessary the procedure presented above, we obtain after a finite[2] number of steps that $\det L'(0, x) \neq 0$ (for all x), and hence the decomposition (B.6) of the lemma is proven. The final set X^0 is obtained as the finite union of the X-sets generated at every step of this procedure. This completes the proof of the lemma.

We are now in a position to prove the theorem.

[2]The process is finite because of the obvious relation, $\det M = \pm\lambda^{\kappa_1+\kappa_2} \det L'$, and the fact that $\det M$ is a rational function, and hence has a finite order pole at $\lambda = \infty$.

PROOF OF THE THEOREM. Because of Lemma B.2, we only need to consider the case of a matrix function $M(\lambda, x)$, rational in λ with coefficients holomorphic in $D_\delta(x_0)$ and such that:

(1) The only possible poles of $M(\lambda, x)$ are $\lambda = 0$ and $\lambda = \infty$. Moreover,
$$\alpha_{0,\infty} := \sup_{x \in D_\delta(x_0)} (\text{order of the pole at } 0, \infty) < \infty.$$

(2) The positions of the zeros of the det $M(\lambda, x)$ in $(\Omega_0 \setminus \omega) \times D_\delta(x_0)$ do not depend on x.

If in addition $M(\lambda, x)$ is holomorphically invertible in $(\Omega_0 \setminus \{0\}) \times D_\delta(x_0)$, then the theorem follows immediately from lemma B.3 (with $X = \emptyset$). Suppose then, that for some $0 \neq \lambda_0 \in \Omega_0 \setminus \omega$, we have $\det M(\lambda_0, x_0) = 0$.

Consider new local coordinates $\zeta = \lambda - \lambda_0$ and define the *polynomial* matrix function
$$\tilde{M}(\zeta, x) := (\zeta + \lambda_0)^{\alpha_0} M(\zeta + \lambda_0).$$
Denote by $\tilde{\Omega}_0$ a small disk in the ζ-plane with center at $\zeta = 0$ such that the matrix function $\tilde{M}(\zeta, x)$ is holomorphically invertible in $(\tilde{\Omega}_0 \setminus \{0\}) \times D_\delta(x_0)$. Applying lemma B.3 to $\tilde{M}(\zeta, x)$ (again with $X = \emptyset$), we obtain the representation

$$M(\lambda, x) = T^{-1}(\lambda, x)(\lambda - \lambda_0)^C \Phi(\lambda, x), \tag{B.17}$$

where $C = \operatorname{diag}(\kappa_1, \kappa_2)$, $T(\lambda, x) \in H^0\Big((\mathbb{C}\setminus\{\lambda_0\}) \times (D_{\delta_0}(x_0)\setminus X^0)\Big)$, and the matrix function $\Phi(\lambda, x)$ is a polynomial in λ whose coefficients are holomorphic in $D_{\delta_0}(x_0)\setminus X^0$ and meromorphic in $D_{\delta_0}(x_0)$. Moreover, by construction, $\det \Phi(\lambda_0, x) \neq 0$, while from the equation (B.17) it follows that $\Phi(\lambda, x)$ is holomorphically invertible in the annular domain ω and its determinant may have zeros in the set $\Omega_0 \setminus \{0\}$ only at the possible zeros of $\det M$ different from λ_0. Noticing that the function $(\lambda/\lambda - \lambda_0)^C$ is holomorphically invertible in $\Omega_\infty \times (D_{\delta_0}(x_0) \setminus X^0)$, we derive from (B.17) the equation
$$M(\lambda, x) = T_1^{-1}(\lambda, x) M_1(\lambda, x),$$
where $M_1(\lambda, x) := \lambda^C \Phi(\lambda, x)$. The function $M_1(\lambda, x)$ has all the properties of the original function $M(\lambda, x)$ except its determinant has one zero less in $\Omega_0 \setminus \omega$. The function $T_1^{-1}(\lambda, x)$ has exactly the properties stated in the theorem. Repeating, if necessary, the above construction for the function $M_1(\lambda, x)$ we can eliminate, after a finite number of steps, all the possible zeros of the det M in $\Omega_0 \setminus \omega$ and hence we arrive to the function $M'(\lambda, x)$ which satisfies the conditions of lemma B.3 (with, generally, a nonempty singularity set X'). Then the theorem again follows from the lemma. □

Bibliography

[Abd] A.S. Abdullaev, On the theory of the second Painlevé equation, *Dokl. Akad. Nauk SSSR*, **273** (1983) No 5, 31–35.

[AC] M. J. Ablowitz and P. A. Clarkson, *Solitons, Nonlinear Evolution Equations and Inverse Scattering*, London Math. Soc. Lecture Notes Series, **149**, Cambridge University Press, Cambridge, 1991

[AS] M.J. Ablowitz and H. Segur, Asymptotic solutions of the Korteweg-de Vries equation, *Stud. Appl. Math.*, **57** (1977) No 1, 13–44.

[Adl] V.E. Adler, Nonlinear chains and Painlevé equations, *Physica D* **73** (1994) 335–351.

[AvM] M. Adler, P. van Moerbeke, Integrals over classical groups, random permutations, Toda and Toeplitz lattices, *Comm. Pure Appl. Math.*, **54** (2001) No 2, 153–205.

[Ai] H. Airault, Rational solutions of Painlevé equations, *Stud. Appl. Math.* **61** (1979) No 1, 31–53.

[AnK1] F.V. Andreev and A.V. Kitaev, Exponentially small corrections to divergent asymptotic expansions of solutions of the fifth Painlevé equation, *Math. Res. Lett.*, **4** (1997) 741–759.

[AnK2] F.V. Andreev and A.V. Kitaev, Connection formulae for asymptotics of the fifth Painlevé transcendent on the real axis, *Nonlinearity*, **13** (2000) No. 5, 1801–1840.

[AnoBol] D.V. Anosov and A. A. Bolibrukh, *The Riemann-Hilbert problem*, Wiesbaden, Braunschweig: Vieweg, 1994.

[ArnIl] V.I. Arnold and Yu. Il'yashenko,*Ordinary differential equations,* Encyclopedia of Mathematical Sciences 1 (Dynamical Systems - I), Springer, Berlin, 1988.

[Bai] J. Baik, Riemann-Hilbert problems for last passage percolation, arXiv, math.PR/0107079.

[BDJ] J. Baik, P. Deift, K. Johansson, On the distribution of the length of the longest increasing subsequence of random permutations, *J. Amer. Math. Soc.*, **12** (1999) No. 4, 1119–1178.

[BKMM] J. Baik, T. Kriecherbauer, K. McLaughlin and P. Miller, Uniform asymptotics for polynomials orthogonal with respect to a general class of discrete weights and universality results for associated ensembles,*Int. Math. Res. Not.* (2003), no. 15, 821-858

[Bal] W. Balser, *Formal power series and linear systems of meromorphic ordinary differential equations*, Springer-Verlag, 2000

[Bar] E. Barouch, B.M. McCoy, C.A. Tracy and T.T. Wu, Zero field susceptibility of the two-dimensional Ising model near T_c, *Phys. Rev. Lett.* **31** (1973) 1409–1411.

[Bas] E.L. Basor, Distribution functions for random variables for ensembles of positive Hermitian matrices, *Comm. Math. Phys.* **188** (1997) 327–350.

[BasT] E.L. Basor and C.A. Tracy, Some problems associated with the asymptotics of τ-functions, Preprint RIMS-845, Kyoto Univ., 1991.

[BCH] A.P. Bassom, P.A. Clarkson and A.C. Hicks, Bäcklund transformations and solution hierarchies for the fourth Painlevé equation, *Atud. Appl. Math.* **95** (1995) 1–71.

[BCLM] A. P. Bassom, P.A. Clarkson, C. K. Law and J. B. McLeod, Application of uniform asymptotics to the second Painlevé transcendent, *Arch. Rat. Mec. Anal.* **143** (1998) 241–271.

[BE] H. Bateman, A. Erdelyi, *Higher Transcendental Functions*, McGraw-Hill, NY, 1953.

[Bea] R. Beals and R.R. Coifman, Scattering and inverse scattering for first order systems, *Comm. Pure Appl. Math.* **37** (1984) 39–90.

[BDT] R. Beals, P. Deift, C. Tomei, *Direct and Inverse Scattering on the Line*, Math. Surveys and Monographs, No 28, Amer. Math. Soc., Providence, 1998.

[BS] R. Beals and D. H. Sattinger, Integrable Systems and Isomonodromy Deformations, *Physica D*, **65** (1993) 17-47

[BIZ] D. Bessis, C. Itzykson, and J. B. Zuber, Quantum field theory techniques in graphical enumeration, *Adv. in Appl. Math.*, **1** (1980) 109–157

[Bir] G.D. Birkhoff, *Collected Mathematical Papers*, **3** Amer. Math. Soc., 1950.

[Bl] P.M. Bleher, Quasi-classical expansions and the problem of quantum chaos, *Lect. Notes Math.* **1496** (1991) 60–89.

[BlE] P. M. Bleher and B. Eynard, Double scaling limit in random matrix models and a nonlinear hierarchy of differential equations, *J. Phys. A: Math. Gen.*, **36** (2003), 3085–3105.

[BlI1] P. Bleher, A. Its, Semiclassical asymptotics of orthogonal polynomials, Riemann-Hilbert problem, and universality in the matrix models, *Ann. of Math.*, **150** (1999) 185–266.

[BlI2] P. Bleher, A. Its, Double scaling limit in the matrix model: the Riemann-Hilbert approach. *Commun. Pure Appl. Math.*, **56** (2003) 433–516.

[Bo1] P. Boalch, Stokes matrices, Poisson Lie groups and Frobenius manifolds, *Invent. math.* **146** (2001) 479–506.

[Bo2] P. Boalch, Symplectic manifolds and isomonodromy deformations, *Adv. in Math.* **163** (2001) 137–205

[Bo3] P. Boalch, *G*-bundles, isomonodromy and quantum Weyl groups, *Int. Math. Res. Not.*, no. 22 (2002) 1129–1166, math.DG/0108152.

[Bo4] P. Boalch, From Klein to Painlevé via Fourier, Laplace and Jimbo, *Proc. Lond. Math. Soc.* (3) **90** (2005) 167-208.

[BobEi1] A. Bobenko and U. Eitner, Bonnet surfaces and Painlevé equations, SFB288 Preprint, No 151, TU Berlin, 1994

[BobEi2] A.I. Bobenko and U. Eitner, Painlevé Equations in the Differential Geometry of Surfaces, *Lecture Notes in Mathematics*, **1753**, Springer-Verlag, Berlin, 2000.

[BobI] A.I. Bobenko and A.R. Its,The Painlevé III Equation and the Iwasawa Decomposition, *Manuscripta Mathematica*, **87**, (1995) 369–377.

[Bol1] A. A. Bolibruch, The Riemann-Hilbert Problem, *Russian Math. Surveys*, **45:2**, 1-47 (1990)

[Bol2] A.A. Bolibruch, The Riemann-Hilbert problem on the complex projective plane, *Matem. Zametki*, **46** (1989) No 3, 118–120.

[Bol3] A.A. Bolibruch, On isomonodromic deformations of Fuchsian systems, *Journ. of Dynamical and Control Systems,* **3** (1997) No 4, 589–604

[Bol4] A.A. Bolibruch, Differential equations of the Fuchs type and holomorphic bundles, Moscow, *MTzNMO*, (2000) (in Russian).

[BolIK] A.A. Bolibruch, A.R. Its, A.A. Kapaev, On the Riemann-Hilbert-Birkhoff inverse monodromy problem and the Painleve equations, *Algebra and Analysis*, **16** (2004), No 1, 121–162.

[Bor1] L.A. Bordag, Painlevé equations and their connection with nonlinear evolution equations, Preprint JINR, F5-80-477, 1980.

[Bor2] L.A. Bordag and A.V.Kitaev, Transformations of the third and fifth Painlevé equations and particular solutions, Preprint JINR, P5-85-740, 1985.

[Br] A. Borodin, Isomonodromy transformations of linear systems of difference equations, *math.CA/0209144*

[Bou] P. Boutroux, Recherches sur les transcendantes de M.Painlevé et l'étude asymptotique des équationes differéntielle du second ordre, *Ann. Sci. Ec. Norm. Super.* (3), **30** (1913) 255–375; **31** (1914) 99–159.

[BreKaz] E. Brézin and V. A.Kazakov, Exactly solvable field theories of closed strings, *Phys. Letts. B* **236** (1990) 144–150.

[Bur1] F.J. Bureau, Differential equations with fixed critical points. I, *Annali di Matematica*, **64** (1964) 229–364; II, ibid. **66** (1964) 1–116.

[Bur2] F.J. Bureau, Équationes differentielles du second ordre en Y et du second degré en $\ddot{Y}$ dont l'integrale générale a ses points critique fixes, *Annali di Matematica*, **91** (1972) 163–281.

[CeVa1] S. Cecotti and C. Vafa, Topological Anti-Topological Fusion, preprint HUTP-91/A031, S1SSA-69/91/EP, 1991.

[CeVa2] S. Cecotti and C. Vafa, Exact results for supersymmetric σ models, *Phys. Rev. Lett.*, **68**, (1992) No. 7, 903–906.

[CeVa3] S. Cecotti and C. Vafa, On classification of $N = 2$ supersymmetric theories, *Comm. Math. Phys.*, **158** (1993) no. 3, 569–644.

[Ch] J. Chazy, Sur les équationes différentielle du troisiéme ordre et d'ordre surérieur dont l'intégrale a ses points critique fixes, *Acta Math.*, **34** (1911) 317–385.

[CG] K. Clancey and I. Gohberg, *Factorization of Matrix Functions and Singular Integral Operators, Operator Theory: Advances and Applications,* **3**, Birkhäuser Verlag, Boston, 1981.

[ClM] P.A. Clarkson and J.B. McLeod, A connection formula for the second Painleve transcendent, *Arch. Rational Mech. Anal.*, **103** (1988) No 2, 97–138.

[ClJP] P. A. Clarkson, N. Joshi and A. Pickering, Bäcklund transformations for the second Painlevé hierarchy: a modified truncation approach, *Inverse Problems* **15** (1999) 175–187.

[CK] T. Clayes and A. B. J. Kijlaars, Universality of the multi-critical double scaling limit in random matrix models, *math-ph/0501074, to appear in Comm. Pure Appl. Math*

[CKV] T. Clayes, A. B. J. Kijlaars and M. Vanlessen, Multi-critical unitary random matrix ensembles and the general Painlevé II equation, *math-ph/0508062*

[dCM] J. des Cloizeaux and M. L. Mehta, Asymptotic behavior of spacing distributions for the eigenvalues of random matrices, *J. Math. Phys.*, **14** (1973) 1648–1650.

[CodL] E.A. Coddington and N. Levinson, *Theory of Ordinary Differential Equations*, McGraw-Hill, New York, 1955.

[Con] R. Conte, edt. *The Painlevé Property. One Century Later.* Volume of CRM series in Mathematical Physics, Springer-Verlag, Berlin, 1999

[ConM] R. Conte and M. Musette, First-degree birational transformations of the Painlevé equations and their contiguity relations, *J. Phys. A: Math. Gen.* **34** (2001) 10507–10522; A truncation for obtaining all the first degree birational transformations of the Painlevé transcendent, *J. Nonlinear Math. Phys.* **9** (2002) 14–28.

[Cos] C.M. Cosgrove, The Painlevé-Gambier class of differential equations, preprint, School of Mathematics and Statistics, University of Sydney, Australia.

[C1] O. Costin, On Borel summation and Stokes phenomena for rank one nonlinear systems of ODE's, *Duke Math. J.*, **93** (1998) No 2, 289–344.

[C2] O. Costin, Correlation between pole location and asymptotic behavior for Painlevé I solutions, MSRI preprint No 1997-094, 1997.

[Dar] G. Darboux, Mémoire sur les équations différentielles algébraiques du premier ordre et du premier degré (Mélanges), *Bull. Sci. Math. 2éme série*, **2** (1878) 60–96, 123–144, 151–200.

[Da] F. David, Loop equations and non-perturbative effects in two-dimensional quantum gravity, *MPLA*, **5**, 13 (1990) 1019–1029

[De1] P.A. Deift, *Orthogonal Polynomials and Random Matrices: A Riemann-Hilbert Approach,* Courant Lecture Notes in Mathematics, New York University, 1999

[De2] P.A. Deift, Integrable systems and combinatorial theory, *Notices of the Ammer. Math. Soc.*, **47** (6) (2000) 631–640.

[DeZ1] P.A. Deift and X. Zhou, A steepest descent method for oscillatory Riemann-Hilbert problems. Asymptotics for the MKdV equation, *Ann. of Math.*, **137** (1993) 295–368.

[DeZ2] P.A. Deift and X. Zhou, Asymptotics for the Painlevé II equation, *Comm. Pure Appl. Math.*, **48** (1995) no 3, 277–337.

[DeZ3] P.A. Deift and X. Zhou, Long-time asymptotics for integrable systems. Higher order theory. *Comm. Math. Phys.*, **165** (1995), 175-191.

[DeZ4] P. Deift and X. Zhou, Perturbation theory for infinite dimensional integrable systems on the line. A case study, *Acta Math.*, **188** (2002) 163–262.

[DeZ5] P. Deift and X. Zhou, A priori L^p estimates for solutions of Riemann-Hilbert problems, *Int. Math. Res. Noties*, **40** (2002) 2121–2154.

[DeIZ] P.A. Deift, A.R. Its and X. Zhou, A Riemann-Hilbert approach to asymptotic problems arising in the theory of random matrix models, and also in the theory of integrable statistical mechanics, *Ann. of Math.*, **146** (1997) 149–235.

[DeIKZ] P. A. Deift, A. R. Its, A. A. Kapaev and X. Zhou, On the algebro-geometric integration of the Schlesinger equations, *Comm. Math. Phys.*, **203** (1999), 613-633.

[DeKMVZ] P. Deift, T. Kricherbauer, K. T.-R. McLaughlin, S. Venakides, and X. Zhou, Uniform asymptotics for polynomials orthogonal with respect to varying exponential weights and applications to universality questions in random matrix theory, *Commun. Pure Appl. Math.*, **52** (1999), 1335–1425.

[DoS] M. Douglas and S. Shenker, Strings in less than one dimension, *Nucl. Phys. B* **335** (1990) 635–654.

[FGZ] Ph. Di Francesco, P. Ginsparg, J. Zinn-Justin, 2D gravity and random matrices, *Phys. Rep. 254*, no. 1-2 (1995) 133 pp.

[D] B.A. Dubrovin, *Geometry and integrability of topological-antitopological fusion*, preprint INFN-8/92-DSF, 1992.

[DM] B. Dubrovin and M. Mazzocco, Monodromy of certain Painleve-VI transcendents and reflection groups, *Invent. Math.*, bf 141 (2000) No 1, 55–147.

[Dy1] F. Dyson, Correlations between eigenvalues of a random matrix, *Comm. Math. Phys.*, **19** (1970) 235–250.

[Dy2] F. Dyson, Fredholm determinants and inverse scattering problems, *Comm. Math. Phys.*, **47** (1976) 171–183.

[Eh] T. Ehrhard, The asymptotics of the Fredholm determinant of the sine kernel on an interval, arXiv:math.FA/0401205

[EM] N.M. Ercolani and K. D. T-R McLaughlin, Asymptotics of the partition function for random matrices via Riemann-Hilbert techniques and applications to graphical enumeration. *Int. Math. Res. Not.* **14** (2003), 755–820.

[Er] N.P. Erugin, Theory of movable singular points for equations of second order, *Differ. Uravn*, **12** (1976) No 4, 387–416; No 5, 579–598.

[FT] L.D. Faddeev, L.A. Takhtadjan, *Hamiltonian methods in the theory of solitons*, Springer Verlag, Berlin, Heidelberg, 1987.

[FLZZ] V. Fateev, S. Lukyanov, A. Zamolodchikov, and Al. Zamolodchikov, Expectation values of local fields in the Bullough-Dodd model and integrable perturbed conformal field theories, *Nuclear Phys. B*, **516** (1998) No. 3, 652–674.

[Fe1] M.V. Fedoryuk, *Asymptotics. Integrals and series*, Moscow, Nauka, 1987.

[Fe2] M.V. Fedoryuk, *Asymptotic Analysis*, Springer-Verlag, New York, 1993.

[FN1] H. Flaschka and A.C. Newell, Monodromy- and spectrum-preserving deformations I, *Comm. Math. Phys.* **76** (1980) 65–116.

[FN2] H. Flaschka, A.C. Newell: The inverse monodromy transform is a canonical transformation *Math.Studies* **61**, North Holland, (1982) 65-91

[FoA1] A.S. Fokas and M.J. Ablowitz, Linearization of the Korteweg-de Vries and Painlevé II equations, *Phys. Rev. Lett.*, **47** (1981) 1096–1100.

[FoA2] A.S. Fokas and M.J. Ablowitz, On a unified approach to transformations and elementary solutions of Painlevé equations, *J. Math. Phys.*, **23** (1982) no. 11, 2033–2042.

[FoA3] A.S. Fokas and M.J. Ablowitz, On the initial value problem of the second Painlevé transcendent, *Comm. Math. Phys.*, **91** (1983) 381–403.

[FoI1] A.S. Fokas and A.R. Its, The Isomonodromy Method and the Painlevé Equations, *in: Important Developments in Soliton Theory,* ed. A.S. Fokas and V.E. Zakharov, Springer-Verlag, 1992.

[FoT] A.S. Fokas and S. Tanveer, A Hele-Shaw problem and the second Painlevé transcendent, *Math. Proc. Camb. Phil. Soc.*, **124** (1998) 169–191.

[FoIK1] A.S. Fokas, A.R. Its, and A.V. Kitaev, Isomonodromic approach in the theory of two-dimensional quantum gravity, *Usp. Matem. Nauk* **45**, 6 (1990) 135–136 (in Russian).

[FoIK2] A.S. Fokas, A.R. Its, and A.V. Kitaev, The matrix models of the two-dimensional quantum gravity and isomonodromi solutions of the discrete Painlevé equations; A.V. Kitaev, Calculations of nonperturbative parameter in matrix model Φ^4; A.R. Its and A.V. Kitaev, Continuous limit for hermitian matrix model Φ^6, in the book: *Zap. Nauch. Semin. LOMI,* **187**, Differential Geometry, Li Groups and Mechanics, **12**, 1991 (in Russian)

[FoIK3] A.S. Fokas, A.R. Its, and A.V. Kitaev, Discrete Painlevé equations and their appearance in quantum gravity, *Commun. Math. Phys.* **142** (1991) 313–344.

[FoIK4] A.S. Fokas, A.R. Its and A.V. Kitaev, The isomonodromy approach to matrix problems in 2D quantum gravity, *Commun. Math. Phys.*, **147** (1992) 395–430.

[FoM] A.S. Fokas and C. Menyuk, Integrability and similarity in transient stimulated Raman scattering, *J. Nonl. Sci.*, **9** (1999) 1–31.

[FoMA] A.S. Fokas, U. Mugan and M.J. Ablowitz, A method of linearization for Painlevé equations: Painlevé IV, V, *Physica D* **30** (1988) 247–283.

[FoY] A.S. Fokas and Y.C. Yortsos, The transformation properties of the sixth Painlevé equation and one-parameter families of solutions, *Lettere Al Nuovo Cimento* **30** (1981) 539–544.

[FoZ] A.S. Fokas and X. Zhou, On the solvability of Painlevé II and IV, *Comm. Math. Phys.*, **144** (1992) 601–622.

[FoMZ] A.S. Fokas, U. Mugan, and X. Zhou, On the solvability of Painlevé I, III and IV, *Inverse Problems*, **8** (1992) 757–785.

[ForW1] P.J. Forrester, N.S. Witte, Application of the τ-function theory of Painlevé equations to random matrices: PIV, PII and the GUE, *Comm. Math. Phys.*, **219** (2001) 357–398.

[ForW2] P.J. Forrester, N.S. Witte, Application of the τ-function theory of Painlevé equations to random matrices: PVI, the JUE, CyUE, cJUE and scaled limits, *math-ph/0204008.*

[Fr] O. Forster, *Lectures on Riemann surfaces*, Springer, 1992.

[Fri] T. Frisch, Spiral waves in nematics and cholesteric liquid crystals, *Physica D* **84** (1995) 601.

[Fu] R. Fuchs, Über lineare homogene Differentialgleichungen zweiter Ordnung mit drei im Endlichen gelegenen wesentlich singulären Stellen, *Math. Annalen*, **63** (1907) 301–321.

[Ga] F.D. Gakhov, *Boundary problems*, Nauka, Moscow, 1977.

[Gam] B. Gambier, Sur les équationes différentielles du second ordre et du premierdegreé dont l'inégrale générale est à points critiques fixes, *Acta. Math. Ann.* **33** (1910) 1–55.

[Gar1] R. Garnier, Sur les équationes différentielles du troisiṁe ordre dont l'intégrale générale est uniforme et sur une classe d'èquationnes nouvelles d'ordre supérieur dont l'intégrale générale a ses points critique fixes, *Ann. Ecole Norm. Super.*, **29** (1912) 1–126.

[Gar2] R. Garnier, Etude de l'intégrale générale de l'équation VI de M.Painlevé, *Ann. Sci. Ecole Norm. Super.*, (3) **34** (1917) 239–353.

[Gar3] R. Garnier, Contribution à l'étude des solutions de l'équatione (V) de M. Painlevé, *J. Math. Pures Appl.*, **46** (1968) 353–413.

[Go] W.W. Golubev, *Vorlesungen über Differentielgleichungen im Komplexen*, Deutsher Verlag des Wiss., Berlin, 1958.

[GJP] P.R. Gordoa, N. Joshi and A. Pickering, Mapping preserving locations of movable poles: a new extension of the truncation method to ordinary differential equations, *Nonlinearity* **12** (1999) 955–968; Mapping preserving locations of movable poles: II. The third and fifth Painlevé equations, *Nonlinearity* **14** (2001) 567–582; Truncation-type methods and Bäcklund transformations for ordinary differential equations: the third and fifth Painlevé equations, *Glasgow Math. J.* **43A** (2001) 23–32.

[GR] I.S. Gradshteyn and I.M. Ryzhik, 5th edn., *Tables of Integrals, Series, and Products*, A. Jeffrey, ed., Academic Press, San Diego, 1994.

[GrNR] B. Grammaticos, F.W. Nijhoff and A. Ramani, Discrete Painlevé equations, in: *The Painlevé Property,* CRM Ser. Math. Phys., Springer, New York, 1999, 413–516.

[GN] P.G. Grinevich and S.P. Novikov, String Equation–2. Physical Solution, *Algebra and Analysis*, **6** (1994) No 3, 118–140.

[Gro1] V.I. Gromak, Algebraic solutions to the third Painlevé equation, *Doklady Acad. Nauk BSSR*, **23** (1979) No 6, 499–502.

[Gro2] V.I. Gromak, On the solutions of the third Painlevé equation, *Differ. Uravnen.* **9** (1973) 2082–2083.

[Gro3] V.I. Gromak, On the theory of Painlevé equations, *Diff. Eq.* **11** (1975) 285–287.

[Gro4] V.I. Gromak, The solutions of the Painlevé's fifth equation, *Diff. Eq.* **12** (1976) 519–521.

[Gro5] V.I. Gromak, Single-parameter families of solutions of Painlevé equations, *Diff. Eqns.* **14** (1978) 1510–1513.

[GroLS] V.I. Gromak, I. Laine, S. Shimomura, *Painleve Differential Equations in the Complex Plane,* Walter de Gruyter, 2002.

[GroL1] V.I. Gromak and N.A. Lukashevich, Special classes of solutions to the Painlevé equations, *Differ. Uravnen.*, **18** (1982) 317–326.

[GroL2] V.I. Gromak and N.A. Lukashevich, *Analytical properties of solutions to the Painlevé equations*, Minsk, 1990.

[GroT] V.I. Gromak and V.V. Tsegel'nik, Functional relations between solutions of equations of P-type, *Differ. Uravn.* **30** (1994) 1118–1124; On the properties of Bäcklund transformations of Painlevé equations, *Differ. Uravn.* **32** (1996) 1018–1023.

[GrMi] D. Gross and A. Migdal, A nonperturbative treatment of two-dimensional quantum gravity, *Nucl. Phys. B* **340** (1990) 333–365.

[Gu] R.C. Gunning, *Lectures on Vector Bundles over Riemann Surfaces*, Mathematical Notes, v.6, Princeton, Princeton University Press, 1967.

[Guz1] D. Guzzetti, The elliptic representation of the general Painleve VI equation, *Comm. Pure Appl. Math.*, **55** (2002) No 10, 1280–1363.

[Guz2] D. Guzzetti, On the critical behavior, the connection problem and the elliptic representation of a Painlevé VI equation, *Math. Phys. Annal. Geom.*, **4**, (2001) No 4, 293–377.

[Ha1] J. Harnad, Dual isomonodromic deformations and moment maps to loop algebras, *Comm. Math. Phys.* **166** (1994), 337–365; preprint CRM-1844 (1993).

[Ha2] J. Harnad and M. A. Wisse, Loop algebra moment maps and Hamiltonian models for the Painlevé transcendants, *Fields Inst. Commun.* **7** (1993), 155-169

[H] P. Hartman, *Ordinary Differential Equations*, Wiley-Interscience Publication, New York, 1964.

[HasM] S.P. Hastings and J.B. McLeod, A boundary value problem associated with the second Painlevé transcendent and the Korteweg-de Vries equation, *Arch. rational Mech. Anal.* **73** (1980) 31–51.

[He] G. Heading, *The method of phase integrals*, McGraw-Hill, New York, 1965.

[HiM] G. Hilfer and C.R. Menyuk, Stimulated Raman scattering in the transient limit, *J. Opt. Soc. Am. B*, **7** (1990) 739–749.

[Hil] E. Hille, *Ordinary Differential Equations in the Complex Domain*, Wiley-Interscience Publication, New York, 1978.

[In] E.L. Ince, *Ordinary Differential Equations*, Dover, New York, 1956.

[I1] A.R. Its, *The Painlevé Transcendents as Nonlinear Special Functions*, Proc. of the Conference, Montreal, September,1990.

[I2] A.R. Its, Asymptotic solution of the nonlinear Schrödinger equation and isomonodromy deformations of systems of linear differential equations, *Sov. Math. Dokl.*, **24**, (1981) 452–456.

[I3] A.R. Its, "Isomonodromic" solutions of equations of zero curvature, *Izv. Akad. Nauk SSSR*, ser. mat., **49** (1985) no. 3, 530–565 (English Transl.: *Math. USSR-Izv.* **26**, no. 3, 497–529, 1986).

[I4] A.R. Its, Connection formulae for the Painlevé transcendents, in: *The Stokes Phenomenon and Hilbert's 16th Problem,* eds. B.L.J. Braaksma, G.K. Immink, and M. van der Put, World Scientific, 1996

[IFoK] A.R. Its, A.S. Fokas and A.A. Kapaev, On the asymptotic analysis of the Painlevé equations via the isomonodromy method, *Nonlinearity*, **7** (1994) 1291–1325.

[IK1] A.R. Its, A.A. Kapaev, The method of isomonodromy deformations and connection formulas for the second Painlevé transcendent, *Izvestia Acad. Nauk SSSR*, ser. matem., **51** (1987) 878–892 (English transl.: *Math. USSR Izvestiya*, **31** (1988) No 1, 193–207.

[IK2] A.R. Its and A.A. Kapaev, Connection formulae for the fourth Painlevé transcendent; Clarkson-McLeod solution, *J. Phys. A: Math. Gen.* **31** (1998) 4073–4113.

[IK3] A.R. Its and A.A. Kapaev, The irreducibility of the second Painlevé equation and the isomonodromy method, in: *Toward the Exact WKB Analysis of Differential Equations, Linear or Non-Linear,* eds. C.J. Howls, T. Kawai and Y. Takei, Kyoto University Press, 2000, 209–222.

[IK4] A.R. Its and A.A. Kapaev, The nonlinear steepest descent approach to the asymptotics of the second Painlevé transcendent in the complex domain, in: *MathPhys Odyssey 2001, Integrable Models and Beyond,* eds. M. Kashiwara and T. Miwa, Birkhäuser Boston, 2002, 273–311.

[IK5] A.R. Its and A.A. Kapaev, Quasi-linear Stokes phenomenon for the second Painlevé transcendent, *Nonlinearity*, **16** (2003) 363–386.

[IN1] A.R. Its and V.Yu. Novokshenov, *The Isomonodromy Deformation Method in the Theory of Painlevé Equations*, Lect. Notes in Math., **1191**, Springer-Verlag, 1986.

[IN2] A.R. Its and V.Yu. Novokshenov, On the effective sufficient conditions for solvability of the inverse monodromy problem for the systems of linear ordinary differential equations *Funct. Anal. Prilozhen.*, **22** (1988) No 3, 25–36.

[IPe] A.R. Its and V.E. Petrov, "Isomonodromic" solutions of the sine-Gordon equation and the time asymptotics of its rapidly decreasing solutions, *Soviet Math. Dokl.* **26** (1982) no. 1, 244–247.

[IKK] A.R. Its, A.A. Kapaev, A.V. Kitaev, The isomonodromic deformation method and special functions, in the book *Problems of Theoretical Physics*, eds. Yu.N. Demkov, Yu.V. Novozhilov, P.P. Pavinsky, Leningrad University, 1988

[IIwS] M.-A. Inaba, K. Iwasaki, M.-H. Saito, Moduli of stable parabolic connections, Riemann-Hilbert correspondence and Geometry of Painlevé equations of type VI, Part I. arXiv: math.AG/0309342

[Iw] K. Iwasaki, An area-preserving action of the modular group on cubic surfaces and the Painlevé VI equation, *Commun. Math. Phys.*, **242** (2003) 185–219.

[IwKSY] K. Iwasaki, H. Kimura, S. Shimura, and M. Yoshida, *From Gauss to Painlevé: a Modern Theory of Special Functions*, Aspects of Mathematics E **16**, Viewag, Braunschweig, 1991.

[JU] K. Ueno, *Proc. Jpn. Acad.* **56A** 97, 103 (1980); M. Jimbo, *Prog. Theor. Phys.* **61** 359 (1979).

[J1] M. Jimbo, Monodromy problem and the boundary condition for some Painlevé equations, *Publ. Res. Inst. Math. Sci.*, v.18, N 3, p.1137-1161, 1982

[JMU] M. Jimbo, T. Miwa, K. Ueno Monodromy preserving deformation of linear ordinary differential equations with rational coefficients, *Physica D* **2** (1980) 306–352;

[JM] M. Jimbo, T. Miwa, Monodromy preserving deformation of linear ordinary differential equations with rational coefficients. II, *Physica D* **2** (1981) 407–448; M. Jimbo, T. Miwa, Monodromy preserving deformation of linear ordinary differential equations with rational coefficients. III, *Physica D* **4** (1981) 26–46.

[JMMS] M. Jimbo, T. Miwa, Y. Môri, M. Sato, Density matrix of impenetrable bose gas and the fifth Painlevé transcendent, *Physica D* **1** (1980) 80–158.

[JS] M. Jimbo, H. Sakai, A q-analog of the sixth Painlevé equation, *Lett. Math. Phys.* **38** (1996), no. 2, 145–154.

[JoK] N. Joshi and A.V. Kitaev, On Boutroux tritronquée solutions of the first Painlevé equation, *Stud. Appl. Math.*, **107** (2001) 253–291.

[JoKr1] N. Joshi and M.D. Kruskal, An asymptotic approach to the connection problem for the first and the second Painlevé equations, *Phys. Lett. A*, **130** (1988) 129–137.

[JoKr2] N. Joshi and M.D. Kruskal, A simple proof that Painlevé equations have no movable essential singularities, *Preprint no. CMA-R06-90*, Centre for Math. Anal., Australian Nat. Univ., 1990.

[KvM] M. Kac and P. van Moerbeke, On an explicitly soluble system of nonlinear differential equations related to certain Toda lattices, *Advances in Math.*, **16** (1975) 160–169.

[KM] K. Kajiwara and T. Masuda, A generalization of determinant formulae for the solutions of Painlevé II and XXXIV equations, *J. Phys. A: Math. Gen.* **32** (1999) 3763–3778.

[KMNOY] K. Kajiwara, T. Masuda M. Noumi, Y. Ohta and Y. Yamada, Determinant formulas for the Toda and discrete Toda equations, *Funk. Ekvac.* **44** (2001) 291–307.

[KO1] K. Kajiwara and Y. Ohta, Determinant structure of the rational solutions for the Painlevé II equation, *J. Math. Phys.* **37** (1996) no. 9, 4693–4704.

[KO2] K. Kajiwara and Y. Ohta, Determinant structure of the rational solutions for the Painlevé IV equation, *J. Phys. A: Math. Gen.* **31** (1998) 2431–2446.

[K1] A.A. Kapaev, Singular Solutions of the Painleve II Equations, in: Lect. Notes in Math, Springer-Verlag, **1191** (1986) 261–283.

[K2] A.A. Kapaev, Asymptotics of solutions of the Painlevé equation of the first kind *Diff. Eqns.* **24** (1989) 1107–1115.

[K3] A.A. Kapaev, Irregular singular point of the second Painlevé function and the nonlinear Stokes phenomenon, *Zap. Nauch. Semin. LOMI*, **187** (1991) 139–170.

[K4] A.A. Kapaev, Scaling limits for the second Painlevé transcendent, *Zap. Nauch. Semin. POMI*, **209** (1994) 60–101.

[K5] A.A. Kapaev, Global asymptotics of the second Painlevé transcendent, *Phys. Lett. A*, **167** (1992) 356–362.

[K6] A.A. Kapaev, Global asymptotics of the fourth Painlevé transcendent, POMI Preprint **5/1996**, http://www.pdmi.ras.ru/preprint/1996/index.html; *Steklov Math. Inst. and IUPUI Preprint* # 96-5, 1996.

[K7] A.A. Kapaev, WKB method for 2×2 systems of linear ordinary differential equations with rational coefficients, Steklov Math. Inst. and IUPUI Preprint # 96-6, 1996; http://www.pdmi.ras.ru/preprint/1996/index.html.

[K8] A.A. Kapaev, *Painlevé transcendents as nonlinear special functions*, Doctor of Sciences Thesis, Steklov Math. Institute, St. Petersburg, 1998.

[K9] A.A. Kapaev, Monodromy approach to the scaling limits in isomonodromy systems, *Theor. Math. Phys.* **137** no. 3 (2003) 1691–1702; (translated from *Teoret. i Matemat. Fizika* **137** no. 3 (2003) 393-407); arXiv: nlin.SI/0211022

[K10] A.A. Kapaev, Quasi-linear Stokes phenomenon for the Hastings-McLeod solution of the second Painlevé equation, arXiv: nlin.SI/0410009

[K11] A.A. Kapaev, Riemann-Hilbert problem approach to the "physical solution" of the Painlevé first equation, arXiv: nlin. SI/0404026

[KH] A.A. Kapaev and E. Hubert, A note on the Lax pairs for Painlevé equations, *J. Phys. A: Math. Gen.* **32** (1999) 8145–8156.

[KK1] A.A. Kapaev and A.V. Kitaev, The limit transition $P2 \to P1$, *Zap. Nauchn. Sem. LOMI*, **187** (1991) 75–87.

[KK2] A.A. Kapaev and A.V. Kitaev, Connection formulae for the first Painlevè transcendent in the complex domain, *Lett. Math. Phys.*, **27** (1993) 243–252.

[KN] A.A. Kapaev and V.Yu Novokshenov, Two-parametric family of the real solutions of the second Painlevé equation, *Sov. Phys. Dokl.* **31**, 719 (1986).

[KaV] V. Katsnelson and D. Volok, Deformations of Fuchsian systems of linear differential equations and the Schlesinger system, *preprint*, 2005.

[KaM] D.J. Kaup and C.R. Menyuk, Model initial-data problem in stimulated Raman scattering, *Phys. Rev. A*, **42** (1990) 1712–1717.

[Ki1] A.V. Kitaev, The justification of asymptotic formulas that can be obtained by the method of isomonodromic deformations, *Zap. Nauchn. Sem. LOMI*, **179** (1989) 101–109; (English Transl.: *J. Soviet Math.* **57** (1991) no. 3, 3131–3135).

[Ki2] A.V. Kitaev, Self-similar solutions of modified nonlinear Schrödinger equation, *Teor. Math. Phys.* **64** (1985) no 3, 347–369.

[Ki3] A.V. Kitaev, The method of isomonodromic deformations and the asymptotics of the solutions of the "complete" third Painlevé equation, *Mat. Sbornik* **134 (176)** (1987) no 3, 421–444 (English Transl.: *Math. USSR-Sb.* **62** (1989) no 2, 421–444).

[Ki4] A.V. Kitaev, The method of isomonodromic deformations for the "degenerate" third Painlevé equation, *Zap. Nauchn. Sem. LOMI*, **161** (1987) 45–53; (English Transl.: *J. Soviet Math.* **46** (1989) no. 5, 2077–2083).

[Ki5] A.V. Kitaev, Asymptotic description of the fourth Painleve equation. Solutions on the Stokes Rays Analogies, *Zap. Nauch. Semin. LOMI*, **169** (1988) 84–112. (Russian)

[Ki6] A. V. Kitaev, On similarity reduction of the three-wave resonant system to the Painlevé equations, *J. Phys. A.: Math. Gen.* **23** (1990) 3543–3553.

[Ki7] A.V. Kitaev, Elliptic asymptotics of the first and second Painlevé transcendents, *Uspekhi Mat. Nauk* **49** (1994) No. 1, 77–140; (English Transl.: *Russian Math. Surveys* **49** (1994) No. 1, 81–150).

[Ki8] A.V. Kitaev, Special functions of the isomonodromy type, *Acta Appl. Math.* **64** (2000) no. 1, 1–32.

[Ki9] A.V. Kitaev, Special functions of the isomonodromy type, rational transformations of spectral parameter, and algebraic solutions of the sixth Painlevé equation, *nlin.SI/0102020*

[Ki10] A.V. Kitaev, Dessins d'enfants, their deformations and algebraic solutions of the sixth Painlevé and Gauss hypergeometric equations, *nlin.SI/0309078*

[KiV] A.V. Kitaev and A.H. Vartanian, Connection formulae for asymptotics of solutions of the degenerate third Painlevé equation. I, *Inverse Problems,* v. 20, N 4, p.1165-1206, 2004.

[Ko1] D.A. Korotkin, Isomonodromic deformations in genus zero and one: algebro-geometric solutions and Schlesinger transformations, *in*: Integrable systems: from classical to quantum (Montreal, QC, 1999), *CRM Proc. Lect. Not.* **26** (2000) 87–104.

[Ko2] D. Korotkin, Matrix Riemann-Hilbert problems related to branched coverings of $\mathbb{C}P^1$, *arXiv: math-ph/0106009*, 2001.

[Ko3] D. Korotkin and H. Samtleben, Quantization of Coset Space σ-Models Coupled to Two-Dimensional Gravity, *Comm. Math. Phys* **190** (1997) 411-457,

[KoMS] D. Korotkin, N. Manojlović and H. Samtleben, Schlesinger transformations for elliptic isomonodromic deformations, *J. Math. Phys.* **41** (2000) no. 5, 3125–3141.

[Kra] I.V. Krasovsky, Gap probability in the spectrum of random matrices and asymptotics of polynomials orthogonal on an arc of the unit circle, *IMRN*, no. 25 (2004), 1249-1272.

[Kri1] I. Krichever, Isomonodromy equations on algebraic curves, canonical transformations and Whitham equations, *hep-th/0112096*

[Kri2] I. Krichever, Analytic theory of difference equations with rational and elliptic coefficients and the Riemann-Hilbert problem, *math-ph/0407018*

[L] L.D. Landau, E.M. Lifshitz, *Theory of elasticity*, Moscow, 1987.

[La] I.A. Lappo-Danolevskii, *An Application of Matrix Functions to the Theory of Systems of Ordinary Differential Equations,* Acad. Sci. SSSR, Moscow, 1957.

[LeM] D. Levi, C.R. Menyuk and P. Winternitz, Self-similarity in transient stimulated Raman scattering, *Phys. Rev. Lett.* **69** (1992) 3048–3051.

[LiS] G. Litvinchuk and T. Spitkovskii, *Factorization of measurable matrix functions*, Birkhäuser Verlag, Basel, 1987

[Lu1] N.A. Lukashevich, Theory of the fourth Painlevé equation, *Diff. Equations* **3** (1967) 395–399.

[Lu2] N.A. Lukashevich, On the theory of the third Painlevé equation, *Diff. Equations* **3** (1967) 994–999.

[Lu3] N.A. Lukashevich, The second Painlevé equation, *Diff. Equations* **7** (1971) 853–854.

[Ma] A. P. Magnus, Painlevé-type differential equations for the recurrence coefficients of semi-classical orthogonal polynomials, *J. Computat. Appl. Math.*, **57** (1995), 215–237.

[Mal] B. Malgrange, La Classification des Connection Irréguliéres a une Variable, in: *Mathématique et Physique, Progress in Mathematics,* **37**, L. Boutet de Monvel, A. Douady, J.-L. Verdier, eds., Birkhäuser, Boston, 381-400 (1983); Sur les Déformations Isomonodromiques, I. Singularités Regulieres, ibid., 401-426 (1983); Sur les Déformations Isomonodromiques, II. Singularités Irrégulieres, ibid., 427-438 (1983).

[Mq] J. Malmquist, Sur les équations différentielles du second ordre, dont l'intégrale générale a ses points critiques fixes, *Ark. Mat. Astr. Fys.* **17** (1922/23) 1–89.

[Mn] S.V. Manakov, Nonlinear Fraunhoffer diffraction, *Zh. Eksp. Teor. Fiz.*, **65** (1973) 1392–1403.

[Mas] L.J. Mason, M.A. Singer, N.M.J. Woodhouse, Tau functions and the twistor theory of integrable systems, *Preprint, Univ. of Oxford*, 1999.

[MOK] T. Masuda, Y. Ohta and K. Kajiwara, A determinant formula for a class of rational solutions of Painlevé Y equation, *Nagoya Math. J.* **168** (2002) 1–25.

[Mc1] B.M. McCoy, Spin systems, statistical mechanics and Painlevé functins, in: *Painlevé Transcendents, Their Asymptotics and Physical Applications,* Eds. D. Levi and P. Winternitz, NATO ASI Series B: Physics v. 278, pp. 377-391, 1992

[McPS] B.M. McCoy, J.H.H. Perk and R.E. Shrock, Time-dependent correlation functions of the transverse Ising chain at the critical magnetic field, *Nucl. Phys.* B, v. 220, p. 35-47, 1983

[McT1] B.M. McCoy and Sh. Tang, Connection formulae for Painlevé V functions, *Physica D* v.19, 42-72, 1986

[McT2] B.M. McCoy and Sh. Tang, Connection formulae for Painlevé V functions: II. The δ-function Bose gas problem, *Physica D* v. 20, 187-216, 1986

[McTW] B.M. McCoy, C.A. Tracy and T.T. Wu, Painlevé functions of the third kind, *J. Math. Phys.* **18** (1977) No 5, 1058–1092.

[Me] C.R. Menyuk, Long distance evolution in transient stimulated Raman scattering, *Phys. Rev. A*, **47** (1993) 2235–2248.

[Mi] K.B. Migler and R.B. Meyer, Spirals in liquid crystals in a rotating magnetic field, *Physica D* **71** (1994) 412.

[Meh] M.L. Metha, *Random Matrices*, Second Edition, Academic Press, San Diego, 1991.

[Mil1] J.W. Miles, On the second Painlevé transcendent, *Proc. R. Soc. Lond.* A **361** (1978) 277–291

[Mil2] J.W. Miles, The second Painlevé transcendent: a nonlinear Airy function, *Mechanics Today* **5** (1980) 297–313 A **361** (1978) 277–291

[MCB] A.E. Milne, P.A. Clarkson and A.P. Bassom, Bäcklund transformations and solution hierarchies for the third Painlevé equation, *Stud. Appl. Math.* **98** (1997) 139–194.

[Miw] T. Miwa, Painlevé property of monodromy preserving deformation equations and the analyticity of τ-functions, *Publ. Res. Inst. Math. Sci.* **17** (1981), 703–712.

[Mo] G. Moore, Geometry of the String Equations, *Commun. Math. Phys.*, **133** (1990), 261–304.

[Mor] G. Moore, Matrix models of 2D gravity and isomonodomic deformations, *Progr. Theoret. Phys. Suppl.*, **102** (1990) 255–285.

[MuFo] U. Mugan, A. S. Fokas, Shlesinger transformations of Painlevé II – V, *J. Math. Phys.* **33** (1992) 2031–2045.

[Mu] U. Mugan and A. Sakka, On the solvability of the Painlevé VI equation, *J. Math. Phys.* **36** (1993) 1284–1298.

[Mur] Y. Murata, Classical solutions of the third Painlevé equation, *Nagoya Math. J.* **139** (1995) 37–65; Rational solutions of the second and fourth Painlevé equations, *Funk. Ekvac.* **28** (1995) 1–58.

[MuSY] M. Murata, H. Sakai and J. Yoneda, Riccati solutions of discrete Painlevé equations with Weyl group symmetry of type $E_8^{(1)}$, *J. Math. Phys.* **44** (2003) no. 3, 1396–1414.

[NiRGO] F.W. Nijhoff, A. Ramani, B. Grammaticos and Y. Ohta, On discrete Painlevé equations associated with the lattice KdV systems and the Painlevé VI equation, *Stud. Appl. Math.*, arXiv: solv-int/9812011.

[NiW] F.W. Nijhoff and A.J. Walker, The discrete and continuous Painlevé VI hierarchy and the Garnier systems, *Glasgow Math. J.* **43A** (2001) 109–123.

[No1] M. Noumi, Affine Weyl group approach to Painlevé equations, *Proceedings of the International Congress of Mathematics*, Vol. III (Beijing, 2002) 497–509, Higher Ed. Press, Beijing, 2002.

[No2] M. Noumi, *Painlevé equations through symmetry*, Translations of Math. Monographs, vol. 223, AMS, Providence, Rhode Island, 2004.

[NoO] M. Noumi and K. Okamoto, Irreducibility of the second and the fourth Painlevé equations, *Funk. Ekvac.* **40** (1997) 139–163.

[NoY1] M. Noumi and Y. Yamada, Symmetries in the fourth Painlevé equations and Okamoto polynomials, *Nagoya Math. J.* **153** (1999) 53–86; Symmetries in Painlevé equations, *Sugaku* **53** (2001) 62–75.

[NoY2] M. Noumi and Y. Yamada, A new Lax pair for the sixth Painlevé equation associated with $\widehat{so}(8)$, in: T. Kawai (ed.) at al., *Microlocal analysis and complex Fourier analysis.* River Edge, NJ: World Scientific, 2002, pp. 238–252.

[NMPZ] S. Novikov, S.V. Manakov, L.P. Pitaevskii and V.E. Zakharov, *Theory of Solitons (the Inverse Scattering Method)*, Consultants Bureau, New York, 1984.

[N1] V.Yu. Novokshenov, Boutroux ansatz for the second Painlevé equation in the complex domain, *Izv. Akad. Nauk SSSR* ser. math., **54** (1990) 1229–1251.

[N2] V.Yu. Novokshenov, On the asymptotics of the general real solution of the Painlevé equation of the third kind, *Sov. Phys. Dokl.*, **30** (1985) 666–668.

[N3] V.Yu. Novokshenov, The method of isomonodromic deformation and the asymptotics of the third Painlevé transcendent, *Funct. Anal. Appl.*, **18** (1984) no. 3, 260–262.

[N4] V.Yu. Novokshenov, Nonlinear Stokes phenomena, *Physica D*, **63** (1993) No. 1, 1–7.

[N5] V.Yu. Novokshenov, Radial-symmetric solution of the cosh-Laplace equation and the distribution of its singularities, *Russian J. Math. Phys.* **5** (1997) No 2, 211–226.

[N6] V.Yu.Novokshenov, A.G.Shagalov, Bound states of the elliptic sine-Gordon equation, *Physica D* **106** (1997) 81–94.

[O] Yu. Ohyama, On the third Painlevé equation of type D_7 and On classical solutions of the third Painlevé equation of type D_7, Preprints 2001.

[Ok1] K. Okamoto, Sur les Feuilletages Associés aux Equations du Second Ordre à Points Critiques Fixes de Painlevé, Espaces des Conditions Initiales, *Japan J. Math.* **5** (1979) 1–79.

[Ok2] K. Okamoto, Polynomial Hamiltonians associated with Painlevé equations. II. Differential equations satisfied by polynomial Hamiltonians, *Proc. Japan Acad. Ser. A Math. Sci.*, **56** (1980) No. 8, 367–371.

[Ok3] K. Okamoto, On the τ-function of the Painlevé equations, *Physica D*, **2** (1981) No. 3, 525–535.

[Ok4] K. Okamoto, Isomonodromic deformation and Painlevé equations and the Garnier system, *J. Fac. Sci. Univ. Tokyo Sect. IA Math.* **33** (1986) 575–618.

[Ok5] K. Okamoto, Studies on the Painlevé equations, I. Sixth Painlevé equation P_{VI}, *Ann. Math. Pure Appl.* **146** (1987) 337–381; Studies on the Painlevé equations, II. Fifth Painlevé equation P_V, *Japan J. Math.* **13** (1987) 47–76; Studies on the Painlevé equations, III. Second and fourth Painlevé equations, *Math. Ann.* **275** (1986) 221–255; Studies on the Painlevé equations, IV. Third Painlevé equation P_{III}, *Funkcial. Ekvac.* **30** (1987) 305–332.

[Ok6] K. Okamoto, Painlevé equations and Dynkin diagrams, *Painlevé Transcendents*, D. Levi, P. Winternitz eds., Plenum Press, New York (1992) 299–313.

[Ol] F.W.J. Olwer, *Asymptotics and special functions*, Academic Press, New York, 1974.

[P] P. Painlevé, Mémoire sur les équations differentielles dont l'integrale générale est uniforme, *Bull. Soc. Math. France* **28** (1900) 201-261.

[Pa] J. Palmer, Zeros of the Jimbo, Miwa, Ueno tau Function, *J. Math. Phys.*, **40:12** (1999) 6638-6681.

[PS] L. Pastur and M. Shcherbina, Universality of the local eigenvalue statistics for a class of unitary invariant random matrix ensembles, *J. Statist. Phys.*, **86** (1997), 109-147.

[Pl] J. Plemelj, *Problems in the sense of Riemann and Klein*, Inter. Publ. New York-Sydney, 1964.

[RGH] A. Ramani, B. Grammaticos and J. Hitarinta, Discrete versions of the Painlevé equations, *Phys. Rev. Lett.* **67** (1991) 1829–1832.

[S] H. Sakai, Rational surfaces associated with affine root systems and geometry of the Painlevé equations, *Comm. Math. Phys.* **220** (2001) 165–229.

[Sch] L. Schlesinger, Über eine Klasse von Differentialsystemen beliebliger Ordnung mit festen kritischen Punkten, *J. für Math.*, **141** (1912) 96–145.

[Se] N. Seiberg and D. Shih, Flux vacua and branes of the minimal superstring, *hep-th/0412315*

[Sh] A.B. Shabat, The third version of dressing method, *Theor. Math. Phys.*, **121** (1999) No 1, 165–176.

[Sha1] A.G.Shagalov, Critical phenomena in large area Josephson junctions with defects, in: *Nonlinear superconducting devices and high-T_c materials,* ed. R.D. Parmentier and N.F. Pedersen, World Scientific, Singapore, 1995, p.125.

[Sha2] A.G. Shagalov, Damped and driven sine-Gordon equation as a model for pattern formation in liquid crystals, *Phys. Lett. A* **199** (1995) 229.

[Si1] Y. Sibuya, Stokes Phenomena, *Bull. Amer. Math. Soc.*, **83** (1977) No 5, 1075–1077.

[Si2] Y. Sibuya, *Linear differential equations in the complex domain: problems of analytic continuation*, Translations of mathematical monographs, **82**, Providence, 1990.

[SY] P.G. Silvestrov and A.S. Yelkhovsky, INP preprint 90-81, Novosibirsk, 1990.

[SL] S. Slavyanov and W. Lay, *Special functions: A united theory based on singularities*, Oxford University Press, 2000.

[So] A. Soshnikov, Universality at the edge spectrum in Wigner random matrices, *Commun. Math. Phys.*, **207** (1999) 697-733.

[Su1] B.I. Suleimanov, On Asymptotics of Regular Solutions for a Special Kind of Painleve V Equation, *Lect. Notes in Math.*, Springer Verlag, **1191** (1986) 230–255.

[Su2] B. I. Suleimanov, The relation between asymptotics properties of the second Painlevé equation in different directions towards infinity, *Diff. Eqns.*, **23** (1987) 569–576.

[Ta1] Y. Takei, On the connection formula for the first Painlevé equation – from the viewpoint of the exact WKB analysis, *Sūrikaisekikenkyūsho Kōkyūroku*, **931** (1995) 70–99.

[Ta2] Y. Takei, An explicit description of the connection formula for the first Painlevé equation, in: *Toward the exact WKB analysis of differential equations, linear or non-linear* (Kyoto, 1998) (Kyoto: Kyoto Univ. Press) 271-296, 2000.

[Tk] T. Takenawa, Weyl group symmetry of type $D_5^{(1)}$ in the q-Painlevé V equation, *Funk. Ekvac.* **46** (2003) 173–186.

[TBAZ] R. Teodorescu, E. Bettelheim, O. Agam, A. Zabrodin, and P. Wiegmann, Normal random matrix ensemble as a growth problem, *hep-th/0401165* (2004).

[TW1] C.A. Tracy and H. Widom, Introduction to random matrices, *Geometric and quantum aspects of integrable systems* (Schreveningen, 1992), 103–130. Lecture Notes in Physics, 424. Springer. Berlin, 1993

[TW2] C.A. Tracy and H. Widom, Level spacing distributions and the Airy kernel, *Comm. Math. Phys.*, **159** (1994) 151–174.

[TW3] C.A. Tracy and H. Widom, Asymptotics of a class of solutions to the cylindrical Toda equations, *Commun. Math. Phys.* **190** (1998) 697–721.

[TW4] C.A. Tracy and H. Widom, On exact solutions to the cylindrical Poisson-Boltzmann equation with applications to polyelectrolytes, *Phys. A*, **244** (1997) no. 1–4, 402–413.

[TW5] C.A. Tracy and H. Widom, Correlation functions, cluster functions, and spacing distributions for random matrices, *J. Statis. Phys.* **92** (1998) 809–835.

[TW6] C.A. Tracy and H. Widom, Random unitary matrices, permutations and Painlevé, *Comm. Math. Phys.*, **207** (1998) 665–685.

[TW7] C.A. Tracy and H. Widom, Universality of the distribution functions of random matrix theory. II, *in Integrable Systems: From Classical to Quantum, CRM Proceedings & Lecture Notes* **26**, eds. J. Harnad, G. Sabidussi, and P. Winternitz, AMS, Providence (2000) 251–264.

[U] M. Ugaglia, On a Poisson structure on the space of Stokes matrices, *Int. Math. Res. Not.*, no 9, (1999) 473–493.

[Um1] H. Umemura, Second proof of the irreducibility of the first differential equation of Painlevé, *Nagoya Math. J.* **117** (1990) 125–171.

[Um2] H. Umemura, The Painlevé equation and classical functions, *Sugaku*, **47** (1995) No. 4, 341–359; (English Transl.: *Sugaku Expositions*, **11** (1998) No. 1, 77–100).

[UmW] H. Umemura and H. Watanabe, Solutions of the second and fourth Painlevé equations. I, *Nagoya Math. J.* **148** (1997) 151–198.

[VT] H.G. Vaidya and C.A. Tracy, One-particle reduced density matrix of impenetrable bosons in one dimension at zero temperature, *Phys. Rev. Lett.* **42** (1979) 3–6.

[Va] V.S. Varadarajan, Linear meromorphic differential equations: a modern point of view, *Bull. Amer. Math. Soc.*, **33** (1996) no. 1, 1–42.

[W1] W. Wasow, *Asymptotic Expansions for Ordinary Differential Equations,* Interscience-Wiley, New York, 1965.

[W2] W. Wasow, *Linear Turning Point Theory,* Applied Mathematical Sciences, Vol. 54, Springer-Verlag, New York, 1985.

[Wa1] H. Watanabe, Solutions of the fifth Painlevé equation. I, *Hokkaido Math. J.* **24** (1995) 231–267.

[Wa2] H. Watanabe, Birational canonical transformations and classical solutions of the sixth Painlevé equation, *Ann. Scuola Norm. Sup. Pisa* **27** (1998) 379–425.

[Wi1] H. Widom, The asymptotics of a continuous analogue of orthogonal polynomials, *J. Approx. Theory,* **77** (1994) 51-64

[WinL] P. Winternitz and D. Levi, eds., *Painlevé transcendents, their asymptotics and physical applications,* NATO ASI Series B, **278**, 1992

[Wit] E. T. Whittaker, On a class of differential equations whose solutions satisfy integral equations, *Proc. Edinb. Math. Soc.* **33** (1914) 14–23.

[Wt] N. S. Witte, New transformations for Painlevé's third transcendent, *Proc. Am. Math. Soc.* **132** (2004) no. 6, 1649–1658.

[WTMTB] T.T. Wu, B.M. McCoy, C.A. Tracy and E. Barouch, Spin-spin correlation functions of the two-dimensional Ising model: Exact theory in the scaling region, *Phys. Rev. B* **13** (1976) 316–374.

[Y] Y. Yamada, Determinant formulas for the τ-functions of the Painlevé equations of type A, *Nagoya Math. J.* **156** (1999) 123–134.

[ZK] V.E. Zakharov, E.A. Kuznetsov and S.L. Musher, Quasiclassic wave collapse, *JETP Lett.*, **41** (1985) No 3, 125–127.

[ZM] V.E. Zakharov and S.V. Manakov, On the asymptotic behavior of the nonlinear wave systems solvable by the inverse scattering method, *Zh. Eksp. Teor. Fiz.,* **71** (1976) 203–215.

[Zh] X. Zhou, The Riemann-Hilbert problem and inverse scattering *SIAM J. Math. Anal.* **20** (1989) No 4, 966–986.

Subject Index

Titles in This Series

128 **Athanassios S. Fokas, Alexander R. Its, Andrei A. Kapaev, and Victor Yu. Novokshenov,** Painlevé transcendents, 2006

127 **Nikolai Chernov and Roberto Markarian,** Chaotic billiards, 2006

126 **Sen-Zhong Huang,** Gradient inequalities, 2006

125 **Joseph A. Cima, Alec L. Matheson, and William T. Ross,** The Cauchy Transform, 2006

124 **Ido Efrat, Editor,** Valuations, orderings, and Milnor K-Theory, 2006

123 **Barbara Fantechi, Lothar Göttsche, Luc Illusie, Steven L. Kleiman, Nitin Nitsure, and Angelo Vistoli,** Fundamental algebraic geometry: Grothendieck's FGA explained, 2005

122 **Antonio Giambruno and Mikhail Zaicev, Editors,** Polynomial identities and asymptotic methods, 2005

121 **Anton Zettl,** Sturm-Liouville theory, 2005

120 **Barry Simon,** Trace ideals and their applications, 2005

119 **Tian Ma and Shouhong Wang,** Geometric theory of incompressible flows with applications to fluid dynamics, 2005

118 **Alexandru Buium,** Arithmetic differential equations, 2005

117 **Volodymyr Nekrashevych,** Self-similar groups, 2005

116 **Alexander Koldobsky,** Fourier analysis in convex geometry, 2005

115 **Carlos Julio Moreno,** Advanced analytic number theory: L-functions, 2005

114 **Gregory F. Lawler,** Conformally invariant processes in the plane, 2005

113 **William G. Dwyer, Philip S. Hirschhorn, Daniel M. Kan, and Jeffrey H. Smith,** Homotopy limit functors on model categories and homotopical categories, 2004

112 **Michael Aschbacher and Stephen D. Smith,** The classification of quasithin groups II. Main theorems: The classification of simple QTKE-groups, 2004

111 **Michael Aschbacher and Stephen D. Smith,** The classification of quasithin groups I. Structure of strongly quasithin K-groups, 2004

110 **Bennett Chow and Dan Knopf,** The Ricci flow: An introduction, 2004

109 **Goro Shimura,** Arithmetic and analytic theories of quadratic forms and Clifford groups, 2004

108 **Michael Farber,** Topology of closed one-forms, 2004

107 **Jens Carsten Jantzen,** Representations of algebraic groups, 2003

106 **Hiroyuki Yoshida,** Absolute CM-periods, 2003

105 **Charalambos D. Aliprantis and Owen Burkinshaw,** Locally solid Riesz spaces with applications to economics, second edition, 2003

104 **Graham Everest, Alf van der Poorten, Igor Shparlinski, and Thomas Ward,** Recurrence sequences, 2003

103 **Octav Cornea, Gregory Lupton, John Oprea, and Daniel Tanré,** Lusternik-Schnirelmann category, 2003

102 **Linda Rass and John Radcliffe,** Spatial deterministic epidemics, 2003

101 **Eli Glasner,** Ergodic theory via joinings, 2003

100 **Peter Duren and Alexander Schuster,** Bergman spaces, 2004

99 **Philip S. Hirschhorn,** Model categories and their localizations, 2003

98 **Victor Guillemin, Viktor Ginzburg, and Yael Karshon,** Moment maps, cobordisms, and Hamiltonian group actions, 2002

97 **V. A. Vassiliev,** Applied Picard-Lefschetz theory, 2002

96 **Martin Markl, Steve Shnider, and Jim Stasheff,** Operads in algebra, topology and physics, 2002